Advances in Intelligent Systems and Computing

Volume 1031

The series "Advances in Intelligent Systems and Computing" contains publications on theory, applications, and design methods of Intelligent Systems and Intelligent Computing. Virtually all disciplines such as engineering, natural sciences, computer and information science, ICT, economics, business, e-commerce, environment, healthcare, life science are covered. The list of topics spans all the areas of modern intelligent systems and computing such as: computational intelligence, soft computing including neural networks, fuzzy systems, evolutionary computing and the fusion of these paradigms, social intelligence, ambient intelligence, computational neuroscience, artificial life, virtual worlds and society, cognitive science and systems, Perception and Vision, DNA and immune based systems, self-organizing and adaptive systems, e-Learning and teaching, human-centered and human-centric computing, recommender systems, intelligent control, robotics and mechatronics including human-machine teaming, knowledge-based paradigms, learning paradigms, machine ethics, intelligent data analysis, knowledge management, intelligent agents, intelligent decision making and support, intelligent network security, trust management, interactive entertainment, Web intelligence and multimedia.

The publications within "Advances in Intelligent Systems and Computing" are primarily proceedings of important conferences, symposia and congresses. They cover significant recent developments in the field, both of a foundational and applicable character. An important characteristic feature of the series is the short publication time and world-wide distribution. This permits a rapid and broad dissemination of research results.

**** Indexing: The books of this series are submitted to ISI Proceedings, EI-Compendex, DBLP, SCOPUS, Google Scholar and Springerlink ****

More information about this series at http://www.springer.com/series/11156

Vipul Jain · Srikanta Patnaik ·
Florin Popențiu Vlădicescu ·
Ishwar K. Sethi
Editors

Recent Trends in Intelligent Computing, Communication and Devices

Proceedings of ICCD 2018

Springer

Editors
Vipul Jain
Victoria Business School
Victoria University of Wellington
Wellington, New Zealand

Florin Popenţiu Vlădicescu
University "Politehnica" of Bucharest
Bucharest, Romania

Srikanta Patnaik
Department of Computer Science
and Engineering, Faculty of Engineering
and Technology
SOA University
Bhubaneswar, Odisha, India

Ishwar K. Sethi
Department of Computer Science
and Engineering
Oakland University
Rochester, MI, USA

ISSN 2194-5357 ISSN 2194-5365 (electronic)
Advances in Intelligent Systems and Computing
ISBN 978-981-13-9405-8 ISBN 978-981-13-9406-5 (eBook)
https://doi.org/10.1007/978-981-13-9406-5

This Springer imprint is published by the registered company Springer Nature Singapore Pte Ltd.
The registered company address is: 152 Beach Road, #21-01/04 Gateway East, Singapore 189721, Singapore

Preface

The integration of the Internet of things (IoT) and cloud computing has led to the evolution of new paradigms as well as their extensive applications. Technically, both are different, but when integrated with each other, together they lead to disruptive and intelligent next-generation devices and technologies with exponential applications in almost all sectors. While cloud computing has changed the software development and enterprise management strategies, IoT has fused both digital and physical worlds together, thus promising new opportunities for research and development.

This conference proceeding covers a collection of recent research works presented at The 4th International Conference on Intelligent Computing, Communication and Devices (ICCD-2018) held at Guangzhou, China, during December 7–9, 2018. The research papers provide a wide coverage to all the aspects of next-generation computing technologies and communication and intelligent devices. The papers are broadly categorized into four major tracks: (i) intelligent computing, (ii) smart sensor and devices, (iii) next-generation communication and networking, and (iv) Industry 4.0 and its applications.

The first track on "intelligent computing" addresses research works on intelligent computing in distributed networks and cloud-based networks. It further includes papers on soft computing, data mining, data exchange systems, data crawling, cluster analysis, context-based adaptive filters, automatic decision support systems, collaborative systems, etc.

The second track, namely "smart sensor and devices," includes state-of-the-art research works on intelligent devices and sensors in WSN-, SDN-, and RFID-based devices, collision avoidance in narrow channels, object detection over networks by smart devices, smart home systems, design structure of smart communities, multi-sensor data fusion, control of unmanned vehicles, and integration of IoT and artificial intelligence for remote diagnosis.

The subsequent track, namely "next-generation communication and networking," covers recent research and trends being developed in various areas such as M2M communications and networks, signal transmissions, signal processing of mechanical vibrations, artificial intelligence-based network security, and fault detection and diagnosis. It further exhibits work on ad hoc networks, radar signals, wideband differential phase shifters, dual-band antenna array, miniaturized antenna units, and their transmission lines.

Last but not least, the track on "Industry 4.0 and its applications" showcases intelligent applications in various industrial sectors to increase flexibility, scalability, and productivity of the sector while coping with various challenges and circumstances. Some of them are development of industrial robotic arms, modeling of assembly system complexity and further applying it to planning horizon problem, adaptive cruise model for controlling trucks, monitoring armored vehicles remotely, preheater air leakage rate measuring for gas boilers, electrification of autonomous transportation vehicles, designing of micro-energy grids for nuclear power plants, optimizing natural gas pipeline designs, etc.

Wellington, New Zealand — Dr. Vipul Jain
Bhubaneswar, India — Prof. Srikanta Patnaik
Bucharest, Romania — Dr. Florin Popenţiu Vlădicescu
Rochester, USA — Prof. Ishwar K. Sethi

Acknowledgements

The contributions covered in this proceeding are the outcome of the contributions from more than one hundred researchers. We are thankful to the authors, paper contributors to this volume, and the departments which support the event a lot.

We are thankful to the Editor-in-Chief of the Springer book series "Advances in Intelligent Systems and Computing," Prof. Janusz Kacprzyk, for his support to bring out the fourth volume of the conference, i.e., ICCD-2018. It is noteworthy to mention here that constant support from the Editor-in-Chief and the members of the publishing house makes the conference fruitful for the third edition.

We would like to extend our heartfelt thanks to Dr. Thomas Ditzinger, Executive Editor, and his Springer publishing team, for their encouragement and support.

We are thankful to Prof. Florin Popenţiu Vlădicescu, "Politehnica" University of Bucharest, for his well-researched keynote address "New Advancements in Computational Intelligence for Software Engineering" and equally thankful to Prof. Dr. Wen-Jun Zhang, University of Saskatchewan, Canada, for his path-breaking talk on "Data Relativity: over 20 year's reflection." We extend our thanks to Prof. Gilbert Yuk Sing Chan, Hong Kong Polytechnic University, for his talk on "The roles of IT for agriculture and aquaculture in nowadays China."

Last but not least, we are thankful to our friend Prof. Dr. Sc. Kazumi Nakamatsu, School of Human Science and Environment, University of Hyogo, Japan; Prof. Dr. Andrew W. H. Ip, Department of Mechanical Engineering, University of Saskatchewan, Canada; Prof. Jinwen Wu, School of Economics and Management, South China Normal University, Guangzhou, China; Prof. Chong Shen, Hainan University, Hainan, China; Prof. Xilong Qu, School of Information Technology and Management, Hunan University of Finance and Economics, Changsha, China; Prof. Jinhui Zeng, College of Electrical and Information Engineering, Hunan University of Technology, Zhuzhou, China; Prof. Kun Zhang, Hainan Tropical Ocean University, Hainan, China; Dr. WU Chun Ho, Jack, Department of Supply Chain and Information Management, Hang Seng University of Hong Kong, Hong Kong; Prof. Yikui Zhai, School of Information and Engineering, Wuyi University,

Jiangmen, China, for their support and guidance. We are also thankful to the experts and reviewers who have worked for this volume despite the veil of their anonymity.

We are also thankful to our academic partners: School of Economics and Management, South China Normal University, Guangzhou, China; Scientific Research Office, Hainan University, Hainan, China; Hunan University of Finance and Economics, Changsha, China; School of Electrical and Information Engineering, Hunan University of Technology, Zhuzhou, China; School of College of Ocean Information Engineering, Hainan Tropical Ocean University, Hainan, China; IRNet International Academic Communication Center, China; Financial Big Data Science and Technology Key Laboratory of Hunan Province, Changsha, China; Hunan Provincial 2011 Collaborative Innovation Center for Development and Utilization of Finance and Economics Big Data Property, Changsha, China; Hunan Province Higher Educational Institutions Key Laboratory "Information Technology and Information Security," Changsha, China; Key Laboratory for Electric Drive Control and Intelligent Equipment of Hunan Province, China; Guangdong Graphic Image Association, China; Guangdong Massive Biometric Information Processing Engineering Technology Research Center; Jiangmen Computer Federation, China; and last but not the least, Interscience Institute of Management & Technology, Bhubaneswar, India.

We look forward to your valued contribution and support to the next edition of the 5th International Conference on Intelligent Computing, Communication and Devices (ICCD-2019), whose venue will be announced shortly.

We are sure that the readers shall get immense benefit and knowledge from the fourth volume of the Intelligent Computing, Communication and Devices.

Contents

Industry 4.0 and Applications

About the Editors

Dr. Vipul Jain is currently working in the area of Operations and Supply Chain Management at Victoria Business School, Victoria University of Wellington, New Zealand. He has also worked as a French Government researcher for the French National Institute for Research in Computer Science and Control at Nancy, France. He is the Editor-in-Chief of the International Journal of Intelligent Enterprise and an Editorial Board Member for seven international journals.

Dr. Srikanta Patnaik is a Professor at the Department of Computer Science and Engineering, Faculty of Engineering and Technology, SOA University, Bhubaneswar, India. He is the Editor-in-Chief of the International Journal of Information and Communication Technology and International Journal of Computational Vision and Robotics, both published by Inderscience, England, and of the book series "Modeling and Optimization in Science and Technology," published by Springer, Germany.

Prof. Florin Popențiu Vlădicescu graduated with a degree in Electronics and Telecommunications from the University POLITEHNICA of Bucharest in 1974 and holds a Ph.D. in Reliability. He has been appointed as a Director of the "UNESCO Chair in Information Technologies Department" at the University of Oradea. Professor Vlădicescu is the founder of the first "UNESCO Chair of Information Engineering" in the UK, established at City University London in 1998. He is also an Advisory Board Member for several international journals, including "Reliability and Risk Analysis: Theory & Applications."

Dr. Ishwar K. Sethi is currently a Professor in the Department of Computer Science and Engineering at Oakland University in Rochester, Michigan, USA. He has served on the editorial boards of several prominent journals including IEEE Transactions Pattern Analysis and Machine Intelligence, and IEEE Multimedia. He was elected an IEEE Fellow in 2001 for his contributions to artificial neural networks and statistical pattern recognition.

Intelligent Computing

Global Stability of a Viral Dynamical Model

Lin Li and Fengyin Gao

Abstract In this paper, a viral (HIV) infection model with the influence coefficient is considered. The global stabilities of the healthy equilibrium and infection equilibrium of the model are studied by constructing the suitable Lyapunov functions. The threshold of the various equilibrium existences is found. When the threshold is not greater than 1, the healthy equilibrium is globally stable on the feasible region, which implies that in-host viral (HIV) dies out eventually; when the threshold is greater than 1, the model has a unique infection equilibrium, which is globally stable in the feasible region; that is, the viral (HIV) persists in the body of the infected individuals, and the concentration of in-host viral (HIV) tends to a positive number.

Keywords Equilibrium · Basic reproduction number · Lyapunov function · Global stability

1 Introduction

Virology studies show that the main target of HIV attacks is the CD4+T cells in the host. Long-term HIV infection causes the depletion of CD4+T cells in the host, leading to acquired immunodeficiency syndrome (AIDS). Since its discovery in the 1980s, AIDS is spreading rapidly around the world, posing a major threat to human health. Many medical and mathematical workers at home and abroad are working on the prevention and control of AIDS. Using mathematical model to analyze the dynamics of virus infection has become one of the hot topics in applied mathematics nowadays. Studies on viral dynamics in the world have been progressing rapidly, and a large number of HIV infection models have been proposed and studied by scholars [1–7], Stability Analysis of Equilibrium Point of Infectious Disease Dynamics Model was studied in [8–10]. In [11], Nowak and Bangham first

L. Li · F. Gao (✉)
Department of Basic Sciences, Air Force Engineering University, Xi'an 710051, China
e-mail: 156171643@qq.com

V. Jain et al. (eds.), *Recent Trends in Intelligent Computing, Communication and Devices*, Advances in Intelligent Systems and Computing 1031,
https://doi.org/10.1007/978-981-13-9406-5_1

introduced the model of CD4+T cells to stimulate cytotoxic T lymphocyte (CTL) immune response. In [12], Nowak and May considered a model which was with self-regulating CTL immune response. The [5] considered models with latent stage and drug therapy. Combined with [12] and [5], this paper will exist in assuming no HIV infection when CTL cell constant input, described in the general function of infected CD4+T cells on CTL cell stimulation, by constructing Lyapunov function to prove the global stability of the model.

2 Model

In 1996, Nowak et al. [11] proposed the dynamic model of HIV infection:

$$\begin{cases} \mathrm{d}x/\mathrm{d}t = \lambda - dx - \beta xv \\ \mathrm{d}y/\mathrm{d}t = \beta xv - ay \\ \mathrm{d}v/\mathrm{d}t = kay - \gamma v \end{cases} \tag{1}$$

where $x = x(t), y = y(t), v = v(t)$, respectively, denote the concentration of uninfected (healthy) CD4+T cells, infected CD4+T cells, and free HIV virus in the infected body at time t. λ is the new CD4+T cell production rate, d is the mortality coefficient of healthy CD4+T cells, β is the infection coefficient of HIV in healthy CD4+T cells, $a(a > d)$ is the mortality coefficient of the infected CD4+T cells, γ is the coefficient of clearance of HIV, and k is the number of viruses produced by the death of an infected CD4+T cell per unit time.

This paper assumes that the change of CTL cell concentration is consistent with the following rules when individuals are not infected with HIV.

$$\mathrm{d}w/\mathrm{d}t = \lambda_1 - d_1 w \tag{2}$$

where $w = w(t)$ denotes the concentration of CTL cells in the infected body at time t, λ_1 is the input rate of CTL cells, and d_1 is the mortality coefficient of CTL cells. Equation (2) means that when an individual is not infected with HIV, CTL cells in the body tend to be in equilibrium. Usually, acute infections (such as acute hepatitis B) are recovered because the immune system of the infected individual plays a role in clearing the virus. In the case of chronic infection, the immune system is not enough to remove the virus, but it can inhibit or control the proliferation of the virus in the body by dissolving the infected cells [13]. CTL cells play an important inhibitory role in HIV infection of CD4+T cells. This paper assumes that when an individual is infected with HIV, the effective coefficient of the infected CD4+T cells on CTL cell reproduction is $f(w)$. So let's say that $f(w)$ satisfies the conditions: $f'(w) \geq 0$ and $[f(w)/w]' \leq 0$, $f'(w) \geq 0$ means that $f(w)$ increases with the concentration of CTL cells. $[f(w)/w]' \leq 0$; it means that the change rate of the influence coefficient decreases with the increase of the concentration of CTL cells [6, 7]. Considered the immune system two forms: $f(w) = c$ and $f(w) = cw$. But the global stability of the model is not analyzed. There $f(w)$ contains two special forms $f(w) = c$ and $f(w) = cw$.

In this paper, an HIV dynamic model with a more general impact coefficient is considered.

$$\begin{cases} \mathrm{d}x/\mathrm{d}t = \lambda - dx - \beta xv \\ \mathrm{d}y/\mathrm{d}t = \beta xv - ay - myw \\ \mathrm{d}v/\mathrm{d}t = kay - \gamma v \\ \mathrm{d}w/\mathrm{d}t = \lambda_1 + f(w)y - d_1 w \end{cases} \tag{3}$$

It is easy to know that

$$R_+^4 = \{(x, y, v, w) : x > 0, y \geq 0, v \geq 0, w > 0\}$$

is positively invariant to the model (3). Further, the set $\Omega = \{(x, y, v, w) \in R_+^4 : x + y \leq \lambda/d, v \leq kaM/\gamma, w \geq 0\}$ is positively invariant to the model (3).

In fact, the first two equations of model (3) gain

$$\mathrm{d}(x + y)/\mathrm{d}t = \lambda - dx - ay - myw.$$

By $a > d$, on the set R_+^4, $\mathrm{d}(x + y)/\mathrm{d}t \leq \lambda - d(x + y)$ is gained, then it follows that $\lim\ \sup_{t\to\infty}(x + y) \leq \lambda/d$. That is, for any given positive number ε, there is a positive number T. So when $t > T$, we have $x + y < \lambda/d + \varepsilon =: M$. When $y < M$, the third equation of the model (3) gains $\mathrm{d}v/\mathrm{d}t \leq kaM - \gamma v$. So

$$\lim \sup_{t\to\infty} v \leq kaM/\gamma.$$

3 The Existence of Equilibria

Calculation shows that the model (3) always has the disease-free equilibrium $E_0(x_0, 0, 0, w_0)$, where $x_0 = \lambda/d, w_0 = \lambda_1/d_1$.

The endemic equilibrium of the model (3) $E^*(x^*, y^*, v^*, w^*)$ is determined by equations

$$\begin{cases} \lambda - dx - \beta xv = 0 \\ \beta xv - ay - myw = 0 \\ kay - \gamma v = 0 \\ \lambda_1 + f(w)y - d_1 w = 0 \end{cases} \tag{4}$$

From the first and third equations of Eqs. (4), we can obtain $x = \lambda/(d + \beta v)$,

$y = \gamma v/ka$. For $v \neq 0$, substituting it into the second and fourth equation gives

$$v = \frac{\lambda ka}{\gamma(a+mw)} - \frac{d}{\beta} =: h_1(w), v = \frac{ka(d_1 w - \lambda_1)}{\gamma f(w)} =: h_2(w).$$

For $v^* > 0$, in the $h_1(w)$, w^* must satisfy $w^* < \lambda\beta ka/md\gamma - a/m$, in the $h_2(w)$, w^* must satisfy $w^* > \lambda_1/d_1$. When $\lambda_1/d_1 < \lambda\beta ka/md\gamma - a/m$, the model (3) has the endemic equilibrium $E^*(x^*, y^*, v^*, w^*)$. It is easy to see when $\lambda_1/d_1 < \lambda\beta ka/md\gamma - a/m$, $f'(w) \geq 0$, and $[f(w)/w]' \leq 0$, $h_1(w)$, is the monotone decreasing function on w, $h_2(w)$, is the monotone increasing function on w. Denote $w_1 = \lambda_1/d_1$, $w_2 = \lambda\beta ka/md\gamma - a/m$, $F(w) = h_1(w) - h_2(w)$.

It is easy to see that $F'(w) \leq 0$, also because

$$F(w_1) = h_1(w_1) - h_2(w_1) = h_1(w_1) > 0$$
$$F(w_2) = h_1(w_2) - h_2(w_2) = -h_2(w_2) < 0$$

So equation $F(w) = 0$ has a unique root w^*, substituting $w = w^*$ into $h_1(w)$ gives v^*, and then gains x^*, y^*. By the method of the next generation matrix proposed in [14], the basic reproduction number of the model (3) is given by

$R_0 = \lambda\beta ka/[\gamma d(m\lambda_1/d_1 + a)]$, that is $R_0 = \beta kax_0/[\gamma(mw_0 + a)]$, and then
$$\lambda_1/d_1 < \lambda\beta ka/md\gamma - a/m \Leftrightarrow R_0 > 1.$$

Theorem 1 The model (3) always has the disease-free equilibrium $E_0(x_0, 0, 0, w_0)$.

When $R_0 > 1$, besides the disease-free equilibrium E_0, the model (3) also has an endemic equilibrium $E^*(x^*, y^*, v^*, w^*)$, where $x_0 = \lambda/d$, $w_0 = \lambda_1/d_1$; $x^* = \lambda/(d + \beta v^*)$, $y^* = \gamma v^*/ka$, $v^* = \lambda ka/\gamma(a + mw^*) - d/\beta$ and w^* is the positive root of equation $h_1(w) = h_2(w)$.

4 The Global Stability of Equilibria

The model (3) has the global stability of equilibria, as follows

Theorem 2 For the model (3), the disease-free equilibrium E_0 is globally stable on the set Ω when $R_0 \leq 1$; the endemic equilibrium E^* is globally stable in the interior of the set Ω when $R_0 > 1$.

Proof We first prove the global stability of the disease-free equilibrium E_0 as $R_0 \leq 1$, since $x_0 = \lambda/d, w_0 = \lambda_1/d_1$, then the model (3) can be equivalent to deformation as

$$\begin{cases} dx/dt = x[\lambda(1/x - 1/x_0) - \beta v] \\ dy/dt = \beta(x - x_0)v - ay - my(w - w_0) \\ \qquad + \beta x_0 v - (a + mw_0)y \\ dv/dt = kay - \gamma v \\ dw/dt = \lambda_1 + f(w)y - d_1 w = f(w)[d_1(w_0 - w)/f(w) + y] \end{cases} \tag{5}$$

Define a Lyapunov function

$$L_1 = \int_{x_0}^{x} \frac{\theta - x_0}{\theta} d\theta + y + \frac{a + mw_0}{ka} v + m \int_{w_0}^{w} \frac{\theta - w_0}{f(\theta)} d\theta,$$

then the derivative of L_1 along solutions of system (5) is given by

$$\begin{aligned} dL_1/dt|_{(5)} &= \lambda(2 - x_0/x - x/x_0) - \{(a + mw_0)\gamma/ka - \beta x_0\}v - md_1(w - w_0)^2/f(w) \\ &= \lambda(2 - x_0/x - x/x_0) - \beta x_0(1/R_0 - 1)v - md_1(w - w_0)^2/f(w). \end{aligned}$$

Therefore, $dL_1/dt|_{(5)} \leq 0$ for $R_0 \leq 1$. The disease-free equilibrium E_0 is globally stable on the set Ω when $R_0 \leq 1$ by LaSalle's invariable principle [15].

Second, we prove that the endemic equilibrium E^* is the global stability. Since x^*, y^*, v^*, w^* satisfy the Eq. (4), the model (3) can be equivalent to deformation as

$$\begin{cases} dx/dt = x[\lambda(1/x - 1/x^*) - \beta(v - v^*)] \\ dy/dt = y[\beta(xv/y - x^*v^*/y^*) - m(w - w^*)] \\ dv/dt = kav(y/v - y^*/v^*) \\ dw/dt = f(w)[\lambda_1(1/f(w) - 1/f(w^*)) + (y - y^*) - d_1(w/f(w) - w^*/f(w^*))] \end{cases} \tag{6}$$

Define a Lyapunov function

$$L_2 = \int_{x^*}^{x} \frac{\theta - x^*}{\theta} d\theta + \int_{y^*}^{y} \frac{\theta - y^*}{\theta} d\theta + \frac{\beta x^* v^*}{kay^*} \int_{v^*}^{v} \frac{\theta - v^*}{\theta} d\theta + m \int_{w^*}^{w} \frac{\theta - w^*}{f(\theta)} d\theta$$

then the derivative of L_2 along solutions of system (6) is given by

$$\begin{aligned} dL_2/dt|_{(6)} &= dx^*(2 - x^*/x - x/x^*) + \beta x^* v^*(3 - x^*/x - yv^*/y^*v - xy^*v/x^*yv^*) \\ &\quad + \lambda_1 m(w - w^*)[1/f(w) - 1/f(w^*)] - md_1(w - w^*)[w/f(w) - w^*/f(w^*)]. \end{aligned}$$

Since $f'(w) \geq 0$, $(w - w^*)[1/f(w) - 1/f(w^*)] \leq 0$; since $[f(w)/w]' \leq 0$, $(w - w^*)[w/f(w) - w^*/f(w^*)] \geq 0$. According to the relation between the arithmetic and the associated geometric means, we have $dL_2/dt|_{(6)} \leq 0$ and the equality holds if and only if $x = x^*, w = w^*, y/y^* = v/v^*$. The model (3) has the largest invariant set on the set

$$\{(x, y, v, w) \in \Omega : x = x^*, w = w^*, y/y^* = v/v^*\},$$

it is the singleton $\{E^*\}$. Therefore, according to LaSalle's invariable principle [15], the endemic equilibrium E^* of the model (3) is globally stable in the feasible region when it exists.

Acknowledgements We are very grateful to the anonymous referees for their valuable comments and helpful suggestions, which led to a great improvement of the original manuscript. This work was supported by the Research Fund of Department of Basic Sciences at Air Force Engineering University (JK2019109).

References

1. Perelson, A.S., Kirschner, D.E., Deboer, R.: Dynamics of HIV infection of CD4+T cells. Math. Biosci. **114**(1), 81–125 (1993)
2. Perelson, A.S., Nelsonp, W.: Mathematical analysis of HIV-1 dynamics in vivo. Siam Rev. **41** (1), 3–44 (1999)
3. Ma, Z.: Mathematical Modelling and Study of Population Ecology. Anhui Education Press, Hefei, China (1996). (in Chinese)
4. Wang, H., Zhu, H.: Stability analysis of the time delay HIV infection system with saturated CTL immune response. Math. Biosic. **27**(2), 274–282 (2012)
5. Mclean, A.R., Emery, V.C., Webster, A., Griffiths, P.D.: Population dynamics of HIV within an individual after treatment with zidovudine. AIDS **5**, 485–489 (1991)
6. Yan, Y., Wang, W.: The global dynamic state of HIV infection model considering CTL immunisation. J. Southwest Univ. Nat. Sci. 33(5), 655–662 (2011)
7. Tarfulea, N., Blink, A., Nelson, R., Turpin, D.: A CTL-inclusive mathematical model for antiretroviral treatment of HIV infection. Int. J. Biomath. **4**(1), 1–22 (2011)
8. Ma, Z., Zhou, Y.: Qualitative and Stability Methods of Ordinary Differential Equations. Science Press (2007)
9. Li, J., Wang, F., Ma, Z.: A global analysis of a class of isolated infectious disease models. J. Eng. Math. **1**(22), 20–24 (2005)
10. Song, X., Li, J.: Global stability of an SEIR epidemic model with nonlinear incidence. J. Eng. Math. **33**(2), 175–183 (2016)
11. Now, K.M.A., Ban Gham, C.R.M.: Population dynamics of immune responses to persistent viruses. Science **272** (5258), 74–79 (1996)
12. Now, K.M.A., May, R.M.: Virus Dynamics: Mathematical Principles of Immunology and Virology, pp. 52–66. Oxford University Press, New York (2000)
13. Perelson, A.S., Nelson, P.W.: Mathematical analysis of HIV-1 dynamics in vivo. SIAM Rev. **4**(1), 3–44 (1999)
14. Van den Driessche, P., Watmough, J.: Reproduction numbers and sub-threshold endemic equilibria for compartmental of disease transmission. Math. Biosci. **180**, 29–48 (2002)
15. Lasalle, J.P.; The stability of dynamical systems. In: Regional Conference Series in Applied Mathematics Philadelphia, SIAM (1976)

Research on Construction of the Software Technology Profession in Vocational Institutes Based on the External Vision

Dengfeng Xiong, Haizhen Zhou and Lihong Huang

Abstract Professional construction of vocational education is the key to determine the quality of talent cultivation and an important symbol to determine the level of institute running. The target and vision of professional construction shall focus on external social market, enterprises, and labor contributions, instead of being limited inside the vocational institutes. Only in this way can the professional construction resources be fully utilized and can the professional construction be turned from the extensive development pattern to economical development pattern. In this paper, through taking the professional construction of software technology as an example, the detailed implementation method and process of professional construction in vocational institutes based on the external vision are illustrated from four aspects such as professional construction thoughts, teaching team construction, on-campus and off-campus practice base construction, and the construction of social service ability.

Keywords Professional construction · Vocational institutes · Social marketing · Talents training · Teaching team

1 Introduction

Vocational education is one of the three major education sectors in our country. Profession is the interface between vocational education and market demands and is the basic education unit for higher vocational institutes to serve the society. Professional construction is the foundation and leading part of the teaching work in higher vocational institutes and, simultaneously, is also the key to determine the quality of talent cultivation and an important symbol to determine the level of

D. Xiong (✉) · H. Zhou · L. Huang
Hunan Software Vocational Institute, Hunan Xiangtan, China
e-mail: 187244191@qq.com

V. Jain et al. (eds.), *Recent Trends in Intelligent Computing, Communication and Devices*, Advances in Intelligent Systems and Computing 1031,
https://doi.org/10.1007/978-981-13-9406-5_2

institute running. As a comprehensive, systematic, and long-term fundamental work for higher vocational institutes, professional construction is not only one of the key links for teaching works of schools to adapt to the market actively and flexibly, but also the concentrated expression of the level of institute running. Therefore, it is of great significance to pay adequate attention to the professional construction in higher vocational institutes in the time of current great development of higher vocational education, especially in the time of current great development of the connotation of higher vocational education.

The professions of vocational institutes are the main source to supply all kinds of public service products of education to the social markets. Comparing with the compulsory education and general education, the public service products of education provided by vocational institutes have strong productivity, directness, and practicality, and therefore, the product attributes of profession determine the destination of professional construction, which shall focus on the contribution made for the external social markets other than the vocational institutes. In recent years, in the process of constructing demonstrative (backbone) vocational institutes in Hunan Province, the department of software engineering in Hunan Software Vocational Institute, has explored and implemented the professional construction mode actively, and in this paper, through taking the professional construction of software technology as an example, the detailed implementation method and process of professional construction in vocational institutes based on the external vision are illustrated from four aspects such as professional construction thoughts, teaching team construction, on-campus and off-campus practice base construction, and the construction of social service ability.

2 Professional Construction Ideas Based on the External Vision

The idea is the basis and guide to the action; therefore, when thinking about the construction and development strategy of software technology profession, we must first make sure the guiding ideology of professional construction. Drucker said [1]: “Professional construction exists for the result, which must be started with the expect result, and must be organized the resources to achieve the results, and should be achieved the goal to product the results, not only on the plan but also on the action.”

According to the opinion on professional construction proposed by Peter Ducker, if the focus of professional construction is limited in the institutes only, for example, if we only concentrate on the number of software training bases and double-qualified teachers in the process of professional construction of software technology, then it means that the professional construction cares about hard-working only, instead of caring about the external contribution and acceptance by outside social market. Obviously, adopting such kind of professional

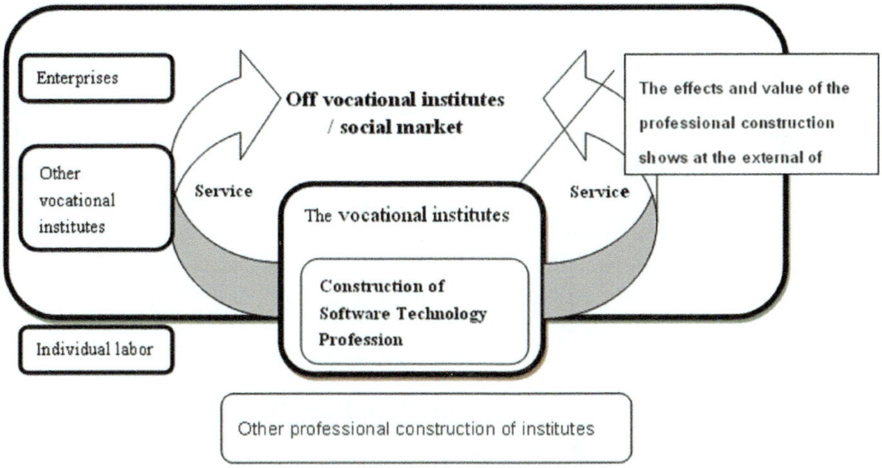

Fig. 1 Principle diagram of the software technology professional construction for the vocational institutes under the external vision

construction, it not only is meaningless but also obeys our original intention to do the professional constructions.

As illustrated in Fig. 1, in the process of the professional construction of software technology, we shall focus on contribution to the social market (the external of vocational institutes) firstly, and ask and answer the following questions in the initial period of the professional construction: (1) What kinds of service can we offer for enterprises? Obviously, the answer is to provide them with qualified personnel and software products to support the regional companies focusing on core industries; (2) What kinds of service can we offer for other vocational institutes? Obviously, the answer is to provide them with guidance and resources of the professional construction, and sharing and optimization of the professional software and hardware resources; (3) What kinds of service can we offer for individual labor? Obviously, the answer is to provide a satisfying job for them.

According to the above description, it is obvious that focusing on the contribution is more important than the construction method. However, from technology to idea, efficiency to achievement, the self-examination of the professional teaching team is often needed. On the basis of self-examination, we proposed the following two questions in the process of professional construction further: (1) Why shall we implement the professional construction? (2) What contributions can the professional construction provide for the social market and other service objects?

The above self-examination questions are not only straightforward but also formal. After obtaining the answers, then we shall think about the specific contents of the professional construction, and the major and value expression for students in the social markets. More importantly, it requires us to take responsibility instead of executing orders to satisfy the school or school leaders only. If we focus on the contribution of professional construction, then we shall not only attach importance to the method but also include the targets and results in the process of professional constructions.

Based on the above ideas of professional construction, in the process of professional construction of the software technology, through focusing on the social market and the external needs outside of the vocational colleges, not only the resources can be allocated optimally, the massive resources investment, which leads to extensive development pattern, can be deserted in the process of the professional construction, but also the waste of resources on the ineffective construction can be avoided effectively. Consequently, it makes the professional development and construction quality having a better promotion space.

3 Versatile Teaching Team Construction Based on the Strengths of Teachers

Teachers are the first resource in vocational colleges, which determine the level and the quality of personnel, and also are the main force for training high-skilled personnel. Constructing high-level and professional teaching team is an important guarantee for improving the teaching quality of higher vocational institutes.

Through visiting and investigation, the authors discover that in the professional construction process of software technology, the teaching team has the following major problems: (1) lack of high-level professional leaders; (2) the experienced of teachers is very single, they lack of actual production experience, and the double type [2] structure unreasonable, part-time teachers lack of stability; (3) the professional of teachers is awareness and lack of cohesion. It is easy to find that this series of problems in most of the questions are put forward based on the teacher individual of the teaching team. Supposing constructing professional teaching team in this foothold, it is necessary to build a goal that puts each teacher in the team to versatile personnel. Obviously, this kind of versatile teacher exists rarely. The introduction of this kind of versatile teacher is very difficult, and it is far harder to cultivate either.

From the perspective of the social market, it is not for one teacher to supply of education provided by the public service product, but the result of a teacher professional team. In the building of professional team, we shall take hold in the team. In the process of building a professional teaching team, the strengths of teacher should be developed, if only to seize weaknesses and the shortcomings of teachers in the process of building and that must build a bad team. Therefore, to achieve the goal, it must be based on the team, with the teachers' institute.

It is easier and effective to construe a versatile teacher team than building a number of teams of individual teachers with versatile teachers. In the process of constructing versatile teaching team of software technology based on giving full play to the strengths of teachers, to the teachers that have wealth of teaching experience and strong research capabilities cannot simply ask them to improve the practical ability. Information systems project manager for the social enterprise recruitment, senior software engineer, or programmer cannot concentrate on their research capacity which requires strengthening teaching.

In the process of actual professional construction of software technology, we combine their strengths, so they are to serve as teachers teach the theory and skills training respectively, and according the ability to develop course materials, based on the work process libraries design case, etc. projects, and achieved good results. After nearly three years of construction, the current team of Software Technology Teaching Software College has become one of the best teaching teams and has won a series of honors and awards, such as: (1) in the college teaching contest, teaching team won team recognition 2–5 times of personal recognition; (2) technology professional teaching team of software is not only the best student evaluation of teaching team, but also gain levels research projects, construction quality up to the best team, among them, the school more than 60% provincial projects, is chaired by the teaching team of software technology professional teachers; (3) technology professional teaching team of software or the entire school more than students in "Challenge Cup" and other kinds of provincial large-scale professional skills in the game the only award or winning the largest number of specialized teaching team.

Therefore, make good use of these advantages which can bring real opportunities for teaching team. Teachers should make full use of the advantages, because it is the only purpose of existence of team. To know that everyone must have many shortcomings and weaknesses, and the shortcomings and weaknesses are almost impossible to change. But we can try to avoid the disadvantage. Our task is to make full use of every teacher's strengths to complete the task.

4 Social Service Ability Construction Based on Demands of Social Markets

Social service is one of the most core functions for vocational institutes. It is the extension of higher education teaching and research functions. Higher vocational college is the most direct reflection and embodiment of core competence. Meaning of social service functions' vocational colleges has broad and narrow sense. Generalized social services refer to the vocational colleges' social functions and roles. It includes training, development of science and technology as well as directly to the social services. Narrow social services, it is to point to in vocational institutes in teaching and scientific research task outside often for social development in various forms, concrete, service nature of the activities. Software technology professional social services' main task is to provide technology applied to regional enterprises and high-skilled personnel training and training. Provide software technology promotion, service and innovation, the implementation of advanced culture and the spread of radiation, has distinct regional and industry characteristics [3].

On the outside of the focus on vocational institutes and under the horizon of social service ability construction, we combined the technology of software professional social market demand, integrated schools and gold overseas investment holding group, Tai Fen technology co., LTD., the United States shipping magnate clothing co, LTD, Changsha general agent of resources, cooperation project implementation of the "school, the student and the enterprise that three win-win mode of cooperation and practice" project, on the basis of independent innovation laboratory, and the premise of area enterprise needs, developed a guessing game version of the English words of a Chinese Odyssey, mobile games, mobile ordering system and other software products. These projects are in vocational institutes' external social market demand and resources as the foundation; with the help of the software industry characteristics and professional advantages, less investment, quick effect, social services, benefit is obvious.

All in all, whether it is a special social service or general social services, promote professional social service ability is the most effective strategy of vocational institutes, must focus on external horizon of the social service demand, also coupled with the professional resources.

5 Construction of School Training Bases Integrated Into Occupation Job Elements and the Construction of the Public Off-Campus Training Bases

The construction of training bases is an important part of the professional construction, which is also necessary to the talents training. The bases of software engineering include the school training bases and the off-campus training bases.

(1) There are mainly three kinds of patterns of the construction of school training bases, including curriculum content directed, the physical environment simulation-directed and working process simulation directed. The bases of curriculum-content- directed advocate designing needs of course practice teaching content, there are usually multiple training rooms, simulated and real equipment, which can meet the professional skills training; The bases of the physical environment simulation-directed are according to the workshop space layout, environment decoration, which reflect a real scene in training; working process simulation model is in accordance with the working process of the enterprise to organize the teaching process design training base, to enable students to gain work experience in practice.
    Curriculum-content-directed model is typical internal vocational training base for college construction methods, and the physical environment simulation mode is a training base construction between internal and external way between the two methods, but the process simulation model focuses on the training base outside the sight of the building vocational colleges underway. By the requirements of software product development, we transform the classroom into a software

research and development, hardware development, technical support and maintenance department, testing department and meeting rooms, and configure the server for each department, material chips, test equipment, production tools, and real-work scenarios in order to achieve work task-oriented teaching reform, where students finish both inside and outside work tasks.

(2) The off-campus training bases give services of post-practice to students. In the selection and construction of the off-campus training bases, enterprise can represent the characteristics of the industry, can provide many internship positions, internships content with the scientific and technological content, but also to consider the students' lives, security, and other issues, the school shall provide teaching equipment and the establishment work classroom.

The construction of vocational training bases in the ways of equipment selection, physical environment, instructional design, and other aspects of the project should reflect the typical tasks of enterprises, should reflect the technical content of the training program, should reflect the development direction of new technologies, should reflect the business professional atmosphere, which is both different from the experimental base research university but also different from the general vocational education training base. Meanwhile, construction of bases must consider the teaching function and the training, vocational skills identification, and application of technology-oriented research and development and other social services outside the school.

With a view to the development and reform process of the social market (off the vocational colleges) training base, we use the training base of the establishment of public schools instead of each individual's training base construction. It allows the service to the profession itself, and other vocational colleges, at the same time, can build a low level to avoid duplication. Plans are made for future construction of a number of skills-converted center which are built by government, businesses, schools, open to the public, all sharing resources, such centers are skills training, technology incubators, which will become the best combination of schools and businesses.

In addition, the construction of school training bases focusing on external sight vocational colleges and public training base can be substantial reform of the professional development. Construction makes further improvement in saving energy and cost, which is the only way that the vocational education can develop rapidly.

Acknowledgements The 12th five-year education scientific planning project of Hunan Province: The study on the quality evaluation mechanism of professional construction in higher vocational colleges based on the "external contribution" of the major, the item number: XJK014QZY003.

References

1. Drucker, P.: Effective Managers. Machinery Industry Press (2009)
2. The Double Type: Is Means the Double Type Teachers, Who Have the Practice and Teaching 2 Experience Meantime
3. Jing: My professional hospital association and regional industrial structure analysis. Panyu Polytechnic College. **2**, 36–39 (2008). http://www.cnki.net/

Numerical Study on the Expanded Mixed Covolume Method for One-Dimensional Sobolev Equation

Na Li

Abstract We continue on studying the expanded mixed covolume method (EMCM) to solve Sobolev equation in this paper. The conservation laws of properties of Sobolev equation are emphasized. By designing several test functions, the accuracy, convergence, and robustness of EMCM are studied numerically. The results show that EMCM is an easy-to-implement technology that provides a precise and robust numerical solution while maintaining conservation.

Keywords Sobolev equation · Mixed covolume method · Conservation laws · Numerical experiment

1 The Expanded Mixed Covolume Method

Nonlinear Sobolev equations are widely used in mathematical physics problems, such as the percolation theory of fluid percolating through rock fissures, the theory of moisture transfer in soil, and the heat conduction problem with many mediums (see [1, 2, 3, 4, 5, 6]). Therefore, many scholars study this kind of equation (see [7, 8]).

In this paper, we will propose an EMCM-based numerical scheme for the initial boundary value problem of the following types of Sobolev equations.

$$\begin{cases} u_t + f(u)_x - \mu u_{xxt} = 0, & (x,t) \in I \times (0,T] \\ u(X_1,t) - u(X_2,t) = 0, & t > 0 \\ u(x,0) = u_0(x), & x \in I \end{cases} \tag{1}$$

N. Li
School of Data Science and Computer Science, Shandong Women's University, Jinan 250300, Shandong, China

N. Li (✉)
School of Mathematics and Physics, University of Science and Technology Beijing, Beijing 100083, China
e-mail: dajiena2002@163.com

V. Jain et al. (eds.), *Recent Trends in Intelligent Computing, Communication and Devices*, Advances in Intelligent Systems and Computing 1031,
https://doi.org/10.1007/978-981-13-9406-5_3

where $I = [X_1, X_2]$. Here, μ is a given positive constant, and $u_0(x)$ is a prescribed function.

1.1 The Semi-discrete Scheme

Let $w = u_t$, $p = w_x$, $q = u_x$, then Eq. (1) is equivalent to the following first-order differential equations with respect to the unknowns u, q, w, and p:

$$u_t = w, \quad q_t = p, \quad w - \mu p_x = -(1+u)q, \quad p - w_x = 0, \quad \forall (x,t) \in I \times (0,T] \tag{2}$$

with initial value $u(x,0) = u_0(x), q(x,0) = u_{0,x}(x), \quad x \in I$.

The weak form can be obtained by multiplying the above equations by the test function v:

$$\begin{aligned}&(u_t, v) = (w, v), (q_t, v) = (p, v), (p, v) + a(w, v) = 0,\\&(w, v) + \mu a(p, v) = -(f_u'(u)q, v), \quad \forall v \in H_0^1,\end{aligned} \tag{3}$$

where $a(u,v) = \sum_{j=1}^{n-1} [-(u, v_x)_{I_j^*} + (u,v)|_{x_{j-\frac{1}{2}}}^{x_{j+\frac{1}{2}}}]$.

(u_h, q_h, w_h, p_h) are used to approximate (u, q, w, p), the expanded mixed covolume scheme is obtained: find $(u_h(\cdot, t), q_h(\cdot, t), w_h(\cdot, t), p_h(\cdot, t)) : [0, T] \to Q_h \times Q_h \times Q_h \times Q_h$, such that

$$\begin{aligned}&(u_{ht}, v_h) = (w_h, v_h), (q_{ht}, v_h) = (p_h, v_h), (p_h, v_h) + a(w_h, v_h) = 0,\\&(w_h, v_h) + \mu a(p_h, v_h) = -(f_u'(u_h)q_h, v_h), \quad \forall v_h \in R_h,\end{aligned} \tag{4}$$

where v_h can take different values in different equations.

Theorem 1 *The expanded mixed covolume scheme has the unique solution.*

Theorem 2 *Suppose (u, qw, p) are the solutions of* (3) *and (u_h, q_h, w_h, p_h) are the solutions of* (4)*, respectively. Assume that $(u, q, w, p) \in H_0^1 \times H_0^1 \times H_0^1 \times H_0^1$, the prior error estimates for the semi-discrete scheme are estimated as follows*:

$$\begin{aligned}&\|u - u_h\|_{L^\infty(0,T;H^0(I))} + \|q - q_h\|_{L^\infty(0,T;H^0(I))}\\&\leq \|(u - u_h)(0)\|_0 + \|(q - q_h)(0)\|_0 + Ch\Big(h\|u\|_{L^1(0,T;H^2(I))}\\&+ h\|q\|_{L^1(0,T;H^2(I))} + \|w\|_{L^1(0,T;H^2(I))} + \|p\|_{L^1(0,T;H^2(I))}\Big)\end{aligned}$$

1.2 The Fully Discrete Scheme

Let $0 = t_0 < t_1 < \cdots < t_N < T$ be a subdivision of the region $[0, T]$, $\Delta t_k = t_k - t_{k-1}$ $(k = 1, 2, \ldots, N)$, and $(u_h^k, q_h^k, w_h^k, p_h^k)$ approximate (u, q, w, p) at t_k. Then, on the basis of the semi-discrete format, using the forward difference instead of the time derivative, the fully discrete scheme is obtained: find $(u_h^k, q_h^k, w_h^k, p_h^k)$ satisfy

$$\begin{cases} \left(\frac{u_h^{k+1}-u_h^k}{\Delta t}, \pi_h^* v_h\right) = \left(w_h^k, \pi_h^* v_h\right), & \forall v_h \in R_h \\ \left(\frac{q_h^{k+1}-q_h^k}{\Delta t}, \pi_h^* v_h\right) = \left(p_h^k, \pi_h^* v_h\right), & \forall v_h \in R_h \\ \left(p_h^k, \pi_h^* v_h\right) + a\left(w_h^k, \pi_h^* v_h\right) = 0, & \forall v_h \in R_h \\ \left(w_h^k, \pi_h^* v_h\right) + \mu a\left(p_h^k, \pi_h^* v_h\right) = -\left(f_u'(u_h^k) q_h^k, \pi_h^* v_h\right), & \forall v_h \in R_h \end{cases} \tag{5}$$

Theorem 3 *Let* $(u_h^k, q_h^k, w_h^k, p_h^k)$ *be the solution of* (5), *other conditions are the same as Theorem* 2, *the prior error for the fully discrete scheme is estimated:*

$$\begin{aligned} &\max_{1\le k\le N}\|u^k - u_h^k\|_0 + \max_{1\le k\le N}\|q^k - q_h^k\|_0 \\ &\le C\Big[\|(u-u_h)(0)\|_0 + \|(q-q_h)(0)\|_0 + h\Big(h\|u\|_{L^\infty(0,T;H^2(I))} + h\|q\|_{L^\infty(0,T;H^2(I))}\Big) \\ &+ \|w\|_{L^\infty(0,T;H^2(I))} + \|p\|_{L^\infty(0,T;H^2(I))}\Big] + \Delta t\left(\left\|\frac{\partial^2 u}{\partial t^2}\right\|_{L^1(0,T;H^0(I))} + \left\|\frac{\partial^2 q}{\partial t^2}\right\|_{L^1(0,T;H^0(I))}\right). \end{aligned}$$

2 Numerical Invariants

In this section, we will discuss the conservation of numerical format on the basis of the EMCM method. For mass, momentum, and energy, we introduce the following three quantities [9].

$$I_1 = \int_{-\infty}^{+\infty} u_h \mathrm{d}x = \sum_{j=1}^{N} \int_{I_j} u_h^k \mathrm{d}x, \tag{6}$$

$$I_2 = \int_{-\infty}^{+\infty} \left[u_h^2 + \mu u_{hx}^2\right] \mathrm{d}x = \sum_{j=1}^{N} \int_{I_j} \left[(u_h^k)^2 + \mu (u_{hx}^k)^2\right] \mathrm{d}x, \tag{7}$$

$$I_3 = \int_{-\infty}^{+\infty} \left[u_h^3 + 3u_h^2\right] \mathrm{d}x = \sum_{j=1}^{N} \int_{I_j} \left[(u_h^k)^3 + 3(u_h^k)^2\right] \mathrm{d}x, \tag{8}$$

We test the conservation of numerical schemes in the simulation by continuously monitoring the fluctuations of the three quantities I_1, I_2, and I_3 over time (see [10, 11, 12, 13, 14, 15]).

2.1 A Single Solitary Wave

In this section, we consider Eq. (1) by using the scheme given

$$\begin{cases} u_t + u_x + uu_x - u_{xxt} = 0, & x \in (-80, 120), \quad t > 0 \\ u(-80, t) = u(120, t) = 0, & t > 0 \\ u(x, 0) = 3\alpha \sec h^2(\beta x), & x \in (-80, 120) \end{cases} \tag{9}$$

where α is a constant. The exact solution is $u(x,t) = 3\alpha \text{sech}^2 [0.5 * \sqrt{\alpha(1+\alpha)}(x-t)]$.

We choose $\alpha = 0.03$, $h = \frac{1}{32}$, $\tau = \frac{1}{512}$. From Fig. 1, we can see u_h and q_h approach u and q very well. The discrete L_2 error norm is defined by $L_2 = [h\sum_{j=1}^{N}\left[\left|u(x_j) - u_h(x_j)\right|^2\right]^{\frac{1}{2}}\sum$.We use this norm to calculate the error between the numerical solution. The error norm is set in Tables 1 and 2. The results show that the error of the method is small and the position of the solution can be predicted accurately, and the movement of the isolated wave is successfully simulated.

In the simulation of isolated wave motion, the conservation of numerical algorithms is verified by monitoring the invariants. The cross-sectional views of the solitary waves (from $t = 0$ to $t = 10$) shown in Figs. 2 and 3 are almost indistinguishable from the numerical solution and the exact solution. The simulations are carried out at $t = 10$, and the results are recorded in Table 3, indicating that the method has good conservation. The invariants are $I_1 = 0.021094114$, $I_2 = 0.072714717$ and $I_3 = 0.003854728$. These changes are less than 10^{-5}, 10^{-3}, 10^{-5}, respectively. The method produces very satisfactory conservation properties.

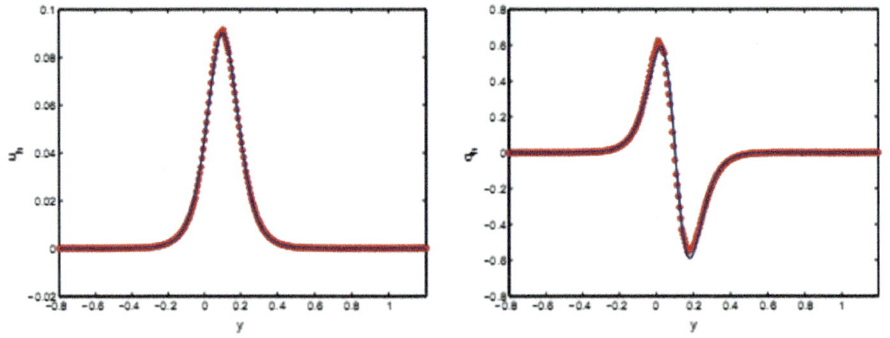

Fig. 1 From the two figures, we can find the differences between the approximate solutions and the exact solutions are satisfactory small at $t = 10$

Table 1 $\|u - u_h\|_{0,h}$

$\tau/h^2 = 1/4$	$t = 2.5$	$t = 5.0$	$t = 7.5$	$t = 10.0$
$h = 1/2^2$	$1.59953846e^{-2}$	$1.96626929e^{-2}$	$2.26687655e^{-2}$	$2.51216877e^{-2}$
$h = 1/2^3$	$5.88255975e^{-3}$	$4.63824084e^{-3}$	$3.48869770e^{-3}$	$5.18384102e^{-3}$
$h = 1/2^4$	$1.24794477e^{-3}$	$1.093244833e^{-3}$	$1.22495810e^{-3}$	$1.49698996e^{-3}$
$h = 1/2^5$	$2.92617878e^{-4}$	$2.44878295e^{-4}$	$2.84272002e^{-4}$	$3.19770283e^{-4}$

Table 2 $\|q - q_h\|_{0,h}$

$\tau/h^2 = 1/4$	$t = 2.5$	$t = 5.0$	$t = 7.5$	$t = 10.0$
$h = 1/2^2$	$4.28431664e^{-2}$	$9.53674716e^{-2}$	$1.53053619e^{-1}$	$2.18724075e^{-1}$
$h = 1/2^3$	$2.10506796e^{-2}$	$3.31288155e^{-2}$	$7.55683036e^{-2}$	$1.15615453e^{-2}$
$h = 1/2^4$	$1.05063608e^{-2}$	$1.95153991e^{-2}$	$2.81032398e^{-2}$	$3.68007007e^{-2}$
$h = 1/2^5$	$5.05371068e^{-3}$	$1.00742452e^{-2}$	$1.50239488e^{-2}$	$1.98651939e^{-2}$

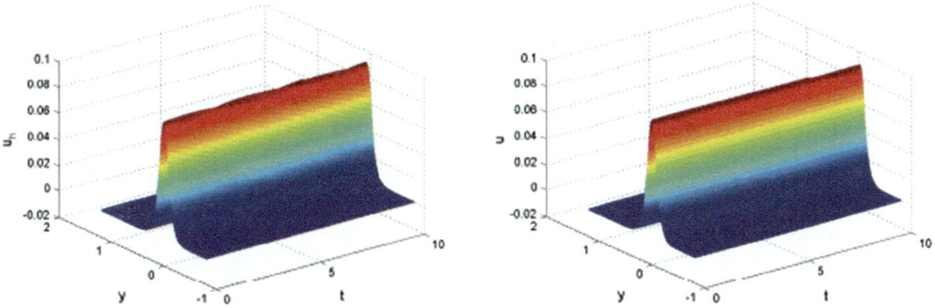

Fig. 2 Motion of u_h and u at various times

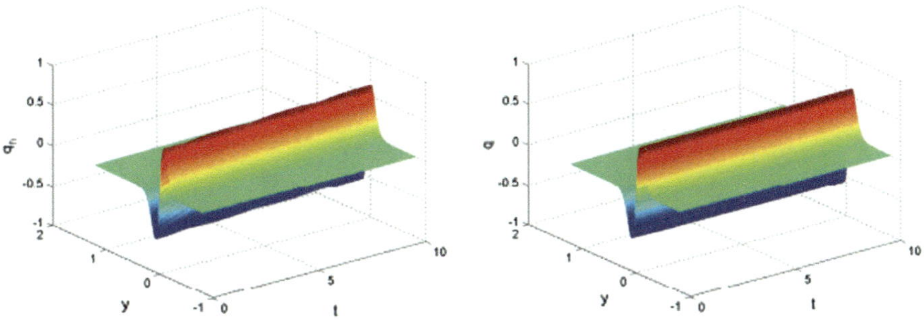

Fig. 3 Motion of q_h and q at various times

Table 3 Invariants and error norms of solitary waves at different time points

Time	I_1	I_2	I_3
$t = 0.0$	0.0210940495713021	0.0725653039178565	0.0038508536573052
$t = 2.5$	0.0210946696514096	0.0726296237060465	0.0038563291753335
$t = 5.0$	0.0210963894822037	0.0726839687502454	0.0038695202055441
$t = 7.5$	0.0210988602228716	0.0727211788998445	0.0038878159110631
$t = 10.0$	0.0211016148536019	0.0727345430265978	0.0039078013035750

2.2 *Interaction of Two Solitary Waves*

Secondly, we consider Eq. (1) with the initial boundary conditions

$$\begin{cases} u(-120,\, t) = u(180,\, t) = 0, & t > 0 \\ u(x,0) = \sum_{j=1}^{2} 3\alpha_j \mathrm{sech}^2\left[0.5 * \sqrt{\alpha_j/(1+\alpha_j)}(x - x_j)\right], \end{cases} \tag{10}$$

It can be seen from the initial condition that there are two solitary waves at the initial moment, one with amplitude $3\alpha_1$ originally located at $x = x_1$ and another with amplitude $3\alpha_2$ at $x = x_2$. Waves with larger amplitudes are known to be faster than those with smaller amplitudes. Therefore, let $\alpha_1 > \alpha_2$ and $x_1 < x_2$ to ensure that the interaction between two solitary waves increases over time. All calculations in the simulation are based on the parameters $\alpha_1 = 0.03$, $x_1 = -18$, $\alpha_2 = 0.01$, $x_2 = -58$, $h = 0.03$ and $\Delta t = 0.1$ in the range $-120 \leq x \leq 180$. The experiments are conducted from $t = 0$ to $t = 10$ to facilitate the interaction, as shown in Fig. 4.

The values of invariants at different times are listed in Table 4. They are about $I_1 = 0.033141406$, $I_2 = 0.070279601$, and $I_3 = 0.005027136$. The changes are less than 10^{-5}, 10^{-2}, 10^{-4}, respectively. The invariant value can basically be considered to remain unchanged while the computer is running.

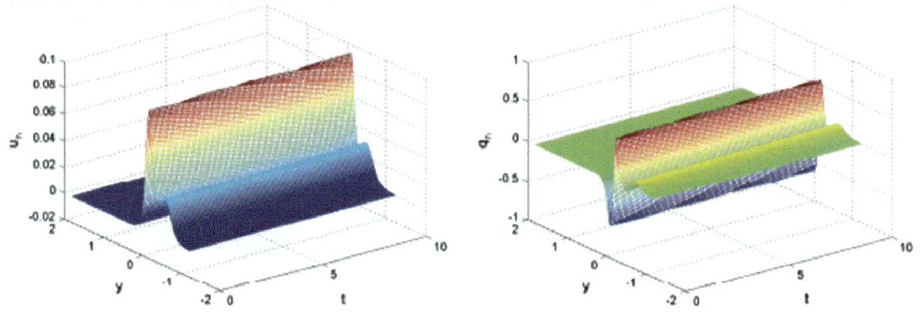

Fig. 4 Motion of u_h and q_h at various time

Table 4 Invariants of two interacting solitary waves at different time points

Time	I_1	I_2	I_3
$t = 0.0$	0.0331285528631741	0.0692198389657512	0.0050153618179368
$t = 2.5$	0.0331348209129624	0.0697001706813762	0.0050252667862336
$t = 5.0$	0.0331410326153673	0.0701749908068718	0.0050430559093074
$t = 7.5$	0.0331472326643209	0.0706370601191356	0.0050661698238232
$t = 10.0$	0.0331533178002679	0.0710794288167175	0.0050911384603003

References

1. Li, N., Gao, F.Z., Zhang, T.D.: An expanded mixed covolume method for Sobolev equation. J. Shandong Univ. (Nat. Sci.) **47**, 34–48 (2012)
2. Li, N., Gao, F.Z., Zhang, T.D.: An expanded mixed finite element method for Sobolev equation. J. Comput. Anal. Appl. (JoCAAA) **15**, 534–543 (2013)
3. Seyler, C.E., Fenstermacler, D.C.: A symmetric regularized long wave equation. Phys. Fluids **27**, 4–7 (1984)
4. Shang, Y.D., Guo, B.L.: Analysis of chebyshev pseudospectral method for multi-dimensional regularized SRLW equations. Appl. Math. Mech. **24**, 1035–1048 (2003)
5. Omarani, K.: The convergence of fully discrete Galerkin approximation for the Benjamin-Bona-Mahony (BBM) equation. Appl. Math. Comput. **180**, 614–621 (2006)
6. Gao, F.Z., Qiu, J.X., Zhang, Q.: Local discontinuous Galerkin finite element method and error estimates for one class of Sobolev equation. J. Sci. Comput. **41**, 436–460 (2009)
7. Bluman, G.W., Kumei, S.: Symmetries and differential equations. Springer, New York (1989)
8. Noether, E.: Invariante variationsprobleme, Nachr. K¨onig. Gesell. Wiss. G¨ottingen Math.-Phys. Kl. Heft 2, 235–257 (1918). English translation in P. S. Laplace, Trait¨e M¨ecanique C¨eleste, vol. 1, Paris, 1798. English translation, Celestial Mechanics, New York (1966)
9. Olver, P.J.: Euler operators and conservation laws of the BBM equation. Math. Proc. Camb. Phil. Soc. **85**, 143–160 (1979)
10. Gardner, L.R.T., Gardner, G.A., Dogan, A.: A least-squares FE scheme for the RLW equation. Commun. Numer. Meth. Eng. **12**, 705–804 (1996)
11. Saki, S.I.: Solitary waves of the splitted RLW equation. Comput. Phs. Commun. **138**, 80–91 (2001)
12. Babuska, I.: Error-bounds for finite element method. Numer. Math. **16**, 322–333 (1970/1971)
13. Brezzi, F.: On the existence, uniqueness and approximation of saddle-point problems arising from Lagrangian mulipliers. Rev. Francaise Automat. Informat. Recherche Operationalle Ser. Rouge, **8**, 129–151 (1974)
14. Falk, R.S., Osborn, J.E.: Error estimates for mixed methods. RAIRO Anal. Numer. **14**, 249–277 (1980)
15. Brezzi, F., Fortin, M.: Mixed and hybrid finite element method. Springer Ser. Comput. Math. **15**

Research on the Control Method of Coal Sample Blanking Based on the BP Neural Network and the PID Algorithm

Faquan Zhang, Baokun Liu, Guofu Wang and Jincai Ye

Abstract Due to the uneven blanking and time-variation characteristics of the coal sample blanking control system, traditional PID algorithm can hardly obtain good performance. A control method of coal sample blanking based on the BP neural network and the PID algorithm is proposed. This method combines the BP neural network and the PID algorithm. Environmental parameters and the number of vibration of coal sample bottle were taken as the input of the BP neural network. Experimental results show that the controller combined with the BP neural network and the PID algorithm has better control characteristics than traditional PID controller. The amount of coal sample dropped is even, and the experiment completion time is more stable.

Keywords BP neural network · PID · Stepper motor · Blanking control

1 Introduction

When enterprises choose coal, the quality of coal needs to be standardized analysis to ensure that the coal used for coking meets the relevant national regulations. The main process of standardized analysis includes three parts: sampling, sample making and test. In the process of test, more accurate coal samples need to be extracted from coal sample bottles. For example, if coulomb titration is used to measure the sulfur content in coal, (0.05 ± 0.005)g of air-dried coal sample that it's particle size less than 0.2 mm needs to be extracted for the experiment.

F. Zhang · B. Liu (✉) · G. Wang · J. Ye
School of Information and Communication, Guilin University of Electronic Technology, Guilin 541004, China
e-mail: 1592344186@qq.com

F. Zhang · G. Wang · J. Ye
Key Lab of Wireless Wideband Communication &Signal Processing of Guangxi, Guilin 541004, China

V. Jain et al. (eds.), *Recent Trends in Intelligent Computing, Communication and Devices*, Advances in Intelligent Systems and Computing 1031,
https://doi.org/10.1007/978-981-13-9406-5_4

After investigation, many enterprises use PID [1] algorithm when extracting coal sample from coal sample bottle. However, due to the uncertainty of coal sample's hardness, humidity and particle size, it is difficult to control the time and accuracy of coal sample extraction. In order to improve the control effect, the traditional PID [1] algorithm needs to be improved. Neural network is an intelligent control algorithm. It does not depend on the quantitative model. It has strong learning ability and is suitable for the control object with higher nonlinear degree. Among them, BP neural network [2] is a mature neural network, which has the ability to approximate arbitrary nonlinear functions and has a good effect on solving non-linear modeling problems. In this paper, BP neural network [2] algorithm is added on the traditional PID [1] algorithm to improve the control effect. The control object of this algorithm is the stepper motor [3]. A small hole is left at the mouth of the sample bottle and the coal sample bottle is inverted. By controlling the motion amplitude of the stepper motor resulting in the coal sample bottle is vibrated, then the coal sample falling into the crucible. The amount of coal sample falling can be precisely controlled by this method.

2 Control Method of Coal Sample Blanking Based on the BP Neural Network and the PID Algorithm

Figure 1 illustrates the structure of coal sample blanking [4] control system based on the BP neural network [2] and the PID [1] algorithm. The control system consists of three parts:

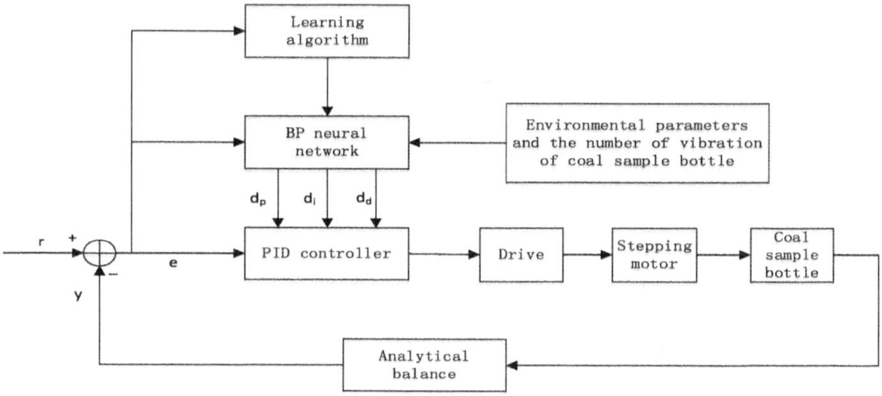

Fig. 1 Structure of coal sample blanking control system based on the BP neural network and the PID algorithm

(1) PID controller: PID algorithm is widely used in the field of industrial process control. Its output value is adjusted according to the error of feedback in real time.
(2) BP neural network: BP neural network algorithm has self-learning function. Its weight is adjusted by the back-propagation learning rule. The output of the BP neural network [2] corresponds to the three adjustable parameters of the PID [1] algorithm: d_p, d_i and d_d.
(3) Actuator: The actuator consists of three parts: drive, stepper motor and coal sample bottles. The coal sample bottle makes different vibration according to the output of the controller, causing the coal powder to fall into the crucible.

2.1 PID Control Algorithm

In this paper, we adopt the algorithm of the classical incremental PID [1]. The expression is:

$$u(d) = u(d-1) + d_p[e(d) - e(d-1)] + d_i e(d) + d_d[e(d) - 2e(d-1) + e(d-2)] \tag{1}$$

$$e(d) = r(d) - y(d) \tag{2}$$

d_p is the proportion coefficient, d_i is the integral coefficient, d_d is the differential coefficient, $u(d)$ is the output of PID [1] controller, $r(d)$ is the expected amount of single fall, $y(d)$ refers to the actual single output, $e(d)$ is the deviation signal of the expected amount of single fall and the actual single output.

2.2 BP Neural Network Algorithm

Single hidden layer BP neural network [2] was used, its structure is shown in Fig. 2. Four neurons were selected in the input layer: $x_1(d) = e(d)$, $x_2(d) = h(d)$, $x_3(d) = t(d)$, $x_4(d) = n(d)$.$e(d)$ is the deviation signal of the expected amount of single fall and the actual single output, $h(d)$ is the current coal sample humidity, $t(d)$ is the external temperature, $n(d)$ is the number of vibration of current coal sample bottle. With the increase of vibration times of coal sample bottles, the coal samples in the coal sample bottles will have different degrees of compaction. Meanwhile, the difference of coal sample humidity will also have a great impact on the accuracy of single blanking amount. In this paper, the environmental parameters (coal sample humidity, external temperature) and the number of vibration of the current coal sample bottle are taken as the input layer parameters.

The hidden parameters have six neurons. Sigmoid function as the activation function of the hidden layer neurons.

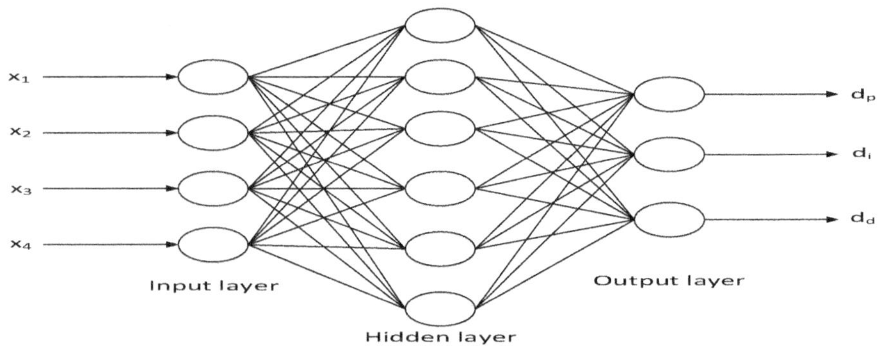

Fig. 2 Structure of the BP neural network

$$g(x) = \frac{1}{1 + e^{-x}} \tag{3}$$

The output parameters of the neural network are three adjustable parameters of d_p, d_i and d_d. So the number of neurons in the output layer is three. Because d_p, d_i and d_d is positive, the activation function of output layer neurons is the linear rectification function ReLU:

$$g(x) = \max(0, x) \tag{4}$$

The input layer of neural network is:

$$O_j^{(1)} = x_j; \quad j = 1, 2, 3, 4 \tag{5}$$

The input parameters and the output parameters of the hidden layer of neural network are:

$$net_i^{(2)}(d) = \sum_{i=1}^{4} w_{ij}^{(2)} O_j^{(1)} \tag{6}$$

$$O_i^{(2)}(d) = g(net_i^{(2)}(d)); \quad i = 1, 2, 3, 4, 5, 6 \tag{7}$$

$w_{ij}^{(2)}$ is the weighting coefficient of the hidden layer, $g(x)$ is the activation function.

The input parameters and the output parameters of the output layer of the neural network are:

$$net_l^{(3)}(d) = \sum_{i=1}^{6} w_{li}^{(3)} O_i^{(2)}(d) \tag{8}$$

$$O_l^{(3)}(d) = g(net_l^{(3)}(d)); \quad l = 1, 2, 3 \tag{9}$$

$$O_1^{(3)}(d) = d_p, O_2^{(3)}(d) = d_i, O_3^{(3)}(d) = d_d \tag{10}$$

The performance index function of the neural network is:

$$E(d) = \frac{1}{2}[r(d) - y(d)]^2 \tag{11}$$

Gradient descent method is used to adjust the weight coefficient of the network. The search is adjusted according to the weight coefficient. A global minimum inertia term that makes the search rapidly converge is added:

$$\Delta w_{li}^{(3)}(d) = -\upsilon \frac{\partial E(d)}{\partial w_{li}^{(3)}} + \beta \Delta w_{li}^{(3)}(d-1) \tag{12}$$

υ is the learning rate, β is the inertia coefficient.

$$\frac{\partial E}{\partial w_{li}^{(3)}} = \frac{\partial E}{\partial y(d)} \frac{\partial y(d)}{\partial \Delta u(d)} \frac{\partial \Delta u(d)}{\partial O_l^{(3)}(d)} \frac{\partial O_l^{(3)}(d)}{\partial net_j^{(3)}(d)} \frac{\partial net_j^{(3)}(d)}{\partial w_{li}^{(3)}(d)} \tag{13}$$

$$\frac{\partial net_j^{(3)}(d)}{\partial w_{li}^{(3)}(d)} = O_l^{(2)}(d) \tag{14}$$

In the case, the $\frac{\partial y(d)}{\partial \Delta u(d)}$ is unknown; it is replaced by the near analog function $\operatorname{sgn}\left(\frac{y(d)}{\partial \Delta u(d)}\right)$. Adjusting the learning rate can compensate the error effects.

After the analysis, the learning algorithm of the output layer's weight parameters is:

$$\Delta w_{ij}(d) = \beta \Delta w_{ij}^{(2)}(d-1) + \upsilon \delta_i^{(2)} O_j^{(1)}(d) \tag{15}$$

$$\delta_i^{(3)} = e(d) \operatorname{sgn}\left(\frac{\partial y(d)}{\partial u(d)}\right) \frac{\partial u(d)}{\partial O_l^{(3)}(d)} g'(net_l^{(3)}(d)) \tag{16}$$

Similarly, the learning algorithm of the hidden layer's weight parameters is:

$$\Delta w_{ij}(d) = \alpha \Delta w_{ij}^{(2)}(d-1) + \eta \delta_i^{(2)} O_j^{(1)}(d) \tag{17}$$

$$\delta_i^{(2)} = g'(net_i^{(2)}(d)) \sum_{l=1}^{3} \delta_l^{(3)} w_{li}^{(3)}(d); \quad i = 1, 2, 3, 4, 5, 6 \tag{18}$$

3 Experimental Verification

The experiment was carried out on the control platform of coal sample discharge. The balance was developed by the Sartorius (Fig. 3).

The highest requirement for the extraction accuracy of coal sample is to extract (0.05 ± 0.005)g coal sample. In this paper, two control algorithms are used for comparison.

It can be seen from Figs. 4 and 5 that the experimental effect control ability based on the traditional PID algorithm is poor. When the coal sample is relatively

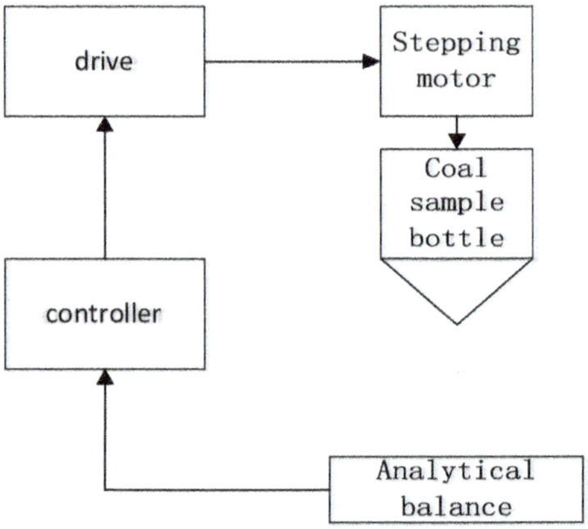

Fig. 3 Experimental equipment of coal sample blanking

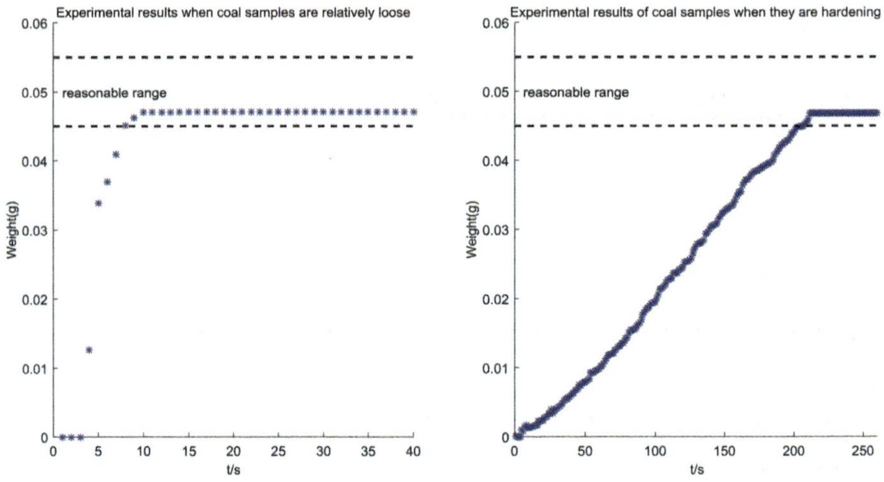

Fig. 4 Experimental results of traditional PID control algorithm

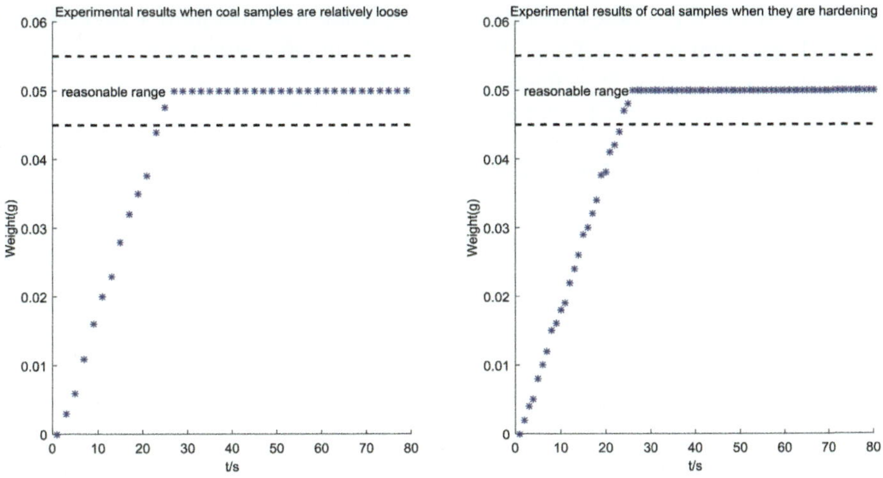

Fig. 5 Experimental results of the BP neural network and the PID algorithm

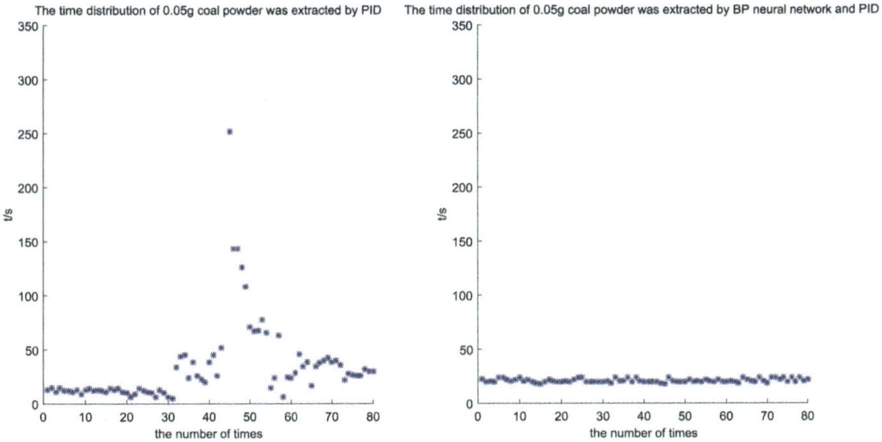

Fig. 6 The completion time distribution of (0.05 ± 0.005) coal sample extracted by the two control algorithms

loose, the single falling coal sample quantity is larger than 0.01 g and the falling coal sample quantity is not even. When coal samples are caked, the amount of coal samples falling in a single time is extremely small, and the task completion time is extended. The experimental results based on the BP neural network [2] and the PID algorithm [1] are relatively good. The single falling coal sample is even, and no single falling coal sample is too large or too small.

As shown in Fig. 6, the experimental completion time distribution based on the PID [1] algorithm is not uniform. When the coal sample is stiff due to vibration, the experiment completion time fluctuates and the maximum time is close to 260 s. However, the experimental completion time of the BP neural network [2] and the PID [1] algorithm was stable, which was concentrated around 20 s, greatly improving the stability of the system.

Acknowledgements I would like to thank the teachers and students in the laboratory for their encouragement and support in the process of completing this paper. The work was partially supported by NSFC under Grant no. 61362020 and 61761009.

References

1. Miao, Z., Song, C., Hu, C., Liu, L.: Applying on BLDCM speed PID controller of optimizing BP neural network based on GA, Lecture Notes in Information Technology, pp. 434–437 (2011)
2. Zhou, W., Shunqing, Xiong: Optimization of bp neural network classifier using genetic algorithm. Adv. Intell. Syst. Comput. **180**, 599–605 (2013)
3. Tom, H.: Stepper motor controller/driver simplifies stepper motor system. Power Electron. Technol. **36**, 24–27 (2010)
4. Al-Momani, E.S., Mayyas, A.T., Rawabdeh, I., Alqudah, R.: Modeling blanking process using multiple regression analysis and artificial neural networks. J. Mater. Eng. Perform. **21**, 1611–1619 (2012)

A Practice on Neural Machine Translation from Indonesian to Chinese

Lin Bai and Wuying Liu

Abstract Machine translation is used to implement the translation between different languages. Neural machine translation is one of the most popular machine translation methods which have made a great progress in recent years especially on universal languages. However, domestic translation software for non-universal languages is limited and also needs improving. In this paper, we carry out a practice on neural machine translation from Indonesian to Chinese. We first build a bilingual corpus of Indonesian and Chinese. After that, we train translation models using neural machine translation method, with traditional statistical machine translation models as baselines. In the later stages of our practice, we develop a Web software, named as Lore Translator, basing on our so-far best translation model. The performance of our model is comparable to previous work.

Keywords Machine translation · Neural machine translation · Non-universal languages · Indonesian · Chinese

1 Introduction

In the era of globalization, contacts between countries are more frequent than before. At the same time, failing to understand a different language becomes an obstacle in communication. Machine translation (MT) is a helpful and efficient way to overcome barricades in communication between different languages. It is a process of using a computer to implement the translation between different languages, that is, converting the source language into target language.

L. Bai · W. Liu (✉)
Laboratory of Language Engineering and Computing, Guangdong University of Foreign Studies, Guangzhou 510420, Guangdong, China
e-mail: wyliu@gdufs.edu.cn

W. Liu
Engineering Research Center for Cyberspace Content Security, Guangdong University of Foreign Studies, Guangzhou 510420, Guangdong, China

V. Jain et al. (eds.), *Recent Trends in Intelligent Computing, Communication and Devices*, Advances in Intelligent Systems and Computing 1031,
https://doi.org/10.1007/978-981-13-9406-5_5

Machine translation has been making great progress these years, especially in the translation between universal languages such as English and French, English and Chinese, etc. However, the number of domestic translation software for non-universal languages is limited.

Meanwhile, with the promotion of the "one belt one road," many Chinese companies have begun to strengthen cooperation with ASEAN countries. As one of the founding countries of ASEAN, Indonesia is also a member of the G20 as well as the largest economy in Southeast Asia. China and Indonesia have established cooperative relations in social infrastructure and trade. Therefore, the demand for translation on Indonesian and Chinese is also increasing. With the help of machine translation, we can reduce the burden on translators and interpreters and sweep some obstacles in cross-language and cross-cultural cooperation, and therefore promote political, economic and cultural exchanges between the two countries.

In recent years, neural machine translation (NMT) has become one of the most popular machine translation methods.

2 Related Works

Neural machine translation is based on deep learning. Compared with the statistical machine translation (SMT) method, the performance of a well-trained NMT model has been significantly improved. NMT has gone through several stages.

In 2013, researchers proposed recurrent continuous translation models for machine translation [1]. This model uses a convolutional neural network (CNN) to encode a given piece of source text into a continuous vector and then uses a recurrent neural network (RNN) as a decoder to convert the vector into target language.

In 2014, long short-term memory (LSTM) was introduced into NMT [2]. In this research, a multilayer LSTM is used to encode source sentence into a fixed-size vector and then decode it into target using another LSTM. The application of LSTM effectively solved the problem of gradient vanishing, which allows the model to capture information over long distance in a sentence.

To settle the problem of generating fixed-length vectors for encoders, researchers introduced attention mechanism into NMT [3]. The attention mechanism allows the neural network to pay more attention to the relevant parts of the input, and pay less attention to the unrelated parts. Since then, the performance of the neural machine translation method has been significantly improved.

Basing on the previous researches, we carry on practice on machine translation from Indonesian to Chinese.

3 Lore Translator

Figure 1 shows the framework of our practice. We first establish a corpus of Indonesian and Chinese. Then we segment Chinese sentences of this corpus in two different ways. One is to take word (namely one Chinese character) as unit of a

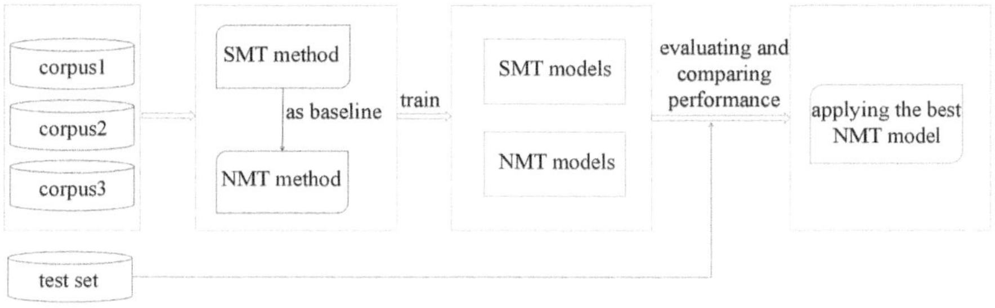

Fig. 1 Framework of experiments

sentence; the other is to take phrase as unit of a sentence. For comparison, we also train SMT models on MOSES with the same training sets. In the later stages of our experiments, we apply our best translation model into a Web application. We hope our efforts could provide some reference for further studies.

4 Experiments

4.1 Corpus

We obtain a bilingual corpus of 2.44 million pairs of Indonesian-Chinese sentences less than 50 words in length as training set. The source language is Indonesian and the target language is Chinese. We choose 20 k pairs of original dataset as development set and 20 k pairs as fixed test set. Each word out of vocabulary is represented by <UNK> token.

Besides, we separate Chinese sentences in two ways. The first is to separate sentences by word (corpus1), the other is to separate them by phrase, namely separated by 1 or more than 2 Chinese characters according to its meaning (corpus2). The training sentences in Indonesian contain 173,097 words. The Chinese sentences separated by word contains 6522 unique words, while sentences separated by phrase contains about 122,472 unique phrases, both including punctuation.

In the later stages of our practice, we focus on improving the performance of NMT model and developing a demo system. We further enlarge our dataset into a bilingual corpus of 3.21 million pairs of Indonesian-Chinese sentences (corpus3). Each sentence has less than 50 words. Chinese sentences in corpus3 are segmented by word. Thus, Indonesian sentences of corpus3 contain 173,216 unique words, and Chinese sentences contain 6599 unique words.

All the information of our training sets can be concluded as in Table 1.

Table 1 Datasets details

Corpus	Size (m)	Segmentation	Indonesian vocab	Chinese vocab
Corpus1	2.44	Word	173,097	6522
Corpus2	2.44	Phrase	173,097	122,472
Corpus3	3.21	Word	173,216	6599

4.2 Training

We conduct our research basing on open-source codebase [4]. This codebase is written in Python, using TensorFlow, a open-source software library. We also train SMT models based on MOSES as baseline, using the same sets of corpus.

For NMT model, we use a four-layer LSTM network with residual connections as well as attention mechanism to train a translation model. We initialize the parameters of neural network with uniform distribution between −0.1 and 0.1. We train our model for about 10 epochs, using mini-batch gradient descent with a batch size of 128. The learning rate is initialized as 1.0. After 1/2 of total train steps, we start halving the learning rate for 5 times before finishing. We apply the same hyper-parameters for different datasets.

The experiments are mainly executed on Tianhe-2. Each training takes about two weeks with above hyper-parameters. Experiments achieved a speed of 0.92 K (both Indonesian and Chinese) words per second with a mini-batch size of 128.

4.3 Experimental Results and Discussion

We use the BLEU score [5] to evaluate the performance of all the translation models we train. The performance of all models measured on test set is listed in Table 2.

The performance of NMT models is comparable to baseline models. Comparison between the results of corpus1 and corpus2 shows that model which is trained using sentences with a smaller unit has a better translating performance. Besides, comparing the performances of models trained on corpus2 and corpus3, we can find that

Table 2 BLEU scores of translation models

Model	Corpus	Translation method	BLEU scores
Model1_SMT	Corpus1	SMT	11.54
Model1_NMT		NMT	12.02
Model2_SMT	Corpus2	SMT	15.28
Model2_NMT		NMT	17.13
Model3_SMT	Corpus3	SMT	19.13
Model3_NMT		NMT	20.30

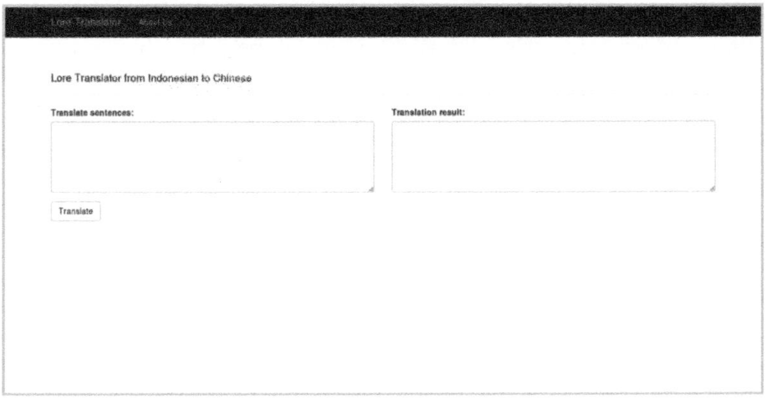

Fig. 2 Homepage of Lore Translator

translation model performs better with a larger training set for both SMT and NMT models.

After manually check the translation results, we found that Lore 1.0 can roughly translate Indonesian sentence into Chinese sentence. However, it performs not well in translations on long sentences and rare words.

4.4 Demo System

We carry on a further practice on applying our best NMT model which got 20.30 BLEU scores. We develop a NMT translator from Indonesian to Chinese translation, named as Lore Translator. This system is developed basing on Flask, a lightweight web application framework written in Python.

Figure 2 shows the homepage of Lore Translation V1.0. It is built as a Web application, and it is easy to use. Users enter sentences in Indonesian and click the "Translate" button, and then they can get corresponding Chinese output.

Lore Translator implements translation from Indonesian to Chinese which could be a reference for other non-universal languages. Besides, Lore Translator can be extended to other non-universal languages translation, given a large-scale parallel corpus of any two languages.

5 Conclusion

In this paper, we carry out a practice on NMT from Indonesian to Chinese. We first build a corpus of Indonesian and Chinese. Then we train models using NMT method. MOSES was used to generate SMT models as baseline. Experiments result shows that model which is trained using sentences with a smaller unit has a better translating performance. Also, translation model performs better with a larger training set for both SMT and NMT models. We also develop a extensible Web application, Lore Translator, basing on our best NMT model.

The experimental data is limited to further illustrate and verify the relation between the increase of training set and the promotion of model performance. Besides, our model still needs improving. Lore 1.0 cannot handle long sentences as well as rare words. Future work will be focused on the further verification of the findings and the improvement of our translation model. Besides, the Web application will be further enhanced in terms of stability and fault tolerance.

We believe improved Lore Translator is prospective. It can be used for assisting the teaching of Indonesian and Chinese, providing technical support for enterprises and governments. We hope our efforts could assist translators and interpreters, as well as providing some reference for further studies in the future.

Acknowledgements The research is supported by the Key Project of State Language Commission of China (No. ZDI135-26), the Natural Science Foundation of Guangdong Province (No. 2018A030313672), the Featured Innovation Project of Guangdong Province (No. 2015KTSCX035), the Bidding Project of Guangdong Provincial Key Laboratory of Philosophy and Social Sciences (No. LEC2017WTKT002), and the Key Project of Guangzhou Key Research Base of Humanities and Social Sciences: Guangzhou Center for Innovative Communication in International Cities (No. 2017-IC-02).

References

1. Kalchbrenner, N.: Recurrent continuous translation models. In: EMNLP (2013)
2. Sutskever, I.: Sequence to sequence learning with neural networks. Neural Inf. Process. Syst., 3104–3112 (2014)
3. Bahdanau, D.: Neural machine translation by jointly learning to align and translate. arXiv preprint arXiv:1409.0473 (2014)
4. Neural Machine Translation (seq2seq) tutorial. https://github.com/tensorflow/nmt. Last accessed 2017
5. Papineni, K., Roukos, S., Ward, T., Zhu, W.J.: BLEU: a method for automatic evaluation of machine translation. In: ACL (2002)

Keyword-Based Indonesian Text Summarization

Jinru Liu and Wuying Liu

Abstract Automatic summarization is a hot research topic in the field of natural language processing. Now the automatic summarization technology is mostly for majority languages such as English and Chinese, while less for the rare languages such as Indonesian. We aim to analyze the development trends of summarization and explore the processing methods for Indonesian summarization. This paper introduces the implementation process and experimental results of Indonesian text summarization based on the keyword frequency extraction method. The experimentation compares the RUGOE-2 result of our keyword-based system with that of the PSKSUMSUM one in Indonesian text summarization. The experimental results show that the keyword-based Indonesian text summarization is more effective.

Keywords Indonesian text summarization · Single-document summarization · Multi-document summarization · Keyword

1 Introduction

With the era of big data as background, information explosion has made the speed of traditional manual processing of documents far behind the growth of the literature. How to improve the efficiency of accessing information is an urgent problem to be solved. In order to solve this problem, people have long thought of using computer technology instead of manual processing of documents, so the concept of automatic summarization was gradually proposed [1]. H.P. Luhn of IBM in the

J. Liu · W. Liu (✉)
Laboratory of Language Engineering and Computing, Guangdong University of Foreign Studies, Guangzhou 510420, Guangdong, China
e-mail: wyliu@gdufs.edu.cn

W. Liu
Engineering Research Center for Cyberspace Content Security, Guangdong University of Foreign Studies, Guangzhou 510420, Guangdong, China

V. Jain et al. (eds.), *Recent Trends in Intelligent Computing, Communication and Devices*, Advances in Intelligent Systems and Computing 1031,
https://doi.org/10.1007/978-981-13-9406-5_6

United States published the first paper on automatic summarization in 1958. As a result, research on the field of automatic summarization has kicked off.

Our research focuses on single-document and multi-document summarization for Indonesian. Although the resources of non-universal languages are scarce, the application requirements of the corresponding languages are very strong. With the implementation of the "Belt and Road Initiative," China's links with ASEAN countries have also strengthened, and the demand for Southeast Asian language processing technologies such as Indonesian has risen sharply. The Indonesian region is also in the forefront of the global number of Internet users. The amount of information on the Internet is growing at an explosive rate every day. This has become an important breakthrough point for China to understand the local conditions and the national situation [2]. Also, our automatic summarization technology has strong applicability, and the research results can be transferred to other non-universal languages in Southeast Asia.

2 Related Work

2.1 The Development of Automatic Summarization

Automatic summarization began in 1958. Luhn published a paper entitled "The Automatic Creation of Literature Abstracts," which pioneered the field of automatic summarization. In 1969, Edmundson applied a variety of features of the text based on Luhn, including word frequency, indicative words or phrases, titles or subtitles, the position of the sentence, etc., and weighted the sentences to extract the abstracts. The research work of multi-document automatic summarization first appeared in the 1980s. At that time, this technology was mainly used for the processing of scientific articles. The application fields and methods were limited, which was not conducive to promotion. The real multi-document abstract research in any field is Since 1997.

2.2 Existing Systems

At present, through continuous research and improvement methods, there are many automatic summarizing systems formed at home and abroad. The more mature systems in China include PSKSUMSUM system, Korean intelligent abstract system, and Uyghur single-document automatic summarization system. There are many document automatic summarization systems such as Newsblaster, WebInEssence system, and NeATS system [3].

The PSKSUMSUM system integrates a variety of different document abstracts, such as centroid-based, subject-based sentences, and methods based on TextRank, supporting single-document, multi-document, and topic-based multi-document summarization. The Korean intelligent summarization system implements techniques such as morphological analysis, syntactic analysis, and semantic analysis of Korean and generates summaries by paragraph segmentation, sentence segmentation, and keyword extraction. The multi-document automatic summarization system Newsblaster is a digest system for the news field. The system uses a clustering method to summarize the same topic news that occurs every day. The WebInEssence system is a personalized web-based multi-document automatic summarization and content recommendation system. The system uses a multi-document abstraction method based on centroids.

3 Keyword Contribution Method

3.1 Overall Architecture

Due to the consideration of real needs, the keyword-based Indonesian text summarization system is mainly divided into three levels: text import layer, text processing layer, and text export layer, focusing on the text processing layer. The text processing layer includes techniques such as text preprocessing, sentence segmentation, word segmentation, statistical probability, and sentence matching.

Starting from the text import layer, when there comes a text, it is read into the memory in the form of a byte stream, and then two steps are performed at the same time: One is to perform sentence segmentation to form a set of original sentences and a set of lowercase sentences; the other is text preprocessing, including phrase-tagging and stopword-filtering. Then enter the text processing layer. The text to be processed are segmented into words, the word frequency in the word set is counted as the probability of the word, and the keywords are sorted according to the word frequency. The sentence weights are calculated according to the number of keywords in the lowercase sentence subset, thereby obtaining a sentence set with weights. By default, sentences with high weights are divided into summary sentences, and the number of summary sentences is determined according to the length of the required summary. Finally, the text export layer matches the original sentence subset with the weighted sentence subset, and outputs the summary sentences in the original order (see Fig. 1).

3.2 Algorithm

The key algorithms of text summarization introduced in this paper include word segmentation, sentence segmentation, and keyword extraction, with emphasis on keyword extraction. Feature selection plays a crucial role in text summarization techniques, and there are many methods. Among them, the surface features contains feature selection based on high-frequency words, feature selection based on

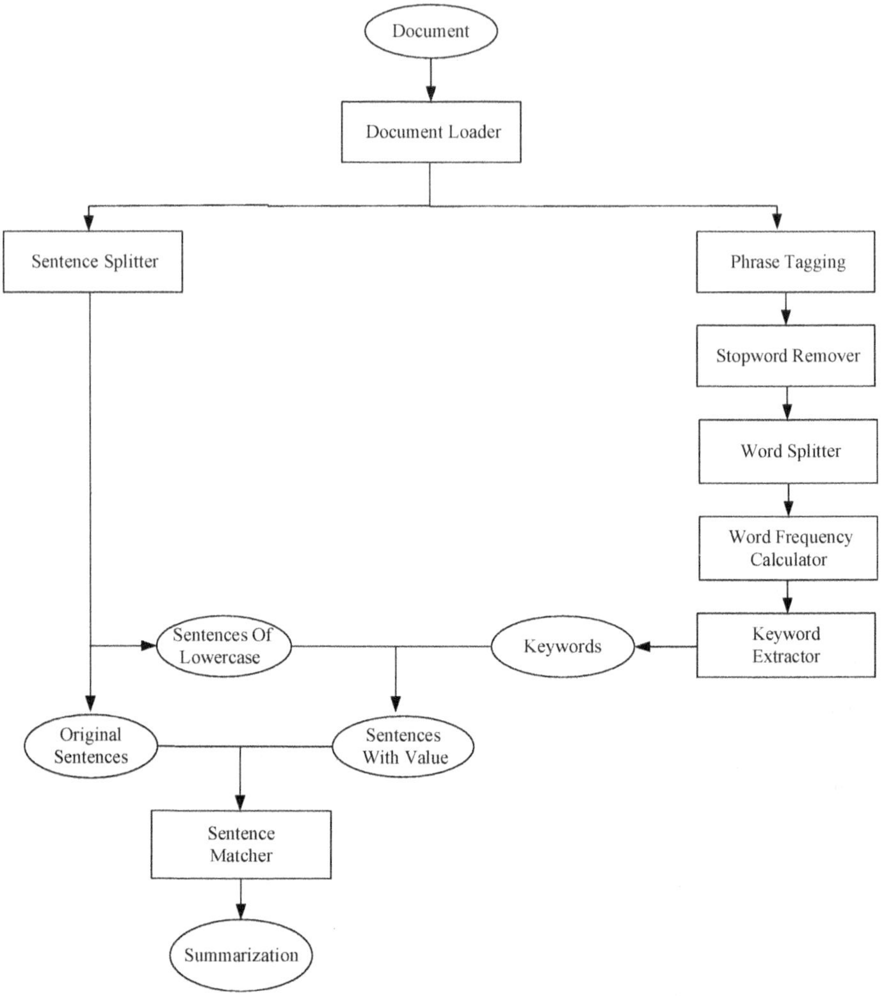

Fig. 1 Overall architecture

sentence position, and feature selection combined with high-frequency words and sentence position. In subsequent studies, more complex features were proposed, such as features based on entities and semantic levels. Entity features include named entities, proximity, associations of entity words, similarity, and logical relationships. Semantic features include rhetorical structure, syntactic features, and semantic concepts. The single-document summarization introduced in this paper adopts the feature selection method based on high-frequency words, that is, the keyword extraction method based on statistics. Treat text as a collection of words, count how often each word appears in a text, and calculate the weight of the keyword accordingly [4].

The strategy is as follows:

- Segmenting the preprocessed text into words;
- Count the frequency of each word, sorted according to the frequency from high to low;
- Extracting the first 25% of the words as keywords;
- Counting the number of keywords as a sentence weight for each sentence in the sentence set;
- A sentence with a high weight is used as a summary sentence [5].

4 Experiment

There are two evaluation methods for automatic summarization, including F value and ROUGE. This paper uses ROUGE to evaluate the performance of summaries. The ROUGE evaluation method compares the text summaries generated by the system with the artificially generated standard digests and evaluates the quality of the summaries by counting the number of basic units (n-grams, word sequences, or word pairs) that overlap between both of them. Improve the stability and robustness of the evaluation system by comparing with multiple standard summaries [6]. The ROUGE formula is as follows:

$$\text{ROUGE-}N = \frac{\sum_{\{S \in \text{ReferenceSummaries}\}} \sum_{\text{gram}_n \in S} \text{Count}_{\text{match}}(\text{gram}_n)}{\sum_{\{S \in \text{ReferenceSummaries}\}} \sum_{\text{gram}_n \in S} \text{Count}(\text{gram}_n)}$$

As shown in the single-document summarization experiment in Table 1, the keyword-based text summarization is compared with the single-document summarization function in the PKUSUMSUM system, and ROUGE-2 is used as the evaluation index [7]. The evaluation corpus is from the Indonesia IPI (Indonesia Publication Index) website. Among them, "Areas" represents the document set in each field; "Documents" represents the number of documents in each field; "+" represents the number of documents based on the keyword-based text summarization which performs better than the PKUSUMSUM system; "−" represents the number of documents based on the PKUSUMSUM system which performs better than the keyword-based text summarization; and "=" represents the equal value of both ROUGE-2.

The Table 2 shows the comparison of one document between ROUGE-1, ROUGE-2, ROUGE-3, and ROUGE-4. Among them, KITS represents the keyword-based Indonesian text summarization, and PKS represents the PKUSUMSUM system.

According to the above experimental results, the performance of keyword-based Indonesian text summarization is better. There are three points in the summary: (1) PSKSUMSUM adopts a method based on centroid processing single-document summarization, which can handle the language of multiple countries, but mainly

Table 1 Experiment results

Areas	Documents	+	−	=
Aerospace engineering	284	61	6	217
Agriculture, biological science, and forestry	1666	951	72	643
Arts and humanities	1025	607	61	357
Astronomy	190	35	0	155
Automotive engineering	620	348	24	248
Biochemistry, genetics, and molecular biology	775	597	40	138
Chemical engineering, chemistry, and bioengineering	1268	622	20	626
Chemistry	1490	790	36	664
Civil engineering, building, construction, and architecture	1573	1058	113	402
Computer science and IT	1670	965	132	573
Control and systems engineering	359	247	34	78
Decision sciences, operations research, and management	1215	282	198	735
Dentistry	776	361	53	362
Earth and planetary sciences	1079	645	63	371
Economics, econometrics and finance	3098	1944	150	1004
Education	3714	2414	194	1106
Electrical and electronics engineering	983	595	82	306
Energy	400	195	21	184
Engineering	882	473	51	358
Environmental science	888	431	35	422
Health professions	1961	876	368	717
Immunology and microbiology	764	630	39	95
Industrial and manufacturing engineering	642	416	39	187
Language, linguistics, communication, and media	551	373	29	149
Law, crime, criminology, and criminal justice	1011	588	79	344
Library and information science	810	432	23	355
Materials science and nanotechnology	702	446	28	228
Mathematics	917	539	58	320
Mechanical engineering	579	320	53	206
Medicine and pharmacology	1582	813	88	681
Neuroscience	228	192	25	11
Physics	763	611	32	120
Public health	1061	565	215	281
Social sciences	1950	1139	113	698
Transportation	503	383	37	83
Veterinary	1670	1456	54	160
Total	39,649	23,400	2665	13,584

Table 2 Compare from ROUGE-1 to ROUGE-4

nROUGE	KITS	PKS
ROUGE-1	0.945	0.670
ROUGE-2	0.849	0.484
ROUGE-3	0.789	0.415
ROUGE-4	0.767	0.375

used to process Chinese and English documents [8]. The key to this experiment is to process the Indonesian. (2) Our system is more targeted. The preprocessing work such as phrase-tagging and stopword-filtering is performed first, so that the generated summaries are of higher quality. (3) The method of ROUGE evaluation is more inclined to the text summarization based on keywords.

5 Conclusion

Keyword-based Indonesian text summarization has performed well on ROUGE evaluation but extracted summarization is not logical enough for reading. Therefore, we consider two aspects to the future work. One is to conduct experimental research on multi-document summarization, and the other is to explore automatic summarization methods combined with generating models [9]. The generated against network was first proposed in 2014, and the idea behind it is two competitive neural network models. A model takes noise as input and generates a sample (so-called a generator). Another model, called the discriminator, receives samples from the generator and training data and must be able to distinguish between the two sources [10]. Most successful applications for generating against networks are in the field of computer vision and applying these techniques to the field of natural language processing is a valuable research direction.

Acknowledgements The research is supported by the Key Project of State Language Commission of China (No. ZDI135-26), the Natural Science Foundation of Guangdong Province (No. 2018A030313672), the Featured Innovation Project of Guangdong Province (No. 2015KTSCX035), the Bidding Project of Guangdong Provincial Key Laboratory of Philosophy and Social Sciences (No. LEC2017WTKT002), and the Key Project of Guangzhou Key Research Base of Humanities and Social Sciences: Guangzhou Center for Innovative Communication in International Cities (No. 2017-IC-02).

References

1. Barzilay, R.: Text Summarization. MIT (2005)
2. Radev, D.R., Fan, W., Zhang, Z.: WebInEssence: A Personalize Web-Based Multi-Document Summarization and Recommendation System (2001)
3. Radev, D.R., Fan, W.: Automatic summarization of search engine hit lists. In: Proceeding, ACL Workshop on Recent Advances in NLP and IR, Hong Kong, October

4. Saggion, H., Gaizauskas, R.: Multi-document summarization by cluster profile relevance and redundancy removal (2004)
5. Evans, D.K., Klavans, J.L.: Columbia Newsblaster: Multilingual News Summarization on the Web (2004)
6. Lloret, E.: Text Summarization: An Overview (2008)
7. Lu, D., Pan, X., Pourdamghani, N., Chang, S.-F., Ji, H., Knight, K.: A Multi-media Approach to Cross-lingual Entity Knowledge Transfer. Computer Science Department, Rensselaer Polytechnic Institute, Information Sciences Institute, University of Southern California, Electrical Engineering Department, Columbia University (2016)
8. Zhang, J., Wang, T., Wan, X.: PKUSUMSUM: A Java Platform for Multilingual Document Summarization. Institute of Computer Science and Technology, Peking University, The MOE Key Laboratory of Computational Linguistic, Peking University (2016)
9. Yao, J.G., Wan, X., Xiao, J.: Recent Advances in Document Summarization (2017)
10. He, Z., Chen, C., Bu, J., Wang, C., Zhang, L., Cai, D., He, X.: Document summarization based on data reconstruction. In: Proceedings of the Twenty-Sixth AAAI Conference on Artificial Intelligence (2012)

Automatic Decision Support for Public Opinion Governance of Urban Public Events

Zhaoxuan Li and Wuying Liu

Abstract Though the process of urbanization in China is accelerating and the size of the cities continues to expand, the urban danger sources are increasing. When public emergencies such as natural disasters, major accidents, environmental hazards, and vandalism occur in the city, people receive information from major media platforms; it is easy to cause rapid spread of public opinion by smart terminals. Therefore, the research on the public opinion governance model of urban public events has great practical significance for ensuring the economic development and social stability. This paper mainly focuses on the prediction, analysis, and governance of the current public sentiment caused by urban public emergencies. Facing different sensational outbreak nodes, the system analyzes the current node's emotional inclination, assists the government to improve their supervisory capability, and proposes rationalized constructive opinions and solutions.

Keywords Automatic decision support · Public opinion governance · Urban public events · Sentiment analysis · Extended 5W model

1 Introduction

At present, there are corresponding researches at home and abroad. Internationally, developed countries have made a lot of investment in the public opinion governance of urban public events in recent years [1] and pay special attention to the research and development of the integrated decision-making support system for interdisciplinary. Automatic decision support is the core of public opinion governance system.

Z. Li · W. Liu
Laboratory of Language Engineering and Computing, Guangdong University of Foreign Studies, Guangzhou, China

W. Liu (✉)
Engineering Research Center for Cyberspace Content Security, Guangdong University of Foreign Studies, Guangzhou 510420, China
e-mail: wyliu@gdufs.edu.cn

V. Jain et al. (eds.), *Recent Trends in Intelligent Computing, Communication and Devices*, Advances in Intelligent Systems and Computing 1031,
https://doi.org/10.1007/978-981-13-9406-5_7

For example, Giuliano Manno used ontology and cloud computing technology to construct a semantic joint cloud system for emergency response [2].

Now, some researchers in China have initially established a monitoring and early warning system in the field of public security. For the first time in the national medium- and long-term science and technology development plan, public safety emergency response was listed as a key research topic. As a core component of the emergency response system, auto decision support system refers to a session system that supports unstructured or semi-structured public opinion governance decisions and facilitates decision-makers to call the analysis decision model and access to the public opinion database.

In summary, the auto decision support system is to establish a rule base, a knowledge base, a model library, and a plan base related to the decision-making theme, and to assist the decision-makers solving the semi-structured or unstructured decision-making problem information system by a human–computer interaction manner.

2 Implementation

In our research program, big data public opinion monitoring and guidance can make monitoring more scientific. By distinguishing big data public opinion data source, it can be better to strengthen the evaluation, monitoring, and guiding role of big data public opinion [3]. Through the monitoring of multimedia platforms, the public opinion guidance mechanism will be improved, and early warning and rapid response mechanisms will be established [4]. Big data public opinion governance support technology mainly includes data mining, intelligence analysis, language processing, and other technologies to analyze data sources and social media, combined with "crowdsourcing" and "cloud computing" technologies to build cloud platform virtual machines. Building a database of public opinion governance plans to promote interdisciplinary and cross-disciplinary information sharing and co-governance [5].

Currently, the big data model that fits the urban public event public opinion governance model is the 5W model, which is What Where When Who Why. Based on the model, we expand the description of the event into name, specific location, specific time, category, level, details, loss, causing casualties, and response measures. We extend the 5W urban public event theory model on mainstream social media platforms for big data detection analysis and discussion, including:

(1) The mainstream social media platforms include WeChat, microblog, Zhihu, etc.
(2) Distinguishing the subjects: offenders, regulators, and communicators. Analyzing their cognitive, emotional, and behavioral trends on social media.
(3) Combining the secondary communication with communication information (from elite to public opinion) and crisis communication theory (two-way symmetric communication mode), making suggestions and opinions on the response of different subjects on social media.

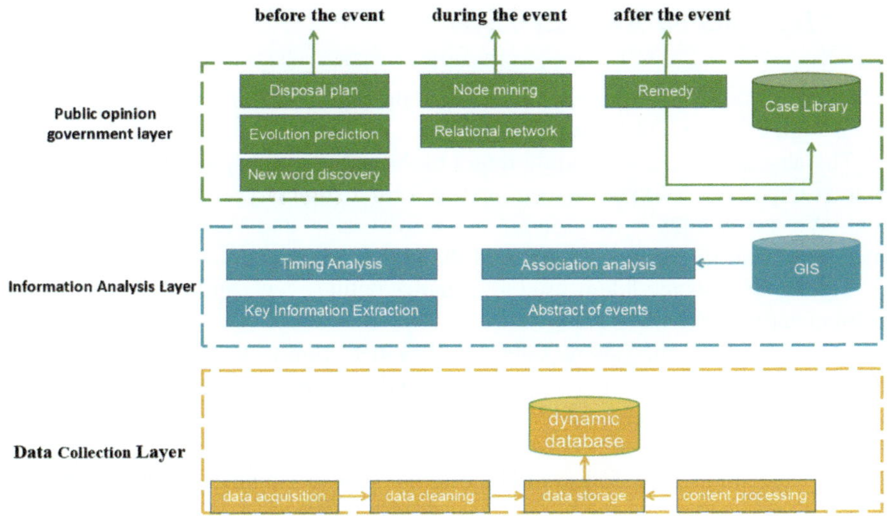

Fig. 1 The overall architecture of auto decision support system

The overall structure of the auto decision support system is shown in Fig. 1. It mainly includes three aspects: data collection layer, information analysis layer, and public opinion governance layer.

The system design goal is mining network emergencies in massive data information. On this basis, the urban public event public opinion governance model is used further to understand the events in depth and to realize the functions of disaster public opinion prediction, multimedia information intervention, and negative public opinion prevention and control. The front end is based on the Bootstrap framework. The after end is based on Flask framework; this framework is suitable for flexible and rapid development. It can also call python third-party modules and deep learning framework to fit the system development.

2.1 Data Collection Layer

The spread of public events usually begins on social media platforms, reports through various news portals, and ferments on mainstream social platforms. This research mainly discusses Baidu news information, based on Sina Weibo hotspot search, narrowing the scope of text mining and improving the accuracy and depth of mining. The information subject is divided into offenders, regulators, and communicators. At present, statistical learning methods have a solid theoretical basis for training text classification standards by training known classification documents. Weigh the advantages and disadvantages of various statistical learning algorithms. This paper uses SVM classifier to classify microblog hot search.

2.2 Information Analysis Layer

The key information extraction module uses the jieba and tf–idf models for Chinese word segmentation and keyword extraction.

The abstract of events module refers to the use of a computer to automatically extract important information from text to form a summary to express the original text. We use word2vec to generate the word vector, and then use the TextRank method to find the central sentence of the text to form a summary.

Timing analysis module is used to visualize public events and provide a basis for subsequent causal analysis (Fig. 2).

The association analysis module uses link detection tasks (LDT) to detect two randomly selected network's news reports to determine if these documents belong to the same topic. We have built a case library that contains a detailed description of historical events in cities according to the extended 5W model presented in this paper. The sample is as follows (The original sample language is Chinese):

Title: 11 · 11 Shanghai supermarket collapse accident
Place: Home Delight Supermarket, No. 138, Xiaxia Road, Zhuqiao Town, Zhuqiao Town, Pudong New Area, Shanghai
Coordinate: 121.752965, 31.183431
Category, level: safety accident, III
Detail: According to the Shanghai Municipal Administration of Work Safety, at 6:53 on the 11th, at the Jiadele Supermarket, No. 138, Xiaxia Road, Zhuqiao Town, Pudong New Area, the attic of the second floor of the house collapsed due to

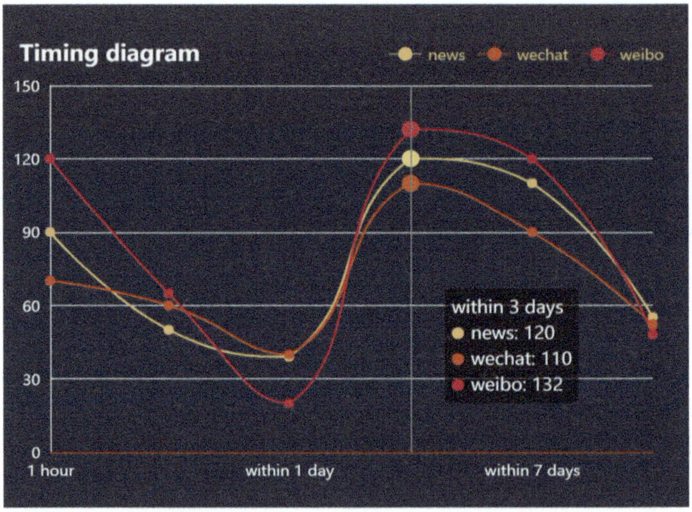

Fig. 2 Timing analysis module visualization

excessive stacking, causing person to be crushed. Seven people have been rescued and two of them have been seriously injured.
Loss: Attic damage
Casualties: Six people were injured and three were killed
Response: After the supermarket accident, Shanghai public security, fire protection, and related rescue experts quickly rushed to the scene to carry out search and rescue work.

2.3 *Public Opinion Government Layer*

The novelty discovery module is able to identify new knowledge that was not previously realized with the support of machine learning algorithms. The evolution prediction module can achieve an unknown prophecy and predict disaster events based on the current information atmosphere. Under the support of the previous two modules, the disposal plan foresight module is implemented for pre-event management. The relationship network module uses the complex network method to analyze the interpersonal relationship network, the human–land relationship network, and the human–institutional relationship network based on the knowledge of graph theory. In this way, the key point blocking module is supported to realize the governance of the event. The remedy module integrates the public opinion information about multi-source media according to the second-level communication and crisis communication theory of communication information and uses the urban disaster event case database to propose timely and reasonable post-remediation measures to form a late-stage urban disaster event response solution.

Governance Before the Event Module. The core of this layer is on the discovery of new words. The conventional word segmentation system divides the text containing new words to pieces, which result is fragmented and which accuracy is reduced. At present, the three factors that determine that the word is a new word are frequency, degree of free use and degree of aggregation. Through the new words obtained for topic detection and tracking to determine if there is a trend of public opinion, to achieve pre-governance, we used the new word discovery algorithm based on information entropy put forward by Matrix67. We extract candidate words that meet the threshold and remove the words that exist on the dictionary as new words.

Governance During the Event Module. In the analysis of the key nodes of traditional social networks, node centrality analysis is a more common method, mainly a degree of centrality, weighted degree centrality, intermediate degree, and near centrality. The HITS algorithm is a fundamental and important algorithm in link analysis. The authority of microblog bloggers is obtained through the HITS algorithm to realize key node mining. In order to alleviate or even eliminate the harm caused by negative public opinion, this paper will emotionally analyze the authoritative microblog bloggers through sentiment analysis. Using deep learning technology to solve the sentiment analysis problem is the current hot research, powerful feature learning capabilities can capture the essential characteristics of data from a small sample set. Zheng Xiao proposed a microblog short text sentiment analysis method based on convolutional memory neural network (CMNN). Using CNN's convolutional layer to extract local features, as the input of the next layer of long-term short-term memory network (LSTM) to better utilize local parallel features, context information and reducing manual annotation to solve microblog essays. The emotional classification of this [6]. Wu Peng proposed to realize the sentiment analysis of emergencies based on the two-way long-term and short-term memory model [7]; we will compare the two methods in the follow-up work.

Governance After the Event Module. As the heat for urban public emergencies gradually declines, the enthusiasm of netizen is no longer high. We will summarize the event process, supplemented by visual means, and present it to the public through web services. At the same time, the information is integrated and processed to realize the automatic incremental case library.

3 Conclusion

The new decision-making model has brought us effective solutions to deal with urban public events. We have initially implemented system prototypes according to system design, but not only that, due to the development of deep learning, researchers have proposed more and more natural language processing methods. We will explore an automatic extraction model based on deep learning and sentiment analysis based on two-way long-term and short-term memory models. Gradually expanding the case base and include as much as possible the reported incidents since the founding of New China, making the system more intelligent.

Acknowledgements The research is supported by the Key Project of State Language Commission of China (No. ZDI135-26), the Natural Science Foundation of Guangdong Province (No. 2018A030313672), the Featured Innovation Project of Guangdong Province (No. 2015KTSCX035), the Bidding Project of Guangdong Provincial Key Laboratory of Philosophy and Social Sciences (No. LEC2017WTKT002), and the Key Project of Guangzhou Key Research Base of Humanities and Social Sciences: Guangzhou Center for Innovative Communication in International Cities (No. 2017-IC-02).

References

1. Huosong, Xia, Huachun, Zhen: Public opinion analysis and decision support study under big data surroundings. Intell. Mag. **34**(2), 1–6 (2015)
2. Manno, G., Smari, W.W., Spalazzi, L., Taccari, G.: A semantic-based federated cloud system for emergency response. Concurr. Comput. Pract. Exper. **11**(8), 12–21 (2015)

3. Xu, Z., Liu, Y., Yen, N., Mei, L., Luo, X., Wei, X., Hu, C.: Crowdsourcing based description od urban emergency events using social media big data. IEEE Trans. Cloud Comput. **99**, 1 (2016)
4. Cieri, C., Maxwell, M., Strassel, S., Tracey, J.: Selection criteria for low resource language programs. In: The 10th International Conference on Language Resources and Evaluation (LREC) Proceedings. ELRA, pp. 4543–4549 (2016)
5. Qian, Zhang: The application of big data in government decision-making in emergencies. J. Northeast Agric. Univ. (Social science edition) **2013**(6), 73–79 (2013)
6. Zheng Xiao Wang Yizhen Yuan Zhixiang Qin Feng.: Sentiment analysis of micro-blog short text based on convolutional memory neural network. J. Electr. Measur. Instru. https://doi.org/10.13382/j.jemi.2018.03.028
7. Wu Peng, Ying Yang, Shen Si: Negative emotions of online users. Anal. Based Bidirect. Long Short-Term Memory. https://doi.org/10.3772/j.issn.1000-0135.2018.08.011

Improved Local Morphology Fitting Active Contour with Weighted Data Term for Vessel Segmentation

Xuan Wang and Kaiqiong Sun

Abstract An improved local morphology fitting active contour model with weighted data term is proposed in this paper for automated segmentation of the vascular tree on 2-D angiogram. In the original local morphology fitting model, morphological fuzzy minimum and maximum opening are adopted to approach the background and vessel object, separately. The structuring elements used in the morphology operator are linear ones, and their scale and orientation are computed from the image. The energy of the active contour model is minimized through a level set framework. This model is robust against both the inhomogeneous background and the initial contour location. However, the same weight coefficient is adopted for object structure. It leads to that bigger vessel structure that will dominate the contour evolution. In this paper, a normalized weight is added to the data term of the local morphology fitting to balance the data energy and encourage the segmentation of small vessel structure. The results on angiogram compared with the original local morphology fitting method are presented.

Keywords Inhomogeneous image · Vessel segmentation · Local image information · Level set method

1 Introduction

Accurate and automated detection of coronary vessel tree in angiogram is a fundamental step in various cardiovascular and cerebrovascular diseases-related medical imaging applications. Vessel segmentation remains a difficult problem because of inhomogeneous background, fuzzy boundary, image noise and complex vessel structure. Existing vessel segmentation methods normally consist of two main components: a local transform process and a global grouping one [1]. These

X. Wang · K. Sun (✉)
School of Mathematics and Computer Science, Wuhan Polytechnic University, Wuhan, China
e-mail: kqsunn@yeah.net

V. Jain et al. (eds.), *Recent Trends in Intelligent Computing, Communication and Devices*, Advances in Intelligent Systems and Computing 1031,
https://doi.org/10.1007/978-981-13-9406-5_8

two components corporately transform image data into segmentation result. Matched filter [2], morphology filter [3], skeleton and intensity ridge detection [4] are included in the local process; the method for grouping includes tracking-based method [5], scanning-based method and active contour models [12, 16]. The threshold [6], region growing [7] and clustering methods [8] usually mentioned in the literature can be treated as a specific combination of the above-mentioned methods.

Active contour models (ACMs) have been widely used in vessel segmentation. The main ACMs are edge-based, local as well as global region-based models. For suffering from weak contrast caused by low imaging quality, early popular edge-based active contour models [9] are adopted in angiogram for robust segmentation with improved edge feature instead of the simple gray gradient, such as the general edge detection [10]. The edge-based methods have small convergence range for only the edge information is used to drive the contour evolution.

In the energy of the global region-based ACMs, the global region information is calculated, combing with the image gray intensity, which constructs the force to guide the curve evolution. They are robust to the initial contour location. However, in the traditional region-based active contour models [11], for the gray intensity in each of the regions, a principle assumption is that the gray intensity fits a Gaussian distribution. Thus, the models can only deal with an image with regions being homogeneous. The assumption derogates from the inhomogeneous intensity property of angiogram. Therefore, many methods used a vessel sensitive feature image in the global region model instead of gray intensity [12].

In contrast, for the local region-based ACMs, the energy consists of the gray statistics from a local neighborhood of the image pixel point. Therefore, for the local region information is used; these models [13, 14] have the advantage in segmenting inhomogeneous images which hamper the global region-based models. Though the problem of intensity inhomogeneity is solved by the local region-based models, a consequent side effect is that the local model brings in initialization sensitivity. Different initial locations may lead to different results. This drawback of the models decreases their practicability. For the above problem with the local model, hybrid models have been proposed that combines global region-based with local region-based energy. For the advantages of both types of information are exploited, the hybrid model provides enhanced performance [15].

For the problem of automatic vascular segmentation on 2-D angiogram, the active contour model with local morphology fitting (LMF) was proposed in [16], which has similar energy in the local region-based model in [13]. The method uses directional morphology measure instead of the isotropic weighted mean of image intensity as a fitting term in the energy of contour. The local statistic for fitting the gray intensity can be computed before evolution and does not depend on the contour. Thus, during iteration of the contour evolution, they do not require to be updated. Thus, the level set evolution of this model is robust against the initial condition. However, the same weight coefficient is adopted for object structure in the energy of the LMF model. For certain smoothing parameter for the model, bigger vessel structure that has high gray contrast to the background has lower energy than the smaller ones. It leads to that bigger vessel structure that will dominate the contour evolution. In this paper, a normalized weight is added to the data term of the energy to balance the data energy and encourage the segmentation of small vessel structure.

2 Related Local Region-Based Models

The gray inhomogeneity is a distinct property of the vessel angiogram, which is difficult for the global region-based method to deal with. To deal with the intensity inhomogeneity problem, the local binary fitting (LBF) model has been designed. In the energy of the model, two fitting functions $f_1(x)/f_2(x)$ are used as the local approximation of the gray intensity of the points located in the two sides of the contour, respectively. The two fitting functions have spatial variation values. The data term in the energy functional of the LBF model [13] is defined as follows:

$$\begin{aligned} E^{\mathrm{LBF}} = \lambda_1 \int\limits_{\Omega} \int\limits_{\mathrm{in}(C)} K_\sigma(x-y)|I(y) - f_1(x)|^2 \mathrm{d}y\mathrm{d}x \\ + \lambda_2 \int\limits_{\Omega} \int\limits_{\mathrm{out}(C)} K_\sigma(x-y)|I(y) - f_2(x)|^2 \mathrm{d}y\mathrm{d}x \end{aligned} \tag{1}$$

where λ_1 and λ_2 are two positive constants. in(C) and out(C) mean inner and outer of the curve C, respectively. $I(x)$ denotes the gray intensity at x point of the original image. $K_\sigma(.)$ denotes a kernel function weighting the square error of the approximate value of $I(y)$ by the fitting values $f_1(x)/f_2(x)$. They correspond to the point in internal and external of the contour, respectively.

The contribution of $I(y)$ to the fitting energy decreases to zero as the point y moves away from the center point x, due to the localization of the kernel function. Therefore, the introduced kernel function results in the intensity $I(y)$ near x dominating the energy. For the local image information is used, the LBF model can deal with the gray inhomogeneity and is widely used for image segmentation. However, because of no global information involved and inefficient force presented in the comparatively smooth region, the segmentation results of the active contour depend on the initialization.

To overcome the above disadvantages of LBF model, the LMF model [16] fits the image gray intensity to the local image estimation which is independent of the contour location. The energy function of the LMF is defined as follows:

$$F(C) = \lambda_1 \int\limits_{\text{in}(C)\cap N(C)} |I(x) - I_{\max}(x)|^2 \mathrm{d}x + \lambda_2 \int\limits_{\text{out}(C)\cap N(C)} |I(x) - I_{\min}(x)|^2 \mathrm{d}x + \mu|C| \tag{2}$$

where $I_{\max}(x)/I_{\min}(x)$ denote the maximum/minimum fuzzy opening. They are also the weighted sum of gray intensity on a directional local neighborhood of a point. Other parameters are the same as in (1). $N(C)$ is the narrow region along C, which allows energy to be defined along the curve C, instead of the entire image area. μ is a normalized parameter to smooth the contour.

3 Proposed Method

With a fixed smooth parameter μ, the evolution of the contour of LMF can be treated as a competition process between the smoothing term and the data term. The large vessel structure usually has a higher gray intensity and bigger fitting error in data tern in (2) than the small vessel structure. Thus, the bigger vessel structure will dominate the contour evolution. To make the fitting error of big and small vessel into similar level, a normalized weight is added to the data term of LMF. Therefore, the energy of the proposed improved model is as follows:

$$F^{\text{NLMF}}(C) = \lambda_1 \int\limits_{\text{in}(C)\cap N(C)} |I(x) - I_{\max}(x)|^2 / I^2(x) \mathrm{d}x + \lambda_2 \int\limits_{\text{out}(C)\cap N(C)} |I(x) - I_{\min}(x)|^2 / I^2(x) \mathrm{d}x + \mu|C| \tag{3}$$

The original fitting error square of the LMF in (2) is divided by the square of the image gray intensity in (3). The added weight coefficient balances the contribution of fitting error from big and small vessel. The above energy can be transformed into a function formula of the level set Φ, and the evolution equation of the level set can be solved.

A regularization term for the level set [16] is added to (3) with the following formula,

$$P(\phi) = \int \frac{1}{2} (|\nabla \phi(x)| - 1)^2 \mathrm{d}x \tag{4}$$

which describes the difference between a signed distance function and the level set function. This term is adopted to preserve the regularity of the level set function Φ.

To minimize the total energy function, the standard gradient descent methods are adopted by the calculus of variations,

$$\begin{aligned}\partial\phi/\partial t = \delta_\varepsilon(\phi)\left\{-\lambda_1(I - I_{\max})^2/I^2 + \lambda_2(I - I_{\min})^2/I^2 + \mu\mathrm{div}\left(\frac{\nabla\phi}{|\nabla\phi|}\right)\right\} \\ + v\left(\nabla^2\phi - \mathrm{div}\left(\frac{\nabla\phi}{|\nabla\phi|}\right)\right)\end{aligned} \quad (5)$$

where $\delta_\varepsilon(.)$ is given by the smooth Dirac function.

4 Results

Figure 1a, d (first column) are two initial contours denoted by a rectangle on angiogram. The segmentation results by the LMF model is presented in the second column in Fig. 1, and those by the proposed method are presented in the last column. From the results, it can be observed that the LMF method obtained satisfactory results at some initial locations, but the segmentation was not comprehensive, and branches of some blood vessels were not completely segmented, as

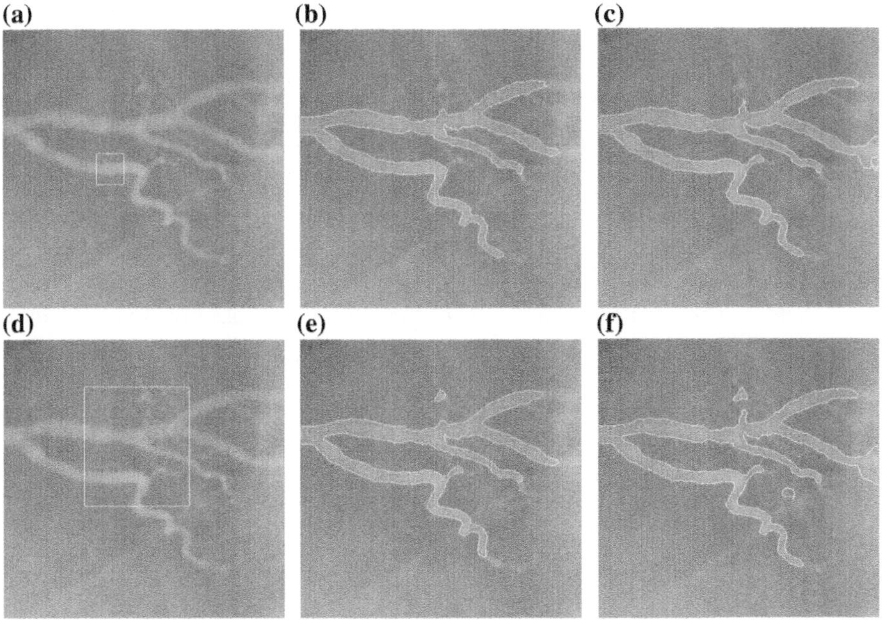

Fig. 1 First column: the initial contours; second column: the segmentation results of the LMF model; third column: the results of proposed method. The smoothing parameter for the both methods $\mu = 0.001 \times 255^2$. The iteration number is 300

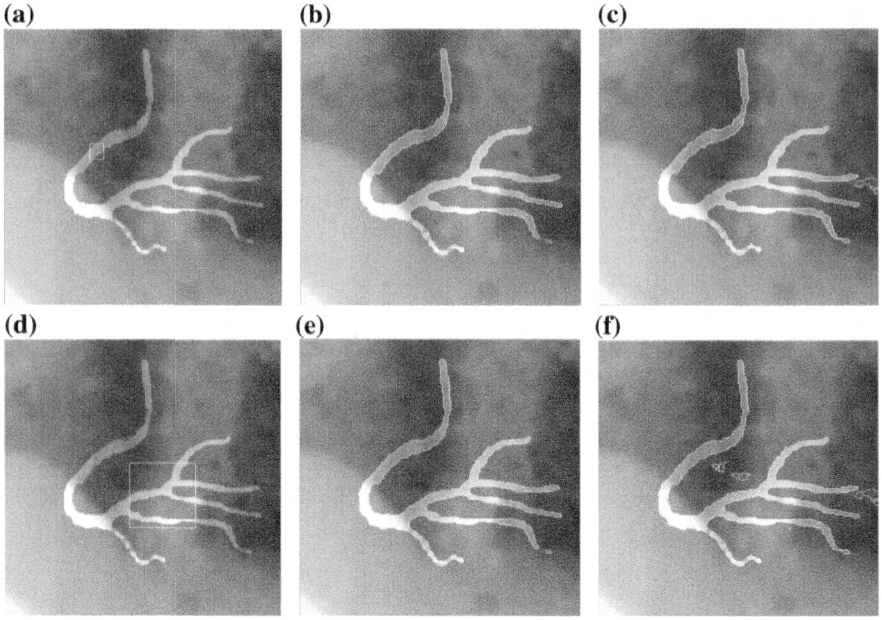

Fig. 2 First column: the initial contours; second column: the segmentation results of the LMF model; third column: the results of proposed method. The smoothing parameter $\mu = 0.001 \times 255^2$ in both of the two methods. The iteration number is 300

shown in Fig. 1b, e. In contrast, the results of the proposed method achieved satisfactory results, presented in Fig. 1c, f, and more details of the vessel structure were segmented with the same smoothing parameter.

Figure 2 presents the results on a synthetic vessel image. It is observed that some ends of the vessel are missed in the result as shown in (b) and (e). However, the missed part is completed with the proposed method shown in (c) and (f).

Figure 3 gives the results of the third group of experiments, which are similar to Figs. 1 and 2. The segmentation results of the LMF model are given in Fig. 3b, e, and those of the proposed method are presented in Fig. 3c, f.

5 Conclusion

An improved active contour model with local morphology fitting is proposed in this paper for vessel segmentation on angiogram. A normalized weight coefficient is added to the data term in the energy of the original LMF model. The weight coefficient makes the fitting error from big and small vessel structure fall into a comparative level. Therefore, neither big vessel nor small vessel will dominate the data energy. Preliminary experiments on several images reveal that improved result

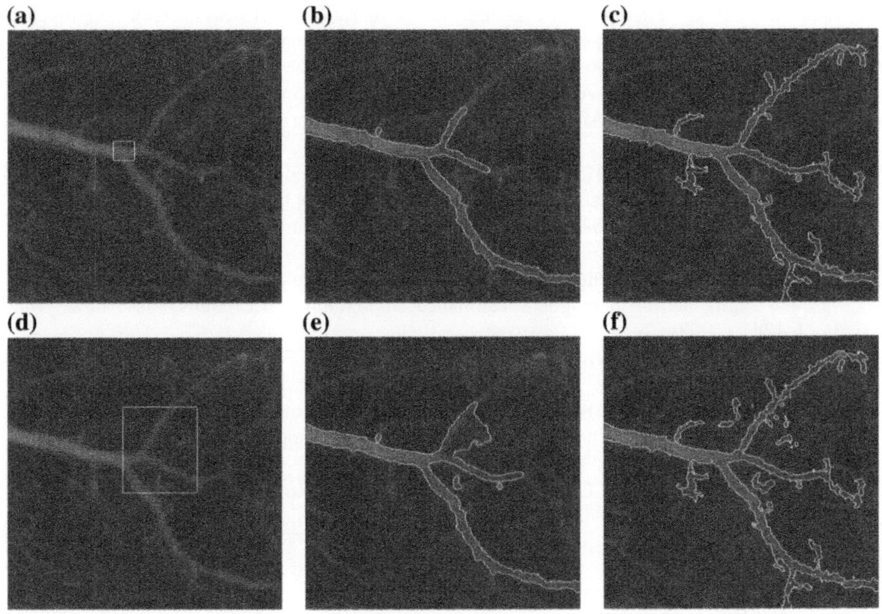

Fig. 3 First column: the initial contours; second column: the segmentation results of the LMF model; third column: the results of proposed method. The smoothing parameter $\mu = 0.001 \times 255^2$ in both of the two methods. The iteration number is 300

can be achieved by the proposed method compared to the original LMF method. The proposed method is required to be combined with the adaptive parameter method and the further work also includes extensive tests on other images from a real application.

References

1. Sun, K.Q.: Development of segmentation methods for vascular angiogram. IETE Tech. Rev. **28**(5), 392–399 (2011)
2. Lu, C.Y., Jing, B.Z, Chan, P.P.K., et al.: Vessel enhancement of low quality fundus image using mathematical morphology and combination of Gabor and matched filter. In: Wavelet Analysis and Pattern Recognition (ICWAPR), 2016 International Conference on. IEEE, pp. 168–173 (2016)
3. Sun, K.Q., Chen, Z., Jiang, S., et al.: Morphological multiscale enhancement, fuzzy filter and watershed for vascular tree extraction in angiogram. J. Med. Syst. **35**(5), 811–824 (2011)
4. Jin, D., Iyer, K.S., Chen, C., et al.: A robust and efficient curve skeletonization algorithm for tree-like objects using minimum cost paths. Pattern Recogn. Lett. **76**, 32–40 (2016)
5. Kalaie, S., Gooya, A.: Vascular tree tracking and bifurcation points detection in retinal images using a hierarchical probabilistic model. Comput. Methods Progr. Biomed. (2017)

6. BahadarKhan, K., Khaliq, A.A., Shahid, M.: A morphological hessian based approach for retinal blood vessels segmentation and denoising using region based OTSU thresholding. PLOS one **11**(7), e0158996 (2016)
7. Kerkeni, A., Benabdallah, A., Manzanera, A., et al.: A coronary artery segmentation method based on multiscale analysis and region growing. Comput. Med. Imag. Graph. **48**, 49–61 (2016)
8. Zeng, S., Wang, X., Cui, H., Zheng, C., Feng, D.: A unified collaborative multikernel fuzzy clustering for multiview data. IEEE Trans. Fuzzy Syst. **26**(3), 1671–1687 (2018)
9. Caselles, V., Kimmel, R., Sapiro, G.: Geodesic active contours. Int. J. Comput. Vis. **22**, 61–79 (1997)
10. Law, M., Chung, A.: Weighted local variance-based edge detection and its application to vascular segmentation in magnetic resonance angiograph. IEEE Trans. Med. Imag. **26**(9), 1224–1241 (2007)
11. Chan, T.F., Vese, L.A.: Active contours without edges. IEEE Trans. Image Process. **20**, 266–277 (2001)
12. Zhao, Y., Rada, L., Chen, K., et al.: Automated vessel segmentation using infinite perimeter active contour model with hybrid region information with application to retinal images. IEEE Trans. Med. Imag. **34**(9), 1797–1807 (2015)
13. Li, C., Kao, C.-Y., Gore, J.C., Ding, Z.: Implicit active contours driven by local binary fitting energy. In: Proceedings of IEEE Conference Computer Vision and Pattern Recognition, vol. 1, pp. 1–7 (2007)
14. Lankton, S., Tannenbaum, A.: Localizing region-based active contours. IEEE Trans. Image Process. **17**(11), 2029–2039 (2008)
15. Sun, K.Q., et al.: Hybrid active contour model for inhomogeneous image segmentation with background estimation. J. Electron. Imag. **27**(02), 1 (2018)
16. Sun, K.Q., Chen, Z., Jiang, S.: Local morphology fitting active contour for automatic vascular segmentation. IEEE Trans. Biomed. Eng. **59**(2), 464–473 (2012)

3D Point Cloud Data Splicing Algorithm Based on Feature Corner Model

Siyong Fu

Abstract Three-dimensional point cloud data splicing technology has always been a research hotspot and it is also a difficulty in reverse engineering, computer vision, pattern recognition, surface quality detection, and photogrammetry. Taking reverse engineering as an example, three-dimensional digital technology is the first link in reverse engineering. In the actual measurement process, due to the limitation of the geometric shape and measurement method of the measured object, the measurement device needs to perform multiple positioning measurements on the object from different viewing angles. Then, the splicing of multiple views is performed on point cloud data measured from different perspectives. The 3D point cloud splicing technology is also called repositioning, registration, or splicing technology on different occasions, whose essence is to coordinate transformation of data point clouds measured under different coordinate systems. The key to the problem is the determination of coordinate transformation parameters (rotation matrix) and translation vectors. Therefore, in the field of engineering, how to carry out non-contact reverse measurement of large-size parts and complex surfaces, how to digitize parts with high efficiency and high precision, and how to effectively realize the three-dimensional reconstruction of large-size and complex parts based on the digitized results of the part surface are key issues. This paper proposes the novel perspective of dealing with the above-mentioned challenges, and the experiment shows its effectiveness.

Keywords Feature points · Corner extraction · Feature vector model · 3D point cloud data · Data mosaic algorithm

---

S. Fu (✉)
Xinyu University, Xinyu 338024, China
e-mail: fusiyong58@163.com

V. Jain et al. (eds.), *Recent Trends in Intelligent Computing, Communication and Devices*, Advances in Intelligent Systems and Computing 1031,
https://doi.org/10.1007/978-981-13-9406-5_9

1 Introduction

With the wide application of the general reverse engineering in three-dimensional design and precision measurement, reverse reconstruction algorithms play an increasingly important role [1]. This project is mainly to study the algorithm for stitching point cloud data in the precision measurement of three-dimensional objects based on reverse engineering to solve the common problems in the traditional test methods, such as complex equipment shape and non-contact measurement in the measurement process. At present, the main algorithm of point cloud data splicing generally finds corresponding points by performing feature extraction in the frequency domain or space domain, such as Harris algorithm based on eigenvalues of autocorrelation matrix and SUSAN algorithm based on graphic edge points and corner point features. Such algorithms are highly targeted and have good splicing effects on specific graphics, but their versatility is not strong [2–4].

Currently, the common solution is to establish the correspondence between point cloud data by using the point cloud feature information that does not depend on three-dimensional rotation and translation transformation [5, 6]. However, the problem is that although the low-dimensional feature description function is simple to calculate, it contains less feature information and is sensitive to noise. The high-dimensional feature function can reflect the feature information well, but the calculation is complicated. Also, the cost of time and space is large, and how to make reasonable feature comparison is a difficult problem. Therefore, in the modeling process, noise detection and removal work are required for the obtained data.

The ICP algorithm calculation is easy and intuitive and allows for the stitching with good accuracy, but the speed of the algorithm and the global optimal convergence depend on, to a large extent, the given initial transformation as well as the estimation of the iterative process for the establishment of a basic relationship. A variety of bold stitching technologies provide the ICP algorithm with a good starting position, and the iterative process establishes the correct corresponding points in order to avoid the trap of the iteration limit as the part of the key to improving the algorithm, determining the algorithm's convergence with the end of the stitching accuracy.

Because the algorithm can use multiple features of the overlapping part of the point cloud data and also adopt an iterative algorithm with low time complexity and space complexity, the algorithm can make full use of the clustering characteristics of the transformation matrix parameter in the multi-dimensional space and can eliminate it. The influence of inaccurate corresponding points on the calculation results lowers the requirement for the accuracy of the corresponding point search algorithm; on the other hand, the problem of a large number of corresponding points in the search result of the initial corresponding point can be solved. Finally, the paper verifies the feasibility and validity of the algorithm through simulation.

2 3D Point Cloud Data Splicing Algorithm Based on Feature Corner Model

2.1 Model Innovation and Data Mining

Corner detection algorithms can be classified into three categories: corner detection based on grayscale images, corner detection based on binary images, and corner detection based on contour curves. Corner detection based on grayscale images can be further divided into three methods based on gradient, template-based, and template-based gradient combination. The template-based method mainly considers the grayscale change of the pixel field, i.e., the change of the image brightness, and the point where the point brightness contrast is large enough to be defined as a corner point. The common template-based corner detection algorithms include Kitchen–Rosenfeld corner detection algorithm, Harris corner detection algorithm, KLT corner detection algorithm, and SUSAN corner detection algorithm. Compared with other corner detection algorithms, SUSAN corner detection algorithm is characterized by a simple algorithm, accurate location, and strong anti-noise ability.

The KLT algorithm has higher quality than the Harris algorithm for detecting corner points, but the KLT algorithm is suitable for situations where the number of corner points is small and the light source is simple. Harris is suitable for situations where the number of corner points is large and the light source is complicated. In addition to the corner detection of a single image, the KLT algorithm and the Harris algorithm have better detection of the corners of the image sequence. Kitchen–Rosenfeld algorithm and SUSAN algorithm are generally not suitable for the corner tracking of the image sequence. For corner detection of a single image, the SUSAN algorithm is much better than Kitchen–Rosenfeld algorithm. However, the Harris algorithm has a smooth part in its implementation formula, so it has strong robustness and is also less sensitive to noise. However, in the actual calculation process, the circular template needs to be discretized, which brings about a large quantization error, thus easily leading to the confusion in the judgment of edge points and corner points. For edge-blurred images, the corner will be lost using small templates, which needs to determine which template is optimal through the dynamic determination.

Regardless of the data obtained by the contact method or the non-contact method, there will be a difference or an error point. These noise points have a great influence on the physical structure and if we do not eliminate the noise points, the shape of the final constructed entity will vary greatly due to the presence of noise points, thus seriously affecting the quality of testing. For the accuracy of the subsequent curve and surface reconstruction, we must perform denoising and smoothing filtering.

The data collected by the laser scanning method is often very intensive, and the data volume is generally in the megabytes, even up to tens of megabytes. Normally, these data will not be directly used for curve or surface reconstruction because it

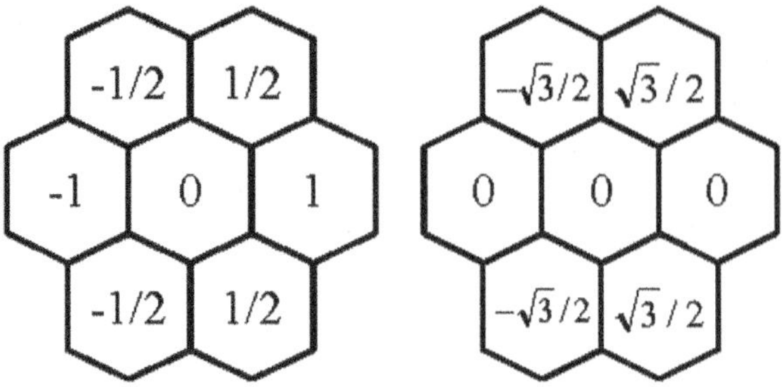

Fig. 1 Measurement matrix

will greatly reduce the computational efficiency and consume more memory. Therefore, it is necessary to compress the "point cloud" data (Fig. 1).

Data segmentation methods are divided into two methods based on measurement-based segmentation and automatic segmentation. Measurement-based segmentation is the process of measurement. The operator divides the contour surface into different subsurfaces according to the physical shape features and marks the contours, holes, groove boundaries, surface ridges, and other features of the surface based on this. The measurement of path planning is performed to provide a great convenience for modeling. The automatic segmentation method has two basic methods based on the edge and surface.

$$d(p, q) = d_e(p, q) + \alpha^2 d_f(p, q) \tag{1}$$

$$d = \sqrt{\left| \|q - p\|^2 - \frac{(q^T p)^2}{q^T q} \right|} \tag{2}$$

$$e = \|q - p\| \tag{3}$$

2.2 The Overall Structure of the System

The overall structure of the basic system, as the mosaic of the three-dimensional point cloud data, uses the coordinates of the corresponding points in the two observed coordinate systems to calculate the transformation relationship between these two observed coordinates, namely the coordinate transformation matrix. After the coordinate transformation matrix is obtained, the point under one of the observation coordinates can be conveniently projected into another observation coordinate system, thereby realizing the three-dimensional graphics splicing.

This system uses the binocular camera to obtain the original point cloud data of the object under different viewing angles and then uses a variety of algorithms to extract features and match the corresponding points on the overlapping parts. Finally, the mathematical model of the graphic transformation is pushed out in the homogeneous coordinates. The Kalman algorithm is introduced to fuse the corresponding points to obtain the transformation matrix of the splicing, so as to realize the splicing and non-contact precision measurement of the 3D point cloud data.

After acquiring a set of corresponding points extracted according to different features, the problem to be solved is that the transformation matrix needs to be solved according to the corresponding points. There are two kinds of transformation relations between two coordinate systems: One is a rotation operation; the other is a translation operation.

The Forstner operator is a well-known point-positioning operator in photogrammetry. The basic idea is to perform the edge of each pixel in the best window case for corner points. Weighted centralization, the position coordinates of the corner points are obtained; for the garden point, the gradient line passing through each pixel in the best window is weighted and centered to obtain the coordinates of the circle center. Therefore, the detailed steps of the model can be then summarized as follows. The SURF algorithm simplifies the DoH in the SIFT algorithm and simplifies the processing of the Gaussian second-order differential template into a box-filtering operation using the integral image. Search for the speckle response extremum in the image scale space pyramid: Compare the response values of the 26 pixel points in the adjacent region between adjacent scales to determine whether they are extreme. After processing these two steps of the response value of the Haar wavelet, the main direction of the core feature point is sought again. SURF algorithm also uses a cumulative method to get the main direction of feature points, but the cumulative range is the sector area centered on the feature points and the angle is targeted.

After processing these two steps of the response value of the Haar wavelet, the main direction of the core feature point is sought again. SURF algorithm also uses the cumulative method to get the main direction of feature points, but the cumulative range is the sector area centered on the feature points and the angle is the target.

After establishing the above-mentioned transformation matrix to solve the mathematical model, each group of corresponding points can calculate a result with clustering characteristics. In the case of fewer corresponding points, people usually use the least squares estimate to calculate. However, in the present system, since the data is often large. The least square algorithm cannot fully utilize the measurement data, which affects the estimation accuracy.

2.3 *Point Cloud Data Splicing Algorithm*

To measure the complete point cloud data of an object, the acquisition device needs to perform multiple measurements on the measured object from different perspectives. For the point cloud data collected from different perspectives, there must be some degree of overlap between the point cloud data at each viewpoint, and the rotation matrix is solved according to the point cloud data of the overlapped portion. Then, the rotation matrix obtained according to the point cloud data collected from different perspectives is converted into a unified coordinate system. At this point, complete point cloud data of the measured object can be obtained, and the entire point cloud data mosaic is completed. Therefore, the technique of novelty can be then summarized as the following aspects.

3D point cloud processing. A discrete point set is selected as the initial point to reconstruct the original surface of the point set. By connecting the scattered data points in the space with the optimized processing method, the optimized triangular mesh 3D model is obtained.

The multi-frame data fusion theory is firstly used to get 3D point cloud data and smooth the 3D point cloud data. Combined with the bilateral filtering theory of depth data, the outlier noise and internal high-frequency noise of three-dimensional point cloud data under white noise jamming are processed, respectively.

On the basis of a neighborhood characteristic measurement point and its immediate neighbors neighborhood point values of the curvature of the figure, in the case of the 3D point cloud data multi-scale space, the curvature of the characteristics in the analysis extracts the 3D point cloud data characteristics of the feature point set and then matches a 3D point cloud data feature point. The matched 3D point cloud data feature point is on a coordinate transformation solver. On this basis, the iterative closest point theory is used to perform the 3D point cloud data registration. On this basis, white noise and 3D point cloud data reconstruction are completed (Fig. 2).

For the algorithm of point cloud data registration, many scholars both at home and abroad have conducted in-depth researches and have also achieved fruitful research results. Among them, the most widely used is the iterative closest point algorithm proposed by Besl. In addition, there are some other mainstream algorithms. The geometrical principles used by these registration algorithms mainly include rigid transformation, projection transformation, radiation transformation, and curve transformation. Although the principle of each registration algorithm is different, the registration process is not much different, which can be mainly divided into: point cloud data extraction, point cloud data matching, error correction processing, and transformation solution.

Fig. 2 Simulated data and the results. **a** The first part of the dragon model. **b** The second part of the dragon model. **c** The result of the splicing of the first part and the second part

3 Conclusion

The first step of the iterative closest point algorithm is to determine an initial iteration value. The selection of the initial iteration value determines the accuracy of the point cloud registration to a great extent. If the selection of the initial iteration value is unreasonable, it is easy to fall into the algorithm, and the local optimum results lead to the final convergence of the entire iterative process. At the same time, the number of executions of the entire iteration process and the search speed of the corresponding point pair are both important factors affecting the efficiency of the iterative closest point algorithm. Therefore, how to select the initial iterative value and how to quickly and accurately find the correspondence between the source point cloud data set and the target point cloud data set are the key research contents in the iterative closest point algorithm.

Acknowledgements Thank the anonymous reviewers for their comments and constructive suggestions. This work was supported by the Science and Technology Research Project of Jiangxi Education Department of China (GJJ181032).

References

1. Poonam, Sharma, M.: A corner feature adaptive neural network model for partial object recognition. In: International Conference on Reliability, INFOCOM Technologies and Optimization. IEEE, pp. 1–6 (2015)
2. Shajina, A.R., Varalakshmi, P.: A novel dual authentication protocol (dap) for multi-owners in cloud computing. Clust. Comput. **20**, 1–17 (2017)
3. Qin, X., Li, S.: Finding scale-invariant corner feature from full-view image based on discrete spherical model. In: International Conference on Systems and Informatics, IEEE, pp. 1914–1918. (2012)
4. Wang, Y., Li, J., Wang, H.H.: Cluster and cloud computing framework for scientific metrology in flow control. Clust. Comput. **1**, 1–10 (2017)
5. Eckhardt, H.D.: Correction to: simple model of corner reflector phenomena. Appl. Opt. **10**(11), 2547 (1971)
6. Olague, G., Hernández, B.: A new accurate and flexible model based multi-corner detector for measurement and recognition. Pattern Recogn. Lett. **26**(1), 27–41 (2005)

A Preliminary Study on Mobile Learning

Dan Zhao

Abstract With the deep integration of information technology and education, especially the rise and deep using of new generation information technologies such as cloud computing, Internet of Things, big data, artificial intelligence, mobile learning are gradually becoming the mainstream of online education. Moreover, it is an indispensable and innovative learning method and also a new model for future learning. By analyzing the concept, characteristics, and composition of mobile learning, we give a feasibility analysis of mobile learning in its own advantages, network environment, professional construction, and requirement for industrial education integration. In addition, we also analyzed the current problems in the development of mobile learning and gave some corresponding solutions. This provides a complete basis for mobile learning in practice of distance education, online education, professional technology improvement, professional construction, etc. and also lays a solid theoretical foundation for the implementation of mobile learning.

Keywords Mobile learning · Characteristics of mobile learning · Feasibility analysis · The composition of mobile learning

1 The Concept, Characteristics, and Composition of Mobile Learning

With the deep integration of information technology and education, especially the rise and deep using of new generation information technologies such as cloud computing, Internet of Things, big data, and artificial intelligence, it will lead to a revolution in learning thinking, teaching methods, and cognitive models [1]. As a new learning mode after E-Learning, mobile learning can better reshape the human

D. Zhao (✉)
School of International Business, Yunnan College of Business Management, Kunming 650106, Yunnan, China
e-mail: tudou-yeah@163.com

V. Jain et al. (eds.), *Recent Trends in Intelligent Computing, Communication and Devices*, Advances in Intelligent Systems and Computing 1031,
https://doi.org/10.1007/978-981-13-9406-5_10

brain and avoid the inherent drawbacks of traditional classroom learning methods [2]. Thus, it has attracted the attention of many researchers and the public.

"The 40th China Internet Development Statistics Report" released by CNNIC shows that as of December 2017, China's mobile Internet users reached 753 million, an increase of 57.34 million compared with the end of 2016. The proportion of netizens using mobile phones to access the Internet has increased from 95.1% in 2016 to 97.5%. The number of online education users in China is 155 million, which is 12.7% higher than December 2016 [3]. Based on online education, mobile learning brings a new experience to learners. Through the combination of mobile computing technologies, mobile learning satisfies the needs of learners to learn and communicate anytime and anywhere and promotes the construction of learners' personal knowledge learning [4]. Mobile learning is gradually becoming the mainstream of online education. At the same time, mobile learning is also considered as a new model for future learning [5].

In general, we call mobile learning M-Learning or M-Education. In technology, it is an integrated product of computer, multimedia, network, and wireless communication [6]. At the same time, relying on wireless mobile communication networks and devices, it enables learners to obtain digital learning information, resources, and services at anytime and anywhere and improve their learning performance through interaction [6]. Mobile learning is an important way to build a learning society where everyone can learn, learn everywhere, and learn from time to time. It has a good development prospect. Its main features are as follows [7]:

- Mobility of learning forms: Learners can be free from the limitations of time, space, and wired networks.
- Wireless of learning device: The learning device should support wireless transmission.
- Interactivity of learning process: Two-way communication based on network communication and mobile computing technology is the basic guarantee for learning interaction.
- Integration of learning technologies: A variety of different learning devices are combined with each other.
- Mix of learning styles: E-learning and traditional classroom instruction are combined with each other.
- Personalization of learning content: Learning content can be customized according to user needs.

Mobile learning is composed of learners, mobile learning devices, Internet, learning platforms or apps, and mobile learning resources [8]. Learners can learn from the platform or app while interacting with teachers or other learners on it. Figure 1 shows the structure of mobile learning.

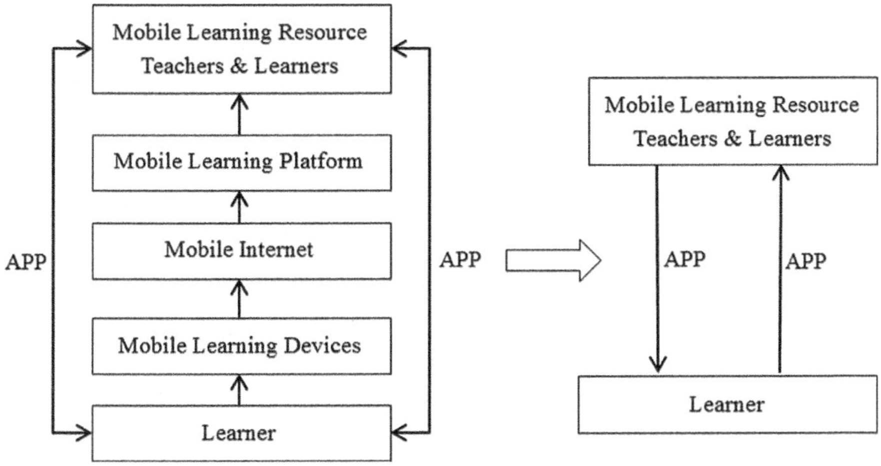

Fig. 1 Structure of mobile learning

2 The Feasibility of Mobile Learning

As a new type of digital learning, mobile learning has many possibilities. We will discuss the feasibility of mobile learning from three aspects: the characteristics of mobile learning, the inherent advantages of the network environment, and the needs of professional construction.

2.1 The Characteristics of Mobile Learning

The characteristics of mobile learning enable learners to freely arrange the time, place and content of learning, and even determine the specific learning time and difficulty according to their current energy and state. It shows the individualization and autonomy of learning [9].

Mobile learning is a flexible and open personalized education [10]. Various learning platforms provide scientific guidance and advice for learners' learning, support and guarantee in spirit, and a variety of learning resources in content. It has changed the boring form of book learning so that more learners are willing to learn actively and attract more adults to participate in continuing education.

In addition, making full use of the advantages of mobile learning can effectively compensate for the shortage of teachers and infrastructure in schools. By utilizing the infinite and open advantages of the network virtual space, we can make full use of the resources of different universities, educational institutions, and platforms so that the learning content can be expanded to a wider range of areas. It can promote educational exchanges and resources sharing in different regions, achieve educational equity, and play an important role in improving the quality of the people and strengthening the national culture [11].

2.2 The Inherent Advantages of the Network Environment

With the introduction of the 4G communication protocol, the performance of DSP and communication chips continues to improve; China's mobile communication network has entered the 4G era. Compared with 3G networks, 4G networks have significant advantages in terms of the frequency band, service, and function [12]. The broadband mobile communication system integrated by multiple functions has more stable quality and lower price, which provides a good network environment for the further development of mobile learning [13].

Mobile learning technology has broken through the teacher-centered one-way knowledge transfer model by making full use of the Internet, portal, WeChat platform, mobile app, and online learning resources based on the cloud education platform [14]. Learners, especially adult learners who need to work and learn at the same time, can make use of convenient mobile devices to make full use of amateur or fragmented time to achieve fragmented learning and persistent learning of knowledge. It realizes the individualized learning of learners and has greatly promoted for lifelong education and comprehensive education.

2.3 The Needs of Professional Construction

Mobile learning is conducive to the development of teaching models such as school and business cooperation, industry, and education integration. The integration of industry and education has promoted the interactive development of the labor employment system and the vocational education system. It has opened up and broadened the channels for training technical personnel, and promoted the development of the modern vocational education system [15].

In terms of professional construction, through the mobile learning platform, learners can master the operation in professional skills and practice while learning. In practice, when difficulties are encountered, learners can solve problems in a timely manner through a mobile learning platform to improve their ability to solve problems independently. At the same time, learners can also communicate with other learners, experts, teachers, etc. through the mobile learning platform to obtain the required professional knowledge quickly. In addition, learners can help other learners solve problems to improve their professionalism so that learners are more in line with the current professional construction needs of composite, innovative, and technical talents.

3 Potential Problems and Solutions in Mobile Learning

Mobile learning can provide learners with self-learning on mobile phones and effectively solve complex problems. However, through questionnaires Yin Lijie [15] and others found that learners who use mobile learning methods use the mobile network to learn, although learners can make full use of leisure time and peacetime to complete the learning, but still have the following problems:

Firstly, in the process of learning, learners cannot fully understand the knowledge gained through mobile learning and have not thought about how to improve the efficiency of mobile learning. This makes the knowledge gained too fragmented and easy to forget. Learners will only organize related knowledge when they complete the study with a task.

Secondly, since the study is mainly based on leisure time, learners cannot effectively manage the duration of mobile learning, and they also do not set long-term or short-term learning goals. In addition, in the learning process, learners are also susceptible to interference from other information, such as advertisements on web pages and prompt information of various types of apps so that learning efficiency cannot be fully guaranteed. Finally, knowledge on the web is relatively fragmented, and its reliability is yet to be verified, while relatively concentrated knowledge, courses, or learning platforms often require additional charges. Therefore, many learners need to spend more time to find and identify effective learning resources, and a lot of time is wasted. This greatly hurts the learner's enthusiasm and does not cultivate the habits to learn by using mobile learning. Therefore, in order to solve these problems, first of all, on the basis of mobile learning, we can use the mobile library, WeChat public number, MOOC platform, mobile learning app, and other different ways to help learners to make their learning plans, learning tasks, and learning objectives at the beginning of learning.

Secondly, in the process of learning, we can add the punching mechanism and help the learner to set the mobile learning time reminder and other functions [16] to assist students in self-discipline and management. At the same time, by adding interactive sections among learners, such as learning communication space, learners can exchange learning experiences and learning gains so that learners can encourage each other to enhance learning motivation and learning enthusiasm.

Finally, the government should accelerate the development and utilization of hardware facilities such as modern information networks, teaching servers, and mobile digital devices. At the same time, the government should make full use of the resources of different universities, educational institutions, and platforms and open up to the learners as free as possible, give full play to the efficient and flexible advantages of mobile education, and continuously improve the learning enthusiasm and enthusiasm of learners. Moreover, it lays the basic foundation for learners to achieve personalized learning, lifelong education, and comprehensive education.

4 Conclusion

With the deep integration of information technology and education, mobile learning as an emerging learning model is constantly changing traditional learning. Learning anywhere, anytime, fragmented learning has gradually become the mainstream of adult education, distance education, and continuing education. How to make full use of mobile learning methods to improve learners' enthusiasm for learning and make learning more personalized, lifelong, and socialized is the direction of continuous reform of learning models in the future. By analyzing the concept, characteristics, and composition of mobile learning, our paper gives the feasibility of mobile learning in its own advantages, network environment, and professional construction needs, as well as the existing problems and solutions in the development of mobile learning. It realizes a comprehensive explanation of mobile learning, provides a complete basis for a specific practice, and lays a solid theoretical foundation for the development and comprehensive implementation of mobile learning.

References

1. Gan, Z.: Blended learning: a new trend in digital learning. In: Higher Education Modernization 2013 International Forum on Higher Education, pp. 439–443 (2013)
2. Jiang, Q., Zhao, W., Li, S.: Personalized adaptive learning research—the new normal of digital learning in the age of big data. China Educ. Technol. **2**, 25–32 (2016)
3. CNNIC.: The 40th China internet development status report [R/OL]. http://www.cnnic.cn/hlwfzyj/hlwxzbg/hlwtjbg/201708/P020170807351923262153.pdf (2017)
4. Zhang, T., Shang, J.: Application status of mobile learning based on information teaching in higher vocational colleges. Netw. Secur. Technol. Appl. **9**, 97–98 (2018)
5. Xie, S.: Problems and strategies in mathematical modeling teaching of higher vocational education in mobile learning environment. Hebei Voc. Educ. **05**, 58–60 (2017)
6. Tong, Y., Zhang, L., Zhang, Y.: Research progress and evaluation of mobile learning at home and abroad. Educ. Voc. **02**, 101–106 (2017)
7. Luo, L., Huang, H.: Innovation of adult education model in colleges and universities based on mobile learning. China Adult Educ. **04**, 31–33 (2017)
8. Li, Z., Xiong, X.: Mobile learning—a new trend in digital learning. Softw. Guide **2**, 10–11 (2007)
9. Niu, Y., Li, H., Jiang, Y.: Research on the innovation of college ideological and political education model based on mobile learning. New Heights **02**, 145–149 (2016)
10. Zhang, L., Wu, W.: Mobile learning—a new chapter in digital learning. J. Inner Mongolia Normal Univ. (Educ. Sci.) (7), 147–149 (2015)
11. Lv, M.: Mobile learning: a new way of learning for continuing education. Ability Wisdom **12** (34), 68, 70 (2017)
12. Liu, D.: Analysis on the strategies of adult english autonomous learning in the mobile learning environment. J. Jilin TV Radio Univ. **09**, 36–38 (2018)
13. Wang, Y.: On mobile learning modes under the informational environment. J. Hanjiang Normal Univ. **6**(37), 7–10 (2017)

14. Hu, D.: Exploration on the construction of talent cultivation system of "modern apprenticeship system" in secondary vocational schools in ethnic areas. Policy Sci. Consult. (12), 32–32 (2016)
15. Yin, L., Chang, Z.: Analysis of the components of mobile internet autonomous learning in higher vocational education. Educ. Modern. **1**(04), 293–295 (2018)
16. Liang, M., Huang, H., Zhang, M.: A survey of students mobile learning on WeChat public platform. J. HuBei Corresp. Univ. **1**(31), 160–162 (2018)

Design and Implementation of Real-Time Inquiry System of the Stock Market

Peiyao Nie, Yulian Wen and Meng Yan

Abstract Nowadays, the development of Internet finance is in full swing. The application of the Internet will certainly play a vital role in the rapid changing stock market, such as inquiring the stock market, analyzing the trend of stock, and investing the stock market. And this is the original intention of the design of this system. The design and implementation of the real-time stock inquiring and displaying system will be introduced explicitly. To be specific, this paper will contain the summary of technologies, demand analysis, system design, system implementation, data storage, system testing, and conclusion. Combined with Hadoop and data visualization technology, the system uses the technologies of JavaEE, AJAX, and JQuery, achieving the inquiry and display of the stock market. Utilizing spiral incremental development, the system is developed iteratively by four phases, which are requirement definition, risk analyzing, engineering achievement, and evaluation.

Keywords Stock query · JavaEE · Data visualization · Hadoop

1 Introduction

Since Chinese reform and opening, the rapid and stable economic increment has been one of the main Chinese features. As one of the most important components, Chinese securities market has a rapid development since the opening of Shanghai and Shenzhen Stock Exchange in 1990s, which becomes an important driving force to promote rapid economic growth.

Securities market has made tremendous contributions to Chinese economic growth [1]. For stock market, which is one of the usual investment channels of

P. Nie (✉)
Sanya University, Sanya 572022, China
e-mail: niepeiyao@126.com

Y. Wen · M. Yan
School of Computer Science & Technology, Shandong University of Finance and Economics, Jinan 250014, China

V. Jain et al. (eds.), *Recent Trends in Intelligent Computing, Communication and Devices*, Advances in Intelligent Systems and Computing 1031,
https://doi.org/10.1007/978-981-13-9406-5_11

securities market, it has been hovering in the low state since the bear market in 2008. But due to the limitation of the investment channels, stock market is still the popular candidate of the investment channels for the public. However, the domestic stock market is in initial stage in general. In other words, there is still much room for improvement, such as relative laws and regulations, operating mechanism, and management system. To better serve the development of the real economy, meet the requirements of the investors and let the stock market play the role of "economic blood"—the two sessions made relative goal clearly in its report in 2016. For its goal, it is to promote the reform of stock market and securities market and relative legal construction, pushing forward the healthy development of multi-level capital market, as well as improving the proportion of direct financing.

With the rapid development of the Internet, the Internet has become an important investment tool for investors. Besides, Internet financial firms have been got extensive attention recent years. It is predicted by massive finance chiefs that the business model of commercial banks would be completely overturned as the rising of Internet financial enterprises [2]. What is more, the model of "Internet and finance" has been gradually accepted and integrated into the daily life of the public. Given all of this, this paper will explore how to play the role of the Internet in the stock market to achieve real-time queries of the stock quotes, providing the convenience of the shareholders.

2 Relative Technologies

2.1 Hadoop

It is a long-term problem to process large-scale data. The reason for this lies in it which relies on the expensive computing resources, such as high-performance computing of distributed data, network computing technology, and so on. Besides, it needs tedious programming to realize the valid segmentation of large-scale data and the rational allocation of computing tasks. Luckily, the development of Hadoop distributed technologies supports an effective solution for these problems [3].

So far, the most common application of the Hadoop platform is the log storage and log analysis. There are extensive direct and relative applications of Hadoop. Hive, which is used by Facebook for log management, is a Hadoop-based data analysis tool. In 2009, 30% non-programming staff in Facebook used SQL supported by Hive to process data. Taobao, which has very powerful searching function within the Web site, also uses Hive to filter the results by searching condition.

Pig, which is as powerful as Hive, is another outstanding framework of data analysis in Hadoop family. It is accomplished by Pig to recommend your potential friends, which is the function in most of the social networking software, especially for Twitter and LinkedIn. For pig, it is usually used for the advanced data process and data analysis. There are abundant applications of Pig as well. First of all, the product recommendation function of Amazon is based on collaborative filtering using Pig. Besides, Taobao is trying to utilize Pig to achieve the product recommendation. Moreover, Tianmao, which belongs to Alibaba, achieves most of its functions depending on the products of Hadoop family. Furthermore, the well-known Internet company, Yahoo, also implements most of its functions depending on Hadoop family, which includes spam analysis and filtering, analysis of user behavior, and model building of user behavior.

2.2 *HBase*

HBase, which is the abbreviation of Hadoop database, is an excellent NoSQL database product of Hadoop family. And its initial idea comes from the thought of three published papers, which is the Bigtable. There are a lot of advantages of HBase. Firstly, there is certainly a big table inside HBase, and its data storage could reach infinitely in theory. Secondly, its insertion and reading speed are not slower than the outstanding products of traditional structured databases, such as Oracle. More importantly, HBase does not limit the number of fields, which is unparalleled by structured databases like Oracle. For Oracle, its performance would be affected seriously when the number of fields is large. But, the complex business logic may not be completed easily when the number of fields is small. However, HBase can nicely make up for this deficiency.

3 Requirement Analysis and Design

3.1 *System Function Analysis*

According to the analysis, the designed system has following functions.

(1) Return potential reminder after inputting stock code or stock name.
(2) Check the stock k-line graph based on stock code or stock name.
(3) Grab domestic and foreign hot news and financial news from the Internet.
(4) Support multitudinous storage of historical stock data.
(5) Support the millisecond-level query of stock data in large data context.
(6) Support multi-terminal access and open rest style interface.
(7) Update and maintain the basic stock information.

3.2 Overall Design of the System

There are five projects in this system, which are stock search—parent, stock search—common, stock search—rest, stock search—portal, and stock search—manager. For stock search—parent, which is the super class of all projects, it supplies the information of all jar package of the whole system and is used for engineering polymerization. For stock search—common, it supplies the projects of the common classes, basic tool classes, and public responding results. For stock search—rest, it is the basic service project and supports most of the functions of the system. For example, the fuzzy query of stock information is based on the http interface supported by stock search—rest. For stock search—portal, it is the front-end interface, which is used for the Web function of PC. Besides, all data are obtained through the business layer request interface without the direct operation of the database. For stock search—manager, it is the back-end management project operated by mangers for viewing the operation information of the system and maintaining the infrastructure information of the system.

3.3 Database Design

In database design, there are several problems that need to be concerned by relational database designers, which are data redundancy, updating anomalies, and integrity constraint [4]. The system mainly uses two kinds of databases, which are MySQL and HBase. For MySQL, it is used to store basic stock information, supporting the operation of basic stock information. For HBase, it is used to store the historical stock information, supporting the storage and querying of big data.

4 Implementation of the Key Modules of the System

4.1 The Real-Time Prompt Module of Fuzzy Query of Stock Information

The detailed process of fuzzy query would be demonstrated explicitly in this part. First of all, the system asks the users to input relative stock codes or Chinese names through the search box of the browser. Please note that the input content could be arbitrary part of the stock codes or names. Then JavaScript does the response to the event by JQuery, using AJAX to send the response to stock search—portal system. After accepting the request, the layer of portal parses the parameters, judging whether the string is in digital format or Chinese character format. According to the string format, the relative parameters would be sent to its corresponding business module. Through http client tool, the business module

Fig. 1 Effect diagram of the fuzzy query of stock information

would call corresponding interfaces in service layer of rest to obtain the stock information. According to accepted data by service layer of rest, its relative service methods would make a fuzzy query with corresponding parameters through SQL statements. Then the retrieved results would be returned to the interface caller and finally to the browser. Browser-side AJAX verdicts whether the response is successful according to the returned data. If the response were successful, the received data would be sent to the relative parts of the search box. If not, the relative wrong prompt would be demonstrated as well. Figure 1 is an example of the searching results.

4.2 News Information Module

The front-end module of the browser will send the request to back-end portal layer based on the asynchronous event mechanism when the home page of the system is requested. After receiving the request, portal layer will call the relative business methods. Then the business method will obtain relative data by calling Baidu API. By parsing the retrieved data, the business layer concatenates the data into a string and returns it to the browser.

There are three asynchronous requests responding to relative interfaces of portal layer, which are requesting financial-related news, requesting domestic hot news, and requesting international hot news. After receiving request, portal layer calls Baidu API to obtain data of news. Then portal layer completes the analysis of JSON string in the background, stitches them into corresponding HTML codes, and passes the HTML codes to the browser. Finally, the browser appends the relative data into corresponding part by JQuery, completing the display of the list of news.

4.3 Query Module of Large Quantities of Historical Data

The platform of this system uses Hadoop 2.4.1 version, running on the operating system of cent OS 6.4. There are seven nodes in the cluster.

In this module, the HBase cluster is built on the basis of the Hadoop cluster. And the stock data table is created in HBase. Besides, the historical stock data in HDFS is imported in HBase by the data migration tool of Sqoop. There are several inquiry functions in this module. Firstly, the cluster and data could be maintained by shell. Secondly, through Java interface, the historical data of the stock could be retrieved according to the stock code and time. For the inquiry system, it is similar to other subsystems except that the database of Hadoop is operated by Spring for Apache Hadoop in the level of database operation. Finally, the data are transferred into the interface layer by business layer, providing the service for other systems.

5 Conclusion

This paper introduces the design and implementation of the stock real-time query system based on Hadoop. The system mainly uses JavaEE platform to build the service-based stock inquiry function by using maven. Combined with big data technology of Hadoop, it realizes the system building of large-scale data storage, providing a good data storage platform. Besides, combined with data visualization technology, the real-time query system is achieved successfully.

Acknowledgements This work was supported by Human and Social Science Project of MOE (15YJAZH042) and Research Project of Teaching Reform of Higher Education in Shandong Province (2015Z058).

References

1. Yu, N.: A theory of the development trend of China's securities industry. Financ. Theor. Pract. **11**, 41–43 (2002)
2. Wang, R., Wang, J.: The SWOT analysis of the internet financial enterprise competition of the commercial banks. Small Medium-sized Enterp. Manag. Technol. Mag. **8**, 62–63 (2014)
3. Cui, J., Li, T., Lan, H.: Design and development of the mas data storage platform based on Hadoop. J. Comput. Res. Dev. **49**(z1), 12–18 (2012)
4. Wang, W.: Initial exploration about the basic principles of database design. J. Yunnan Natl. Univ. (Nat. Sci. Ed.) **9**(4), (2000)

An Improved Feature Selection Algorithm for Fault Level Identification

Weiwei Pan

Abstract In fault level identification, there are ordinal structures between different levels, and some features have the monotonic dependency with decision. In this paper, we propose a feature selection algorithm for fault level identification based on ordinal classification. First, we design a new feature evaluation function to evaluate the quality of features based on ordinal rough set. Second, combining with the search strategy of genetic algorithm (GA), an improved feature selection algorithm is proposed. Finally, the proposed feature selection algorithm is employed to crack level identification. Experimental results show that the proposed algorithm not only can reduce the feature dimension but also improve the accuracy of identification.

Keywords Fault level identification · Feature selection · Ordinal classification

1 Introduction

Fault diagnosis technology has attracted further research in the past decade. Different from the traditional fault diagnosis, fault level identification does not pay attention to the fault types, but focus on identifying the different severity information of the fault under the same fault type. Fault severity levels can be represented as "slight fault", "moderate fault", "severe fault" and so on. There are ordinal structures between different severity levels. In the field of pattern recognition, the fault level identification can be considered as ordinal classification (also called ranking or ordinal regression) [1, 2]. In addition, some fault features have monotonic relationship with different severity levels; such features are called monotonic features. The monotonic features reflect the monotonic trend with different severity levels, which can provide straightforward and simple diagnostic information for fault level identification. However, most features do not have the monotonic

---

W. Pan (✉)
School of Applied Mathematics, Xiamen University of Technology, Xiamen, China
e-mail: panweiwei@xmut.edu.cn

V. Jain et al. (eds.), *Recent Trends in Intelligent Computing, Communication and Devices*, Advances in Intelligent Systems and Computing 1031,
https://doi.org/10.1007/978-981-13-9406-5_12

relationship with the severity levels, which are called non-monotonic features. Therefore, how to use and represent the potential ordinal information in the data is the first issue for fault level identification. Compared with fault identification and fault classification, fault level identification is more difficult.

The existing methods for fault level identification can be mainly grouped into two categories: one method is generating an effective indicator which varies monotonically with the severity levels; the other method is applying the classification techniques for classifying the fault severity levels automatically [3–5]. The monotonic indicator can reflect the development trend of the severity levels, which provide a straightforward and intuitive way for fault level identification. However, these proposed methods neglected the existing monotonic information between some features and different levels.

Feature selection is regarded as an important preprocessing step in machine learning and pattern recognition. There are many benefits to apply feature selection algorithms, such as improving classification performance and reducing computation time by deleting many irrelevant features. In recent years, for ordinal classification, the problem of feature selection has been discussed and studied [6–8]. However, most of these proposed algorithms are designed under the assumption of all features having the monotonic relationship with decision. As we known, in many real-world applications, supposing all features having the monotonic trend with decision may not be accurate. No much attention has been paid to this special problem up to now.

The aim of this paper is to propose a feature selection algorithm for fault level identification. As we know, in fault severity level identification, there are ordinal structures between different levels and some features change monotonically with the severity levels. To take these special characteristics into consideration, we propose a new feature evaluation function. Then the proposed feature evaluation function is combined with the search strategy of genetic algorithm (GA), we propose a feature selection algorithm. At the same time, the genetic algorithm (GA) is used to decide the partition of non-monotonic and monotonic features based on the original feature set.

Some basic concepts of ordinal classification are given in Sect. 2. In Sect. 3, we design an improved feature selection for ordinal classification. In Sect. 4, we apply the proposed algorithm to gear crack level identification and show the experimental results. The conclusion is presented in Sect. 5.

2 Basic Concepts of Ordinal Classification

For ordinal classification, the existing techniques try to analyze and represent the ordinal information. This section firstly gives a review of ordinal classification.

Let $\mathrm{DT} = \langle U, A, D \rangle$ be an information table, where $U = \{x_1, \ldots, x_N\}$ denotes a set of samples, $A = \{a_1, a_2, \ldots, a_J\}$ is a non-empty finite set of attributes which characterize the samples, and $D = \{d_1, \ldots, d_K\}$ is a set of decisions. If there is ordinal relationship between different class's values, assuming $d_1 < d_2 < \cdots < d_K$, we can say DT $= \langle U, A, D\rangle$ is an ordinal classification system.

From the ordinal relations of $D = \{d_1, \ldots, d_K\}$, we can derive two nested preference decision structures: $d_i^{\leq} = \cup_{j \leq i} d_j$ and $d_i^{\geq} = \cup_{j \geq i} d_j$. Here, $d_i^{\leq}$ denotes a preference subset of samples whose decision is equal or less than d_i; $d_i^{\geq}$ denotes a subset of samples whose decision is not smaller than d_i.

Let $v(x_i, a)$ be the value of sample x_i on attribute $a \in A$. If $v(x_i, a) \leq v(x_j, a)$, we say that x_i is no better than x_j regarding to attribute a, denoted by $x_i \leq_a x_j$. Furthermore, for any $a \in B$, if $v(x_i, a) \leq v(x_j, a)$, we say $x_i \leq_B x_j$.

Monotonic classification is a special ordinal classification task. Let f be an information function: $f: U \rightarrow D$, the monotonicity constraints between features and decision can be described as: $\forall x_i, x_j \in U$, $\forall B \subseteq A$, if $x_i \leq_B x_j$, the inequality $f(x_i) \leq f(x_j)$ holds. We call this classification function as a monotonic classification function and DT $= \langle U, A, D\rangle$ is a monotonic classification. In monotonic classification, the monotonicity constraints should be taken into considered as the basic assumption.

3 Feature Selection Algorithm for Ordinal Classification

As we known, monotonic dependency is very sensitive under the case of existing noisy information. This problem lays in the existing feature selection algorithms assuming that all the features are monotonic. In ordinal classification task, this assumption may not be accurate. In most cases, only partial features have the monotonic dependency with the decision. The non-monotonic features getting a small value of feature quality when using monotonic feature evaluation functions may be also useful for classification or prediction. So, a new feature evaluation function needs to be developed.

Definition 1 Given an ordinal information system $\langle U, A, D\rangle$, we define two subsets of features: $A_{nm} \subseteq A$ and $A_m \subseteq A$. Let A_{nm} and A_m represent two subsets of non-monotonic and monotonic features, respectively. We have that $A_{nm} \cup A_m = A$ and $A_{nm} \cap A_m = \Phi$.

Definition 2 Given an ordinal information system $\langle U, A, D\rangle$, for $B \subseteq A$, B can be represented with the union of two subsets B_{nm} and B_m; B_{nm} and B_m are non-monotonic and monotonic subset of B, respectively.

Here, $B \subseteq A$ means $B_{nm} \subseteq A_{nm}$ and $B_m \subseteq A_m$.

Definition 3 Given an ordinal information system $\langle U, A, D\rangle$, $B \subseteq A$, for $x_i \in U$, we define the following set:

$$[x_i]_B^{\sim <} = \{x_j \in U | x_i \sim_{B_{nm}} x_j\} \cap \{x_j \in U | x_i <_{B_m} x_j\};$$
$$[x_i]_B^{\sim >} = \{x_j \in U | x_i \sim_{B_{nm}} x_j\} \cap \{x_j \in U | x_i >_{B_m} x_j\}.$$

For ordinal classification, we need a strategy for generating hybrid monotonic relation from data.

Definition 4 Given an ordinal information system $\langle U, A, D \rangle$, $B \subseteq A$, let $R^{\sim}$ denotes the fuzzy relation induced by the non-monotonic features B_{nm}, $R^{<}$ and $R^{>}$ denote the fuzzy preference relation induced by monotonic features B_m, respectively. Then the hybrid monotonic relation can defined as $R^{\sim <} = \min\{R^{\sim}, R^{<}\}$ and $R^{\sim >} = \min\{R^{\sim}, R^{>}\}$.

Definition 5 Given an ordinal information system $\langle U, A, D \rangle$, the memberships of object x belong to the hybrid lower and hybrid upper approximations of $d_i^{\leq}$ and $d_i^{\geq}$ with respect to B are computed as follows:

(1) hybrid lower approximation:

$$\underline{R^{\sim <}} d_i^{\leq}(x) = \inf_{u \in U} \max\{1 - R^{\sim <}(u, x), d_i^{\leq}(x)\};$$
$$\underline{R^{>}} d_i^{\geq}(x) = \inf_{u \in U} \max\{1 - R^{>}(u, x), d_i^{\geq}(x)\};$$

(2) hybrid upper approximation:

$$\overline{R^{\sim <}} d_i^{\leq}(x) = \sup_{u \in U} \max\{1 - R^{\sim <}(x, u), d_i^{\leq}(x)\};$$
$$\overline{R^{\sim >}} d_i^{\geq}(x) = \sup_{u \in U} \max\{1 - R^{\sim >}(x, u), d_i^{\geq}(x)\}.$$

Definition 6 Given an ordinal information system $\langle U, A, D \rangle$, the hybrid lower and upper monotonic dependencies of D with respect to B are computed by:

(1) hybrid lower dependency:

$$r_B^{\sim <}(D) = \frac{\sum_i \sum_{x \in d_i^{\leq}} \underline{R^{\sim <}} d_i^{\leq}(x)}{\sum_i |d_i^{\leq}|} \tag{1}$$

(2) hybrid upper dependency:

$$r_B^{\sim >}(D) = \frac{\sum_i \sum_{x \in d_i^{\geq}} \underline{R^{\sim >}} d_i^{\geq}(x)}{\sum_i |d_i^{\geq}|} \tag{2}$$

If $r_B^{\sim <}(D) = 1$ or $r_B^{\sim >}(D)$, then we say decision is upward or downward consistent on features B, respectively. It can be seen that $0 \leq r_B^{\sim <}(D) \leq 1$ and $0 \leq r_B^{\sim >}(D) \leq 1$.

We following combine the hybrid dependency with the search strategy of genetic algorithm (GA) to search the optimal feature subset. Before feature selection, it deserves to point out that the partition of non-monotonic and monotonic features is significant for the performances of the proposed feature selection. So in this paper, we use GA to decide the proportion of monotonic features for the original feature set.

4 Feature Selection and Classification Results

In this paper, the synthetic data is collected from an experimental system, refer to [3]. The experiment has been conducted on gear crack with three damage levels: baseline (no crack), slight (tooth crack with levels 25%), moderate (tooth crack with levels 50%). The data set includes 108 samples and 52 features. The detail information of the 52 features is available in reference [8].

We firstly discuss the monotonicity between each feature and decision. Rank mutual information has proved its effectiveness for reflecting the monotonic consistency in ordinal classification [7, 9]. The rank mutual information of each feature with decision is shown in Fig. 1. We can see that some features have higher values of rank mutual information, which demonstrate these features have monotonic relevance with decision, and others do not have such relationships. On the other hand, this result proves that only some features have the monotonic relationships with decision in real-world applications.

According to the rank mutual information of each feature, we randomly select some features; F27 and F48 have the higher monotonic consistency, and the monotonic dependence of F13 and F24 are nearly close to 0, respectively. The scatter plots of the above four selected features are shown in Fig. 2. It can be seen that F27 and F48 change monotonically with fault severity levels clearly. The other two features (F13 and F24) have not shown the monotonic relationship.

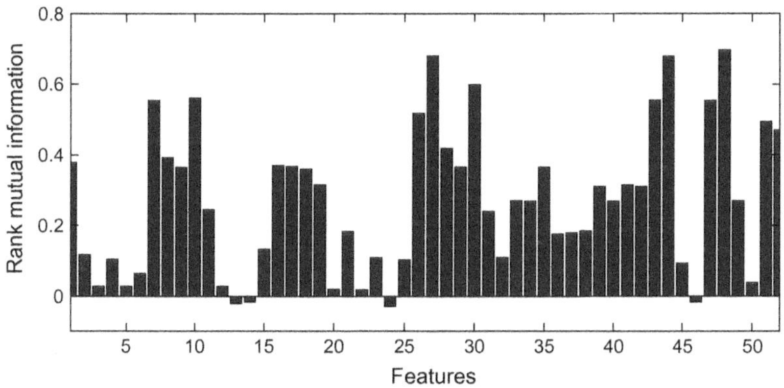

Fig. 1 Feature evaluation using rank mutual information

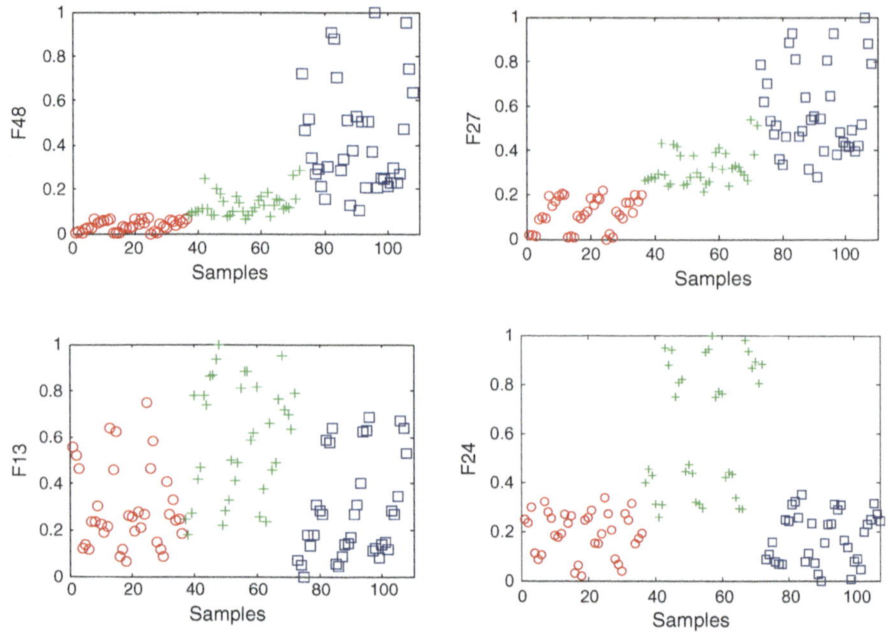

Fig. 2 Scatter plot of some monotonic features and non-monotonic features

The original feature space has 52 features. As we know, there are some irrelevant features, which do not have relevant information with the fault severity. In this section, we firstly employ the proposed feature selection algorithm to get a feature subset. Then some classification algorithms are introduced for validating the classification performance of feature subsets. Now, we compare the selected features, where hybrid monotonic dependency and monotonic dependency are used as evaluation function. The selected features are shown in Table 1.

In what follows, to test the performance of feature selection, five classifiers are selected as the classifiers: ordinal class classifier (OCC) [10], rank entropy-based decision tree (REDT) [9], rank tree (RT) [11] are three ordinal classifiers, and C4.5 and KNN ($K = 1$) are two non-ordinal classifiers. Based on ten-fold cross-validation, mean absolute error (MAE) is used to evaluate the classification performance. Table 2 shows the mean absolute error (MAE) computed with different features. Comparing the results, we can derive that the combination of hybrid monotonic dependency with genetic algorithm (GA) can reach better performance

Table 1 Feature selected with different evaluation functions

	Monotonic dependency				Hybrid monotonic dependency					
Features selected	7	25	30	48	13	25	27	30	44	48

Table 2 Mean absolute error of the feature subsets

Classifier	Hybrid monotonic dependency	Monotonic dependency
OCC	0.0237	0.0935
REDT	0.1167	0.1167
RT	0.0333	0.0361
C4.5	0.0237	0.0300
KNN	0.0062	0.0432

than integration of monotonic dependency with genetic algorithm (GA), which can demonstrate the proposed algorithm is effective for fault level identification.

5 Conclusions

This paper designs a feature selection algorithm for ordinal classification. First, we introduce a new concept of hybrid monotonic dependency for ordinal classification. Second, we combine this measure with GA search strategy to search an optimal subset. Finally, the proposed feature selection algorithm is applied to gear crack level identification. The results show that the proposed method not only can reduce the features size, but also improve the classification performance.

Acknowledgements This work is supported by Youth Project of Fujian Province of China (Grant No. JAT160350).

References

1. Ben-David, A., Sterling, L., Pao, Y.H.: Learning and Classification of Monotonic Ordinal Concepts. Blackwell Publishers (1989)
2. Gutiérrez, P.A., Pérez-Ortiz, M., Sánchez-Monedero, J., et al.: Ordinal regression methods: survey and experimental study. IEEE Trans. Knowl. Data Eng. **28**, 127–146 (2016)
3. Lei, Y., Zuo, M.: Gear crack level identification based on weighted K nearest neighbor classification algorithm. Mech. Syst. Signal Process. **23**(5), 1535–1547 (2009)
4. Zhao, X., Zuo, M.J., Patel, T.H.: Generating an indicator for pump impeller damage using half and full spectra, fuzzy preference-based rough sets and PCA. Meas. Sci. Technol. **23**(4), 45607–45616 (2012)
5. Zhang, L., Huang, W., Xiong, G.: Assessment of rolling element bearing fault severity using multi-scale entropy. J. Vib. Shock **33**(9), 185–189 (2014)
6. Hu, Q., Pan, W., Zhang, L., et al.: Feature selection for monotonic classification. IEEE Trans. Fuzzy Syst. **20**(1), 69–81 (2012)
7. Pan, W., Hu, Q.: An improved feature selection algorithm for ordinal classification. IEICE Trans. Fundam. Electron. Commun. Comput. Sci. **99**(12), 2266–2274 (2016)
8. Pan, W., He, H.: An ordinal rough set model based on fuzzy covering for fault level identification. J. Intell. Fuzzy Syst. **33**(5), 2979–2985 (2017)

9. Hu, Q., Che, X., Zhang, L.: Rank entropy-based decision trees for monotonic classification. IEEE Trans. Knowl. Data Eng. **24**, 2052–2064 (2012)
10. Frank, E., Hall, M.: A simple approach to ordinal classification. In: Proceedings of the 12th European Conference on Machine Learning, pp. 145–156 (2001)
11. Xia, F., Zhang, W., Li, F., et al.: Ranking with decision tree. Knowl. Inf. Syst. **17**(3), 381–395 (2008)

Computing Model of Musical Multiple Perception Based on Memory Mapping Perception Inversion

Yumeng He and Ping He

Abstract Music pattern description and model establishment are the research hotspots of music cognition and computing musicology. The purpose of this paper is to propose a music fuzzy perception computing model based on fuzzy sets and perceptual learning. Firstly, according to the inversion principle of relational mapping, human music cognition is regarded as the neural network of memory mapping and perception inversion (MMPI), and a conceptual model is established for description. Secondly, based on the analysis of music fuzzy perception characteristics, the fuzzy perceptual feature index (FPCI) of music elements is defined. Finally, a music cognitive learning model based on fuzzy perception features is proposed. Research shows that human music cognition is a learning process based on fuzzy perception, which is a fusion process between music memory and reality cognition. The principle of testing this fusion is to optimize the fuzzy perception of the music.

Keywords Musical computing model · Fuzzy perception · Music mapping perception inversion (MMPI) · Musical multiple perception

1 Introduction

Music pattern description and model establishment are the research hotspots of music cognition and computing musicology. The purpose of this paper is to propose a music fuzzy perception computing model based on fuzzy sets and perceptual learning. Firstly, according to the inversion principle of relational mapping, human music cognition is regarded as the neural network of memory mapping and perception inversion (MMPI), and a conceptual model is established for description.

Y. He
School of Music, Liaoning Normal University, Dalian 116029, China

P. He (✉)
Liaoning Police College, Dalian 116036, China
e-mail: heping2000@163.com

V. Jain et al. (eds.), *Recent Trends in Intelligent Computing, Communication and Devices*, Advances in Intelligent Systems and Computing 1031,
https://doi.org/10.1007/978-981-13-9406-5_13

Secondly, based on the analysis of music fuzzy perception characteristics, the fuzzy perceptual feature index (FPCI) of music elements is defined. Finally, a music cognitive learning model based on fuzzy perception features is proposed. Research shows that human music cognition is a learning process based on fuzzy perception, which is a fusion process between music memory and reality cognition. The principle of testing this fusion is to optimize the fuzzy perception of the music.

Music perception is an attractive issue in the field of music cognition and intelligent computing. In many research literatures, the measurement of music perception and the study of relevance are involved [1–3]. It is unclear which factors determine this correlation. However, it can be confirmed that the musical ability is based on the perceptual ability of music training. What is the perceived ability of music learning? How to evaluate music perception? These problems have become a research hotspot in music cognition, music computing, and music intelligent system design. In the past 20 years, the research on music computing mode and music intelligence system has been rapidly developed. The research involves two aspects, one is the computer-based formal description of music, and the other is the analysis of music perception characteristics based on the category of music cognition.

The structure of this paper is as follows: Sect. 2 presents the fuzzy perception of music as well as the fusion framework of two modes: memory mapping and perception inversion. Section 3 gives the music cognitive model based on the music characteristics of fuzzy perception. Section 4 presents the result of the experiment. Section 5 is the conclusion of this paper.

2 Fuzzy Perception of Music

2.1 Fuzziness of Music

Music is the art of sound, and the perception of this sound feature is fuzzy, and the perception of this fuzzy determines the human's musical ability. People get their cognition of music through their perception of musical elements. For example, the height of the sound, the length of the sound, the strength of the sound, and the timbre are the basic elements of music. The perception of these basic musical elements is fuzzy, and different people have different degrees of perception. Therefore, fuzzy set methods can be used to describe the perception of these musical elements. For the height of the sound, it can be expressed in very high, high, higher, normal, lower, low, and very low. For the strength of sound, it can be very strong, strong, stronger, general, weaker, weak, very weak to express, and so on [4]. In the perception of these musical elements, there is no definite measurement standard, but the musical ability can be used to obtain the hearing ability of the ambiguity of these musical elements, which is based on the perceptual ability of music element memory.

2.2 Memory Mapping and Perception Inversion (MMPI)

There is a music intelligence system with memory and perception in the human brain. The function of this intelligent system is to get the recognition of music through the mapping and inversion process of memory and perception. Specifically, a mapping is realized from memory to perception, and a new memory is obtained through the process of perceptual music, and an inversion is completed. This paper refers to a process of music mapping and inversion to realize a music computing. In this way, after a limited number of music memories and perceptual computing, a certain music cognition model is obtained [5, 6].

In musical computing, suppose that there exist a group of mode $S = (X, R)$ and $S^* = (X^*, R^*)$ on perception mode set X and its memory mode set X^* as well as perception mode relationship R, memory mode relationship R^*, and, suppose there is a perception mode $x \in X$ in music mode structure S, it is an undetermined mode in the musical intelligent system under the memory mapping M, that is $M : x^* \to x(x^* \in X^*, x \in X)$. In the musical cognition process, M is a memory mapping that it not only known music mode x^* into x but also music relationship (or algorithm) R^* of X^* into R of X. Therefore, known music mode structure $S^* = (X^*, R^*)$ can be mapped into perception mode structure $S = (X, R)$ by M. The framework of this model is shown by Fig. 1.

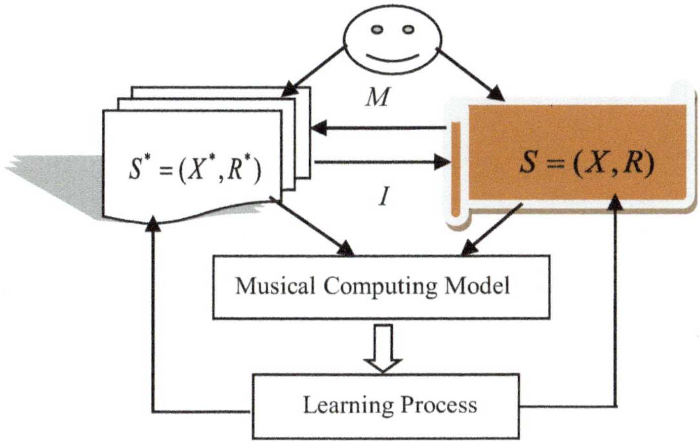

Fig. 1 The fusion framework of memory and perception

3 Music Intelligence Based on Fuzzy Perception

3.1 Musical Feature Index

The so-called music perception of human beings is based on the comprehensive judgment results of music knowledge, music experience, and music intuition through the mapping-inversion process, and this judgment is fuzzy. The music perception of the memory mappings $p(x) = p(x|x^*) \cup \bar{p}(x|x^*)$ is a fuzzy perception process [7]. This process illustrates the basic characteristics of human brain music perception. Although the formal expression given in [8] has certain limitations, it has certain significance for exploring the perceptual characteristics of human brain music.

Therefore, any musical element in the perceptual field has a one-to-one correspondence with its fuzzy perceptual set. It includes two aspects: First, the degree of perceptibility of music is determined by the fuzzy perceptual and non-fuzzy perceptual features of the musical element. Secondly, the cognitive ability of music can be measured by the perceived feature index. In other words, the same musical features will have different musical cognitions under different perception angles [9]. The formal definition is given as follows.

Definition 3.1 Let $S = (X, R)$ be a space of music elements, $S^*(R^*, X^*)$ be a space of music memory, and there exists a mapping M with fuzzy perception $p(x|x^*)$ and non-fuzzy perception $\bar{p}(x|x^*)$ based on $S^*(X^*, R^*)$, then $I(x) = \{ < x, p(x|x^*), \bar{p}(x|x^*) > | x \in X, x^* \in X^* \}$ be called a music perception feature index (MPFI), where $0 \leq p(x|x^*) - \bar{p}(x|x^*) \leq 1$, without loss of generality, we have

$$I(x) = \begin{cases} 1 & p(x|x^*) = 1, \bar{p}(x|x^*) = 0 \\ 0 & p(x|x^*) = 0, \bar{p}(x|x^*) = 1 \\ \frac{1}{2}\{1 + [p(x|x^*) - \bar{p}(x|x^*)]\} & 0 < p(x|x^*) - \bar{p}(x|x^*) < 1 \\ 0.5 & p(x|x^*) = \bar{p}(x|x^*) \end{cases}.$$

The definition with nature comes to following:

(1) If $p(x|x^*) = 1$, then $\bar{p}(x|x^*) = 0$, $I(x) = 1$ that is $S = (X, R)$ with the degree of the maximum fuzzy perception.
(2) If $p(x|x^*) = 0$, then $\bar{p}(x|x^*) = 1$, $I(x) = 0$, there is not music judgment for $S = (X, R)$.
(3) If $0 < p(x|x^*) - \bar{p}(x|x^*) < 1$, then $S = (X, R)$ with fuzzy perception $I(x)$ ($I(x) \in (0, 1)$.
(4) If $p(x|x^*) = \bar{p}(x|x^*)$, then $S = (X, R)$ with maximum hesitate degree.

The music perception feature index is a measure of music perception and a computing model that describes the intelligent features of music perception. It is not difficult to find through this computing model that the intelligent behavior of music

Table 1 Type's distribution of MMPI

	MMPI					
	M	*C*	*I*	CC	LC	DC
Knowledge (*K*)	(k, m)	(k, c)	(k, i)	(k, cc)	(k, lc)	(k, dc)
Experience (*E*)	(e, m)	(e, c)	(e, i)	(e, cc)	(e, lc)	(e, dc)
Intuition (*I*)	(i, m)	(i, c)	(i, i)	(i, cc)	(i, lc)	(i, dc)

perception is formed in the choice of fuzzy perception and non-fuzzy perception of music elements [10].

3.2 The Basic Type of Music Perception

In the process of perceiving music, there are three types to perceive the musical form elements such as melody, timbre, and tonality of the music: (1) knowledge-based memory-perception, (2) experience-based memory-perception, and (3) intuition-based memory-perception. In other words, different musical perceptions will have different perceptual characteristics and different musical cognitive abilities [11]. From the perspective of MMPI research, the first type can be determined as a full memory-perception feature (*M*), the second type as a comparative memory-perception feature (*C*), and the third type as an inspiration memory-perception feature (*I*). In addition, from the relationship between music perception and cognition, it can be determined that the music cognition obtained by the first type of music perception is conceptual cognition (CC), the music cognition obtained by the second type of music perception is learning cognition (LC), and the music cognition obtained by the third type of music perception is the discovery of cognition (DC) as shown in Table 1.

4 Experimental Results

According to the music perception computing model proposed in this paper, an experimental analysis of the music perception ability of 120 students who studied music was conducted at Dalian Qimeng Music School. Select the violin concerto "Liang Zhu" to test students' perception and cognition of this music work. The experimental results show that (1) 35 students (29%) are based on music knowledge memory to obtain music perception. Among the students who rely on knowledge memory, 22 students (63%) have satisfactory fuzzy perception ability, and 12 of them have been trained in music perception, which can enhance the non-fuzzy perception of some musical elements to fuzzy perception. (2) There are 28 students (23%) who are based on intuition to gain musical perception. Among students who rely on intuition, 8 students (29%) have strong musical talent. (3) 57 students (48%) are based on experience to get music perception. The music experience refers to the influence of music practice and music environment. Musical perception through music experience accounts for a large proportion of students engaged in music learning, which shows that music experience plays an important role in music perception.

The experimental results of this study show that music perception is a kind of intelligence of human beings. Through the MMPI process, not only can students discover the characteristics of students' musical talents, but also cultivate music memory-perception ability. Moreover, the role of the music perception computing model is to be able to effectively correct the music fuzzy perception and the music non-fuzzy perception.

5 Conclusions

This paper proposes a new concept—memory mapping and perceptual inversion (MMPI) from the perspective of the relationship between music memory and perception. Based on the fuzzy perception and non-fuzzy perception of music elements, a music perception computing model is established. Studies showed that the music memory-perception feature is an intelligent feature of the human brain, with three types: knowledge memory-perception, experience memory-perception, and intuition memory-perception. Different people have different types of music memory and perception. Therefore, music perception and cognitive learning and training should be designed according to different types.

Acknowledgements This paper is supported by the Liaoning Education Science Planning Fund (JG18DB290). In the process of experimental analysis, we got the help of the teachers and students of Liaoning Normal University and Dalian Qimeng Music School. The experimental data in this paper was provided by Wang Xin and Zhang Jiahui. Thank you very much for their help.

References

1. Homenda, Wladyslaw, Jastrzebska, Agnieszka, Pedrycz, Witold: Multicriteria decision making inspired by human cognitive processes. Appl. Math. Comput. **290**(1), 392–411 (2016)
2. Trainor, L.J., Unrau, A.: Development of Pitch and Music Perception, vol. 42, pp. 223–254 Springer Handbook of Auditory Research (2012)
3. Camurri, A., De Poli, G., Rocchesso, D.: A taxonomy for sound and music computing. Comput. Music J. **19**, 4–5 (1995)
4. Perlovsky, Leonid, Ilin, Roman: Mathematical model of embodied symbols: cognition and perceptual symbol system. J. Behav. Brain Sci. **2**, 195–220 (2012)
5. Chen, P., Zhao, L., Xin, Z., Qiang, Y.: A scheme of MIDI music emotion classification based on fuzzy theme extraction and neural network. In: International Conference on Computing Intelligence, pp. 323–326 (2017)

6. Kiendl, H., Kiseliova, T., Rambinintsoa, Y.: Use of fuzzy strategies for interpretation of music. J. Mult.-Valued Log. Soft Comput. **12**, 221–265 (2006)
7. Heister, H.W.: Invariance and Variance of Motives: A Model of Musical Logic and/as Fuzzy Logic, vol. 273, pp. 423–450. Springer, Berlin, Heidelberg (2012)
8. Corbin, J.C., Reyna, V.F., Weldon, R.B., Brainerd, C.J.: How reasoning, judgment, and decision making are colored by gist-based intuition: a fuzzy-trace theory approach. J. Appl. Res. Mem. Cognit. **4**(4), 344–355 (2015)
9. Kiendl, H., Kiseliova, T., Rambinintsoa, Y.: Fuzzy interpretation of music. Lect. Notes Comput. Sci. **44**, 756–764 (2004)
10. Ping, He: Design of interactive learning system based on intuition concept space. J. Comput. **5**(3), 478–487 (2010)
11. Yang, Y.H., Liu, C.C., Chen, H.H.: Music emotion classification: a fuzzy approach. In: ACM International Conference on Multimedia, pp. 81–84 (2006)

A Natural Immersive Closed-Loop Interaction Method for Human–Robot “Rock–Paper–Scissors” Game

Xvjun Yuan, Shan Dai and Yeyang Fang

Abstract The existing Kinect somatosensory games and other human–computer interaction are lack of natural interaction experience. We made research on a natural immersive closed-loop interaction method for human–robot “rock–paper–scissors” game. Kinect was used to make somatosensory dynamic recognition. Mechanical arm was used to make gestures to interact with human and to show the result of the game. System through the development of Modbus master and Modbus slave based on Arduino to control the mechanical arm, and used the multithread concurrent to achieve a better real-time experience. The PC was programmed to determine the outcome and record the result. The natural immersive closed-loop interaction method has the advantages of natural interaction, fast response, and strong expansibility.

Keywords Human–robot interaction · Kinect recognition · Mechanical arm

1 Introduction

The human–robot interaction is a research hotspot in recent years. As a recognition device with good performance, Kinect is widely used in the related studies. Many researchers made research on the recognition based on Kinect. Skeleton is used to track human action [1] and measure shoulder joint angles [2]. What’s more, Kinect shows a great potential in hand gesture recognition [3]. Based on these studies,

X. Yuan
Shanghai Cohere Electronics Technology Co., Ltd., Shanghai 201101, China

S. Dai (✉)
School of Mechanical Engineering, Shanghai Jiao Tong University, Shanghai 200240, China
e-mail: daishankey@sjtu.edu.cn

Y. Fang
University of Michigan, Ann Arbor 48109, USA

V. Jain et al. (eds.), *Recent Trends in Intelligent Computing, Communication and Devices*, Advances in Intelligent Systems and Computing 1031,
https://doi.org/10.1007/978-981-13-9406-5_14

many researchers achieved a lot in human–robot interaction. Gesture-based human–robot interface is developed [4]. However, few researchers have focus on the human–robot entertainment system. And, researchers usually study on letting robots make response to specific human gestures.

In addition, somatosensory game has applied to many areas in recent years. Researchers have studied somatosensory game's application in rehabilitation [5], and some researchers have managed to develop somatosensory games with portable devices [6]. However, despite the broad application of somatosensory games, few studies were made to improve the experience of somatosensory games. During the game, users usually interact with screen, which is not immersive and naturally interactive.

This paper presents a natural immersive closed-loop interaction method for human–robot "rock–paper–scissors" game. A human–robot game system including mechanical arm and Kinect is built based on the research. The mechanical arm was programmed to show gestures of its own and thus to play games with human. Kinect is used to recognize player's gesture.

2 System Description

Interactivity, real-time, and immersion are the keys to improve interaction experience. This system is aimed at the above key points. Through the judgment of the 'rock-paper-scissors' game result to achieve human-robot closed-loop interaction.

Figure 1 shows the framework of the natural immersive interaction system. PC sends signals to Modbus master. Modbus master sends corresponding signals to Modbus slave and driving the steering gear in the mechanical arm to make the appropriate gestures. Meanwhile, PC receives the dynamic gesture recognition from Kinect. Players only need to interact with the mechanical arm to play the game, and they can also know the result of the game easily from the gestures after the game..

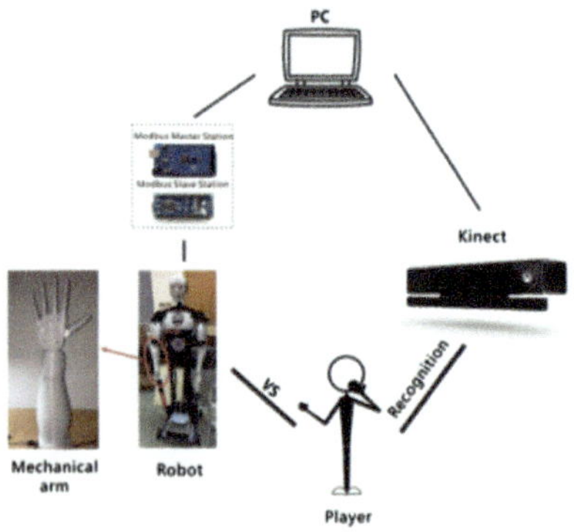

Fig. 1 Framework of the interaction system

2.1 The Natural Interaction Implementation

Through the construction of robot to achieve the natural interaction. In the testing "rock–paper–scissors" game system, a mechanical arm is built. The mechanical arm is response for showing countdown before the game, PC's gesture during the game and the gestures representing win-or-lose result after the game.

It can be manufactured easily with a 3D printer. The fingers of mechanical arm have two gestures which can be either straight or bent, corresponding to the open or close of the fingers. In addition, the wrist also contains a degree of freedom of rotation. The five fingers and wrist are controlled by Arduino development board.

The mechanical arm uses wire to control the fingers' position. Wires are pulled by steering gears. Interphalangeal joints are connected through rotation shafts, to realize mutual rotation, as shown in Fig. 2. The wires are fixed to the fingertips, passing through the finger joints, then passing through the wrist, arm, and finally connected to the steering gear. When the steering gear is turned to a certain angle, one end of the wires will produce corresponding tension to achieve the fingers' bending or straightening gestures. Figure 3 shows the overall structure of mechanical arm.

To match the robot, a recognition system based on Kinect is built. Players can interact with the device in natural ways like gestures, movements, and voice. The mechanical arm is designed to show the gestures of PC program while Kinect is used to identify the player's gesture dynamically.

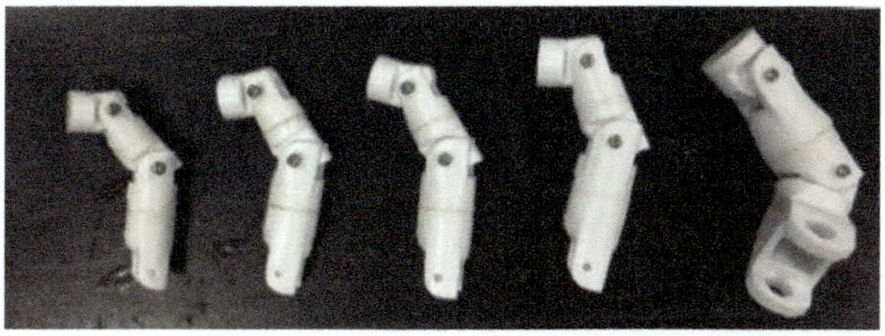

Fig. 2 Structure of interphalangeal joints

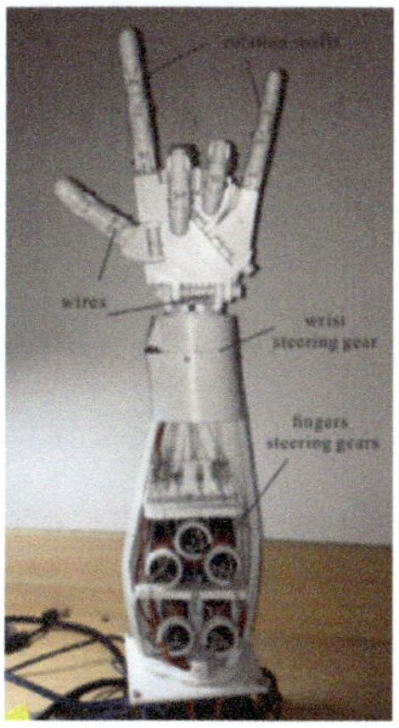

Fig. 3 Overall structure of mechanical arm

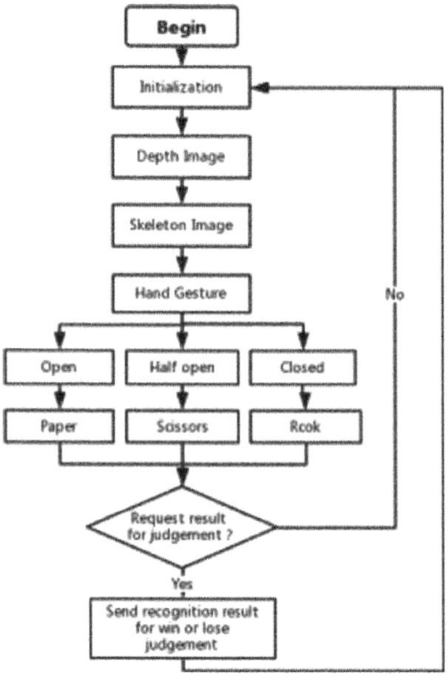

Fig. 4 Flowchart of recognition process

Using cameras on the Kinect to achieve the recognition of the player's gesture. Figure 4 shows the flowchart of recognition process. First, a depth image is used to identify human body from the environment. Then, the skeleton of human is recognized, which helps to locate the position of the hand of player. Kinect is able to identify the hand "open, half open, closed"—three states, respectively, corresponding to "paper, scissors, rock"—three gestures. Figure 5 shows the identified image. The recognition process keeps running cyclically and dynamically. The recognition result is dynamically changing with the change of human gesture. When the win-or-lose judgment program requests the recognition result, it will send related information and then go back to the loop.

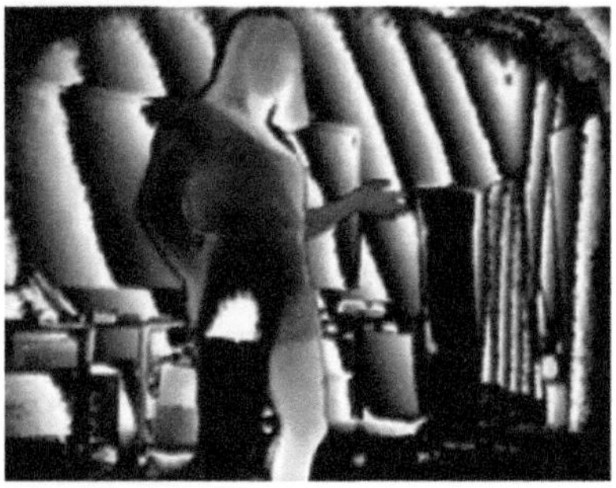

(a) The depth image

(b) The skeleton image

Fig. 5 Identified images

2.2 The Real-Time Performance Implementation

A multithread concurrent PC program is developed to improve the real-time performance of the system. The program is developed in Qt. In the Qt program, two threads are defined. One of the threads is used to perform the program of image generation, to identify the gesture, and use OpenCV to generate the depth image and the human skeleton image. Another thread is used to display GUI, send action signals to robot through serial communication, and receive the information of Kinect human gesture recognition result. This thread is also responsible for the result judgment. In this way to realize the Kinect gesture real-time recognition.

In the testing "rock–paper–scissors" game, usually players make the final gesture after the countdown ends. Since the Kinect recognition results change as the player's gesture changes, it is necessary to select the appropriate moment to obtain the Kinect recognition result for win-or-lose judgment to get the correct game result.

In human versus robot games, the countdown process is carried out by the mechanical arm gestures. When pressed "Game Begin" button, the mechanical arm shows the "3, 2, 1" countdown gestures first, then shows one of the "rock, paper, scissors" gesture randomly. Player only needs to make a gesture at the end of the countdown. At this moment, the GUI thread obtains the Kinect recognition result from the recognition thread and uses it to make the win-or-lose judgment.

2.3 The Immersion Mode Implementation

The immersion mode is mainly achieved by allowing players to play games with robot instead of screen. Arduino development boards supporting Modbus protocol were developed to control the robot.

The steering gear and motors in the robot are controlled by Arduino series development boards. Modbus is adopted as data transition protocol between development boards. A Arduino Mega development board is programmed to be the Modbus master and many Arduino Nano development boards are programmed to be Modbus slaves. Figure 6 shows the software framework in the robot. The PC sends different action instructions to the Modbus master via the USB serial port. The master responds to it and sends a corresponding signal to the slaves. TTL-485 module, RS485 hub, 485-TTL module are used as a splitter. In this way, one master can connect to plenty of the slave. The slaves respond to the master's signal, run the associated program, and drive the steering gears to specific angle.

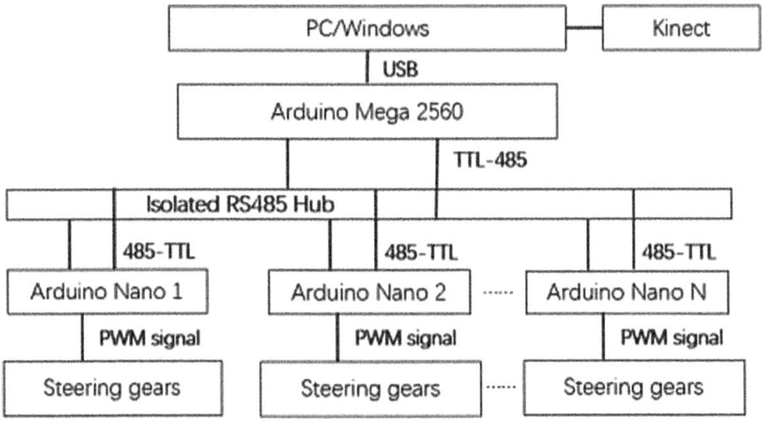

Fig. 6 Software framework of the system

3 The Closed-Loop Characteristics

Based on the natural immersive closed-loop interaction method for human–robot, a testing interaction system running "rock–paper–scissors" game is developed. After the countdown, the player and the mechanical arm will make the gesture at the same time. Comparing both gestures to judge the result.

Figure 7 shows the GUI interface and Fig. 8 shows the flowchart of the testing game system. The player clicks the "Open Serial" bottom to complete the connection between PC and Modbus master. Clicking the "Game Begin" bottom to activate the game. The mechanical arm makes the countdown gestures. At the end of countdown, Kinect recognizes the player's gesture and the mechanical arm shows the gestures randomly generated by PC. Kinect recognized gesture and the mechanical arm gesture are displayed in corresponding text box, respectively. PC makes the win-or-lose judgment and shows in the "Game Result" text box and the cumulative results are recorded as well. Meanwhile, the mechanical arm does

Fig. 7 GUI interface

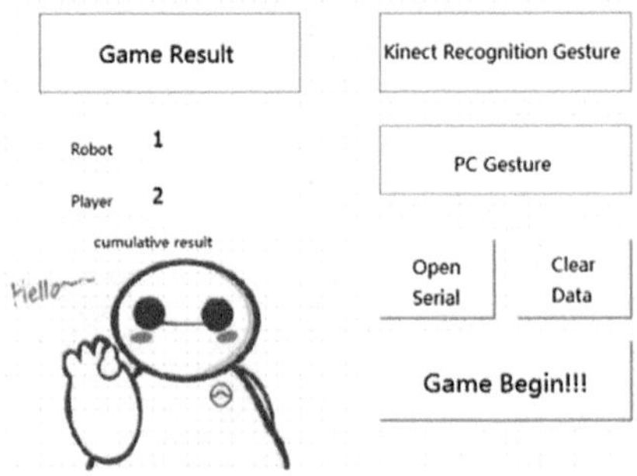

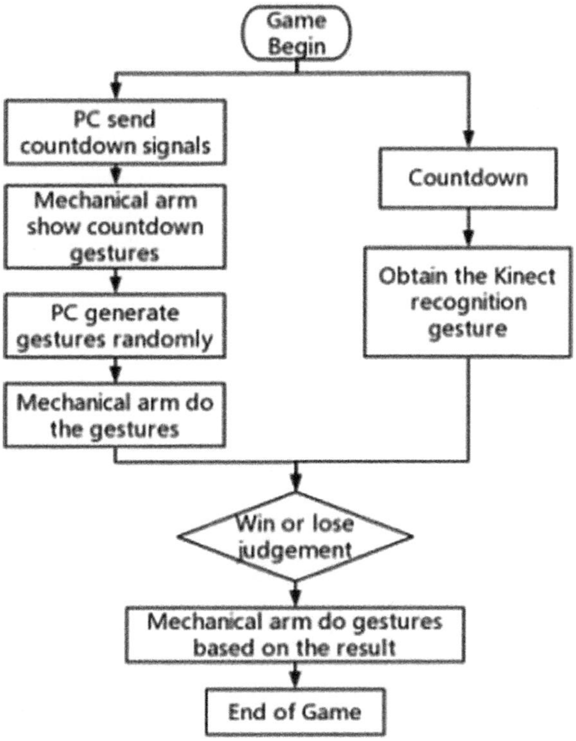

Fig. 8 Flowchart of the system

gestures based on the result. It shows "paper" gestures for a tie, rotating its wrist for player's loss and thumbs up for player's winning. And, the player can clear the cumulative results by clicking "clear data" button.

4 System Testing and Evaluation

Accuracy and real-time performance are two important factors for the method discussed above. The testing game system may appear to have problems like mechanical arm response delay, identification error, and judgment error, and thus affecting the player's experience. Five different people are selected to perform the test. Each player plays the game 50 times and Table 1 shows the testing data and Fig. 9 shows the result.

The result shows that the mechanical arm response, the recognition, and the judgment system performed well. For the Kinect recognition accuracy, sometimes recognition error occurs. The situations of error are studied and we found that mostly the wrong result is "stone", corresponding to the first gesture. We analyzed the game process and realized that it is because the player did not make the gestures

Table 1 The testing data of the system

Number	Times for fast response	Percentage for fast response (%)	Times for correctly recognition	Percentage for correctly recognition (%)	Times for correctly judgement	Percentage for correctly judgement (%)
1	47	94	48	96	48	100
2	49	98	46	92	48	100
3	48	96	47	94	47	100
4	47	94	48	96	46	100
5	48	98	49	98	49	100

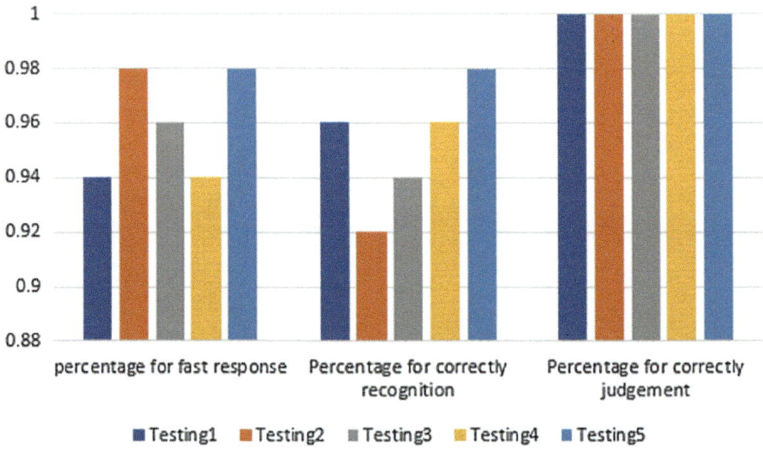

Fig. 9 Testing result of the game system

in time. When PC began to get the recognition result, the player still remains the "stone" gesture during the countdown period and changed it after PC has already got the recognition result. This problem can be solved by delaying another 0.5 s after the countdown and then obtain the recognition result. When the identification is correct, the judgment is correct at the rate of 100%. It shows that the judgment algorithm is running well.

5 Conclusion

This paper studies a natural immersive closed-loop interaction method for human–robot. A testing "rock–paper–scissors" game system based on the method was built. The innovation of the study lies in the study of the interactivity, real-time, and immersion of the human–robot interaction method. Using Kinect and mechanical arm to naturally interact with players. The innovative design of the mechanical arm countdown process allows player to make gesture at the right moment and combined with multithread concurrent program ensures the real-time recognition. The development of the Modbus master and slaves reduce the wiring complexity and the response time of the robot.

A natural immersive closed-loop interaction method uses Kinect to make dynamical recognition, uses robot to make naturally interaction and uses Modbus protocol to make the system simple and convenient for further development. The same technology can be applied in the development of various human–robot interactive games, and the design of the guide robot and traffic-command robot, etc.

References

1. Papadopoulos, G.T., Axenopoulos, A., Daras, P.: Real-time skeleton-tracking-based human action recognition using kinect data. In: International Conference on Multimedia Modeling, pp. 473–483 (2014)
2. Huber, M.E., Seitz, A.L., Leeser, M., Sternad, D.: Validity and reliability of Kinect skeleton for measuring shoulder joint angles: a feasibility study. Physiotherapy **101**, 389–393 (2015)
3. Ren, Z., Yuan, J., Meng, J., Zhang, Z.: Robust part-based hand gesture recognition using kinect sensor. IEEE Trans. Multimedia **15**(5), 1110–1120 (2013)
4. Qian, K., Yang, H., Niu, J.: Developing a gesture based remote human-robot interaction system using kinect. Int. J. Smart Home **7**(4), 203–208 (2013)
5. Chou, C.H., Tsai, C.Y., Fang, S.J., Hsiao, Y.C., Chen, H.L.: A home rehabilitation system combined with somatosensory games for stroke patients. ICIC Express Lett., **73**(B), 1005–1010 (2013)
6. Pan, M.-S., Hsu, W.-C., Liu, C.-H., Huang, K.-C., Cheng, C.-F.: InstantGaming: playing somatosensory games using smartwatches and portable devices. In: IEEE International Conference on Applied System Innovation, pp. 1072–1074 (2017)

Capsule Network-Based Facial Expression Recognition Method for a Humanoid Robot

Jingru Zhang and Nanfeng Xiao

Abstract Compared to the classical convolutional neural network (CNN), the capsule net Hinton put forward can use fewer network layers to achieve the classification tasks very well and arrive at the convergence with a faster speed. The principle of the capsule net is based on the CNN, and it is just that the neuron form is converted from the scalar to the vector, which is a capsule, and then chooses the suitable capsule for the final output through the dynamic routing method (Sabour in Dynamic routing between capsules, [1]). In this paper, on the basis of the capsule net, use deconvolution to restore images and optimize the error between original images and restored images. The classical facial emotions database named Cohn-Kanade Database Plus (CK+) that is processed through Data Augmentation is used to conduct experiments. Lately, the classification results are combined with the NAO robot. The NAO robot is able to visualize the emotion by changing its eyes colors and speaking the results, achieving the purpose of combining theory with practice.

Keywords Capsule · Convolutional neural network (CNN) · Facial expression · NAO robot

1 Introduction

With the development of various technologies and theories, more and more work can be done by robots. Along with the rapid development of artificial intelligence, robots also become smarter [2]. Facial recognition will be an integral part of a service robot if it is to perform functions such as emotional substitution. This function needs the basis of emotion classification model such as the hidden Markov model, which can transform the state of emotion from a calm state to the

---

J. Zhang (✉) · N. Xiao
School of Computer Science and Engineering, South China University of Technology, Guangzhou, China
e-mail: 305109567@qq.com

V. Jain et al. (eds.), *Recent Trends in Intelligent Computing, Communication and Devices*, Advances in Intelligent Systems and Computing 1031,
https://doi.org/10.1007/978-981-13-9406-5_15

corresponding emotional state under the stimulus through a series of calculation of transition of probability matrix [3].

Nowadays, various kinds of CNNs are proliferating, and their abilities are varied such as Inception v3 and ResNet-50 and so on. The idea of the capsule net is based on cognitive neurologic and CNN, put forward by Hinton, who wants to realize the recognition process of human eyes. Hinton noticed an interesting fact that most current neural anatomy studies support (most of the mammals, especially primates) that in cerebral cortex there exists micro-column structures called Cortical mini-column, which contain hundreds of neurons and have an internal delamination [4]. That structure can help humans learn many attributes of one object at the same time such as the spatial relation and the shape.

At present, the development of service robot is changing and the request is high. If the service robot masters the ability of recognizing facial expressions, it will improve the quality of service of the robot greatly [5]. To solve related problems, the experiment is based on the NAO robot in the laboratory, combining with the capsule net that realizes facial emotions recognition.

2 Advantages and Properties of the Capsule Net

2.1 The Differences

Compared with other classical CNNs, the most important change is that the input form and the output form are changed from the scalar characteristics to the vector that is called the capsule, using the dynamic routing based on the similar attention mechanism instead of the max-pooling layer. The reason that why replace max-pooling by dynamic routing is that max-pooling will abandon the relationship of the spatial composition and retain the prominent information, therefore the trained model does not know the spatial relation between noise and mouth. If some images contain the noise, the eyes, and the mouth without organizing like a face, they are still recognized as faces. But the capsule net can store the spatial relation information in vectors and recognize these images that organize mouth and eyes out of order are not images of faces. This is the biggest difference between the capsule net and CNN.

2.2 The Properties of the Capsule Net

There are some problems using the capsule replacing the traditional convolution layer. First one is about how to achieve the capsule architecture. The images will become feature maps after the first layer's process. Each element of one feature map is a scalar but that can become a vector after conducting multiple convolutions of the images. Each time a convolution produces a bunch of identically structured feature maps. Combine these feature maps and then each element of one feature map becomes a vector. The vector is called a capsule. Another problem is how to use activation function for the capsule. In the capsule net, the calculation process of activation function is replaced by the linear combination and the dynamic routing, and the linear combination is defined as [1]

$$\hat{u}_{j|i} = W_{ij} \cdot u_i \tag{1}$$

where the u_i represents the capsule and $\widehat{u}_{j|i}$ is just an intermediate result. Then, the dynamic routing uses $\widehat{u}_{j|i}$ to do the calculation [1]

$$s_j = \sum_i c_{ij} \cdot \hat{u}_{j|i} \tag{2}$$

where c_{ij} is the result of b_{ij} after Softmax processing. The b_{ij} represents the conformity and is initialized to 0 matrix. Because Softmax can make the weight distribution "sharp", only a few values of c_{ij} are relatively high. Then, the conformity of capsules as output on the third layer will become differently. When finishing three iterations, b_{ij} will be updated and the highest b_{ij} of the third layer's capsule is selected, which means the dynamic routing finding the optimal way to process the images. b_{ij} will be updated by

$$b_{ij} = b_{ij} + \hat{u}_{j|i} \cdot v_j \tag{3}$$

And v_j is calculated by [1]

$$v_j = \frac{\|s_j\|}{1 + \|s_j\|^2} \cdot \frac{s_j}{\|s_j\|} \tag{4}$$

The characteristic of Eq. (4) is that the range is monotonically increasing between 0 and 1, and it rewards the longer vector and punishes the smaller vector. This equation plays the role of activation function. Finally, one innovation of this paper is using the error between original images and deconvolution-reconstructed images instead of the calculation of the edge detection error in SVM [6], which gets better results than original capsule net.

3 CK+ Dataset and NAO Robot Experiments

3.1 CK+ Dataset

Some experiments based on textured 3D video technique performed on BU-4DFE database gave an excellent 94.34% recognition rate [7]. As for still images, JAFFE database mainly collects ten Japanese women's facial expressions according to the

Fig. 1 CK+ dataset

instructions in the experimental environment. Although some papers have conducted experiments on JAFFE database and got high classification accuracy [8], this dataset contains fewer images, the amount of which is only about 200 images in total. The generalization and applicability of that are unconvinced. CK+ dataset is a classic and popular facial expression dataset, the amount of which is moderate, including seven emotions made by 123 individuals [9]. Figure 1 shows some examples. Moreover, the quality of each image is very high and emotion features are obvious, which is beneficial to the capsule net for classification. In this paper, the experiments are carried out on the sorted CK+ dataset.

However, the amount of CK+ database is still a little small to use for the capsule net. In order to avoid under-fitting, CK+ database is expanded by rotating, amplifying and adding with the salt and pepper noise. Then, the expand database is 20 times larger than the original dataset and there are over 10,000 images, which is another innovation of this paper.

3.2 NAO Robot

The NAO robot is developed by Aldebaran Robotics company, which possesses a dull appearance, with many functions such as human face recognition, speech recognition, language expression, and various other anthropomorphic limb motions that many service robots should have. NAO robot's intelligent ability relies mainly on its wide variety of the high-performance sensors and the programmable embedded processors using Linux. Above all, it can be programmed on many platforms, such as Windows, Mac OS, and Linux, to control the interaction between different modules of the robot using C++ or python language [10]. Apart from its various sensors, the RGB full-color LEDs are installed on the eyes, the ears, the feet, and so on. The color and the strength of LEDs can be adjusted through the functions called using the advanced language (Fig. 2).

3.3 Experiments

All of the experiments in this paper are conducted under the Ubuntu 16.04 system and the experimental environment of python 2.7 was built by using gtx1070 8g

Fig. 2 NAO robot

memory graphics and 16g memory. The expanded CK+ database is divided into training set and testing set, and training set accounts for 90%, testing set accounts for 10%.

The capsule net adopts a larger convolution kernel whose size is 11×11 because the facial expression image contains much information, and it needs a wider view to obtain more abundant information and the dimension of the capsule net's vectors and the final output is deepened to 16 and 24, respectively. In this way, the network can represent the characterization of the complex image information better. Considering the limit of memory resource allocation, the total training cycle is adjusted to 30 times and the batch size is set to 32. In addition, it is about the image reconstruction model. In the original text, the full connection layer is used to restore images and the error between restored images and original images can become smaller and smaller until the capsule net converges. However, the images of the CK+ database are complex and containing much detailed information. Use the full connection layer to restore images will result in a number of explosions of parameters and exhausting the resource of memory sometimes. This paper adopted the deconvolution reconstruction method for the image deconvolution operation which reduces the parameter numbers. The final result is increased by 3%. Figure 3 shows the test accuracy of capsule net with full connection and deconvolution of each cycle.

In order to verify the outstanding performance of the capsule net, other famous models are used to do the same experiment on the CK+ database, including Inception v3, ResNet-50, AlexNet, and AC-GAN (Auxiliary Classifier Generative Adversarial Nets) [11]. AC-GAN can generate new images by deconvolution and complete the process of classification [12], which offers a good idea that uses deconvolution to restore images in the capsule net. Considering models should be adjusted to the CK+ database, the class numbers of the last full connection layer are changed to 7. Figure 4 shows the test accuracy of each cycle in the expanded CK + dataset of various models, and it can be intuitively seen from Fig. 4 that capsule net converges the fastest.

Table 1 makes a summary about the testing results of various models. It shows that capsule net gets the highest testing accuracy and converges fastest.

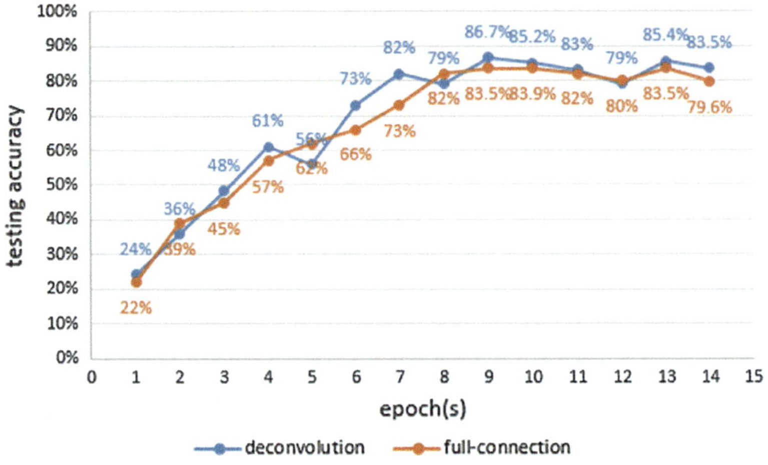

Fig. 3 Accuracies of capsule net with full connection and deconvolution

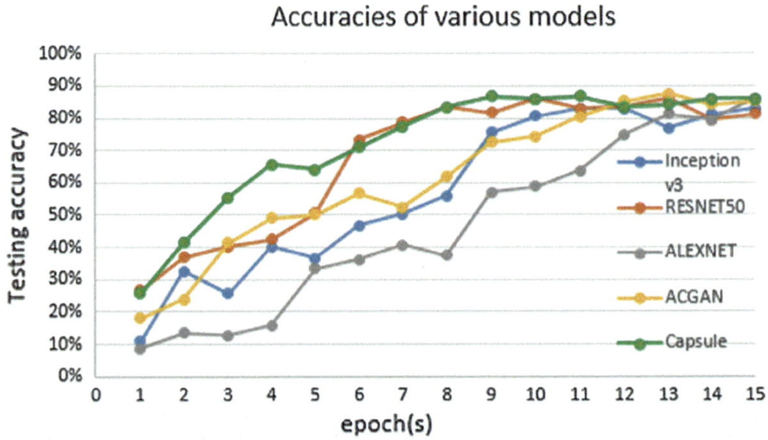

Fig. 4 Testing results of various models

Table 1 Comparison of testing results

Network	Accuracy (%)	Convergence (epochs)
Capsule net (full connection)	83.5	9
Capsule net (deconvolution)	86.7	9
Inception v3	82.9	11
ResNet-50	85.9	10
AlexNet	84.9	15
AC-GAN	85	14

Fig. 5 Eyes turn red

Fig. 6 Eyes turn green

The NAO robot's LED lights in the eyes and microphone in the mouth are used to complete interwork. While taking a picture by camera on the forehead, the NAO robot reports a result according to the trained capsule net. Then, the NAO's eyes will turn green and "correctly identified" is spoken via the microphone. When all tests are completed, the NAO robot reports the final test accuracy. The specific results are as follows (Figs. 5 and 6).

4 Conclusions

In the dataset, a popular CK+ dataset is selected for the experiment. In order to meet the experiment requirements, it is modified and adjusted accordingly. On the network structure, the capsule net theory and design idea are kept, but some of the super parameters are adjusted and the error calculation's formula is replaced by the method of deconvolution, improving the overall network performance. As for the application, the NAO robot can own facial expression recognition function based on the capsule net training model, which can express the recognition results vividly by changing the colors of the LED lights in the eyes and speaking like humans.

There are still some improvements for the experiments. That can be concluded as three aspects: (1) If hardware is allowed, the dimension of the capsule can be increased so that the capsule net can represent the complex images better. That may improve the accuracy of image recognition and classification. (2) The complex images require a deeper vector dimension to represent, and the added dimension will undoubtedly causes taking much longer time to calculate the dynamic routing, which will increase the training burden, decrease the network performance and even result in program breaking down. If there is an improvement on the dynamic routing scheme, it will greatly enhance the generality and training speed of the capsule net. (3) On the application, many complex functions of the NAO robot can be implemented so as to achieve richer emotional expression such as adding a gesture or the head actions to cooperate on the voice and the color change of eyes. Realizing those requires a deeper understanding of the control parameters and usage methods of the NAO robot.

Acknowledgements This work was supported by the National Natural Science Foundation of China under Grant [project no. 61573145], the Public Research and Capacity Building of Guangdong Province under Grant [project no. 2014B010104001], and the Basic and Applied Basic Research of Guangdong Province under Grant [project no. 2015A 03030 8018], and the authors greatly thank these grants.

References

1. Sabour, S., Frosst, N., Hinton, G.E.: Dynamic Routing Between Capsules (2017)
2. Wang, T., Tao, Y., Chen, Y.: Research status and development trends of the service robotic technology. Sci. Sin. (Informationis) **42**(09), 1049–1066 (2012)
3. Xie, L., Wen, H., Xiao, N.: Affective computing model based on HMM for home-service robot. Comput. Eng. Des. **33**(01), 322–327 (2012)
4. Hinton, G.E., Krizhevsky, A., Wang, S.D. (2011, June). Transforming auto-encoders. In: International Conference on Artificial Neural Networks, pp. 44–51. Springer, Berlin, Heidelberg
5. Li, R., Liang, L., Zhao, L.: Research on service robot oriented facial expression recognition based on vision. Mechatronics **16**(02), 43–48 (2010)
6. Osuna, E., Freund, R., Girosi, F.: Training svm: An application to face detection (1997)
7. Mishra, B., Fernandes, S.L., Abhishek, K., et al.: Facial expression recognition using feature based techniques and model based techniques: A survey. In: International Conference on Electronics and Communication Systems. IEEE, pp. 589–594 (2015)
8. Hamester, D., Barros, P., Wermter, S.: Face expression recognition with a 2-channel Convolutional Neural Network. In: International Joint Conference on Neural Networks. IEEE, pp. 1–8 (2015)
9. Lucey, P., Cohn, J.F., Kanade, T., et al.: The extended Cohn-Kanade dataset (CK+): a complete dataset for action unit and emotion-specified expression. In: Computer Vision and Pattern Recognition Workshops. IEEE, pp. 94–101 (2010)
10. NAO: Creating Interactive Robots. Robot Tech. Appl. (03), 63–65 (2014)

11. Odena, A., Olah, C., Shlens, J.: Conditional image synthesis with auxiliary classifier GANs (2016)
12. Springenberg, J.T.: Unsupervised and semi-supervised learning with categorical generative adversarial networks. Comput. Sci. (2015)

Equipment Maintenance Mode Decision Based on Fuzzy Multi-attribute Decision Method

Hongtu Cai, Yuwen Liu, Hao Wu, Pengfei Ma and Ancheng Hu

Abstract Aiming at the problem which can't use a same standard to measure various affect factors in equipment maintenance, establish equipment maintenance mode decision model on the basis of fuzzy multi-attribute by using fuzzy theory and multi-attribute decision method. And it is analyzed and verified by the example of equipment maintenance support, and the result shows that combat readiness, system security, and mission success of the equipment can be improved effectively by the method.

Keywords Maintenance mode · Decision · Fuzzy theory · Multi-attribute

1 Introduction

Equipment maintenance mode decision is the process of selecting the best maintenance mode [1] according to its health status and according to certain optimization criteria, such as maintenance cost, downtime, and system availability. So it is necessary to consider the system availability, fault risk, operational safety, performance reliability, and maintenance cost-effectiveness ratio and so on when determining equipment maintenance mode [2]. Because of the fuzziness and uncertainty of these factors, there is no unified standard to measure these factors at present [3]. Therefore, fuzzy multi-attribute decision method is needed when making decisions.

H. Cai (✉) · Y. Liu · H. Wu · P. Ma · A. Hu
Army Artillery and Air Defense College, Hefei 230031, China
e-mail: sypycht@126.com

V. Jain et al. (eds.), *Recent Trends in Intelligent Computing, Communication and Devices*, Advances in Intelligent Systems and Computing 1031,
https://doi.org/10.1007/978-981-13-9406-5_16

2 Fuzzy Multi-attribute Decision Model

Fuzzy multi-attribute decision (FMAD) is a method developed on the basis of multi-attribute decision [4–7]. It can be defined as a maintenance mode decision set $S = \{S_1, S_2, \ldots, S_n\}$, a corresponding attribute set $u = \{u_1, u_2, \ldots, u_m\}$, and a weight set representing the relative importance of attributes $\omega = \{\omega_1, \omega_2, \ldots, \omega_m\}$. The description of attribute and weight value can be expressed qualitatively or quantitatively. All attribute indexes and weights described qualitatively are expressed as fuzzy subsets or numbers in decision space. The matrix $\widetilde{M}$ composed of fuzzy index values can be described as follows.

$$\widetilde{M} = \begin{bmatrix} \widetilde{m}_{11} & \widetilde{m}_{12} & \cdots & \widetilde{m}_{1n} \\ \widetilde{m}_{21} & \widetilde{m}_{22} & \cdots & \widetilde{m}_{2n} \\ \vdots & \vdots & & \vdots \\ \widetilde{m}_{k1} & \widetilde{m}_{k2} & \cdots & \widetilde{m}_{kn} \end{bmatrix} \quad (1)$$

For weight vector $\widetilde{\omega}$ and $\widetilde{M}$, using generalized fuzzy synthesis operators to transformation operations and obtaining fuzzy decision vectors are $\widetilde{P}$; that is

$$\widetilde{P} = \widetilde{\omega} \odot \widetilde{M} = (\widetilde{p}_1, \widetilde{p}_2, \ldots, \widetilde{p}_m) \quad (2)$$

The elements $\widetilde{p}_1, \widetilde{p}_2, \ldots, \widetilde{p}_m$ in fuzzy decision vectors $\widetilde{P}$ are sorted according to the ranking method of fuzzy sets. The best maintenance strategy can be selected from maintenance strategy $S = \{S_1, S_2, \ldots, S_n\}$, that is $S_{\max}$.

3 Fuzzy Multi-attribute Decision Process

Usually, the main process of fuzzy multi-attribute decision involves two stages [8]. First, determine the fuzzy weights of each attribute and the fuzzy index value of the maintenance mode on this attribute. And determine the appropriate fuzzy operator to merge the fuzzy weights and fuzzy index values into fuzzy utility values which can represent the value of maintenance mode. Secondly, the fuzzy utility values of maintenance methods are compared according to the fuzzy set ranking method. And take the max fuzzy utility value as the final decision result. The basic process of modeling is shown as in Fig. 1. The specific steps are as follows.

Step 1: Establish maintenance decision sets and attribute sets.
Step 2: Relative importance levels are determined by system maintenance support personnel and system designer on system maintenance mode decision sets and attribute sets.

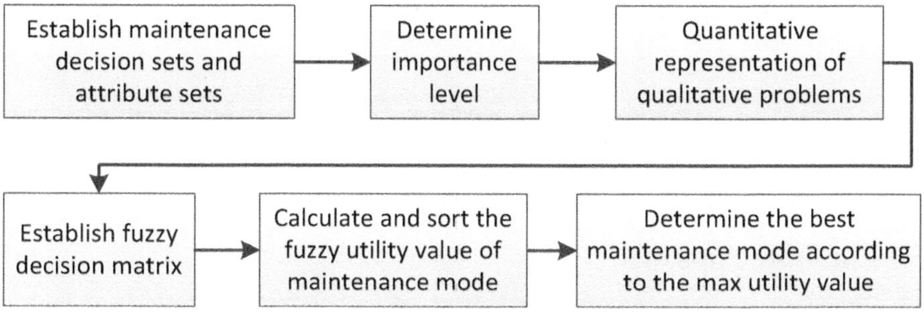

Fig. 1 Basic process of maintenance mode decision modeling

Step 3: Use trapezoidal fuzzy membership function curve, and transform the qualitative description into quantitative index expressed by L-R trapezium fuzzy figure.
Step 4: Establish fuzzy matrix.
Step 5: Calculate the fuzzy avail value of each maintenance mode. And sort the maintenance mode according to the result.
Step 6: Determine the best maintenance mode according to the max utility value.

4 Fuzzy Multi-attribute Decision Method

Fuzzy multi-attribute decision methods mainly include fuzzy weighted average decision, fuzzy optimistic decision, fuzzy pessimistic decision, fuzzy optimistic–pessimistic combination decision, and fuzzy compromise decision. Because maintenance mode decision of artillery command information system needs to consider the influence of all factors on the decision result, fuzzy weighted average decision method can be adopted. The mathematical expression is

$$S^* = \left\{ (S_k | k \in I), \cup (S_k) = \max \left\{ \sum_{j=1}^{n} \widetilde{\omega}_j \cdot \widetilde{x}_{ij} \right\} \right\} \tag{3}$$

If both exact and fuzzy concepts can be represented by L-R trapezoidal fuzzy figures in decision problems, then fuzzy utility function can be described in the form of simple weighted averaging. That is,

$$\widetilde{A}_i = \frac{\sum_{j=1}^{n} \widetilde{\omega}_j \widetilde{x}_{ij}}{\sum_{j=1}^{n} \widetilde{\omega}_j} \tag{4}$$

where $\widetilde{\omega}_j, \widetilde{x}_{ij}$ is L-R trapezium fuzzy figure.

The biggest maintenance mode of fuzzy utility value is the best maintenance mode. That is,

$$A_{\max} = \max\left\{\widetilde{A}_i\right\}, \quad i = 1, 2, \ldots, m \tag{5}$$

5 Example Analysis

Taking a certain artillery command information system as an example, determine three maintenance methods according to the characteristics of system maintenance support. They are early overhaul; preventive maintenance and continuous monitoring are S_1, S_2, S_3, respectively. Therefore, a maintenance decision set $S = \{S_1, S_2, S_3\}$ is obtained. Through comprehensive analysis, select fault risk, system availability, total maintenance cost, operational reliability and operational stability as maintenance mode decision indicators on the basis of integrity and economy. Represented in terms of u_1, u_2, u_3, u_4, and u_5 respectively, and get maintenance method attribute set $u = \{u_1, u_2, u_3, u_4, u_5\}$. Failure risk and maintenance cost are cost indicators, system availability, operational reliability, and work stability are indicators of efficiency. For a cost index, the better the attribute value is the smaller. For the benefit index, it is totally opposite to the cost-type index attribute. The better attribute value is the bigger. The importance of each attribute is divided into five grades. The evaluation grades of failure risk, maintenance cost ratio, and system availability are higher, high, general, low, and lower. The evaluation grades of operational reliability and working stability are better, good, general, poor, and worse. The evaluation grades of the attribute importance are more important, important, general, unimportant, and less important. Maintenance method attributes are comprehensively evaluated by system maintenance support personnel and system designers, and the evaluation result of decision attribute of maintenance method is shown in Table 1.

It is necessary to transform the related qualitative problem into the form of quantitative indicators in order to make artillery command information system state maintenance mode decision more scientific and accurate. Because the qualitative description has some fuzziness, the qualitative description is transformed into the

Table 1 Evaluation result of decision attribute sets

Maintenance decision	Decision attribute				
	u_1	u_2	u_3	u_4	u_5
S_1	Lower	Higher	Higher	Better	Better
S_2	General	General	General	Good	Good
S_3	Higher	Lower	Lower	General	General
ω	General	More importance	General	More importance	importance

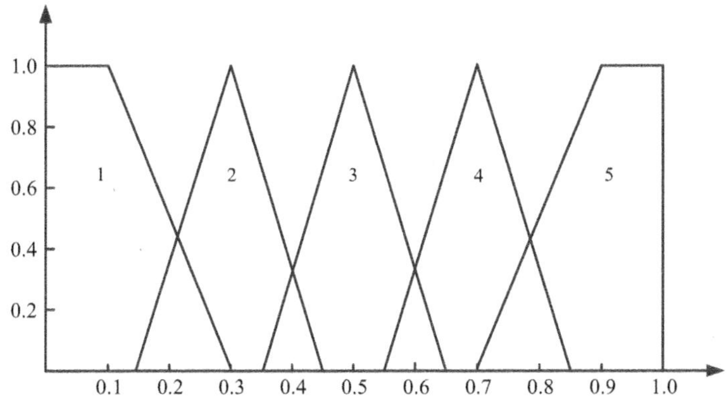

Fig. 2 Fuzzy number membership function curve

quantitative index expressed by L-R trapezoidal fuzzy number by the form of trapezoidal fuzzy number membership function curve, as shown in Fig. 2. And the correspondence between them is shown in Table 2.

According to Tables 1 and 2, establish method decision fuzzy matrix, as shown in Table 3.

Table 2 Correspondence between qualitative description and trapezium fuzzy figures

Figure	Trapezium fuzzy figure	u_1	u_2	u_3	u_4	u_5	ω
1	(0, 0.1; 0, 0.2)	Higher	Lower	Higher	Worse	Worse	Unimportant
2	(0.3, 0.3; 0.15, 0.15)	High	Low	High	Poor	Poor	Less important
3	(0.5, 0.5; 0.15, 0.15)	General	General	General	General	General	General
4	(0.7, 0.7; 0.15, 0.15)	Low	High	Low	Good	Good	Important
5	(0.9, 1; 0.2, 0)	Lower	Higher	Lower	Better	Better	More important

Table 3 Maintenance method decision fuzzy matrix

Mode	u_1	u_2	u_3	u_4	u_5
S_1	(0.9, 1; 0.2, 0)	(0.9, 1; 0.2, 0)	(0, 0.1; 0, 0.2)	(0.9, 1; 0.2, 0)	(0.9, 1; 0.2, 0)
S_2	(0.5, 0.5; 0.15, 0.15)	(0.5, 0.5; 0.15, 0.15)	(0.5, 0.5; 0.15, 0.15)	(0.7, 0.7; 0.15, 0.15)	(0.7, 0.7; 0.15, 0.15)
S_3	(0, 0.1; 0, 0.2)	(0, 0.1; 0, 0.2)	(0.9, 1; 0.2, 0)	(0.5, 0.5; 0.15, 0.15)	(0.5, 0.5; 015, 0.15)
ω	(0.5, 0.5; 0.15, 0.15)	(0.7, 0.7; 0.15, 0.15)	(0.5, 0.5; 0.15, 0.15)	(0.7, 0.7; 0.15, 0.15)	(0.9, 10.2, 0)

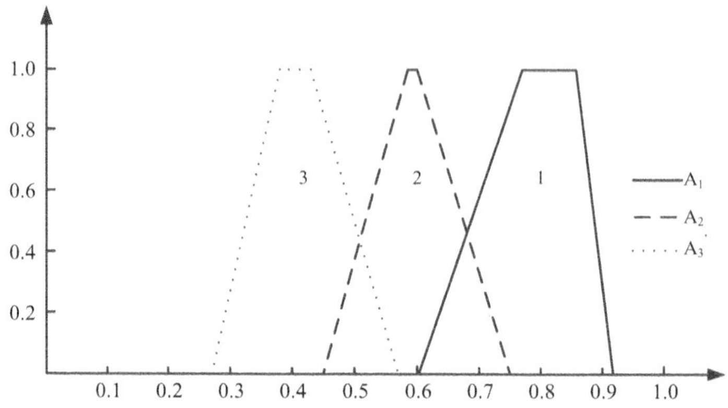

Fig. 3 Fuzzy utility value function curve of each maintenance mode

The method of fuzzy weighted average is used to solve the calculation, and obtaining the value of fuzzy avail of each strategy as follows.

$$\tilde{A}_1 = \sum_{j=1}^{5} \tilde{\omega}_j \tilde{x}_{1j} / \sum_{j=1}^{5} \tilde{\omega}_j = (0.7636, 0.8676; 0.1625, 0.05)$$

$$\tilde{A}_2 = \sum_{j=1}^{5} \tilde{\omega}_j \tilde{x}_{2j} / \sum_{j=1}^{5} \tilde{\omega}_j = (0.597, 0.6; 0.15, 0.15)$$

$$\tilde{A}_3 = \sum_{j=1}^{5} \tilde{\omega}_j \tilde{x}_{3j} / \sum_{j=1}^{5} \tilde{\omega}_j = (0.3788, 0.4324; 0.1031, 0.1375)$$

The fuzzy utility value membership function curve of each maintenance mode is shown in Fig. 3.

According to Fig. 3, the order of maintenance mode is $\tilde{A}_1 > \tilde{A}_2 > \tilde{A}_3$, and the mode S_1 is best. Therefore, artillery command information system should be overhauled ahead of schedule.

6 Conclusion

There are many evaluation indexes which affect equipment maintenance mode decision, and the evaluation of these indexes cannot be measured by means of rigorous evaluation criteria. Therefore, it needs to use fuzzy multi-attribute decision model to deal with multiple evaluation indexes of each maintenance mode by fuzzy weighted average and then get the comprehensive evaluation results of the maintenance mode. Case analysis shows that it is scientific and reasonable to use fuzzy multi-attribute decision model to determine maintenance mode.

References

1. Liang, R., Ma, L.: Research on the civil aircraft condition-based maintenance strategy based on PHM. Aviat. Maint. Eng. **5**, 46–48 (2017)
2. Moya, G.E.J., Veliz, Y.Z.: A dynamic decision making method with discrimination of alternatives using associative aggregation operators. IEEE Lat. Am. Trans. **14**(10), 4310–4317 (2016)
3. Ren, P., Xu, Z., Hao, Z.: Hesitant fuzzy thermodynamic method for emergency decision making based on prospect theory. IEEE Trans. Cybern. **47**(9), 2531–2543 (2017)
4. Xu, Z.S., Xia, M.M.: Hesitant fuzzy entropy and cross-entropy and their use in multi-attribute decision-making. Int. J. Intell. Syst. **27**, 799–822 (2012)
5. Farhadinia, B.: A novel method of ranking hesitant fuzzy values for multiple attribute decision-making problems. Int. J. Intell. Syst. **28**, 752–767 (2013)
6. Xu, Z.S., Zhang, X.L.: Hesitant fuzzy multi-attribute decision making based on TOPSIS with incomplete weight information. Knowl. Based Syst. **52**, 53–64 (2013)
7. Wei, G.W.: Gray relational analysis method for intuitionistic fuzzy multiple attribute decision making. Expert Syst. Appl. **38**, 11671–11677 (2011)
8. Zhou, J., Wang, Z.: Method of dynamic hybrid multi-attribute decision based on triangular fuzzy numbers. J. Wuhan Univ. Technol. Inf. Manage. Eng. **36**(1), 121–129 (2014)

A Method for Facial Kinship Verification Based on Deep Learning

Hao Zhou, Xiongdong Qiu, Huajian Cong, Haiyan Wu, Baiping Wang, Fuli Guo, Hao Li and Lianshui Wang

Abstract The analysis of facial information is always an important hotspot issue in the field of computer vision and pattern recognition and kinship verification by facial image is a challenging problem. Facial kinship verification has wide application range and important research value not only in the field of biometrics analysis but also in the social fields, such as analysis of mining social network data, searching work for scattered family members and so on. At present, with the development of computer vision especially the deep learning and metric learning, face recognition has made great achievements in recent years. In this thesis, a framework for kinship verification based on deep learning is proposed. Comparing with the current research methods that focus on metric learning, we use a deep learning network model to replace the two processes of feature extraction and metric learning. The effectiveness of the method is verified in KinFaceW datasets and TSKinFace datasets. The experimental results show that the accuracy is about 91% in KinFaceW datasets, and about 89.5% in TSKinFace datasets.

Keywords Facial kinship verification · Deep learning · Metric learning · Deep residual network

1 Introduction

In recent years, a large number of face recognition verification methods have been proposed by researchers and articles. The performance of related methods in LFW [1] datasets has reached above 99.8% of the accuracy that surpassing the level of human beings. The kinship recognition based on features extracted from facial images is the further application of facial identification recognition. In general, face recognition is usually used to analyze and compare the feature information of the same individual. This process is called authentication. Compared with authentication, the kinship

H. Zhou · X. Qiu · H. Cong · H. Wu · B. Wang · F. Guo · H. Li · L. Wang (✉)
Yinfeng Gene Technology Co. Ltd., Jinan, China
e-mail: wanglianshui@yfyxjy.com

V. Jain et al. (eds.), *Recent Trends in Intelligent Computing, Communication and Devices*, Advances in Intelligent Systems and Computing 1031,
https://doi.org/10.1007/978-981-13-9406-5_17

recognition needs to judge the relationship between different individuals based on their each facial information. Generally, kinship is defined as the relationship between two individuals with overlapping genetic biological correlations. Thus, there are four different types of kinship, namely father–son (F–S) relationship, father–daughter (F–D) relationship, mother–son (M–S) relationship and mother–daughter (M–D) relationship. Recently, more and more researchers believe that kinship verification has been a new challenge to be faced in the field of facial recognition and analysis [2–4]. Compared with face recognition, kinship analysis needs to analyze multiple samples with different ages and sex. In addition, due to the photographing angle and illumination, it is difficult to keep robustness by using the traditional feature extraction methods based on artificial design. In order to solve the problem that how to compare the similarity between individuals, a great deal of research work has been focused on the calculation of similarity distance between feature descriptors, namely metric learning. The thesis [5–7] summarized and introduced the role of metric learning in face recognition and facial kinship verification and proposed their respective models or methods to solve the problem of kinship verification.

In recent years, deep learning theory has made great progress in the field of image processing and face recognition. We propose a method to design a model from feature extraction and calculation to recognition based on deep learning. The first half of the network uses two or more channels to compute the input image pairs to obtain their feature vector, and the latter half of the model calculates the distance between different feature vectors, outputs the final recognition result according to the distance value. According to the recognition and labels that calibrated in advance, the parameters of network update by backpropagation. After a certain number of iterative training, we obtain the final model. The whole progress realizes an end-to-end training and recognition from input image pair to output result.

The contributions of this paper are summarized as:

We design a deep learning model for kinship verification. For other current methods, the feature extraction and metrics learning process are separated. For example, the work of feature extraction is training a model on large-scale datasets, and then updates the metric learning model on the kinship datasets. The model proposed in the thesis begins with the training on the kinship datasets. We test the mean-square error (MSE) first for calculating the similarity between feature vectors, then use the absolute value of difference and full-connect layer as output which is used for training and updating the parameters of the network.

We design an elastic channel network model for kinship verification. There are two types of kinship classification, one is one-to-one mode, which can classify the relationship between a child and its father (mother). The other is the *N*-to-*N* mode, that is, two samples of parents and their kid are compared at the same time (one-to-*N*), and even more than one child sample is compared with their parents (*N*-to-*N*). The elastic channel network can change number of channels in the model during train and test process for compatibility with different modes.

2 Method

2.1 Channel Backbone Network

The deep convolution network can extract from low-level features to high-level features from images by a large number of convolution operations of different kernel sizes. Generally, the more layers of the network, the more features that can be obtained, and the features have more information. However, in practice, the depth of network cannot be increased blindly, which leads to the dispersion and explosion of gradient information in the network, that making network cannot be trained correctly. It can be solved effectively by the use of regularization technology. But the new problem is introduced at the same time, and with the increase of layers, the performance of network decreases. In the paper [8, 9], a residual learning structure is proposed. As shown in Fig. 1a, a shortcut connection is used to send low-level features directly to the next layer of the network to solve the problem of deep network optimization.

The backbone network proposed in the paper is mainly constructed by two structures shown in Fig. 1, and the structure of ResBlock_B is constructed by ResBlock_A. We build a 48-layers backbone network by overlaying the ResBlock_B.

Fig. 1 **a** ResBlock_A, **b** ResBlocks_B

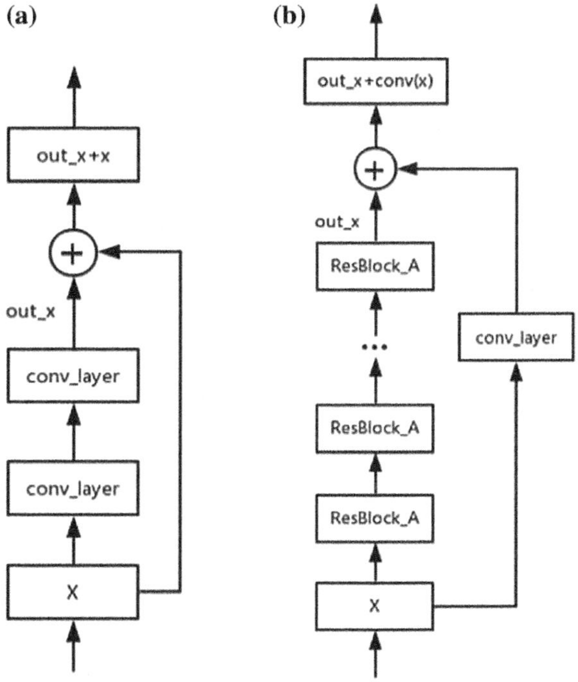

2.2 *Multi-channel Model*

2.2.1 Binary-Channel Model

The deep learning networks used for a traditional computer vision problem, such as image classification, object detection and image segmentation, are usually accepted only one image as input. Even in the case of processing video sequence, the video is decomposed into image sequences first, then the network is fed with pieces of images in proper time. This means that the network processes a single sample image at the same time. However, two or more sample images are needed to be processed simultaneously in order to get the final result of kinship verification. To tackle this problem, we need to design a network model that contains two or more channels, as shown in Fig. 2. Taking the binary-channel model as an example, it supports four types of kinship validation. Model is fed with a pair of sample images simultaneously, and its output determines whether there is a relationship between them. The DCNN+Res module in Fig. 2 is the backbone network mentioned in 2.1 with a Global Average Pool layer (GAP) added. The GAP layer can compress a feature map into a value, which greatly simplifies the computation and parameters the of network. We design and try two different models. The model A calculates a mean-square difference (MSE) for the output of two channels. The model B

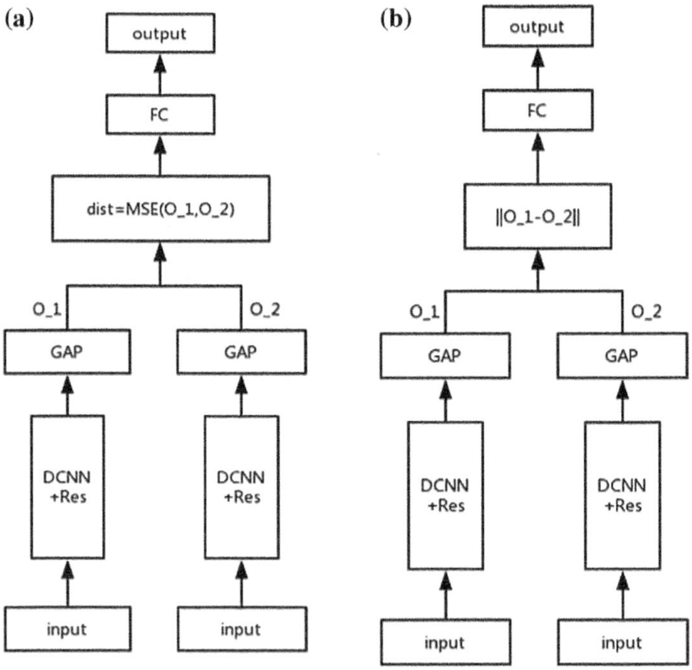

Fig. 2 **a** Binary_channel_model_A, **b** binary_channel_model_B

calculates the absolute difference (AD) of each dimension for two channel outputs. The results of the two models are finally calculated through a full-connection layer. After sorting the final output, the final recognition result can be obtained. 0 means no relationship, 1 means there is a kinship.

2.2.2 Elastic Multi-channel Model

In fact, in addition to one-to-one recognition, one-to-*N* or *N*-to-*N* problem is needed to be solved. In this case, photos of both parents and children are provided, and the model outputs the result that there is whether a relationship between parents and children. Compared with one-to-one recognition, only the photo of mother or father and a child's photo provided, the feature information of both parents and children can be compared at the same time. In genetics, the character of children is inherited from both parents. Similarly, the facial traits are inherited from both parents. Compared with one-to-one, the *N*-to-*N* mode can better reflect the relevance between them.

Therefore, we propose an elastic multi-channel model to deal with one-to-*N* problem, that is, the model is fed freed pictures simultaneously. As shown in Fig. 3, compared with the binary-channel model in Sect. 2.2.1, a new channel-selection module, namely ch_choose is added, which can decide how many channels needed to be opened up to handle the current problem. Figure 3 shows that thc current problem needs three channels, and channel ch_SD and channel ch_MF are used to feed image tensors of child and parents, respectively. After that, we obtain the outputs of three channels, then, the AD is calculated between O_2 and O_1 or O_3. Next, concatenating them to obtain a feature, finally, the output of the model is through a FC layer calculation. If the model is fed with one-to-one image pair, model with two channels is enough after analysis of ch_choose module. The following model is consistent with the one in Fig. 2. That can be compatible with one-to-one mode and one-to-*N* mode and achieve an elastic matching between model and problem.

2.3 Training

The training and validation samples are obtained from two public kinship datasets KinFaceW datasets and TSKinFace datasets. KinFaceW datasets are divided into two parts, KinFace-I and KinFace-II [6], that are collected from the public photos or celebrities. For each person in these two datasets, the face image of his/her parent or child was also collected. There are four kinship relations in the two datasets: F–S, F–D, M–S and M–D. In KinFaceW-I, there are 156, 134, 116 and 127 pairs of kinship images, and each aligned 64 $\times$ 64 image is used for feature extraction. The TSKinFace datasets [10] contain three different types, such as father–mother–daughter (FM–D) relationship, father–mother–son (FM–S) relationship and

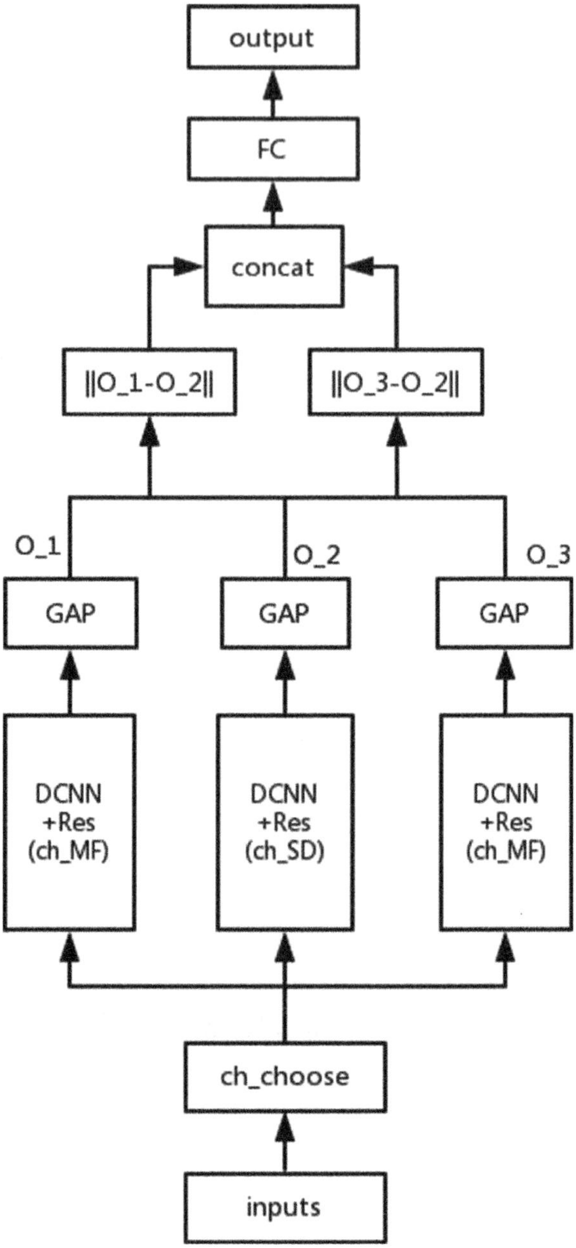

Fig. 3 Elastic multi-channel model

father–mother–son–daughter (FM–SD) relationship. All sample images in the datasets are harvested from the Internet based on knowledge of public figures family and photo-sharing social network such as Flickr.com. Family facial images are collected based on the face region and eye position automatically and, respectively, estimated by the face detector [11] and eye localization [12]. Training labels are classified into two categories, that is, 0—no relationship and 1—there is kinship. The positive sample group is provided by the datasets, and the negative samples group is randomly assembled according to different family photos in the datasets. For KinFaceW datasets, combine KinFace-I and KinFace-II datasets first, and select 382 samples as a validation set randomly, and the remaining 2684 samples as the training set. Similarly, for TSKinFace datasets, 273 and 213 samples are randomly selected from FM–D and FM–S group as the validation set. The implementation of task is used by pytorch0.40 framework. The optimization process uses Adam first, with an initial learning rate of 1e-4 using a cross-entropy loss function while the learning rate is attenuated by 2% every 20 epochs. After 50 epochs, we continue training the network for 450 epochs using SGD with a momentum of 0.9. With every iteration, half of the training sets are randomly selected as positive samples and others as negative samples.

3 Experiments and Results

Since, the method of face alignment and the size of face segmentation that used by KinFaceW and TSKinFace datasets are different. We test two sets of models on the datasets, respectively.

Firstly, in the KinFaceW datasets, we select half of the validation set as negative samples, and the other half as positive samples just like the training stage. Because of the randomness, the positive and negative samples are different every time. So we repeat the test experiment 50 times and take the average value, and the recognition results based on four relationships are shown in Table 1. Model_A corresponds to the binary_model_A structure in Fig. 2, while Model_B corresponds to the binary_model_B structure in Fig. 2. After training respectively, the results show that the model with AD has better performance and less computation compared with the MSE. The benchmark alignment algorithm uses SVM. First, the image is stretched into sequence vectors, and then the model is trained with

Table 1 Results in the KinFaceW datasets

	F–D (%)	F–S (%)	M–D (%)	M–S (%)	Average (%)
SVM	54.52	55.36	53.23	53.40	54.13
DDMML	86.91	81.45	82.30	85.00	83.91
Model_A	92.40	89.40	89.89	87.83	89.88
Model_B	90.52	89.80	93.51	92.39	91.56

Table 2 Results in the TSKinFace datasets

	FM–D (%)	FM–S (%)	Total (%)
DDMML	87.18	89.20	88.07
Model	89.25	89.67	89.51

sequence vectors by SVM. DDMML is a metric learning model which supports multiple features proposed in paper [7]. LBP, DSIFT, HOG, LPQ are used to compute features to obtained multiple feature vectors, and then DDMML is used to get the final result.

Similar to the KinFaceW datasets, positive set and negative set samples of the TSKinFace datasets are selected randomly with 50% probability, and we repeat experiment tests 50 times to obtain the final result by calculating the average value. The model structure used is shown in Fig. 3, and the results are shown in Table 2.

Through the test above, the method proposed in the paper does not require training separately in feature extraction and metric learning compared with the existing methods. It simplifies the training process and improves the recognition accuracy. The whole framework can complete the training of model after an end-to-end training. The elastic channel model can enhance the compatibility of model, and it is compatible with many different recognition problems for kinship verification only by changing the training label.

4 Conclusion

In the thesis, a deep learning model is proposed for kinship verification, and a channel choose module is designed, which can match the current training or testing tasks flexibly by changing the num of the channel. After the experiments on KinFaceW datasets and TSKinFace datasets, 91 and 89.5% of the recognition accuracy are achieved, respectively. However, there are only two types of labels for TSKinFace datasets. Later, we consider increasing the types of categories, such as parent–child, father-no mother–child, no father–mother–child, etc. At present, the model cannot support N-to-N mode and needs to be updated to make it compatible with N-to-N mode. At present, the training and verification sets used are standard datasets, and an image preprocessing process needs to be added in the future, which can be compatible with ordinary photos or images from different datasets with a unified face segmentation method.

Acknowledgements This research is based on work supported by the Yinfeng Gene Technology Co. Ltd. We thank all families who took part in this study. Thanks to the data shared by researchers on the Internet.

References

1. Learned-Miller, E., Huang, G.B, Roychowdhury, A., et al.: Labeled faces in the wild: a survey. In: Advances in Face Detection and Facial Image Analysis. Springer International Publishing (2016)
2. Fang, R., Tang, K.D., Snavely, N., et al.: Towards computational models of kinship verification. In: IEEE International Conference on Image Processing, pp. 1577–1580. IEEE (2010)
3. Xia, S., Shao, M., Luo, J., et al.: Understanding kin relationships in a photo. IEEE Trans. Multimedia **14**(4), 1046–1056 (2012)
4. Lu, J., Zhou, X., Tan, Y.P., et al.: Neighborhood repulsed metric learning for kinship verification. IEEE Trans. Pattern Anal. Mach. Intell. **36**(2), 331–345 (2014)
5. Hu, J., Lu, J., Tan, Y.P.: Discriminative deep metric learning for face verification in the wild. In: Computer Vision and Pattern Recognition, pp. 1875–1882. IEEE (2014)
6. Hu, J., Lu, J., Tan, Y.P.: Deep transfer metric learning. In: IEEE Conference on Computer Vision and Pattern Recognition, pp. 325–333. IEEE Computer Society (2015)
7. Lu, J., Hu, J., Tan, Y.P.: Discriminative deep metric learning for face and kinship verification. IEEE Trans. Image Process. **26**(9), 4269–4282 (2017)
8. He, K., Zhang, X., Ren, S., et al.: Deep Residual Learning for Image Recognition, pp. 770–778 (2015)
9. He, K., Zhang, X., Ren, S., et al.: Identity mappings in deep residual networks. In: Computer Vision—ECCV 2016, pp. 630–645. Springer International Publishing (2016)
10. Qin, X., Tan, X., Chen, S.: Tri-subject kinship verification: understanding the core of A family. IEEE Trans. Multimedia **17**(10), 1855–1867 (2015)
11. Viola, P., Jones, M.J.: Robust real-time face detection. Int. J. Comput. Vision (IJCV) **57**(2), 137–154 (2004)
12. Tan, X., Song, F., Zhou, Z.H., et al.: Enhanced pictorial structures for precise eye localization under incontrolled conditions, In: 2009 IEEE Conference on Computer Vision and Pattern Recognition, pp. 1621–1628. IEEE (2009)

A Domain-Adapting Word Representation Method for Word Clustering

Wanting Zhou, Hanbin Wang, Hongguang Sun and Tieli Sun

Abstract Extracting key information from texts is a goal of natural language processing (NLP) field. A few keywords could prompt the main idea of the text, and the complete vocabulary information is richer, but not easy to organize. This paper proposes a word representation method based on frequently co-occurring entropy (FCE) and fuzzy bag-of-words model (FBoW), named frequently co-occurring entropy and fuzzy bag-of-words model (FCE-FBW). This method is used to cluster the words of different domains and integrate similar words together. These word clusters can be useful for tasks such as building knowledge-based domain repositories. FCE is used to pick out the generalizable features. FBoW supports the description of the same word by multiple dimensions. This paper combines the two models and proposes FCE-FBW method. It provides good performance.

Keywords Word representation · Word clustering · Frequently co-occurring entropy · Fuzzy bag-of-words model

1 Introduction

Large-scale text data contains a wealth of information. This method is used to cluster the words of different domains and integrate similar words together. Users or product suppliers need to know not only which brands or products the reviews primarily refer to, but also what attributes of the product are mentioned. But when

W. Zhou · H. Wang · H. Sun · T. Sun (✉)
School of Information Science and Technology, Northeast Normal University, Changchun 130117, China
e-mail: suntl@nenu.edu.cn

H. Sun
e-mail: sunhg889@nenu.edu.cn

T. Sun
College of Humanities and Science, Northeast Normal University, Changchun 130117, China

V. Jain et al. (eds.), *Recent Trends in Intelligent Computing, Communication and Devices*, Advances in Intelligent Systems and Computing 1031,
https://doi.org/10.1007/978-981-13-9406-5_18

people write comments, they may use very different words, even if they talk about the same property of the same product. For example, "white", "surface", "cover", "handprint", and "paint", are all may be used to comment the appearance of laptops. If someone wants to grasp the feel of clients exactly, these words could not be ignoring. Because the words are massive, it makes sense to bring together semantically similar words. In order to obtain word clusters with high discrimination and rich information, this paper proposes a word representation method FCE-FBW, which combines FCE [1] algorithm and FBoW [2] model. The method considers the special meaning of the word in a specific domain, as well as the co-occurrence information of words.

Some words have different meanings in different domains. For example, "Apple" refers to a cell phone brand in the cell phone domain and a fruit in the plant domain. FCE can utilize text sets that contain multiple categories to find the words have general meanings in all categories and treat them as valid features adapt to different domains.

The position between words also reflects the level of association between them. Word2vec [3, 4] uses a sliding window and one-hot encoding to express words into vectors. The vectors can be utilized to calculate cosine values of words.

The main idea of FCE-FBW is as follows: assume there are two words w_1, w_2 and a general word set $T = \{t_1, t_2, \ldots t_v\}$, when w_1 and w_2 are all have similar cosine values with the words in T, then w_1 and w_2 could be considered as similar words. They can be clustered in the same cluster.

The details of FCE-FBW will be described below. Section 2 will introduce the technologies related to FCE-FBW. Section 3 will discuss the details our proposed method and Sect. 4 will show the results of experiments and comparative analysis.

2 Related Work

Similar to most feature extraction approaches, FCE takes advantage of the presence of words. When words appear more in a certain domain and appear less in other domains, they can be regarded as domain-related features. Otherwise, the words are general features if they have close frequency in all domains. Therefore, when there are two or more categories, the general features can be screened out. Tan et al. [1] used FCE for feature extraction of unlabeled documents. The categories are treated as different domains.

FCE is an algorithm based on statistical features and does not consider expressing words as vectors. We combine it with word vector technique. The methods for obtaining word vectors can be roughly divided into two kinds. One is based on counting and matrix decomposition, for instance, Latent Semantic Analysis (LSA) [5]. Another is based on local context windows, such as skip-gram and CBOW of Mikolov [3].

The FBoW model is based on word2vec, and its core idea is as follows: assume a word and a word list $S = \{s_1, s_2, \ldots s_v\}$, the words in S are synonyms, moreover,

w is related to most of the words in S, then w could be regarded as a synonym of words in S. Even if w is not adjacent to the position of a word s_i in the corpus, their meanings can be considered similar. This model was originally used for document representation. Compared with traditional bag-of-words (BoW) model, FBoW can find sentences with different words but similar semantics. We improved the original FBoW matrix to adapt the task of word representation.

3 Method

3.1 Frequently Co-occurring Entropy

This paper on the use of FCE is different from Tan's [1] work. This part takes two different domains as examples to illustrate our method. Assuming that there are two categories of document sets a and b, all the words in a and b are compiled into a vocabulary V, which size is v, and the words in V are represented by w. Then, the FCE value of w can be calculated as follows:

$$f_w = \log_2\left(\frac{P_a(w) \cdot P_b(w)}{|P_a(w) - P_b(w)| + \beta}\right) \quad (1)$$

where β is the parameter to avoid the denominator being 0, set to 0.0001. $P(w)$ represents the probability of w, and the calculation method is shown in Formula (2):

$$P(w) = \frac{N_w + \alpha}{|D| + 2 \cdot \alpha} \quad (2)$$

N_w in Formula (2) is the frequency of w, $|D|$ is the number of documents of one domain, and α is a parameter set to avoid overflow. In experiments, the value is 0.0001.

Based on the FCE values and threshold t, some general features are needed to pick after get FCE values of words.

$$G = \{g_1, g_2, \ldots, g_m | g \in V, \quad f_w(g_i) > t\} \quad (3)$$

where g denotes the selected general features that satisfy the condition that the FCE values are greater than t. The threshold t can be freely set according to the data of different corpus. If the data sets of the two categories involved in the operation are roughly balanced, then the G set contains those words more frequently in both domains. Next, these words can be used to construct the FBoW matrix.

3.2 *Fuzzy Bag-of-Words Model*

Word vectors can be used to calculate the semantic similarity between words. The algorithm used by Word2vec is cosine similarity.

$$S(\mathbf{w_1}, \mathbf{w_2}) = \cos(\mathbf{w_1}, \mathbf{w_2}) = \frac{\mathbf{w_1} \cdot \mathbf{w_2}}{||\mathbf{w_1}|| \cdot ||\mathbf{w_2}||} \tag{4}$$

$\mathbf{w}$ in Formula (4) represents the word embedding vector of word w.

Word2vec calculates semantic similarity through co-occurrence. Sometimes, the words obtained are not synonyms, but related words. In this case, the relevance of words in each domain to the general features can be calculated, and the word vectors of different domains will be acquired.

Formula (5) gives the method for calculating the similarity between common features and other words:

$$\mathrm{sim}_{g_i}(w) = \begin{cases} \cos(w_{g_i}, w_w) & \text{if} \quad \cos(w_{g_i}, w_w) > 0, \\ 0 & \text{otherwise,} \end{cases} \tag{5}$$

where g represents a generic feature in G; w represents a word in one of the two domains a or b. If the similarity value is not greater than 0, the value is set to 0.

After calculating the similarity between all words in V and the general features, the fuzzy word matrix H is obtained as Formula (6):

$$H = \begin{bmatrix} \mathrm{sim}_{g_1}(w_1) & \mathrm{sim}_{g_1}(w_2) & \mathrm{sim}_{g_1}(w_3) & \ldots & \mathrm{sim}_{g_1}(w_v) \\ \mathrm{sim}_{g_2}(w_1) & \mathrm{sim}_{g_2}(w_2) & \mathrm{sim}_{g_2}(w_3) & \ldots & \mathrm{sim}_{g_2}(w_v) \\ \ldots & \ldots & \ldots & \ldots & \ldots \\ \mathrm{sim}_{g_m}(w_1) & \mathrm{sim}_{g_m}(w_2) & \mathrm{sim}_{g_m}(w_3) & \ldots & \mathrm{sim}_{g_m}(w_v) \end{bmatrix} \tag{6}$$

where m represents the size of G. In our work, each column vector is treated as a word vector of w_i. According to the parameter settings of Word2vec, vectors of words which frequency less than 5 are not calculated, so H does not contain words with less than 5 occurs actually. The H matrix is used for word clustering operations.

The K-means clustering method is used to cluster the words to obtain word clusters, which can be used to sort out words of different domains.

4 Experimental Results

This section discusses the experimental results and their analysis. The scikit-learn tool was used in K-means clustering experiment [6]. As comparison, this tool was used to cluster and evaluate the word vectors obtained by Word2vec. Word2vec's word vector dimension is set to the common 200 dimensions.

We chose English corpus Amazon_6 [7] as experimental data. The NLTK tool [8] was used to perform root restoration and to remove the stop words. Amazon_6 contains 6 classes. They are cameras, mobile phone, TVs, laptops, video surveillance, and tablets. The contents of Amazon_6 are reviews of these electronic products. We extracted the comment texts from the contents. Due to the imbalance of data, four categories with close texts were chosen and divided into two groups for combined experiments. The t value was set to −5.0.

The first experiment utilized the cameras and mobile phone texts. These two classes were combined to extract general features, and then, found the FBoW matrix. In the result of each method, a part of a cluster of words is intercepted for display. The criteria for selection are around similar topics, in order to see the difference between the clustering results of FCE-FBW and word2vec. Since there are many words to be clustered, even if 50–300 cluster centers are selected, each cluster may still have tens to hundreds of words. Intuitive results of some word clustering of 80 cluster centers are shown in Table 1.

In the results of the cameras category, the cluster of words is about photographic effects. In the mobile phone category, the cluster of words is used for display revolves around the brand of the phone. The words found by FCE-FBW are more relevant to the topic. Word2vec tends to bring together words with the same root, but more words are less relevant to the topic.

Silhouette coefficient [9] was chosen to evaluate the experimental results. The overall evaluation for word clustering results is shown in Table 2 where four decimal places are retained according to the principle of rounding.

Table 1 Examples of word clustering results of cameras and mobile phone

Cameras	FCE-FBW	Black effect subject soft spot distort distract dist dark bright pixel bloom grainy blur dim background viewscreen temp noisy horizon
	Word2vec	Airbrush alga area areas artifact awash backdrop background backlight blend blind blob blocky bloom blot blotch blotchy blowout
Phone	FCE-FBW	Galaxy galazy motorol nok nokia handset phones motorola pre mobile carrier moto ericcson rumo ericson sidekick lastest black
	Word2vec	Abolv accsess acro againt alcatel amol androind anyoth appletouch aroudn backflip balckberry bb bberry berry blackberrry blackberry

Table 2 Word clustering results evaluation of cameras and mobile phone

Classes	Cameras		Mobile phone	
Number of cluster centers	FCE-FBW	Word2vec	FCE-FBW	Word2vec
50	0.0598	−0.0126	0.0837	−0.0251
80	0.0537	−0.0262	0.0584	−0.0406
100	0.0440	−0.0266	0.0559	−0.0357
200	0.0285	−0.0358	0.0361	−0.0448
300	0.0282	−0.0389	0.0322	−0.0415

Table 3 Examples of word clustering results for TVs and laptops

TVs	FCE-FBW	Sound jack surround stero suround speak pump loud bass radio mut volum aud ampl speakers hiss stereo headphon klipsch subwoof
	Word2vec	Abx acoust acoustimass aiw altec amp ampl appsgam arc audio audioengin audiophil audyssey aux auxil auxy avr bitstream bois
Laptops	FCE-FBW	Loud speak sound stereo quiet speakers output audio hiss nois volum micro phon mic headphone altec lans hear blast headset
	Word2vec	Altec amp ampl analog arc aud audio audiophil auxy bass blar boom bos bud buzz clack crank db deaf decibel dialog dolby drown dts

Table 4 Evaluation of word clustering results of TVs and laptops

Classes	TVs		Laptops	
Number of cluster centers	FCE-FBW	Word2vec	FCE-FBW	Word2vec
50	0.0444	−0.0352	0.0599	−0.0427
80	0.0440	−0.0432	0.0556	−0.0399
100	0.0403	−0.0463	0.0559	−0.0359
200	0.0383	−0.0476	0.0504	−0.0389
300	0.0354	−0.0455	0.0429	−0.0297

Due to the actual demand, it is easier to subdivide the semantically similar words by setting a larger cluster center number. Although fewer cluster centers can make the contour coefficients larger, more different words will be clustered together. Therefore, this paper still tests more cluster centers. The evaluation of FCE-FBW is close to 0, but its effect is still better than clustering with Word2vec's word vector alone.

The second set of experiments was a combined experiment of two categories of TVs and laptops. Table 3 lists the clusters of words on the topic of "sound". Both TV and computer have the function of sound, so both classes have this topic cluster.

Even on the same topic, different fields of corpus have different words. The words in the cluster of words aggregated by the FCE-FBW method are semantically close, and there are fewer noise words. Table 4 gives a comparison of the contour coefficients of this experiment, with a 10% increase, which is more than the previous experiment.

5 Conclusion

In summary, this paper proposes a method for word vector representation based on frequently co-occurring entropy and fuzzy bag-of-words model for word clustering tasks. Domain-related clusters of words can be obtained by training word vectors using only a smaller corpus. Different clusters of words surround different attributes and topics. This method would be helpful for establishing domain-related vocabulary and fine-grained tendency analysis. It can also make the machine find more abundant keywords according to different levels of labeled data sets. Experiments have shown its effective. In the follow-up work, we can build a domain knowledge base with the keywords found already. Through the experiments of more data sets, the knowledge base is continuously enriched and applied to tasks such as sentiment analysis.

Acknowledgments This paper was sponsored by Jilin Provincial Science and Technology Department of China (Grant No. 20170204002GX), and Jilin Province Development and Reform Commission of China (Grant No. 2014Y056). We would like to thank the organizations for their support.

References

1. Tan, S., Cheng, X., Wang. Y., Xu, H.: Adapting Naive Bayes to domain adaptation for sentiment analysis. Adv. Inf. Retr. **5478**, 337–349 (2009)
2. Zhao, R., Mao, K.: Fuzzy bag-of-words model for document representation, IEEE Trans. Fuzzy Syst. **14**, 8
3. Mikolov, T., Sutskever, I., Chen, K., et al.: Distributed representations of words and phrases and their compositionality. Adv. Neural Inf. Process. Syst. 3111–3119 (2013)
4. Mikolov, T., Chen, K., Corrado, G., et al.: Efficient estimation of word representations in vector space. arXiv preprint arXiv **1301**, 3781 (2013)
5. Deerwester, S., Dumais, S.T., Furnas, G.W., Landauer, T.K., Harshman, R.: Indexing by latent semantic analysis. J Am Soc Inf Sci **41** (1990)
6. http://scikit-learn.org/stable/
7. Wang, H., Lu, Y., Zhai, C.: Latent aspect rating analysis without aspect keyword supervision. In: Proceedings of the 17th ACM SIGKDD international conference on Knowledge discovery and data mining, pp. 618–626. ACM (2011)
8. http://www.nltk.org/
9. Rousseeuw, P.J.: Silhouettes: a graphical aid to the interpretation and validation of cluster analysis. J. Comput. Appl. Math. **20**, 53–65 (1987)

Fitting Complex Nonlinear Function with Belief Rule Base

Xilang Tang, Mingqing Xiao, Bin Hu and Chunqing Gao

Abstract A new approach is proposed to fit complex nonlinear function by using BRB system in this paper. BRB system is a knowledge-based system instead of a black box, but its parameters can be trained with an optimization framework. Therefore, it combines the advantage of the knowledge-based method and data-driven method, so it can not only overcome the inaccuracy of expert knowledge but also overcome the overfitting problem of data-driven method. An experiment is implemented to compare the performance of BRB system and neural network when fitting a nonlinear system, and the results indicate that BRB system performs much better than neural network when training data is insufficient.

Keywords Nonlinear function · Belief rule base · Overfitting problem · Neural network

1 Introduction

Fitting nonlinear function plays an important role in the field of prediction, and it has been one of the fundamental problems which are concerned by mathematic researchers and engineers at present. The prevailing approach is to use neural networks, and it has been proved to be very efficient as a data-driven method [1]. However, the neural network is a black box, whose internal working mechanism is beyond human understanding. In many practical applications, it is difficult to collect complete historical dataset [2], and overfitting problem may occur when using the data-driven method [3].

In this paper, a new approach is proposed to fit complex nonlinear function by using belief rule base (BRB) system. BRB system is proposed by Yang, extended

X. Tang (✉) · M. Xiao · C. Gao
Air Force Engineering University, Xi'an 710051, China
e-mail: tangxilang@sina.com

B. Hu
China Mobile Communications Corporation, Xiangtan, China

V. Jain et al. (eds.), *Recent Trends in Intelligent Computing, Communication and Devices*, Advances in Intelligent Systems and Computing 1031,
https://doi.org/10.1007/978-981-13-9406-5_19

from the conventional rule-based system, and its inference methodology is based on the evidential reasoning (RIMER) approach [4]. BRB system is a semi-quantitative method which uses both numerical data and human judgmental information, and its parameters can be trained with an optimization framework [5]. Therefore, using BRB system to fit nonlinear function cannot only overcome the inaccuracy of expert knowledge but also overcome the problem of overfitting.

1.1 The Basic BRB Model

A belief rule is extended from traditional IF-THEN rule with all possible results associated with belief degree. A BRB consists of a set of belief rules, which can be represented as follow [4],

$$\begin{aligned} &R_k: \\ &\text{IF}\, x_1 \text{ is } A_1^k \wedge x_2 \text{ is } A_2^k \wedge \ldots \wedge x_{T_k} \text{ is } A_{T_k}^k \\ &\text{THEN} \left\{ (D_1, \beta_{1,k}), (D_2, \beta_{2,k}), \ldots, (D_N, \beta_{N,k}) \right\}, \left(\sum_{i=1}^{N} \beta_{i,k} \le 1 \right), \\ &\text{with rule weight } \theta_k \text{ and attribute weights } \delta_{1k}, \delta_{2k}, \ldots, \delta_{T_k k}, k \in \{1, \ldots, L\} \end{aligned} \tag{1}$$

where $A_i^k \quad (i = 1, 2, \ldots, T_k)$ refers to the reference value of x_i, which denotes to the ith antecedent attribute in the kth rule; $\beta_{j,k} \quad (j = 1, 2, \ldots, N)$ is the belief degree assessed to D_j which denotes the jth consequent in the kth rule. If $\sum_{i=1}^{N} \beta_{i,k} = 1$, the kth rule is complete; or else, it is incomplete. θ_k is the relative weight of kth rule. $\delta_{ik} \quad (i = 1, 2, \ldots, T_k)$ is the relative weight of the ith antecedent attribute in the kth rule. L is the number of all belief rules that are used in the BRB. And T is the total number of antecedent attributes that are used in the rule base. In addition, "$\wedge$" is a logical connective to represent the "AND" operator.

1.2 BRB Inference Using ER Approach

If the parameter model for a BRB is known, and input vector is available, then the final result or output can be calculated based on ER analytical algorithms [4], by aggregating all the rules activated by the actual input vector x^*, represented as,

$$O(y) = f(x^*; Q) = \left\{ (D_j, \beta_j), \quad j = 1, 2, \ldots, N \right\} \tag{2}$$

where β_j is the belief degree calculated by ER analytical algorithm which is assessed to consequent D_j. Note that β_j is a function of input vector x^* and the parameter vector Q.

The specific calculation process includes two steps: activation weight calculation and belief degree calculation.

The activation weight refers to the degree of activation of the kth rule, represented by ω_k, which can be calculated by,

$$\omega_k = \frac{\theta_k \prod_{i=1}^{T_k} \left(\alpha_{i,j}^k\right)^{\bar{\delta}_i}}{\sum_{l=1}^{L} \left[\theta_l \prod_{i=1}^{T_l} \left(\alpha_{i,j}^l\right)^{\bar{\delta}_i}\right]} \quad \text{and} \quad \bar{\delta}_i = \frac{\delta_i}{\max\limits_{i=1,2,\ldots T_k}\{\delta_i\}} \tag{3}$$

where $\alpha_{i,j}^k$ $(i = 1, 2, \ldots, T_k)$, called individual matching degree, represents the degree of belief which the input for the ith antecedent attribute belongs to its jth referential value $A_{i,j}^k$ in the kth rule. Therefore, the input of the BRB system is not a set of single values for each antecedent attribute but the individual matching degree.

The belief degree β_j can be calculated by,

$$\beta_j = \frac{\mu * \left[\prod_{k=1}^{L} \left(\omega_k \beta_{j,k} + 1 - \omega_k \sum\limits_{i=1}^{N} \beta_{i,k}\right) - \prod_{k=1}^{L} \left(1 - \omega_k \sum\limits_{i=1}^{N} \beta_{i,k}\right)\right]}{1 - \mu^* \left[\prod_{k=1}^{L} (1 - \omega_k)\right]}, \tag{4}$$

$$j = 1, 2, \ldots, N$$

where

$$\mu = \left[\sum_{j=1}^{N} \prod_{k=1}^{L} \left(\omega_k \beta_{j,k} + 1 - \omega_k \sum_{i=1}^{N} \beta_{i,k}\right) - (N-1) \prod_{k=1}^{L} \left(1 - \omega_k \sum_{i=1}^{N} \beta_{i,k}\right)\right]^{-1} \tag{5}$$

2 Constructing BRB Model for Nonlinear Function

2.1 Nonlinear Function

The characteristics of a complex nonlinear function are the nonlinear relationship between the output variable y and the input variables $x = \{x_1, \ldots, x_n\}$. A nonlinear system can be represented as follows,

$$y = f(x; Q) \tag{6}$$

where $f(\cdot)$ is a nonlinear system model, which means BRB model in this paper, and Q denotes the vector of parameters of the model. Note that the input and output here are quantitative variables, but the input and output of BRB are the match degree or belief degree, so a transformation technique for the quantitative variable is required.

2.2 *Transformation Techniques for Quantitative Data*

There is an important technique, proposed by Yang [4], which allows us transform the quantitative input data into the form of belief degree, described as follows.

Suppose that $x = \{x_1, \ldots, x_n\}$ is a set of quantitative input, and $A_i = \{A_{ij}; j = 1, \ldots, J_i = |A_{ij}|\}$ $(i = 1, \ldots, n)$ is a set of referential values for antecedent attribute. Without loss of generality, $A_{i(j+1)}$ is assumed to be larger than A_{ij}.

Then, an input variable x_i can be represented as,

$$S(x_i) = \{(A_{ij}, \alpha_{ij}); \quad j = 1, \ldots, J_i\} \tag{7}$$

where

$$\alpha_{ij} = \frac{A_{i(j+1)} - x_i}{A_{i(j+1)} - A_{ij}} \tag{8}$$

$\alpha_{i(j+1)} = 1 - \alpha_{ij}$ if $A_{ij} < x_i < A_{i(j+1)}$, and $\alpha_{ik} = 0$ for $k = 1, \ldots, J_i, k \neq j, j+1$

And the output is numerical, and the output of belief degree can be transformed as,

$$y = \sum_{j=1}^{N} u(D_j)\beta_j \tag{9}$$

where $u(D_j)$ is the utility of an individual consequent D_j.

3 Parameter Training for BRB

Because the parameters of a BRB are determined subjectively by experts, it is impossible to ensure that all parameters are accurate for a large-scale rule base. Additionally, a change in parameters will influence the performance of a BRB. Yang [5] proposed a generic learning framework to train BRB parameters and

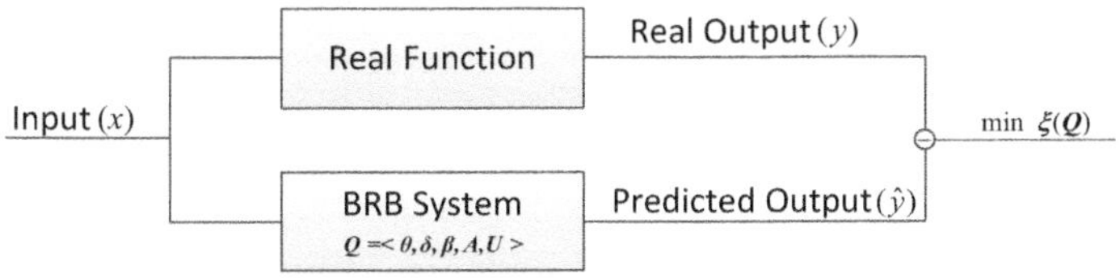

Fig. 1 Framework of training BRB parameter

abstracts the learning process as a nonlinear optimization problem, whose optimal target is to minimize the difference between estimated outputs of BRB system and observed outputs.

Figure 1 shows the framework of training BRB parameter, where x is the input vector, y is the real output generated by function, $\hat{y}$ denotes the predicted output generated by BRB system, and the parameters of BRB is represented as $Q = <\theta, \delta, \beta, A, U>$. And the target of the optimization is minimizing the error between real output y and predicted output $\hat{y}$, which is denoted by $\xi(Q)$. There are several methods to represent $\xi(Q)$, such as mean square error (MSE) and mean average error (MAP). In this paper, MSE is chosen as the optimization target, and therefore, $\xi(Q)$ can be calculated as follow,

$$\xi(Q) = \frac{1}{n}\sqrt{\sum_{i=1}^{n} (\hat{y}_i - y_i)^2} \tag{10}$$

Note that the parameters vector Q must satisfy some constraints for a BRB model [2], listed as Eq. (11).

4 Experiments Study

In this section, we implement an experiment to verify the performance of the BRB system for fitting complex nonlinear functions, and we will also use the BP neural network method for comparison.

$$\begin{aligned}
&0 \le \beta_{j,k} \le 1, \quad j = 1, \ldots, N; \quad k = 1, \ldots, L \\
&\sum_{j=1}^{N} \beta_{j,k} = 1 \\
&0 \le \theta_k \le 1, \quad k = 1, \ldots, L \\
&0 \le \delta_i \le 1, \quad i = 1, \ldots, T \\
&A_{i,j} < A_{i,j+1}, \quad i = 1, \ldots, T; \quad j = 1, \ldots, J_i \\
&u(D_i) < u(D_j) \quad \text{if i<j}, \quad i,j = 1, \ldots, N
\end{aligned} \tag{11}$$

A nonlinear function is used to generate the training dataset represented as follows,

$$\begin{cases} y_1 = (x_1 - 2)x_2, & x_1 \in (2,4), \quad x_2 \in (0,2) \\ y_2 = (x_3 + 1)x_4, & x_3 \in (-1,0), \quad x_4 \in (0,1) \\ y_3 = e^{y_1} + e^{y_2} + \frac{\sin(x_5)}{x_5}, & x_5 \in (0,\pi) \end{cases} \tag{12}$$

In Eq. (12), x_1, x_2, x_3, x_4, x_5 are inputs, and y is the output. To ensure the generality of training data, the training data is generated randomly. The method of generating the training data is as follows: An input x_i is generated randomly from a uniform distribution of its value range, for example, $x_1 \sim U([2,4])$. And other inputs are generated with the same method. Then, calculate the output y with Eq. (12), and thus, a new data pair (x, y) is obtained.

To verify the performance of initial BRB, 20 data pairs are generated as training data randomly using the method mentioned above.

Figure 2 shows the comparison of training data, network output, and trained BRB output, and it can be seen that the outputs of trained BRB are much closer to the training data. And the error of trained BRB has dropped to a very low level, which almost equals to the neural network.

Only comparing the training data with the predicted outputs cannot present the true performance of BRB system or neural network because the parameters are trained on a set of the training dataset, but they are often applied to make predictions on new data points in a real application. So, another new 20 data pairs are generated as testing data to verify the performance of the BRB system and neural network. Figure 2a shows the comparison of testing data, network output, and trained BRB output with 20 training data pairs, and it is obvious that the trained BRB system performs much better than the neural network. This is because the training data is insufficient, and serious overfitting problem occurred in the neural

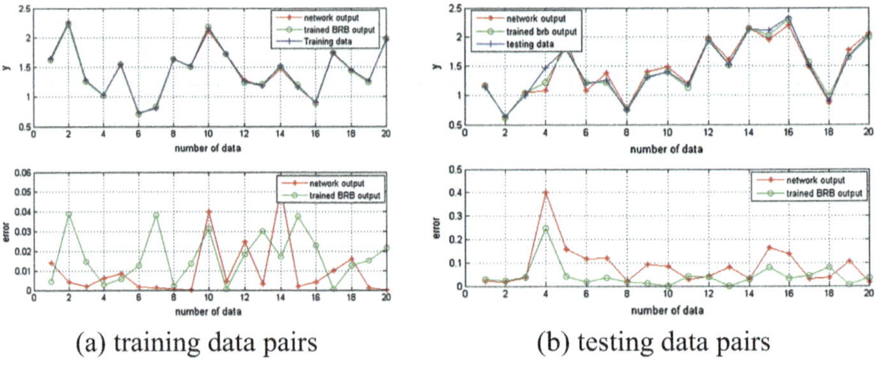

Fig. 2 Comparison of network output and trained BRB output

network. However, the BRB system has no such problem because it combines the knowledge of experts with the data-driven method, and the knowledge of experts plays an important role to predict the outputs discussed before, which can protect the system against overfitting. In contrary, the neural network is a complete black box, and its parameter model is decided by the training process totally, so it is easy has overfitting problem when training data is insufficient.

5 Conclusion

A new method to fit nonlinear function by using BRB system is proposed in this paper, and an experiment is implemented to validate the advantage of BRB system for fitting nonlinear function. Through the comparison with a neural network, it shows that when the training data is insufficient, the performance of BRB is much better than the neural network.

References

1. Ye, H., Yang, L., Liu, X.: Optimizing weight and threshold of BP neural network using SFLA: applications to nonlinear function fitting. In: International Conference on Emerging Intelligent Data & Web Technologies. pp. 211–214. IEEE, (2013)
2. Kay, H., Rineer, B., Kuiper, B.: Semi-quantitative system identification. Artif. Intell. **119**(1), 103–140 (2000)
3. Dietterich, T.: Overfitting and undercomputing in machine learning. ACM Comput. Surv. **27** (3), 326–327 (1995)
4. Yang, J.B., Liu, J., Wang, J., Sii, H.S., Wang, H.W.: A belief rule-base inference methodology using the evidential reasoning approach–RIMER. IEEE Trans. Syst. Man Cybernet. Part A **36** (2), 266–285 (2006)
5. Yang, J.B., Liu, J., Xu, D.L., Wang, J., Wang, H.W.: Optimization models for training belief rule based systems. IEEE Trans. Syst. Man Cybernet–Part A **37**(4), 569–585 (2007)

Approximate Kernel Regression Based on Distributed ADMM Algorithm

Lina Sun, Wenfeng Jing, Cheng Zhang and Haizhen Zhu

Abstract Aiming at the kernel regression of large-scale data, in this paper, we propose a distributed ADMM algorithm based on the Spark platform. It is difficult to calculate and store the kernel matrix of large-scale data. Thus, the Nystrom sampling method is utilized to approximate the kernel matrix, which is applied in solving the kernel regression problem. To verify the effectiveness of the algorithm, we performed numerical experiments on the Spark big data platform. The experimental results show that, given accuracy and computational cost, when the sampling ratio is 2–5%, the kernel matrix reaches the most reasonable approximation degree. The approximate kernel matrix method can solve the problem that the true kernel cannot tackle. Additionally, the approximate kernel regression could be utilized to deal with large-scale data problems, where the computational cost can be greatly reduced and the ideal accuracy can be obtained.

Keywords Kernel regression · ADMM · Distribution algorithm · Spark

1 Introduction

Regression is an important supervised learning problem in machine learning that researches the dependence of variables and is applied to predict unknown data. In practical problems, the relationship between the variables of the data becomes so complicated that it denies description of the relationship between samples. Different from linear regression, the kernel regression [1] uses the correlation between data to

L. Sun · W. Jing (✉)
School of Mathematics and Statistics, Xi'an JiaoTong University, Xi'an 710049, China
e-mail: wfjing@xjtu.edu.cn

C. Zhang
China Railway First Survey and Design Institute Group, Xi'an 710043, China

H. Zhu
ATS Lab, Air Force Engineering University, Xi'an 710038, China

V. Jain et al. (eds.), *Recent Trends in Intelligent Computing, Communication and Devices*, Advances in Intelligent Systems and Computing 1031,
https://doi.org/10.1007/978-981-13-9406-5_20

optimize the model. Kernel regression methods are widely used in image reconstruction [2], image denoising [3], industrial measurement, and prediction [4, 5].

In the context of big data, for m-dimensional data with samples, the kernel matrix size of the data is. For massive computing problems [6], simply improving the computing performance of a single machine can no longer meet the demand for the rapid growing scale of data. The data set is decomposed into a number of small data sets that are convenient to process according to a certain method or strategy. The same model and algorithm are carried out in parallel with the decomposed small data sets, and then, the final result is obtained by integrating the result of small data sets [7]. Big data regression methods have been developed in recent years. Qing [8] designed a efficient parallel extreme learning machine for regression based on MapReduce framework. In [9], a parallel ELM algorithm on the distributed computing MapReduce platform is designed to solve the regression problem, which can process very large-scaled data sets and yields good performance.

In this paper, inspired by the strategy of 'divide and conquer' [10], combining approximate kernel matrix with the kernel regression, we proposed a distributed ADMM algorithm. Moreover, based on Spark platform, the ADMM algorithm is used to solve the approximate kernel regression problem. The effectiveness of the proposed algorithm is verified using large-scale data.

The remaining of this paper is organized as follows: In Sect. 2, we provide a brief review of the kernel regression and ADMM algorithm. Additionally, we introduce a method to construct an appropriate kernel matrix and a distributed ADMM algorithm for kernel regression. Three types of data are utilized to verify the effectiveness of the proposed algorithm in Sect. 3. The conclusion is given at the end of this paper.

2 Method

2.1 Kernel Regression

In practical problems, the relationship between the independent variables and the dependent variables of the data is complicated. The linear regression model cannot describe the relationship between the data well. By adding the kernel function to regression the data with complex relationships. Suppose that the data set $D = \{(x_i, y_i)\}_{i=1}^{N}, (x_i, y_i) \in X \times Y$ is independent distributed samples. K is a kernel, and $K: X \times Y \rightarrow R$, which means that the kernel regression is designed to find a function that reflects the characteristics of the data set well.

$$f_\alpha(x) = \sum_{i=1}^{N} \alpha_i K(x, x_i) \tag{1}$$

where $a_i, i = 1, \ldots, N$ is parameter of regression.

The purpose of the data regression is to find the function $f_\alpha(x)$ of a given type so that the value $f_\alpha(x_1), \ldots f_\alpha(x_N)$ is closest to the observation $y_1, \ldots, y_N$ under certain measure, that is, the sum of the squares of the deviations at each point is minimized.

$$\begin{aligned}\alpha^* &= \arg\min_\alpha L(f_\alpha(x), y)\\ &= \arg\min_\alpha \sum_{i=1}^{N} (y_i - \hat{y}_i)^2 = \arg\min_\alpha \sum_{i=1}^{N}\left(y_i - \sum_{j=1}^{N} \alpha_j K(x_j, x_i)\right)^2\end{aligned}$$

The kernel regression problem can be summarized as an optimization problem as follows:

$$\arg\min_x \frac{1}{2}\sum_{i=1}^{N} ||K_i x - y_i||_2 + \lambda||\mathrm{x}||_1, \quad i = 1, 2, \ldots, N \tag{2}$$

where K is the kernel, x is parameter of regression function, y is the label, and λ is penalty parameter.

2.2 ADMM Algorithm

The ADMM (Alternating Direction Method of Multipliers) algorithm, developed in the 1970s, is a convex optimization algorithm, which is suitable for distributed computing. The original idea of the ADMM algorithm was proposed by Gabay [11]. The ADMM method combines the decomposability of the dual ascending method and the weak convergence of the multiplier method. It is suitable for solving the distributed convex optimization problem in dealing with big data problems.

Given the following problem:

$$\begin{aligned}&\text{minimize } f(x) + g(y)\\ &\text{subject to } Ax + By = c\end{aligned} \tag{3}$$

where $x \in R^n, y \in R^m$, and $A \in R^{p\times n}, B \in R^{p\times m}, c \in R^p$. Suppose f and g are convex functions, the augmented Lagrangian function of the above formula is:

$$L_\rho(x, y, p) = f(x) + g(y) + p^{\mathrm{T}}(Ax + By - c) + (\rho/2)||Ax + By - c||_2^2 \tag{4}$$

Then, the ADMM algorithm is iterated according to the dual rise method

$$x^{k+1} := \arg\min_x L_\rho(x, y^k, p^k) \tag{5.1}$$

$$y^{k+1} := \arg\min_y L_\rho(x^{k+1}, y, p^k) \tag{5.2}$$

$$p^{k+1} := p^k + \rho(Ax^{k+1} + By^{k+1} - c) \tag{5.3}$$

In (4) $\rho > 0$, this algorithm is similar to the dual ascending method and the multiplier method, which is divided into minimization step (5.1), minimization step (5.2), and dual variable update (5.3). Utilizing the multiplier method, the step size is equal to the augmented Lagrangian parameter ρ when the dual variable is updated.

Usually, in order to facilitate the combination of linear terms and quadratic terms in the augmented Lagrangian function and scaling the dual variables, suppose $r = Ax + By - c$, and we can get:

$$\begin{aligned} p^T r + (\rho/2)||r||_2^2 &= (\rho/2)||r + (1/\rho)p||_2^2 - (1/2\rho)||p||_2^2 \\ &= (\rho/2)||r + u||_2^2 - (\rho/2)||u||_2^2 \end{aligned} \tag{6}$$

In (6) $u = (1/\rho)p$ is called scaled dual variable. Using the scaling dual variable, the following ADMM form can be obtained:

$$x^{k+1} := \arg\min_x \left(f(x) + (\rho/2)||Ax + By^k - c + u^k||_2^2\right) \tag{7.1}$$

$$y^{k+1} := \arg\min_y \left(g(y) + (\rho/2)||Ax^{k+1} + By - c + u^k||_2^2\right) \tag{7.2}$$

$$u^{k+1} := u^k + Ax^{k+1} + By^{k+1} - c \tag{7.3}$$

The first form of ADMM (5.1)–(5.3) is ADMM without scaling, and the second form (7.1)–(7.3) derives from the use of dual variables. The scaled form is equivalent to the former two forms and it is simpler compared to the other forms, so it is adopted in our method.

2.3 Kernel Regression Based on ADMM Algorithm

The ADMM algorithm is used in solving the kernel regression problem. We can get

$$\begin{cases} x^{k+1} = \text{soft}\left(x^k - \tau K^*\left(Kx^k - y - p^k/\beta\right), \tau/\beta\right) \\ p^{k+1} = p^k - \gamma\beta\left(Kx^{k+1} - y\right) \end{cases} \tag{8}$$

where K is kernel of regression, K^* is conjugate transposed matrix of K, p is multiplier, and τ, β, γ is given parameter. This paper adds the acceleration method

[12] and the adaptive step size [13]. The detailed iterative steps of the kernel regression are shown in Algorithm 1.

Algorithm 1 Acceleration of ADMM to solve kernel regression
Input: The kernel matrix and label for samples $[K, y] = \{[K_i, y_i]\}_{i=1}^{N}$, penalty parameter λ, parameter τ, β, η and iteration step Maxi;
Output: Kernel regression parameter Z

(1) Divide the entire data into N blocks $\{X_1, \ldots, X_N\}$. The ith block of data is allocated to the ith machine;
(2) Initialize the variable: set $x, p = 0$, residual res $= -b$, $t$new $= 1$, iterative data initialization is $k = 0$;
(3) for $k = 1$:*Maxi*
(4) $t = t$new; $x$old $= x$; resold $=$ old; $p$old $= p$
(5) Calculate the step size, gradient and update the variable x on the Master

$$\text{stepsize} = \eta/\tau$$

$$g = A^{\mathrm{T}}(\text{res} - p/\eta), x = \text{soft}(z - g \cdot \text{stepsize}, \lambda/\tau)$$

(6) Calculate the residuals res in parallel on each slaver and update the multipliers P

$$\text{res} = Ax - b, \quad p = p\text{old} - \gamma\eta \cdot \text{res}$$

(7) Perform adaptive step adjustment on the master according to the following formula

$$\begin{aligned} hx &= \tau \cdot \|x\text{old} - x\|^2 \\ hp &= (2 - \gamma) \cdot \eta \cdot \|\text{res}\|^2 \\ hc &= 2\eta \cdot \text{res}^{\mathrm{T}}(\text{resold} - \text{res}) \\ h &= hx + hp + hc \\ hr &= \eta \cdot hx + r \cdot \eta^2 \cdot \|\text{res}\|^2 \end{aligned}$$

If $h < hr$ so $\tau = 2\tau$, else $\tau = 0.5\tau$
(8) Iteratively accelerates at the master according to the following formula

$$\begin{aligned} t\text{new} &= 0.5 \times \left(1 + \sqrt{4t^2 + 1}\right) \\ z &= x + \tfrac{t-1}{t\text{new}} \times (x - x\text{old}) \end{aligned}$$

(9) Broadcast variables x, z to each slaver compute node to complete the iteration

2.4 Approximate Kernel Matrix

At present, the kernel method has been successfully applied to various practical problems of nonlinear structures, such as support vector machine, kernel regression, and principal component analysis. The kernel function is an important index to measure the similarity between data samples. In the application of manifold learning [14] and dimensionality reduction [15], the eigenvectors of the kernel matrix reveal the true structure of the data and the low-dimensional data flow to some extent. For the data set of n samples, the size of the kernel matrix is $n \times n$. The quadratic space complexity and cubic time complexity of the kernel matrix pose difficulty in processing big data. The spectrum of the kernel function is reduced, and the original matrix is reduced. Performing low-rank approximation is an effective way to solve this problem. We use an efficient low-rank approximation kernel matrix method [16], which mainly uses random SVD decomposition combined with the Nystrom method to perform low-rank approximation of the kernel matrix.

Algorithm 2 D-R Nystrom algorithm
Input: Oversampling parameter p, power parameter q, rank k;
Output: Orthogonal matrix U, diagonal matrix Λ;

1. Divide all data points into N blocks $\{X_1, \ldots, X_N\}$;
2. for the slaver $i = 1, 2, \ldots, N$
3. Assign the ith block data to the ith node;
4. Sampling data points to obtain a sampling matrix S, calculate the kernel matrix W of the samples, and decompose the matrix W using k-rank SVD, so we will get $W \approx \tilde{U}\Lambda\tilde{U}^{\mathrm{T}}$;
5. Broadcast the sampling matrix S to each node and calculate the kernel submatrix C_i of X_i and S;
6. Calculate $U_i \leftarrow C_i\tilde{U}\Lambda^{+}$ in slavers;
7. $U \leftarrow [U_1, \ldots, U_N]$, combine U_i from slavers

3 Experiments

All the experiments in this paper are conducted in the National Engineering Laboratory of Big Data Algorithm and Analysis Technology of Xi'an JiaoTong University. The detailed information of the big data platform is shown in Table 1.

The big data processing framework Hadoop and Spark are deployed on the cluster. The development environment is Eclipse. Spark platform is used to perform simulation experiments on artificial generated data. This experiment uses three

Table 1 Hardware configuration for experimental platform

Name	Number	Configuration detailed
Master	1	Ubuntu Linux, Memory 300 G, 12 kernels
Slaver	27	Ubuntu Linux, Memory 240 G, 12 kernels

different attributes of the data [16] to test the approximate kernel regression algorithm, where the relative error and iteration time between the real and the approximate kernel are compared. The relationship is used to verify the validity of the approximate kernel regression. For the three different properties of the function, five sets of data of different scales are used to conduct experiments.

Experiment 1: Approximate kernel regression experiments on data (1)–(3) and kernel regression experiments using real nuclei, comparing the relative errors of the two experiments, verifying the validity of approximate kernel regression. In this experiment, we set the parameter of Gaussian kernel function as $\tau = 0.3$, sample number as $m = 1000$, and rank as $k = 600$. Setting penalty parameters to $\lambda = 1e - 4$, $\eta = \tau = 10/||b||_2, \gamma = 1$ in the approximate kernel regression. Kernel regression experiments were performed on the Spark platform. The real kernel matrix was calculated on MATLAB and then uploaded to HDFS.

The experimental results are shown in Tables 2, 3, and 4. N_Train and N_Test mean the dimensions of train data and test data. And TKRE, AKRE, and AKE mean the error of true kernel regression, error of approximate kernel regression, and error of kernel approximation, respectively.

Experiment 2: The data (3) is subjected to approximate kernel regression experiments based on different sample numbers, and the relationship between the number of samples and the relative error is explored; the number of samples is 500, 1000, and 2000, respectively. The experimental results of the three types of data are shown in Fig. 1.

Table 2 Results of kernel regression experiments on Wendland data

N_Train	N_Test	Sample	TKRE	AKRE	AKE
4000	500	1000	4.6e−03	5.1e−03	0.0701
10,000	1000	1000	2.1e−03	2.6e−03	0.0726
20,000	2000	1000	2.2e−03	2.7e−03	0.0816

Table 3 Results of kernel regression experiments on Sinc data

N_Train	N_Test	Sample	TKRE	AKRE	AKAKE
4000	500	1000	1.9e−03	4.0e−03	0.0732
10,000	1000	1000	1.0e−03	2.3e−03	0.0101
20,000	2000	1000	1.1e−03	4.6e−03	0.0805

Table 4 Results of kernel regression experiments on Cosc data

N_Train	N_Test	Sample	TKRE	AKRE	AKE
4000	500	1000	0.0102	0.0653	0.0723
10,000	1000	1000	8.7e−03	0.0291	0.0754
20,000	2000	1000	9.1e−03	0.0565	0.0978

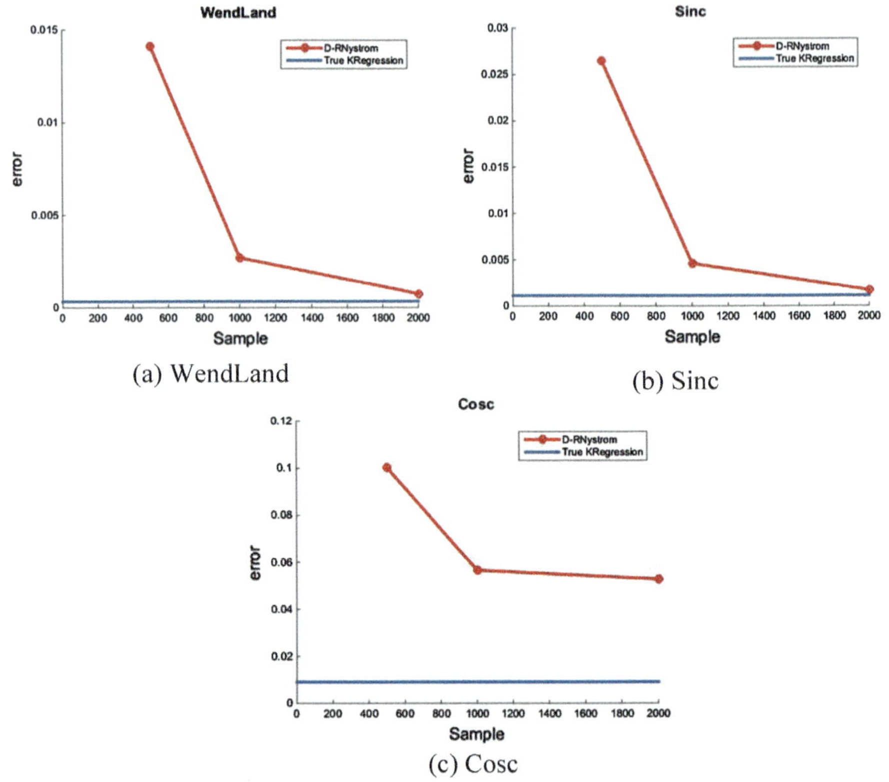

Fig. 1 The relationship between number of samples and kernel regression error

After the above experiments, it can be seen that with the increase in the number of samples of the original data, we can obtain more accurate kernel matrix approximation, but the overhead will become larger. Taking the accuracy and cost into consideration, the kernel matrix of the sampling number would obtain the most reasonable approximation degree. For data with a very large number of samples, calculating the kernel matrix directly requires very large overhead and using the approximate kernel matrix can achieve both the ideal accuracy and the computational overhead.

4 Conclusion

In this paper, the problem of kernel regression in the context of big data is studied, and an efficient distributed ADMM algorithm is proposed. The problem of calculating cost of large-scale sample kernel matrix is solved by approximate kernel matrix method. When dealing with the big data kernel regression problem, the distributed ADMM algorithm is used and the adaptive step as well as acceleration algorithm is used to accelerate the iterative kernel regression algorithm. Furthermore, we conducted several experiments using three types of data. The experimental results show that, given accuracy and computational cost, when the sampling ratio is 2–5%, the kernel matrix reaches the most reasonable approximation degree. The approximate kernel matrix method can solve the problem that the true kernel cannot tackle. Additionally, the approximate kernel regression could be utilized to deal with large-scale data problems, where the computational cost can be greatly reduced and the ideal accuracy can be obtained, showing that the proposed distributed ADMM algorithm yields good performance on approximate kernel regression.

Acknowledgements Lina Sun thanks to the NSFC for its support under grant 11690010 and grant 11631013 as well as the support from National Engineering Laboratory for Big Data Analysis. And the authors thanks to the reviewers for their constructive comments.

References

1. Zhang, Y., Duchi, J.C., Wainwright, M.J.: Divide and conquer kernel ridge regression: a distributed algorithm with minimax optimal rates. J. Mach. Learn. Res. **30**(1), 592–617 (2013)
2. Feng, Q., et al.: Center-based weighted kernel linear regression for image classification. In: IEEE international conference on image processing IEEE, 3630–3634 (2015)
3. Deng, X.G., Tian, X.M.: Kernel regression modeling method based on feature vector selection. Control Eng. China **17**(4), 517–520 (2010)
4. Härdle, W., Vieu, P.: Kernel regression smoothing of time series. J. Time **13**(3), 209–232 (2010)
5. Yang, Y., et al.: Accurate, fast and scalable kernel ridge regression on parallel and distributed systems (2018)
6. Afonso, M.V.: Fast image recovery using variable splitting and constrained optimization. IEEE Trans. Image Process. **19**(9), 2345–2356 (2010)
7. Zhang, L.S., Liu, H.Y., Lei, D.J.: MapReduce-based parallel linear regression for face recognition. Appl. Mech. Mater. **556–562**(11), 2628–2632 (2014)
8. He, Q., et al.: Parallel extreme learning machine for regression based on MapReduce. Neurocomputing **102**(2), 52–58 (2013)
9. Chen, J., et al.: MR-ELM: a MapReduce-based framework for large-scale ELM training in big data era. Neural Comput. Appl. **27**(1), 101–110 (2016)
10. Asanovic, K., et al.: A view of the parallel computing landscape. Commun. ACM **52**(10), 56–67 (2009)
11. Beck, A., Teboulle, M.: A fast iterative shrinkage-thresholding algorithm for linear inverse problems. Siam J. Imaging Sci. **2**(1), 183–202 (2009)
12. Daubechies, I., Defrise, M., Mol, C.D.: An iterative thresholding algorithm for linear inverse problems with a sparsity constraint. Commun. Pure Appl. Math. **57**(11), 1413–1457 (2010)
13. Xiao, Y., Xia, L., Zhang, W.: Face recognition with supervised spectral regression and multiple kernel SVM. In: International Conference on Advanced Computer Control IEEE, pp. 343–346 (2010)
14. Lin, Y.Y., Liu, T.L., Fuh, C.S.: Multiple kernel learning for dimensionality reduction. IEEE Trans. Pattern Anal. Mach. Intell. **33**(6), 1147–1160 (2011)

15. Li, M., et al.: Large-scale Nyström kernel matrix approximation using randomized SVD. IEEE Trans. Neural Netw. Learn. Syst. **26**(1), 152–164 (2014)
16. Xu, C., et al.: On the feasibility of distributed kernel regression for big data. IEEE Trans. Knowl. Data Eng. **28**(11), 3041–3052 (2016)

Detection for Mixed-Characters Based on Machine Learning

Liang Han, Shuai Zou, Dengke He and Wen Jing Zhou

Abstract We propose a new method based on a machine learning technique to detect the mixed-characters on the surface of all kinds of cables applied in industry. Firstly, the images of these characters are captured by a high precision CCD, and then, the captured images need to be preprocessed, normalized, and divided into multiple images. Each image includes a single character. Finally, the training set image is sorted and optimized. We establish a convolutional neural network for the characters recognition, and its parameters are improved based on character features. The average recognition rate of mixed-character is 92.6%. Experimental results show a recognition rate based on machine learning could be higher than the one by using other algorithms.

Keywords Mixed-characters · Machine learning · Convolutional neural networks

1 Introduction

As an important material in modern industrial production, the cable will have a certain application space in the industry in the future, and its surface quality greatly affects the performance of the final product. In recent years, the detection technology based on machine vision [1] has been widely used because of the rapid development of digital image processing and computer technology. Using machine learning to detect cable surface character detection is more effective than traditional characters, the cable surface character detection technology [2] in this paper is to use the camera to get object images, and the visual inspection is a combination of computer vision and image processing technology. We can obtain image and various feature parameters after image processing, the parameters we need to

L. Han (✉) · S. Zou · D. He · W. J. Zhou
Department of Precision Mechanical Engineering, Shanghai University, Shanghai 200072, China
e-mail: hlswj@shu.edu.cn

V. Jain et al. (eds.), *Recent Trends in Intelligent Computing, Communication and Devices*, Advances in Intelligent Systems and Computing 1031,
https://doi.org/10.1007/978-981-13-9406-5_21

process, judge, and make decisions, the features of results and output control signals which are high accuracy, non-contact and high degree of automation [3]. Therefore, the study of the cable surface character detection system and the development of intelligent online detection equipment are of great significance for promoting the automation and intelligence of domestic factories and Industry 4.0.

The cable printing character mostly adopts the method of artificial visual inspection at present, which has low intelligence, low efficiency, and high error. With the rise of machine vision, its stability, intelligence, and accuracy make it successfully applied in many industrial fields and form many detection methods [4, 5]. In the field of character recognition, the neural network has the advantages of self-organization, self-study, and self-adaptability. In this paper, a method based on machine vision [6–8] is proposed to detect the mixed-characters on the surface of different cables, while a convolutional neural network for feature recognition of the printed characters is also established. During experimental work, the character images from several different cables are captured by using a high precision CCD. A liquid lens also is used to match CCD in the image capturing system. The process of all images includes preprocessing, segment, normalization, and feature extraction. Enough images also are collected to avoid overfitting.

2 The Composition of the Detection System

In order to capture some good quality images, an excellent imaging system is built shown in Fig. 1. A CCD (Manta G-504) with 2.2 μm pixel size is adopted, and a liquid lens (C-C-39NO-160-R12) also is used to match with. Here, this liquid lens has a focal length with 16 mm that means CCD always captures the focus plane. Hence, that also ensures the captured images are clear and good quality. A two-dimensional motorized translation stage (GCD-040201M) is used to move the cable because the detected cables keep moving fast in the industry.

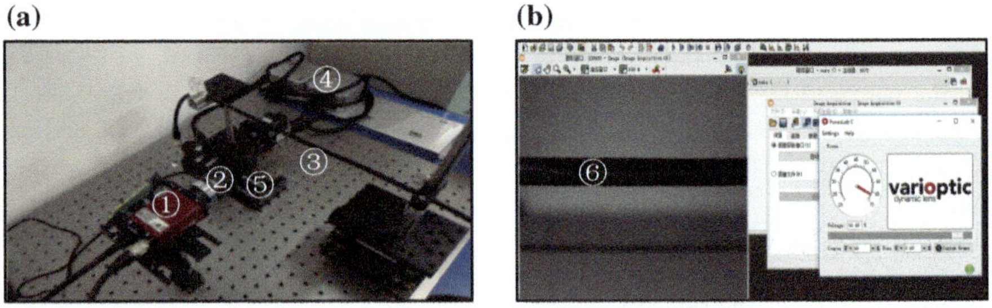

Fig. 1 Photographs of the image capturing system and HALCON interface. **a** Image capturing system; **b** HALCON interface. 1—CCD; 2—liquid lens; 3—cable with characters; 4—motor; 5—motorized translation stages; and 6—captured character image

3 Character Preprocessing

Because of the problem of the illumination or angle, it is inevitable to have noise and uneven illumination when the cable character image is collected. Therefore, the preprocessing of characters is crucial to the segmentation and recognition of characters.

3.1 Binarization

In order to separate the target character from the background area in the image, binarization should be carried out. Considering the obvious difference between the character and the background grayscale value, we use directly experience value as the threshold value. The process of binarization is shown in Fig. 2.

3.2 Character Segmentation

The convolutional neural network can only take a single number as the sample during training and detection, so the multiple characters after binarization need to be segmented, which can be separated by the white background among the characters [9]. The specific steps are:

(1) Determine the left and right width of the character. We use the binarization of the character images, take one column in turn, and scan it from top to bottom when the first black pixel is encountered and recorded its column number which is the left-end position of the character and continue scanning until we encounter a column that has no black pixels, which is the right end of the character. These two column numbers record the approximate width of the character.

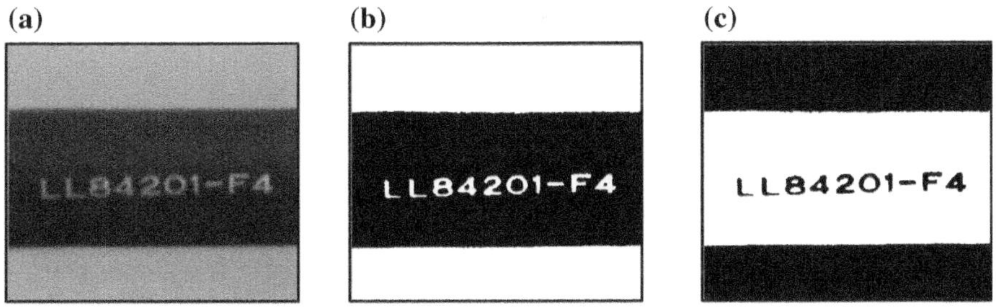

Fig. 2 Schematic diagram of binarization process; **a** original image of the string character; **b** binarization of image; and **c** inverse of image

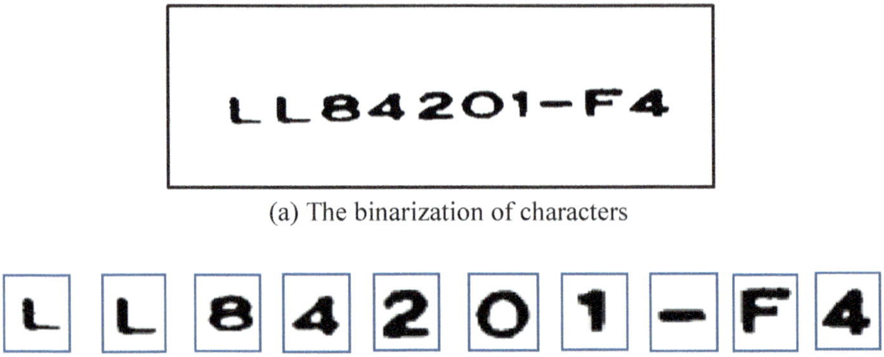

(a) The binarization of characters

(b) The segmentation of character

Fig. 3 Binarization and segmentation of characters

(2) Determine the height range of characters in the image. In the width range obtained in step 1, we scan line by line from top to bottom when we encounter the first black pixel, and we need record its line number, which is the upper position of the character; then, continue to scan until you encounter a line with no black pixels, the lower end of the character. The results are shown in Fig. 3.

3.3 The Character Normalization

Although the collection size of the cable character image can be kept consistent, the general pixel is too high which is not conducive to training and testing. The uniform size of the character image is required in training, testing, and recognition. Therefore, it is necessary to normalize the character image to make the character images has the same size. In this paper, the size of the segmented character image is set to 28 * 28. The algorithm determines the target width and height of the character at first and then calculates the scaling factor in the horizontal and vertical directions, and it can not only zoom the image of the same pixel but also complete the pixel mapping for characters of different sizes to achieve scaling.

4 Data Set Production

Before the design of the network, the production of data sets is very important. A good data set can improve the efficiency of the experiment and the accuracy of recognition. Through algorithm verification, directly training 26 uppercase letters, 10 handwritten numbers, and special characters cannot be achieved. Expected effect, but the improvement can effectively improve the recognition rate.

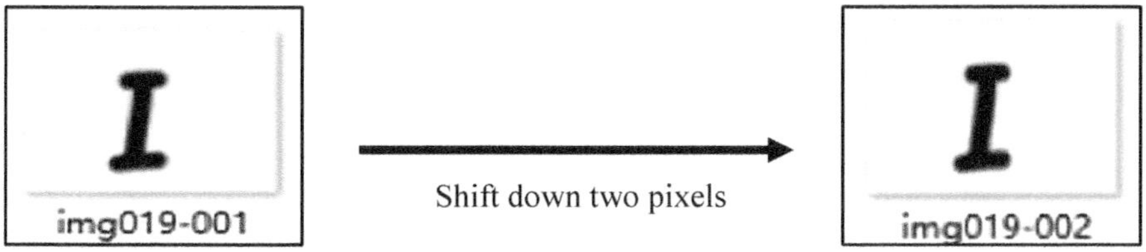

Fig. 4 Standardization of character

Because some characters and numbers are very similar, for example, the number 0 and the letter O are not different when printed, so the similarity between the numbers 0 and O is extremely high, and the uppercase letter I and the similarity of the number 1 is also extremely high, and the classification and recognition is extremely difficult. Therefore, the uppercase letter O and the number 0 can be classified into one class during training so that there are 37 classifications in total, and the average recognition rate can be achieved under the two-layer convolution 0.92. And because the recognition is relatively singular, standard characters are used in the production of data sets for specific similar characters. Under the premise of standard, the corresponding number of data sets are achieved by rotation and displacement, so when the training set is produced, the controversial numbers or standardization of characters, such as the recognition rate, can be improved, as shown in Fig. 4.

In this paper, we mainly focus to recognize the handwritten uppercase English letters from A to Z, ten numbers from 0 to 9 and the special characters as "–" and "/". The handwritten English letters used in this paper are handwritten with the drawing board that comes with the Windows operating system and saved as a monochrome bitmap in PNG format, which is also a binary image, also called a binary image. There are 2775 sample images (37 groups) in this paper, of which 1850 are training samples and 925 are test samples. Figure 5 shows a partial data set.

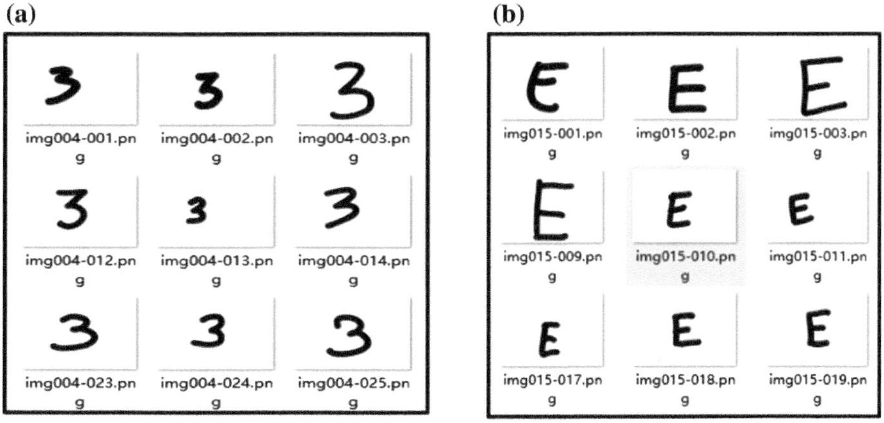

Fig. 5 Partial 28 × 28 pixel data set

5 Neural Network Design

Designing a neural network that recognizes 37 English capital letters, numbers, and special characters means that whenever the trained network has an input that represents a letter or number, the network can correctly indicate the letter or number at the output. The neural network training should be supervised, training 37 sets of letters representing the letters, numbers, and special characters of the one-dimensional vector, which can correspond to the specific positions of the output terminals 1–37. For character recognition, a convolutional neural network is used in this paper. The model mainly includes an input layer, two layers of convolution and pooling layers, full connection layer, and output layer.

The overall structure is shown in Fig. 6.

6 Experimental Results and Analysis

The character detection simulation system designed in this paper mainly includes three modules: input picture, normalization, and recognition. Although they keep independent of each other in algorithm design, they are interdependent in function. The result of the previous module is the input of the latter module, and the performance of the entire system is closely related to each module. For the sample data of the cable, we use the hand-drawn character data set as a sample set, and the character data collected by the actual cable is used as a test set. A total of 1850 training samples and 925 test data, the model cycle training 500 times, each time randomly crawling 55 data points in the training data for training. In order to compare the recognition effect of different neural network algorithms, the convolutional, the LVQ [10], and BP [11] are performed for same images. Table 1 shows the recognition rate of 37 characters by different neural networks and the corresponding parameters. Hence, the convolution is more efficient and accurate than LVQ neural networks and BP neural networks.

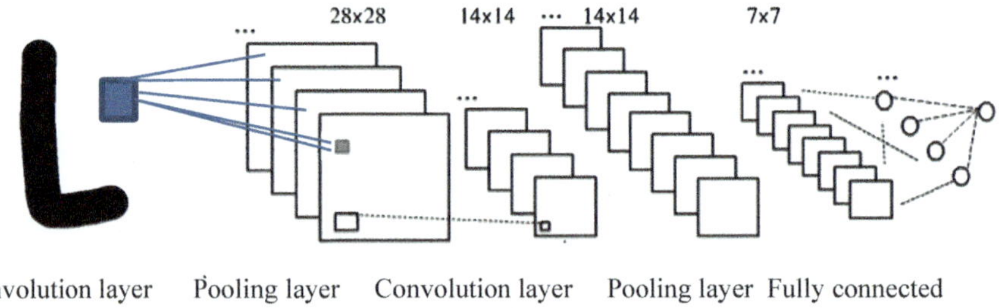

Fig. 6 Overall structure of the training model

Table 1 Accuracy of character recognition

Network type	The training sample	The test sample	Wrong number	Vector	Recognition rate (%)
BP	1850	925	275	0.1	70.2
LVQ	1850	925	237	0.1	74.4
Convolutional	1850	925	68	0.1	92.6

7 Conclusion

Convolutional neural networks are widely used in image recognition, but it is rare to use cable character recognition. Experiments show that the convolutional neural network cannot only perform feature extraction but also complete classification function. After enough data set samples being trained, the convolutional neural network has better recognition and fault tolerance. Through the identification system of this paper, it shows that the contrast rate of the convolutional neural network is significantly improved compared with BP neural network and LVQ neural network. The next step is to improve the convergence algorithm, optimize the network parameters, and further improve the character detection precision and speed.

References

1. Davies, E.R.: Machine Vision: Theory, Algorithms, Practicalities, pp. 1–16. Elsevier (2004)
2. Steger, C., Ulrich, M., Wiedemann, C.: Machine vision algorithms and application. Illumination **5**(2.1), 1 (2008)
3. Zhou, W.J.: Research and Application of Online Idle Speed Detection and Precise Control Problem Based on Machine Vision. Shanghai University, Shanghai (2014). (In Chinese)
4. Wang, Y., Xia, R.Z., Chen, H.L.: Milling machine vision inspection system design. J. Test Technol. **29**(3), 235–240 (2015). (In Chinese)
5. Wu, H.H., Zhang, X.M.: Pre-blow detection algorithm for multi-pin type placement devices. J. Electron. Meas. Technol. **25**(11), 998–1005 (2011). (In Chinese)
6. Xu, Y.S., Gu, J.H., Tao, Z.: Handwritten character recognition based on improved BP neural network. Commun. Technol. **44**(5), 106–109 (2011). (In Chinese)
7. Steger, C., Urich, M., Wiedemann, C.: Machine Vision Algorithms and Applications, Bilingual edn. In: Yang, S.S., Wu, D.J., Duan, D.S. (trans.). Tsinghua University Press, Beijing (2011)
8. Zhang, W., Wang, Y.P., Xue, G.X.: Digital Image Processing and Machine Vision. People's Posts and Telecommunications Press, Beijing (2012). (In Chinese)
9. Wu, H.H., Zeng, X.R., Lai, Y.J., Wu, F.P.: Machine vision based harness connector character detection. Test Technol. **32**(2), 174–179 (2018). (In Chinese)
10. Li, Y.J.: Handwritten English Letter Recognition Based on LVQ Neural Network. Guangdong University of Technology, Guangdong (2008). (In Chinese)
11. Gao, L.: Handwritten English Letter Recognition Based on BP Neural Network. North University, Shanxi (2009). (In Chinese)

Research on 3D Terminal Rendering Technology Based on Power Equipment Business Features

Gang Wang, Xiaodong Zhang, Chengzhi Zhu, He Wang, Lin Peng and Zhansheng Hou

Abstract At present, the traditional grid 3D model is huge, and the running software and hardware environment is complex. It is difficult to apply across computer platforms, the model rendering efficiency is low, and the mobile terminal is difficult to load and run. The core of 3D engine is the real-time high-speed rendering technology of 3D model. The traditional 3D model needs a lot of post-rendering after modeling. It needs to calculate and process the material, texture, and illumination of the model, which consumes a lot of computing resources and labor cost. This paper innovatively adopts the rendering algorithm based on grid business characteristics and realizes the high speed of 3D power equipment model on mobile terminal equipment through the application of modules and algorithms such as material management, particle system, bulletin board, HDR environment, and illumination. Rendering solves the problem of high cost of post-production rendering of traditional 3D models.

Keywords Electric equipment · 3D terminal rendering · Business characteristics

1 Introduction

At present, the traditional grid 3D model is huge, and the running software and hardware environment is complex. It is difficult to apply across computer platforms, the model rendering efficiency is low, and the mobile terminal is difficult to load

G. Wang (✉) · H. Wang · L. Peng · Z. Hou
Global Energy Interconnection Research Institute Co., Ltd., Nanjing 210003, Jiangsu, China
e-mail: wanggang@geiri.sgcc.com.cn

G. Wang · H. Wang · L. Peng · Z. Hou
State Grid Key Laboratory of Information & Network Security, Nanjing 210003, China

X. Zhang
State Grid Corporation of China, Beijing 100031, China

C. Zhu
State Grid Zhejiang Electric Power Co., Ltd., Hangzhou 310007, China

V. Jain et al. (eds.), *Recent Trends in Intelligent Computing, Communication and Devices*, Advances in Intelligent Systems and Computing 1031,
https://doi.org/10.1007/978-981-13-9406-5_22

and run. The core of the 3D engine is the real-time high-speed rendering technology of the 3D model. The traditional 3D model needs to be post-secondary rendering. It needs to calculate the material, texture, illumination, etc., which consumes a large amount of computing resources and labor costs, and because the computing power of the terminal is relatively weak, how to realize the real-time high-speed rendering of the 3D model for the terminal and grid business application requirements under the condition of limited computing resources is necessary. Therefore, it is necessary to study the 3D lightweight engine for the grid business scenario and the 3D model file compression technology.

Computer 3D modeling rendering is a science that uses the powerful computing power of a computer to simulate various visual phenomena in real life [1]. This emerging science has a history of nearly 50 years. In the early stage of development, due to the limitation of computer hardware capabilities, the main research direction focused on how to simplify the physical mathematical model so that the computer-generated computational complexity is not too good [2]. As hardware capabilities improve, scientists and researchers are beginning to focus on how to create realistic graphics. From the 1970s to the present, rendering 3D modeling is a hot and difficult problem in the field of graphics [3]. The difficulty of 3D image rendering is not only reflected in the complexity of the scene and the number of triangular patches that need to be rendered [4]. At the same time, due to the large number of scene objects, a large number of scene scheduling algorithms are involved, and other aspects such as scene modeling and illumination are also faced [5]. Shadows, interactive movements, and other issues affect the rendering of authenticity.

2 3D Model High-Speed Rendering Technology

In the entire 3D model reconstruction system, the rendering of a single 3D model is the basis of the entire system. In general, there are two main ways to render a 3D model. One is a 3D model rendering based on a polygon mesh, and the other is a 3D model rendering based on points. The former belongs to the traditional rendering mode, the technology development is relatively mature, and it is superior in the rendering of the fine model. The latter belongs to the new model rendering method that has emerged in recent years. It has the significant advantages of low storage space and fast speed, and has become a hot spot in the field of augmented reality in recent years [6].

Model rendering based on polygon mesh refers to the method of rendering the model with a polygon mesh as the rendering unit. The advantages of using polygon mesh rendering are as follows: First, the geometry of the polygon is relatively regular, and the geometric information of each face can be directly used to perform very intuitive mapping of texture and material. The second is that the polygon mesh retains the connection information between the nodes when it is stored. Therefore, no matter whether it is enlarged or stretched, it will not have a large degree of distortion, which can meet the high-precision browsing requirements. The third is that when the surface with a highly discontinuous surface is rendered, such as a tree, the polygon mesh has a better rendering effect because it can be infinitely subdivided.

Point-based model rendering refers to the method in which the input model is modeled in the form of a sample point set for model rendering. The "point cloud" that the scanner obtains by scanning the 3D model may first be chunked and need to be merged to obtain a complete model point cloud. Second, there may be disturbing points or redundant points in the scanning process, which need to be removed when the rendering is actually started. Of course, the fully processed point cloud can also be rendered directly as a rendering unit. Oil splash technology is widely used in point-based 3D model rendering. It refers to mapping a point to multiple pixels on the screen, and the color of one pixel point will be the weighted average of the colors of all the points that affect it. Point-based rendering has many advantages: First, the point is a simpler rendering element than the polygon mesh, with low redundancy, so the structure is simple, and the storage space requirements are low. Second, point-based rendering is faster. An algorithm that uses points as a rendering element does not require polygon clipping, scan conversion, texture mapping, or bump mapping, so point-based rendering is fast. The second is that we only determine the attributes of a certain rule area around the point according to the attributes of a certain point, such as normal vector, color, and illumination, so the rendering speed is faster. Third, point-based models are more suitable for objects that represent organic shapes, such as fluids, smoke, and the like, than models represented by polygonal meshes.

3 3D Model Real-Time High-Speed Rendering Technology for Terminal and Grid Business Application Requirements

The traditional 3D model needs a lot of post-rendering after modeling. It needs to calculate and process the material, texture, illumination, etc., of the model, which consumes a large amount of computing resources and labor costs. This paper innovatively adopts the characteristics based on grid business. The rendering algorithm realizes the high-speed rendering of the 3D power device model on the mobile terminal device through the application of modules and algorithms such as material management, particle system, bulletin board, HDR environment, and illumination, and solves the cost of the traditional 3D model post-rendering.

In this paper, the open-source model rendering algorithm is used to realize the real-time high-speed rendering of the 3D model. In order to facilitate the user's use, the model engine's overall interface module is based on WebGL development. The user can conveniently use various functions of the system on the interface, and the design will interface. Separation from the system improves the reusability of the system code. The interface can be designed on WebGL, as well as MFC or even a GUI system. At the bottom of the system is the graphics engine, which provides basic functionality support for the system, and the system can be added to the engine through plug-ins to provide additional functional support for the engine.

The rendering module is the most critical part of the system. Traditional web pages usually contain HTML, CSS, and JS files. WebGL-based applications also include 3D model data and shader source code for display on the basis of traditional web applications. The source code of these 3D model data and shader is submitted to the WebGL rendering pipeline through JavaScript calling WebGL API. The rendering pipeline calls GPU resources to realize 3D visualization and fast rendering in the web environment. As shown in Fig. 1, the rendering module design of this paper is divided into five parts: material management module, particle system module, bulletin board, HDR environment, and lighting system. The material management module is responsible for the management and rendering of the grid model material according to the grid business characteristics; the particle system includes a three-dimensional particle system, which can produce grid device voltage, current, device status display, etc.; the bulletin board module can make the model always face the camera, positive for the viewpoint; the HDR environment can provide automatic analysis and adjustment of the scene light source, providing environment indoor and outdoor environment such as power transformation and transmission; the illumination system adopts the physical illumination algorithm based on energy balance, which is the best performance and effect on the mobile end. Through the application of the rendering module, the engine can realize high-speed rendering of the grid model based on the business characteristics, improve the realistic degree of the model, reflect the characteristics of the grid business, and finally achieve the rendering and optimization of the model more stably.

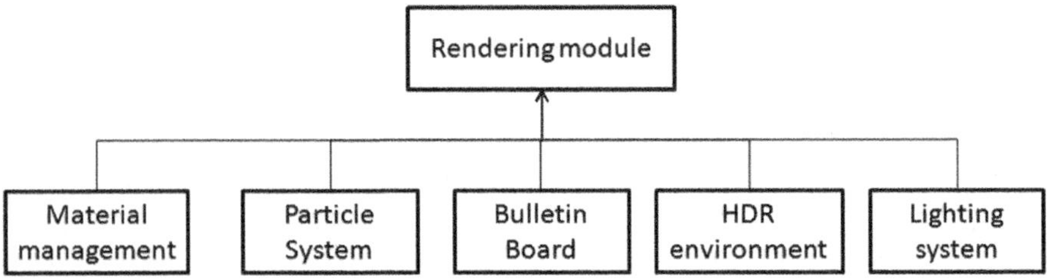

Fig. 1 3D model rendering module architecture design

3.1 Scene Organization

The rendering organization model uses a dynamic LOD algorithm based on a quadtree. Taking the grid environment as an example, there is an environment where there is an object at the camera to query the object. To find out where the small circle is, divide the environment into four large blocks 1, 2, 3, and 4, and continue to divide each block into A, B, C, and D. Then, when you want to query which piece of the small ball, first traverse the first level, find it in the first block, then traverse the second level, and finally find the search in C. This is the analysis of a typical quadtree. Let us look at the dynamic LOD algorithm based on quadtree. The basic idea of the quadtree LOD algorithm is to divide the environment into four and recursively render each mesh. Then, each grid is judged according to the determination condition: whether the highest precision is reached and whether it is within the field of view. If the highest precision is reached, or if it is not in the field of view, it will exit. The recursive mesh is then continued on the eligible mesh. In the simplification of the quadtree, it is more important to establish a reasonable node evaluation mechanism to determine whether the current grid meets the requirements. Since each mesh is square, it can be easily rendered by simply selecting two triangles. However, when such a rendering method differs by more than one time from the adjacent precision levels, the environment rendering will have a significant crack problem. A variety of methods to eliminate cracks are needed to repair.

3.2 Rendering Process

After the user completes the organization of the entire scene, enter the entire rendering stage. First, through the Start Render function in the engine rendering module Root, you can start to enter the rendering, first call the Render One Frame, then enter the Begin Scene function, and start drawing each frame. In the drawing of each frame, first enter the Scene Manager, call the update scene graph function update according to the grid business scenario requirements, in the update function, it will traverse all the scene nodes, and call the update function corresponding to each scene node in combination with the grid business logic features. After completing the update of all the nodes in the whole scene, the model with some details of the grid details will be refined, the nodes with no significant requirements for the detail requirements will be normalized, and then all the elements that need to be rendered on the root scene node will be rendered. Send it to Render Order, then call the corresponding cropping and sorting algorithm to select the visible object, and officially enter the rendering process of the Render System. The process in the Render System is similar to the rendering process of the underlying API, mainly including the setting of the pipeline state, the implementation of GPU programming, generally refers to the Begin Frame () into the rendering, through End Frame () to end the rendering of a frame.

3.3 3D Engine Cross-Platform Technology

The 3D engine overall module of this project is based on WebGL technology. Web Graphics Library (WebGL) is a 3D drawing protocol. This drawing technology standard allows JavaScript and OpenGL ES 2.0 to be combined by adding a JavaScript binding of OpenGL ES 2.0. WebGL can provide hardware 3D accelerated rendering for HTML5 Canvas, so that web developers can use system graphics to more smoothly display 3D scenes and models in the browser, as well as create complex navigation and data visualization. The WebGL technology standard eliminates the hassle of developing web-specific rendering plug-ins, can be used to create website pages with complex 3D structures, and can even be used to design 3D web games. It has many features such as cross-platform and lightweight, so WEBGL is adopted as the technical architecture of the 3D model engine, which has the characteristics of 3D cross-platform. The cross-platform support of the 3D model lightweight engine under terminals and platforms such as PC, Web, Android, and H5 is realized. This project is oriented to the grid business scenario, combined with various 3D application browsers, augmented reality devices, and other features, based on WebGL technology to carry out cross-platform research of 3D engine, and research on model compression and rendering technology based on power business characteristics.

4 Conclusions

In this paper, the three-dimensional terminal rendering technology based on the characteristics of power equipment business is studied. Under the condition of limited computing resources, the real-time high-speed rendering of the 3D model is realized to the terminal and grid business application requirements, and the 3D model library construction and power service 3D for the future power grid are unified. Interaction provides the underlying support. Through the subsequent 3D lightweight engine-based grid intelligent guidance analysis and analysis of interactive system development, it can serve the grid training drill, design, manufacturing, construction, operation, and maintenance operation guidance and analysis interaction, improve grid training effect, improve on-site operation and decision-making, analyze efficiency, accuracy of field device fault diagnosis, and provide on-site work quality.

Acknowledgements This work was financially supported by the science and technology project to State Grid Corporation "Research on 3D Lightweight Engine Technology for Power Grid Service Scenarios".

References

1. Verbree, E., Jansen, F.W.: A multi-view VR interface for 3D GIS. Comput. Gr. **23**, 497–506 (1999)
2. Dougher, D.: Genesis 3D Engine Reference, Tools, and API Function Manuals. China Electric Power Press (2004)
3. Tulip, J., Bekkema, J., Nesbitt, K.: Multi-threaded game engine design. In: Proceedings of the 3rd Australasian Conference on Interactive Entertainment, p. 9. Murdoch University (2006)
4. Jarosz, W., Jensen, H.W., Donner, C.: Advanced global illumination using photon mapping. In: ACM SIGGRAPH 2008 Classes, ACM, p. 2 (2008)
5. Novfik, J., Dachsbacher, C.: Rasterized bounding volume hierarchies. Comput. Gr. Forum **31** (2pt2), 403–412 (2012)
6. Hapala, M., Hawan, V.: Review: Kd-tree traversal algorithms for ray tracing. Comput. Gr. Forum **30**(1), 199–213 (2011)

Community Detection Based on Improved Bee Evolutionary Genetic Algorithm

Shijin Zhang, Sheng Zhang, Jibiao Tian, Zhiqiang Wu and Weikai Dai

Abstract To solve the problem of unsatisfactory optimization ability and requiring prior knowledge in the community detection of the complex networks, an improved bee evolutionary genetic algorithm (IBEGA) was proposed by combining the bee evolutionary genetic algorithm. Firstly, the algorithm took modularity as the fitness function, combined the improved character encoding method with the corresponding genetic operators, and automatically acquired the optimal community number and the community detection solution without the prior knowledge. Then, by utilizing the local information of the network topology structure in initialization population, crossover operation and mutation operation, the search space was compressed, the optimization ability and convergence speed was improved and introducing the number of random population inversely proportional to the number of iterations to improve the exploration ability, robustness, and accuracy of the algorithm. Finally, the proposed IBEGA was simulated on real networks and synthetic networks. The results show that compared with other classical algorithms for community discovery and similar intelligent algorithms, the algorithm has the advantages of high accuracy, which shows that the algorithm is feasible and effective.

Keywords Complex network · Community detection · Genetic algorithm · Modularity

1 Introduction

Nowadays, the network is used in research fields more and more widely. Many systems can be described by network in nature, such as computer networks, social networks, transportation networks, and human disease gene networks. Community structure is a basic feature of the network, and the whole network consists of several

S. Zhang · S. Zhang (✉) · J. Tian · Z. Wu · W. Dai
School of Information Engineering, Nanchang Hangkong University, Nanchang 330063, China
e-mail: zwxzs168@126.com

V. Jain et al. (eds.), *Recent Trends in Intelligent Computing, Communication and Devices*, Advances in Intelligent Systems and Computing 1031,
https://doi.org/10.1007/978-981-13-9406-5_23

communities. Community is a collection of nodes with the same characteristics; the connections between nodes within the community are relatively tight, while the node connections between different communities are relatively sparse [1]. Community detection is a process of finding a community structure for a given network, and it can reveal the interaction between nodes in the network, which is a very meaningful study. Many researchers try to explore the network structure from different angles. The algorithm of community detection can be divided into based on graph segmentation algorithm [2, 3], clustering algorithm [4, 5], based on network dynamics characteristics algorithm [6, 7] and optimization algorithm based on the objective function [8, 9] etc. Among them, the optimization algorithm based on the objective function can divide different types of algorithms according to different objective functions or optimization methods, and the optimization algorithm with the modularity [10] function as the objective function is widely used. The modularity function is proposed by Newman and Girvan. It is a measure of the difference between a network in the algorithm community and the random network. Since the random network nonexistent community structure, when the difference (Q value) is larger, the more obvious the community structure, the better the community division.

Yang et al. [11] proposed a multi-objective ant colony algorithm for community detection and solved the local optimal problem of ant colony algorithm; Shakya et al. [9] applied fuzzy genetic algorithm to social networks to improve the overall performance of the algorithm; He et al. [12] proposed a fast simulated annealing algorithm by improving simulated annealing to improve efficiency while ensuring accuracy; Gong et al. [13] used a multi-objective immune algorithm with both modularity and standard mutual information as objective functions. Tang et al. [14] combined the bat algorithm with the genetic algorithm and proposed an adaptive evolutionary bat algorithm to improve the accuracy. With the continuous development of intelligent optimization algorithm, the algorithm is used to solve the problem of community detection in the network, and can be summed up between them in common, which has the similar basic ideas. Firstly, the corresponding fitness function and encoding method are selected according to the problem. Then, the parameters of the algorithm are set. Finally, different optimization operators are adopted to approximate the optimal solution continuously. However, there is room for improvement in the accuracy and operational efficiency of the algorithm. Meng et al. [15] proposed a new intelligent algorithm bee evolutionary genetic algorithm, compared with other genetic algorithms, the algorithm has the characteristics of small calculation and high precision. In the framework of the bee evolutionary genetic algorithm, combined with the appropriate coding method, this paper proposes an improved bee evolutionary genetic algorithm, which is applied to the community discovery field of the network. First, using the improved character-based coding method, the population is initialized by means of label propagation; then, the number of external populations in the bee evolutionary genetic algorithm is changed in a floating manner, and the crossover operator and the mutation operator are used to update the population for achieving community segmentation of the network.

2 Algorithm Structure and Algorithm Description

The bee population is composed of drones and queen bee. The algorithm only considers the case where the population of bees is constant at N. The initial population is $A(0) = \{m_0^1, m_0^2, \ldots, m_0^{N-1} | Queen\}$, among $m_0^1, m_0^2, \ldots, m_0^{N-1}$ is drones, Queen is the queen bee. Breeding the next generation of population $A(1) = \{m_1^1, m_1^2, \ldots, m_1^{N-1} | Queen\}$. After continuous breeding, when the bees are inherited to the population after the t generation, $A(\mathrm{t}) = \{m_t^1, m_t^2, \ldots, m_t^{N-1} | Queen\}$. Continuously multiplying from the first generation of populations, the bee population will generate a population sequence $\{A(0), A(1), A(2), \ldots, A(\mathrm{t}), \ldots\}$, and the purpose is to be able to get an optimal individual. This is an iterative problem, the key is how iterative algorithm from generation t to $t+1$ on behalf of the population. BEGA algorithm is the basic idea: queen bee and drone cross a certain probability, their offspring mutate individual by individual variation mutation probability obtained, through competitive way to redefine the queen bee, access to new populations. The detailed evolution process of the algorithm, firstly, select $N\gamma/2 (0 \leq \gamma \leq 1)$ male bee individuals for the population $A(t)$ of the tth generation by the selection algorithm (the roulette algorithm). In addition, $(1 - \gamma)N/2$ external drones are randomly generated to prevent local optimum. The $N/2$ drones selected above are paired with the queen bee Queen; after the crossover, N new individuals are generated, which are recorded as the offspring $B(t)$, and then, the offspring $C(t+1)$ is obtained through genetic variation. The best individual in the offspring $C(t+1)$ is labeled Queen_new, and the fitness value is the evaluation standard. The individual with the highest fitness value is the queen bee. If the value of *Queen* fitness is the largest, then *Queen* is the queen and will replace the worst individual in $C(t+1)$; otherwise, it replaces only the worst individual in $C(t+1)$, and the queen is Queen_new; after confirming the new queen, the offspring $C(t+1)$ is evolved into $A(t+1)$.

There are two advantages to the bee evolutionary genetic algorithm compared with the genetic algorithm (GA). First, the optimal individual of the population enters the offspring, which enhances the mining capacity of the algorithm and improves the convergence speed. The second is to introduce new populations (random drones) in the process of evolution, improve the exploration ability of the algorithm, prevent the algorithm from premature convergence, fall into local optimum, and improve the accuracy of the algorithm. Therefore, the BEGA algorithm has the characteristics of fast convergence, good robustness, and high precision.

3 Improved Bee Evolutionary Genetic Algorithm

Based on the evolutionary genetic algorithm of honeybee, this paper proposes the improved bee evolutionary genetic algorithm (IBEGA) based on the idea of the literature [16] and uses the algorithm to solve community detection problems.

3.1 Fitness Function

In the study of community partitioning of complex networks, modularity function is commonly used to evaluate algorithms for various community discovery classes. Since the modularity function is simple to solve and easy to understand, the fitness function of the IBEGA algorithm selects the modularity function. Its function expression is as follows:

$$Q = \frac{1}{2m}\sum_{i,j}^{n}\left[\left(a_{ij} - \frac{k_i k_j}{2m}\right)c(\delta_i, \delta_j)\right] \tag{1}$$

where v_i and v_j are used to represent different nodes in the network; n is the total number of network nodes; m is the total number of sides of the network; a_{ij} is the adjacency matrix element of the graph; k_i and k_j are the degree of nodes v_i and v_j; δ_i and δ_j are the community numbers for nodes v_i and v_j; if $\delta_i = \delta_j$, then, $c(\delta_i, \delta_j) = 1$; else, $c(\delta_i, \delta_j) = 0$. The modularity function Q value represents the difference between the network and a random network under a certain community division; the greater its value, the better the community is divided. For real-world networks, the Q value is generally in the interval (0.3, 0.7). By finding the best Q values of the modularity, and thus to find the optimal network partitioning scheme.

3.2 Encoding

In the paper, the coding method is improved based on character encoding [17], in order to better conform to the structure of the community. Detailed coding method, a network has n nodes, each individual's gene consists of the community number divided by n nodes and the number of communities divided by the network, that is, the dimension of each individual is $n + 1$. For bee population $A(\mathrm{t}) = \{m_t^1, m_t^2, \ldots, m_t^{N-1} | Queen\}$, the individual's gene encodes $m_t^i = \{r_1, r_2, \ldots, r_n | k\}$, where individual m_t^i represents the ith drone in the tth generation, r_i is the community number to which the ith node is divided, and k is the number of network divided communities.

3.3 Population Initialization

In the above coding method, the number k of communities is randomly generated, in order to take the k value more reasonably; refer to the literature [18]. Then, the value of k ranges from $[k_{\min}, k_{\max}]$, where $k_{\min} = 2$, $k_{\max} = n$, and n represent the number of nodes in the network. In the process of individual population initialization, first, the number of communities is randomly generated; then, according to the number of communities, the community number to which each node belongs is randomly assigned; finally, the community information of the node and the number of communities are combined into one chromosome. Considering the characteristics of the community structure, the node and the neighboring nodes are in the same community as a high probability event, and the initialization of the individual is optimized according to this feature; randomly select the α_n gene (α is the model parameter, in the simulation $\alpha = 0.4$), and change the community number of the neighbor node of the selected node (gene) to the same community. The initialization of the population is to repeat the individual initialization N times, and the population with the number N is obtained.

3.4 Crossover Operator

The crossover operator is the main way for genetic algorithms to generate new individuals. The genes of two individuals cross each other to generate new genes. The offspring combines the characteristics of the parents and produces diverse and potential optimal individuals. This paper uses the one-way crossover method and combines the coding method as the crossover operator of IBEGA. The specific operation process is described as follows:

Step 1: The drones and randomly generated drones selected by the selection algorithm form a cross drone population M.
Step 2: The *Queen* in the population is taken as the source individual *src*, and the drone individual is selected from the drone population M in turn as the target individual *dst*.
Step 3: Randomly select a gene position of the source individual *src*. The value of the gene position (the community number of the node) is represented by *ks*, and then find all the gene positions in the source individual whose value is *ks*, and the set of the components is represented by the symbol K.
Step 4: The number of community divisions of the target individual *dst* is expressed as k; if $ks \leq k$, the value of the genetic position K in the target individual *dst* is changed to *ks*; if $ks > k$, the value of the genetic position K in the target individual *dst* is changed to a random integer in the range $[1, k]$.
Step 5: Adjust the genetic values of the new individuals after the crossover to be consistent with the actual number of communities and avoid community numbers with rounds.
Step 6: Take *Queen* as the target individual, and select the individuals selected from the drone M as the source individuals, and repeat Step 3 to Step 5.

In the crossover process, the number of cross drone populations is $N/2$; that is, the number of crossings is $N/2$, and N new individuals are obtained. The above crossover operation is different from the traditional crossover, it is not blindly crossing the genetic positions, but according to the information of the network nodes, and then purposefully intersecting each other, which not only enhances the diversity of the population, but also effectively improves the convergence speed of the algorithm.

3.5 Mutation Operator

For targeted mutations, based on the information of the network nodes and using heuristic mutation operations, the randomly selected gene positions are mutated to the community number of most of the neighbor nodes of the node (gene). If there is more than one community number, then one of them is randomly selected as the variance value. This mutation method effectively narrows the search space, which not only reduces the meaningless search, but also improves the convergence speed of the algorithm.

3.6 Selection Operator

The queen bee selects some of the drones for mating according to probability. For a population size *N*, the selection probability $m_t^i(i = 1, 2, \ldots, N - 1)$ of the individual p_i is defined as follows:

$$p_i = \frac{f_i}{\sum\limits_{j=1}^{N} f_j} \quad (2)$$

where f_i is the fitness of the individual *i*, and for individuals with negative *Q* values, their fitness is changed to 0. The advantage of this selection operator is that individuals with higher fitness values have a higher probability of being selected.

3.7 Optimization of Random Population Size

In the bee evolutionary genetic algorithm, the number of randomly selected external drone populations is $(1 - \gamma)N/2$, where γ is the internal population coefficient; once γ is determined, the number of external drones randomly selected during evolution will remain unchanged. However, as the optimal iteration of the algorithm is more and more close to the optimal individual, the significance of introducing a new random external drone will be reduced or even unnecessary. In order to solve this problem, the paper adopts the random population size floating

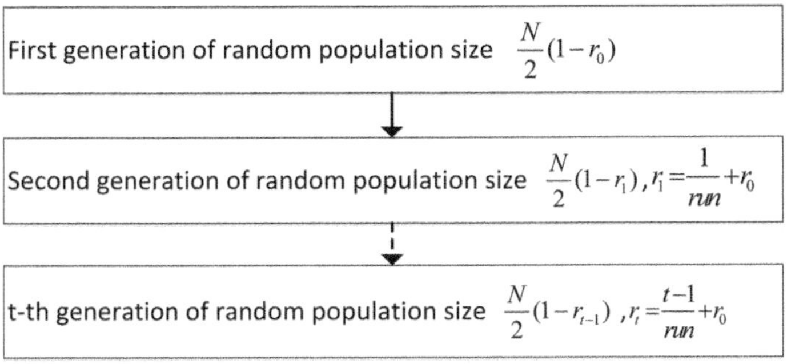

Fig. 1 Random population size stage adjustment process

adjustment strategy, and the number of randomly selected external drones is inversely proportional to the number of iterations. That is, as the number of iterations increases, the number of random external drones decreases continuously. This is the most direct, simplest control method. In the algorithm, let the number of iterations be *run* and the parameters of the random external population size parameter be initialized to r_0, and the adjustment process in the evolution process is shown in Fig. 1.

Improved algorithm description is as follows:

Step 1: Randomly generate an initial population *A*(0) of size *N*, and initialize the algebra $t = 1$ of the initial population to the parameter random population coefficient r_0, the mutation probability P_m, and the crossover probability P_c.
Step 2: Calculate the fitness of each individual in the initial population *A*(0), and select the individual with the highest fitness value as *Queen*, and the fitness value is *f_queen*.
Step 3: If the stop condition is met, the algorithm outputs the result and stops running; otherwise, continues.
Step 4: Set $t : t = t + 1$.
Step 5: Use the roulette algorithm to select $r_{t-1}N/2$ ($r_{t-1} = (t-1)/run$, *run* is the number of iterations) individuals from $A(t-1)$.
Step 6: Randomly generate $(1 - r_{t-1})N/2$ individuals.
Step 7: The $N/2$ individuals obtained from Step 5 and Step 6 are successively cross-calculated with the queen bee *Queen* to obtain the progeny population n, which is denoted as *B*(*t*).
Step 8: After the mutation operation, the population *C*(*t*) is obtained from the population *B*(*t*).
Step 9: Calculate the fitness of each individual in the population *C*(*t*), and select the individual with the highest fitness value as the queen bee candidate Queen_new, and the value of the fitness is recorded as *f_queen*_new.
Step 10: Select a new generation of queen bees according to the value of fitness; if *f_queen_new* > *f_queen*, then Queen_new is a queen *Queen*; otherwise, *Queen* does not change.

Step 11: The new population $A(t+1)$ is generated, and the queen of $A(t)$ is replaced by the individual with the smallest fitness value in the population $C(t+1)$.
Step 12: Turn to Step 3.

4 Simulation and Analysis

In order to verify the effectiveness of IBEGA for community detection, experiments were conducted on artificially synthesized networks and real social network data, respectively. There are many classical algorithms and new algorithms emerging in community detection of complex networks. This paper selects GN [19], LPA [20], and IGA [21] to compare with IBEGA. The comparison indicators between the algorithms have the modularity Q value, the accuracy (divide the correct node to the ratio of all nodes), and the normalized mutual information (NMI) [22], which proves the reasonable effectiveness of IBEGA found in the community. The initial parameters of IBEGA in the experiment are given in Table 1.

4.1 Community Detection of the LFR Benchmark Network

In this paper, the artificial synthetic network used is the LFR benchmark. Since the node degree and community size of the network are adjustable and conform to the power law distribution, the generated network is closer to the real network. The parameter settings of its network are given in Table 2. The parameter μ is the ratio of the number of node connections outside the community of any node to the degree of the node. The parameter is used to obtain a network with different community structures to test the impact of the community structure on the algorithm.

The NMI is the degree of similarity between the community partitioning result and the real community partitioning according to the calculation algorithm and selected as the evaluation. If the community partitioning result of a network has two types, A and B, C is a confusion matrix, and its element C_{ij} is the number of nodes

Table 1 IBEGA parameters and values

Parameter name	Parameter meaning	Parameter value
N	Population size	200
run	Iterative maximum	200
t	Evolutionary initial algebra	1
r_0	Random population ratio	0.2
P_c	Crossover probability	0.4
P_m	Mutation probability	0.01

Table 2 LFR benchmark parameters and values

Parameter name	Parameter meaning	Parameter value
N	Number of network nodes	100
k	Degree average of nodes	10
maxk	Maximum degree of node	30
minc	Minimum number of nodes in the community	20
maxc	Maximum number of nodes in the community	40
μ	The mixing ratio	0.0–0.5

common between the *i* community in *A* and the *j* community in *B*. The definition of NMI (*A*, *B*) is as follows:

$$\mathrm{NMI}(A,B)=\frac{-2\sum_{i=1}^{C_A}\sum_{j=1}^{C_B}C_{ij}\cdot\log\left(\frac{C_{ij}\cdot N}{C_{i}\cdot C_{j}}\right)}{\sum_{i=1}^{C_A}C_{i}\cdot\log\left(\frac{C_{i\cdot}}{N}\right)+\sum_{j=1}^{C_B}C_{ij}\cdot\log\left(\frac{C_{\cdot j}}{N}\right)} \tag{3}$$

where $C_A(C_B)$ is the number of communities divided by $A(B)$, $C_{i\cdot}$ is the sum of the *i*th row of confusion matrix *C*, $C_{\cdot j}$ is the sum of column *j*, and *N* is the number of network nodes. If $A = B$, NMI = 1, *A* and *B* have the same result; if NMI = 0, the division results of *A* and *B* are completely opposite. The value of NMI is [0, 1]. When the *B* is the result of real community partitioning and *A* is the result of the algorithm, the larger the value of NMI is, the closer the result of *A* partition is to the real result, the better the algorithm is.

The IBEGA and the GN, LPA, and IGA perform community detection on the above-mentioned synthetic network, and the division results are compared in the NMI, as shown in Fig. 2. It can be seen from the figure that the GN, IBEGA, and

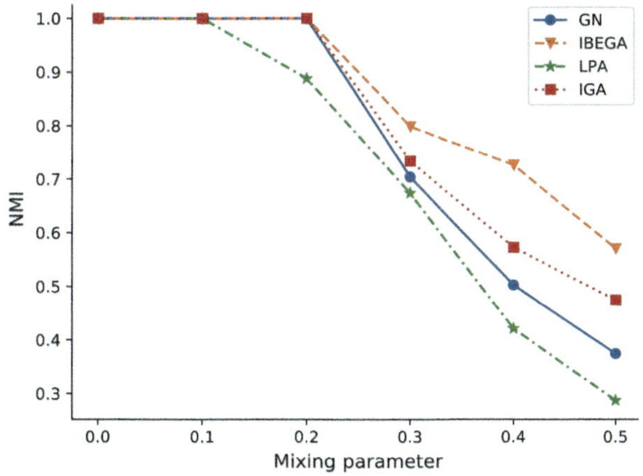

Fig. 2 NMI precision comparison of four algorithms on the LFR benchmark

IGA can be accurately divided by community when the mixing ratio μ is 0, 0.1, and 0.2; IBEGA has the highest NMI value when μ is taken at 0.3, 0.4, and 0.5. The results show that IBEGA also has a good network structure analysis for networks with fuzzy community structures.

4.2 Community Detection of the Karate Club Network

The karate club network is a social network built by Zachary to observe an American university karate club. The members of the club are represented by nodes, and the friendship between the members of the club is represented by the edges between the nodes, so the network can be represented by 34 nodes and 78 edges. During the observation process, the club's managers and coaches disagreed and eventually split into two subgroups centered on club managers and club coaches.

IBEGA divides the karate club network into four communities, and the results are shown in Fig. 3. Communities are represented by different colors and shapes with a modularity value of 0.419. The network is split into two communities, and its division results are shown in Fig. 4, the modularity value is 0.380. IBEGA divides node 8 into the community of club managers, while in reality node 8 is the community that is divided into coaches, by observing the edge relationship between node 8 and nodes 0, 2, 28, 30, 32, 33, node 0 and 2 belongs to the manager community, nodes 28, 30, 32, 33 belong to the coaching community, and the node 8 is divided into the coaching community to be more in line with the community discovery. IBEGA divides node 8 into the group where the coach is located, which is consistent with the sparse connection between the communities. The characteristics of the nodes in the community are closely connected. Nodes 0 and 32 are the club coaches and club managers, while nodes 0 and 32 are also important nodes in

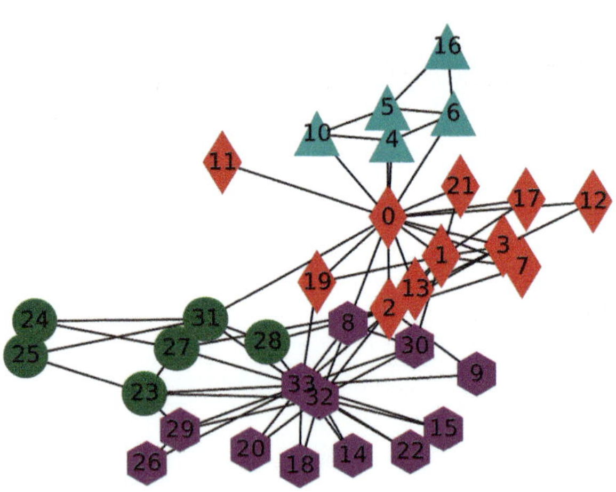

Fig. 3 IBEGA divides the results of karate

Fig. 4 karate split results

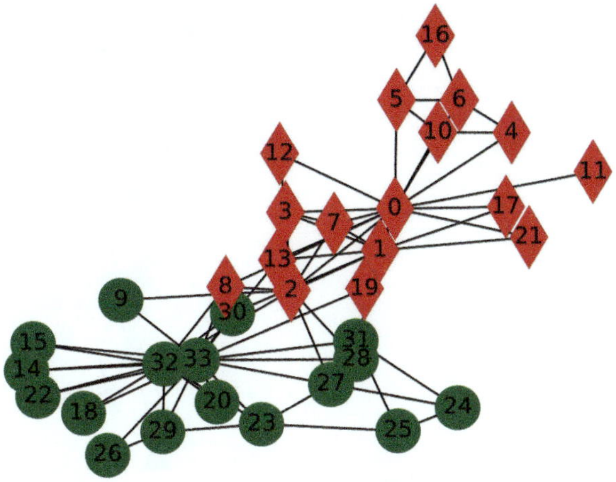

Table 3 Four algorithms for karate network experimental results

Algorithm	Q value	Accuracy	NMI
GN	0.401	0.941	0.514
LPA	0.354	0.912	0.677
IGA	0.402	0.941	1.00
IBEGA	0.420	0.971	0.619

the network, which also means that the club splits into two subgroups. Compared with the results of real community division, IBEGA has increased the value of the modularity, and further divided the real community, and divided the two subgroups into two communities.

IBEGA has a better partitioning effect on karate network. The classic community discovery algorithms GN and LPA and the same type of genetic algorithm IGA are selected. The results of these three algorithms are compared with IBEGA. The results are given in Table 3. The Q value, correct rate division, and NMI of the four algorithms are averaged over 20 runs. It can be concluded from the table that IBEGA has the highest module q value and correct division rate, and the segmentation result is improved, compared with similar algorithms and classical algorithms.

4.3 Community Detection of the dolphin Network

The dolphin dataset is a network of dolphins that Lusseau et al. used to observe the exchange of 62 dolphins in the Doubtful Sound Channel in New Zealand for seven years. The dolphin can be represented by network nodes, and frequent contact between dolphins is represented by edges. Therefore, the dolphin network can be

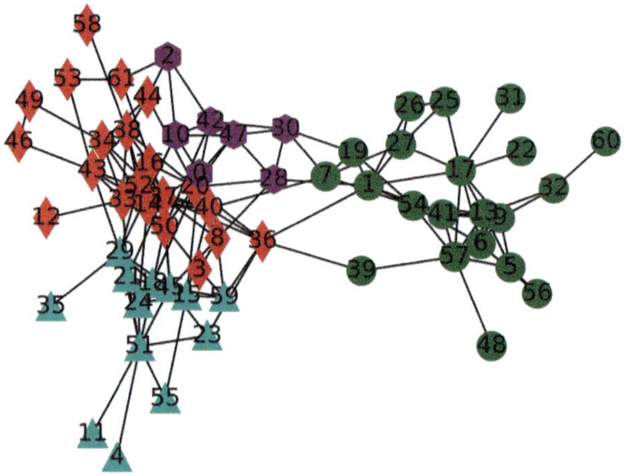

Fig. 5 IBEGA divides the results of dolphin

represented by 62 nodes and 159 edges. IBEGA divides the dolphin network into four communities. The communities are represented by different node shapes and colors. The results of community segmentation are shown in Fig. 5, and the modularity value is 0.527. In the course of research, Lusseau et al. divided the dolphins into two communities according to gender. As shown in Fig. 6, the green circular node cluster on the left is male dolphins, and the red diamond cluster on the right is female dolphins. By comparison, IBEGA divides node 39 into errors. It can be seen from the figure that the neighbor nodes of the node 39 are only 36 and 41, which belong to different genders, and the nodes 39 are difficult to divide because of less information. The overall division is better. In addition to the gender-differentiated community, IBEGA has divided the female dolphins into three small communities. The connections between the small communities are relatively sparse, and the links are closely related.

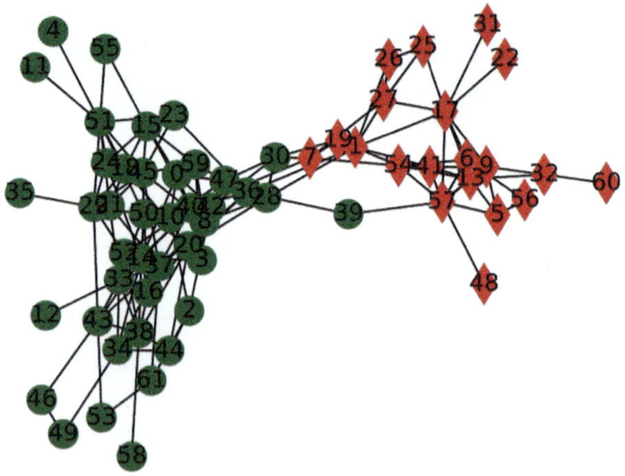

Fig. 6 dolphin split results

Table 4 Four algorithms for dolphin network experimental results

Algorithm	Q value	Accuracy	NMI
GN	0.519	0.942	0.55
LPA	0.511	0.955	0.52
IGA	0.519	0.984	0.64
IBEGA	0.527	0.984	0.56

The results of the evaluation of dolphin network partitioning by the four algorithms are given in Table 4. The values, the correct rate partition, and the NMI are averaged over 20 runs. IBEGA has the highest module value, and the correct division rate ranks first with IGA, which is second only to IGA.

5 Conclusion

For most intelligent algorithms, there is a problem of poor search ability. The paper proposes a complex network community partitioning algorithm based on improved bee evolutionary genetic algorithm, in the case where the prior knowledge is not required, using the improved character-based coding method, and combined with the population initialization, the use of floating external populations, crossover operators and mutation operators. The network can divide the community in the best number of communities and get the optimal partition plan. In the process of algorithm searching the solution space to approximate the optimal space of the problem, the algorithm adopts heuristic method in population initialization, crossover operator and mutation operator, on the one hand, the convergence speed of the algorithm is improved, on the other hand, the robustness of the algorithm is enhanced. Based on the evolutionary genetic characteristics of bees, in the drones of the queen bee, the external population is increased to effectively increase the diversity of the population. The above four aspects of work enable IBEGA to effectively discover the community structure of complex networks. The experimental analysis of IBEGA shows that the algorithm can divide the optimal community scheme and improve the quality of community division while automatically obtaining the optimal number of communities.

Acknowledgments This work is partially supported by the National Natural Science Foundation of China (61661037, 61162002), the Foundation of Jiangxi Provincial Department of Education (GJJ170575), and the Jiangxi province graduate innovation special foundation (YC2018-S370).

References

1. Fortunato, S.: Community detection in graphs. Phys. Rep. **486**(3), 75–174 (2009)
2. Kernighan, B.W., Lin, S.: An efficient heuristic procedure for partitioning graphs. Bell Syst. Tech. J. **49**(2), 291–307 (2014)
3. Suaris, P.R., Kedem, G.: An algorithm for quadrisection and its application to standard cell placement. IEEE Trans. circ. syst. **35**(3), 294–303 (1988)
4. Newman, M.E., Girvan, M.: Finding and evaluating community structure in networks. Phys. Rev. E Stat. Nonlin. Soft Matter Phys. **69**(2), 026113 (2004)
5. Krishnapuram, R., Keller, J.M.: A possibilistic approach to clustering. IEEE Trans. Fuzzy Syst. **1**(2), 98–110 (2002)
6. Van Dongen, S.: Graph clustering by flow simulation. Phd Thesis, University of Utrecht (2000)
7. Rosvall, M., Bergstrom, C.T.: Maps of random walks on complex networks reveal community structure. Proc. Natl. Acad. Sci. U.S.A. **105**(4), 1118–1123 (2008)
8. Cai, Q., Gong, M., Ma, L., et al.: Greedy discrete particle swarm optimization for large-scale social network clustering. Inf. Sci. Int. J, **316**(C), 503–516 (2015)
9. Shakya, H.K., Singh, K., Biswas, B.: An efficient genetic algorithm for fuzzy community detection in social network (2017)
10. Tasgin, M., Herdagdelen, A., Bingol, H.: Community detection in complex networks using genetic algorithms. Corr **2005**(3120), 1067–1068 (2006)
11. Yang, N., Hong-Juan, L.V., Chen, T.: Multi target of community detection algorithm based on ant colony optimization. Comput. Technol. Autom. (2015)
12. He, J., Chen, D., Sun, C.A.: Fast simulated annealing strategy for community detection in complex networks. In: IEEE International Conference on Computer and Communications. IEEE, pp. 2380–2384 (2017)
13. Gong, M.G., Zhang, L.J., Ma, J.J., et al.: Community detection in dynamic social networks based on multiobjective immune algorithm. J. Comput. Sci. Technol. **27**(3), 455–467 (2012)
14. Tang, C.W., Li, Y., Duan, Q.Y.: Research on community detection in complex networks based on self-adaptive evolution bat algorithm. J. Cent. S. Univ (Sci. Technol.) **1**, 109–117 (2018)
15. Meng, W., Han, X.D., Hong, B.R.: Bee evolutionary genetic algorithm. Acta Electronica Sinica **34**(7), 1294–1300 (2006)
16. Huang, M., Ji, B., Liang, X.: An improved bee evolutionary genetic algorithm. In: IEEE International Conference on Intelligent Computing and Intelligent Systems. IEEE, pp. 372–374 (2010)
17. Cao, Y.C., Tian, S.L., Shao, Y.B.: Community detection in complex networks based on immune genetic algorithm. J. Comput. Appl. **33**(11), 3129–3133 (2013)
18. Zhou, S.B., Xu, Z.Y., Tang, X.Q.: New method for determining optimal number of clusters in K-means clustering algorithm. Comput. Eng. Appl. **46**(16), 27–31 (2010)
19. Girvan, M., Newman, M.E.J.: Community structure in social and biological networks. Proc. Natl. Acad. Sci. U.S.A **99**(12), 7821–7826 (2002)
20. Zhu, X., Ghahramani, Z.: Learning from labeled and unlabeled data. Tech. Rep. **3175**(2004), 237–244 (2002)
21. Deng, K., Zhang, J.P., Yang, J.: Community detection in complex networks using an improved genetic algorithm. J. Harbin Eng. Univ. **11**, 1438–1444 (2013)
22. Danon, L., Díazguilera, A., Duch, J., et al.: Comparing community structure identification. J. Stat. Mech. **2005**(09), 09008 (2005)

Named Entity Recognition for Chinese Management Case Texts

Suhui Liu, Xiaodong Zhang and Xinhao Zhan

Abstract Extracting named entity especially organization name from Chinese management case texts is a challenging task due to the lack of labeled data and difficulty in identifying diversified forms of entity names. In this paper, a semi-supervised learning method based on bidirectional long short-term memory and conditional random field (BI-LSTM-CRF) model was proposed. This method has a bootstrapped framework which automatically learns complex text features from a small number of seed sets with the BI-LSTM-CRF model and then updates the seed sets after evaluating the recognition results according to the rule base. It stops iterating when the precision of the model comes to convergence. The experimental results show that the accuracy of the proposed method reaches 89.2%, which outperforms other semi-supervised learning models.

Keywords Chinese management case texts · Named entity recognition · Semi-supervised learning method · BI-LSTM-CRF model

1 Introduction

Chinese management case study has developed rapidly since 2005, and the Internet especially has produced massive management case texts recently [1]. Analyzing the numerous texts manually seems to be impossible. Therefore, automatic case text analysis with text mining technology is necessary for knowledge discovery. Named entity recognition (NER) is the first step of text mining and also the basis of further study on the relationship among entities. This paper focuses on the NER task in Chinese management case texts. Compared with other management texts such as news and monographs, management case texts are more colloquial. A lot of abbreviations and alternative name of organization are used. Furthermore, the

S. Liu · X. Zhang (✉) · X. Zhan
Donlinks School of Economics and Management, University of Science and Technology Beijing, Beijing 100083, China
e-mail: xdzhang@management.ustb.edu.cn

V. Jain et al. (eds.), *Recent Trends in Intelligent Computing, Communication and Devices*, Advances in Intelligent Systems and Computing 1031,
https://doi.org/10.1007/978-981-13-9406-5_24

names of emerging entities such as *Tick Tok* and *Headline* are not as regular as the traditional ones with common suffixes, which make the task more difficult.

Wang et al. [2] once constructed a rule-based method to identify company names in Chinese financial news texts. It did not work well in open test data because making rules to cover all forms of company abbreviations was difficult. Machine learning method performs better than rule-based method in NER task, and it mainly includes the maximum entropy model [3], the support vector machine (SVM) model [4], the hidden Markov model (HMM) [5], and conditional random field (CRF) model [6]. But the challenges are to construct appropriate text feature and prepare annotated training data. Since the distributed representation of words (word embedding) and the language model based on recurrent neural networks (RNNs) [7] are proposed, the deep learning method shows the advantages in sequence labeling. Li et al. [8] used the RNN model to extract biomedical entities, and the results showed that it performed better than the CRF model. Rekia et al. [9] solved the long-distance dependence problem in CCG sequence labeling problem with bidirectional long short-term memory (Bi-LSTM) model. Mourad [10] added a CRF layer to the Bi-LSTM output layer, and the accuracy was greatly improved. However, this method also requires sufficient annotated data like machine learning method. The semi-supervised method based on pattern learning is the main way to solve such problems, which usually has an iterative framework, starts from a seed set and pattern base, and matches similar sentence patterns or entities in unlabeled text. Different semi-supervised methods have different ways to build pattern base. Etzioni [11] et al. constructed pattern rules to match entities according to the syntax and semantic lexical features, and Dang and Aizawa [12] has used the dependency syntax structure. Stevenson and Greenwood [13] has used external resources WordNet to evaluate the newly created patterns. But the shortages of this method are the low accuracy and the dependency on manual construction of pattern base.

This paper proposes a semi-supervised learning method based on Bi-LSTM-CRF model to solve the dependency on the annotated corpus and the problem of high accuracy and recall simultaneously. The method has an iterative bootstrapped framework to continuously learn new entities, and the Bi-LSTM-CRF model can automatically extract the text features around entity. The result showed that this method outperformed the deep learning model with general corpus and the StanfordCoreNLP.

2 Semi-supervised Learning Method

2.1 Bootstrapped Framework

The framework of the proposed method is shown in Fig. 1, which can be mainly divided into three parts. The key techniques in each part will be described in the next sections.

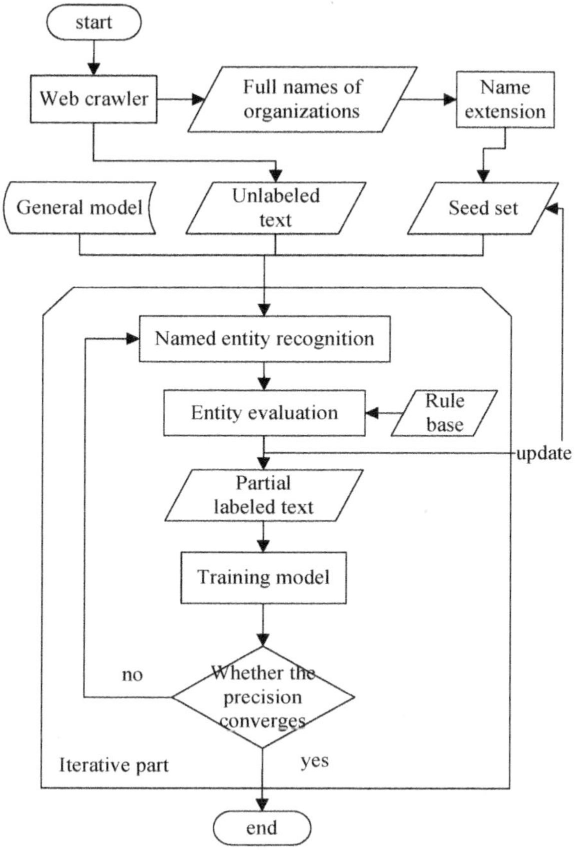

Fig. 1 Framework of the method

(1) Generating the initial seed set

Firstly, we used a Web crawler to fetch Chinese management case texts as unlabeled corpus from China Management Case Sharing Center (http://www.cmcc-dut.cn). Some of the Web pages contain the full name of the background organization which can be collected into the initial seed set. Secondly, we expanded the names with expanding rules to get the initial seed set.

(2) Initializing the iterative modules

The unlabeled corpus in part (1) was automatically matched and annotated with the initial seed set to generate partly labeled corpus. Then, the initial training data was obtained by merging the above corpus with MSRA-labeled corpus. After training the BI-LSTM-CRF model, we extracted named entities from the unlabeled corpus in part (1) to generate candidate entities.

(3) Iterating and updating

The candidate entities were evaluated according to the rule base, and the correct entities were appended to the seed set. Then, the part (2) was repeatedly processed with new seed set until the model's accuracy and recall came to convergence.

2.2 BI-LSTM-CRF Model

BI-LSTM-CRF model is more suitable for annotating long sentences with complex structure because it can make better use of sentence-level semantic features. BIO tag scheme was used in this paper. The structure of the BI-LSTM-CRF model is shown in Fig. 2, which can be divided into four parts.

The first part is character embedding layer; every Chinese character was pre-trained to get the vector representation x_t with the word2vec tool.

The second part is bidirectional long short-term memory (Bi-LSTM) layer, which automatically extracts abstract and semantic features of sentences. The output of the current position is closely related to the input x_t and the forward and backward hiding states $\overrightarrow{h_{t-1}}$, $\overleftarrow{h_{t-1}}$; that is, the output takes full account of the context information. The LSTM model has the input gate i_t, forgetting gate f_t, and output gate o_t to optionally let information through and designs the memory cell state c_t to avoid the problems of gradient disappearance and long-term dependency in standard RNNs. So that the important long-distance information in the sequence can be transmitted effectively along the neural network. The calculation formulas are as follows:

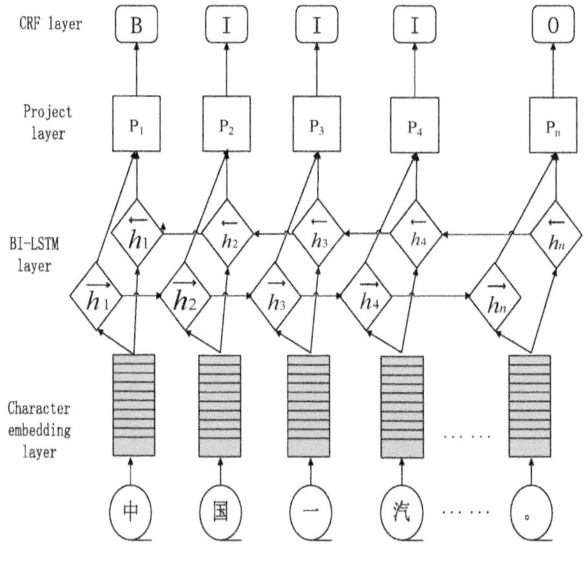

Fig. 2 Structure of Bi-LSTM-CRF model

$$i_t = \text{sigmoid}(W_{xi}x_t + W_{hi}h_{t-1} + b_i) \tag{1}$$

$$f_t = \text{sigmoid}(W_{xf}x_t + W_{hf}h_{t-1} + b_f) \tag{2}$$

$$o_t = \text{sigmoid}(W_{xo}x_t + W_{ho}h_{t-1} + b_o) \tag{3}$$

$$c_t = f_t \cdot c_{t-1} + i_t \cdot \tanh(W_{xc}x_t + W_{hc}h_{t-1} + b_c) \tag{4}$$

$$h_t = o_t \cdot \tanh(c_t) \tag{5}$$

where W_{xi}, W_{hi}, W_{xf}, W_{hf}, W_{xo}, W_{ho} denote the weight matrices and b_i, b_f, b_o, b_c denote the bias vectors.

The third part is the project layer, and it maps the output $[\overrightarrow{h_t}, \overleftarrow{h_t}]$ of Bi-LSTM layer into the m-dimensional vector called P_t. Each dimension represents a label, and m is the number of labels. If we normalize P_t with a softmax layer, the value of each dimension in P_t will represent the probability of classifying x_t into that label. However, in that way, we ignore the dependence between outputs of Bi-LSTM layer, so we replace the softmax layer with a linear-chain conditional random field (CRF) layer.

The fourth part is the linear-chain conditional random field (CRF) layer, which considers the transfer probability between labels. The output of this layer is the label sequence with the highest probability.

2.3 *Evaluating Rule*

In order to avoid semantic drift of entity categories after iterations, the results must be selected to ensure that the named entities entering the seed set are completely correct. By observing the results of named entity recognition, it is found that the misidentified entities can be mainly classified into three types: (1) misidentified entity boundary words, (2) incorrect component of entity's full name, and (3) generalized names.

The first two can be distinguished by establishing rules after the statistical analysis of word length and part of speech of the correct entity in the seed set as well as the analysis of the composition features of the misidentified entities. However, generalized names are not easy to be distinguished since they often contain the same suffixes as correct organization names. In order to ensure the recall and iteration efficiency of the model, we solve this problem by manual checking.

3 Experiment

3.1 Dataset

Eight hundred case texts were crawled from China Management Case Sharing Center (www.cmcc-dut.cn), and 444 organization full names were collected. About 13,000 sentences (1.28 million words) were extracted as unlabeled text. A total of 592 organization names were involved in the initial seed set after expansion. Another 1200 sentences (1.8 million words) were selected as test sets after manual annotation.

3.2 Results and Analysis

Six iterations of named entity recognition experiments were carried out. The results of each iteration are given in Table 1.

In the first five iterations, the precision, recall, and F score increased obviously. However, in the sixth iteration, the recall was still improved, but the accuracy and F score decreased. This may be caused by the model over-fitting after too many iterations, so we should pay attention to the number of iterations.

The size of the seed set had been extended from 592 words to 2234 words after six iterations. Abbreviations and alternative names accounted for about 30–40% of the total, and new media names and other emerging institutional entity names about 10%. The result proves that there exist many different forms of organization names in management case texts, and the research of this paper is indeed necessary.

The results in Table 2 show that the recall of the model trained with open corpus is relatively low because the model cannot effectively identify the abbreviations and alternative names. Adding a small number of annotated management case texts to the training data leads to a significant increase in recall and precision. Therefore, for supervised learning methods, domain consistency between training set and test set is very important to the results. Our model proposed in this paper starting from a small number of seed sets achieved the precision and recall of 89.15 and 83.26%,

Table 1 Results of six iterations

	Precision (%)	Recall (%)	F score (%)
First iteration	82.56	66.64	73.75
Second iteration	87.71	68.94	77.20
Third iteration	88.72	75.70	81.69
Fourth iteration	89.11	78.87	83.68
Fifth iteration	89.15	83.26	86.10
Sixth iteration	87.17	84.76	85.95

Table 2 Comparison with other models

Model	Precision (%)	Recall (%)	F score (%)
StanfordCoreNLP tool	74.82	34.66	47.38
Bi-LSTM-CRF+ open annotated data	71.48	46.48	56.33
Bi-LSTM-CRF+ domain annotated data	81.34	68.44	74.33
Bi-LSTM-CRF+ bootstrapped framework	89.15	83.26	86.10

respectively, after six iterations, which are close to the level of named entity recognition in general domain. It is proved that our method can recognize named entities without annotated data effectively.

4 Conclusion and Discussion

In this paper, we focused on the difficulty in extracting organization names in management case texts and proposed a semi-supervised method based on Bi-LSTM-CRF model to solve it. This method combines the advantages of depth learning method in sequence labeling, such as no manual feature definition and high accuracy, with the advantage of semi-supervised learning method in less dependence on annotated corpus. After the iterative training, we can obtain the annotated texts and the entity dictionary, which has an important impetus for the establishment of the management domain corpus. The training time of each iteration is a bit long, so how to improve the learning efficiency of the model can be further studied.

Acknowledgements The authors gratefully acknowledge the financial support provided by the National Natural Science Foundation of China under Grant No. 71871018.

References

1. Wenchen, Guo, Rong, Dai, Shaosheng, Sun: Research on management case study in China: actuality and prospect. Chin. J. Manag. **13**(5), 664–670 (2016)
2. Ning, Wang, Ruifang, Ge, Chunfa, Yuan, et al.: Company name identification in Chinese financial domain. J. Chin. Inf. Process. **16**(2), 1–6 (2002)
3. Saha, S.K., Sarkar, S., Mitra, P.: Feature selection techniques for maximum entropy based biomedical named entity recognition. J. Biomed. Inform. **42**(5), 905–911 (2009)
4. Lee, K., Hwang, Y., Kim, S., et al.: Biomedical named entity recognition using two-phase model based on SVMs. J. Biomed. Inform. **37**(6), 436–447 (2004)
5. Hongkui, Yu., Huaping, Zhang, Qun, Liu, et al.: Chinese named entity identification using cascaded hidden Markov model. J. Commun.cations **27**(2), 87–94 (2006)
6. Junsheng, Zhou, Xinyu, Dai, Cunyan, Yin, et al.: Automatic recognition of chinese organization name based on cascaded conditional random fields. Acta Electronica Sinica **34**(5), 804–809 (2006)

7. Bengio, Y., Ducharme, R., Vincent, P., et al.: A neural probabilistic language model. J. Mach. Learn. Res. **3**(6), 1137–1155 (2003)
8. Li, L., Jin, L., Jiang, Z., et al.: Biomedical named entity recognition based on extended recurrent neural networks. In: IEEE International Conference on Bioinfonnatics and Biomedicine, pp. 649–652 (2015)
9. Kadari, R., Zhang, Y.U., Zhang, W., et al.: CCG supertagging with bidirectional long short-term memory networks. Nat. Lang. Eng. **24**(1), 77–90 (2018)
10. Gridach, M.: Character-level neural network for biomedical named entity recognition. J. Biomed. Inform. **70**, 85–91 (2017)
11. Etzioni, O., Cafarella, M., Downey, D., et al.: Unsupervised named-entity extraction from the Web: an experimental study. Artif. Intell. **165**(1), 91–134 (2005)
12. Dang, V.B., Aizawa, A.: Multi-class named entity recognition via bootstrapping with dependency tree-based patterns. In: 12th Pacific-Asia Conference on Knowledge Discovery and Data Mining, pp. 76–87 (2008)
13. Stevenson, M., Greenwood, M.A.: A semantic approach to IE pattern induction. In: Proceedings of the 43rd Annual Meeting on Association for Computational Linguistics (ACL'05), pp. 379–386 (2005)

Application of Virtual Simulation Platform in Basic Medical Teaching

Dong Lin, Qin Zhao, Haiyun Luan and Yudong Hou

Abstract In the context of educational informationization, virtual simulation technology has been applied in medical experimental teaching due to its openness, sharing, interactivity, and low cost. This paper researched the function of the virtual simulation experimental platform and focused on the construction concept, content, function, and teaching characteristics of the platform. The basic medical virtual simulation experimental platform provides a new way for the cultivation of medical students' clinical skills and also opens up new ideas for improving the informationization of medical education.

Keywords Informationization · Virtual simulation · Basic medicine · Education

1 Introduction

Medicine is a highly practical subject. Experimental teaching plays an irreplaceable role in cultivating the practical ability and medical thinking of medical students [1]. At present, during the operation of experimental teaching, there are some problems and deficiencies due to some conditions [2]: (1) The experimental content is single and the comprehensive experiments are insufficient. More importantly, it is not compatible with the rapid development of new technologies and new theories in the medical field. (2) The experimental teaching often adopts the traditional teaching method, and the teaching methods are single. It is difficult to stimulate the enthusiasm of the students in the experimental class. (3) The experimental teaching conditions are limited. For example, high pathogenic bacteria and highly toxic or radioactive substances are difficult to achieve; some biochemical experiments

D. Lin (✉) · H. Luan · Y. Hou
Experimental Teaching Management Center, Binzhou Medical University, Yantai 264003, China
e-mail: lindong3503@163.com

Q. Zhao
College of Food Engineering, Ludong University, Yantai 264025, China

V. Jain et al. (eds.), *Recent Trends in Intelligent Computing, Communication and Devices*, Advances in Intelligent Systems and Computing 1031,
https://doi.org/10.1007/978-981-13-9406-5_25

require expensive instruments and reagents, which limits the implementation of the experiment.

Based on the above reasons, the practical ability and medical thinking of medical students are lacking, which have become a serious constraint bottleneck for medical personnel training [3]. In recent years, virtual simulation technology has been applied to the experimental teaching in medical colleges due to its advantages of interactivity, security, openness, and sharing. It provides new ideas and means for the cultivation of medical students [4, 5].

2 Significance of Basic Medical Virtual Simulation Experimental Platform

Virtual simulation teaching is an important part of educational informationization. Virtual simulation technology refers to the construction of various virtual experimental environments on a computer to simulate real experimental scenarios. Experimenters can use virtual instruments and equipment to perform virtual operations on experimental animals, specimens, or human organs as in the real experiment. And then, complete a variety of scheduled experimental projects and gain an immersive learning experience [6–8]. Virtual simulation technology has changed the traditional experimental teaching mode, making the learning forms of students more flexible [9, 10].

2.1 Reduce the Cost of Experiment

Virtual simulation experiments can be used for the experimental projects, which need expensive consumables and equipment, such as complex experiments, dangerous experiments, consumption of large numbers of animals, and cadaveric experiments. This method can not only improve the quality of experiment, but also effectively reduce the cost.

2.2 Provide an Open Experimental Environment

Virtual simulation experimental platform is not limited by time and space. Students can enter the virtual laboratory to perform virtual experiment at any time and any place. Students can develop their own experimental topics and organize their own experimental content, so that medical students can change from passive learning to active learning under the guidance of experimental teachers.

2.3 *Enhance Classroom Fun*

Through virtual simulation technology, a highly simulated experimental environment is generated on the computer, which enables students to experience the feeling of being in the realistic and vivid operation. This can greatly stimulate students' experimental interest and improve their curiosity and fun.

3 Construction of Basic Medical Virtual Simulation Platform

According to the requirements of professional training for medical students, our research group have built a virtual simulation platform for basic medicine. The construction of the platform follows two major concepts: (1) Based on the opening, online and sharing characteristics of virtual simulation experiments, the experimental projects were enriched and the operational opportunities of students increased. Focus on solving the following problems: not being able to carry out extensively, not being able to repeatedly train, and poorly performing actual training. (2) Breaking the limitations of experimental content according to subjects and curriculums, we designed systematic virtual experiment which is based on organs, tissues, and diseases system. The experimental content is consistent with the modern medical teaching rules and development trends. The basic medical virtual simulation platform can provide a great support for cultivating the high-level medical students. The basic medical virtual simulation experimental platform consists of six sub-platforms, covering the main experimental courses of basic medicine (Fig. 1).

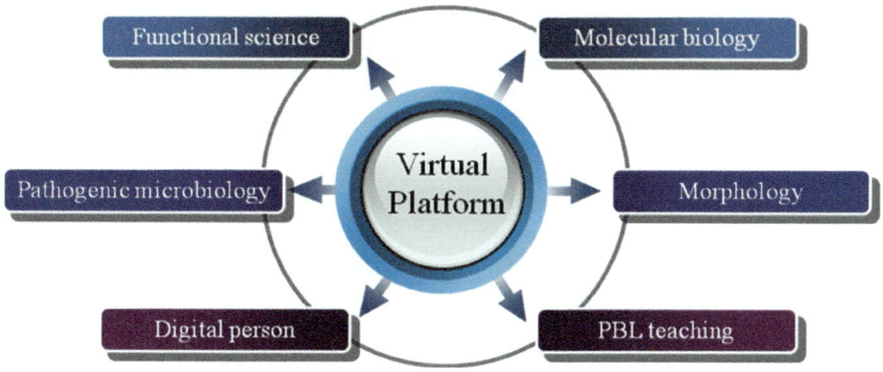

Fig. 1 Basic medical virtual simulation experimental platform

3.1 Functionality Virtual Simulation Experiment Platform

On this platform, students can complete the experimental project through using human–computer interaction virtual simulation operations; at the same time, the animation and video exhibit the operation step of the instruments and experimental animal. The experimental project mainly includes large-scale comprehensive experiments, and each experimental project usually contains multiple sub-experiment modules. Students can grasp relevant experimental skills through virtual operations. This method can effectively train the ability of the medical students to find problems, analyze problems, and solve problems. In real experiments, it is impossible for each student to complete multiple experiments due to the limitations of teaching resources. Through virtual experiments, students are given the more opportunity to increase their practice, so consolidate and expand the knowledge in classroom.

3.2 Molecular Biology Virtual Simulation Experimental Platform

On this platform, the equipment, tools, and biochemical reagents used in the experiment are put into the virtual scene by establishing a simulation laboratory scene. The equipment parameters and experimental data can also be simulated inputs. Students can perform various experimental operations in a virtual scene. The molecular biology virtual simulation experiment can effectively remove the tedious experimental preparation work and reduce the use of expensive experimental materials. At the same time, the virtual experimental resources can be repeatedly used to improve the students' operational skills.

3.3 Pathogenic Microbiology Virtual Simulation Experimental Platform

The platform mainly includes high-risk experimental projects that are difficult to achieve under realistic conditions and some teaching specimens that are difficult to collect. By constructing a virtual realistic experimental environment and experimental objects on a computer, the microscopic world under the microscope is vividly displayed, which can enhance students' experimental interest and stimulate learning enthusiasm.

3.4 Digital Morphology Experiment Platform

The platform mainly integrated digital sectioning modules such as histological sectioning system, pathological sectioning system, gross pathology system, parasitic system, microbial system, and digital embryo system. Students can observe various sections through the network or microscopic digital interactive teaching system, and interact with teachers in real time. Through the interactive teaching and learning of digital sections, students can deepen their understanding of relevant knowledge.

3.5 Digital Human Experimental Platform

The platform is based on the digital human anatomy system developed by the Chinese Digital Human Body Database. It can display the cross section, coronal plane, and sagittal section of the human body and can use a virtual cutter for layer selection. While the students are performing autopsy, the same group of personnel can observe the structural images of the corresponding parts or organs through the system. By using the digital human anatomy system, the deficiencies of the corpses and teaching specimens can be compensated to some extent. At the same time, the experimental class becomes vivid and the teaching effect is improved.

3.6 PBL Teaching Platform

The platform effectively combines PBL teaching with virtual simulation technology. According to the teaching philosophy of PBL, the standardized patient mode is used to simulate the whole process of patient visit, physical examination, laboratory-assisted detection, diagnosis, case analysis, and discussion. This method can effectively cultivate students' clinical thinking and team awareness. This method can resolve some difficulties in the actual teaching process, such as limited number of PBL teaching instructors and insufficient teaching space.

4 Characteristics of Virtual Simulation Teaching

The virtual simulation experimental teaching mode is based on the Internet, which is very different from the traditional face-to-face classroom teaching. It breaks the limitations of traditional experimental time and space. Students can conduct virtual experimental operations at any time and anywhere. It solves the limitations of experimental sites and experimental time during traditional experimental teaching.

In addition, it can flexibly provide different experimental content for different students. Teachers can arrange the corresponding experimental content and learning process according to the students' proficiency and mastery of knowledge.

4.1 Student Experiment

Students can perform experimental operations on the virtual simulation experimental platform according to the arrangement of the teacher, or independently select experimental items according to their own conditions. After the students completed the experiment, the experimental result was submitted online, then the teacher could comment it in real time.

4.2 Teacher Management

The teacher can manage the process of experimental teaching through the management platform, such as grouping of students, setting of experimental content, and real-time checking of the virtual experiments operating. The background scoring system can intelligently evaluate the quality of the completion of the experimental operation and automatically give the experimental results.

4.3 Online Interaction

In the process of operating virtual experiments, students can interact with teachers through the Q&A platform. The teacher can observe the progress of the students' experimental operation in real time and find out the problems that occurred during the experiment and correct them.

Acknowledgments This work was financially supported by Shandong Provincial Medicine and Health Technology Development Plan (Grant No. 2017WSB290215) and Binzhou Medical University Teaching Reform and Research Project (Grant No. JYKT201601).

References

1. Schifferdecker, K.E., Reed, V.A.: Using mixed methods research in medical education: basic guidelines for researchers. Med. Educ. **43**, 637–644 (2010)
2. Denadai, R., Saad-Hossne, R., Todelo, A.P., Kirylko, L., Souto, L.R.: Low-fidelity bench models for basic surgical skills training during undergraduate medical education. Revista Do Colégio Brasileiro De Cirurgiões. **41**, 137–145 (2014)

3. Worthington, R., Becker, P., Hays, G.R.: Appraising the profile of ethics, law and professionalism in basic medical education. Investigación En Educación Médica **3**(12), 209–213 (2014)
4. Jonathan, A., Stevens, J., Peter, K.: The relationship between presence and performance in virtual simulation training. Open J Modell. Simul. **3**, 41–48 (2015)
5. Ragan, E., Sowndararajan, A., Kopper, R., Bowman, D.: The effects of higher levels of immersion on procedure memorization performance and implications for educational virtual environments. Presence **19**, 527–543 (2010)
6. Dalgarno, B., Lee, M.: What are the learning affordances of 3-D virtual environments british. J. Educ. Technol. **41**, 10–32 (2010)
7. Bailenson, J., Yee, N., Blascovich, J., Beall, A., Lundblad, N., Jin, M.: The use of immersive virtual reality in the learning sciences: digital transformations of teachers, students and social context. J. Learn. Sci. **17**, 102–141 (2008)
8. Wolbrink, T.A., Kissoon, N., Burns, J.P.: The development of an internet-based knowledge exchange platform for pediatric critical care clinicians worldwide. Pediatr. Critic. Care Med. **15**, 197–205 (2013)
9. Evans, K.H., Daines, W., Tsui, J., Strehlow, M., Maggio, P., Shieh, L.: Septris: a novel, mobile, online, simulation game that improves sepsis recognition and management. Acad. Med. **90**, 180–184 (2015)
10. Diehl, L.A., Souza, R.M., Alves, J.B., Gordan, P.A., Esteves, R.Z., Jorge, M.L., Coelho, I.C.: A serious game to teach insulin therapy to primary care physicians: design of the game and a randomized controlled trial for educational validation. JMIR Res. Protoc **2**, e5 (2013)

Impacts of Features and Tagging Schemes on Chunking

Xiaofeng Liu

Abstract Text chunking, also known as shallow parsing, is an important task in natural language processing, and very useful for other tasks. By means of discriminate machine learning methods and extensive experiments, this paper investigates the impacts of different tagging schemes and feature types on chunking efficiency and effectiveness on corpora with different chunk specifications and languages. We find out that it costs more time for training and tagging with the machine learning method with more features and more fine-grained tagging schemes on all the corpora. Nevertheless, the tagging time is less affected by them. It is also revealed from our investigation that the method with more features and more fine-grained tagging schemes has better performance, but the chunk specification of corpus may have impacts on the choice.

Keywords Shallow parsing · Chunking · Natural language processing

1 Introduction

Text chunking, also known as shallow parsing, is an important task in natural language processing, and very useful for other tasks, such as spelling check [1], machine translation [2], anaphora resolution [3], semantic role tagging [4], named entity recognition [5], opinion mining [6] and information extraction [7], and so on.

There has been much work on chunking, and a range of machine learning frameworks are used for solving this task, e.g., SVM [8], HMM [9], memory-based learning [10], transformation-based learning [11], CRF [12], MaxEnt [13], and so forth. The machine-learning-based methods, especially supervised ones, have many advantages, and they also outperform the rule-based ones, achieving state-of-the-art and expressive performance.

X. Liu (✉)
School of Software Engineering, Huazhong University of Science and Technology, Wuhan, China
e-mail: kiddenliu@sina.com

V. Jain et al. (eds.), *Recent Trends in Intelligent Computing, Communication and Devices*, Advances in Intelligent Systems and Computing 1031,
https://doi.org/10.1007/978-981-13-9406-5_26

The supervised machine learning approaches to chunking take training samples as input and learn chunking models based on the features extracted from samples. Specifically, the training samples are tagged with various schemes, and for these tagged samples machine learning frameworks mainly adopt two types of features, that is, word- and part-of-speech-related features.

Although with respect to two types of features paper [14] compared the effectiveness of various machine learning frameworks for chunking on Chinese Penn Treebank, there is no work on the impacts of tagging schemes and features on chunking for different languages and chunk specifications. In this paper, we adopt the MaxEnt machine learning framework, a discriminate model, to deal with chunking. Within this framework, three types of tagging schemes as well as two types of features are used to train different chunking models on three corpora, which have two languages, i.e., English and Chinese, and are annotated with different chunk specifications.

The rest of paper is organized as follows: In Sect. 2, we briefly describe chunking, present two types of features and three types of tagging schemes, and introduce the MaxEnt-based chunking method. In Sect. 3, we give experimental settings. Experimental results and corresponding discussions are presented in Sect. 4. We conclude the paper in Sect. 5.

2 Chunking, Features, and Tagging Schemes

Given chunk set $XP = \{x_1, x_2, \ldots, x_m\}$ and sentence $S = (s_1, s_2, \ldots, s_n) = s_{1.n}$, where s_i is a word with part-of-speech or not, a chunk in S is denoted as $(x, s_{p..q})$, where $x \in XP$, $1 \le p \le q \le n$, $s_{p..q}$, is a subsequence of S, which means the words from position p to q in S form a chunk x. Given a sentence S, a chunker maps it into a chunk sequence $C(S) = (c_1, c_2, \ldots, c_t)$, where $c_i = (x_i, s_{p..q})$, $c_{i+1} = (x_{i+1}, s_{q+1..r})$, $x_i \in XP$, $x_{i+1} \in XP$, $1 \le i \le t-1$, $1 \le p \le q < r \le n$.

For a chunk $(x, s_{p..q})$, we can tag it with different schemes. The first-scheme BIO tags the first word in the chunk with B_x, another words, if exist, with I_x. If a word is not a chunk word, it is tagged as O, a special chunk type. The second-scheme BIEO tags the first word in the chunk with B_x, the last word with E_x, and the middle word with I_x. For a word not in a chunk, it is tagged as O. The third-scheme BIEOS tags words like BIEO except that it tags a word with S if the word is the only word of a chunk.

In order to learn a chunker, two types of features can be exploited in machine learning. The first type of features is mainly from words in a chunk, which is denoted as W. The second type of feature is mainly from words as well as their parts-of-speech, which is denoted as W + P. The feature templates used in this paper for W and W + P are shown in Table 1, where w_n and p_n denote word and part-of-speech, respectively, and n denotes their position, i.e., $n = 0$ is the current position.

Table 1 Feature template for W and W + P

W	W + P
$w_n, -2 \leq n \leq 2$	$w_n, -2 \leq n \leq 2$
$w_{n-1}w_n, -1 \leq n \leq 2$	$w_{n-1}w_n, -1 \leq n \leq 2$
$w_{-1}w_1$	$p_n, -2 \leq n \leq 2$
	$p_{n-1}p_n, -1 \leq n \leq 2$
	$p_{n-2}p_{n-1}p_n, 0 \leq n \leq 2$
	$p_{-1}p_1, w_0p_n, -2 \leq n \leq 2$

Table 2 Combinations of tagging schemes and features

	W	W + P
BIO	W + BIO	W + P + BIO
BIEO	W + BIEO	W + P + BIEO
BIEOS	W + BIEOS	W + P + BIEOS

With a chunk training corpus, we can have six configurations to create a chunker based on the combinations of tagging schemes and feature types. The six configurations are illustrated in Table 2. As an example, method W + BIO means that tag scheme BIO and feature type W are used in a chunker.

In this paper, machine-learning-based method, specifically MaxEnt-based method, is used to obtain a chunker. The chunk sequence $C(S)$ of sentence S is as follows.

$$C(S) = \arg\max_{c} P(c|S) = \arg\max_{c} \prod_{i=1}^{n} P(c_i|h_i) \tag{1}$$

where $P(c_i|h_i)$ is the probability of chunk output c_i of word s_i given history or context h_i.

Let $F = \{f_i | 1 \leq i \leq K\}$ be the feature set, and λ_i be the weight of feature f_i, then

$$P(c_i|h_i) = \frac{\prod_{j=1}^{K} e^{\lambda_j f_j(c_i,t_i)}}{\sum_c \prod_{j=1}^{K} e^{\lambda_j f_j(c,t_i)}} \tag{2}$$

During the training phase, we use GIS algorithm to compute the weight parameters $\{\lambda_i\}$ of model, and in the phase of decoding, beam search algorithm is exploited to find the most probable chunk output sequence.

3 Experimental Settings

In order to comprehensively investigate the impacts of tagging schemes and features on chunking, various experiments are conducted on three corpora, which have different chunk specifications and languages. The first one is the CoNLL 2002

Table 3 Statistics for 3 corpora

	CTB	CoNLL	CCL
# of characters	829,546	943,791	269,611
# of words	508,620	211,727	162,649
# of word types	37,382	19,109	17,486
# of POS	37	78	38
# of chunks	240,116	106,978	29,787
# of chunk types	16	11	13
Avg. length of chunk	1.973	1.718	2.264

English chunking corpus, denoted as CoNLL, which is a well-known benchmark for chunking and generated from English PTB. The second one is the CCL Chinese chunking corpus, denoted as CCL, which is created via tagging a Chinese POS corpus in paper [15]. The third one is generated from Chinese PTB by ourself, denoted as CTB, in which we define a chunk as a phrase constituent that has no other phrase constituent as its child. The statistics of three corpora are shown in Table 3.

There are three metrics used for effectiveness in experiments, i.e., precision P, recall R, and $F1$ measure. Precision P is the percentage of all chunks generated by system that are correct. Recall R is the percentage of all chunks in corpus that are generated by system correctly. $F1$ measure is the harmonic mean of precision P and recall R. We adopt training and tagging time as efficiency metrics.

In experiments, all codes are written in Java and run in JRE7. The implementation of MaxEnt is based on OpenNLP1.7, which set cutoff to 3 and the number of iterations to 100. All programs run on a Legend V480 notebook with 4 GB RAM, i5 2.6 GHz CPU and 64 bit Windows 8.

4 Experimental Result and Analysis

In order to examine the impacts of tagging schemes and features, we conduct tenfold cross-validation on each corpus, obtaining average training and tagging times for efficiency, and average P, R, and $F1$ for effectiveness. They are shown in Tables 4, 5, 6, and 7, 8, 9, respectively.

Although the chunk specifications and languages of three corpora are different, it is obvious that with the same tagging scheme, the methods with features W + P

Table 4 Efficiency for CoNLL corpus (s)

	W + BIO	W + BIEO	W + BIEOS	W + P + BIO	W + P + BIEO	W + P + BIEOS
Training time	42.12	52.41	63.57	92.19	113.62	120.09
Tagging time	7.23	10.15	13.52	9.18	12.97	16.77

Table 5 Efficiency for CTB corpus (s)

	W + BIO	W + BIEO	W + BIEOS	W + P + BIO	W + P + BIEO	W + P + BIEOS
Training time	150.33	195.12	219.98	305.86	342.99	372.66
Tagging time	26.15	28.65	36.26	31.59	32.45	40.38

Table 6 Efficiency for CCL corpus (s)

	W + BIO	W + BIEO	W + BIEOS	W + P + BIO	W + P + BIEO	W + P + BIEOS
Training time	38.56	47.78	48.74	83.22	96.25	96.29
Tagging time	5.89	6.77	8.78	6.99	9.18	10.13

Table 7 Effectiveness for CoNLL corpus

	W + BIO	W + BIEO	W + BIEOS	W + P + BIO	W + P + BIEO	W + P + BIEOS
P	85.81	89.86	91.66	93.44	**95.74**	94.59
R	85.22	89.88	91.97	93.31	**95.81**	95.01
*F*1	85.51	89.87	91.87	93.38	**95.77**	94.79

Table 8 Effectiveness for CTB corpus

	W + BIO	W + BIEO	W + BIEOS	W + P + BIO	W + P + BIEO	W + P + BIEOS
P	67.49	76.18	86.47	88.84	95.87	**96.05**
R	66.99	76.08	86.11	89.53	95.73	**95.88**
*F*1	67.24	76.13	86.29	89.19	95.81	**95.97**

Table 9 Effectiveness for CCL corpus

	W + BIO	W + BIEO	W + BIEOS	W + P + BIO	W + P + BIEO	W + P + BIEOS
P	70.79	74.21	75.29	77.62	79.52	**80.21**
R	59.69	61.46	62.66	70.95	71.81	**72.54**
*F*1	64.77	67.24	68.39	74.13	75.47	**76.18**

have longer training and tagging times than ones with features W. The reason is that features W + P will generate more features than features W for the same sample. Nevertheless, the tagging times are less affected by features than training times, and they are just increased by around 20%.

With the same features template, the methods with fine-grained tagging schemes have longer training and tagging times. In a summary, on all corpora training times are more affected by tagging schemes and features than tagging times.

In Tables 7, 8, and 9, the bold figures denote the best effectiveness for each corpus. Among three corpora, method W + P + BIEOS achieves the best effectiveness on two Chinese corpora, i.e., CTB and CCL, while method W + P + BIEO obtains the best effectiveness on English corpus, i.e., CoNLL, and it is slightly better than method W + P + BIEOS. This is because, according to the summary in Table 3, two Chinese corpora have longer average chunk length than English corpora CTB, and fine-grained tagging scheme BIEOS can provide more helpful position information in generated features, which helps the chunker make decision.

5 Conclusion and Future Work

By conducting experiments on three chunk corpora, which have various chunk specifications and languages, we investigate the impacts of tagging schemes and features on chunking efficiency and effectiveness. It is shown from experimental results that more features and fine-grained tagging schemes lead to longer training and testing time, while tagging times are less affected by them. Meanwhile, more features and fine-grained tagging schemes mean better effectiveness. However, the actual impacts on effectiveness are relevant to the average chunk length in chunk specification. In the future, we will further investigate the impacts of tagging schemes and features in other machine learning frameworks.

References

1. Mizumoto, T., Nagata, R.: Analyzing the impact of spelling errors on POS-tagging and chunking in learner English. In: Proceedings of the 4th Workshop on Natural Language Processing Techniques for Educational Applications (2017)
2. Ishiwatari, S., et al.: Chunk-based decoder for neural machine translation. In: Proceedings of ACL (2017)
3. Yeh, C.-L., Chen, Y.-C.: Zero anaphora resolution in Chinese with shallow parsing. J. Chin. Lang. Comput. **1**, 41–56 (2007)
4. Hacioglu, K., et al.: Semantic role labeling by tagging syntactic chunks. In: Proceedings of the Eighth Conference on Computational Natural Language Learning (2004)
5. Zhou, G.D., Su, J.: Named entity recognition using an HMM-based chunk tagger. In: Proceedings of ACL (2002)
6. Ghosh, S., Tonelli, S., Johansson, R.: Mining fine-grained opinion expressions with shallow parsing. In: Proceedings of the International Conference Recent Advances in Natural Language Processing (2013)
7. Meyers, A., et al.: Jargon-term extraction by chunking. In: Proceedings of the COLING Workshop on Synchronic and Diachronic Approaches to Analyzing Technical Language (2014)
8. Kudo, T., Matsumoto, Y.: Chunking with support vector machines. In: Proceedings of NAACL (2001)
9. Li, H., Webster, J.J., Kit, C., et al.: Transductive HMM based Chinese text chunking. In: Proceedings of IEEE NLPKE (2003)

10. Sang, E.F.T.K.: Memory-based shallow parsing. JMLR **2**(3), 559–594 (2002)
11. Ramshaw, L., Marcus, M.: Text chunking using transformation-based learning. In: Proceedings of the Third Workshop on Very Large Corpora (1995)
12. Sha, F., Pereira, F.: Shallow parsing with conditional random fields. In: Proceedings of HLT-NAACL (2003)
13. Wu, S.-H., Shih, C.-W., Wu, C.-W., et al.: Applying maximum entropy to robust Chinese shallow parsing. In: Proceedings of ROCLING (2005)
14. Chen, W., Zhang, Y., Isahara, H.: An empirical study of Chinese chunking. In: Proceedings of the COLING/ACL (2006)
15. Xu, R., Lu, Q., Li, Y., et al.: The construction of a Chinese shallow treebank. In: Proceedings of ACL (2004)

A Generic Stiffness Measurement Method for a 3-DOF Cable-Driven Joint Module

Kaisheng Yang, Guilin Yang, Silu Chen, Zaojun Fang, Yi Wang, Lefeng Gu and Tianjiang Zheng

Abstract A cable-driven module with passive spherical joint is proposed as a fundamental building block for modular cable-driven manipulators to produce compliant and intrinsically safe motions, which are suitable for the human–robot co-existing applications. Stiffness measurement is an important issue for the cable-driven manipulators. In this work, a generic measurement method for the cable-driven modules and manipulators is proposed. The proposed method does not require to measure cable tensions, making it be easier to implement than the model-based method. The effectiveness of the proposed method is validated by a comprehensive simulation.

Keywords Cable-driven manipulator · Redundant actuation · Stiffness measurement · Force/torque sensor

1 Introduction

Cable-driven manipulators (CDMs) have attracted significant attention over the past decades. They utilize cables (tendons) to drive the end-effector, instead of the conventional rigid links. They have advantages of light weight, large workspace, intrinsically safe motion. Now CDMs are applied in fast pick-and-place operations, positioning systems, aircraft test, rehabilitation, and so on [1–5]. Previous works on

K. Yang · G. Yang (✉) · S. Chen · Z. Fang · Y. Wang · L. Gu · T. Zheng
Zhejiang Key Lab of Robotics and Intelligent Manufacturing Equipment Technology, Ningbo Institute of Material Technology and Engineering, Chinese Academy of Sciences (CAS), 315201 Ningbo, China
e-mail: glyang@nimte.ac.cn

K. Yang · Y. Wang · L. Gu
University of Chinese Academy of Sciences (CAS), 100049 Beijing, China

K. Yang
Zhejiang Marine Development Research Institute, 316021 Zhoushan, China

V. Jain et al. (eds.), *Recent Trends in Intelligent Computing, Communication and Devices*, Advances in Intelligent Systems and Computing 1031,
https://doi.org/10.1007/978-981-13-9406-5_27

CDMs have focused on the design, kinematics, workspace analysis [6–9], while stiffness measurement is another important issue for the CDMs.

The model-based measurement method is widely employed in cable-driven manipulators [10–12]. But it needs to measure the cable tensions. Thus, a generic stiffness measurement method is required, which is easy to implement and has high accuracy.

In this work, a 6-cable-driven-spherical-module (6-CSJM) is presented as a fundamental building block for modular CDMs. We study the stiffness model of the joint module and propose a generic stiffness measurement method, which only requires to measure the external load and the pose of the module. A comprehensive simulation is designed to validate the proposed method. The results show that this method is effective for the cable-driven module with high measurement accuracy. Remarkably, this method can be extended to the modular cable-driven manipulators.

2 Structure and Kinetostatics of 6-CSJM

2.1 Structure of 6-CSJM

The 6-CSJM consists of a moving-platform and a base connected by a spherical joint. When the module is at home pose, the moving-platform and the base are parallel to each other, as shown in Fig. 1. Six cables are utilized in the module from six holes A_i on the moving-platform to $B_j (i, j = 1, 2, \ldots, 6)$ on the base. Geometrically, $A_2A_3 = A_4A_5 = A_6A_1 = 0.100$ m, $B_1B_2 = B_3B_4 = B_5B_6 = 0.130$ m, $A_1A_2 =$

Fig. 1 Home pose of 6-CSJM

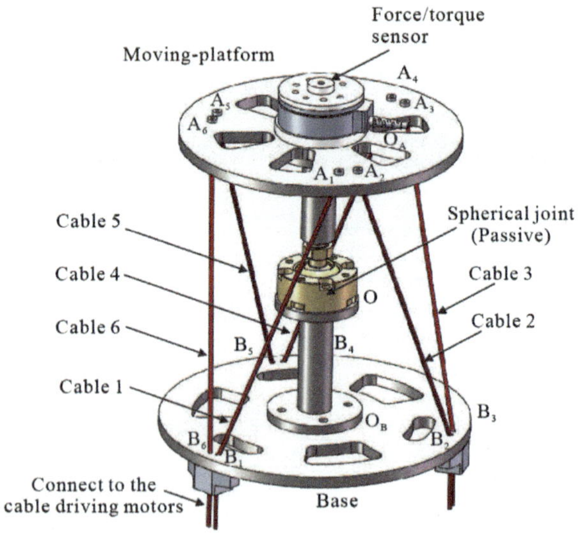

$A_3A_4 = A_5A_6 = 0.005\,\text{m}$ and $B_2B_3 = B_4B_5 = B_6B_1 = 0.005$ m. O, O_A and O_B are the centers of the passive spherical joint, moving-platform and base plate, respectively, in which $OO_A = 0.080\,\text{m}$ and $OO_B = 0.080\,\text{m}$.

2.2 Kinetostatics of 6-CSJM

In this 6-CSJM, the moving-platform realizes rotation motion about the spherical joint. We set a frame $\{B\}$ fixed on the base, named base frame, and a frame $\{A\}$ fixed on the moving-platform, named moving frame. As shown in Fig. 2, $Z_B \perp \odot B_1B_3B_5$, $X_B \perp B_1B_2$, $Y_B \perp \Omega$, and $Z_A \perp \odot A_1A_3A_5$, $X_A \perp A_4A_5$, where $\odot$ represents the plane generated by the points. Thus, these two frames coincide with each other when the module is at home pose.

The pose of the moving-platform with respect to the base can be described by a rotation matrix $R \in SO(3)$. $\mathbf{R}(t)$ can be given by an exponential form

$$\mathbf{R}(t) = e^{\hat{\zeta}(t)} = e^{\hat{\boldsymbol{\sigma}}_1\zeta_1(t) + \hat{\boldsymbol{\sigma}}_2\zeta_2(t) + \hat{\boldsymbol{\sigma}}_3\zeta_3(t)}, \tag{1}$$

where $\hat{\zeta}$ is the element of Lie algebra $so(3)$, and $(\hat{\bullet})$ is defined below

$$\zeta = \begin{pmatrix} \zeta_1 \\ \zeta_2 \\ \zeta_3 \end{pmatrix} \to \hat{\zeta} = \begin{pmatrix} 0 & -\zeta_3 & \zeta_2 \\ \zeta_3 & 0 & -\zeta_1 \\ -\zeta_2 & \zeta_1 & 0 \end{pmatrix}. \tag{2}$$

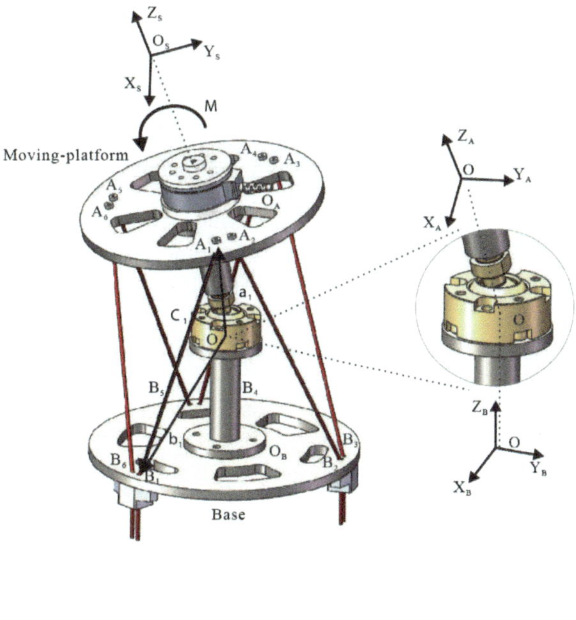

Fig. 2 Arbitrary pose of 6-CSJM

Here $\boldsymbol{\sigma}_1 = (1,0,0)^T$, $\boldsymbol{\sigma}_2 = (0,1,0)^T$ and $\boldsymbol{\sigma}_3 = (0,0,1)^T$ are the basis for vector space $so(3)$, representing the axes of the frame $\{B\}$, X_B, Y_B, and Z_B, respectively.

Denote $\mathbf{u}_i = \frac{\mathbf{b}_i - \mathbf{a}_i}{\|\mathbf{b}_i - \mathbf{a}_i\|}$ as the unit direction vector of the ith cable, where $\mathbf{a}_i$ and $\mathbf{b}_i$ are the position vectors of point A_i and B_i, respectively, then the tension vector of the ith cable $\mathbf{t}_i = t_i\mathbf{u}_i$ with $t_i = \|\mathbf{t}_i\|$. Define $\mathbf{T} = (\, t_1 \quad t_2 \quad \ldots \quad t_6 \,)^T$, the following equilibrium equation is obtained

$$\boldsymbol{\tau} = \sum_{i=1}^{6} \mathbf{a}_i \times \mathbf{t}_i = \sum_{i=1}^{6} (\mathbf{a}_i \times \mathbf{u}_i) t_i = \mathbf{J}^T\mathbf{T}, \tag{3}$$

where $\mathbf{J} = (\, \mathbf{a}_1 \times \mathbf{u}_1 \quad \mathbf{a}_2 \times \mathbf{u}_2 \quad \ldots \quad \mathbf{a}_6 \times \mathbf{u}_6 \,)^T$.

3 Stiffness Measurement Method for 6-CSJM

The stiffness $\mathbf{K} \in \mathbb{R}^{3\times 3}$ is defined by $\mathrm{d}\boldsymbol{\tau} = \mathbf{K}\mathrm{d}\boldsymbol{\zeta}$ in frame $\{A\}$. Based on the kinetostatics analysis, the stiffness $\mathbf{K}$ can be derived as following

$$\mathbf{K} = \mathbf{G} - \mathbf{J}^T\mathbf{K}_{\mathrm{diag}}\mathbf{J} = \mathbf{K}_g + \mathbf{K}_c, \tag{4}$$

where $\mathbf{K}_{\mathrm{diag}} = \mathrm{diag}\{\, k_1 \quad k_2 \quad \ldots \quad k_6 \,\}$ with the stiffness of the ith cable k_i, and $\mathbf{G} = \left(\frac{\partial \mathbf{J}^T}{\partial \varsigma_1}\mathbf{T} \quad \frac{\partial \mathbf{J}^T}{\partial \varsigma_2}\mathbf{T} \quad \frac{\partial \mathbf{J}^T}{\partial \varsigma_3}\mathbf{T} \right)$. Here $\mathbf{K}_c = -\mathbf{J}^T\mathbf{K}_{\mathrm{diag}}\mathbf{J}$ represents the stiffness caused by the elongation of the cable and $\mathbf{K}_g = \mathbf{G}$ represents the stiffness caused by the change of the geometry of the 6-CSJM.

According to the stiffness model (4), the model-based method requires to measure the cable tensions, which is not easy for implementation. In this work, a generic measurement method is proposed to measure the stiffness of 6-CSJM.

The pose $\mathbf{R}$ of the 6-CSJM can be measured by the Laser Tracker. The corresponding twist of $\mathbf{R}$ is computed by $\boldsymbol{\zeta} = \log(\mathbf{R})$. The external wrench exerting on the moving-platform $\mathbf{W}^S = \begin{pmatrix} \mathbf{f}^S \\ \boldsymbol{\tau}^S \end{pmatrix}$ can be measured directly by the force/torque sensor, which is described in the sensor frame $\{S\}$. While $\mathbf{W}^A$ is the external wrench described in the moving frame $\{A\}$, it can be obtained by the transformation $\mathbf{W}^A = \mathbf{T}_{AS}\mathbf{W}^S$. Here, $\mathbf{T}_{AS} = \begin{pmatrix} \mathbf{I}_{3\times 3} & \mathbf{0} \\ \hat{\mathbf{p}}^S & \mathbf{I}_{3\times 3} \end{pmatrix}$ is the transformation matrix of frame $\{S\}$ with respect to $\{A\}$, where $\mathbf{p}^S$ is the position vector of the origin of frame $\{S\}$ with respect to frame $\{A\}$, described in frame $\{S\}$.

N sets of data $\{\Delta\boldsymbol{\tau}^A, \Delta\boldsymbol{\zeta}\}$ can be obtained from N measurements. Denoting $\boldsymbol{\Gamma} = (\Delta\boldsymbol{\tau}_1^A \quad \Delta\boldsymbol{\tau}_2^A \quad \ldots \quad \Delta\boldsymbol{\tau}_N^A)$ and $\boldsymbol{\Lambda} = (\Delta\boldsymbol{\zeta}_1 \quad \Delta\boldsymbol{\zeta}_2 \quad \ldots \quad \Delta\boldsymbol{\zeta}_N)$, the measured stiffness of 6-CSJM $\mathbf{K}_{\text{mea}}$ can be computed by the following equation

$$\mathbf{K}_{\text{mea}} = \boldsymbol{\Gamma}\boldsymbol{\Lambda}^+ \tag{5}$$

4 Simulation of Measurement

4.1 Condition of 6-CSJM for Simulation

A comprehensive simulation is designed for the 6-CSJM to validate the proposed stiffness measurement method. The pose $\mathbf{R}_0$ for measurement is given by $\zeta_0 = (0.087, 0.140, 0.122)^T$, the external load $\boldsymbol{\tau}_0^S = (0.31, 1.72, 0.14)^T$ Nm, and the cable tension vector $\mathbf{T}_0 = (12, 18, 20, 15, 30, 28)^T$ N.

In order to evaluate the effectiveness of the stiffness measurement method, the theoretical stiffness of 6-CSJM given by (4) is taken as the standard value. Furthermore, a parameter η is defined as following to evaluate the error of the measured stiffness $\mathbf{K}_{\text{mea}}$ comparing with the standard stiffness $\mathbf{K}_{\text{std}}$, i.e.,

$$\eta = \frac{||\mathbf{K}_{\text{std}} - \mathbf{K}_{\text{mea}}||_F}{||\mathbf{K}_{\text{std}}||_F} \tag{6}$$

where $||\bullet||_F$ represents the Frobenius norms of the matrix.

4.2 Simulation of the Stiffness Measurement

For any measurement, the moving-platform moves from pose $\mathbf{R}_0$ to a nearby pose $\mathbf{R}_n$. A deformation index $\Sigma_{\Delta\zeta} = |\Delta\zeta_1| + |\Delta\zeta_2| + |\Delta\zeta_3|$ is defined to evaluate the deformation amplitude. Here, we set 40 cases with different deformation indexes to figure out the influence of the deformation amplitude on the stiffness measurement.

The errors of the Laser Tracker are 0.01° for the angle and $15 + 6d$ m for the distance d m ($d \leq 5$). The errors of the force/torque sensor are listed in Table 1.

Table 1 Measurement error of the force/torque sensor

	f_x	f_y	f_z	τ_x	τ_y	τ_z
Error (%)	1.25	1.00	0.75	1.25	1.50	1.25

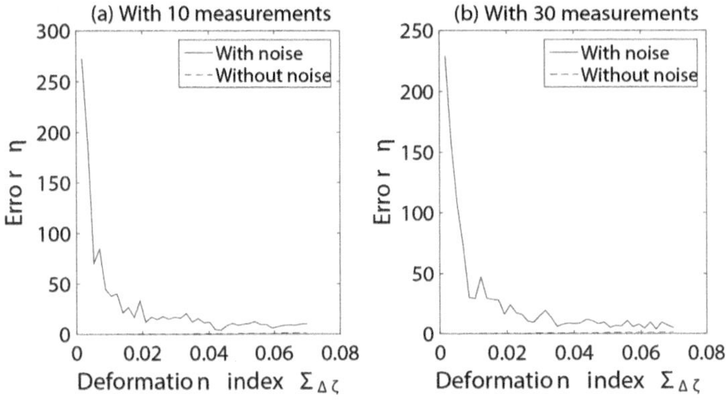

Fig. 3 Error of the stiffness measurement at pose $\mathbf{R}_0$

In this simulation, we produce the noise by using uniform distribution randomly and add them to the external loads and the poses.

Furthermore, we implement 10 measurements and 30 measurements in the simulation to figure out the influence of the measurement times on the measurement results. The results of the simulation are shown in Fig. 3.

4.3 Discussion

As shown in Fig. 3, the measurement accuracy is low when the deformation is small. The reason is that, when the deformation is small, the noise is equivalent or even large than the value of the deformation and the change of the load. The noise will affect the measurement significantly. As the deformation goes large, the measurement accuracy starts to converge and trends to stable, and the method has enough high measurement accuracy. The measurement times do not significantly affect the trend of accuracy, comparing the situation of 10 measurements and 30 measurements.

5 Conclusion

In this work, a generic stiffness measurement method is proposed for 6-CSJM, which is utilized as a fundamental building block for modular CDMs. The stiffness model is complicated and the model-based stiffness measurement method requires to measure the cable tension directly. Thus, a generic method is presented to measure the stiffness. A comprehensive simulation is designed to validate the proposed method, considering the deformation amplitude, measurement times, and the noise. The result shows that the proposed measurement method is effective for the cable-driven module. Now we are building the experimental platform for the 6-CSJM and modular CDMs. The experimental validation for the proposed method will be implemented in our future work.

Acknowledgements This work is supported by the National Natural Science Foundation of China (Code: 51475448 and 51705510) and Qianjiang Talent Project (Code: QJD1602033).

References

1. Dallej, T., Gouttefarde, M., Andreff, N., et al.: Towards vision-based control of cable-driven parallel robots. In: Proceedings IEEE/RSJ International Conference on Intelligent Robots and Systems, San Francisco, USA, 25–30 September 2011, pp. 2855–2860
2. Lafourcade, P., Llibre, M.: Design of a parallel wire-driven manipulator for wind tunnels. In: Proceedings Workshop on Fundamental Issues and Future Research Directions for Parallel Mechanisms and Manipulators, Quebec City, Canada, 3–4 October 2002, pp. 187–194
3. Deschenes, J.-D., Lambert, P., Gosselin, C., et al.: A cable-driven parallel mechanism for capturing object appearance from multiple viewpoints. In: Proceedings Sixth International Conference on 3-D Digital Imaging and Modeling, Montreal, Canada, 21–23 August 2007, pp. 367–374
4. Suzuki, Y., Kuwahara, H., Ohnishi, K., et al.: Development and verification of tendon-driven rotary actuator for haptics with flexible actuators and a PE line. In: Proceedings 11th IEEE International Workshop on advanced motion control, Nagaoka, Japan, 21–24 March 2010, pp. 484–489
5. Mustafa, S.K., Yang, G., Yeo, S.H., et al.: Optimal design of a bio-inspired anthropocentric shoulder rehabilitator. Appl. Bionics Biomech. **3**(3), 199–208 (2006, June)
6. Lim, W.B., Yeo, S.H., Yang, G., et al.: Design and analysis of a cable-driven manipulator with variable stiffness. In: Proceedings IEEE/ASME International Conference on Robotics and Automation, Karlsruhe, Gemany, 6–10 May 2013, pp. 4519–4524
7. Lim, W.B., Yeo, S.H., Yang, G., et al.: Kinematic analysis and design optimization of a cable-driven universal joint module. In: Proceedings IEEE/ASME International Conference on Advanced Intelligent Mechatronics, Suntec Convention and Exhibition Center, Singapore, 14–17 July 2009, pp. 1933–1938
8. Mustafa, S.K., Yeo, S.H., Pham, C.B., et al.: A biologically-inspired anthropocentric shoulder joint rehabilitator: workspace analysis and optimization. In: Proceedings IEEE International Conference on Mechatronics and Automation, Niagara Falls, Canada, 29 July–1 August 2005, pp. 1045–1050
9. Lum, G.Z., Mustafa, S.K., Lim, H.R., et al.: Design and motion control of a cable-driven dexterous robotic arm. In: Proceedings IEEE Conference on Sustainable Utilization and Development in Engineering and Technology, Kuala Lumpur, Malaysia, 20–21 November 2010, pp. 106–111
10. Franklin, D.W., Osu, R., Burdet, E., et al.: Adaptation to stable and unstable dynamics model. J. Neurophysiol. **90**(5), 3270–3282 (2003)
11. Perreault, E.J., Kirsch, R.F., Crago, P.E.: Multijoint dynamics and postural stability of the human arm. Exp. Brain Res. **157**(4), 507–517 (2004)
12. Balasubramanian, R., Howe, R.D., Matsuoka, Y.: Task performance is prioritized over energy reduction. IEEE Trans. Biomed. Eng. **56**(5), 1310–1317 (2009)

Research on Data-Driven Fault Diagnosis Technology of Cloud Test

Weijie Kang, Jiyang Xiao and Xiaoruo Kong

Abstract A data-driven cloud test fault diagnosis method is proposed for the current testing system based on cloud computing, which has a low utilization rate of test data and fails to give full play to the operation and storage capacity of cloud computing. Firstly, the initial fuzzy reasoning fault diagnosis method is constructed based on expert knowledge and system parameters. Secondly, GSA is used to optimize the model based on historical data. Finally, the simulation platform is used for experimental verification. The results show that the system can effectively improve the utilization rate of cloud test data and achieve more accurate fault diagnosis.

Keywords Cloud computing · Data-driven · Fault diagnosis · GSA · Cloud test

1 Introduction

The automatic test system based on cloud computing solves the information interaction problem of the traditional auto-test system. Cloud testing system has nearly unlimited data storage and processing capacity [1]. How to make good use of these massive test data for more effective fault diagnosis is an important research direction of cloud test system [2].

Aiming at the problems of the low utilization rate of test data and poor adaptability of the model, a data-driven cloud fault diagnosis method is proposed in this paper. Firstly, the initial fuzzy reasoning fault diagnosis method is constructed based on expert knowledge and system parameters. Secondly, Gravitational Search

W. Kang
Aeronautics Engineering College, Air Force Engineering University, Xi'an, China

J. Xiao (✉)
ATC and Navigation College, Air Force Engineering University, Xi'an, China
e-mail: 905132738@qq.com

X. Kong
Journalism School, Fudan University, Shanghai, China

V. Jain et al. (eds.), *Recent Trends in Intelligent Computing, Communication and Devices*, Advances in Intelligent Systems and Computing 1031,
https://doi.org/10.1007/978-981-13-9406-5_28

Algorithm (GSA) is used to optimize the model based on historical data. Finally, the simulation platform is used for experimental verification. The results show that the system can effectively improve data utilization and achieve more accurate fault diagnosis.

2 Fault Diagnosis Based on Fuzzy Reasoning

In the cloud system, the test and diagnosis data are processed centrally by the cloud [3], which makes the cloud test platform gather massive test and diagnosis data, which puts forward new requirements for the real-time response of the diagnosis method and data mining ability [4]. Based on this, a fault diagnosis method based on fuzzy reasoning is proposed.

2.1 Fault Separation and Correlation Matrix

The traditional fault diagnosis methods are mostly based on the interpretation of test data by means of distance or angle [5, 6]. However, due to the diversity of test signal types, large differences in test data types and magnitude, and the different fault characterization degree of the same signal, traditional similarity calculation methods, such as angle, distance and coefficient matrix similarity, are difficult to accurately diagnose the fault in the actual application process. By using data mining and fuzzy reasoning, the above problems can be effectively overcome, and the fault diagnosis can be made efficiently.

In the process of fault diagnosis, a good diagnostic method should be able to distinguish between normal data and fault data to a greater extent.

Definition 1 (*Fault Isolation (FI)*) When the working state of the device under test is abnormal or fails, its test data will deviate from the measured data under normal operation, and the size of such deviation is defined as fault isolation. Based on historical data and test examples, D_0 represented the theoretical work data, D_{fmin} represented the measured minimum deviation fault data, and D_{nmax} represented the maximum deviation of normal work data, then the fault separation degree is shown in Eq. (1):

$$\beta_{k,j} = \frac{(D_{\text{fmin}} - D_0) - (D_{\text{nmax}} - D_0)}{(D_{\text{nmax}} - D_0)} \tag{1}$$

Definition 2 (*Fault Correlation Matrix (FCM)*) The abnormal working state of the device under test, will often cause a chain reaction, leading to an abnormal state of multi-channel test signals. The corresponding relationship between standardized test signal value and fault type of the device under test is defined as a fault correlation matrix as shown in Eq. (2).

$$R = \begin{bmatrix} sp_{1,1} & \cdots & sp_{1,s} \\ \vdots & \ddots & \vdots \\ sp_{p,1} & \cdots & sp_{s,p} \end{bmatrix} \quad (2)$$

where the corresponding row represents the signal type and the column represents the fault type. $sp_{i,j}$ represented the correlation of the signal value of i to the fault type of j. $sp_{i,j}$ is defined as fault correlation; the specific calculation method is given in the next section.

2.2 Fault Diagnosis Based on Fuzzy Reasoning

In the cloud test system, the fault diagnosis method based on fuzzy reasoning is generally divided into three stages: (1) establishing the initial fuzzy reasoning fault diagnosis model based on system parameters and expert knowledge; (2) using intelligent algorithms based on massive historical data. Optimize the key parameters in the model; (3) Substitute the relevant data of a test instance into the model to obtain the fault diagnosis result. The diagnostic results are then added to the historical database to continuously update the inference model. The diagnostic process is shown in Fig. 1.

In actual fault detection and diagnosis, the degree of deviation of the signal from the standard value can be divided into multiple hazard levels according to expert knowledge and system parameters.

F is the fault diagnosis vector and f is the fuzzy inference rule base. After obtaining the fault deviation amount and the fault correlation matrix, the fault diagnosis method is combined with the fuzzy reasoning method as follows:

$$F = S_{kf}R = f^{\circ}S_kR = f^{\circ}(S_k - S_0)R \quad (3)$$

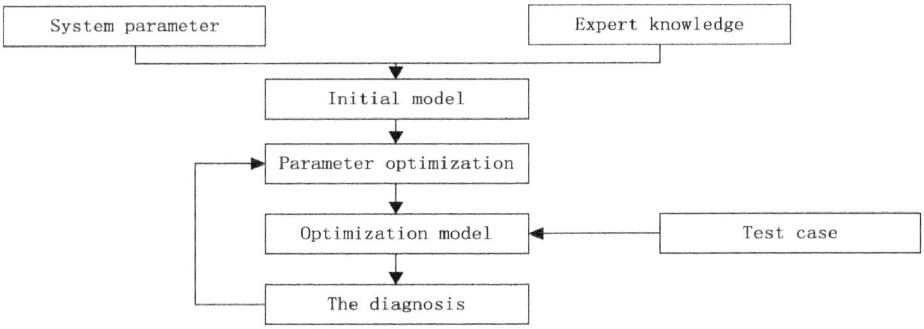

Fig. 1 Fault diagnosis process

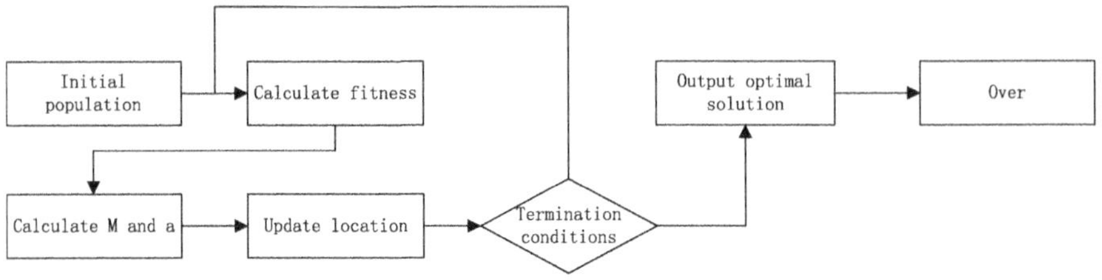

Fig. 2 Algorithm flow chart

Among them, S_k is the test instance, S_0 is the standard signal, and R is the fault correlation matrix. The fault diagnosis can be effectively realized by comparing the size of its components or the set parameters.

2.3 Parameter Optimization Based on GSA

Based on the fuzzy evaluation system established by expert knowledge, there are initial errors and attenuation errors that are difficult to eliminate. Therefore, with the accumulation of measured data, it is necessary to optimize the evaluation system in combination with historical data. In this paper, the GSA is used to optimize the key parameters of the fuzzy evaluation system. The specific implementation process is shown in Fig. 2.

(1) Input the fault diagnosis system of initial fuzzy reasoning.
(2) Set the discrete offset, substitute historical data, and give the objective function.
(3) Using a discrete differential evolution algorithm for parameter optimization.

3 Simulation Experiment and Result Analysis

3.1 Experimental Data

The experimental data in this paper comes from the electronic components of a certain type of missile launcher. For common faults, see the literature [7], as shown in Table 1.

The structure of the missile launching device is relatively complicated, and its fault type and test signal types are various. It is difficult to intuitively express the mapping relationship between its signal and fault by means of functions and the like. Therefore, the key to simplifying the test process is to select a small number of test signals to characterize typical equipment failures. This paper analyzes the common faults of the electronic components of a missile launcher and selects the following test signal types.

Table 1 Fault type

	*F*0	*F*1	*F*2	*F*3	*F*4	*F*5
Fault type	Normal	Voltage regulator failure	Audio amplifier failure	Bit marker unlocking module fault	Ignition control module malfunction	Fault of servo control system of bit marker

Based on a certain type of private cloud test platform test task instance, 2000 normal data of a certain type of fault, 2000 fault data, and 1000 test data are randomly selected for model verification.

3.2 Fuzzy Inference Model Optimization Experiment

After obtaining the initial fuzzy inference model based on system parameters and expert knowledge, the GSA is used to optimize the model.

Set the discrete offset to 20% of the values of the two adjacent intervals. For the four-level classification system, the input variable has a dimension of 6, and the GSA has a population size (**NP**) of 200 and a maximum genetic algebra of 100. The constant **G0** is 100, and the variation **a** is 20. The model is optimized based on the MATLAB 2016 platform. The classification result of fault data is shown in Fig. 3.

3.3 Diagnostic Model Comparison Experiment

In the troubleshooting process, it is often necessary to compare the test data of the current test instance with the standard data. The traditional similarity comparison methods are mainly divided into ratio, angle, distance, etc., and the distance method is the most widely used. In order to verify the effectiveness of the proposed method, it is compared with commonly used Euclidean distance, standardized Euclidean distance, and Mahalanobis distance.

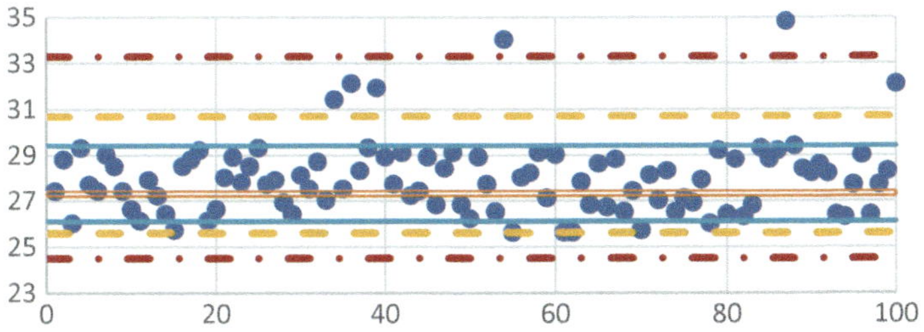

Fig. 3 The classification results

3.3.1 Diagnostic Model Comparison Experiment

Select different data types to verify the above methods. At first, the separation degree of data of different orders of magnitude is calculated, so that the adaptability of each algorithm can be estimated. The results are shown in Fig. 4.

It can be seen from the above table that the traditional Euclidean distance is difficult to be used for fault diagnosis of multi-channel signals. The Markov distance still has good fault resolution in the small sample interval, but with the sample increasing its fault separation ability is significantly reduced. The similarity calculation method based on fuzzy reasoning has better fault separation ability in the interval of different sample sets, and is suitable for the characteristics of massive test data of cloud test system.

3.3.2 Accuracy Comparison

In the actual fault diagnosis, higher requirements are put forward for the accuracy of fault diagnosis. A good fault diagnosis system should have a low false alarm rate, that is, a high fault diagnosis accuracy rate. Under the premise of known normal data and fault data, fault diagnosis problem is transformed into the classification problem. By comparing the test instance data with the standard data, the corresponding fault type is determined. By using traditional classification methods, such as BP neural network, KNN, and PSO, fault diagnosis was conducted on 100 groups of test data, and the number of correct classification is shown in Table 2.

According to the above table, the classification performance of BP neural network is good, but the training process is complex. The KNN algorithm is the

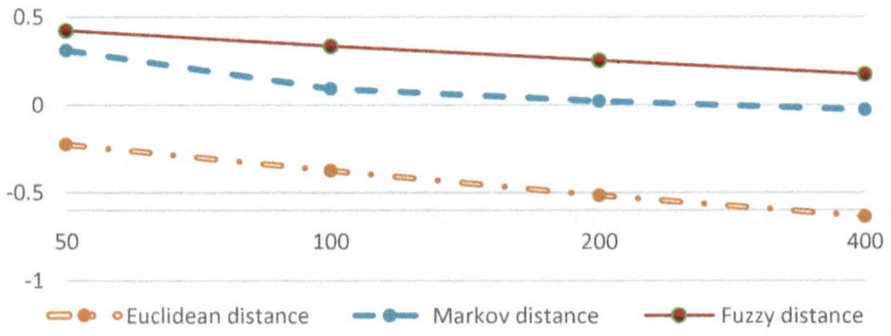

Fig. 4 Separation degree of each algorithm

Table 2 Fault diagnosis accuracy rate

	*F*0	*F*1	*F*2	*F*3	*F*4	*F*5	All
BP	88	87	84	90	87	80	86.00
KNN	85	85	83	86	82	81	83.67
PSO	87	82	79	84	86	89	84.50
FR	87	91	90	86	90	89	88.83

simplest but has poor classification effect. The complexity of SVM is between the two and the performance is relatively good. The fuzzy reasoning method is simple and has the best classification effect.

To sum up, the fault diagnosis algorithm based on fuzzy reasoning is superior to the traditional similarity calculation method in adaptability, slightly superior to the traditional method in accuracy, simple in implementation, and can realize more efficient and accurate fault diagnosis of UUTs.

4 Conclusion

This paper proposes a fault diagnosis method based on fuzzy inference. The method is suitable for cloud test mass storage and processing test data requirements, effectively improves the fault diagnosis accuracy of the cloud test system, and is applied as a key technology to a cloud test system. The data-driven cloud test fault diagnosis method proposed in this paper expands the new method to improve equipment testing efficiency and guarantee level, which has important theoretical significance and practical value.

Acknowledgements This work does not have any fund support.

References

1. Xiao, M.Q., Zhao, Y, Huihui, X, Tang, X, Deng, J.: Exploration of cloud computing and its application in the field of testing. J. Air Force Eng. Univ. (Nat. Sci. Ed.) **16**(01), 50–55 (2015)
2. Xiao, M.Q., Zhao, Y., Zhao, X.: Concept and application exploration of cloud testing. Comput. Meas. Control **24**(01), 1–3 + 11 (2016)
3. Zhao, Y.: Research on cloud test resource virtualization and its optimal scheduling method. Air Force Engineering University (2018)
4. Guo, R., Zhao, X.: Development trend of automatic test system. Overseas Electron. Meas. Technol. **33**(06), 1–4 (2014)
5. Zhou, Y., Jing, B., Zhang, J., Zhou, H.: Application model research of aircraft fault prediction and health management. Comput. Meas. Control **19**(09), 2061–2063 + 2101 (2011)
6. Xin, L., Zhou, Y., Kong, Q., Zhao, Y.: Study on fault prediction of aerial equipment based on markov distance. Comput. Meas. Control **22**(07), 2052–2054 + 2058 (2014)
7. Zhang, J., Feng, J., Li, Q., Lu, Q.: Study on fault diagnosis method of parallel test system based on fuzzy clustering. Comput. Meas. Control **17**(01), 30–32 (2009)

Android Malware Detection Model Based on LightGBM

Guangyu Wang and Zhijing Liu

Abstract Android malware detection is an important research area against Android malware apps. In this paper, we propose an Android malware detection model based on LightGBM. The model consists of a new feature selection method, which contains Chi2 and ExtraTrees, and a LightGBM classifier method. A corpus of 2000 malware and equal number of benign samples are prepared to verify the model. Finally, an experiment is designed to test the model accuracy and training time. The results show high model accuracy (about 96.4%) and a heavy reduction in training time as compared to existing models.

Keywords Android malware · Static detection · LightGBM · Feature selection

1 Introduction

Android malware detection is becoming more and more challenging everyday since Android has become the most popular mobile operating system. According to '***China Mobile Security Ecology Research Report 2018***' [1] from Qihoo 360, there are 2,831,000 new malicious samples of Android platform appeared in the first six months of 2018, and nearly 16,000 new mobile phone malicious samples were intercepted every day. Figure 1 shows the each month's additions and interceptions of malicious samples in the first half of 2018.

Many research works have been proposed in this area. Naser Periavian [2] proposed a method based on machine learning using the combination of permission and API calls as features. Yerima et al. [3] presented an Android detection model based on Bayesian classification. Milosevic et al. [4] provided a permission-based approach and source code-based approach to analyze the mobile applications. Hou et al. [5] proposed a Linux kernel system call graph-based deep learning framework

G. Wang (✉) · Z. Liu
School of Computer Science and Technology, Xidian University, Shaanxi, 710071 Xian, China
e-mail: 1783538369@qq.com

V. Jain et al. (eds.), *Recent Trends in Intelligent Computing, Communication and Devices*, Advances in Intelligent Systems and Computing 1031,
https://doi.org/10.1007/978-981-13-9406-5_29

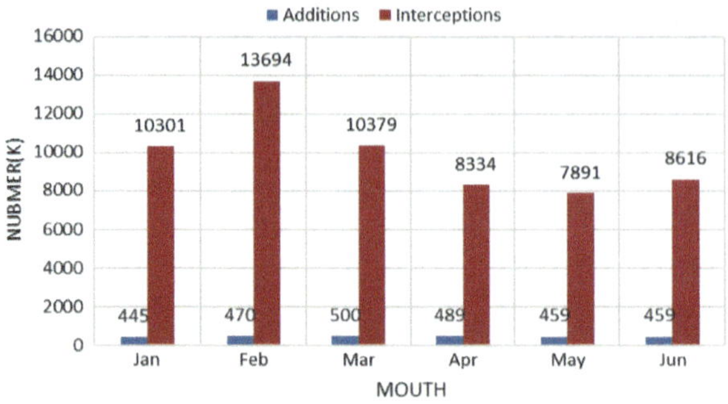

Fig. 1 Additions and interceptions of malware

for malware detection. Feng [6] introduced a method based on information flow analysis to detect malware.

The major contributions of our work are (1) collection of 2000 benign samples from Baidu app store and Google app store and 2000 malicious samples from www.virusshare.com, (2) extracted four features namely 'permissions,' 'actions,' 'categories,' and 'SO file names' to build the feature vector, (3) prepared a feature selection model based on Chi2, that selects 500 of features having high rank in Chi2, and ExtraTrees, (4) the classifier is implemented in Python and designed using LightGBM that is effective and provides high accuracy [7], and (5) the result is compared with the existing works and shown that our model outperforms other models.

The rest of the paper is organized as given below. Section 2 describes the concepts of Chi2, ExtraTrees, and LightGBM. In Sect. 3, we design an experiment to test the validity of the model we made using different lengths of feature vectors. At the end, we summarize the work.

2 Methodology

Figure 2 depicts the proposed detection model that contains feature extraction, feature selection, and classification.

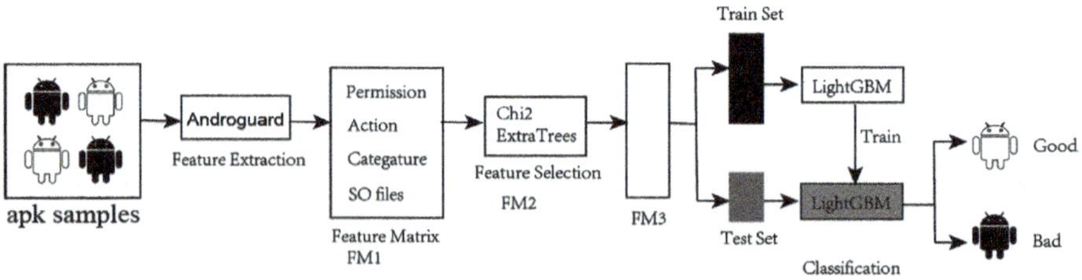

Fig. 2 Full process of detection model

2.1 Feature Extraction

There are two types of features, static and dynamic. Static feature detection is the most common option to build a feature vector, whereas dynamic feature needs sandbox to simulate the environment and it often takes too long time. Therefore, we choose static feature for our model. The purpose of 'permission label' feature is to protect Android users' privacy is found in AndroidMainfest.xml. The feature 'action' is used to add an action to intent filter, which contained in intent filter. Category label aims to add a category to intent filter, which is also included in intent filter as action. The feature 'SO file' is a compiled library file that performs similar function as in Windows DLL. These four features are used to build the feature vector. Androguard [8] is a Python-based decompile tool to analyze Apk files. It is an open-source project that provides many useful functions to get information of Apks.

With the help of Androguard, we can extract features we need to build our feature vector easily. There are three steps to get a feature vector from an Apk:

(1) Analysis of Apk files with Androguard gets information about features (permission, action, category, SO files) and is saved as Feature_Set.
(2) Compare every Apk information with Feature_Set. The value one (1) means vector has this element, and zero (0) indicates an Apk does not have this element. Save the result is saved as Feature_Vector.
(3) Get all samples' Feature_Vector, store it in as a matrix (FM_1), and then save it as a CSV file for further use.

2.2 Feature Selection

The size of the matrix of the features created above is 4000 × 16,287. The size of this matrix needed to be shrunken (feature selection) using either a filter way or a wrapper way.

As the feature list is too long, we introduce a two-step method of feature selection that consists of Chi2 selection and ExtraTrees selection. Both the methods are provided by scikit-learn [9] (www.scikit-learn.com). Figure 3 presents the complete process of feature selection.

Chi2 is a filter-type method. In this method, every feature's grade of Chi2 is calculated and the top 500 features are selected, which is the stored as a new matrix (FM_2) of size 4000 × 500 and saved in a new CSV file.

Unlike random forest, ExtraTrees use all original training set samples as each tree's training set and pick an random value to split the feature. Consequently, we choose it as the feature selection method. According to the importance of every feature, we get 73 features, whose importance is greater than zero. Moreover, save this new matrix FM_3 of size 4000 × 73, as a new CSV file.

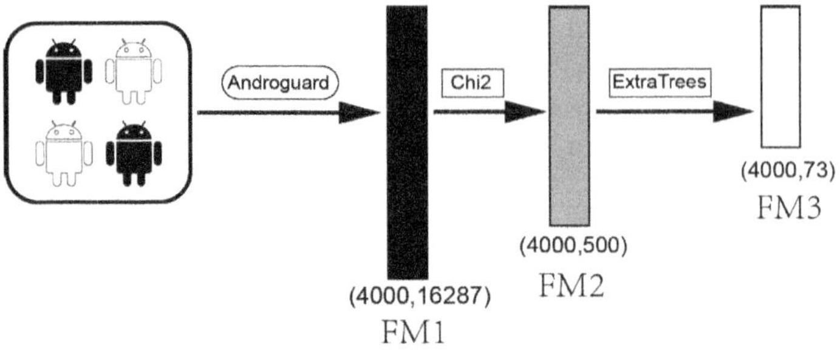

Fig. 3 Process of feature selection

2.3 Classification Model

GBDT is the most popular machine learning algorithm in recent years. They have many improved versions of it, such as LightGBM, XGBoost [10]. XGBoost introduced by Tianqi Chen has high accuracy rate, but sometimes the time required is higher than LightGBM.

LightGBM is an open-source framework for machine learning developed by Microsoft. Unlike other pre-sort-based boosting algorithms, LightGBM uses histogram-based algorithms for decision tree learning that can reduce the time complexity. LightGBM also uses Gradient-based One-Side Sampling (GOSS), that excludes instances with little gradients, and Exclusive Feature Bundling (EFB), which can shrink the size of features. Therefore, LightGBM can reduce the time of execution.

3 Evaluation

3.1 Experimental Setup

We run this model on a laptop, whose configurations are shown in Table 1.

2000 benign samples are downloaded from Baidu app store and Google app store again 2000 malware samples are downloaded from Virusshare [11] (www.virusshare.com). We analyze those samples with Androguard to get all the

Table 1 Configurations

Items	Configuration
CPU	Core i7-7700HQ (2.80 GHz)
Memory	16 GB (DDR4 2400 MHz)
Hard drive	1 TB (5400 r/min)
Python	2.7
Androguard	3.1.0
Scikit-learn	0.19.1

features we need. Then, the model is tested with all the samples features. We use all those samples with Androguard and build an auto-Python-based detection model.

3.2 Model Evaluation

500 features are selected using Chi2 features' selection and 73 features using ExtraTrees features' selection. The matrix is used as the input to the classifier. Testing of the classifier is performed with different length of the feature vector. The length of the feature vector starts with ten and increased by five each time. We then use tenfold cross-validation for ten times. At each step, 90% of samples are used as training set and the rest as validation samples. The average value of the result is calculated to test the accuracy of the proposed model along with the time required by each test.

Considering malware as MAL, and benign as BEN, we set $N_{\text{BEN}\to\text{BEN}}$ as the number of good samples classified as good, and $N_{\text{MAL}\to\text{MAL}}$ indicates the number of bad samples classified as bad. $N_{\text{MAL}\to\text{BEN}}$ represents the number of bad samples those have been classified as good samples, and $N_{\text{BEN}\to\text{MAL}}$ indicates the number of good samples those have been classified as bad samples. Therefore, the accuracy is:

$$\text{Acc} = \frac{N_{\text{BEN}\to\text{BEN}} + N_{\text{MAL}\to\text{MAL}}}{N_{\text{BEN}\to\text{BEN}} + N_{\text{MAL}\to\text{MAL}} + N_{\text{BEN}\to\text{MAL}} + N_{\text{MAL}\to\text{BEN}}} \tag{1}$$

Time requirement contains training cost and prediction cost. Figure 5 introduces the time requirements of both LightGBM and XGBoost.

We tested our model and XGBoost together and compared their result. Figures 4 and 5 show the time requirement and accuracy of LightGBM versus XGBoost. As we can see from those two figures, LightGBM requires less time than XGBoost in training process and has higher accuracy than XGBoost when the feature vector is longer than 30, and the accuracy of LightGBM is 96.4% when the feature vector

Fig. 4 Accuracy

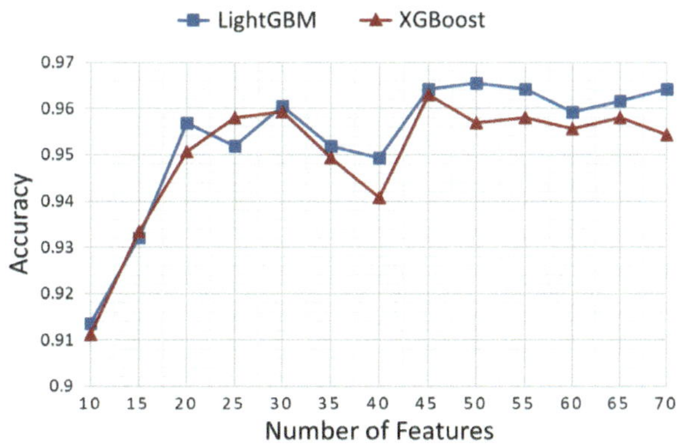

Fig. 5 Time cost

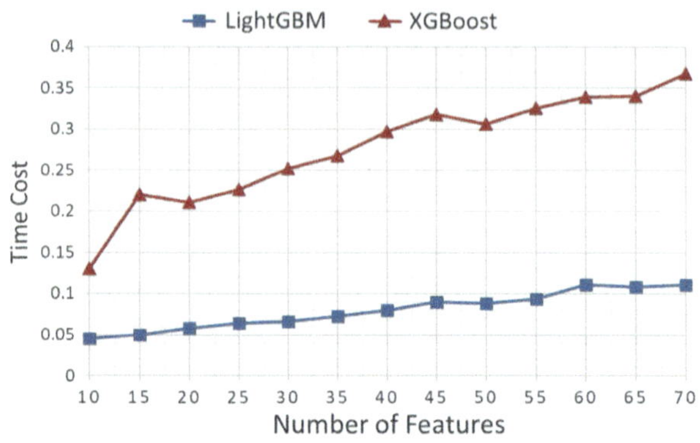

Table 2 Results from LightGBM with different types of feature selection methods

Combination	Feature-length	Time cost (s)	Accuracy (%)
Chi2 (500), ExtraTrees	50	0.085	96.41
Chi2 (500)	500	0.256	94.93
ExtraTrees	927	0.275	96.52
No selection method	16287	2.602	96.17

length is 50. Figure 5 shows that the time requirement of LightGBM is much less than XGBoost. As the length of the feature vector is longer than 50, the time requirement of LightGBM is about a quarter of XGBoost's. Therefore, we choose LightGBM as the classification algorithm and set the feature vector length to 50, and the sample matrix is of size 4000×50.

We further test LightGBM with different types of feature selection methods that are combination of Chi2 and ExtraTrees, Chi2 without ExtraTrees, ExtraTrees without Chi2, and no selection method. Table 2 shows the feature-length, time cost, and accuracy of those different types of combinations. 'Chi2 (number)' is the feature-length kept after the Chi2 selection.

As we can see from the table, our feature selection can shrink the size of the feature vector and reduce the time requirement while holding the accuracy rate at the same time.

4 Conclusion

In this paper, we introduce an Android malware detection model with feature selection and LightGBM. At first, four types of features are extracted from Apk samples, and then the size of the matrix is shrunken using Chi2 and ExtraTrees. Finally, two boosting algorithms XGBoost and LightGBM are used for testing. According to the result found, the accuracy of LightGBM is as good as XGBoost and sometimes even better. However, it spends less time than XGBoost on training as well as in prediction. Therefore, we conclude that the LightGBM performs better than XGBoost in this experiment.

In the future, we will modify this model by collecting more Apk samples for the experiment as LightGBM works better on a larger sample set. Currently, the advantage of time requirement is not that obvious. Also, more feature selection algorithms may be considered to obtain better performance.

References

1. Qihoo 360 Technology Co, Ltd. China Mobile Security Ecology Research Report 2018 [EB/OL]. http://zt.360.cn/1101061855.php?dtid=1101061451&did=491398428
2. Machine Learning for Android Malware Detection Using Permission and API Calls. In: 2013 IEEE 25th International Conference on Tools with Artificial Intelligence. IEEE (2014)
3. Yerima, SY., Sezer, S., Mcwilliams, G., et al.: A new android malware detection approach using bayesian classification (2016)
4. Milosevic, N., Dehghantanha, A., Choo, K.K.R.: Machine learning aided android malware classification. Comput Electri. Eng. **61** (2017)
5. Hou, S., Saas, A., Chen, L., et al.: Deep4MalDroid: a deep learning framework for android malware detection based on linux kernel system call graphs. In: IEEE/WiC/ACM International Conference on Web Intelligence Workshops, pp. 104–111. IEEE (2017)
6. Feng, S., Vecchio, J.D., Mohaisen, A., et al.: Android malware detection using complex-flows. In: IEEE, International Conference on Distributed Computing Systems, pp. 2430–2437. IEEE, (2017)
7. Ke, G., Meng, Q., Finley, T., Wang, T., Chen, W., Ma, W., Ye, Q., Liu, T.Y.: LightGBM: a highly efficient gradient boosting decision tree. Adv. Neural Info. Process. Syst. **30**, 3149–3157 (2017)
8. Androguard, "An apk analysis tool" [EB/OL]. https://code.google.com/p/androguard/
9. Scikit-learn, "Machine Learning Framework in Python" [EB/OL].http://scikit-learn.org/stable/modules/classes.html
10. Chen, T., He, T., Benesty, M., et al.: xgboost: extreme gradient boosting (2016)
11. VirusShare, A repository of malware samples[EB/OL].https://virusshare.com/

ID3-Based Classification of College Students' Physical Fitness Data

Gang Ma, Liumei Zhang and Shoubang Li

Abstract This paper uses the ID3 algorithm to analyze and extract the classification rules hidden in the original data of the physical fitness test of junior college students in mobile app platform database of "Running Shida". These classification rules are highly consistent with the actual data in the database and are highly consistent with the results of individual survey of students. The forecasting conclusion of these classification rules is of great significance for quickly and scientifically determining students' physique, putting forward reasonable suggestions for sports training and promoting the reform of "integration in-and-outside class" teaching mode of physical education in colleges and universities.

Keywords Data mining · Classification analysis · Physical fitness testing · Decision tree · ID3 algorithm

1 Introduction

The physical health level of contemporary college students is not only related to the healthy growth and happy life of individuals, but also to the health quality of the whole nation and the quality of personnel training in China. According to the results of the research on adolescents' physique in China, the declining physique level has attracted the attention of many ministries and colleges. Hu Jingchao and Wang Li studied the original data of students' physical fitness test in Henan University of Technology and found that physical fitness can be promoted from the flexibility, speed and endurance [1]. Zhang Chonglin et al. utilized an association rule to

---

G. Ma (✉) · L. Zhang
School of Computer Science, Xi'an Shiyou University, Xi'an, Shaanxi, China
e-mail: jhlmg@xsyu.edu.cn

S. Li
Department of Physical Education, Xi'an Shiyou University, Xi'an, Shaanxi, China

V. Jain et al. (eds.), *Recent Trends in Intelligent Computing, Communication and Devices*, Advances in Intelligent Systems and Computing 1031,
https://doi.org/10.1007/978-981-13-9406-5_30

analyze the data of physique test of faculty and staff in a university in Shanghai and drew corresponding conclusions. It was suggested that in order to promote the physique health of faculty and staff, priority should be given to developing their cardiopulmonary function and improving the strength of upper limbs and trunk [2]. Liu Xin and Yang Sujin adopt an array-based Apriori algorithm to mine and analyze the physical fitness test items of college students, find out the relationship between the test items and judge the rationality of the test items [3]. Zhao Changhong and Wang Lin made a comparative study on the data of the physical fitness test of male and female students in Northwest University for Nationalities. They suggested that the physical fitness test of female students should strengthen the training of standing long jump and male students' pull-up, which has important application value for improving the comprehensive quality of physical function of students in colleges for nationalities [4].

In view of the serious situation of the continuous decline of college students' physical fitness level, the majority of sports educators are making efforts to study and have achieved some results. However, most of these research results remain in the shallow statistics and analysis of individual test data, or only the correlation analysis of some project data, which cannot reveal the internal relationship of all test items, nor point out the classification rules of each test item, so it cannot be aimed at the actual physical condition of contemporary college students, it is unable to put forward comprehensive, scientific and training suggestions.

As an important part of the reform of the "integration of physical education in-and-outside class" teaching mode in colleges and universities, college students' physical fitness testing mode and evaluation mechanism have been highly valued by scholars and physical education in recent years. In our previous work, we have adopted the ID3 algorithm for mining hidden classification rules from mass students' physical constitution evaluation and sports training result data [5]. Based on that, in this paper, we use the original data of college students' physical fitness test, which was accumulated three years by mobile app of "Running Shida." Then, it uses ID3 algorithm to analyze and extract the classification rules of all test items in the dataset, analyzes and excavates the same type of physical fitness characteristics, common nature and characteristic knowledge of individual differences in different physiques. The results have a direct guiding role for different students to take specific training and training measures on individual differences and improve individual physical quality.

2 Classification Analysis

Classification is the analysis process of classifying a group of data with common attributes (a group of training sample data) according to the value of their attributes (learning through data mining classification algorithm), and searching for the rules of each class through the analysis results. The classification rules are used as the basis for future data classification, and the results of future data classification are predicted [5]. Classification is mainly used for prediction. Its purpose is to find a set of models or functions that can describe the typical features of data sets, so as to identify the attribution or category of unknown data [6].

Decision tree is one of the important construction methods of classification model. It is a directed acyclic graphics (DAG) based on machine learning, which consists of root node, internal node and leaf node. For the original training tuple dataset, attribute selection measurement is the most critical problem in constructing decision tree, and the final decision tree is the most favorable for the generation of classification rules [7–9]. Attribute selection metric is a criterion for selection classification. The attributes with the highest information gain and maximum information gain rate should be selected as the attributes selection of the current classification [6].

ID3 algorithm is one of the typical algorithms for decision tree learning. The algorithm uses the maximum information gain and maximum information gain rate as the criteria for selecting attributes at all levels of nodes in the decision tree. When testing on each non-leaf node, the largest category information of the tested example can be obtained. After using this attribute, the set of examples can be divided into subsets. The entropy of the system can be minimized [10].

3 Application of ID3 Algorithm in Mining of College Students' Fitness Data

In this paper, the physical test data of the third-year students of Xi'an Shiyou University in 2018 are selected as statistical data sets R_1 and R_2. Tables 1 and 2 show the sample example of the data set. Before the classification, the original data sets are preprocessed by sorting and cleaning. Then, the results of all test items are converted according to the test standards. The corresponding training sets R_1' and R_2' of statistical marker data are derived, as shown in Tables 3 and 4.

From the description of the ID3 algorithm, it can be seen that the core idea is the selection of attributes. The decision tree is constructed by the ID3 algorithm. The data of physical fitness test are classified and analyzed. Finally, the influence of

Table 1 College student (female) physical fitness test statistical data set (R_1)

ID	Height (cm)	Weight (Kg)	FVC (ml)	Sit-reach (count)	Stand-leap (m)	50 M (s)	800 M (s)	Sit-up (count)	Body mass
S00	168.0	60.0	2336	18	150.00	9.9	4′00″	45	72.7
S01	175.0	77.0	3489	24	170.00	8.9	4′00″	44	80.7
…	…	…	…	…	…	…	…	…	…
S99	165.0	54.0	2370	13	180.00	8.6	3′45″	45	78.3

Table 2 College student (male) physical fitness test statistical data set (R_2)

ID	Height (cm)	Weight (Kg)	FVC (ml)	Sit-reach (count)	Stand-leap (m)	50 M (s)	1000 M (s)	Pull-up (count)	Body mass
S00	170.0	55.0	4055	2	227.00	7.4	4′00″	2	67.5
S01	174.0	94.0	3800	11	170.00	8.4	4′20″	2	54.5
…	…	…	…	…	…	…	…	…	…
S99	174.0	60.0	3500	13	222.00	7.4	3′43″	5	70.5

Table 3 College student (female) physical fitness test statistical class tag data set (R_1')

ID	Weight index	FVC (ml)	Sit-reach (count)	Stand-leap (m)	50 M (s)	800 M (s)	Sit-up (count)	Body mass
S00	Normal	Pass	Pass	Fail	Pass	Pass	Pass	Pass
S01	Overweight	Excellent	Excellent	Pass	Pass	Pass	Pass	Good
…	…	…	…	…	…	…	…	…
S99	Normal	Pass	Pass	Pass	Pass	Pass	Pass	Pass

Table 4 College student (male) physical fitness test statistical class tag data set(R_2')

ID	Weight index	FVC (ml)	Sit-reach (count)	Stand-leap (m)	50 M (s)	1000 M (s)	Pull-up (count)	Body mass
S00	Normal	Pass	Fail	Pass	Pass	Pass	Fail	Pass
S01	Overweight	Pass	Pass	Fail	Pass	Pass	Fail	Fail
…	…	…	…	…	…	…	…	…
S99	Normal	Pass	Pass	Pass	Pass	Pass	Fail	Pass

other attributes on the results of their BMI attributes is obtained. In this paper, the attributes are selected according to the following manner. Taking body mass as the class label attribute, 100 items were sampled and recorded, i.e., 100 discrete values of attribute ID, four discrete values of weight index (malnutrition, normal, overweight, obesity); FVC as the vital capacity; Sit-Reach; Stand-Leap as the standing long jump; 50 M as the 50 m running; 1000 M (male) as the 1000 m running; 800 M (female) as the 800 m running; Pull-up (male) as the pull-up; Sit-up as the sit-up, which consist of numeral discrete values, class label attributes body mass has four discrete values (fail, pass, good, excellent).

According to the definition of information gain by DT classical algorithm, for the training set of statistical labeled data for female college students' physique test in Table 3, the calculation process and results are as follows:

$$\mathrm{Info}(R_1') = -\frac{19}{100}\log_2\frac{19}{100} - \frac{64}{100}\log_2\frac{64}{100} - \frac{12}{100}\log_2\frac{12}{100} - \frac{5}{100}\log_2\frac{5}{100} = 1.4505$$

$$\mathrm{Info}_{\mathrm{Weight-Index}}(R_1') = \frac{7}{100}\left(-\frac{4}{7}\log_2\frac{4}{7} - \frac{2}{7}\log_2\frac{2}{7} - \frac{1}{7}\log_2\frac{1}{7} - \frac{0}{7}\log_2\frac{0}{7}\right)$$
$$+ \frac{64}{100}\left(-\frac{6}{64}\log_2\frac{6}{64} - \frac{44}{64}\log_2\frac{44}{64} - \frac{10}{64}\log_2\frac{10}{64} - \frac{4}{64}\log_2\frac{4}{64}\right)$$
$$+ \frac{17}{100}\left(-\frac{3}{17}\log_2\frac{3}{17} - \frac{10}{17}\log_2\frac{10}{17} - \frac{3}{17}\log_2\frac{3}{17} - \frac{1}{17}\log_2\frac{1}{17}\right)$$
$$+ \frac{12}{100}\left(-\frac{3}{12}\log_2\frac{3}{12} - \frac{7}{12}\log_2\frac{7}{12} - \frac{2}{12}\log_2\frac{2}{12} - \frac{0}{12}\log_2\frac{0}{12}\right) = 1.4007$$

Thus,

$$\begin{aligned}\mathrm{Gain}_{\mathrm{Weight-Index}}(R_1') &= \mathrm{Info}(R_1') - \mathrm{Info}_{\mathrm{Weight-Index}}(R_1') = 1.4505 - 1.4007 \\ &= 0.0498\end{aligned}$$

Similarly, computations are available as follows:

$$\mathrm{Gain}_{\mathrm{FVC}}(R_1') = \mathrm{Info}(R_1') - \mathrm{Info}_{\mathrm{FVC}}(R_1') = 1.4505 - 1.1376 = 0.3129$$
$$\mathrm{Gain}_{\mathrm{Sit-Reach}}(R_1') = \mathrm{Info}(R_1') - \mathrm{Info}_{\mathrm{Sit-Reach}}(R_1') = 1.4505 - 1.3627 = 0.0878$$
$$\mathrm{Gain}_{\mathrm{Stand-Leap}}(R_1') = \mathrm{Info}(R_1') - \mathrm{Info}_{\mathrm{Stand-Leap}}(R_1') = 1.4505 - 1.0195 = 0.4310$$
$$\mathrm{Gain}_{\mathrm{50M}}(R_1') = \mathrm{Info}(R_1') - \mathrm{Info}_{\mathrm{50M}}(R_1') = 1.4505 - 0.8631 = 0.5874$$
$$\mathrm{Gain}_{\mathrm{800\,M}}(R_1') = \mathrm{Info}(R_1') - \mathrm{Info}_{\mathrm{800M}}(R_1') = 1.4505 - 0.9502 = 0.5003$$
$$\mathrm{Gain}_{\mathrm{Sit-up}}(R_1') = \mathrm{Info}(R_1') - \mathrm{Info}_{\mathrm{Sit-up}}(R_1') = 1.4505 - 1.3860 = 0.0645$$

In fact, the measurement of information gain tends to have more output attributes. For the training set of statistical marker data for female in Table 3 and the training set of statistical marker data for male in Table 4, the ID attribute has the largest attribute value (which can be calculated by $\mathrm{Info}_{\mathrm{ID}}(R) = 0$, thus $\mathrm{Gain}_{\mathrm{ID}} = \mathrm{Info}(R)$). But, it is obvious that classifying by using ID attributes is meaningless to solve problems.

By comparing the information gains of weight index, FVC, Sit-Reach, Stand-Leap, 50 M, 800 M and Sit-up, which are the attributes of body weight, vital capacity, standing long jump, sitting-up and so on, the attribute 800 M with the highest information gains are selected as the root nodes of the decision tree. Using the same computing process, we can get the subsequent decision tree nodes and finally generate the decision tree, as shown in Fig. 1.

For the decision tree in Fig. 1, the classification rules are extracted along the path from the root node to each leaf node as follows:

Rule_1: IF 800 M = fail AND 50 M = fail AND Stand-Leap = fail Then Body-Mass = fail

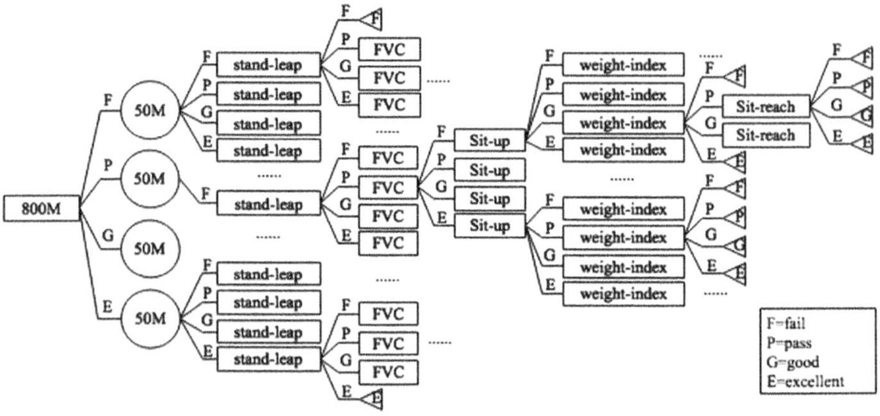

Fig. 1 Decision tree of college student(female) physical fitness test statistical class tag

……

Rule_m: *IF 800 M = pass AND 50 M = fail AND Stand-Leap = pass AND FVC = fail AND Sit-up = good AND Weight-Index = fail Then Body-Mass = fail*

Rule_m + 1: *IF 800 M = pass AND 50 M = fail AND Stand-Leap = pass AND FVC = fail AND Sit-up = good AND Weight-Index = pass AND Sit-Reach = fail Then Body-Mass = fail*

Rule_m + 2:IF 800 M = pass AND 50 M = fail AND Stand-Leap = pass AND FVC = fail AND Sit-up = good AND Weight-Index = pass AND Sit-Reach = pass Then Body-Mass = fail

Rule_m + 3: *IF 800 M = pass AND 50 M = fail AND Stand-Leap = pass AND FVC = fail AND Sit-up = good AND Weight-Index = pass AND Sit-Reach = good Then Body-Mass = pass*

Rule_m + 4: *IF 800 M = pass AND 50 M = fail AND Stand-Leap = pass AND FVC = fail AND Sit-up = good AND Weight-Index = pass AND Sit-Reach = excellent Then Body-Mass = pass*

……

Rule_s: *IF 800 M = pass AND 50 M = fail AND Stand-Leap = pass AND FVC = fail AND Sit-up = good AND Weight-Index = excellent Then Body-Mass = good*

……

Rule_t: *IF 800 M = pass AND 50 M = fail AND Stand-Leap = pass AND FVC = excellent AND Sit-up = pass AND Weight-Index = fail Then Body-Mass = pass*

Rule_t + 1: *IF 800 M = pass AND 50 M = fail AND Stand-Leap = pass AND FVC = excellent AND Sit-up = pass AND Weight-Index = pass Then Body-Mass = pass*

Rule_t + 2: IF 800 M = pass AND 50 M = fail AND Stand-Leap = pass AND FVC = excellent AND Sit-up = pass AND Weight-Index = good Then Body-Mass = pass

Rule_t + 3: IF 800 M = pass AND 50 M = fail AND Stand-Leap = pass AND FVC = excellent AND Sit-up = pass AND Weight-Index = excellent Then Body-Mass = good

……

Rule_w: IF 800 M = excellent AND 50 M = excellent AND Stand-Leap = excellent Then Body-Mass = excellent

According to the same principle and algorithm, the information gain of seven attributes of body mass index such as the weight index, FVC, Sit-Reach, Stand-Leap, 50 M, 1000 M and Pull-up can be obtained for the statistical labeled data training set R_2' in Table 4. In which, Pull-up with the highest information gain is selected as the root node of the decision tree. This article is limited in length and will not be repeated here.

4 Conclusion

In view of the present situation of the continuous decline of the physical fitness level of contemporary college students, combined with the specific requirements of the reform of the teaching mode of "integration of physical education in-and-outside class" for the physical fitness test mode and evaluation mechanism of college students, this paper adopts the ID3 algorithm to classify the physical fitness data of college students' physical test. Therefore, the decision tree extract and classification rules are generated for male and female students. Through comparison, it is found that these classification rules are highly consistent with the actual data in the database and highly consistent with the students' individual survey. Therefore, the prediction conclusion of these classification rules can quickly and scientifically determine the individual physique of each student, so as to classify the students with different physiques, and put forward reasonable suggestions for sports training, which has very high reference value and promotion for students' physical exercise and school sports teaching reform.

References

1. Hu, J.C., Wang, L.: Application research on college students' physical health test with data mining. J. Jilin Sport Univ. **33**(3), 8–11 (2017)
2. Zhang, C.-L., Yu, L.J., Wu, W.B.: The application of data mining technology of association rules in physical fitness testing analysis. J. Shanghai Univ. Sport **36**(2), 42–44 (2012)

3. Liu, Xin, Yang, Su-jin: The application of an array-based association rule mining algorithm to the physical test data analysis. J. Shandong Univ. Technol. (Natural Science Edition) **25**(5), 55–58 (2011)
4. Zhao, C.H., Wang, L.: Analysis and application of college students' physical fitness test data-a case study of Northwest University for nationalities. J. Lanzhou Univ. Arts Sci. (Natural Sciences) **31**(3), 108–112 (2017)
5. Zhang, Q., You, K., Ma, G.: Application of ID3 Algorithm in exercise prescription. Electrical Power Systems and Computers, pp. 669–675. Springer, Berlin, Heidelberg (2011)
6. Xu, J.P.: Data Warehouse and Decision Support System. Science Press, Beijing, China (2005)
7. Ma, G.: Application of data mining technology in hospital management information system. Xi'an Shiyou University (2008)
8. Su, X.N., Yang, J.-L., et al.: Data Warehouse and Data Mining. Tsinghua University Press, Beijing, China (2006)
9. Ma, G., Liu, T.S., Li, J.: Application study of ID3 algorithm in mining of doctor classification regulations. China Acad. J. Electron. Publishing House. **16**(11), 79–81 (2008)
10. Gehrke, J., Ganti, V., Ramakrishnan, R., Loh, WY.: BOAT: optimistic decision tree construction. In: Proceedings of the 1999 ACM SIGMOD International Conference on Management of Data, Philadelphia, Pennsylvania (1999)

Influences of R&D Input on Brand Value Based on Coupling Threshold Regression Analysis

Duan Qi

Abstract Based on evaluation data of brand value in our country from 2013 to 2017, we will explore the influences of enterprises' R&D input on brand value through coupling test model and threshold regression model of panel, and the coupling of enterprises innovation system is deemed as the threshold variable. Results of empirical research show that there are obvious dual-threshold effects between R&D input and brand value in three lag phases. When the coupling is less than 0.620, R&D input is negatively correlated or uncorrelated with the brand value; when the coupling is more than 0.620, R&D input will greatly facilitate the brand value.

Keywords Brand value · Regional innovation system · Threshold regression

1 Introduction

In the era of economic globalization, brand has become the most important factor in global economy and technology competition. The brand value determines status of different countries in global industrial value chain. Therefore, the focus of international economic competition is to promote the competitive force of brand of a country, and technology-oriented international competition is the important method for all countries to promote comprehensive strength and international status [1]. Enterprise innovation system is the important factor which influences the brand value as well as important force to increase enterprise revenue. Establishment of enterprise innovation system is helpful to gather innovation resources and attract innovation talents. In this case, enterprises will realize innovation in technology and knowledge by virtue of various resources and services [2, 3], thus increasing enterprise revenue and promoting brand value of enterprises. With the coming of knowledge economy, technological innovation has become an important engine for

D. Qi (✉)
China National Institute of Standardization, Beijing, China
e-mail: duanqi@cnis.gov.cn

V. Jain et al. (eds.), *Recent Trends in Intelligent Computing, Communication and Devices*, Advances in Intelligent Systems and Computing 1031,
https://doi.org/10.1007/978-981-13-9406-5_31

economic development of contemporary enterprises. Research and Development (R&D) is the important indicator of technological innovation, R&D input has decisive effects on technological innovation, and it is also of great significance for enterprises' technological progress and promotion of brand value [4, 5]. Thus, to research function mechanism of R&D input for enterprise innovation system, further explore and analyze relation and degree of different influences brought for enterprises' brand value by R&D input, and to put forward solutions and suggestions are of practical significance for technological progress of innovation system in our country and promotion of brand value.

2 Literature Review

Research on influences of R&D input on brand value is to research the relation between innovation input and performance output from aspect of "input–output". According to previous researches in relation to these two factors, R&D input is not positively or negatively correlated with enterprise performance in the current phase but reflects its obvious hysteretic nature [6–8]. In the third year of lag phase, these two factors show obvious positive correlation [9]. Besides, the interaction between two factors is influenced by some other control variables such as financing system [10] and development phase of enterprises [11].

Based on previous researches, the thesis is aimed at exploring influences of the third relation variable (coupling of enterprise innovation system) on interaction relation of R&D input and brand value. Research is aimed at finding out the nonlinear interaction relationship between R&D input and enterprises' brand value by focusing on changes in coupling of enterprise innovation system. Based on brand value evaluation and data in our country from 2013 to 2017, the panel regression model of Hansen is adopted to detect and analyze whether influences of R&D input into enterprise innovation system in different coupling stages on brand value have threshold effects.

3 Research Design

3.1 Variable Selection

In the thesis, coupling of enterprise innovation system is deemed as threshold variable. The coupling reflects degree of mutual influences between subjects or main factors of enterprise innovation system. From aspect of synergetic theory, the synergistic effects of various system parameters are the key to turn the whole system from unordered to orderly status, and coupling is the measurement of synergistic effects. The coupling of enterprise innovation system refers to the innovation subjects in enterprise innovation system, namely R&D center, post-doctorate R&D base, national key laboratories, etc. With the guidance of government policy and strategy, the innovation subjects will utilize resources provided by government and financial institutions, etc., and closely cooperate to ensure enterprise innovation system is in dynamic and benign synergy. From aspect of system theory, enterprise innovation system is divided into innovation subsystems and innovation support subsystems. Innovation subsystem consists of two dimensionalities such as innovation input and innovation output in which innovation input is measured by R&D input, expenditure of new product development, technology acquisition, and transformation cost, and innovation output is measured by sales income of new products and number of patents. Innovation support subsystem consists of two dimensionalities such as government support and market support in which government support is measured by proportion of amount of government in internal expenditure of R&D fund and scientific and technological expenditure in total financial expenditure, and market support is measured by transaction volume of technology market. The method of factor analysis is adopted to verify structure validity of indicator system, and variable coefficient is used to determine weight of various indicators. Besides, coupling measurement model is used to calculate coupling of each enterprise innovation system.

R&D input mainly consists of enterprises' input in personnel, finance, materials, information, and ideas. In all these items, it is easy to acquire the information about input into personnel and finance which is objective, quantifiable, and measurable. In most of previous researches, R&D input is mainly measured according to expenditure and personnel input into R&D. R&D expenditure input refers to the expenditure input for research and development. It is the foundation of technological innovation and directly reflects various enterprises' support of research and development as well as other innovation activities. R&D personnel input refers to personnel who are engaged in research and development and other innovation activities, and it is the powerful pusher and basic element of technological innovation activities. In general, scholars will express R&D expenditure input, R&D personnel input with R&D expenditure, R&D expenditure input rate and full-time equivalent of R&D personnel. Nevertheless, R&D input shows hysteretic nature in influencing brand value. In other words, R&D input may have no influences on value in the current phase but has influences of varying degrees on value in a period in the future. In the thesis, the variable used to express brand value is the calculated value of brand value, and it can be deemed as the product caused by R&D input. Moreover, the increase or decrease of value can be also deemed as the product after commercialization of research and development innovation. Besides, to further correct the hysteretic nature, R&D expenditure input in previous three phases is deemed as the core variable to explain quantity of granted patent and sales income of new products in the thesis.

In terms of control variable selection, the scale and structure of enterprise innovation system as well as capacity of government support will also influence brand value of enterprises. As for industry scale, frequently selected indicators include operating income, total number of employees, total assets, etc. In the thesis, the total asset of enterprises is deemed as the variable to express scale of enterprise innovation system. Besides, number of enterprises in the market is often used by scholars to measure market structure. According to division of market structure from aspect of economics, the number of enterprises reflects fierce degree of market competition and it is vital for research and development of enterprises.

3.2 Model Establishment

Enterprise innovation system coupling in the research is selected as threshold variable. Based on the measurement of coupling by establishing measurement model of enterprise innovation system coupling, establishing threshold regression model explores influence of R&D input on enterprise brand value in different coupling phases of enterprise innovation system.

3.2.1 Coupling Measurement Model Establishment

(1) Efficiency function of enterprise innovation system

Suppose that variable $u_i(i = 1, 2, \ldots, n)$ is order parameter of enterprise innovation system, namely innovation subsystem and innovation support subsystem. u_{ij} is the j-th index of the i-th-order parameter, and the value is $X_{ij}(i = 1, 2, \ldots, n)$. A_{ij} and A_{ij} are upper limit value and lower limit values of order parameter, and then, efficiency function is:

$$u_{ij} = \begin{cases} \dfrac{X_{ij} - B_{ij}}{A_{ij} - B_{ij}}, u_{ij} \text{ has positive efficiency} \\ \dfrac{A_{ij} - X_{ij}}{A_{ij} - B_{ij}}, u_{ij} \text{ has negative efficiency} \end{cases} \quad (1)$$

where u_{ij} is efficiency contribution of variable X_{ij} to system. u_{ij} reflects satisfaction degree of all indexes reaching objectives. $u_{ij} \in [0, 1]$, and it is more satisfied if the value is closer to 1.

Linear weighting method is adopted to calculate contribution degree of all indexes to system in innovation subsystem and innovation support subsystem, as shown below:

$$u_i = \sum_{j=1}^{m} w_{ij}\mu_{ij} \quad (2)$$

In the equation, u_i is contribution of subsystem to degree of order of overall system, m is number of index of subsystem, and w_{ij} is weight of all indexes, and then:

$$\sum_{i=1}^{n}\sum_{j=1}^{m} w_{ij} = 1 \quad (3)$$

Analytic hierarchy process method, entropy evaluation method, or variable coefficient method can be adopted to determine w_{ij}. As entropy evaluation method and variable coefficient method are objective assignment methods, weight index is calculated according to objective data fully, which avoids deviation caused by subjective factors, and calculation of variable coefficient method is simpler compared with that of entropy evaluation method, so variable coefficient method is adopted in the research to determine all index weights. Steps of determining index weights are by variable coefficient method:

① is used to calculate variable coefficient θ of all indexes.

$$\theta_{ij} = \frac{\sigma_{ij}}{\bar{x}_{ij}}, \quad i = 1, 2, \ldots, n, \quad j = 1, 2, \ldots, m \quad (4)$$

where $\bar{x}_{ij}$ and σ_{ij} denote average value and standard deviation of observed index of evaluation indexes, respectively. If ratio between standard deviation and average value of the index and is high, respectively, namely the variable coefficient is higher, it shows that difference shown by different enterprises in the evaluation index is also greater. Therefore, the evaluation index should be given a higher weight; on the contrary, if the variable coefficient is lower, then difference shown by enterprises is also smaller and weight is smaller correspondingly.

② normalizes variable coefficient θ and weights of all indexes are obtained.

$$w_{ij} = \frac{\theta_{ij}}{\sum_{j=1}^{m} \theta_{ij}}, \quad i = 1, 2, \ldots, n, \quad j = 1, 2, \ldots, m \quad (5)$$

(2) Measurement model of enterprise innovation system coupling

Coupling calculation of enterprise innovation system refers to capacity coupling concept and capacity coupling coefficient model in physics. The research promotes it to coupling measurement model of multiple systems, which reflects overall efficiency of enterprise innovation system coupling. Considering practical significance of enterprise innovation coupling, coupling measurement model is indicated as:

$$C' = \sqrt{CT} \quad (6)$$

where C' is coupling of enterprise innovation system, C is coupling in physics, and T is comprehensive subsystem regulation index of enterprise innovation system, which reflects overall synergy effect of enterprise innovation system. The calculation equation is as follows:

$$C = n^n \sqrt{(u_1 u_2 \ldots u_n) / \prod (u_i + u_j)} \tag{6}$$

$$T = au_1 + bu_2 \tag{7}$$

where a and b are undermined coefficients and should be determined after suggestions of industry experts of the enterprise are obtained.

3.2.2 Threshold Regression Model Establishment

Dual-threshold regression model is established according to threshold regression model and research contents of the paper:

$$y_{it} = \varphi_1 x_{it} I(q_{it} \leq \gamma_1) + \varphi_2 x_{it} I(\gamma_1 < q_{it} \leq \gamma_2) + \varphi_3 x_{it} I(q_{it} > \gamma_2) + \varphi' x'_{it} + u_i + e_{it} \tag{8}$$

where y_{it} is enterprise brand value;
x_{it} is R&D fund input of three phases in advance;

Control variable x'_{it} is total output value of enterprises, the number of enterprise R&D centers involving in R&D activities, expenditures for science and technology in financial allocations, R&D fund input of the current phase, number of enterprise, and so on;

q_{it} is threshold variable;
I is indicative function;
φ_1, φ_2, φ_3, and φ' are parameters to be estimated.

4 Empirical Research

In the study, the data from 2013 to 2017 of more than 3,242 enterprises were collected for the empirical research. All the data derive from the annual evaluation activities targeting China's brand value. All the data were processed on the standard basis prior to the development of threshold regression, aiming to eliminate dimensional differences.

4.1 Measurement of the Coupling of Enterprise Innovation System

The coupling of enterprise innovation system was measured based on the established coupling measurement model for regional innovation system and as described in Eqs. (1) to (7), and some examples of measurement results are shown in (Table 1).

4.2 Results of Threshold Regression Taking Brand Value as Dependent Variable and Relevant Analysis

The threshold regression was carried out by taking brand value as dependent variable, R&D fund input of 3 phases in advance as core explanatory variable, the coupling of enterprise innovation system as threshold variable, and total output value of enterprises, R&D fund input of the current phase, the number of enterprise R&D centers involving in R&D activities and expenditures for science and technology in financial allocations as control variables. Single-threshold model, dual-threshold model, and triple-threshold model were subject to inspection separately. Table 2 shows the results for threshold value estimation and threshold effect inspection with brand value as dependent variable.

Table 1 Examples of measurement results of enterprise innovation system coupling

	Year				
	2013	2014	2015	2016	2017
Enterprise 1	0.499	0.505	0.506	0.499	0.537
Enterprise 2	0.345	0.352	0.382	0.381	0.417
……	……	……	……	……	……
Enterprise 3242	0.371	0.384	0.364	0.356	0.387

Table 2 Results for threshold value estimation and threshold effect inspection with brand value as dependent variable

	Estimated threshold value	*P*	Number of BS	95% confidence interval
Single-threshold model	0.662	0.05	300	[0.620, 0.662]
Dual-threshold model	0.349 0.620	0.000	300	[0.274, 0.483] [0.593, 0.620]
Triple-threshold model	0.257	0.840	300	[0.257, 0.479]

Note P value is the result originating from repeated sampling for 300 times by means of Bootstrap method

Table 3 Results for threshold model coefficient and relevant inspection with brand value as dependent variable

Single threshold			Dual threshold		
	Coef	*P*		Coef	*P*
Total output value of enterprises	0.219	0.149	Total output value of enterprises	0.700	0.000
Number of R&D institutes	0.205	0.004	Number of R&D institutes	0.291	0.000
R&D input of the current phase	−0.381	0.051	R&D input of the current phase	−0.640	0.001
Ste	0.450	0.000	Ste	0.530	0.000
R&D input of 3 phases in advance (innovation system coupling <=0.662)	0.192	0.288	R&D input of 3 phases in advance (innovation system coupling <=0.349)	−1.085	0.004
R&D input of 3 phases in advance (innovation system coupling >0.662)	0.412	0.011	R&D input of 3 phases in advance (0.349 < innovation system coupling <=0.620)	−0.185	0.314
			R&D input of 3 phases in advance (innovation system coupling > 0.620)	0.303	0.042
Cons	−0.604	0.000	Cons	−0.931	0.000

The estimated threshold value of single-threshold model is 0.662, while 0.349 and 0.620 for dual-threshold model. P value of single-threshold model and dual-threshold model is 0.05 and 0.000, respectively. These two models are significant at the significance level of 0.05 while triple-threshold model is not significant. Thus, the study makes an analysis based on dual-threshold model. Table 3 shows the results for threshold model coefficient and relevant inspection with brand value as dependent variable.

Taking the coupling of enterprise innovation system as threshold variable, the nonlinear relationship between R&D input of three phases in advance and brand value is obvious. When the coupling of the enterprise innovation system is less than 0.349, the corresponding *P* value is 0.004. R&D input of three phases is negatively correlated with the brand value at the significance level of 0.05. For 1 unit of increase in internal R&D input of three phases in advance, the brand value will decline by 1.085 units. It indicates that R&D input has not been effectively utilized, which is inseparable from the effectiveness of cooperation among and management of the main body of the enterprise innovation system, namely the R&D center, the key laboratory, and center for post-doctoral studies. When the coupling of the enterprise innovation system ranges from 0.349 to 0.620, R&D input of three phases is still negatively correlated with the brand value. However, the significance level under such condition is 0.314, meaning the negative correlation is not significant. When the coupling of the enterprise innovation system exceeds 0.620,

R&D input of three phases is positively correlated with the brand value to a great extent at the significance level of 0.05. For 1 unit of increase in internal R&D input of three phases in advance, the brand value will climb by 0.303 units. Such significant positive correlation indicates that the main body of enterprise innovation system featuring high coupling is capable of utilizing R&D funds in a more effective way to facilitate innovation in technological capability during the process of technological innovation, thus realizing the improvement of enterprise profitability. It is obvious that the brand value realizes technology-driven growth under such condition. At the significance level of 0.05, all control variables, namely the total output value, R&D input of the current phase, the number of enterprise R&D centers involving in R&D activities, and expenditures for science and technology in financial allocations, are positively correlated with the brand value, facilitating the improvement of the brand value.

In terms of dual-threshold model with brand value as dependent variable, statistics was made concerning the threshold value distribution covering over 2,000 enterprises participating in brand value evaluation activities from 2013 to 2017. It is found that enterprise innovation system with coupling less than 1 threshold value of 0.349, between two threshold values of 0.349 and 0.620, and greater than two threshold values of 0.620 accounts for 38.7, 41.9 and 19.4%, respectively.

5 Conclusion

In the study, brand value is taken as dependent variable, the coupling of enterprise innovation system as threshold variable, and total output value of enterprises, the number of enterprise R&D centers involving in R&D activities, expenditures for science and technology in financial allocations, and R&D fund input of the current phase as control variables. The sample data are collected from the evaluation results of enterprises participating in brand value evaluation activities from 2013 to 2017. Threshold regression of data is conducted by software of Stata 13.0, and the conclusion is as follow:

(1) There is no constant negative correlation between R&D input and brand value. When the coupling of enterprise innovation system is less than 0.349, R&D input is negatively correlated with brand value; when the coupling of enterprise innovation system is greater than 0.620, there is strong positive correlation between R&D input and brand value at the significance level of 5%.
(2) There are relatively large proportions of enterprises with the innovation system where R&D input is not positively correlated with brand value. Enterprises whose R&D input of three phases in advance is positively correlated with brand value only accounts for 19.4%.

According to the results of empirical research, suggestions raised in the study are as follows:

(1) In the light that there is not constant negative correlation between R&D input of the enterprise innovation system and brand value, all enterprises are encouraged to take the initiative to implement the reform in the management model and organizational structure of R&D center, laboratory, and center for post-doctoral studies, develop strategies for technological innovation and industrial commercialization, optimize resource allocation, improve the efficiency of R&D input, strengthen industry-university-research alliance and further promote the leading role of technological innovation in their profitability;
(2) Enterprises with innovation system featuring relatively low coupling is advised to take actions to strengthen the cooperation and interaction among the main body, facilitate the exchanges and communication among R&D personnel, foster a spirit of innovation featuring opening and cooperation, and mitigate the negative impacts R&D input has on brand value;
(3) The government is advised to give full play to its macro-control function, optimize the input object and quota of innovation resources by allocating resources to the main body or fields of innovation with highest efficiency to the greatest extent, and further facilitate prominent enterprises to improve their capability in technological innovation and innovation system.

Acknowledgements This research was financially supported by China Central Public-interest Scientific Institution Basal Research Fund (Grant No.: 552018Y-5930) and National Key Technology Support Program (Grant No.: 2015BAK46B02 & 2015BAK46B03-3).

References

1. Liu, H.: Research on regional R&D efficiency and its influencing factors in China—an empirical analysis based on stochastic Frontier function. Sci. Res. **29**(4), 549–556 (2011)
2. Xiangbin, W., Guangdong, Y., Weizhong, Y.: Research on the game mechanism of dynamic evolution of regional innovation system. Res. Manag. **11**(33), 1–8 (2011)
3. Wei, C., Zhijun, F., Xin, K., et al.: Research on coordination development measurement and evaluation of regional innovation system—based on the perspective of duality theory. Sci. Sci. Res. **2**(29), 306–313 (2011)
4. Du, L., Miao, C.: The influence of R&D input on the technological innovation performance of regional innovation system—an empirical analysis based on panel data in eastern region. J. Anhui Univ. Sci. Technol. (Soc. Sci. Ed) **7**, 36–40 (2015)
5. Wan, Y.: Regional R&D performance and its relationship with economic growth—based on the perspective of literature research. J. Technol. Econ. Manag. **1**, 118–122 (2014)
6. Lu, Y., Wang, C.: Research on the Impact of R&D investment on the performance of listed companies—taking manufacturing and information technology as examples. Sci. Technol. Manag. Res. **5**, 122–127 (2011)

7. Zhao, Y., Xu, M.: Research on the Impact of R&D investment on firm performance—based on panel data of listed companies in the yangtze river delta from 2006 to 2010. Sci. Technol. Manag. Res. **12**, 95–98 (2013)
8. He, D.: Research on the relationship between financing constraints, R&D investment and enterprises—an empirical analysis of listed companies based on communication and related equipment manufacturing industry. Sci. Technol. Ind. **5**, 38–43 (2015)
9. Zheng, X.D., Huang, Y., Zhuang, Y.: Research on the impact of R&d investment on corporate performance—an empirical analysis based on listed companies in communication and related equipment manufacturing industry. Sci. Technol. Ind. **5**, 38–43 (2015)
10. Karjalainen, P.: R&D investments: the effects of efferent financial environments on firm profitability. J. Multinat. Financial Manag. **2**, 79–93 (2008)
11. Liang, L., Jin, Y., Zhao, N.: Research on the relationship between R&D input and firm performance based on enterprise life cycle—experience data from listed companies. Sci. Technol. Manag. **31**(12), 11–35 (2012)

Global Analysis of a Class of SIRS Models with Constant Input and Nonlinear Infectivity

Fei Wang

Abstract A SIRS epidemic model is established for the case of nonlinear infection rate, constant immigration and exponential birth, and the immunity acquired by individuals disappeared after a period of time. The threshold conditions for the existence of the equilibrium point are discussed. With the help of Hurwitz criterion and Bendixson theorem, Lyapunov function is used to discuss the global asymptotic stability of positive equilibrium point.

Keywords Nonlinear infection · Equilibrium · Asymptotic stability

1 Introduction

The SIR model assumes that the infected individuals will acquire permanent immunity after recovery. This model is commonly used in the transmission of viral hepatitis, varicella, measles, and AIDS. However, the immunity gained by many diseases will disappear after a period of time. The SIRS model describes this situation. Research literature [1–5] on models with nonlinear infectious force has gained practical value.

The infection coefficient $\beta(N)$, meanwhile, the incidence [6–10] of the disease is $\beta(N)SI$combined with constant immigration and exponential birth, and unit time lost immunity ratio μ, the establishment of SIRS nonlinear infectious disease model. Then we study the global asymptotic behavior of its dynamics.

F. Wang (✉)
Department of Applied Mathematics and Military Cryptography,
Air Force Engineering University, 710051 Xi'an, Shaanxi, China
e-mail: wangfeishuxue@163.com

V. Jain et al. (eds.), *Recent Trends in Intelligent Computing, Communication and Devices*, Advances in Intelligent Systems and Computing 1031,
https://doi.org/10.1007/978-981-13-9406-5_32

2 Model Description

Assuming that the infection coefficient β is a function of the total population N, satisfying $\beta(N) > 0, \beta'(N) \leq 0, \ (N\beta(N))' \geq 0$; A is the total number of people moving from outside, which $a > 0, b \geq 0, c \geq 0$ and $a + b + c = 1$.

S input susceptible person, I infected and R Proportion of Moving out, respectively. Meanwhile $d > 0, \alpha > 0, k > 0 \, (k < d)$ are natural mortality, morbidity, and birth rate, respectively. In addition, $\mu > 0$ the proportion of immunity lost per unit time $r > 0$ indicates recovery rate, and assuming that newborn babies are all susceptible. Based on the above assumptions, the following models can be established

$$\begin{cases} \frac{\mathrm{d}s}{\mathrm{d}_t} = kN + aA + \mu R - \beta(N)SI - \mathrm{d}S \\ \frac{\mathrm{d}_I}{\mathrm{d}_t} = bA + \beta(N)SI - (d + \alpha + r)I \\ \frac{\mathrm{d}_R}{\mathrm{d}_t} = cA + rI - (\mu + d)R \end{cases} \tag{1}$$

Let $N = S + I + R$, the total population equation is

$$\frac{\mathrm{d}N}{\mathrm{d}t} = A - (d - k)N - \alpha I \tag{2}$$

Putting $S = N - I - R$ into the second equations of the system (1) available equivalent system

$$\begin{cases} \frac{\mathrm{d}_I}{\mathrm{d}_t} = bA + I[\beta(N)(N - I - R) - (d + \alpha + r)] \\ \frac{\mathrm{d}_R}{\mathrm{d}_t} = cA + rI - (\mu + d)R \\ \frac{\mathrm{d}_N}{\mathrm{d}_t} = A - (d - k)N - \alpha I \end{cases} \tag{3}$$

3 Model Analysis

In this paper, a model is researched for those inputting with no infectious diseases, that is $\mathbf{b} = \mathbf{0}$. At this point, the system (3) is deformed into

$$\begin{cases} \frac{\mathrm{d}_I}{\mathrm{d}_t} = I[\beta(N)(N - I - R) - (d + \alpha + r)] \\ \frac{\mathrm{d}_R}{\mathrm{d}_t} = cA + rI - (\mu + d)R \\ \frac{\mathrm{d}_N}{\mathrm{d}_t} = A - (d - k)N - \alpha I \end{cases} \tag{4}$$

Lemma

(1) If $I(0) = 0$, then $t > 0, I(t) \equiv 0, \lim_{t \to +\infty} R(t) = \frac{cA}{\mu + d}$;

(2) If $I(0) > 0$, then $I(t) > 0, \Omega = \left\{(I, R, N) \in R_+^3 \,|I + R < N \le \frac{A}{d-k}\right\}$ is a positive invariant set of (4);

(3) The system (4) always has a disease-free equilibrium $P_0\left(0, \frac{cA}{\mu+d}, \frac{A}{d-k}\right)$ and the basic reproductive number is $R_0 = \beta\left(\frac{A}{d-k}\right)\frac{A}{d-k}\frac{(\mu+d)+c(k-d)}{(\mu+d)(\alpha+d+r)}$.

Theorem 1 If $R_0 > 1$, there is a unique endemic equilibrium of system (4) $P_1(I_1, R_1, N_1)$, which $I_1 = \frac{A+(k-d)N_1}{\alpha}, R_1 = \frac{(\alpha c + r)A + r(k-d)N_1}{\alpha(\mu+d)}$, N_1 is the only positive root between $\left(0, \frac{A}{d-k}\right)$ of (5).

$$\begin{aligned} &\beta(N)\{[(\mu+d)(\alpha-k+d) - r(k-d)]N - (\mu+d+\alpha c+r)A\} \\ &\quad = \alpha(\mu+d)(d+\alpha+r) \end{aligned} \tag{5}$$

Proof Let

$$\begin{aligned} \varphi(N) = {} & \beta(N)\{[(\mu+d)(\alpha-k+d) - r(k-d)]N \\ & -(\mu+d+\alpha c+r)A\} - \alpha(\mu+d)(d+a+r) \end{aligned}$$

then $\varphi'(N) = (\beta(N)N)'[(\mu+d)(\alpha-k+d) \; -r(k-d)] - \beta'(N)(\mu+d+\alpha c+r)A$, according to $\beta(N) > 0, \beta'(N) \le 0$ and $(N\beta(N))' \ge 0$, obtaining $\varphi'(N) \ge 0$, that is $\varphi(N)$ is monotonous increasing.

Meanwhile $\varphi(0) = -\alpha(\mu+d)(d+\alpha+r) < 0, \varphi\left(\frac{A}{d-k}\right) > 0$, therefore, (5) is a unique positive root in $\left(0, \frac{A}{d-k}\right)$.

3.1 *Local Stability of Equilibrium Point*

Theorem 2 If $R_0 < 1$, it is asymptotically stable at disease-free equilibrium point P_0; if $R_0 > 1$, the equilibrium point P_0 is instability, endemic equilibrium point $P_1(I_1, R_1, N_1)$ is locally asymptotic stability.

Proof The Jacobi matrix of the system (4) at the equilibrium point P_0 is

$$J(P_0) = \begin{bmatrix} \beta\left(\frac{A}{d-k}\right)\left(\frac{A}{d-k} - \frac{cA}{\mu+d}\right) - (d+\alpha+r) & 0 & 0 \\ r & -(\mu+d) & 0 \\ -\alpha & 0 & k-d \end{bmatrix}$$

And the three characteristic roots of $J(P_0)$ are respectively:

$$\lambda_1 = \beta\left(\frac{A}{d-k}\right)\left(\frac{A}{d-k} - \frac{cA}{\mu+d}\right)$$
$$- (d+\alpha+r) = (\alpha+d+r)(R_0 - 1);$$
$$\lambda_2 = -(\alpha+d); \lambda_3 = k-d.$$

Because of $\alpha > 0, d > 0$, when $R_0 < 1$ the disease-free equilibrium P_0 is locally asymptotic stability; then $R_0 > 1$ the equilibrium point P_0 is unstable.

The Jacobi matrix of the system (4) at the equilibrium point P_1 is

$$J(P_1) = \begin{bmatrix} -I_1\beta(N_1) & -I_1\beta(N_1) & -I_1\left[\frac{\beta'(N_1)}{\beta(N_1)}(d+\alpha+r)+\beta(N_1)\right] \\ -r & -(\mu+d) & 0 \\ -\alpha & 0 & k-d \end{bmatrix}$$

The characteristic equation of $J(P_1)$ is $\lambda^3 + m\lambda^2 + n\lambda + p = 0$, which $m = [\mu + d + I\beta(N) + d - k]$;

$$n = [I\beta(N)(\mu+d+r) + (d-k)(\mu+d$$
$$+ I\beta(N)) + \alpha I\left(\frac{\beta'(N)}{\beta(N)}(d+r+\alpha) + \beta(N)\right)\Bigg];$$
$$p = I\beta(N)(\mu+d+r)(d-k)$$
$$+ \alpha I(\mu+d)\left[\frac{\beta'(N)}{\beta(N)}(d+r+\alpha) + \beta(N) = 0\right]$$

because of $m > 0, n > 0, p > 0$,

$$mn - p = I\beta(N)(\mu+d+r)(\mu+d+I\beta(N)))$$
$$+ (d-k)(\mu+d+I\beta(N)+d-k)(\mu+d+I\beta(N))$$
$$+ \alpha I(I\beta(N)+d-k).\left[\frac{\beta'(N)}{\beta(N)}(d+r+\alpha) + \beta(N)\right] > 0,$$

According to the Routh-Hurwitz discriminant method, when $R_0 > 1$ endemic equilibrium point $P_1(I_1, R_1, N_1)$ is locally asymptotic stability.

3.2 Global Stability of Equilibrium Points

Theorem 3 If $R_0 < 1$, the disease-free equilibrium P_0 is global asymptotical stability in the region Ω, when $R_0 > 1$ the positive equilibrium point $P_1(I_1, R_1, N_1)$ is globally asymptotically stable.

Proof If $\lim_{t \to +\infty} I(t) = 0$, the limit system of system (4) is

$$\begin{cases} \frac{dR}{d_t} = cA - (\mu + d) = P(R, N) \\ \frac{dN}{d_t} = A - (d - k)N = Q(R, N) \end{cases} \quad (6)$$

The positive invariant set of system (6) is $\Omega_0 = \{(R,\ N)|R \geq 0, R < N \leq \frac{A}{d-k}\}$, and $\frac{\partial P(R,N)}{\partial R} = -(\mu + d)$, $\frac{\partial Q(R,N)}{\partial N} = -(d - k)$, that is $\frac{\partial P(R,N)}{\partial R} = -(\mu + d) - (d - k) < 0$.

Because of the Bendixson rule, there is no periodic solution of system (6) in Ω_0. So it is globally asymptotically stable at the only equilibrium point P_0.

On the other hand, if $R_0 > 1$, the endemic equilibrium $P_1(I_1, R_1, N_1)$ is globally asymptotically stable.

Definition: Lyapunov function

$$V(I, R, N) = \left(I - I_1 - I_! \ln \frac{I}{I_1}\right) + \frac{\beta(N_1)(R - R_1)^2}{2r} + \frac{\beta(N_1)(N - N_1)^2}{2\alpha},$$

Then there is

$$\begin{aligned} \left.\frac{dV}{dt}\right|_{(4)} &= -\beta(N_1)(I - I_1)^2 - \frac{\beta(N_1)(\mu + d)}{r}(R - R_1)^2 \\ &\quad - \frac{\beta(N_1)(d - k)}{\alpha}(N - N_1)^2 \\ &\quad + (\beta(N) - \beta(N_1))(N - I - R)(I - I_1) \end{aligned}$$

because of $d > k, \beta(N) > 0, \beta'(N) \leq 0$, so $\left.\frac{dv}{d_t}\right|_{(4)}$ is negative definite about (I_1, R_1, N_1). The positive equilibrium point of system (4) is globally asymptotically stable.

4 Concluding Remarks

In this paper, the SIRS model of infectious diseases with constant immigration and exponential birth and disappearance of immunity acquired by individuals after a period of time is analyzed dynamically. The results show that there are always disease-free equilibrium points, and $R_0 < 1$ the system is globally asymptotically stable; there is only endemic equilibrium point and $R_0 > 1$ the system is globally asymptotically stable. Because immigrants may bring infected persons, it will be necessary to further study the infectious disease models.

References

1. Wang, L.: Global analysis of two types of epidemic models with nonlinear incidence rate. J. Eng. Math. **22**(4), 640–644 (2005)
2. Ruan, S., Wang, W.: Dynamical behavior of an epidemic model with a nonlinear incidence rate. Diff Equs **188**(1), 135–163 (2003)
3. Cheng, X., Hu, Z.: A class of epidemic models with nonlinear infectious diseases. Pract. Cogn. Math. **39**(7), 118–122 (2009)
4. Wang, L., Li, J.: Qualitative analysis of a class of SEIS epidemic model with nonlinear infection rate. Appl. Math. Mech. **27**(5), 591–596 (2006)
5. Lin, Z.Z., Dong, L., Li, X.: A class of SIRI epidemic model with constant input and vertical transmission. Pract. Cogn. Math. **41**(5), 156–164 (2011)
6. Zhang, J., en Ma, Z.: Global analysis of SEI epidemic models with constant input in each compartment. J. Xi'an Jiaotong Univ. **37**(6), 653–656 (2003)
7. Korobeinikov, A.: Lyapunov functions and global stability for SIR and SIRS epidemiological models with non-linear transmission. Bull. Math. Biol. **30**, 615–625 (2006)
8. Cai, L., Guo, S., Wang, S.: Analysis of an extended HIV/AIDS epidemic model with treatment. Appl. Math. Comput. **236**, 621–627 (2014)
9. Cai, L., Li, X., Ghosh, M., Guo, B.: Stability analysis of an HIV/AIDS epidemic model with treatment. J. Comput. Appl. Math. **229**, 313–323 (2009)
10. Hyman, J.M., Li, J.: The reproductive number for an HIV model with differential infectivity and staged progression. Linear Algebra Appl. **398**, 101–116 (2005)

A Novel Method for Touchless Palmprint ROI Extraction via Skin Color Analysis

Qin Li, Hong Lai and Jane You

Abstract In order to solve the security problems caused by smart mobile terminals, this work designed an effective ROI extraction scheme in order to extract the ROI of touchless palmprint. The proposed method is based on image difference technology, skin color model, and simple interactive operations, which can eliminate environmental influences such as illumination, accurately implement palmprint segmentation, and realize live detection; the ROI obtained by this scheme is invariant to rotation, translation, and scaling transformations.

Keywords Smart mobile terminal · Touchless palmprint · ROI extraction
Skin color analysis

1 Introduction

In recent years, smart mobile terminals such as smartphones and iPads have become more and more popular in daily life. At the same time when people enjoy the convenience brought from these mobile devices, the fraud and identity theft arising from which has also increased rapidly. Especially with the promotion of mobile Internet applications such as mobile banking and e-commerce, the security issues that constrain its development have become more prominent. It can be said that if the security problems of smart terminals can be solved properly, applications such as mobile banking will realize huge and further development.

Biometric identification technology can complement or replace traditional identity authentication and improve the security of the mobile Internet. Compared to other biometric identification, touchless palmprint recognition has many advantages, including security, ease of use, applicability, hygiene.

Q. Li · H. Lai (✉)
School of Engineering, Shenzhen Institute of Information Technology, Shenzhen, China
e-mail: laih@sziit.edu.cn

J. You
Department of Computing, The Hong Kong Polytechnic University, Hung Hom, Hong Kong

V. Jain et al. (eds.), *Recent Trends in Intelligent Computing, Communication and Devices*, Advances in Intelligent Systems and Computing 1031,
https://doi.org/10.1007/978-981-13-9406-5_33

Even the touchless palmprint recognition has been studied since the year 2008 [1], the accuracy is still not satisfied. Different from the traditional palmprint recognition system [2], the position between the palm and the acquisition device cannot be fixed, which caused the rotation, translation, and scaling of the palm image. Further, the lighting condition cannot be controlled during the image acquisition. These problems caused the difficulty of touchless palmprint recognition.

The first problem of touchless palmprint recognition is the extraction of the region of interest (ROI). At present, the ROI extraction method is majorly classified into two types: One type is the central square area in the palm [3–7], and another type is the inner inscribed circle area in the palm [8–10].

This work follows the idea of the inner inscribed circle area. This type of ROI is invariant to rotation, translation, and scaling. However, because the lighting condition cannot be controlled during the acquisition, the palm image segmentation is difficult. We tried to solve this problem by skin color analysis. By designing a skin color model specially used for touchless palmprint, we obtained satisfied ROI.

2 Method

In order to solve the problem specifically in smart terminals, we developed an Android app to capture the palm image, see Fig. 1.

In order to extract the ROI of touchless palmprint, we designed an effective ROI extraction scheme based on image difference from simple interactive operations and a skin color model established from test samples, which can eliminate environmental influences such as illumination, accurately implement palmprint segmentation and live detection, extract the maximum inscribed circle in the segmented

Fig. 1 Touchless palmprint acquisition

palmprint image, and get the ROI irrelevant to the rotation and translation transformations. The main steps of the scheme are as follows:

(1) Remind the user to make a fist and then spread the palm to get two palm images, see Fig. 2a–b.
(2) Convert P_1 and P_2 to grayscale images P_1^g and P_2^g, see Fig. 2c–d, and use Gaussian filter to perform smoothing and then generate the difference image D of them. D is defined by Eq. (1), where * represents convolution, (x, y) represents the coordinates in the image, $g_\sigma(x,y) = \exp[-(x^2+y^2)/2\sigma^2]$ represents a Gaussian function, and the difference image is shown in Fig. 2e.

$$D(x,y) = \left|g_\sigma(x,y) * P_2^g(x,y) - g_\sigma(x,y) * P_1^g(x,y)\right| \quad (1)$$

(3) To conduct double threshold segmentation of the difference image, we set a low threshold t_l and a high threshold $t_h = 2t_l$. The angiogram obtained by the high threshold is used as a seed, and the angiogram obtained by the low threshold is used as a mask to perform morphological reconstruction, and the obtained segmentation result is used as a palmprint pre-segmented image P_{seg}^1. See Fig. 2f.
(4) Mark the palmprint area in the training image and convert to YC_bC_r chromaticity space, and establish a skin color model based on training samples P_S^{Train}, see Eq. (2); adopt palmprint pre-segmented image P_{seg}^1 to mark the palmprint area in the current test sample and convert to YC_bC_r chromaticity space, and establish a skin color model based on training samples P_S^{Test}, see Eq. (3). Where $x = \begin{bmatrix} C_b & C_r \end{bmatrix}^T$ represents the value of sample pixel in the YC_bC_r chrominance space, $m = E(x)$ represents the mean of the sample in the chromaticity space, $C = E\left[(x-m)(x-m)^T\right]$ represents the covariance matrix of sample chromaticity distribution. Use the mean $P_S = \left(P_S^{\mathrm{Train}} + P_S^{\mathrm{Test}}\right)/2$ of P_S^{Train} and P_S^{Test} as the final skin color model, and the skin color similarity is defined as $P_S^{\mathrm{Simility}} = P_S^i(C_b,C_r)/\,\mathrm{MAX}_i\left(P_S^i(C_b,C_r)\right)$, where i represents the pixel index in the current test samples. Conduct double threshold segmentation on P_S^{Simility}, and use morphological dilation to fill small holes inside to obtain palmprint segmentation images of the current test samples. See Fig. 2g.

$$P_s^{\mathrm{Train}}(C_b,C_r) = \exp\left[-0.5(x-m_{\mathrm{Train}})^T C_{\mathrm{Train}}^{-1}(x-m_{\mathrm{Train}})\right] \quad (2)$$

$$P_s^{\mathrm{Test}}(C_b,C_r) = \exp\left[-0.5(x-m_{\mathrm{test}})^T C_{\mathrm{test}}^{-1}(x-m_{\mathrm{test}})\right] \quad (3)$$

(5) Calculate the center of gravity of palmprint segmentation image P_{seg}^2 as a preliminary estimate of incenter. Start from the center of gravity to search for the maximum radius and determine the final center and radius, and map them to the original image to obtain the ROI, see Fig. 2h.

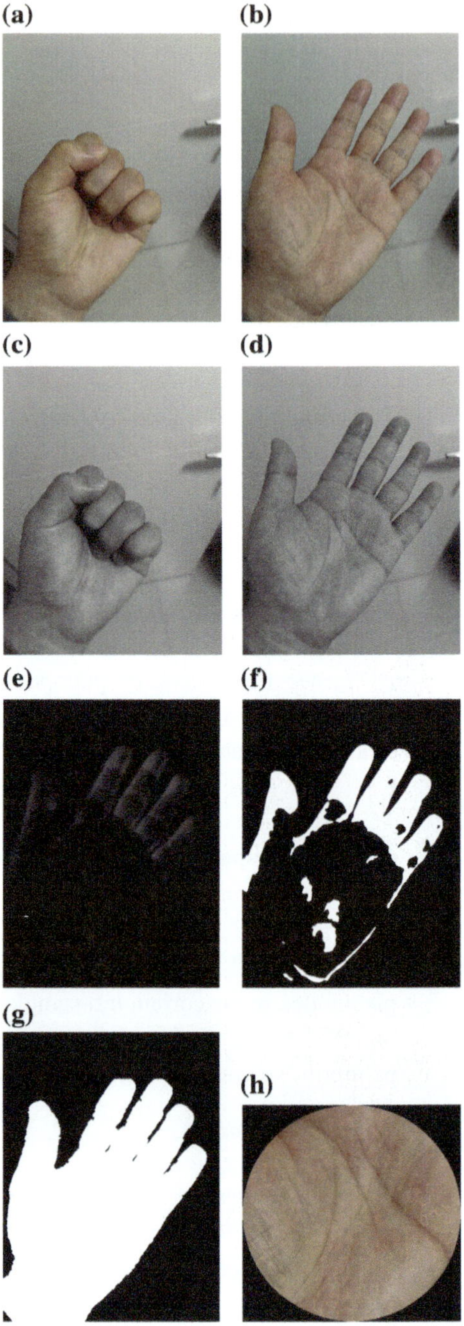

Fig. 2 Touchless palmprint ROI extraction

3 Experiments

We used two datasets to evaluate the performance of the proposed scheme: IndiaCM and MobilePalm. The IndiaCM includes palm images from 167 people, and each one has 8 images. The MobilePalm dataset was built using our self-designed Android app, which includes palm images from 53 people, and each one has 42 images. Images of the MobilePalm dataset were captured without any positioning constraint. Images of the IndiaCM dataset were captured in relatively fixed position with the help of some small poles. We applied the proposed scheme on IndiaCM dataset without the difference image calculation part. Figure 3 shows some examples of the extraction results. In order to evaluate the stability of the ROI, we hand-labeled the ROI of the two datasets and divided the distance between the hand-labeled center and extracted center by the diagonal length of the ROI image. The results are recorded in Table 1. We can see that the stability error is small.

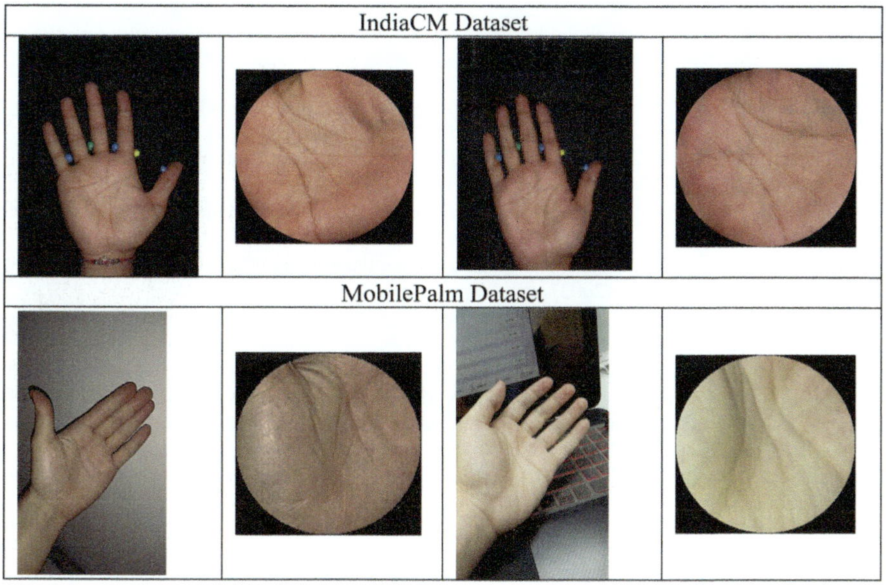

Fig. 3 Extraction examples

Table 1 ROI extraction stability of the proposed method

Dataset	Stability error
IndiaCM	0.015
MobilePalm	0.028

4 Conclusion

This work designed an effective ROI extraction scheme in order to extract the ROI of touchless palmprint. The proposed method is based on image difference technology, skin color model, and simple interactive operations, which can eliminate environmental influences such as illumination, accurately implement palmprint segmentation, and realize live detection; the ROI obtained by this scheme is invariant to rotation, translation, and scaling transformations.

Acknowledgements This work was supported by the Shenzhen Fundamental Research Fund under Grant No. JCYJ20160530141902978.

References

1. Goh, M.K.O., Tee, C., & Teoh, A.B.J.: Touch-less palm-print biometrics: novel design and implementation. Image Vision Comput. (2008)
2. Zhang, D., Kong, W., You, J., et al.: Online palmprint identification. IEEE Trans. Pattern Anal. Mach. Intell. **25**(9): 1041–1050 (2003)
3. Zhang, X.F., Zhang, Z.L., Xie, H.: Research and realization of palmprint ROI segmentation algorithm. Comput. Sci. **43**(11), 170–173 (2016)
4. Han, C., Cheng, H., Lin, C., et al.: Personal authentication using palm-print features. Pattern Recogn. **36**(2), 371–381 (2003)
5. Li, W., Zhang, D., Xu, Z.: Palmprint identification by Fourier transform. Int. J. Pattern Recogn. Artif. Intell. **16**(4), 417–432
6. Ramli, D.A., Ibrahim, S.: Evaluation on palm-print ROI selection techniques for smart phone based touch-less biometric system. Am. Acad. Sch. Res. J. (2013)
7. Puranik, A., Patil, R., Patil, V., Rane, M.: Touch less, camera based palm print recognition. Int. J. Appl. Res. Stud. (2013)
8. Li, W.-X., Xia, S.-X., Zhang, D.-P., et al.: A new palmprint identification method using bi-directional matching based on major line features. J. Comput. Res. Dev. **41**(6), 996–1002 (2004)
9. Wang, Y.-X., Ruan, Q.-Q.: A new preprocessing method of palmprint. J. Image Graph. **13**(6), 1115–1122 (2008)
10. Liu, G., Zhang, J.: Research on extraction algorithm of palm ROI based on maximum intrinsic circle. Int. Symp. Parallel Architect. Algorithm Program. (2017)

Face Detection Based on YOLOv3

Chong Li, Rong Wang, Jinze Li and Linyu Fei

Abstract Face detection is the precondition of various research fields, involving face recognition, face identification, face expression analysis, etc. The existing object detection methods, whether two-stage methods or one-stage ones, expect to balance speed and accuracy. Meanwhile, YOLOv3, as a popular object detection algorithm, has gained obvious advantages in both speed and accuracy. Thus, we migrated YOLOv3 to the face detection area and made some improvements to adjust it to the face detection problem, including changing the detection layer to detect smaller faces, choosing the Softmax as the loss function instead of the logistic classifier to maximize the difference of inter-class features, and decreasing the dimension of features on detection layers to improve the speed. This method was trained on the WIDER FACE database and the CelebA database and tested on the FDDB database. Experimental results have shown that the face detection method based on YOLOv3 has obtained great performance on small face, speed, and accuracy for the face detection task.

Keywords Face detection · YOLOv3 · Softmax

1 Introduction

Recently, facial analytics is becoming more and more notable in computer vision, artificial intelligence, and so on. Face detection, as the basis and precondition of face recognition, face identification, face expression analysis, etc., has gotten much attention from more and more researchers.

As the development of Deep Learning, on which the face detection methods based has surpassed traditional ones towards performance on speed, accuracy, transportability, etc.

C. Li (✉) · R. Wang · J. Li · L. Fei
College of Information Technology and Network Security, People's Public Security University of China, Beijing, China
e-mail: lichong7564@163.com

V. Jain et al. (eds.), *Recent Trends in Intelligent Computing, Communication and Devices*, Advances in Intelligent Systems and Computing 1031,
https://doi.org/10.1007/978-981-13-9406-5_34

Traditional ones are in reference to face detection algorithms based on designed features, like Haar [1], integrogram [2], histogram of oriented gradient (HOG) [3], etc., which are less representable and time-consuming.

Then, face detection methods have gotten great breakthrough with the object detectors based on modern convolutional neural networks (CNNs), including R-CNN [4–6], SSH [7], SSD [8], etc. Meanwhile, many researchers have proposed a list of approaches to raise the accuracy and speed of face detection. The cascade CNN [9], a cascade architecture built on CNNs, has shown powerful discriminative capability. Based on this cascade architecture, the MTCNN [10] that uses a deep cascaded multitask framework to improve face detection performance was proposed.

YOLOv3 is an object detector and developed from You Only Look Once (YOLO). YOLO [11] was first introduced in 2015 for object detection, which has shown its superiority on VOC2012 database [12], offering an easy trade-off between speed and accuracy. The system frames detection as a regression issue and output bounding boxes and class probabilities. YOLO is a universal, real-time object detection system with good general representations and is also easy to be constructed. However, YOLO cannot perform well when two objects are too close or there is a wide variety of aspect ratios, which makes numerous localization errors, leading to poor coordinate accuracy and low recall. To alleviate this problem, YOLO9000 has been proposed [13], which introduces anchor boxes to predict bounding boxes so that different input sizes can be adapted well. Though YOLO9000 model has good performance, the main issue with this approach is that it cannot handle with overlapping effectively. To mitigate the above issue and weigh speed and accuracy better, the YOLOv3 was proposed [14]. YOLOv3 runs significantly faster than many other detection methods with comparable performance.

As YOLOv3 has gained obvious advantages in both speed and accuracy on object detection, we migrate YOLOv3 to the face detection area and make some improvements to adjust it to the face detection task. Firstly, we use the backbone of YOLOv3, named Darknet-53 and change the detection layers to detect smaller faces effectively; Secondly, we choose the Softmax loss as the loss function instead of the logistic loss to maximize the difference of inter-class features; Finally, we decrease the dimension of features on detection layers to improve the speed.

In this paper, we presented a face detection method based on YOLOv3. The rest structure is organized as follows: Sect. 2 describes our method particularly. Section 3 illustrates the result and analysis of experiment. Finally, Sect. 4 summarizes the paper and explores future research.

2 Proposed Method

Our face detection approach is based on YOLOv3. YOLOv3 detects objects on three different scales {8,16,32}. For the first detection layer, the high-resolution and low-level features are gotten after some convolutional layers. For the second detection layer, the features are the combination of the 2* up-sampled features from the first detection layer and the features from the earlier network (with the same size) through a residual skip connection. Similarly, for the third detection layer, the low-resolution and high-level features are the combination of the 2* up-sampled features from the second detection layer and the much earlier network.

2.1 Feature Map

YOLOv3 extracts features from different network layers, using a similar concept to the feature pyramid networks (FPNs) [15]. The first detection layer features are high-resolution and low-level, containing great semantic information, which are benefit for the object location. In contrast, as for the low-level and high-resolution third detection layer features, they are fundamental for the detection of small objects. Thus, to improve the performance on small face detection, which is one of the most changeling difficulties on face detection task, we change the third detection layer by a 4* up-sampled layer instead of 2* and a residual skip layer toward much earlier network to get a feature map with higher resolution and lower level. Then, we get the detection scales of {8,16,64}. The framework for face detection is shown in Fig. 1.

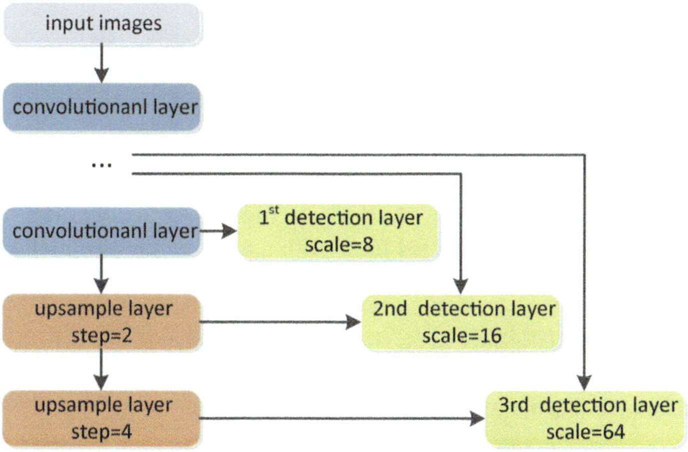

Fig. 1 Framework for face detection based on YOLOv3

Meanwhile, in this paper, we decrease the dimension of detection layer features from 75 to 18 when it comes to the speed of our method.

2.2 *Loss Function*

Object detection is a multilabel classification task and should handle overlapping labels in some complex fields, so YOLOv3 dispenses the Softmax loss function which assumes that each object has exactly one class. However, as for the face detection task, prediction results are either 'face' or 'not face,' which means the Softmax loss function is feasible in the face detection domain. Moreover, the Softmax loss could maximize the difference of inter-class features. Thus, we choose the Softmax loss as the loss function of our method, which is defined as follows:

$$L = -\frac{1}{M}\sum_{i=1}^{M}\log\left(\frac{e^{wy_i \cdot x_i + by_i}}{\sum_{j=1}^{c} e^{w_j \cdot x_i + b_j}}\right) \tag{1}$$

where M is the number of input images; x_i is the ith input image; c is the number of classes; w and b are the weights and bias, respectively.

3 Experiment

In this section, we train our CNNs on two benchmark databases: the recently released WIDER FACE database [16] and the CelebA database [17], and test it on the FDDB benchmark database [18].

3.1 *Databases*

WIDER FACE: WIDER FACE database is used for face detection, containing 32,203 images and 393,703 labeled faces with a high degree of variability in scale, pose, and occlusion, which makes it to be one of the most challenging public face databases.

CelebA: CelebA is a large-scale face attributes database, in which the images have various poses and messy background. It includes more than 200,000 images, and each one has been annotated with 40 attributions.

FDDB: FDDB is a database of face regions for unconstrained face detection task, which contains 5171 annotations for faces and 2845 images. FDDB includes color and grayscale images, and these faces present a variety of states like occlusion, rare poses, low resolution, and out of focus.

3.2 Training

We train our face detection model on the WIDER FACE database and the CelebA database by Python. Firstly, we combine the YOLO format databases and labels from WIDER FACE and CelebA database. After that, we adjust the parameters in darknet/cfg/yolov3.cfg as follows. For detecting smaller face, the input image is adapted to 1024 × 768 and the up-sample step on the third detection layer is set to 4. Considering the training time efficiency, we only set the dimension of features on detection layer to 18 and the epoch to 50K, which guarantees good results in a few hours in the 1080Ti GPU configuration environment. So, we do not need to increase the number of iterations. Furthermore, the initial learning rate is reduced to 0.0001 and declines separately again at 20K, 40K, and 45K iterations. Some qualitative detection results on training databases are shown in Fig. 2.

3.3 Results and Evaluation

We test our face detection model based on YOLOv3 trained before and the MTCNN method mentioned in Sect. 1 on the FDDB benchmark database. The results on the database FDDB are shown as Figs. 3, 4, and Table 1.

Our detection model has a good performance on FDDB database. As shown in Fig. 3, our method performs well on small face detection. As shown in Fig. 4 and Table 1, with 100 error samples, it can achieve 85% accuracy, and the highest accuracy can even reach 93.9%. It can be seen that our face detection model has good adaptability and robustness to face detection in complex environments.

Fig. 2 Face detection results by YOLOv3 on **a** WIDER FACE and **b** CelebA

Fig. 3 Face detection results by YOLOv3 on FDDB (test)

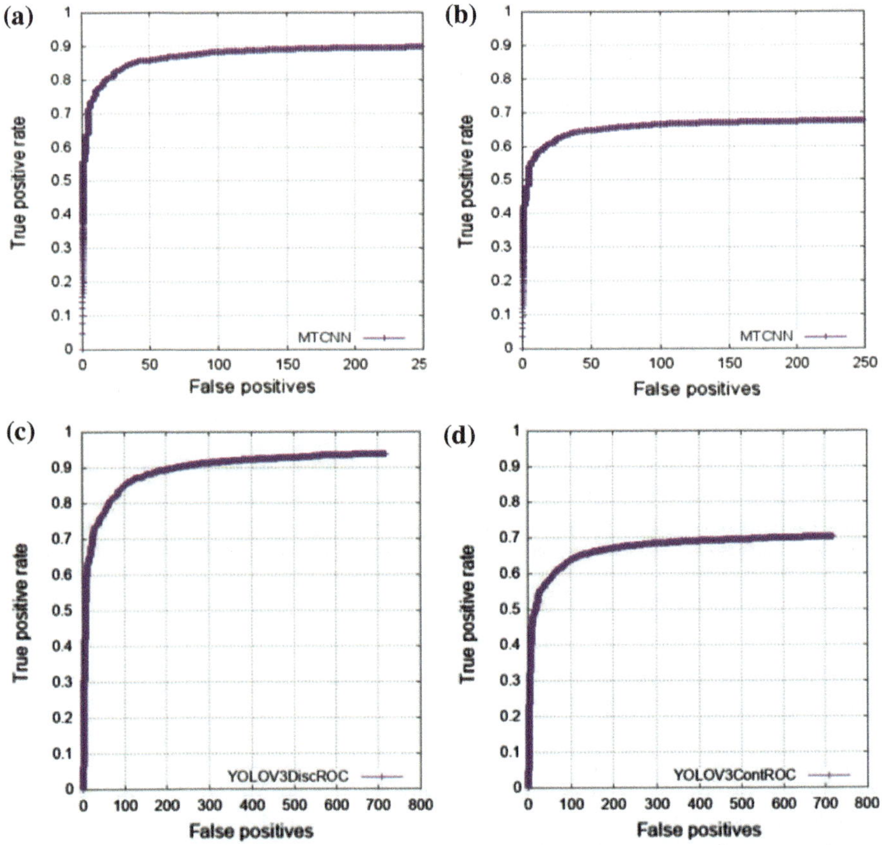

Fig. 4 Comparison of the ROC curves of two methods: **a**, **b** MTCCN and **c**, **d** YOLOv3

Table 1 Performance of two methods on FDDB

Method	Databases	Acc %
MTCNN	FDDB	90.2
YOLOv3	FDDB	93.9

4 Conclusions

We put forward a face detection method based on YOLOv3: we changed the third detection layer to detect smaller faces, chose the Softmax as the loss function to increase the difference of inter-class features, and decreased the dimension of features on detection layers to improve the speed. We trained our method on the WIDER FACE database and the CelebA database and tested it on the FDDB benchmark database. The results of experiment suggested that our approach has gained great performance on small face detection and is by far one of the most balanced face detection networks for speed and accuracy. Due to the integration of a variety of advanced methods like FPN and ResNet, it gives us a significant performance boost. Subsequently, we will try larger face databases and other CNNs models to improve the performance and applicability of the face detection task.

Acknowledgements Our paper is supported by the National Key Research and Development Plan (Grant No. 2016YFC0801005) and the Basic Research Fund Project of People's Public Security University of China (Grant No. 2018JKF617).

References

1. Viola, P., Way, O.M., Jones, M.J.: Robust real-time face detection. Int. J. Comput. Vision **57**(2), 137–154 (2004)
2. P. Viola and M. Jones. Rapid object detection using a boosted cascade of simple features. Comput. Vision Pattern Recogn. (CVPR), 511–518 (2001)
3. Forsyth, D.: Object detection with discriminatively trained part-based models. IEEE Trans. Pattern Anal. Mach. Intell. **32**(9), 1627–1645 (2010)
4. Girshick, R., Donahue, J., Darrell, T., et al.: Rich feature hierarchies for accurate object detection and semantic segmentation. In: IEEE Conference on Computer Vision and Pattern Recognition, pp. 580–587. IEEE Computer Society (2014)
5. Girshick, R.: Fast R-CNN. Comput. Sci. (2015)
6. Ren, S., He, K., Girshick, R., et al.: Faster R-CNN: towards real-time object detection with region proposal networks (2015)
7. Najibi, M., Samangouei, P., Chellappa, R., et al.: SSH: Single stage headless face detector. In: IEEE International Conference on Computer Vision, pp. 4885–4894. IEEE Computer Society (2017)
8. Liu, W., Anguelov, D., Erhan, D., et al.: SSD: Single Shot MultiBox Detector. In: European Conference on Computer Vision, pp. 21–37. Springer, Cham (2016)
9. Li, H., Lin, Z., Shen, X., et al.: A convolutional neural network cascade for face detection. In: 2015 IEEE Conference on Computer Vision and Pattern Recognition (CVPR). IEEE Computer Society (2015)

10. Zhang, K., Zhang, Z., Li, Z., et al.: Joint face detection and alignment using multitask cascaded convolutional networks. IEEE Signal Process. Lett. **23**(10), 1499–1503 (2016)
11. Redmon, J., Divvala, S., Girshick, R., et al.: You only look once: unified, real-time object detection (2015)
12. Everingham, M., Gool, L.V., Williams, C.K.I., et al.: The Pascal Visual Object Classes (VOC) challenge. Int. J. Comput. Vision **88**(2), 303–338 (2010)
13. Redmon, J., Farhadi, A.: YOLO9000: better, faster, stronger. In: 2017 IEEE Conference on Computer Vision and Pattern Recognition (CVPR), pp. 6517–6525 (2017)
14. Redmon J, Farhadi A.: YOLOv3: an incremental improvement (2018)
15. Lin, T.Y., Dollár, P., Girshick, R., et al.: Feature pyramid networks for object detection (2016)
16. http://mmlab.ie.cuhk.edu.hk/projects/WIDERFace/
17. http://mmlab.ie.cuhk.edu.hk/projects/CelebA.html
18. http://vis-www.cs.umass.edu/fddb.html

Research on Computational Thinking Ability Training and Blended Learning

Chong Shen and Kun Zhang

Abstract The highest level of computational thinking training is to improve students' ability to use computational thinking to innovate. Constructing a scientific teaching mode of computational thinking, teaching guidance of computational thinking in blended learning is deeply excavated. Developing students' thinking, students' ability of autonomic learning and flexible application is strengthened. Training of computational thinking ability is combined with cultivation of innovative talents. New ideas and new ways for the country to train innovative talents are provided. In this study, computational thinking and blended learning are first analyzed and summarized, and the common learning theories are summarized. It is supported by the practice of solving problems by using computational thinking in daily life. The characteristics of learners, the current situation of learning resources, and learning environment in blended learning are analyzed. On this basis, a hybrid learning model based on computational thinking is constructed. Through the analysis and research of computational thinking ability training and blended learning case practice, the effectiveness and feasibility of blended learning based on computational thinking are summarized.

Keywords Computational thinking · Blended learning · Innovative talents · Learning theory

---

C. Shen · K. Zhang
School of Information and communication Engineering, Hainan University, Haikou, Hainan 570228, China

K. Zhang (✉)
Education Center of MTA, Hainan Tropical Ocean University, Sanya, Hainan 572022, China
e-mail: kunzhang@hainu.edu.cn

V. Jain et al. (eds.), *Recent Trends in Intelligent Computing, Communication and Devices*, Advances in Intelligent Systems and Computing 1031,
https://doi.org/10.1007/978-981-13-9406-5_35

1 Introduction

With the rapid development of computer technology, it affects not only our life, way of thinking, and habits, but also profoundly our thinking ability [1]. At present, the concept of computing is penetrating into many fields such as cosmology, physics, biology, and even social sciences. Computing has not only become a universal method for people to understand nature, life, thinking, and society, but also is trying to become a new world outlook. Facing various complex problems of environment, ecology, energy, security, economy, and politics encountered by countries all over the world, it is a general trend to cultivate cross-disciplinary thinking, highly rational and objective, problem-solving-oriented talents [2, 3].

2 Computational Thinking and Blended Learning

2.1 Computational Thinking

The first mention of computational thinking is a concept put forward by Professor Zhou Yizhen of Carnegie University in 2006. "Computational thinking is a way of using basic concepts of computer science to solve problems, design systems and understand human behavior. It covers a wide range of thinking tools in the field of computer science." Computational thinking is not only a concept and thought adapted to computer science, but also a visual angle widely used in work, study, life, organization, and analysis of problems. Computational thinking thinks that when we encounter problems, we need to consider whether we can formulate problems in order to solve them by computer? Computational thinkers can understand problems by collecting and analyzing data; it can not only solve individual problems, but remove details, generalize abstractions, and find patterns. So as to solve all problems of the same kind, they can also formulate steps to solve problems. If possible, a simulation model will be set up to test and debug the solution. The six stages of computational thinking include:

Put forward questions; these problems can be solved by computer or other tools;
Organize and analyze problems according to logic;
Through abstract express data, such as modeling and simulation;
Use algorithm thinking (sequential steps) to automate the solution;
Identify, analyze, and implement possible solutions to achieve the most effective combination of resources and steps;
Summarize the problem-solving process and migrate to different problems.

Through these six stages, we can find a solution to the problem. In the form of program, it can be executed on the computer [4, 5]. In the form of process or regulation, it can be carried out by people. That is to say, computational thinking is a way of thinking based on the concept of computer science. It is not limited to computers. In the final analysis, the computer is only a tool; the greatness of this tool is that it has promoted people to develop the way of thinking [6]. Nowadays, people try to solve problems by using computer thinking in many disciplines. When people put forward the difficult problem that can be easily solved by computer or through big data analysis to explore the internal law, it shows that they are using computer thinking to think. Computational thinking has led to the development of computational biology, computer chemistry and other fields, as well as new technologies that can be applied in literature, social research, and art [7].

2.2 Blended Learning

Blended learning is a new learning method in the information age. After experiencing the upsurge of digital learning, the drawbacks of online learning have gradually emerged, and the advantages of traditional classroom teaching have been paid attention to [8]. Therefore, blended learning is conducive to make full use of the advantages of online learning and traditional learning, and promoting better learning of students [9].

The purpose of blended learning is to apply information technology, meet students' needs, and improve learning efficiency. The traditional classroom teaching adopts the "one-size-fits-all" mode; students can only learn the fixed content according to the rhythm of teachers. The blended learning paves the way for students of "What to study," "How to study," and "Why to study." Those differences can make students feel comfortable of choosing their preferred pattern to start their studies. First of all, blended learning is student-centered so that students are guided by their own learning objectives and fully grasp the dominant power of learning. Secondly, blended learning generally adopts a variety of learning modes, such as classroom teaching, online learning, project collaboration, group discussion, and so on. Students can switch between different modes according to their own needs or teachers' suggestions and flexibly arrange learning tasks [10]. In terms of learning content, on the one hand, students can collect information or read relevant textbooks on the Internet by themselves; on the other hand, the use of information technology can expand the way of knowledge display, and students can carry out text, pictures, audio, video, and other forms of learning, which is conducive to the smooth acquisition of knowledge. In addition, blended learning is not restricted by place and time. Students can learn from their own place. Blended learning integrates technology and education, considers students' individual needs, and provides students with full, diverse, and flexible choices [11].

Pictured here are three levels of blended learning (Fig. 1).

The first level: online and offline mixing

The mixing of online and offline is the traditional learning environment and online learning environment, as well as the large-scale online open course platform and other learning environment. It covers all kinds of learning resources, such as text, sound, picture, image, video, animation, and so on so that learners can find and

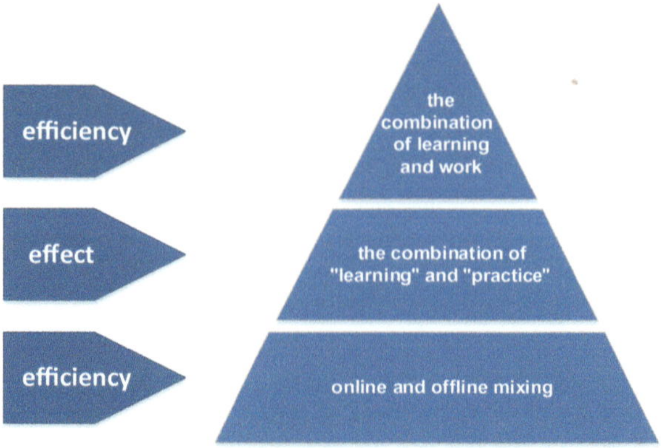

Fig. 1 Three levels of blended learning

solve problems efficiently, and achieve the goal of expanding transfer and internalizing knowledge. The advantage of online learning is that intelligent interaction is not limited by time and place, and the learning resources provided are better than the traditional learning environment.

The second level: the combination of "learning" and "practice"

The combination of "learning" and "practice" is the real meaning of blended learning. It is a higher learning goal to apply learning content to practice through "practice." Usually, we equate "learning" with learning. But the "practice" was completely omitted. The vast majority of face-to-face or online learning are just "learning," not real learning. In fact, designing the combination of "learning" and "practice" is the most effective blended learning. This blended learning is actually the simplest design. After "learning," arrange certain learning tasks. Let the learners complete "practice" through their own practical activities, and really learn something in the process of "practice."

The third level: the combination of learning and work.

Blended learning, which combines learning and work, is also known as "embedded learning" or "action learning." Rather than a learning method, it is a learning realm. In a sense, work is learning. This level is mainly applied to management training. The effect of managers in promoting this level is often reflected in some measures to promote employees' work summary, experience sharing, business innovation, and so on.

2.3 Computational Thinking and Blended Learning Model

According to the four factors of the learning process of blended learning mode based on computational thinking (instructor, learner, blended learning, and computational thinking), we constructed a blended learning model based on computational thinking, as shown in Fig. 2.

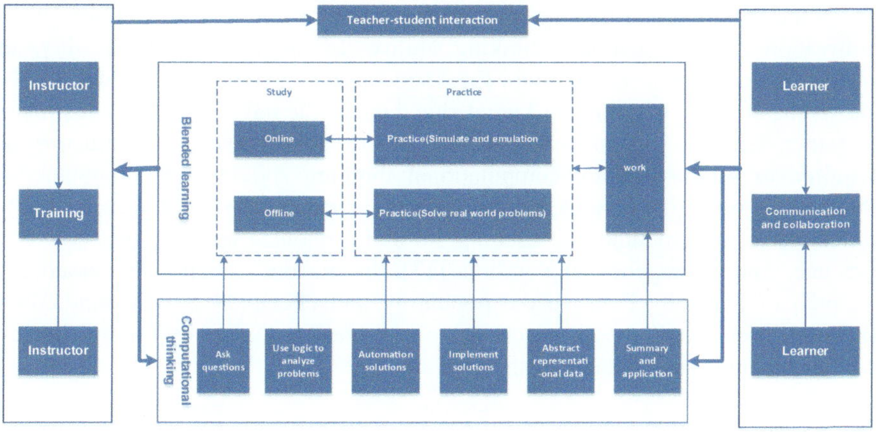

Fig. 2 Blended learning model based on computational thinking

In blended learning, the first level of online and offline learning environment is linked to the organic integration of learning resources. Based on the superior blended learning environment and learning resources, the instructor is guided by the concept of computational thinking and based on the method of computational thinking. Using various teaching methods and computational thinking methods to assist and guide learners to carry out effective learning. So learners can naturally use computational thinking methods to effectively construct knowledge and improve their ability to acquire knowledge, and at the same time improve their computational thinking ability.

Under the guidance, assistance, and supervision of instructors, learners can solve problems in a mixed learning environment and a computational thinking way. Learners need to ask questions in the process of "learning" in the second-level mode of blended learning. Then, a series of computational logics, such as recursive thinking, parallel processing, concern analysis (SOC), abstraction and decomposition, worst-case recovery, and heuristic reasoning, are used to find solutions. In the process of "practice," the feasibility scheme of automation is worked out and implemented through simulation, which is also the process of practice. When learners internalize their knowledge and master computational thinking methods, they can construct knowledge networks and learning frameworks independently through the acquired knowledge experience and methods and complete the communication and collaboration between learners [12]. At the same time, they can use computational thinking methods to improve their problem-solving ability and innovation ability. In the process of blended learning based on computational thinking, all the learning links are developed on the basis of computational thinking under the guidance of the concept of computational thinking. Through the continuous application of computational thinking methods in work, the aim of effective training of learners' computational thinking ability is achieved.

Mutual learning and training among instructors are of great significance to the cultivation of computational thinking ability. In the process of teaching, the instructor does not simply explain the knowledge of computational thinking, but shows the thinking of computer knowledge through the refinement of the instructor, and then lets the students produce their desire for knowledge. Instructors are required to seriously study computational thinking and research computational thinking. In preparing lessons, carefully design teaching contents and cases and try different teaching methods in teaching, so as to innovate teaching methods, bold practice, and courageous exploration [13]. In practice, small class practice is adopted to gradually improve the popularity and effectiveness of computational thinking in college computer teaching, so as to realize the steady improvement of college students' computational thinking ability.

3 The Importance of Computational Thinking and Blended Learning

Computational thinking is the quality of thinking that each of us should possess. It exists in every field of people's life. Before, people had a narrow understanding of computational thinking. Most people think that computational thinking means the thinking of computers. Only those who often use or specialize in computers understand or need to master this ability. In fact, while teaching students knowledge, teachers should not only teach students the ability to use knowledge flexibly, but also let them learn to use computational thinking flexibly. Teachers should pay attention not only to the imparting of knowledge, but also to the cultivation of quality, which is also the content of quality education that experts have always emphasized. That is, teachers should teach students "fishing methods" instead of "fish." This is also the main difference between our country and other countries. In our country, some scholars suggest that a special course for explaining computational thinking should be set up when college students enter the university so that students can learn and master computational thinking systematically, and achieve the ability of flexible application of this thinking to solve practical problems in practical application.

With the development of computer miniaturization and intelligentization in life, computer has been closely related to people's daily life. With the rapid development of communication technology and the emergence of the Internet of things, our life is inseparable from computers [14]. We should cultivate computational thinking, understand it from the perspective of computer creators, better use it, and make it play a greater value. In computational thinking, we can deal with problems by means of recursion, subtraction, embedding, transformation, simulation, and so on. We can decompose complex problems into simple ones. We can hand over complex and boring calculations to computers, and people can solve those problems that can be solved. At the same time, we can combine simple programs and systems to get complex systems to play a greater role.

4 Research and Application of Computational Thinking Training and Blended Learning

The learning of information technology course is an important way to cultivate computational thinking. In the classroom, students should not only master programming language, but also further understand a series of concepts and methods of computational thinking, such as recursion, abstraction, formalization, and so on. When encountering practical problems, students use computational thinking to reason and analyze so that they can find more effective and efficient ways to solve practical problems [15].

As a way to solve problems, computational thinking is not only applied in design procedures. Migrating to other courses, computational thinking can help you analyze problems, select tools, form automated solutions, select optimal solutions, and form generic solutions.

In the literature or history classes, students use spreadsheets and book word frequency statistics to make statistics of common points in language use. In mathematics class, students solve problems with educational resources and show them how to build models based on collected data to predict how long it will take for mobile phones to be full of electricity. In science class, students use computer simulation program to build computer simulation models for water cycle, ecosystem, and other scientific systems and carry out experiments. When necessary, students can also modify code and establish simulation system to learn scientific knowledge.

4.1 Case Analysis

How to sort things on the Internet? We can notice that the files obtained from the computer are arranged in sequence. Files can be sorted alphabetically. Data can be sorted according to time or file size, which facilitates users' observation and analysis of data. For this reason, computer scientists spend a lot of time looking for a good way to sort computers.

In the process of learning computer basic courses in universities, the problem-solving process of "ten different random natural numbers from small to large order" in learning basic knowledge of programming is analyzed. Under the guidance of instructors, learners have learned that the key of sorting algorithm is the correct completion of comparison and exchange, and the use of recursive thinking and separation of concerns in computational thinking. We can decompose the ranking problem of ten different random natural numbers and change the focus from the comparison and ranking of ten different random numbers to the size comparison and ranking problem of two different random natural numbers. Here, we will use a computational thinking to sort quickly, and we will sort six numbers. The rules are as follows:

Fig. 3 The sorting process of 6 random numbers

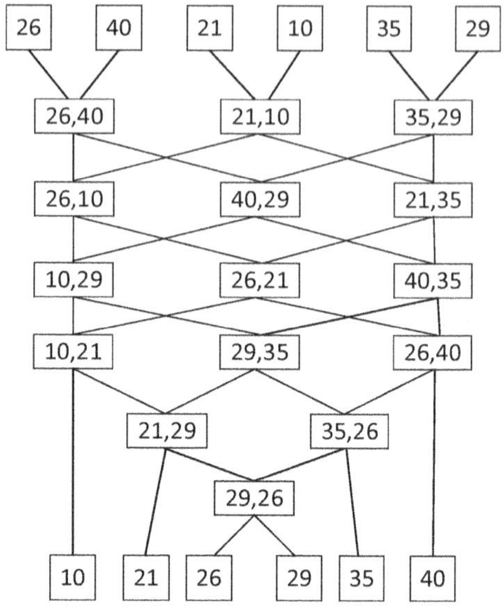

Divide 6 numbers into 3 groups at random;
In comparison with same group number, the small numbers go to the left and the big numbers go to the right;
The three sets of figures are compared at the same time, that is to execute three simultaneous comparisons;
Continue comparing until you get the final result. Get the sequence from left to right ascending order (Fig. 3).

Unlike traditional sorting, this case can be compared with three groups at a time, which greatly saves running time. Instructors guide learners to use computational thinking to think about problems. With the initial solution, learners can use computer programming to simulate, so as to achieve the process of "learning" and "practice" in mixed learning.

Through the analysis of the above cases, we can easily find that the use of computational thinking to decompose problems and build mathematical models to solve problems, giving full play to learners' subjective initiative in learning. It can effectively promote learners' knowledge construction, optimize learning effect, and improve learners' problem-solving ability and computational thinking ability.

5 Summary

The blended learning model based on computational thinking meets the learning needs of learners to a certain extent, effectively optimizes the learning effect, and promotes the cultivation of learners' computational thinking ability. Mixed learning based on computational thinking makes the resources of teaching and learning sustainable development, which conforms to the trend of education and social development, and is an effective new learning mode to adapt to social development. To some extent, the research on the mixed learning model of cultivating computational thinking ability provides help and guidance for optimizing the learning effect of college students and developing the cultivation of their computational thinking ability. At the same time, it provides a theoretical basis for the construction of three-dimensional blended learning environment and resources. Enrich the related research in the field of blended learning, and provide effective support for training college students' computational thinking ability.

Acknowledgements This research was financially supported by the 2016 Hainan Education Science Research Topics of the 13th Five-Year Plan (No. QJY13516030).

References

1. Cui, L., Song, Q.X.: Computational Thinking oriented multidimensional blending and flipped teaching mode—taking the course of "Database Principle" as a case in point. J. Panzhihua Univ. **34**(2), 100–105 (2013)
2. Li, Y.J.: Research on the reform of computer basic teaching in colleges and universities under the mode of mixed teaching. Comput. Knowl. Technol. **13**(1), 128–129 (2017)
3. Gong, P.Z., Yang, H.P.: The cultivation of computational thinking in computer basic teaching. China Univ. Teach. **5**, 51–54 (2012)
4. Zhang, K., Wang, H.F., Chen, X.Y.: Research on teaching reform of computer course based on computational thinking. In: DEStech Transactions on Social Science, Education and Human Science, 2017 International Conference on Education Innovation and Economic Management (EIEM2017), pp. 109–113
5. Zhang, K., Chen, X.Y., Wang, H.F.: Research on the mixed-learning model and the innovative talent cultivation mechanism based on computational thinking. In: International Conference on Intelligent Computing, Communication & Devices (ICICCD2017), Shenzhen, Guangdong, China, pp. 59–66, 9–10 Dec 2017
6. Zhang, K., Shen, C., Huang, M.X., Wang, H.F., Li, H.W., Gao, Q.: Interrupt protection control of anti-interference nodes in network based on band sampling decision filter modulation. Cluster Comput. (2018) https://doi.org/10.1007/s10586-018-2131-1
7. Shi, Y., Chen, S.B.: Construction and implementation of a mixed teaching evaluation index system for "Computer Introduction" of computational thinking. Educ. Teach. Forum. (5), 209–210 (2016)
8. Zhang, K., Shen, C., Wang, H.F., Gao, Q., Li, H.W., Li, N.: An improved three-dimensional location algorithm and simulation of AOA and TDOA based on wave interference sensors. BoletinTecnico/Technical Bull. **55**(19), 211–219 (2017)
9. Jiang, X.H., Xue, H.R., Liu, X., Li, Y., Gao, X.J., Ji, C.: Research on MOOC teaching mode of computer basic course based on computational thinking. J. Inner Mongolia Agric. Univ. (Soc. Sci. Ed.). **18**(6), 106–111 (2016)
10. Zhang, K., Shen, C., Wang, H.F., Gao, Q., Wang, C.X., Feng, X.M.: Desing of ship medical rescue communication system based on MIMO precise positioning. Indian J. Pharm. Sci, **80**(S1), 42–43 (2018)

11. Zhang, K., Wang, H.X., Wang, H.F., Li, Z.: The time series prediction algorithm of enrollment based on the hesitancy degree correction rule of vague sets. ICIC Express Lett. **9**(5), 1311–1318 (2015)
12. Li, H.M.: Research and reflection on the reform of computer basic teaching based on mixed teaching mode. China Comput. Commun. (21), 250–251 (2016)
13. Zhan, D.C., Nie, L.S.: Computational thinking and the basic thinking of university computer course reform. China Univ. Teach. (2), 56–60 (2013)
14. Zheng, L.K., Guan, S.Y., Guo, D.: The research on the college computer basis teaching under the guidance of computational thinking. Heilongjiang Res. High. Educ. (2), 174–176 (2016)
15. Zhao, W.D., Zhao, C.N., Zhang, L.: Teaching reform of programming language course based on computational thinking. Comput. Educ. (12), 28–30 + 34 (2016)

Research on Architecture Design of Aerospace Simulation System Integrating Cloud and Edge Computing

Zhou Jun, Zhao Yang, Shi Zijun and Liang Lei

Abstract The increasing demand of super large-scale spacecraft simulation and spacecraft multi-disciplinary collaborative simulation brings great pressure to the simulation system architecture. In order to satisfy the current complex simulation requirements, combined with cloud computing architecture and edge computing architecture, a new and efficient simulation architecture composed of core cloud, edge cloud, and user terminal is designed. The core cloud manages and monitors the edge cloud; the edge cloud allocates virtual resources and completes simulation; and user terminals provide users with portals for describing tasks and viewing results. The simulation system architecture integrating cloud and edge computing has the advantages of high resource utilization, quick service response, strong robustness, and simple promotion in other research fields. And it is great significance to improve simulation efficiency and conserve computing resources.

Keywords Architecture design · Core cloud · Edge cloud

1 Introduction

With the spacecraft system simulation changing from traditional single-target simulation to multi-disciplinary and collaborative simulation, the simulation scale becomes more and more huge, and the simulation conditions become more and more complex [1]. A construction of multi-disciplinary collaboration design system for aerocraft has been proposed by Wang Jincheng et al., realizing the standardization, flowing process, and synergy of each professional design and simulation work [2]. And Wang Peng has established a space environment elements modeling and simulation system based on high-level architecture [3]. However, like most of the current simulation system construction, this system construction only focuses on the realization of functions without considering the utilization efficiency of

Z. Jun · Z. Yang (✉) · S. Zijun · L. Lei
Harbin Institute of Technology, 150001 Harbin, P.R.China
e-mail: yangzhao@hit.edu.cn

V. Jain et al. (eds.), *Recent Trends in Intelligent Computing, Communication and Devices*, Advances in Intelligent Systems and Computing 1031,
https://doi.org/10.1007/978-981-13-9406-5_36

computing resources. The spacecraft system in complex structure normally requires a large number of models, such as rigid and flexible structural models, motion mechanism models, load models, solar sail models, and antenna models. Besides, simulation tasks can involve many discipline fields, such as attitude and orbit dynamics simulation, structural vibration characteristic simulation, and thermal field simulation. Nowadays, the simulation system architecture, such as distributed simulation architecture, that is used to solve this kind of problems, cannot bring the performance of computers into full play. And the simulation system can only perform one simulation task at a time with a lot of time waste, resulting in low research efficiency.

To improve resource utilization efficiency, cloud computing architecture has been extensively used. By cloud computing architecture, the virtualization technology is further expanded to extend the application and service of data, saving resources for users in the fields of computing power and spatial storage [4]. In addition, with the development of Internet of things, the research of edge computing architecture has been carried out to relieve the pressure of network bandwidth and data center [5]. Therefore, integrating cloud computing architecture and edge computing architecture, an imaginative and efficient simulation architecture is built with high resource utilization and reliable information security to support large complex spacecraft system multi-task co-simulation.

2 The System Architecture Design

The simulation system architecture integrating cloud and edge computing is a service-oriented hierarchical architecture, as shown in Fig. 1.

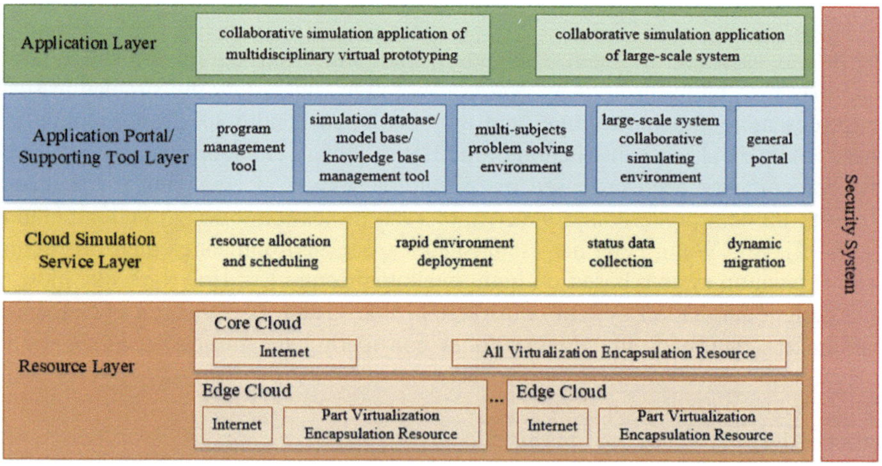

Fig. 1 System architecture of the simulation system

Resource layer, including the core cloud resource and the edge cloud resource, provides kinds of simulation resources for the simulation system operation, such as Internet resource, virtual modeling and simulating resource, and software resource. The core cloud resource contains all resources, but edge cloud resource just contains the resources used by this edge cloud users before.

Cloud simulation service layer provides various services to remain the system normally operating, including resource allocation and scheduling, rapid environment deployment, status data collection, and dynamic migration. Through these services, the edge cloud manages all virtual machines, while the core cloud manages all edge clouds.

Application portal/Supporting tool layer provides portal in form of browsers and desktop or supporting tool for users to carry out simulation activities, including program management tool, simulation database/model base/knowledge base management tool, multi-subjects problem-solving environment, large-scale system collaborative simulating environment, and general portal.

Application layer includes various simulation applications, such as collaborative simulation application of multi-disciplinary virtual prototyping and collaborative simulation application of large-scale system.

Security system provides safe and reliable cloud simulation environment. In order to protect users' data security, the system assigns different privileges to users through user authentication.

3 The Physical Architecture Design

Based on the traditional cloud simulation architecture, introducing the theory of edge computing, the advanced simulation system architecture integrating cloud and edge computing is established. This architecture can improve simulation efficiency and reduce computation cost to meet the demands of the simulation of large complex spacecraft system which has the property of high cooperativity and intensive computing. This architecture is composed of core cloud, edge cloud, and user terminal, as shown in Fig. 2.

3.1 The Core Cloud

The core cloud is a special cloud simulation system aiming at managing and monitoring the use of the resource and services in every edge cloud. The core cloud is connected with the edge cloud through the backbone network; meanwhile, it combines the core resource manager of the core cloud to realize the large complex spacecraft system multi-task co-simulation based on multi-edge clouds joint scheduling.

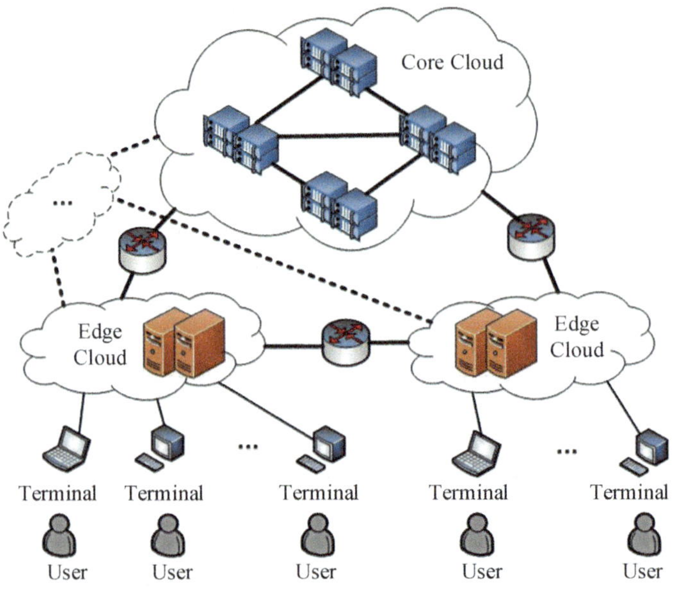

Fig. 2 Physical architecture of the simulation system

3.2 The Edge Cloud

The edge cloud is composed of server nodes distributed in the same area for specifically handling the service requests of users in this area as well as quickly and flexibly providing cloud simulation resource to users. The edge clouds are connected by backbone network, so users can log into the nearest edge cloud through the network and use the services provided by the edge cloud nearby. On the one hand, the edge cloud is responsible for processing the data stream between the core cloud and user terminals, such as compression, encryption, and decryption, and it makes use of the correlation between communication data to reduce network overhead as well as latency in order to ensure the quality of cloud simulation services; on the other hand, the edge cloud stores the common data needed by terminals accessing to use cloud simulation services.

3.3 The User Terminal

The user terminal is a request entity of services related to the cloud simulation. The main functions of the terminal are submitting user's service request as input and displaying service results as output. Therefore, the user terminal does not need great ability in computing and storing, avoiding in occupying excessive operating resources of user computers.

3.4 The Collaboration Between Each Node

Based on the architecture of the simulation system integrating cloud and edge computing, nodes cooperate with each other to provide cloud simulation services ensuring the quality of service for users under the coordination of core cloud. And the interaction of data flow between each node in the network is shown in Fig. 3.

Firstly, user sends the simulation task information through the local network with high bandwidth and low delay to the edge cloud which he has logged in. Secondly, the edge cloud appraises whether it has the conditions and capabilities to independently accomplish the accepted simulation tasks. If the edge cloud cannot provide the services by itself, the co-simulation request will be sent to the core cloud. Then the core cloud establishes the co-simulation process based on interface protocol, after checking permissions of the request edge and other edges involved in the simulation task. Thirdly, each edge cloud joined in the simulation allocates virtual resources based on tasks description and completes the rapid deployment of the simulation environment through the image resource. Finally, the result of the simulation is collected and saved in the edge cloud containing the accessing terminal. And the edge cloud returns the result data to the user terminal for users to view. Besides, the core cloud monitors resource usage for each edge cloud and simulation progress for each simulation task on each edge cloud.

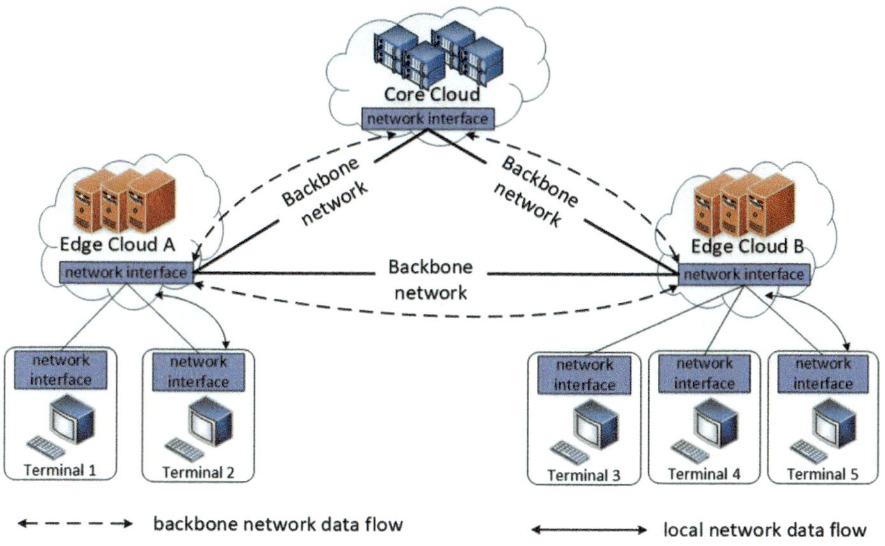

Fig. 3 Data flow between each node in the network

3.5 Characteristics of the System Architecture

The architecture of the simulation system integrating cloud and edge computing combines the advantages of centralized cloud simulation system and distributed simulation system, effectively avoiding each architecture's shortcomings. This innovative architecture inherits the advantages of centralized network structure, so it has the advantages of single resource composition (each edge cloud is composed of different kinds of resources, but it provides a unified resource representation for the core cloud upward), simple resource management, high degree of resource aggregation, and simple resource collaboration service.

Meanwhile, this architecture avoids the high dependence of the network and the inadequate use of existing resources. Compared with the traditional distributed network structure, the edge cloud simulation network structure solves such problems as the independence of internal network, the decentralization of service resources, the difficulties of resource management, and the trouble of resource collaboration service. At the same time, it still retains the advantages of traditional distributed network, such as high utilization rate of existing resources, high extendibility of system architecture, high robustness of service, and low network dependence.

4 Conclusion

The simulation system architecture integrating cloud and edge computing is a new and efficient simulation framework to support large complex spacecraft system multi-task co-simulation. The system architecture consists of five layers to provide powerful, efficient, and secure simulation services for users. In addition, the physical architecture is composed of the core cloud, edge cloud, and the user terminal with high resource utilization, quick service response, and strong robustness. This simulation system architecture is not only suitable for the aerospace field, but also for other research fields that require plenty of computing resources and a high degree of synergy, such as large and complex mechanical system design, intelligent traffic control.

Acknowledgements Author is extremely grateful to Prof. Zhang Weizhe (HIT, Harbin, P.R.China) for his extremely valuable help with aspects of cloud computing.

References

1. Cao, J.: Status, challenges and perspectives of aero—engine simulation technology. J. Propul. Technol. **39**(5), 961–970 (2018)
2. Wang, J., Zhou, Y., et al.: Construction and application of multidisciplinary collaboration design system for aerocraft. Intell. Manuf. **1**, 82–85 (2018)
3. Wang, P.: Space environment elements modeling and simulation based on HLA, doctoral thesis. Information Engineering University. (2006)
4. Baomin, X.U., Xuguang, N.I.: Development trend and key technical progress of cloud computing. Bull. Chin. Acad. Sci. **30**(2), 170–180 (2015)
5. Zhao, Z., Liu, F., et al.: Edge computing: platforms, applications and challenge. J. Comput. Res. Dev. **55**(2), 327–337 (2018)

Multi-person Collaborative Interaction Algorithm and Application Based on HoloLens

Chunfeng Xu, Yange Wang, Wei Quan and He Yang

Abstract In order to solve the concurrency conflict problem generated by multiple HoloLens in the process of collaborative interaction, a multi-person collaborative interaction algorithm is proposed. The algorithm adopts a three-dimensional scene information sharing algorithm and a multi-user collaborative interaction concurrent control strategy, so that different users share the same scene information, and all the user information in the scene is updated and synchronized. Solve the concurrency conflict problem in the process of multi-person collaborative interaction. A multi-person collaborative interaction system based on HoloLens is designed, which supports multiple HoloLens users to interact interactively in the same scene. The results show that the system achieves the cooperative interaction of multiple HoloLens users in augmented reality environment. Compared with the previous human–computer interaction platform, not only the interactive environment is more realistic, but also the interaction is more natural and the user experience is higher.

Keywords HoloLens · Internet · Multi-person collaboration · Natural interaction · Concurrency control

1 Introduction

With the progress of science and technology, large, medium and small kinds of augmented reality devices have appeared, such as Blade 3000, DAQRI smart helmet, SmartEye glass and HoloLens. HoloLens [1, 2] is one of the mainstream AR glasses. Today, HoloLens-based multi-user interaction has been studied in various fields. In the field of collaborative creation, it can solve the problem of duplicate work caused by different opinions of many people and improve the efficiency of the whole creation. In the game area, you can increase user's game

C. Xu · Y. Wang · W. Quan (✉) · H. Yang
School of Computer Science and Technology, Changchun University of Science and Technology, Changchun 130022, Jilin, China
e-mail: quanwei@cust.edu.cn

V. Jain et al. (eds.), *Recent Trends in Intelligent Computing, Communication and Devices*, Advances in Intelligent Systems and Computing 1031,
https://doi.org/10.1007/978-981-13-9406-5_37

experience. In the field of medicine, many doctors can cooperate to complete the same operation. Therefore, multi-person collaborative interaction based on HoloLens has become a hot research topic.

Hanna [3] believes that HoloLens can be used for autopsy and remote pathological explanations during the examination and other users can have remote access. However, these users can only synchronize the scenes in the main HoloLens, and the two sides cannot simultaneously control the scene. Chen [4] implements remote collaboration between a single HoloLens user and a tablet or PC. However, this method cannot realize the natural interaction based on gesture between multi-users. It can only be watched on tablet or PC or interact with mouse and keyboard. The interaction mode is very inconvenient.

The multi-person cooperative interaction algorithm based on HoloLens can solve the above problems well. Multiple HoloLens users can cooperate with each other naturally and solve the concurrent conflicts in the process of collaborative interaction, so that users can have a good interactive experience.

2 Multi-person Collaborative Interaction Algorithm

2.1 Collaborative Interaction Control Model

The collaborative interaction control model is divided into three levels: collaborative control layer, interaction sharing layer and user application layer. As shown in Fig. 1, it is a cooperative interaction control model which enables multiple people to cooperate and interact with each other. From the perspective of computer network architecture, the server side is the collaborative control layer, and the ID information and operation information of all clients constitute an interactive sharing layer.

The interactive sharing layer is composed of individual information and operation information of multiple clients, as shown in Fig. 2. This layer mainly realizes the interaction of individual users with the autonomous information of virtual objects. This part of the interaction needs to realize the control of the virtual object in the three-dimensional environment through human–computer interaction and provide operational data for the evolution module. All current parameters and strategies will be controlled by the control module. All information will be stored in the current user's local database, and the data transmission between the interface and the collaborative control layer will be realized.

The collaborative control layer is mainly composed of a server. As shown in Fig. 3, the layer is mainly based on the data information provided by each client of the interaction sharing layer, and the processing of the layer is performed to realize the cooperative operation and information sharing among the users. The collaborative control unit obtains information through the interface of the collaborative control layer and the interaction sharing layer, and processes user operation information of each interface through coordination, control and the like to avoid conflicts. All processed user information is stored in the global database.

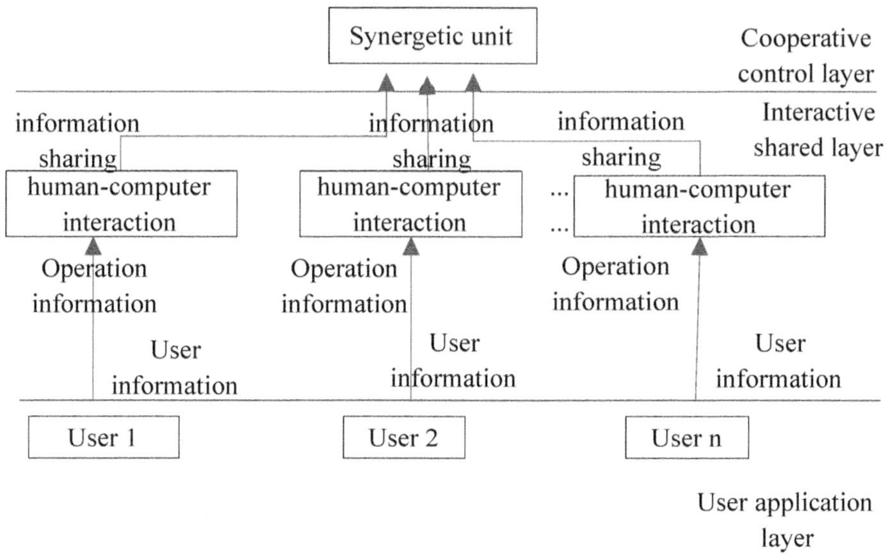

Fig. 1 Multi-person cooperative interaction control model

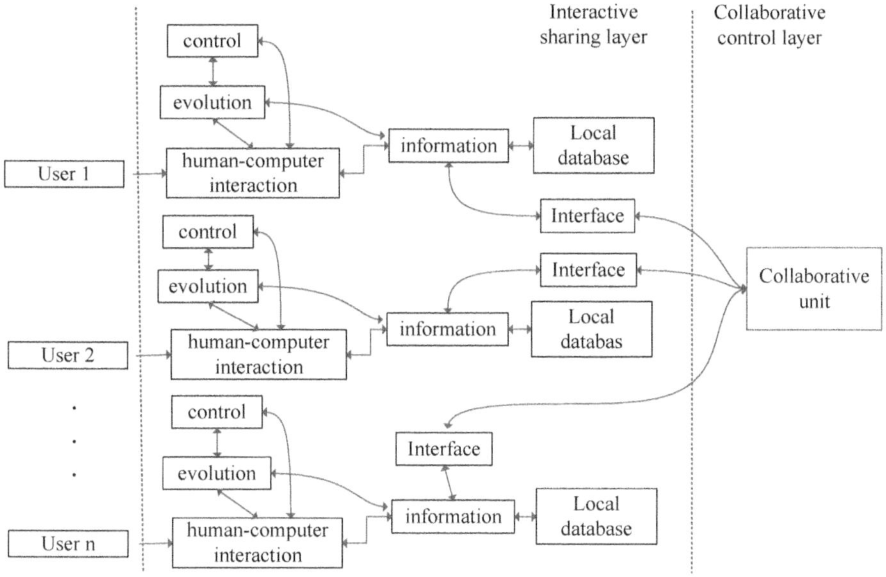

Fig. 2 Interactive sharing layer

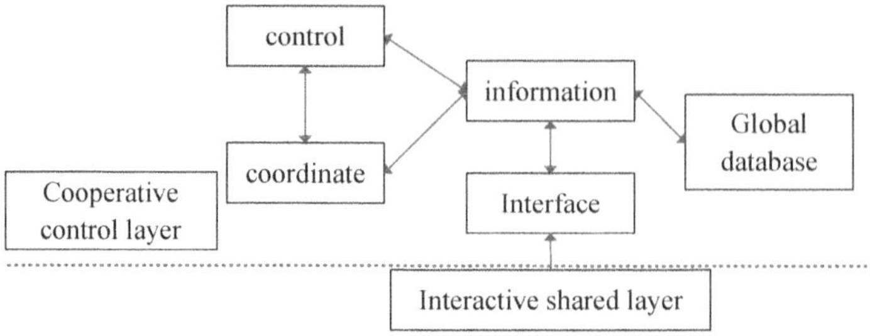

Fig. 3 Cooperative control layer

2.2 3D Scene Information Sharing Algorithm

In an augmented reality environment, multiple clients can share the same AR scenario through real-time network transmission and ensure that all client information in the scenario is updated and synchronized. Therefore, you first need to create an index information base for the scene and store the identity information and operation information of all clients in the scene in turn. The newly added client can update and synchronize data by accessing the index information database already established in the scenario. Figure 4 shows the scenario consistency implementation process.

The state information in the scene is changing all the time. The operation of the client to the scene, the departure of the client, the addition of the new client, the creation and destruction of virtual objects, the change of object position and the deformation of objects will all have an impact on the current scene. Therefore, it is necessary to constantly refresh the scene in order to achieve information

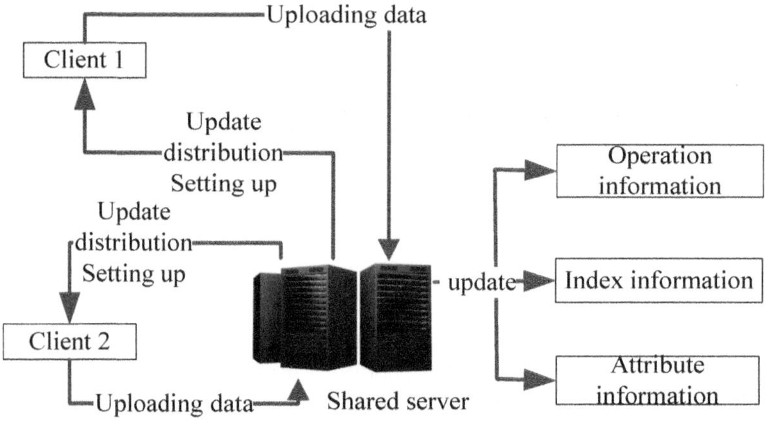

Fig. 4 Scenario consistency implementation process

synchronization between different clients. Figure 5 shows the specific implementation process of the refresh operation in the scene.

The basic idea of scene refreshing is to achieve the update and synchronization of all client information in the scene. When a client joins the scene, you should first determine whether there are other clients before the client. If the first one of the client is added to the scenario, it is the primary client and needs to allocate memory for it and build an index library to store all the index information in the scene.

If there are multiple clients in the room at this time, and each client operates on the scene, the index information of these clients must be stored in the server. Then for the newly added clients, the information of these clients needs to be loaded and synchronized. By looping through the idea of the index information base to locate the ID information of all clients in the scene, the server will update all the index information updates to the newly added client according to the ID information and update the data in the room at this time.

In addition to updating the scene information to the newly added client, it also needs to allocate memory and index in the index library. The index information allocation is to allocate a memory space for each client joining the scene and to

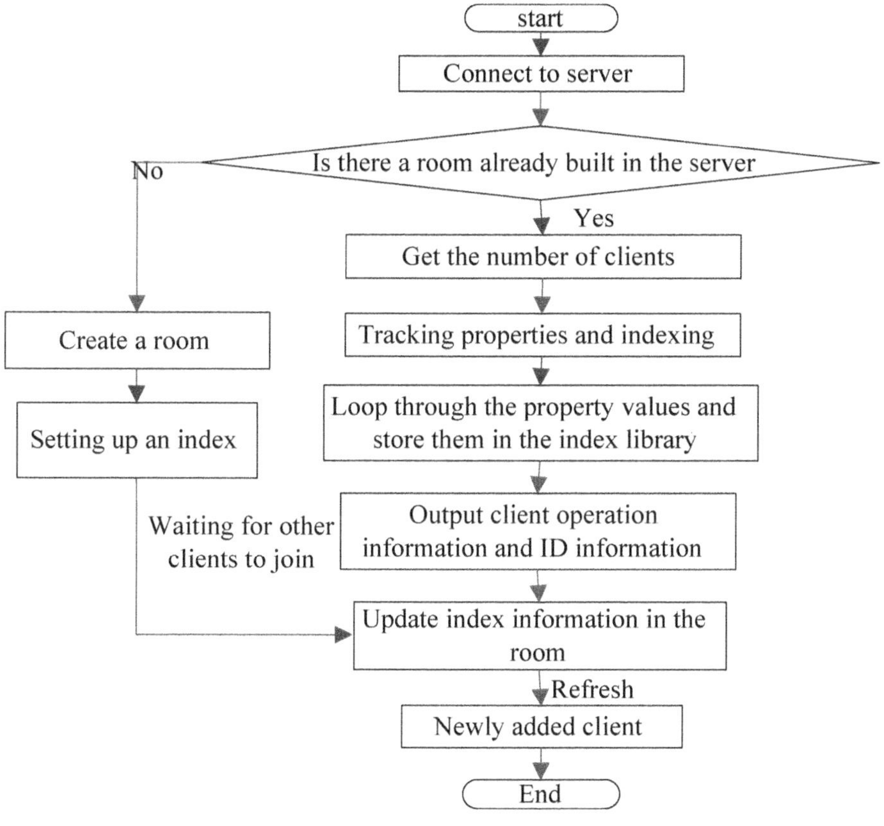

Fig. 5 Specific implementation process of scene refresh

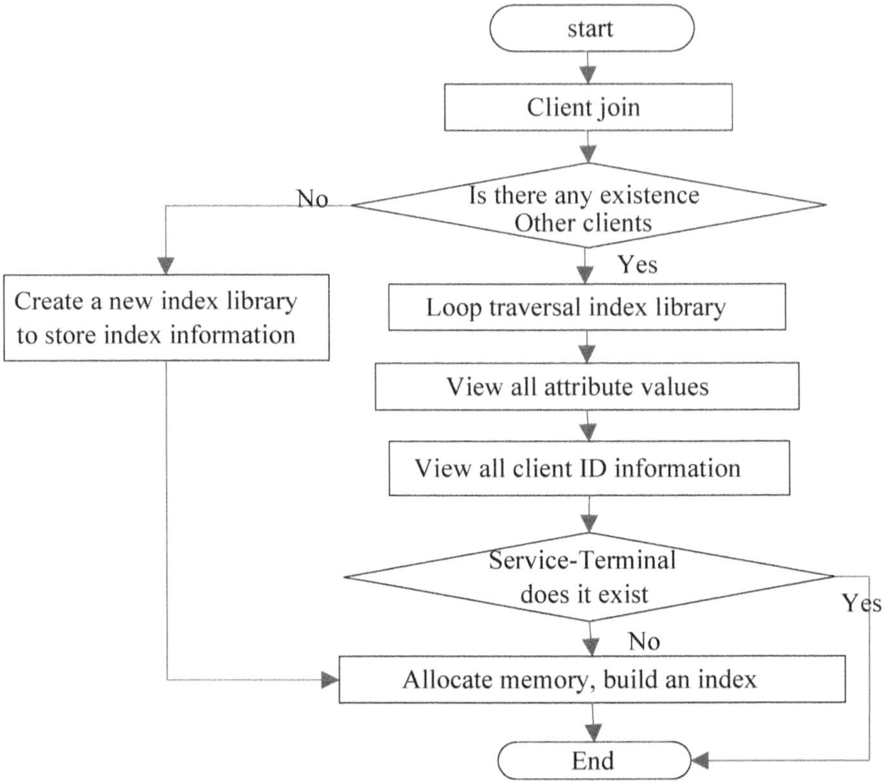

Fig. 6 Specific implementation process of index information distribution

store index information of the client and the objects in the scene. According to the allocated index information, the specific operation process of each client to the scene can be determined. The specific implementation process of index information distribution is shown in Fig. 6.

The allocation of index information is actually to obtain the client ID information that does not exist in the index library by looping through the index library. The server allocates memory and indexes of the client based on the output ID information.

2.3 Multi-user Collaborative Interaction Concurrency Control Strategy

The multi-user collaborative interaction technology in the augmented reality environment mainly realizes the coordinated control of the three-dimensional scene by users in different geographical locations through the Internet. Therefore, it is necessary to deal with the concurrency control problem between users and to perform conflict detection and complete concurrency control. Figure 7 shows a multi-user concurrency control solution.

Concurrency conflict occurs when multiple users simultaneously issue an operation request to a server for a three-dimensional virtual scene. This can be solved by using the concurrency control algorithm. The specific implementation steps of the algorithm are as follows:

(1) One of the users in the 3D scene sends a scene control request to the server, and the server automatically adds it to the application queue.
(2) Check whether there is any request information of other users in the current queue. If it does not exist, you can directly operate the 3D virtual object and upload the operation result to the server. If there is a conflict in the current queue, proceed to the next step.
(3) It is detected that there are multiple users' application information in the current queue at the same time. The server will assign control rights according to the priority of these users. The high priority first has control rights over the three-dimensional virtual scene. The other users enter the waiting queue and re-detect the conflicts and continue to compare the priorities. If these users have the same priority, the control rights will be allocated according to the order in which each user applies for the three-dimensional scene.
(4) The control user uploads the operation result to the server, and the server broadcasts all the operation information of the user to other users in the scene and releases the resources occupied by the user.

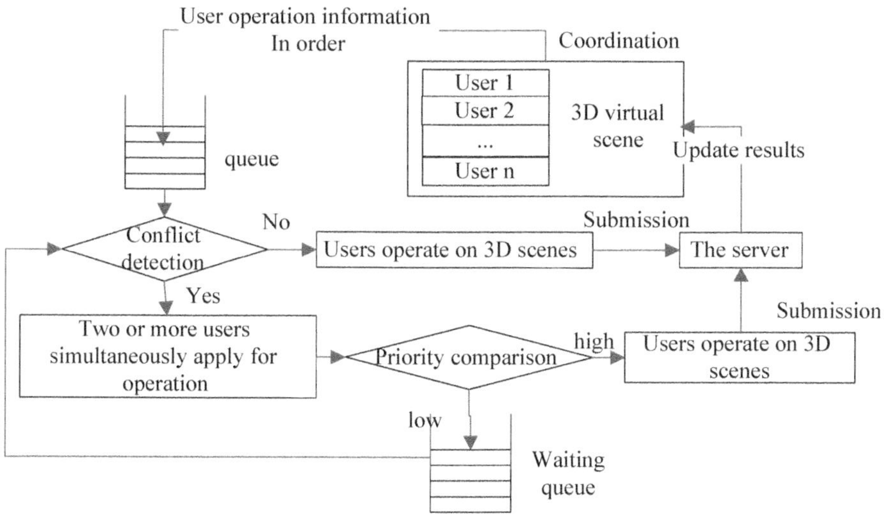

Fig. 7 Multi-user concurrency control solution

This method solves the conflict problem of multi-user cooperative operation in augmented reality environment. By creating multi-threads and setting the priority level of each thread, the order of each applicant's control of three-dimensional scene is determined.

3 Design and Implementation of Augmented Reality Multi-player Cooperative Interaction

3.1 System Structure

The design of this system is to build a virtual scene using the Unity3D platform to realize multi-person collaborative interaction based on HoloLens. The built-up scene is released through the UWP platform, the successful solution is debugged by using VS2017, the created virtual scenes are separately deployed to the two HoloLens devices, and the sharing experience is started. Figure 8 is the basic structure of the system.

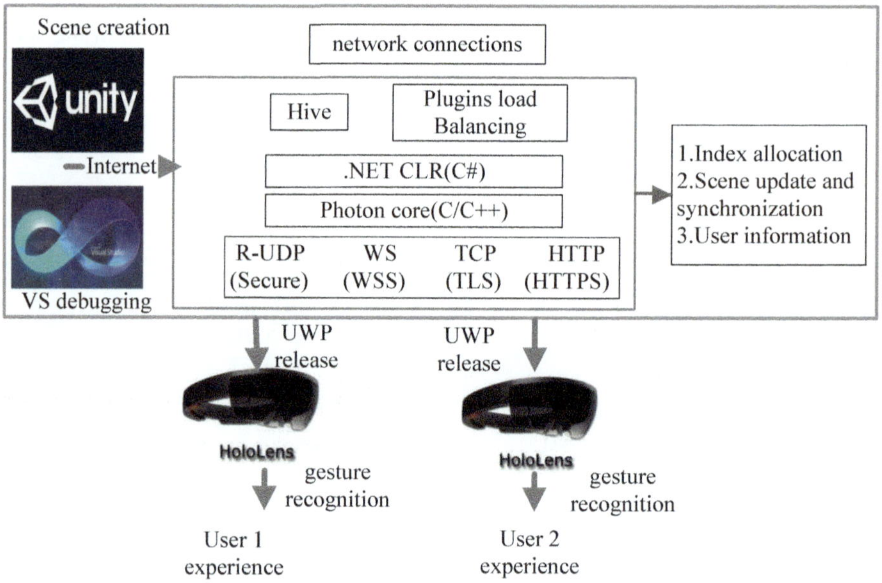

Fig. 8 System structure

3.2 Experimental Results and Analysis

The main implementation process of the multi-player collaborative interaction system in the augmented reality environment is described in detail above. In this system, a connection through the Internet is required to obtain a sharing experience between multiple users, and the same reference point is established for all users, and multiple devices share the same reference frame, so that each user can see the same augmented reality scene.

In the same shared environment, in order to obtain a better interactive experience between users, each user can be given a unique avatar in the scene, which can find each other in the shared scene and see the current status of each user position. Figure 9 shows the user's avatar selection perspective. Each user can use the gesture to select different avatars before entering the 3D main scene, and then enter the main scene. Figure 10 shows the three-dimensional main scene that the user sees after confirming the avatar. After entering the main scene, each HoloLens user can see a three-dimensional castle and a dedicated avatar floating above other users.

Figure 11 is a multi-person collaborative interaction picture taken by myself in the laboratory, which has not been made public. Figure 11a–c show three effects of collaborative interaction between two HoloLens users in the system. In the three-dimensional shared scene, users can interact with other users through gestures and fire artillery shells in all directions. Users can attack each other, fire artillery shells at the castle in front of each other. Each user in the scene can see other users to the scene operation.

Fig. 9 Head selection perspective

Fig. 10 3D main scene

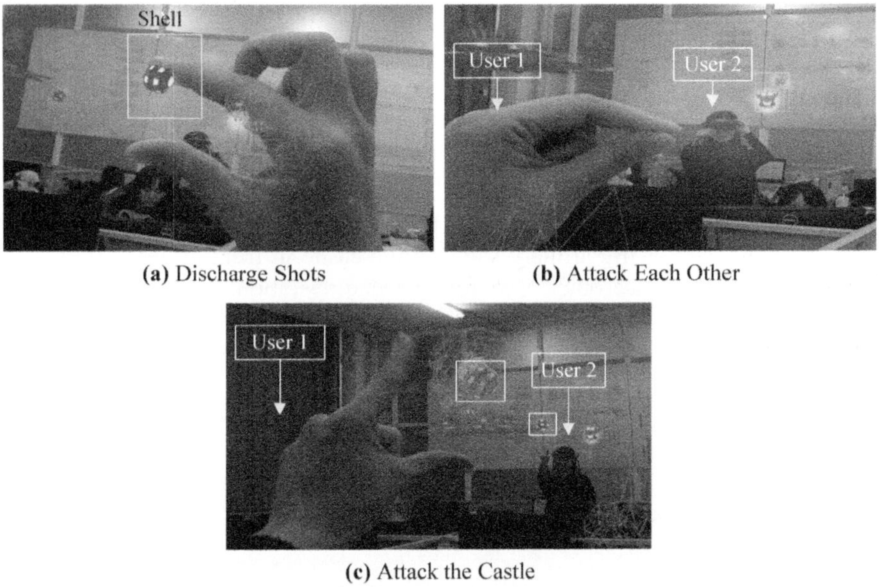

Fig. 11 Two HoloLens users' collaborative interaction diagram (This picture was taken by myself and was not disclosed to the public.)

3.3 Experimental Comparison

In order to verify the superiority of this paper, the related literatures at home and abroad in recent years are selected, which are compared with the paper in terms of working environment, number of users, coordination and interaction. Table 1 compares the functions of this paper with the existing platforms at home and abroad. This paper implements multi-person collaborative interaction based on HoloLens. It can be seen from Table 1 that domestic and foreign scholars' research

Table 1 Comparison of existing platform functions

Literature comparison	Working environment	Amount of users	Collaboration	Interactive mode
This paper	Augmented reality	Multi-user	Yes	Natural interaction of human hands
Literature [3]	Augmented reality	Multi-user	Yes	Main client gesture interaction, other user viewing
Literature [4]	Augmented reality	Multi-user	Yes	Gesture interaction with mouse and keyboard
Literature [5]	Virtual reality	Multi-user	Yes	Handle interaction
Literature [6]	Virtual reality	Multi-user	Yes	Mouse keyboard
Literature [7]	Virtual reality	Multi-user	Yes	Haptic interaction
Literature [8]	Virtual reality	Multi-user	Yes	Interaction and screen interaction
Literature [9]	PC platform	Multi-user	Yes	Keyboard and mouse
Literature [10]	PC platform	Multi-user	Yes	Machine display
Literature [11]	PC platform	Multi-user	Yes	Body interaction
Literature [12]	PC platform	Multi-user	Yes	Mouse and keyboard interaction
Literature [13]	Augmented reality	Single user	\	Speech interaction
Literature [14]	Augmented reality	Single user	\	Body interaction

on multi-person collaborative interaction based on PC is very common. Most of the interaction methods use traditional mouse and keyboard or physical movement. Compared with this article, it is not only inconvenient for interaction, but also very heavy. Some research has been carried out under the virtual reality platform, but the interactive environment is closed, and the interactive mode uses the handle interaction or the traditional keyboard and mouse method. Compared with this article, not only the interactive environment is not real, but also the interaction is not natural. Compared with previous HoloLens multi-player collaboration, this paper achieves multiple HoloLens users interacting simultaneously. In addition, compared with single-user human–computer interaction in augmented reality environment, not only can users communicate with each other, but also the interactive experience is greatly improved.

4 Conclusion

This paper is mainly about multi-person collaborative natural interaction based on HoloLens. Real-time update, sharing and interaction of 3D scenes are realized by multi-person collaborative interaction algorithm. Multi-user collaborative interaction concurrency control strategy is used to realize coordinated control of 3D virtual scenes by users, avoiding concurrency conflicts generated during interaction. The designed multi-person collaborative interaction system realizes the collaborative interaction of multiple HoloLens users in the same three-dimensional scene. By using natural and comfortable interaction mode, not only the sense of reality is stronger, but also the interactive experience brought to users is more realistic.

Acknowledgements This work is supported by the Technology Program of Jilin Province, China (No.20170203003GX, No.20170203004GX and No.20180201069GX). Thanks to the reviewers for their comments and suggestions.

References

1. Chen, A.D., Lin, S.J.: Discussion: mixed reality with HoloLens: where virtual reality meets augmented reality in the operating room. Plast. Reconstr. Surg. **140**(5), 1071–1072 (2017)
2. Furlan, R.: The future of augmented reality: HoloLens—Microsoft's AR headset shines despite rough edges (Resources_Tools and Toys). IEEE Spectr. **56**(6), 21 (2016)
3. Hanna, M.G., Ahmed, I., Nine, J., et al.: Augmented reality technology using microsoft HoloLens in anatomic pathology. Arch. Pathol. Lab. Med. **142**(5), 638–644 (2018)
4. Chen, H., Lee, A.S., Swift, M., et al.: 3D collaboration method over HoloLens™ and Skype™ end points. In: Proceedings of the 3rd International Workshop on Immersive Media Experiences, pp. 27–30. ACM (2015)
5. Yuan, S., Chen, B., Yi, C., Xu, B.: Research and implementation of immersive multi-user collaborative interaction technology in virtual geographic environment. J. Geo-inf. Sci. **20**(8), 1055–1063 (2018)
6. Lanlan, L.: Virtual reality based collaborative product appearance design under human-computer interaction. Mod. Electron. Tech. **41**(7), 111–114 (2018)
7. Huang, Y., Gao, X., Zhou, H., et al.: Design and implementation of distributed collaborative virtual assembly system with three-dimensional force feedback. In: 2012 IEEE 2nd International Conference on Cloud Computing and Intelligent Systems (CCIS), pp. 941–945. IEEE, 2 (2012)
8. Liao, Y., Liu, Y., Wei, X.: Multiplayer collaborative training system based on mobile AR innovative interaction technology. In: International Conference on Virtual Reality and Visualization, pp. 81–85. IEEE (2015)
9. Xingwang, H., Peng, S., Rui, H., et al.: Smart TV human-computer interaction system based on multi-screen collaboration. Comput. Appl. Softw. **33**(11), 49–52 (2016)
10. Shengqing, L., Yimin, C., Chen, H., et al.: A network-based parallel rendering and cross-platform synchronous display system. Comput. Appl. Softw. **34**(10), 113–117 (2017)
11. Bo-Wen, S., Jia-Liang, Z., Ya-Fei, C., et al.: Multiplayer online virtual experiment system based on Kinect somatosensory interaction. Comput. Sci. **43**(9), 284–288 (2016)
12. Qiang, X.U., Jian-Hua, S.: Research on wearable computing in the multi-collaborative soft system. J. Univ. Electron. Sci. Technol. China **s1**, 57–60 (2010)

13. Hamidia, M., Zenati, N., Belghit, H., et al.: Voice interaction using Gaussian mixture models for augmented reality applications. In: International Conference on Electrical Engineering, pp. 1–4. IEEE (2016)
14. Sun, G., Qiu, C., Qi, Q., et al. The implementation of a live interactive augmented reality game creative system. In: International Conference on Multimedia Information Networking & Security. IEEE Computer Society, pp. 440–443 (2013)

Design and Implementation of the Context-Based Adaptive Filtering System for Sensitive Words

Jun Yu, Qingfeng Wei, Changshou Luo and Junfeng Zhang

Abstract Semantics for language application differ among different contexts. Sensitive words are also affected by application scenarios and application timeliness. The sensitivity of words will also change in different scenarios and timeliness. This study sets the context as political context, agricultural context and common context. The path research of acquiring method of sensitive words is studied and a set of sensitive word library is established based on self-learning, through the implementation of the system. Moreover, the context-based detecting method of sensitive words is established.

Keywords Sensitive words · Context · Path research · Detecting method

1 Introduction

At present, many enterprises generally prefer self-built database of sensitive words and use the extraction algorithm of sensitive words to detect and filter users' information, aiming to achieve the filtering of commonly used sensitive information. However, the disadvantage is that the filtering of sensitive information is affected by sensitive word library and extraction algorithm of sensitive words, so it needs to be supported by multiple sets of corresponding sensitive word library in different application scenarios. At the same time, this kind of detecting method is relatively static, without the ability of changing based on the context. Information

J. Yu · Q. Wei · C. Luo (✉)
Institute of Agricultural Information and Economics, Beijing Academy of Agriculture and Forestry, Beijing 100079, China
e-mail: Luocs@agri.ac.cn

J. Yu · Q. Wei · C. Luo
Beijing Research Center of Engineering Technology on Rural Distance Information Service, Beijing 100079, China

J. Zhang
Beijing Academy of Agricultural and Forestry Sciences, Beijing 100079, China

V. Jain et al. (eds.), *Recent Trends in Intelligent Computing, Communication and Devices*, Advances in Intelligent Systems and Computing 1031,
https://doi.org/10.1007/978-981-13-9406-5_38

filtering is relatively mechanical. In order to adapt to the multi-business environment system better, we propose the design and implementation of context-based adaptive filtering system for sensitive words.

2 Design Scheme

As we know, different contexts create different semantics for language application. Sensitive words are also affected by application scenarios and application timeliness. The sensitivity of words will also change in different scenarios and timeliness. This study sets the context as political context, agricultural context and common context. The path research of acquiring method of sensitive words is studied and a set of sensitive word library is established based on self-learning, through the implementation of the system. Moreover, the context-based detecting method of sensitive words is established.

2.1 Design Principles

The principle of context-based word filtering is as follows. Adopting the natural language processing technology, automatic data crawling and analysis of the current application environment are established. Based on the analysis of context word library, automatic sensitive category identification of application environment through clustering technology is built. Depending on the classification relationship of sensitive words, sensitive words in content can be filtered.

Construction principle of classification relationship of sensitive words: The machine self-learning technology is used to crawl the application scenario data, and the keyword extraction technology of natural language processing is used for automatic learning to generate the keyword library of context.

2.2 Key Technology

The implementation of the system mainly relies on the study of natural language processing technology, and the key technology framework involved is shown in Fig. 1.

The research method is mainly used in data pre-processing in the data preparation stage. Specifically speaking, data crawling technology is used for automatic acquisition of corpus to provide data support for system automatic learning. Natural language processing technology is used to classify the acquired data. In this system, NLPIR is adapted. The ontology object is used to provide a basis for data set classification. Machine learning technology is mainly used for feature extraction of

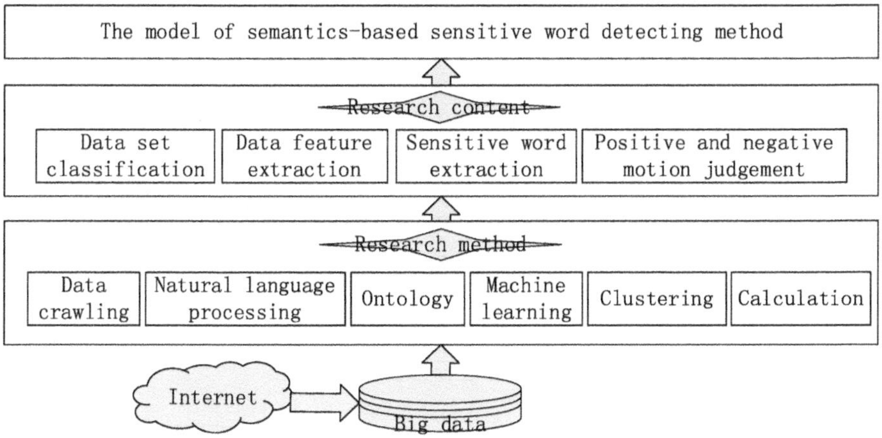

Fig. 1 Technology framework

corpus. Clustering technology is used for automatic classification on the basis of feature extraction results. Calculation mainly provides algorithm support for software functions.

The research content is used to solve the technical problems that need to be solved in dynamic application. Data set classification is used for automatic classification in the acquired context. Data feature extraction is used for the acquisition of keywords in context. Sensitive word extraction is to extract sensitive words through sensitivity analysis of keywords in context. The positive and negative emotion judgment is used for the positive and negative judgment of the sensitive words extracted from the relative context, which provides the basis for whether to shield or not.

2.3 *Overall System Framework*

The whole operation system is divided into five layers: The data pre-processing layer is mainly used to crawl scenario data automatically based on application scenarios, extract scenario feature after data cleaning and build context feature database. At the same time, the sensitive word database and sensitive word labelling classification database are built to collect, label and store the data of sensitive words, which are realized by ‘context-based knowledge management system of sensitive words’. Secondly, the database construction layer mainly designs the database according to the knowledge data relationship, which provides knowledge data and knowledge database support for the system construction. The data analysis layer uses various algorithms to calculate the weight feature of knowledge data to provide knowledge analysis and extraction for the system application. The interface

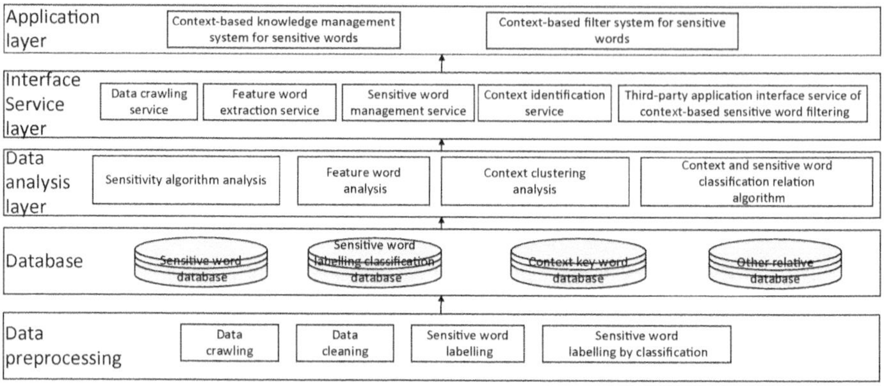

Fig. 2 Overall system framework

service layer provides unified data application and management interface for various data knowledge service and management, realizing the standardization of interface and data format. The application layer includes 'context-based knowledge management system for sensitive words' and 'context-based filter system for sensitive words'. The former is used to build the relationship between knowledge and knowledge, while the latter is based on examples of implementation of interface scenario application (Fig. 2).

2.4 *Technical Route*

In the first phase of the development process, data crawling is carried out, which mainly realizes the classification selection of data crawling websites and the automatic extraction ability of data crawling algorithms for websites developed by different technologies. In the second phase, data are classified and cleaned up, which mainly deals with the cleaning of irrelevant data elements and the extraction of key information in the text. In the third phase, based on natural language technology, the key feature extraction training of key information extracted is carried out. So that the coarse-grained sensitive word list is established. The fourth phase is to improve precision of sensitive words by artificial interpretation. At the same time, taking this as a sample, the parameters of training network are adjusted. In the fifth phase, the algorithm of sensitivity detection is implemented to realize the core code programming of semantics-based sensitive word detecting method. The sixth phase is to develop the sensitivity detecting system and the third-party common interface to mainly achieve the implementation of distributed architecture application. The seventh phase is the deployment and release of the platform, achieving the possibility of putting it into actual operation. The figure of technical route is as follows (Fig. 3).

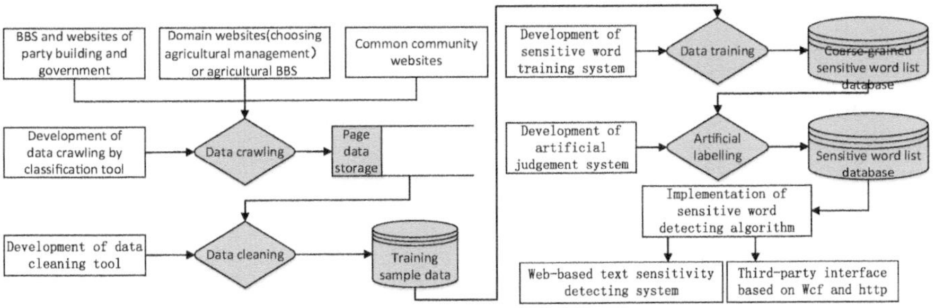

Fig. 3 Figure of technical route

3 Implementation of the System

Based on Visual Studio 2017, the system is developed with C# along with JavaScript. The database adopts Oracle12C as the data storage platform and ensures the integrity of platform storage knowledge by writing SQL database trigger mechanism.

3.1 Context-Based Knowledge Management System for Sensitive Words

The system realizes unified standard interface through WCF technology and adopts intelligent C/S mode to realize thin client development. The system mainly involves the following functional modules: user management module, context construction module, sensitive word management module and context-sensitive word mapping module. The detailed management module diagram is shown in Fig. 4, and the software interface is shown in Fig. 5.

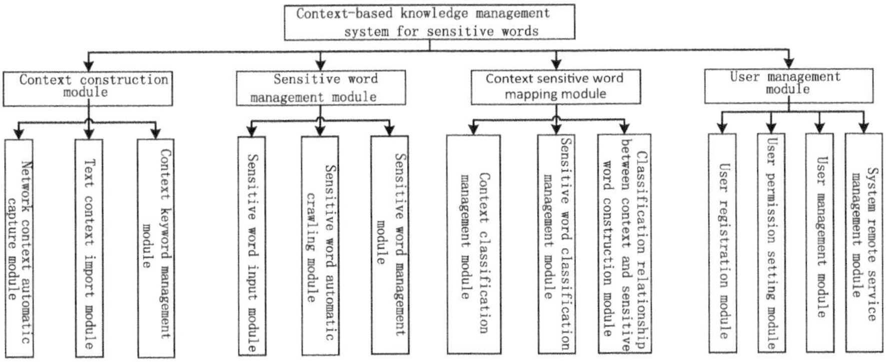

Fig. 4 Functional modules

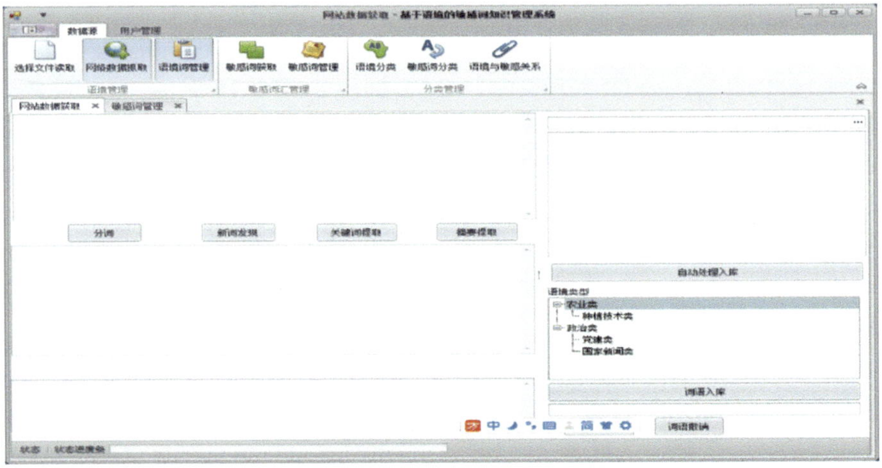

Fig. 5 Software interface

3.2 *Implementation of Context-Based Filter System for Sensitive Words*

Context-based filter system for sensitive words is an application built on the knowledge base of sensitive words. Based on autonomous learning of different knowledge bases, it can be dynamically used in different network interactive systems. The processing flow chart is shown in Fig. 6.

3.3 *Description of Main Relevant Implementation Methods*

3.3.1 Data Acquisition

The data source mainly relies on the Internet. According to the set classification, relevant websites are crawled to the local site as the initial data source. Political

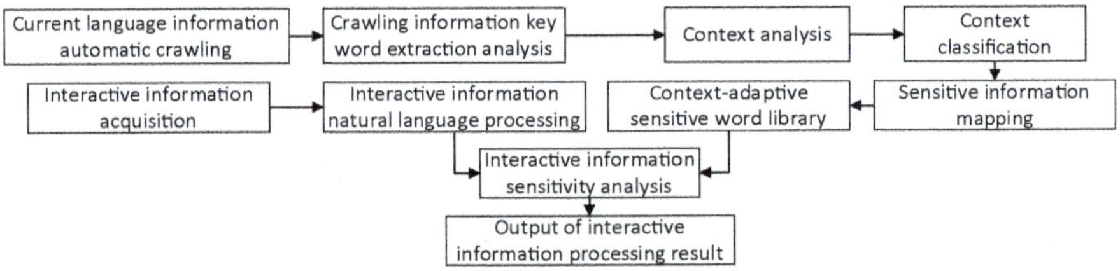

Fig. 6 Context-based processing flow for sensitive words

words mainly come from websites and BBS of party building and government. For domain-sensitive words, we choose agricultural field to build sensitive words, and mainly choose agricultural management website and agricultural management BBS for data collection. For generic-sensitive words, we use BBS websites with high activity, such as community websites, tianya BBS, etc. The number of pages crawled per sample is not less than 10,000.

3.3.2 Crawling of Sensitive Word Data Network

The main purpose of network crawling is to download web pages on the Internet to form a local image backup of web content. The automatic crawling module of network is to realize automatic crawling of website and big data acquisition. Due to different data acquisition targets, different crawling rules must be made according to different target in order to obtain reasonable and useful data sources. The schematic diagram is as follows (Fig. 7).

3.3.3 Construction of Sensitive Word Library

Sensitive word library is used to record sensitive words and their weights, which is the basis of sensitive words detection. The extracting method of keywords is as follows. This paper is based on the calculation method of generalized Jaccard coefficient. On the basis of this calculation method, irrelevant data are added to carry out the seeking-the-same operation, while the result obtained is added to carry out the seeking-difference operation. So that it further improve the precision of sensitive data, and finally sensitive word list of this classification is obtained. On this basis, an artificial labelling system for sensitive words is developed. The basic

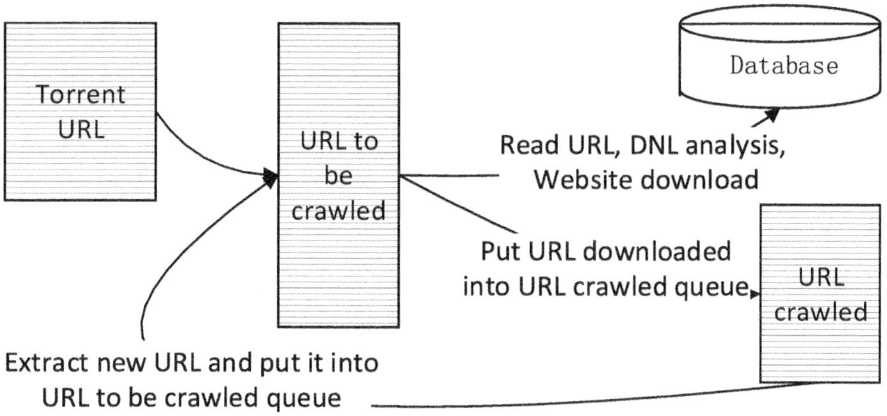

Fig. 7 Schematic diagram of crawling

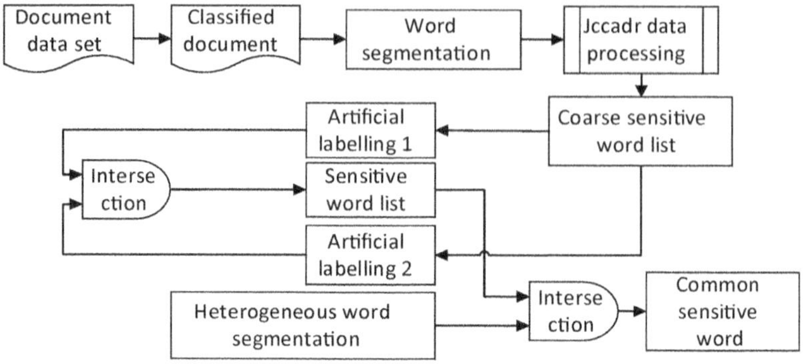

Fig. 8 Construction process of sensitive word library

word list was labelled by different types of people, the labelled results are statistically counted, and the weight of sensitive words is obtained according to the labelled word frequency and the standardized operation. The specific process is shown in Fig. 8.

3.3.4 Research on Detecting Method of Sensitive Words

In mobile Internet, in order to obtain economic benefits or achieve certain political purposes, some people often use some methods to interfere with the keywords in the information to avoid filtering. As Chinese characters have their own complexity compared to English, the common detecting methods of sensitive words for Chinese text information are as follows. Replace keywords with pinyin or the first letter. Such as 'Falun' or 'Gong' instead of 'Falun Gong'. Replace with traditional characters. Separate keywords with special symbols such as 'Falun ? Gong' separate the Chinese characters into pieces. This study adopts the semantics-based sensitive word detecting method. Specifically speaking, firstly, the input information is pre-processed to remove special characters. Afterwards, semantic segmentation is carried out. The words segmented are added into the word order list and the words in the list will be de-sensitized. If the list contains sensitive words, label them; while if not, non-sensitive words will be removed, and the adjacent words will be combined, and then the consent words will be extended. If there are no sensitive words in the list, the combined single words will be combined with pinyin, and then they are matched with sensitive homophones through the pinyin semantic synthesis with sensitive words. If there are, it will be considered that there are sensitive words. The specific flow chart is shown in Figs. 9 and 10.

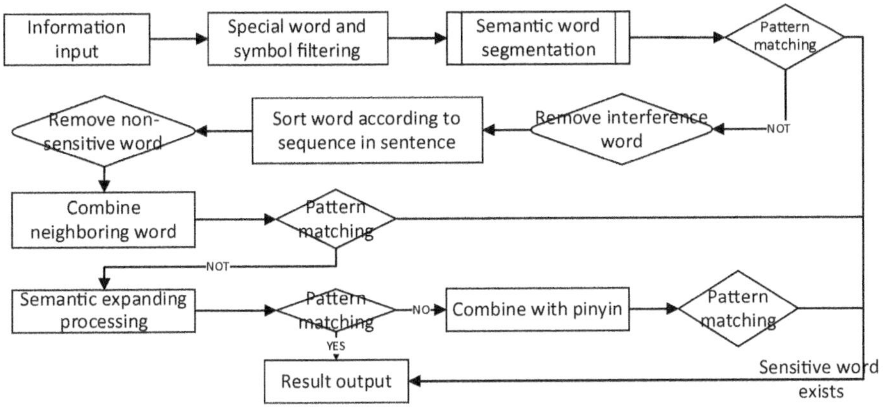

Fig. 9 Process of detecting sensitive words

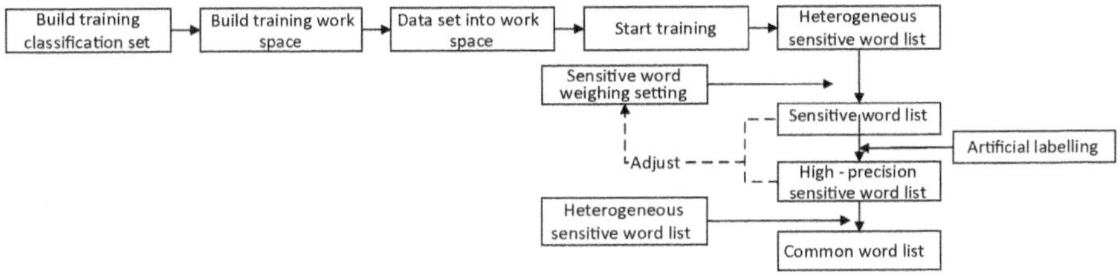

Fig. 10 Process of detecting sensitive words

3.3.5 Sensitive Word Training

The training system of sensitive words carries out natural semantic analysis and processing based on the cleaned text data, completes calculation, collection, storage and management of sensitive words and enriches the list of sensitive words. The operation process is as follows.

4 Conclusion

Context-based adaptive filtering system relies on the classification of basic knowledge data. Therefore, in the application of the system, attention should be paid to the studying, organizing and labelling of basic knowledge. The more the classifications of basic context, the more specialized the application is, and the richer the knowledge data is, the stronger and more accurate the sensitivity detection is. At the same time, the more the sensitive words are learned and labelled, the more accurate the information sensitivity is in the application. However, the disadvantage is that the accuracy of feature extraction and sensitive word labelling based on machine learning is also dependent on interference from artificial labelling, which cannot be fully automatic. Thus, the improvement of sensitivity algorithm and detecting method is needed.

References

1. Lin, X., Xia, Y., Guo, J., Yu, X.: Sensitive file detection method based on CNN. Jisuanji Yu Xiandaihua **7**, 29–32 (2018)
2. Li, Y., Pan, Q., Yang, T.: Sensitive information recognition based on short text sentiment analysis. J. Xi'an Jiaotong Univ. **50**(9), 80–84 (2016)
3. Xu, J., Luo, Z.: Zhang L (2016) Application of semantic extension approach in sensitive data identification. Modern Electronics Technique **39**(12), 80–82 (2016)
4. Collobert, R., Weston, J., Botton, L., et al.: Natural language processing (almost) from scratch. J. Mach. Learn. Res. **12**(1), 2493–2537 (2011)
5. Shen Y, He X, Gao J et al (2014) Learning semantic representations using convolutional neural net-works for web search. In: Proceedings of the 23rd Inter-national Conference on World Wide Web pp. 373–374 (2014)
6. Mikolov T, Karafiát M, Burget L, et al. Recurrent neural network based language model. In: Proceedings of the 11th Annual Conference of the International Speech Communication Association pp. 1045–1048 (2010)
7. Hinton, G.E.: Salakhyurdinov reducing the dimensionality of data with neural networks. Science **313**(5786), 504–507 (2006)

Data Crawling and Cluster Analysis of Online Reviews in Xi'an Catering Industry

Jie Kong and Meng Ren

Abstract Nowadays, online reviews on the social network play an important role in catering industry. However, the acquisition and analysis of massive review data propose challenges for researchers and practitioners. In this study, a web crawler program is designed to collect the data of reviews of restaurants in Xi'an, China, from a famous Chinese social network, then the features of customers are visualized, and the segmentation of restaurants is analyzed by unsupervised learning. The contribution of this study could help catering researchers and practitioners better understand the features of customers and restaurants.

Keywords Web crawler · *k*-means · Data mining · Restaurant

1 Introduction

With the rapid development and wide application of the Internet, people's lifestyles, such as shopping, catering, entertainment, and transportation, have undergone tremendous changes in recent years. More and more people are accustomed to share their opinions and experiences with each other through online reviews on the social networks. In the catering industry, these reviews can not only help customers better understand the quality of service provided by restaurants, but also contain valuable information about the features of the customers. Therefore, how to obtain the data of online reviews and extract the valuable information efficiently and precisely has become an important task for catering business.

Web crawler, also known as web spider, is a solution to obtain the online review of restaurants. It is a computer program or script that automatically fetches the World Wide Web information according to certain rules [1]. Because of its efficiency, automatic, and accuracy in obtaining online data, it is widely used for online information collecting in recent years. To extract valuable knowledge and infor-

J. Kong (✉) · M. Ren
School of Computer Science, Xi'an Shiyou University, Xi'an, China
e-mail: jkong@xsyu.edu.cn

V. Jain et al. (eds.), *Recent Trends in Intelligent Computing, Communication and Devices*, Advances in Intelligent Systems and Computing 1031,
https://doi.org/10.1007/978-981-13-9406-5_39

mation from the data generated by daily business activities, data mining techniques are getting popular by scholars of computer science and management [2]. It could provide insightful analysis ability for the massive and complicated data that is difficult for humans to understand. This study tries to investigate the characteristics of customers and restaurants from the online reviews of restaurants in Xi'an, China. A web crawler program is designed to collect the data of reviews from Dazhong Dianping (www.dianping.com), a famous social network for catering and entertainment in China. On the basis of data crawling, customer's characteristics and restaurant's segmentation are analyzed by data visualization and k-means algorithm. The work of this study could help researchers design more targeted data acquisition and analysis solutions for specific Web sites, as well as provide comprehensible customer and restaurants characteristics for catering practitioners.

2 Related Works

Previous researches about the data mining on the restaurant reviews have shown great interests in the customer's sentiment analysis. For instance, review documents are collected by restaurant search sites and analyzed by improved Naïve Bayes to understand the sentiments of the reviews [3]. Yan et al. investigate customer revisit intention by text mining from online reviews [4]. The online reviews have also been collected to understand customer satisfaction [5], multi-dimensional sentiment [6], and tourist dining preference [7] of the restaurant through text mining. Some researchers try to investigate the restaurant online reviews by supervised and unsupervised learning methods. For example, Zhang et al. perform sentiment analysis on online restaurant reviews with the help of Naïve Bayes and SVM [8]. Claypo and Jaiyen identify the positive and negative groups of Thai restaurants by using unsupervised learning algorithm on customer reviews [9]. However, to the best of our knowledge, there are only a few unsupervised learning-related studies which have been proposed in recent years, which could be applied to identify the market segmentation or the clustering of customers.

Besides, most of the existed studies did not elaborate on the design and implementation of web crawler programs, which makes it difficult for other researchers to get a reference when obtaining online reviews. The data sources of these studies are mainly from foreign social networks, such as Yelp [10] and TripAdvisor [11], which cannot accurately reflect the characteristics of China's domestic catering industry.

For the above-mentioned limitations, this study tries to specifically introduce the design of web crawler for a famous social network of catering and entertainment in China. After the data collection, an unsupervised learning algorithm is performed to analyze the characteristics of different restaurant segments.

3 The Implementation of Web Crawler

The web crawler for Dazhong Dianping is implemented on the basis of the Scrapy, which is a web crawling framework written by Python [12]. The Scrapy could be used to extract data from Web sites efficiently through its APIs. The component of the Scrapy used in the web crawler is listed in Table 1.

The web crawler implemented in this study is based on the above-mentioned components of the Scrapy through its APIs. The workflow of the web crawler is shown in Fig. 1. The obtained items include the restaurant information, such as ID, name, rank, address, number of reviews, category, score of taste, score of environment, and score of service, as well as the customer information, such as user ID, user name, city, and birthday. The Spider is a class to obtain data from a Web site, so as to generate items. It is implemented by inheriting the scrapy.Spider class. Data storage is set in Item pipeline. In this study, SQL server is adopted as the database. The process_item() function stores the data crawled from the web page into the linked database tables through SQL statements, while the close_spdier() function is used to close the database connection.

4 Data Segmentation

In this study, the data segmentation analysis is performed by *k*-means, which is a classical unsupervised learning algorithm. Given the sample set $D = \{x_1, x_2, \ldots, x_m\}$, *k*-means algorithm minimizes the square error of the clusters obtained by segmentation.

Table 1 Scrapy components applied in this study

Name of component	Introduction
Scrapy engine	It handles data flow of the whole system and triggers transactions
Scheduler	It is used to receive requests from the Scrapy engine, press the requests into the queue, and return when the engine requests again
Downloader	Used to download the contents of Web sites and return the contents to the Spider
Spider	Developing rules to parse specific domain names or web pages, and write classes for analyzing responses and extracting items. Each spider is responsible for handling a specific Web site
Item pipeline	Responsible for the processing of items extracted by the Spider from the web pages. Its main task is to clean, verify, and store data
Downloader middleware	It is a hook framework between Scrapy engine and Downloader, deals with requests and responses between the two components

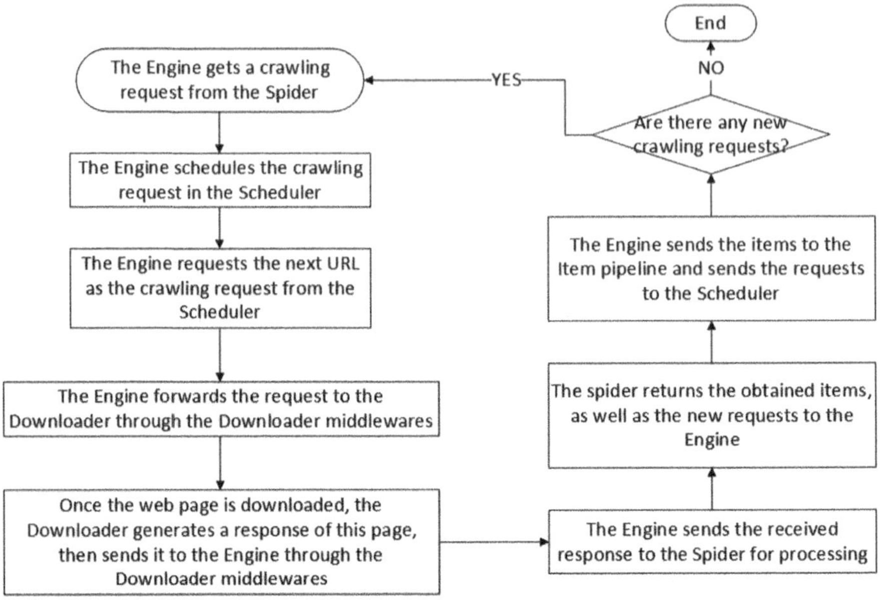

Fig. 1 Workflow of the web crawler

$$E = \sum_{i=1}^{k} \sum_{x \in c_i} \|x - \mu_i\|^2, \quad (1)$$

where $\mu_i = 1/|C_i| \cdot \sum_{x \in c_i} x$ is the mean vector of cluster C_i. This equation indicates the closeness of samples to the mean vector within a cluster. The smaller the value of E, the higher the similarity of samples in the cluster. Due to the limitation of space, the k-means algorithm is not specifically introduced in this study. The detailed information about this algorithm could be found in [13].

5 Experiment Results

The demographic data of customers (namely reviewers) on Dazhong Dianping is collected by web crawler and then divided into two groups: the local customers and the outsiders. The statistical interpretations of some attributes of customers in the two groups are shown in Fig. 2. Due to the privacy limitation, customers' demographical information displayed on the Web site only includes the age, the relationship status, the constellation, the gender, etc. It can be observed from Fig. 2a, b that the largest number of local customers was born in 1985–1994, while the

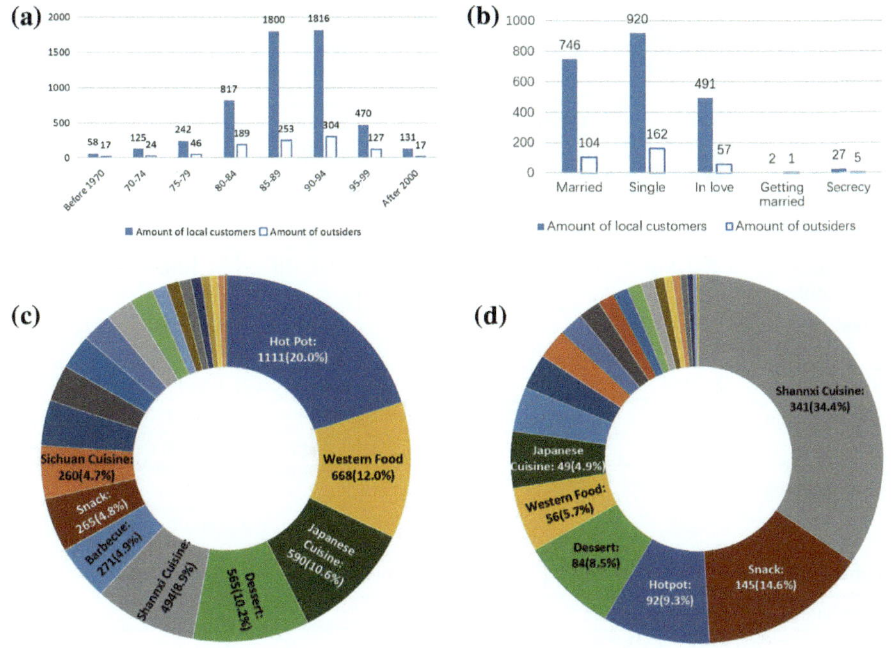

Fig. 2 **a** Age distribution of the local customers and outsiders. **b** Relationship status distribution of the local customers and outsiders. **c** Review amount of local customers on different types of restaurants. **d** Review amount of outsiders on different types of restaurants

outsiders in 1980–1994. For both of local customers and outsiders, the largest number of people is single, followed by married people and those in love.

Through the amount of reviews, it could be inferred that which kinds of foods or restaurants are more popular. Figure 2c indicates that hot pot, Western food, Japanese cuisine, dessert, and Shaanxi cuisine are the top five popular foods for local customers. Figure 2d shows that the food preferences of outsiders are quite different from those of local customers. Shaanxi cuisine is the most popular food for outsiders, followed by snack, hot pot, dessert, and Western food.

The web crawler has totally obtained the information of 753 restaurants in Xi'an. To present the illustrative analysis result, clustering analysis of edge data is performed for the top 100 and bottom 100 restaurants in overall score, respectively. When $k = 6$, the clustering result of the top 100 restaurants is shown in Table 2. When $k = 4$, the clustering result of the bottom 100 restaurants is shown in Table 3.

In these two tables, the values of attributes in each column belong to the centroid in each cluster. The "score" is the overall rating of a restaurant, with a full score of 5. The "Taste," "Environment," and "Service" reflect the score of restaurant in some aspects, with a full score of 10. It could be found that the top 100 restaurants are concentrated in busy commercial districts, while the bottom 100 restaurants may located in both busy commercial districts, such as Bell tower/Drum tower, Qujiang,

Table 2 Clustering result of the top 100 restaurants

Attributes	Cluster 1	Cluster 2	Cluster 3	Cluster 4	Cluster 5	Cluster 6
Score	4.5	4.5	4	4.5	4.5	5
Commercial district	Bell tower/ Drum tower	Qujiang	Xiaozhai	Beilin District	Minleyuan	Xiaozhai
# of reviews	174	22	27	113	61	731
Per capital expenditure	25 RMB	83 RMB	54 RMB	57 RMB	71 RMB	83 RMB
Food type	Snack	Hot pot	Hot pot	Western food	Hot pot	Hot pot
Taste	8.5364	8.675	8.0857	8.45	8.5846	9.0371
Environment	8.9545	8.5	8.5095	8.65	8.7077	9.0657
Service	8.8182	8.2625	8.1619	8.8417	8.6923	9.1086

Table 3 Clustering result of the bottom 100 restaurants

Attributes	Cluster 1	Cluster 2	Cluster 3	Cluster 4
Score	4.5	4	4	4
Commercial district	Weiqu	Bell tower/Drum tower	Qujiang	Weiyang Road
# of reviews	38	52	815	1105
Per capital expenditure	58 RMB	15 RMB	45 RMB	65 RMB
Food type	Hot pot	Snack	Hot pot	Guangdong cuisine
Taste	7.52	8.1457	8.397	7.9889
Environment	7.56	7.7486	7.8939	8.3593
Service	7.24	7.6886	7.6758	7.9556

and ordinary districts, such as Weiqu and Weiyang Road. In particular, the hot pot restaurants in Weiqu and Qujiang, as well as the Guangdong cuisine restaurants on Weiyang Road, need attention due to the features in clusters 1, 3, and 4 of the bottom 100 restaurants. Generally, hot pot restaurants are more popular in Xi'an, which account for the majority of the six clusters in top 100 restaurants and half of four clusters in bottom 100 restaurants.

6 Conclusion

To investigate the characteristics of customers and restaurants in Xi'an, China, from the online review on Dazhong Dianping, a web crawler is implemented. On the basis of the obtained data by web crawler, clustering analysis of the restaurants is performed by *k*-means algorithm. In the experiment of this study, some customer features are visualized and clusters for the top 100 and bottom 100 restaurants are presented, which could help catering researchers and practitioners better understand the features of customers and restaurants in Xi'an.

Acknowledgements This work is supported by the Science Research Project of Shaanxi Provincial Department of Education (Grant No: 17JK0614) and the Youth Innovation Fund of Xian Shiyou University (Grant No: 2013BS025).

References

1. Udapure, T.V., Kale, R.D., Dharmik, R.C.: Study of web crawler and its different types. IOSR J. Comput. Eng. **16**(1), 01–05 (2014)
2. Li, J., Xu, L., Tang, L., Wang, S., Li, L.: Big data in tourism research: a literature review. Tour. Manag. **68**, 301–323 (2018)
3. Kang, H., Yoo, S.J., Han, D.: Senti-lexicon and improved Naïve Bayes algorithms for sentiment analysis of restaurant reviews. Expert Syst. Appl. **39**(5), 6000–6010 (2012)
4. Yan, X., Wang, J., Chau, M.: Customer revisit intention to restaurants: evidence from online reviews. Inf. Syst. Front. **17**(3), 645–657 (2015)
5. Berezina, K., Bilgihan, A., Cobanoglu, C., Okumus, F.: Understanding satisfied and dissatisfied hotel customers: text mining of online hotel reviews. J. Hospitality Mark. Manag. **25**(1), 1–24 (2016)
6. Gan, Q., Ferns, B.H., Yu, Y., Jin, L.: A text mining and multidimensional sentiment analysis of online restaurant reviews. J. Qual. Assur. Hosp. Tourism **18**(4), 465–492 (2017)
7. Vu, H.Q., Li, G., Law, R., Zhang, Y.: Exploring tourist dining preferences based on restaurant reviews. J. Travel Res. https://doi.org/10.1177/0047287517744672 (2017)
8. Zhang, Z., Ye, Q., Zhang, Z., Li, Y.: Sentiment classification of Internet restaurant reviews written in Cantonese. Expert Syst. Appl. **38**(6), 7674–7682 (2011)
9. Claypo, N., & Jaiyen, S.: Opinion mining for thai restaurant reviews using K-Means clustering and MRF feature selection. In: 7th International Conference on Knowledge and Smart Technology (KST), pp. 105–108 (2015)
10. Nakayama, M., Wan, Y.: Is culture of origin associated with more expressions? An analysis of yelp reviews on Japanese restaurants. Tour. Manag. **66**, 329–338 (2018)
11. Zhang, H.Y., Ji, P., Wang, J.Q., & Chen, X.H.: A novel decision support model for satisfactory restaurants utilizing social information: a case study of TripAdvisor.com. Tour. Manag., **59**, 281–297 (2017)
12. Xie, D.X., Xia, W.F.: Design and implementation of the topic-focused crawler based on scrapy. Adv. Mater. Res. **850–851**, 4 (2013)
13. Jain, A.K.: Data clustering: 50 years beyond K-means. Pattern Recogn. Lett. **31**(8), 651–666 (2010)

A Preliminary Study on the Assessment of Restrictedness in High Functioning Autism

Zhong Zhao, Xiaobin Zhang, Xinyao Hu, Xiaolan Cao, Jianping Lu and Xingda Qu

Abstract Restricted and repetitive behavior is one of the two cardinal symptoms of autism, and it might find its root in the neuropsychological trait of "insistence on sameness". Conceived from the perspective of embodied psychology, our study aimed to investigate whether "restrictedness" might be embodied in the patient's motor activity. Four patients with high functioning autism (ASD) and five participants of typical development (TD) aged between 6 and 13 years were recruited to perform a horizontal left–right movement with their dominant hand. Instructions were to perform as complex movement as possible by demonstrating variant amplitude and frequency. Entropy of movement amplitude, frequency, and velocity was calculated as indices of movement "restrictedness". Results demonstrated that the velocity entropy was significantly lower in ASD than in TD (both $p < 0.01$), indicating a higher level of restrictedness in ASD's movement. Given the fact that current evaluation of RRB is questionnaire-based, which might be biased by subjective factors of the evaluator, our finding sheds light on potential objective assessment of restrictedness in autism. A discussion on the comparison between our approach and recent method on automatic RRB pattern recognition was formulated.

Keywords Autism · Restrictedness and repetitive behavior · Entropy · Behavioral assessment

Z. Zhao · X. Zhang · X. Hu · X. Qu
Institute of Human Factors and Ergonomics, Shenzhen University, Shenzhen, China

X. Cao · J. Lu (✉)
Department of Child Psychiatry of Shenzhen Kangning Hospital, Shenzhen Mental Health Center, Shenzhen, China
e-mail: szlujianping@126.com

V. Jain et al. (eds.), *Recent Trends in Intelligent Computing, Communication and Devices*, Advances in Intelligent Systems and Computing 1031,
https://doi.org/10.1007/978-981-13-9406-5_40

1 Introduction

According to DSM-V, restricted repetitive behavior (RRB) is one of the two cardinal symptoms of autism spectrum disorder [1]. Patients with ASD exhibit RRBs such as body rocking, hand flapping, circumscribed interest, and ritualistic behavior. RRBs have been found to impair social communication, quality of life, and many other aspects of ASD patients.

The current evaluation of RRB is achieved by the informant's evaluation, which is widely criticized as having questionable accuracy. Therefore, a more objective assessment tool on RRBs is required. Regardless of the differences in overt manifestations, extant literature postulated that RRBs might find its root in the neuropsychological trait of restrictedness—a feature that is resistant to change [2]. From the perspective of embodied psychology, human behaviors play a vital role in revealing the inner psychological trait [3]. Accordingly, it is reasonable to assume that the behavioral assessment would approach the measurement the restricted trait in ASD.

In addition, a recent study conducted by Słowiński et al. [4] inspired us to explore the restrictedness in ASD through behavioral assessment. The objective of their research was to investigate the existence of individual motor signature—a personalized motor feature that differentiates individuals. Healthy participants were instructed to perform one-dimensional movement as complex (with ever-changing amplitude and frequency) as possible. Their results evidenced that the velocity profile effectively captured the subtle motor differences between individuals. This finding enlightened us on the search of the autistic motor signature with the same motor task.

In the present study, we employed the same motor task to investigate whether restrictedness is embodied in the patient's movement. It was hypothesized that due to the restricted trait, the overall kinematic complexity would be lower in ASD patients. We computed the entropy of amplitude, frequency, and velocity as indices of the flexibility and complexity of movement. We specifically hypothesized that the entropy of amplitude, frequency, and velocity would be significantly lower in ASD.

2 Method

2.1 Participants

Four patients with high functioning autism (ASD) and five participants of typical development (TD) aged between 6 and 13 years were recruited to perform a horizontal left–right movement with their dominant hand. Patients with other clinical conditions such as mental retardation and ADHD were not included. Written informed consent was signed by the participant's caregivers. IQ was administrated with Raven's Advanced Progressive Matrices to assure that ASD had an average non-verbal IQ level. The experimental protocol conformed the Declaration of Helsinki and followed the ethical guidelines of Shenzhen University.

2.2 Apparatus

The experimental apparatus incorporated a computer (Lenovo Legion R720-15IKBN), a Leap Motion device (Leap Motion Inc.), two sticks, and a string (Fig. 1). The LeapMotion served the purpose of recording the hand movement of the participants at a sampling frequency over 200 Hz. The string was tied to the two sticks in order to allow participants to perform hand movements above it. The distance between the two sticks was 60 cm.

2.3 Experimental Procedure

Participants were required to perform one-dimensional movement within the recording area of LeapMotion. The instruction was to perform as complex movement as possible. Prior to the real experiment, the experimenter behaviorally demonstrated that the simple movement referred to periodic movements with mono-amplitude and frequency, whereas complex movement to unpredictable movements with variant amplitude and frequency. Participants had several practice

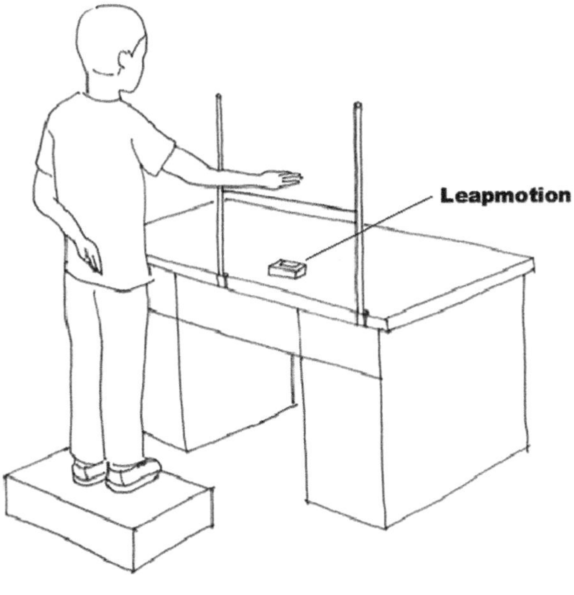

Fig. 1 Experimental setup

trials to assure that they fully understood the instructions. In order to avoid falsely registering the movement of the subdominant hand, participants were explicitly told to keep their subdominant hand behind their back. A solid mat was offered to make sure that the dominant hand could move naturally and comfortably above the string. Three trials of movement task were recorded with each trial lasting 60 s. Participants were required not to withdraw their hand out of the recording zone or to put the subdominant hand in it. A trial was repeated if any experimental rules were violated (e.g., hand withdraw from recording zone, stop hand moving voluntarily). A break of 2–5 min was arranged between two consecutive trials to reduce fatigue.

2.4 Data Analysis

Data were pre-processed with MATLAB (2017a). The original time series of the palm position was processed with interpolation by means of piecewise cubic hermite interpolating polynomial method and filtering with a second-order low-pass Butterworth filter (5 Hz cut-off). Afterward, the amplitude of each single oscillation within a trial, frequency spectrum, and velocity time series could be derived from the processed position time series. As for the amplitude, since each trial of movement was composed of multiple oscillations, the amplitude of a single oscillation was computed as the distance between two endpoints (where velocity equaled 0). In this way, a set of amplitude values could be obtained by taking into account the amplitude values of all oscillations. The oscillating frequency spectrum was achieved by applying the fast Fourier transform, and the velocity time series was computed as the first derivative of the corresponding position time series.

In the calculation of entropy, we first set the threshold ranges for amplitude, frequency, and velocity values to be incorporated into entropy computation. The range for amplitude entropy was 0–60 cm, frequency 0–5 Hz, and velocity −3 and 3 m/s. Values out of these ranges were discarded from further analysis. Afterward, we used a normalized histogram with 101 equally distant bins [4] within the range to compute the probability of each bins. Then, the Shannon entropy was calculated as:

$$\text{Entropy} = -\sum_{i=1}^{n} p(x_i) * \log 2p(x_i)$$

In the above equation, n equals 101 in the present study, and $p(x_i)$ denotes the probability of the *i*-th bin.

2.5 *Statistical Analysis*

Statistical analysis was conducted on R by implementing linear mixed-effects models (LMEMs). The dependent variables were entropy of amplitude, frequency, and velocity. We entered autism, age, and the participant's gender as fixed factors. The random factors were participant and the order of the experimental trials. Maximal random-effects structure was included justified by the experimental design and assumptions [5]. In all the LMEMs, we specify random slopes for the by-subject effect of age. The lme4 [6] package for *R* was used to perform LMEMs, and pairwise comparisons were conducted by implementing the Lsmeans [7] package on the significant predictor.

3 Results

Results demonstrated that autism was a significant predictor of velocity entropy ($p < 0.01$). Pairwise comparison showed that the velocity entropy was significantly higher in TD than ASD ($p < 0.01$). As for the entropy of amplitude and frequency, results failed to show that autism was a significant predictor of either of these two variables (both $p > 0.1$) (Fig. 2).

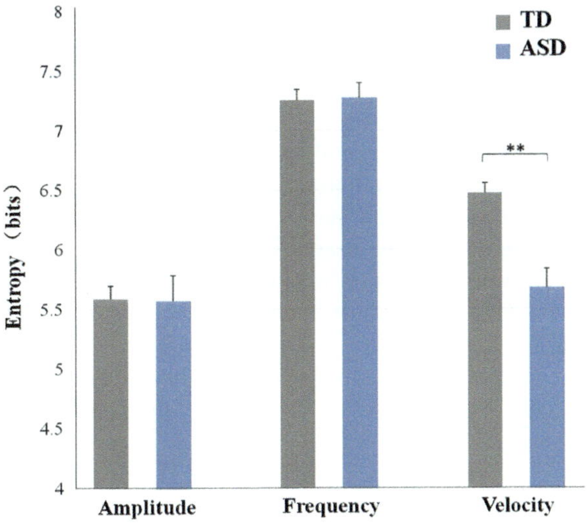

Fig. 2 Comparisons on entropy of amplitude, frequency, and velocity between ASD and TD. Error bars represent standard errors. ** means $p < 0.01$

4 Discussion

The present study investigated whether the restrictedness is embodied in the movement of patients with high functioning autism. The initial hypothesis was fulfilled by showing that the velocity entropy was significantly lowered in ASD as compared to the TD group. The result that the velocity entropy in ASD was lower than TD was consistent with the finding by Słowiński et al. [4], who evidenced the existence of individual motor signature in the velocity profile. It was implied that information resides in the velocity profile that could help distinguish ASD from TD and could be further utilized for ASD screening and diagnosis.

In order to objectively measure RRBs, a couple of recent scientific attempts were tried to develop motion pattern recognition algorithms to automatically detect repeated behaviors with the aid of accelerometers [8, 9]. For instance, Goodwin et al. [9] utilized accelerometers attached to the body of six ASD patients to detect repetitive behaviors, such as hand flapping or body rocking. The most obvious advantage of their method over ours is that their method enables the detection of RRBs in natural environment, whereas our method was confined to be used in experimental or testing settings. In addition, their method could be utilized to detect RRBs in infants, and our approach could only be employed to detect restrictedness in patients who are able to perform the motor task, and obviously, it could not be used with infants. However, the detection of RRBs in natural setting also determines that their method might help less to screen autism out of healthy people or to detect the severity of ASD. On one hand, RRB is not a unique characteristic of ASD, it could also be found in healthy individuals (e.g., leg swinging). The detection of RRBs in natural environment is not sufficient to confirm the decision of ASD diagnosis. On the contrary, our method could potentially be used to screen ASD out of TD by showing that the movement was more restricted in ASD than TD. On the other hand, since human behavior depends on both inner psychological trait and environmental factors, RRBs captured in natural environment could be hardly attributed to pure restricted trait of ASD. Chances are that mild ASD patients exhibit more RRBs than severe ASD if mild ASD is more frequently exposed to RRB-inducing environments. As for our method, although only high functioning ASD was tested, it is worth being examined that the restrictedness in severe ASD would be significantly higher than in mild ASD when performing the movement task.

In sum, this is a preliminary study on the objective measure of restrictedness in high functioning ASD. Our finding sheds light on the possibility of measuring the restricted trait in autism through behavioral assessment. Due to the profound heterogeneity among ASD patients, however, we reckon that more patients should be recruited to confirm our finding.

Acknowledgements The study was financially supported by the SZU funding project (#85303-00000130), Science and Technology Innovation Committee of Shenzhen (No. JCYJ20160429185235132), and Sanming project of medicine in Shenzhen (No. SZSM201612079).

References

1. American Psychiatric Association: DSM 5 [Internet]. Am. J. Psychiatry (2013)
2. Kim, S.J., et al.: A quantitative association study of SLC25A12 and restricted repetitive behavior traits in autism spectrum disorders. Mol. Autism **2**(1), 8 (2011)
3. Niedenthal, P.M., et al.: Embodiment in attitudes, social perception, and emotion. Pers. Soc. Psychol. Rev. **9**(3), 184–211 (2005)
4. Słowiński, P., et al.: Dynamic similarity promotes interpersonal coordination in joint action. J. Roy. Soc. Interface. **13**(116) (2016)
5. Barr, D.J., et al.: Random effects structure for confirmatory hypothesis testing: keep it maximal. J. Mem. Lang. **68**(3) (2013)
6. Bates, D., et al., Fitting linear mixed-effects models using lme4. J. Stat. Softw. **67** (2015)
7. Lenth, R.V.: Least-squares means: the R package lsmeans, 2016. **69**(1), 33 (2016)
8. Rad, N.M., et al.: Deep learning for automatic stereotypical motor movement detection using wearable sensors in autism spectrum disorders. Sig. Process. **144**, 180–191 (2018)
9. Goodwin, M.S., et al.: Automated detection of stereotypical motor movements. J. Autism Dev. Disord. **41**(6), 770–782 (2011)

Investigation and Analysis of University Libraries' Participation in the Construction and Service of Think Tanks

Yongxin Qu, Nan Guan, Changwei Huang and Zihan Xu

Abstract University think tanks have become an important force for new think tank with Chinese characteristics. Through the statistical analysis of the questionnaire survey on the participation of university libraries in the construction and service of think tanks, this paper holds that the opportunities and challenges of university libraries' participation in the construction and service of think tanks coexist, and puts forward some enlightenments for university libraries' services of think tanks.

Keywords University library · University think tank · Information service

1 Introduction

Since the 18th National Congress, the Party and the state have attached great importance to the construction of think tanks. President Xi Jinping has repeatedly stressed the need to speed up the construction of new think tanks with Chinese characteristics. Chinese think tanks have ushered in a period of rapid development. Because of the advantages of concentration of high-end talents, complete disciplines, abundant basic research strength, extensive academic and external exchanges, university think tanks have been paid special attention to. Think tanks in universities have become the new force in the construction in China. Because information plays an important leading role in the construction of think tanks, university libraries need to extend information services to the fields of management and decision making. Libraries need to relay out their business systems and take the construction of think tanks as their new business growth points and service commanding heights.

Y. Qu (✉) · C. Huang · Z. Xu
Harbin University of Commerce, Harbin, China
e-mail: quxin75@163.com

N. Guan
School of Electrical and Information, Northeast Agricultural University, Harbin, China

V. Jain et al. (eds.), *Recent Trends in Intelligent Computing, Communication and Devices*, Advances in Intelligent Systems and Computing 1031,
https://doi.org/10.1007/978-981-13-9406-5_41

In recent years, more and more scholars are concerned about the libraries' participation in the construction of think tanks. By searching CNKI and using "Library + Think Tank Construction" as a subject, 380 documents were retrieved, including 362 periodical papers, 11 master papers, 4 newspapers, and 3 conference papers. The years of 2016 and 2017 have become the focus of research; 116 papers were published in 2016 and 154 papers in 2017. The core journals such as "Library and Information Work," "Library," "Library Forum," "Modern Information," and "Library Work and Research" rank first. There are 127 papers about university libraries' participation in the construction of think tanks, accounting for 33.4%. In practice, more and more university libraries have participated in the construction of think tanks. In view of this situation, the author has issued a questionnaire on the think tank services of university libraries through the questionnaire star system.

2 Investigation

This survey was completed from March to April, 2018. Through Library and Information Group, Library Transformation and Reform Group, Library Planning and Evaluation Statistical Discussion Group, and National Financial and Economic University Library Group, the author released the questionnaire "University Libraries Serve as Think Tanks" such as Table 1, and finally received the valid data submitted by 344 university libraries in China. The sampling is more reasonable and has a certain representativeness.

Table 1 Questionnaire

No.	Contents	Options	Subtotal	Proportion (%)
1	Do you think university libraries can serve as think tanks?	Yes	320	93.02
		No	12	3.49
		Don't know	12	3.49
2	Has your library served as a think tank?	Yes	160	46.51
		No	184	53.49
3	What advantages do you think university libraries have as think tanks?	Literature information resources	316	91.86
		Subject services	296	86.05
		Information analysis	284	82.56
		Talent people	208	60.47
4	The relationship you think between think tanks and institutional repositories	The same	108	31.4
		Two fields	236	68.6
5	The focus of university libraries on think tank services	Scientific research advantages and competitiveness	332	96.5
		Talent selection and analysis	228	66.28
		University evaluation	196	56.98

(continued)

Table 1 (continued)

No.	Contents	Options	Subtotal	Proportion (%)
6	What are the functions of think tanks?	Public policy	288	83.72
		Strategic planning	320	93.02
		Public cognition	236	68.6
7	What are the support of think tanks?	Policy needs	272	79.07
		Academic research	292	84.88
		Database and platform	308	89.53
		Literature resources	296	86.05
		Talent people	260	75.58

3 Summary and Enlightenment

Through the statistical analysis of the questionnaire, we believe that the opportunities and challenges for university libraries to participate in the construction and service of think tanks coexist.

3.1 Opportunities

University think tanks have become an important force in building new think tanks with Chinese characteristics. In the first batch of 25 national high-end think tank construction pilot units announced in 2016, seven universities including Peking University and Tsinghua University were selected. In February 2018, the Think Tank Research Center of Shanghai Academy of Social Sciences released the "China Think Tank Report 2017," which includes the National Development Research Institute of Peking University, the National Development and Strategic Research Institute of Renmin University, the National Situation Research Institute of Tsinghua University, the Chinese Research Institute of Fudan University, the Research Institute of International Law of Wuhan University and the Research Institute of Guangdong, Hong Kong and Macao of Zhongshan University. The nine think tanks were selected as the top forty influential groups. In the CTTI (China think tank index) 2017 and 2018 of the 489 source think tanks, 255 think tanks in colleges and universities, accounting for 52%. In June 2017, 20 key think tanks were published in Heilongjiang Province, of which 13 were college think tanks, accounting for 65%. At present, the country is implementing the "double first-class" development strategies. Universities need to make scientific evaluation of universities, disciplines, and specialties to support decision making, which provides a good opportunity for university libraries to participate in the service and construction of think tanks.

In recent years, a number of academic seminars have been held by the Research and Publication Center of Library and Information Magazine and Guangming Daily in conjunction with the Research and Evaluation Center of Think Tank of Nanjing University. The participants include experts and scholars from think tanks and library and information institution. In February 2016, "Theory and Practice of Think Tanks," co-sponsored by the Documentation and Information Center of the Chinese Academy of Science and Nanjing University, started its publication [1] 46, which accelerated the organic integration of think tanks and library and information institution. At present, some university libraries are still at the stage of theoretical discussion and wait-and-see in terms of think tank services. There are not many domestic university libraries that really carry out think tank services while the situation abroad is optimistic. Through literature research, the typical cases of think tank services in university libraries at home and abroad are summarized as shown in Table 2.

Table 2 A list of typical cases of think tanks in university libraries

Libraries	Serve for (Institution or enterprise)	Contents
Harvard University	Bay Center	Think tank librarians and space services, resource retrieval and lending services, reference and academic guidance, special collections and other special services
	David Asia-Europe and Russia Research Center	
Stanford University	Hoover Center	
Princeton University	Science and Global Security Center	
Hong Kong Polytech University	Academic Think Tank	Displaying academic achievements and archives
Peking University	Research Center of Higher International Education of Peking University	Providing data support for the formulation of education policy in China
Nanjing University	South China Sea Research and Collaborative Innovation Center	Information collections on South China Sea issues
Xiamen University	Southeast Asian Research Center	Information collections on Southeast Asia and overseas Chinese
Zhejiang Normal University	African Research Institute	Establishment of a database of African research resources
Inner Mongolia University	Mongolian Research Center	Maintaining Mongolian Information Network
Jilin Tonghua Normal University	Koguryo and Northeast Ethnic Research Center	Strengthening the Construction of Koguryo Research Database
Jilin University	Northeast Asia Research Institute of Jilin University	Collection of Northeast Asian Literature

3.2 Challenges

Think tanks were well known in the West in the 1970s. Since the 1990s, they have become a hot topic in international academic researches. At the beginning of the twenty-first century, they have gradually attracted the attention of the Chinese people. There is a big gap in the concept and operation of think tanks both at home and abroad. Foreign think tanks generally operate independently and are not subject to outside interference. They have absolutely dominant information resources and top information service personnel. Forecasting and judging are mostly based on their own first-hand information, which is more objective. Foreign think tanks have their own reliable sources of information and data, and generally have their own libraries and information networks. Libraries and information institutions are important guarantee institutions for decision-making information support of think tanks. They have a rich collection of books, periodicals, newspapers and various documents, data and information. They are important support for foreign think tanks, and their service ability is beyond doubt. Most think tanks in our country have official nature, most of which are financed by financial appropriations. Most of the members, after guessing the intention of the government, make a judgment according to government documents and internal references that the government should not be guilty at least. Libraries and information institutions, including libraries, are only their subsidiary departments and have not received the attention that think tanks deserve.

Nowadays, there are still some misunderstandings about the relationship between libraries and think tanks. Some think libraries and think tanks have nothing to do with libraries, and they think libraries should just serve the readers well. Others are blindly optimistic that libraries can build think tanks and become think tanks.

3.3 Enlightenment

At present, libraries and think tanks lack effective communication. Libraries do not understand the operation mechanism and the needs of think tanks. Think tank organizations still remain in the traditional category of "borrowing and returning books," lack of understanding and recognition to library reference services and subject services, do not believe that libraries have the ability to serve think tank, while think tank institutions are more willing to pay and cooperate with database manufacturers. Moreover, the information resources needed by think tanks have their own particularities, including scientific data, social data, economic data, and industrial data. They require more objective data, factual data, objective environment, and international comparison. They need to do in-depth data mining and analysis.

The information needs of university think tanks have six basic characteristics: timeliness, comprehensiveness, reality, authority, continuity, and discipline. There is a certain gap between the information services of university libraries and the information demands of university think tanks, which is mainly reflected in the imperfect information service mechanism of university libraries, the weak pertinence of literature and information resources, the insufficient depth of information service contents, the insufficient information service abilities, the backward information service methods and means, and the backward service concepts, etc. University libraries have a long way to go to support the construction and service of think tanks.

4 Conclusion

University libraries should base their services on the reality of their own libraries and should not blindly follow the trend, rash and aggressive. They should start from the foundation, and on the basis of the original reference and subject services, further do a good job in knowledge consultation and then develop into policy consultation, so that they can truly play the role of serving think tanks. The services of think tanks in university libraries should be promoted at different levels. Some libraries with leading business, abundant resources, and complete functions can directly serve their own think tanks in universities. Other libraries can first carry out subject services based on the idea of think tanks and provide decision-making services for scientific research, subject construction, and talent introduction in schools. We must continue to accumulate experience for the think tanks in the future.

Acknowledgements This thesis is a stage achievement of the project (No. 17TQB060) of the Heilongjiang Province Society Scientific Fund. and the project (No. 17XN088) of Harbin University of Commerce. Many people have contributed to it. This thesis will never be accomplished without all the invaluable contributions selflessly from my group members. It is my greatest pleasure and honor working with my group. I'm deeply grateful to them.

References

1. Chu, J., Tang, G.: The libraries and think tanks. Libr. Inf. Serv. (01), 1–7 (2018)
2. Coburn, M.M.: Competitive Technical Intelligence: A Guide to Design, Analysis, and Action. American Chemical Society, Washington, DC; Oxford University Press, New York (1999)
3. Savioz, P.: Technology Intelligence: Concept Design and Implementation in Technology-based SME's. Basingstoke Palgrave Macmillan (2004)
4. April, K., Bessa, J.A.: Critique of the strategic competitive intelligence process within a global energy multinational. Probl. Perspect. Manag. **4**(2), 86–99 (2006)

SAR Target Recognition Via 2DPCA and Weighted Sparse Representation

Yue Zhao, Yulong Qiao and Xiaoyong Men

Abstract Target recognition about synthetic aperture radar (SAR) image using weight sparse representation-based classification (WSRC) is one of the research hotspots in recent years. For test samples, the weight coefficient of a training sample is computed by the distance or degree of correlation between the test sample and the training sample in WSRC. Different from the traditional PCA method, the image matrix needs to be converted to a vector before feature extraction, and 2DPCA is on the basis of 2D image matrices instead of 1D vectors. 2DPCA and classification method based on weighted sparse representation are combined to identify the synthetic SAR image. We improve the typical WSRC to make it more suitable for our framework. Our proposed method in this article uses the data set of moving target location and recognition (MSTAR), also compared with other existing methods. The experiment results prove that this method is superior to the existing classifier and have certain effectiveness and robustness.

Keywords Target recognition · Weighted sparse representation · 2DPCA

1 Introduction

Synthetic aperture radar (SAR) is a radar monitoring equipment device proposed and developed in the 1950s. Its imaging operations are not limited by light, climate and clouds [1]. SAR's distinctive advantages make it have high practical value in the fields of surveying and mapping projects, national defense, disaster prevention, geological exploration, agriculture and forestry. The MSTAR database is a public database of current SAR target recognition performance assessments. It plays a very important role in the current research of target recognition based on SAR images.

Y. Zhao (✉) · Y. Qiao · X. Men
School of Information and Communication Engineering, Harbin Engineering University, Harbin, China
e-mail: zhaoyue1994@hrbeu.edu.cn

V. Jain et al. (eds.), *Recent Trends in Intelligent Computing, Communication and Devices*, Advances in Intelligent Systems and Computing 1031,
https://doi.org/10.1007/978-981-13-9406-5_42

Based on the fact that SAR images contain both targets and background clutter, different background clutters will affect the recognition results. Therefore, many researchers preprocess the original images. In the literature [2], the target parts and the shadow part of each SAR image are extracted separately by applying different threshold targets, and then the Chebyshev moment is calculated for feature extraction. In the literature [3], it used logarithmic transformation and adaptive threshold segmentation to complete image preprocessing and feature extraction-based PCA.

A variety of different feature extraction methods are introduced in SAR image target recognition. They include some different methods such as LDA, PCA and ICA techniques which also usually appear in remote sensing image classification and face recognition [4–6]. Moreover, the PCA method has the highest recognition rate among several other methods. However, when PCA is used to extract image features, the 2D image matrix must be pre-converted into one-dimensional image vector [7]. In the literature [7], a two-dimensional PCA (2DPCA) is proposed to solve this problem by extracting image features, which can evaluate covariance matrix more accurately and reduce the computational cost.

First, the sparse representation-based classifications (SRCs) are introduced [8] to achieve face recognition. Later, it was used as a classification containing three categories of MSTAR targets (BMP2, BTR70 and T72) in the literature. At the same time, it verifies that SRC has good classification performance on SVM. However, the traditional SRC method does not consider the intrinsic relation between the test samples and the training samples. When the training samples are quite different from the test samples, it cannot achieve good experimental results. It has been proved that the training samples which bear much more resemblance to the test samples are usually more significant on behalf of the test samples in SRC [9]. To overcome this difficulty, the literature [10] makes a new development of SRC, called WSRC.

Based on the above advantages, many researchers use WSRC classifiers to perform SAR target recognition work. Karine proposed a method combining weighted statistical dictionary with Weibull distribution to complete SAR image classification [11]. And Zhou proposed an SAR image recognition method employing monogenic scale selection on the basis of weighted multitask joint sparse representation [12]. In this paper, a method of SAR target recognition using vector correlation distance extracted by 2DPCA feature as the weight in WSRC is selected, and good experimental results are obtained.

2 Proposed Method of This Article

2.1 *Two-Dimensional PCA*

We consider image X as the random matrix with the size of $m * n$, and let U denote an n-dimensional unit column vector. This method projects X onto U by linear transformation as follows.

$$Y = XU \tag{1}$$

where the size of Y is $m * n$. In 2DPCA, the total dispersion of projection matrix Y is used as the criterion function $J(U)$ to determine the direction of the best projection U.

$$J(U) = \mathrm{tr}(S_U) \tag{2}$$

where S_U is the covariance matrix of Y and $\mathrm{tr}(S_U)$ is the trace of matrix S_U. The equation of S_U is shown as:

$$S_U = E\left\{[x - E(x)]^{\mathrm{T}}[x - E(x)]\right\}U \tag{3}$$

The covariance matrix G of the image is defined as:

$$G = E\left\{[x - E(x)]^{\mathrm{T}}[x - E(x)]\right\} \tag{4}$$

where G is a non-negative positive definite matrix of $n * n$. If there are M test samples, $x_i(i = 1, 2, \ldots, M)$. We find the average matrix as:

$$G = \frac{1}{M}\sum_{i=1}^{M}(x_i - u)^{\mathrm{T}}(x_i - u) \tag{5}$$

Each column vector in U that makes $J(U)$ obtain the maximum value is the eigenvector corresponding to r non-zero eigenvalues of the image covariance matrix G. Generally speaking, only one optimal projection axis can not meet the actual needs. Therefore, the first p larger eigenvalues are selected as follows: $\lambda_1 > \lambda_2 > \cdots > \lambda_p$, corresponding eigenvectors $U_1, U_2, \ldots, U_p$, so that $J(U)$ can be maximized and the eigenvectors are orthogonal to each other. The characteristic space U formed according to the above theory is shown as:

$$U = [U_1, U_2, \ldots, U_p] \tag{6}$$

Finally, the projection matrix U is substituted into Eq. (1) to obtain the feature matrix of the sample X.

2.2 The Weighted Sparse Representation-Based Classification

In the traditional SRC algorithm, X_i represents the training sample of the ith class, and l_i is the number of samples. The training samples of all categories (a total of K classes) are combined into a matrix of $A = [X_1, \ldots, X_i, \ldots, X_k]$. For any test sample, it can be expressed in the following linear form:

$$y = Aw \tag{7}$$

Therefore, Formula (1) has no unique solution. For the sake of a unique solution, **we** need to add a constraint that the coefficient w is as sparse as possible. Therefore, Formula (7) evolves into the following optimal problem. For a given learning dictionary A, the mathematical model for sparse representation of A is shown as:

$$w' = \arg\min\|w\|_0 \tag{8}$$

where w' is a sparse matrix. In general, the solution of the sparse representation can be transformed into the problem shown in Eq. (9):

$$w' = \arg\min\|w\|_1, \quad \text{s.t. } \|y - Aw\|_2^2 \leq \varepsilon \tag{9}$$

At last, classifier uses the distance from the test sample as the judgment basis of the category. After solving the appropriate optimal coefficient w', the residual is determined by Formula (10) to get the corresponding category.

$$y = \arg\min\|y - X_i w_i\|_2 \tag{10}$$

WSRC algorithm includes two steps. First, we figure out the distance between the training sample and the given test sample. Second, determine the weight of the training sample according to the distance information. Then, calculate SRC employing weighted training samples.

We adopt the correlation distance between the test sample and the training sample that has been reduced by PCA to compute the distance between the samples. That is, given the training sample x and the test sample y, the correlation coefficient of them is:

$$\rho_{xy} = \frac{\operatorname{cov}(x, y)}{\sqrt{D(x)}\sqrt{D(y)}} = \frac{E((x - Ex)(y - Ey))}{\sqrt{D(x)}\sqrt{D(y)}} \tag{11}$$

where the range of correlation coefficient is [−1, 1]. And then, the distance between them is as follows:

$$D_{xy} = 1 - \rho_{xy} \tag{12}$$

And then, we choose the reciprocal of the correlation distance representing the weight of the training sample.

After obtaining the weight of each training sample, we can acquire a new training set $A' = [X'_1, \ldots, X'_i, \ldots, X'_k]$. The training sample of the kth class is:

$$X'_k = [w_{k1}x_{k1}, w_{k2}x_{k2}, \ldots, w_{kn}x_{kn}] \tag{13}$$

At last, we take test samples into Eq. (13) and get the reconstructed residuals to complete the classification.

3 Experiments and Discussion

These experimental data are based on the measured SAR ground target data published by the MSTAR program. The sensor for collecting the data set is a high-resolution spotlight SAR with a resolution of 0.3 m × 0.3 m. The radar operates in the X band, using the HH polarization mode. The collected data are preprocessed to extract slice images with size of 128 * 128 pixels that contain various targets. In this experiment, ten most widely used targets (BTR70, BMP2, T72, 2S1, BRDM2, BTR60, D7, T62, ZIL13 and ZSU234) are selected. The types and quantity of samples in different depression views are shown in Table 1.

Since the location, rotation and inhomogeneous scattering of the target will affect the performance of feature extraction and classification algorithm, we need to preprocess the original image before the experiment begins. The target image is cut into 64 * 64 pixel images based on each target center point. And then, the normalization of the amplitude of the image makes the average amplitude of each pixel of the image zero, and the standard deviation is 1.

To prove that the weighted distance algorithm is suitable for the features extracted by 2DPCA method, we compare the correlation distance selected in this paper with other distance algorithms, including Jaccard coefficient, Gaussian distance, Chebyshev distance, Mahalanobis distance, histogram distance and correlation distance. The experimental data are listed in Table 2.

From Table 2, it can be seen that the method using correlation distance performs much better than other distance methods, which has verified the superiority of weighted sparse representation. The method mentioned in this article can make the average accurate recognition rate of the relevant distance reach 99.18%, which is better than other distance algorithms (2.46, 2.19, 1.38, 0.83 and 0.79%, respectively). To sum up, we conclude that weight sparse representation has a higher recognition rate compared with other methods.

Table 1 Data set descriptions

	BMP2	T72	BTR70	BTR60	2S1	BRDM2	D7	T62	ZIL131	ZSU234
Training	233	232	233	256	299	298	299	299	299	299
Testing	195	196	196	195	274	274	274	273	274	274

Table 2 Recognition results of ten classes

	Jaccard	Gaussian	Chebyshev	Mahalanobis	Histogram	Correlation
Accuracy (%)	96.72	96.99	97.80	98.35	98.39	99.18

Comparing the experimental results of the WSRC method based on scale selection with the other four classifiers, the superiority of the algorithm we proposed is proved. To analyze the misclassification of targets, the confusion matrices of research methods are displayed in Fig. 1.

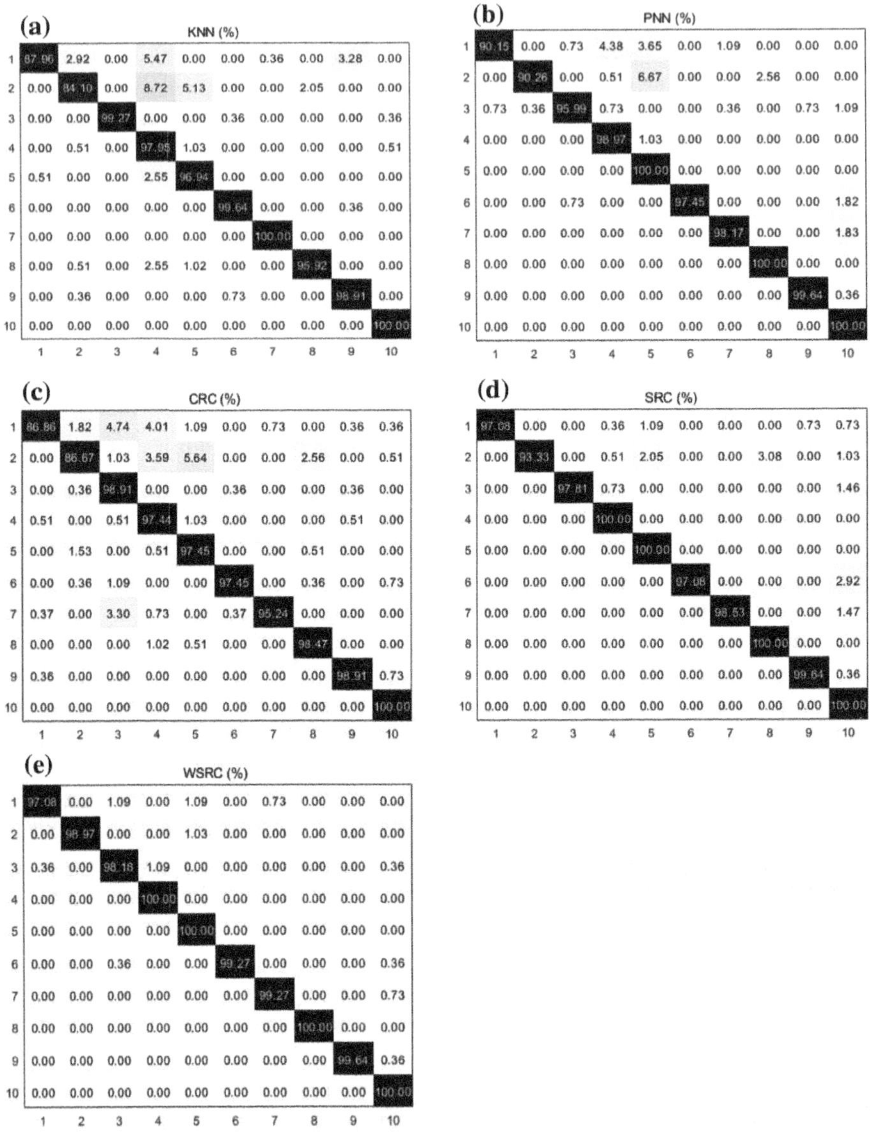

Fig. 1 Confusion matrices belonging to different classifiers. From left to right, the accuracy rate is as follows: **a** KNN-96.37%, **b** PNN-97.03%, **c** CRC-95.84%, **d** SRC-98.35% and **e** WSRC-99.18%

The results presented in Fig. 1 present that the new method proposed in this article is superior to all other methods considering the recognition accuracy about ten-class targets. Specifically, the classifier selected in this article fetches a high recognition accuracy of 99.18%. Its performance is better than KNN, PNN, CRC and SRC, which are 2.81, 2.15, 3.34 and 0.83, respectively. The experimental results in MSTAR database prove that using correlation distance as the weight of training samples in WSRC can improve the recognition rate obviously.

4 Summary

In this paper, a method combining 2DPCA with WSRC is put forward for SAR target configuration recognition. A preprocessing of the MSTAR images is employed to unify the format of the images, reduce the noise and improve the property of the classifiers. The proposed classification method can well model SAR images and extract the basic features of SAR images. More importantly, we have improved the existing method of calculating the distance between the training samples and the test samples in the classifier. The correlation distance between samples is used to evaluate the weight of samples. The recognition rate of our method is also verified through experiments compared with other classifiers, including KNN, PNN, SRC and CRC. Experiments show that our method can achieve greater classification results.

Acknowledgements This work is subsidized by National Natural Science Foundation of China under Grant 61871142. The authors would be sincerely grateful for your valuable revise opinion and suggestions in improving the technical presentation.

References

1. Tait, P.: Introduction to Radar Target Recognition, p. 432. Institution of Electrical Engineers (2005)
2. Bolourchi, P., Demirel, H., Uysal, S.: Target recognition in SAR images using radial Chebyshev moments. SIViP **11**(6), 1–8 (2017)
3. Yu, M., Dong, G., Fan, H., et al.: Remote sensing SAR target recognition via local sparse representation of multi-manifold regularized low-rank approximation. Remote Sens. **10**(2) (2018)
4. Mettam, G.R., Adams, L.B.: How to prepare an electronic version of your article. In: Jones, B. S., Smith, R.Z. (eds.) Introduction to the Electronic Age, pp. 281–304. E-Publishing Inc., New York (1999)
5. Mishra, A.K.: Validation of PCA and LDA for SAR ATR. In: TENCON 2008–2008 IEEE Region 10 Conference, pp. 1–6 (2008)
6. Zhang, X., Chen, C.H.: A new independent component analysis (ICA) method and its application to SAR images. In: Neural Networks for Signal Processing XI, 2001. Proceedings of the 2001 IEEE Signal Processing Society Workshop, pp. 283–292. IEEE (2001)

7. Yang, J., Zhang, D., Frangi, A.F., et al.: Two-dimensional PCA: a new approach to appearance-based face representation and recognition. IEEE Trans. Pattern Anal. Mach. Intell. **26**(1), 131–137 (2004)
8. Dugelay, J.L.: Improved combination of LBP and sparse representation based classification (SRC) for face recognition. In: IEEE International Conference on Multimedia and Expo, pp. 1–6. IEEE Computer Society (2011)
9. Fan, Z., Ni, M., Zhu, Q., et al.: Weighted sparse representation for face recognition. Neurocomputing **151**(1), 304–309 (2015)
10. Lu, C.Y., Min, H., Gui, J., et al.: Face recognition via weighted sparse representation. J. Vis. Commun. Image Represent. **24**(2), 111–116 (2013)
11. Karine, A., Toumi, A., Khenchaf, A., et al.: Target recognition in radar images using weighted statistical dictionary-based sparse representation. IEEE Geosci. Remote Sens. Lett. **99**, 1–5 (2017)
12. Zhou, Z., Wang, M., Cao, Z., et al.: SAR image recognition with monogenic scale selection-based weighted multi-task joint sparse representation. Remote Sens. **10**(4), 504 (2018)

Research on the Intelligent Unmanned Vehicle Measurement System

Bing Zhang, Lirong Liu and Wenji Zhao

Abstract With the expansion of cities and the development of modern life, the traditional vehicle-mounted measurement control system has been difficult to meet the measurement demand in some cases. In order to meet the requirements of vehicle-mounted measurement intelligence and solve the problems of complicated and refined measurement tasks, an intelligent unmanned vehicle-mounted measurement system is proposed. This paper is based on artificial intelligence technology, unmanned vehicle as a platform, the control system modular design, realized unmanned vehicle measurement. This paper presents part of the data obtained from the current manned vehicle measurement, briefly introduces the control system, and discusses the advantages of the intelligent unmanned vehicle measurement system. This study provides some suggestions for the development of intelligent unmanned vehicle measurement equipment.

Keywords Vehicle-mounted measurement · Intelligence · Artificial intelligence · Unmanned vehicle · Control system · Unmanned vehicle measurement

1 Introduction

Since it was proposed in 1956, artificial intelligence [1, 2] has been continuously deepening and expanding in the application field and subject development, and has made a series of achievements, mainly in the fields of speech recognition [3, 4], machine vision [5, 6], natural language [7], and so on. Pedro Domingos divided artificial intelligence into symbolic school, connective school, evolutionary school, Bayesian school, analogy school, and self-taught school according to algorithms.

B. Zhang (✉) · L. Liu · W. Zhao
College of Resource Environment and Tourism, Capital Normal University,
Beijing 100048, China
e-mail: zhangbing_1989as@sina.cn

L. Liu
Satellite Surveying and Mapping Application Center, NASG, Beijing 100048, China

V. Jain et al. (eds.), *Recent Trends in Intelligent Computing, Communication and Devices*, Advances in Intelligent Systems and Computing 1031,
https://doi.org/10.1007/978-981-13-9406-5_43

He elaborated on how each type of algorithm operates and the difference between algorithm characteristics and final results. Unmanned control system [8, 9] refers to the control system that includes control device, communication station for information transmission, equipment mobility, information storage, object detection, and other supporting devices applied to a particular function. Driverless vehicles [10] are wheeled mobile robots that obtain road information through onboard sensors, analyze and process computer vision information acquired by machine vision technology, and automatically control the vehicle to reach the specified target. Vehicle-mounted measurement system [11] is a measurement system that uses vehicles as a platform to carry measurement equipment to obtain urban information. Intelligent unmanned vehicle measurement system is an intelligent device that integrates and synchronously controls the previous measurement sensors [12, 13] on the basis of driverless cars, responds to the periodic mutual data analysis results of monitoring targets by the data center, and obtains targeted data to pay attention to the changes of monitoring targets in the later stage.

2 Chinese and International Researches

In 1914, the outbreak of the First World War promoted the development of automobile types and technologies. And then, it goes from Earle age, the Sloan model, the age of technological progress, the age of diversity, and a period of low prices to the globalization of cars. In the 1980s, emerging electronic technologies [14] contributed to the emergence of intelligent control systems, making unmanned driving a possibility. In the 1970s, Britain, the USA, Germany, and other countries began the research of driverless cars. China began the research in the 1980s. The first driverless car in China was developed in 1992. There are different definitions for the understanding of intelligent systems in each period. The intelligence in the manual age is steam replacing human, followed by electricity replacing steam, followed by the information age. Now it is artificial intelligence under the environment of big data [15], but the ultimate development purpose is only one, that is, to liberate human. At present, researches on artificial intelligence include brain simulation, symbol processing, character number method, statistical method, and integration method. The learning method is mainly machine learning.

3 System Framework

Intelligent unmanned vehicle-mounted measurement system is divided into two framework systems: main control system and secondary control system. Master control is mainly the supervision and control of real-time coordination scheduling between vehicles and measurement equipment by ground user center, which can

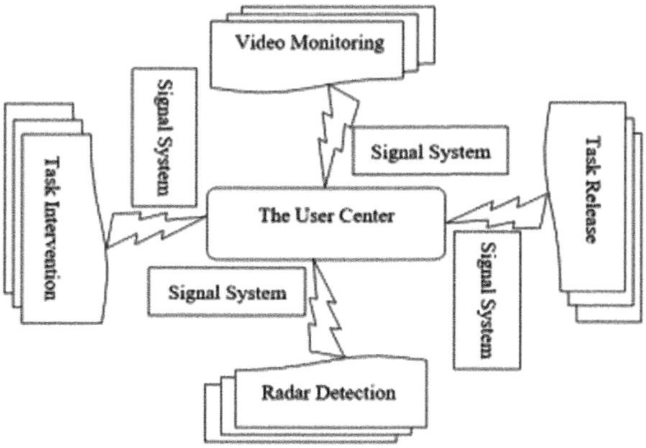

Fig. 1 Main control framework system

intervene in measurement tasks, modify measurement tasks, or issue new tasks. The main control system is shown in Fig. 1.

The secondary control system mainly refers to the intelligent unmanned vehicle-mounted platform control device system, including the start and stop of unmanned vehicle according to the wake up of the user center, self-inspection of the operating functions of each module of the system, command reception and response, route planning, task planning, platform movement and target judgment, environmental judgment, and sub-device regulation. The secondary control system automatically responds and wakes up according to the tasks accepted by the user center, performs the measurement tasks, and decides whether to mobilize the auxiliary measurement module according to the actual situation. The secondary control system is shown in Fig. 2.

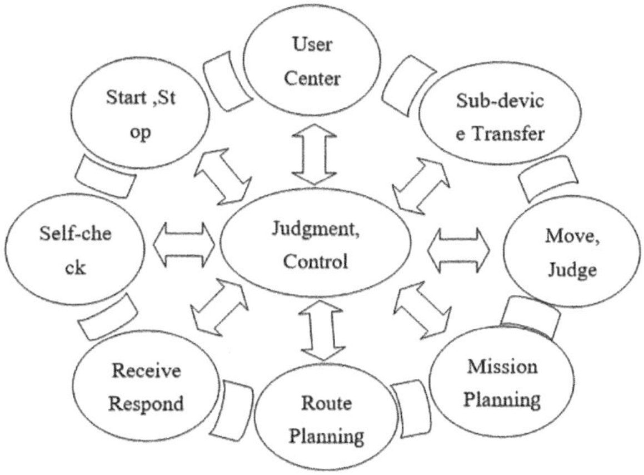

Fig. 2 Framework system of secondary control

4 Some Results of Vehicle-Mounted Mobile Measurement Displaying

At present, there are two kinds of data available for mobile measurement, one is laser data [16], and the other is panoramic data [17]. Baidu, Alibaba, and other companies in China have applied the data of laser and video to driverless experiments, and achieved some results. This paper mainly presents laser and panoramic data for surveying and mapping [18] and geographic information system industries. Laser data can be used for object model construction, and panoramic data can be used for target recognition. The laser data are shown in Figs. 3 and 4.

Panoramic data of the measurement industry are photographs taken by panoramic lens mounted on the vehicle-mounted platform. The main picture shown here is the road surface view. The panorama is shown in Fig. 5.

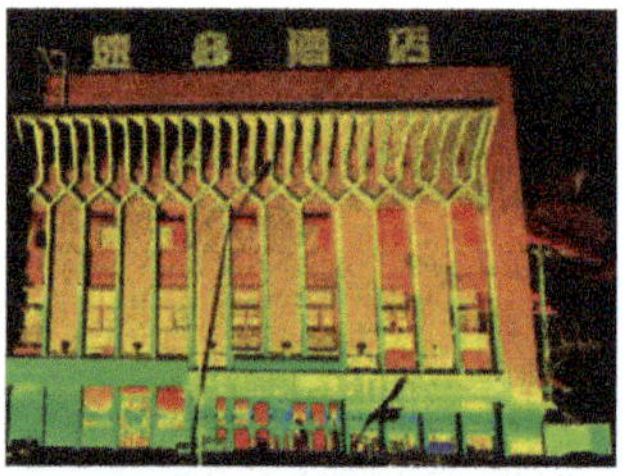

Fig. 3 Building model

Fig. 4 Road model

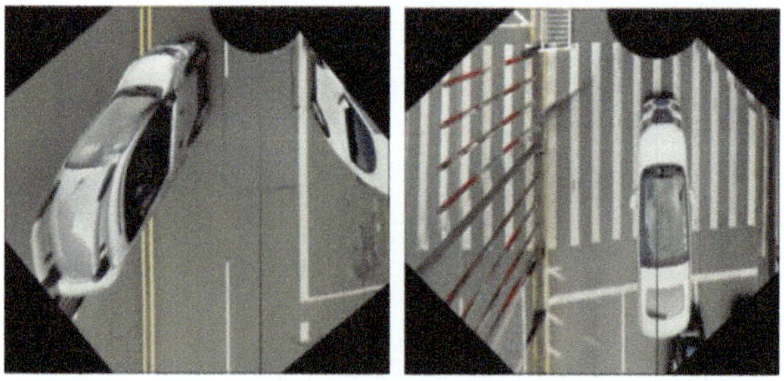

Fig. 5 Panorama

5 Conclusion

Based on the current practical situation of onboard measurement equipment, this paper proposes an intelligent unmanned onboard measurement system from the perspective of practical exploration, which has the following characteristics:

(1) Replace the current manned driving with unmanned to achieve intelligent movement of the platform, greatly reducing human control.
(2) Modular and independent operation of the control system automatically completes the measurement tasks. At the same time, the main control module selectively decides to mobilize the secondary control module to assist in completing the measurement tasks according to the actual situation.
(3) The integration of measurement, control, and artificial intelligence technology is combined to realize intelligent unmanned measurement.

References

1. Zhou, Z.-H., Wu, J., Tang, W.: Corrigendum to "Ensembling neural networks: many could be better than all" [Artificial Intelligence 137 (1–2) (2002) 239–263]. Artif. Intell. **137**(1), 239–263 (2002)
2. Ramos, C., Augusto, J.C., Shapiro, D.: Ambient Intelligence—the next step for artificial intelligence. IEEE Intell. Syst. **23**(2), 15–18 (2008)
3. Ververidis, D., Kotropoulos, C.: Emotional speech recognition: resources, features, and methods. Speech Commun. **48**(9), 1162–1181 (2006)
4. Hinton, G., Deng, L., Yu, D., et al.: Deep neural networks for acoustic modeling in speech recognition: the shared views of four research groups. IEEE Signal Process. Mag. **29**(6), 82–97 (2012)
5. Inigo, R.M.: Application of machine vision to traffic monitoring and control. IEEE Trans. Veh. Technol. **38**(3), 112–122 (2002)
6. Paliwal, J., Visen, N.S., Jayas, D.S., et al.: Cereal grain and dockage identification using machine vision. Biosyst Eng **85**(1), 51–57 (2003)
7. Pennebaker, J.W., Mehl, M.R., Niederhoffer, K.G.: Psychological aspects of natural language use: our words, our selves. Annu. Rev. Psychol. **54**(1), 547–577 (2003)
8. Cai, G., Chen, B.M., Dong, X., et al.: Design and implementation of a robust and nonlinear flight control system for an unmanned helicopter. Mechatronics **21**(5), 803–820 (2011)
9. Du, H., Fan, G., Yi, J.: Autonomous takeoff control system design for unmanned seaplanes. Ocean Eng. **85**(3), 21–31 (2014)
10. Xia, Y., Pu, F., Fu, M., et al.: Modeling and compound control for unmanned turret system with coupling. IEEE Trans. Industr. Electron. **63**(9), 5794–5803 (2016)
11. Conejero, J.A., Jordán, C., Sanabria-Codesal, E.: An algorithm for self-organization of driverless vehicles of a car-rental service. Nonlinear Dyn. **84**(1), 107–114 (2016)
12. Chang, T.H., Hsu, C.S., Wang, C., et al.: Onboard measurement and warning module for irregular vehicle behavior. IEEE Trans. Intell. Transp. Syst. **9**(3), 501–513 (2008)
13. Wang, Q., Terzis, A., Szalay, A.: A novel soil measuring wireless sensor network. IEEE, 412–415 (2010)

14. Howlader, M.M.R., Selvaganapathy, P.R., Deen, M.J., et al.: Nanobonding technology toward electronic, fluidic, and photonic systems integration. IEEE J. Sel. Top. Quantum Electron. **17**(3), 689–703 (2011)
15. Zaharia, M., Xin, R.S., Wendell, P., et al.: Apache Spark: a unified engine for big data processing. Commun. ACM **59**(11), 56–65 (2016)
16. Brovelli, M.A., Cannata, M.: Digital terrain model reconstruction in urban areas from airborne laser scanning data: the method and an example for Pavia (northern Italy). Comput. Geosci. **30**(4), 325–331 (2004)
17. Zheng, J.Y., Tsuji, S.: Panorama representation for route recognition by a mobile robot. Int. J. Comput. Vision **9**(1), 55–76 (1992)
18. Pires, A., Chaminé, H.I., Piqueiro, F., et al.: Combining coastal geoscience mapping and photogrammetric surveying in maritime environments (Northwestern Iberian Peninsula): focus on methodology. Environ. Earth Sci. **75**(3), 196 (2016)

Design and Implementation of Smart Classroom System Based on Internet of Things Technology

Qian Zhu

Abstract The smart classroom system based on the IOT technology which integrates classroom monitoring, equipment control, and other functions together to realize intelligent management has become an important direction for the development of teaching management. This paper analyzes the functional requirements and basic framework of the IOT system for smart classroom, expounds the design scheme of the system, and designs the overall architecture, network architecture, and functional modules in detail. Finally, the feasibility of the system is implemented and verified.

Keywords Internet of things · Smart classrooms · Intelligent control

1 Introduction

Smart classroom is a kind of high-power classroom [1], which integrates high-tech software and hardware equipment, and makes full use of the IOT application technology to realize smart classroom management. In the context of the rapid development of the IOT, enhancing the management level of classrooms and improving the utilization of classroom resources to meet the diverse needs of indoor activities such as lectures have become a key issue in the informatization and modernization of classroom management [2]. Building smart classroom IOT application system has a very important application value.

Q. Zhu (✉)
City College of Wuhan University of Science and Technology, Wuhan 430083, China
e-mail: zhuqian722@163.com

V. Jain et al. (eds.), *Recent Trends in Intelligent Computing, Communication and Devices*, Advances in Intelligent Systems and Computing 1031,
https://doi.org/10.1007/978-981-13-9406-5_44

2 Analysis of IOT System in Smart Classroom

The "intelligent" design of smart classroom mainly includes: optimized presentation of teaching contents, convenient acquisition of learning resources, deep interaction of classroom teaching, situational awareness and detection, classroom layout, and electrical management. When researching and developing the IOT system for smart classrooms, it is necessary to build a wireless communication network [3] and control logic, give full play to the functions of existing equipment, and design the IOT application platform with smart classroom management as the core. The basic architecture is shown in Fig. 1.

3 Design of IOT System in Smart Classroom

3.1 System Structure Design

Functions of IOT application system in smart classroom are interconnected through control logic, and sufficient system application interfaces and expansion interfaces are reserved, so that the system has reasonable, extensible, and multi-application system architecture, as shown in Fig. 2.

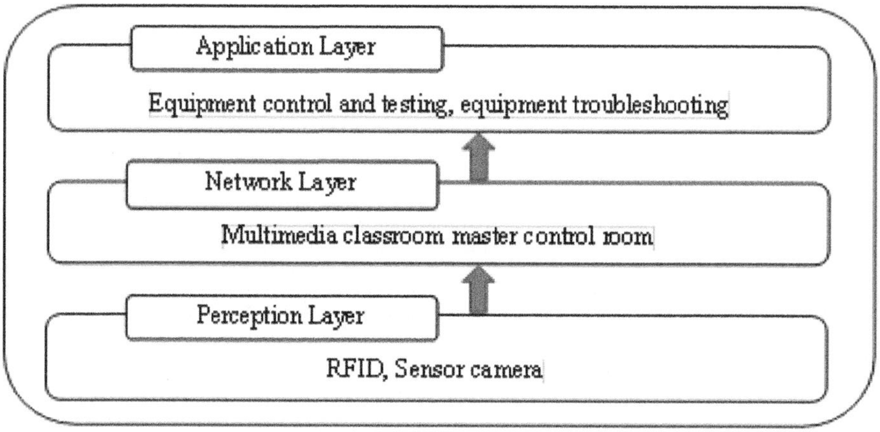

Fig. 1 Basic architecture

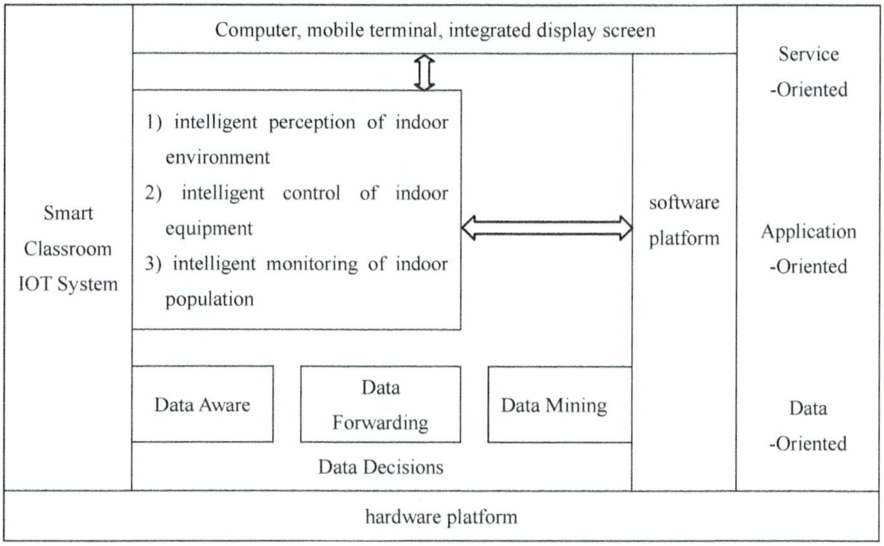

Fig. 2 System structure

3.2 System Network Architecture Design

In the IOT application platform, data forwarding network architecture is the main platform supported by data services [4], as shown in Fig. 3.

Smart classrooms are connected with each other. Different classrooms can be connected to campus server nodes by star network or wireless base station or wireless base station provided by the Internet mobile operators. Distributed network structure communication is adopted.

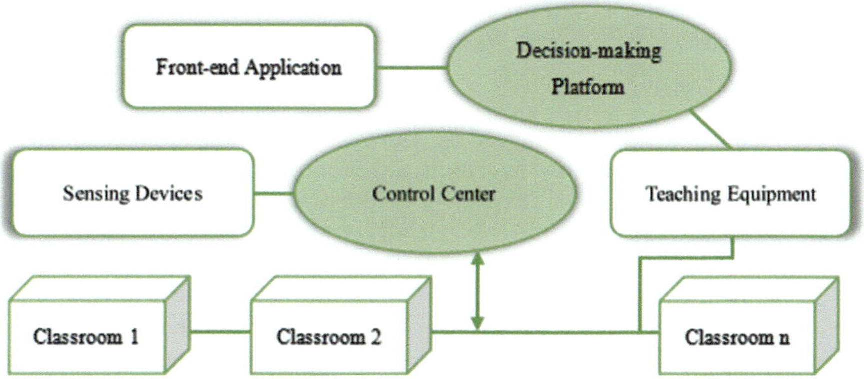

Fig. 3 Data network architecture

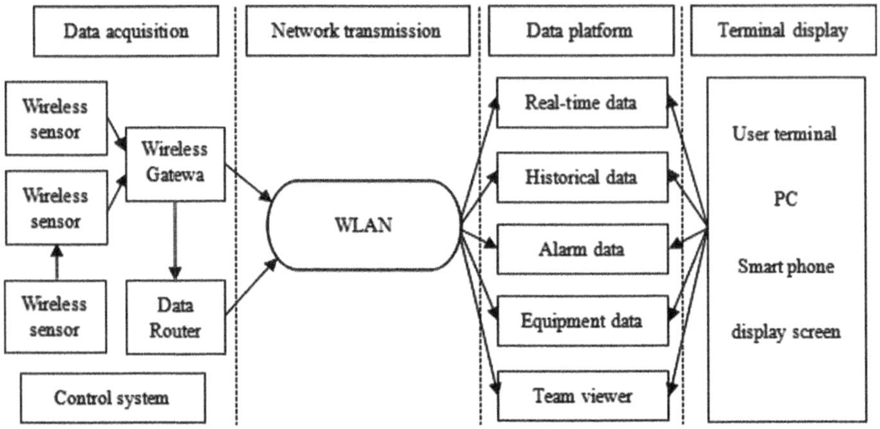

Fig. 4 Overall network architecture

The overall network architecture of the environmental intelligent sensing module can be divided into four parts: field data collection, network transmission, intelligent data processing platform and remote control, as shown in Fig. 4.

3.3 System Function Module Design

1. Indoor environment intelligent perception module

The intelligent sensing module of indoor environment is the core function of data source and linkage control of the smart classroom IOT application system. Various types of environmental data monitoring sensors are deployed to collect and store data such as temperature, humidity, and light intensity. The wireless networking and communication technologies such as Wi-Fi and Zigbee are adopted to provide a basis for real-time monitoring of indoor environment and intelligent control of indoor equipment.

2. LED display module

The LED display module is composed of LED panel, which is installed on the top of the teacher's blackboard. It is used to display the course name, professional class, teacher, attendance rate, and environmental data collected by sensors in the classroom.

3. Teaching module

The teaching module is composed of a touch projector integrated with built-in electronic whiteboard function, power amplifier, speaker, wireless microphone, enclosure, and other supporting control software. Instead of traditional blackboard teaching, touch projectors with built-in electronic whiteboard function can be used to realize dust-free teaching and protect the health of teachers and students [5]. The computer can be operated on the projection screen to realize interactive classroom teaching for teachers and students.

4 Implementation of IOT System in Smart Classroom

4.1 The Main Technologies for Implementing the System

1. Wireless communication network and campus network

The IOT application system is based on classroom and uses wireless networking technology and communication technology to perceive data for the physical layer hardware platform and forward the indoor equipment working status signal to the server.

2. Monitor the indoor integrated hardware platform of the classroom

Through the computer or smartphone, users can access the real-time information set of the classroom internal environment through the wireless network or mobile communication network. They can also conduct multiple types of in-depth inquiry and control of classroom information and its status through the Internet and remote terminals.

3. Intelligent control based on perceptual data decision

The IOT application system can intelligently control the indoor equipment according to the real-time status parameters of the classroom internal environment and the alarm threshold or level required by the user during indoor activities or teaching.

4.2 Android IOS Implementation Environment

A mobile terminal management platform is designed based on the Android platform [6], and its main function is to provide classroom information for users to query and manage. Enter the classroom number to query the usage of the classroom, occupancy information, etc., to facilitate users to obtain classroom information or classroom allocation management. The structure is shown in Fig. 5.

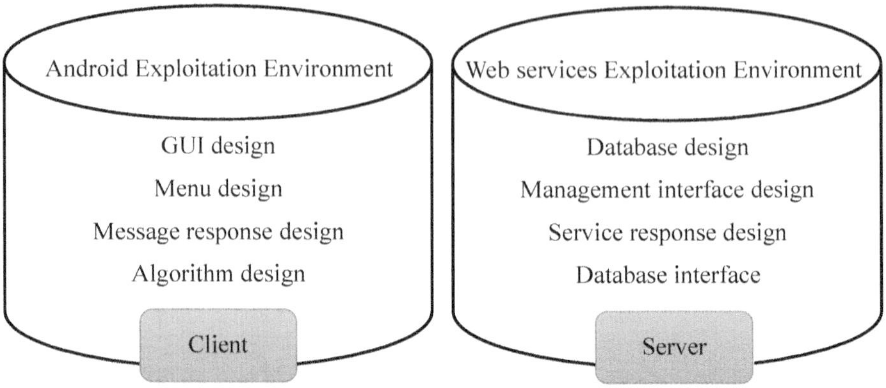

Fig. 5 Mobile terminal architecture

4.3 Realization of System Function

According to the design, the following specific functions are realized:

1. Indoor environment intelligent perception

The real-time data is given by computer, mobile terminal, or sensor LCD screen, which is used to perceive much real-time environmental information in classroom. Data traceability, historical comparison, and information query function are realized on the software platform. In addition, users are allowed to design and input the threshold value of various parameters in the indoor environment or the boundary value of different levels for quality assessment and warning.

2. Intelligent control of indoor lighting equipment

The indoor light intensity is measured by the photoresistor, amplified by the amplifier, then sampled by the single-chip A/D converter to obtain the light intensity data, and combined with the clock module to distinguish the influence of the classroom lighting on it.

3. Intelligent control of indoor door and window equipment

The real-time status detection and intelligent control of indoor doors, windows and curtains. On the one hand, actively open or close according to the data collected by the light intensity sensor; on the other hand, users or administrators can set the threshold value of intelligent control according to the actual situation.

4. Intelligent control of multimedia teaching equipment

The real-time status detection and intelligent control of the indoor multimedia teaching equipment, on the one hand, actively open or close according to the teaching plan; on the other hand, users or administrators can set and control according to the scene situation. In addition, the page layout, content updating, and dynamic adjustment of display system such as display screen or projector in multimedia teaching equipment need real-time monitoring, feedback, and response according to the real state of mobile data and decision-making conclusions.

5. Classroom statistics

Using the camera to collect image information and analyze it to detect the human flow, the accurate statistics of multiple people entering and exiting the classroom at the same time is realized.

5 Conclusion

Based on the analysis of the research background and system requirements of smart classroom system, this paper presents the functional requirements and design schemes of the system. From the main technical aspects of system implementation, the paper presents the technical realization and relevance of wireless communication network and campus network, classroom indoor integrated hardware platform, and intelligent perception control. According to the need, the experimental environment is built and the system is implemented. Due to the limited time and self-level problems, this paper only completed a preliminary study on the implementation. There are still many areas for improvement, such as transplanting more Web functions to the mobile end, accessing more applications, so as to better for teachers and students.

Acknowledgements This paper is supported by Hubei Provincial Education Department of Scientific Research Project of B2017420, Hubei Provincial Education Department of Humanities and Social Science Research Project of 16G251 and City College of Wuhan University of Science and Technology Teaching Research Project of 2019CYYBJY005.

References

1. Lang, Z., Zhe, L., Hu, X.: Design and implementation of a smart classroom scheme based on the Internet of things. Wirel. Commun. Technol. **5**(4), 56 (2014)
2. Xun, Y., Can, D., Feng, T.: Design of lighting control system for smart classroom. Internet Things Technol. **8**(10), 89–91 (2014)
3. Wenyi, X., Hongxing, Y.: Application of Internet of things in the construction of smart campus. Inf. Technol. **4**, 63–65 (2013)
4. Ronghua, H.: The concept and characteristics of smart classroom. Open Educ. Study **19**(2), 22–27 (2012)
5. Yazhen, Z.: Review and prospect of smart classroom research at home and abroad. Dev. Educ. Study **1**, 81 (2014)
6. Guanglin, L., Jun, L.: Intelligent campus application and practice based on the internet of things. Logistics Technol. **9**, 31 (2012)

Research on Heterogeneous Data Exchange Technology Based on Shadow Table

Hua-li Zhang, Fan Yang, Hua-yong Yang and Wei Jiang

Abstract In the process of continuous development of enterprise informatization, with the expansion of old applications and the increasing of new applications, users will face the problem of data exchange between different hardware platforms, different network environments and different databases. Due to the coexistence of multiple application modes, the problems of data exchange between multiple systems are not standardized, network data sharing is not easy, data synchronization is not guaranteed, and data security is not guaranteed, which results in the phenomenon of "information island" and brings great inconvenience to users. The functions provided by traditional system software or tool software can not solve the problem of heterogeneous data exchange well. The research and implementation of a flexible, efficient, concise and transparent heterogeneous data exchange system is very necessary. This paper plans to study and implement a user configurable heterogeneous data exchange system which can shield complex interfaces between various data systems and has high security and reliability.

Keywords Relational database · Data exchange · Heterogeneous data · Shadow table

1 Introduction

With the development of information technology and network technology, various application systems have been developed one after another. The development time, development environment, operation platform and specific business functions of these application systems are different. They can not communicate and exchange data with each other, which makes it difficult to share data among systems. A series of problems, such as data islands [1, 2], the need to spend a lot of human resources

H. Zhang (✉) · F. Yang · H. Yang · W. Jiang
City College, Wuhan University of Science and Technology, Wuhan 430083, China
e-mail: zhanghuali@foxmail.com

V. Jain et al. (eds.), *Recent Trends in Intelligent Computing, Communication and Devices*, Advances in Intelligent Systems and Computing 1031,
https://doi.org/10.1007/978-981-13-9406-5_45

to find data, and with the further development of computers, these problems are becoming more and more serious.

Many users have long been concerned that data sharing and data exchange [3] are hindered by the heterogeneity of various database management systems and operating environment. In order to solve this problem, database manufacturers and software developers have been exploring and researching constantly, and many data exchange solutions have been put forward. And technology, the following technical channels occupy the mainstream position at present.

- Data exchange tool

At present, some database manufacturers and application software developers have expanded and developed the functions of various database management systems [4, 5]. In their database management systems, they have provided their own data exchange tools.

- Data exchange middleware

Middleware is an interface software between client and server [6, 7]. It mainly solves the data interoperability problem of database application systems in heterogeneous environments.

- Intermediate data method

The solution of intermediate data method is that both sides of heterogeneous data exchange use a common data format document to exchange heterogeneous data [8, 9].

This paper designs and implements an overall framework of heterogeneous data exchange system, which includes data update acquisition module, data update analysis and reconstruction module, data update encryption and decryption and transmission module, and data update analysis module.

2 General Design of Heterogeneous Data Exchange System

In heterogeneous data exchange system, multiple database nodes can provide data synchronization service, that is, the data of each data exchange node is equal to each other and has the same function. Therefore, in the design of the exchange system, these data exchange nodes and data exchange servers exchange data to realize the exchange and synchronization of multi-node data. The structure of the system is shown in Fig. 1.

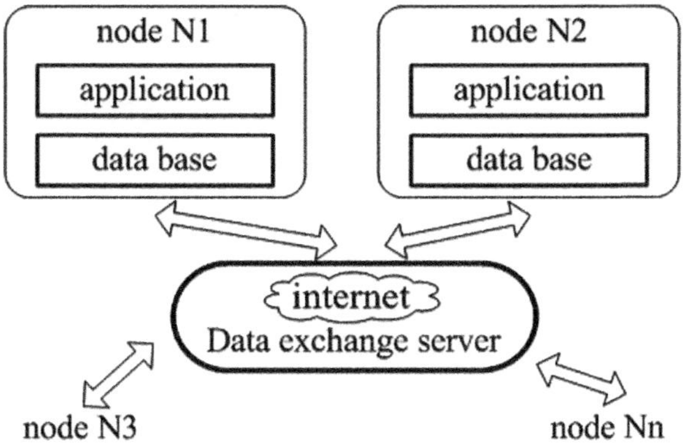

Fig. 1 System structure diagram

3 Data Change Capture

Data change capture technology refers to the technology of obtaining data changes in a database system by certain means. The change of data includes the increase of new data, the deletion of original data, and the modification of original data.

3.1 Comparison of Data Capture Technology

Common data change capture technologies are based on snapshot, trigger, log, timestamp, API, shadow table, change trajectory, etc. [10–15].

As can be seen from Table 1, shadow table method is a very suitable method for data change capture in heterogeneous data exchange system, because it is based on net change update mode, has wide application scope, occupies less resources, does not depend on heterogeneous database system and has high portability.

Table 1 Comparison of data capture technology

Capture method	Update mode	Dependence
Snapshot method	Complete copy	Nothing
Flip-flop method	Incremental modification	Trigger
Log method	Incremental modification	Log
Time stamp method	Incremental modification	Time stamp
API method	Incremental modification	Middleware
Shadow table method	Only change	None
Change trajectory method	Only change	Trigger

3.2 Shadow Table Method

At the beginning, a Shadow table *S* is established for replicating the object table *T*, that is to say, a copy is made at that time, and then net change information can be obtained by comparing the contents of the current *T* and *S* tables at an appropriate time. The shadow table method is suitable for solving heterogeneous database replication, for storage space and management.

(1) The operation information of the intermediate process is lost, which can not provide enough control information;
(2) When judging the operation time, FCLD (First Change Last Detect) phenomenon will occur;
(3) Because each acquisition needs to scan the entire *T* and *S* tables, the acquisition efficiency is relatively low.

4 Data Exchange Design

The heterogeneous data exchange system usually determines the data exchange according to the structure of each field of the target data system.

4.1 Data Change Acquisition

In this process, because of the heterogeneous structure of the source data system and the destination data system, there is no one-to-one correspondence between the tables and fields of the source data system and the destination data system. On the contrary, there will be a "many-to-many" relationship: several fields in several tables of the source data system correspond to the destination data. Multiple fields in multiple tables in the system, as shown in Fig. 2 [16, 17].

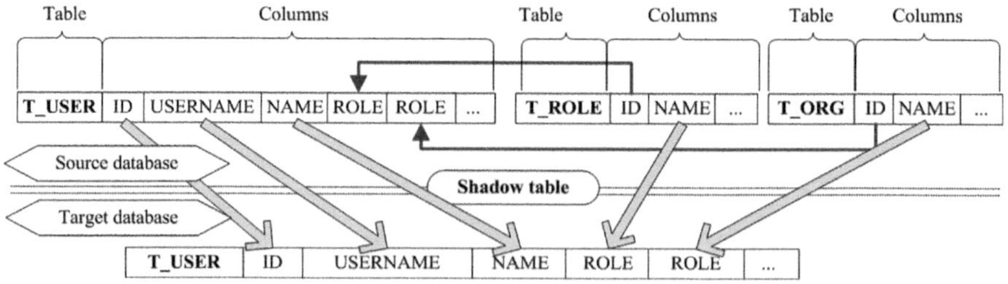

Fig. 2 Data synchronization process in heterogeneous tables

When grabbing data in source data system, some data grabbing depends on the fields of other tables. Media tables are tables that are used to connect with other tables to indirectly obtain the required fields.

4.2 Data Change Analysis

The main function of the database update acquisition module is to extract the corresponding database data according to the needs of the target database system. The process of data update and acquisition can be divided into two steps: monitoring the changes of database tables and generating the update result set according to the changes of database tables. Data change analysis is shown in Fig. 3.

A snapshot of a table at *T*1 time is taken as a *T*1 shadow table, and *T*1 is exported to the local temporary text file *FILE_TEMP1*. A snapshot of a table at *T*2 time is taken as a control table *T*2, and *T*2 is exported to the local temporary text file *FILE_TEMP2*. By comparing the differences between *FILE_TEMP1* and *FILE_TEMP2* through text difference module, two difference files *FILE_DIFF*− and *FILE_DIFF*+ are obtained. The change of database can be obtained by analyzing two different files.

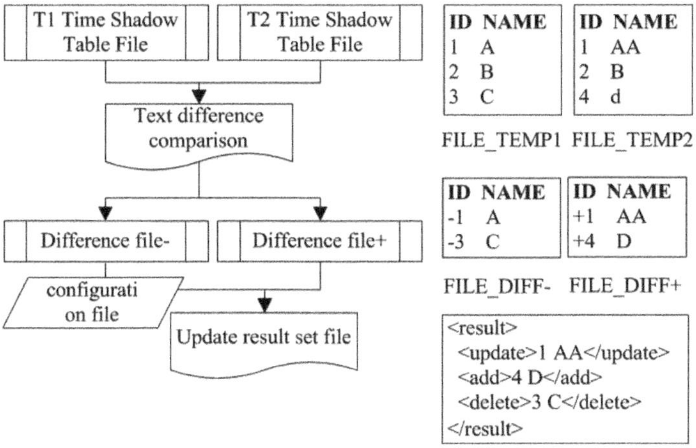

Fig. 3 Data change analysis process

5 Design of Data Updating Sequence Algorithms

This paper uses the topological sorting algorithm to determine the import order of the target data system. To put it simply, a partial order on a set is used to obtain a total order on that set. This operation is called topological ordering. Definitions of partial order and total order in discrete mathematics [18]: If the relation on set X is reflexive, antisymmetric and transitive, then R is a partial order relation on set X. Let R be a partial order on set X. If for each x, y belongs to X and must have xRy or yRx, then R is a total order relation on set X [19, 20].

In this paper, the author first converts the principal-external code relationship of the target database system into a directed graph and stores it with a two-dimensional matrix. After topological sorting, the import order of the target database system is obtained, as shown in Fig. 4.

With the deepening of enterprise informatization, more and more collaboration and communication activities will inevitably put forward higher and higher requirements for heterogeneous data exchange systems. Based on the analysis of the characteristics and related technologies of heterogeneous data systems, a complete design scheme of heterogeneous data exchange is proposed, and a point-to-point data exchange engine is constructed. The technology studied in this paper can provide a good data exchange platform for the information construction of all parts, departments and enterprises. However, heterogeneous data exchange is a very complex task. Due to the limited level and time of research and implementation of heterogeneous data exchange engine, only part of the work has been done in this paper.

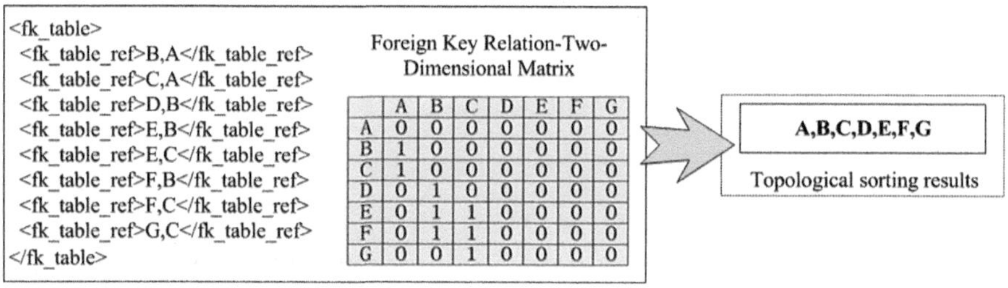

Fig. 4 Sequential topology algorithms for data insertion

Acknowledgements This research was financially supported by the Science and Technology Research Program of Hubei Provincial Department of Education (B2017424).

References

1. Garrido, A.L., Sangiao, S., Cardiel, O.: Improving the generation of infoboxes from data silos through machine learning and the use of semantic repositories. Int. J. Artif. Intell. Tools **26**(5), 1760022 (2017)
2. Wibowo, M., Sulaiman, S., Shamsuddin, S.M.: Machine learning in data lake for combining data silos (2017)
3. Tyagi, H., Viswanath, P, Watanabe, S.: Interactive communication for data exchange. IEEE Trans. Inf. Theory (99):1–1 (2016)
4. Simas, F., Barros, R., Salvador, L., et al.: A data exchange tool based on ontology for emergency response systems. In: Research Conference on Metadata and Semantics Research. Springer, Cham, pp. 74–79 (2017)
5. Waagen, J.: Learning sites, free open source communication and data-exchange tools for archaeological fieldwork and education. Comput. Appl. Quant. Methods Archaeol. (2012)
6. Wang, Y.: Data exchange middleware for Oracle and Sql Server database. Railway Comput. Appl. (2016)
7. Preden, J.S., Llinas, J., Motus, L.: Middleware for exchange and validation of context data and information. Context-enhanced information fusion. Springer International Publishing (2016)
8. Cramer, M.J.: Management of intermediate data spills during the shuffle phase of a map-reduce job (2017)
9. Tang, Z., Zhang, X., Li, K., et al.: An intermediate data placement algorithm for load balancing in Spark computing environment. Future Gener. Comput. Syst. **78** (2016)
10. Domingosalvany, A.: The science of real-time data capture: self-reports in health research. J. Epidemiol. Commun. Health **62**(5), 471 (2008)
11. Stanescu, L., Brezovan, M., Burdescu, D.D.: Automatic mapping of MySQL databases to NoSQL MongoDB. Comput. Sci. Inf. Syst. (2016)
12. Gamper, J., Jensen, C.S.: Extending the Kernel of a Relational DBMS with Comprehensive Support for Sequenced Temporal Queries[J]. ACM Trans. Database Syst. **41**(4), 26 (2016)
13. Tan, D.P., Li, L., Zhu, Y.L., et al.: An Embedded cloud database service method for distributed industry monitoring. IEEE Trans. Ind. Inf. (99):1–1 (2017)
14. Balis, B., Bubak, M., Harezlak, D., et al.: Towards an operational database for real-time environmental monitoring and early warning systems. Procedia Comput. Sci. **108**, 2250–2259 (2017)
15. Xia, Y., Chen, J., Lu, X., et al.: Big traffic data processing framework for intelligent monitoring and recording systems. Neurocomputing **181**(C), 139–146
16. Ye, Y., Li, T., Adjeroh, D.: A survey on malware detection using data mining techniques. Acm. Comput. Surv. **50**(3), 1–40 (2016)
17. Wang, Y., Wen, J., Fang, W., et al.: Research on incremental heterogeneous database synchronization update based on web service. In: International Conference on Computational Intelligence and Communication Networks, pp. 1415–1419. IEEE (2016)
18. Yan, Y.U., Science, S.O.: Analysis on teaching points about topological sort in data structure. J. Sci. Teach. Coll. Univ. (2017)
19. Loach, J.C., Wang, J.: Optimizing the learning order of chinese characters using a novel topological sort algorithm. PLoS ONE **11**(10), e0163623 (2016)
20. Hu, Q., Office, A.A., University, S.A.: Study on the relationship of course dependency based on topological sorting in the background of credit system. Guangdong Chem. Ind. (2017)

Simulation of Gain Effect of Solid-State Impact Ionization Multipliers

Yu Geng, Qin Li and Wan Qiu

Abstract A simulation study of the gain effect of solid-state impact ionization multipliers (SIMs) is reported. The SIM device separates the absorption and multiplication regions and is an alternative to avalanche photodiodes (APDs) used in LiDARs. The effects of geometric structures on electric field and potential distributions are simulated and studied. The optimized geometric structure increases the gain to around 270 at the voltage of 60 V. Based on this simulation, an optimized structure is proposed for future device fabrication.

Keywords APD · Simulation · Impact ionization multiplier · LiDAR · Autonomous driving

1 Introduction

Avalanche photodiodes (APDs) have an internal gain mechanism through avalanche multiplication and have been widely used in optical communications, imaging, single-photon detections and automotive applications [1, 2]. APDs are highly preferred for autonomous driving LiDARs due to their high gain, high bandwidth, and low noise.

Most LiDARs use 905-nm lasers with Si APDs which are inexpensive. Si has the advantage of large electron to hole ionization coefficient ratio (k), which leads to APDs with large bandwidth and low noise [1, 3]. But this wavelength can penetrate eyes so the power of 905 nm lasers is strictly regulated. One alternative approach is to use 1550-nm lasers that are much safer for eyes. This makes 1550-nm LiDARs to use higher power and detect farther objects. However, Si APDs cannot be used in the wavelength range larger than 1000 nm [4]. Hence, III-V materials such as InP

Y. Geng (✉) · Q. Li
School of Engineering, Shenzhen Institute of Information Technology, Shenzhen, China
e-mail: gengy@sziit.edu.cn

W. Qiu
Shenzhen Zhaoyang Institute of Information Technology, Shenzhen, China

V. Jain et al. (eds.), *Recent Trends in Intelligent Computing, Communication and Devices*, Advances in Intelligent Systems and Computing 1031,
https://doi.org/10.1007/978-981-13-9406-5_46

and InGaAs were widely used instead. Though this material system has the advantage of high absorption ratio, it has the disadvantage of small k [5]. Many types of APD with complicated structures were developed, such as SACM APDs and APDs with submicron scaling of the multiplication. However, these structures make APDs hard to design and fabricate [1–6].

It is ideal that Si can be used in multiplication layers and other materials with high absorption ratio, such as InP, be used in absorption layers. This led to the solid-state impact ionization multiplier (SIM) [7, 8]. This completely separated the absorption and multiplication regions. This separation makes SIM easier to be optimized without affecting the absorption efficiency [7, 8]. Various geometric structures of SIM were simulated and optimized.

2 Design

The original design of the SIM is based on the device structure of Lee [8] as shown in Fig. 1. The SIM is on a p^- Si wafer with a doping concentration of $1 \times 10^{15}\ \text{cm}^{-3}$.

The geometric structures determine the electric field distribution of the SIM. Optimized design of these geometric structures helps to create higher gain under a certain voltage.

3 Simulation

The SIM is simulated by Silvaco. The original design is shown in Fig. 1. The widths of the anode and the cathode are both set to be 5 μm, while the distance between them (d) is set as 10 μm. The width of SiO_2 insulation region (w) is 5 μm,

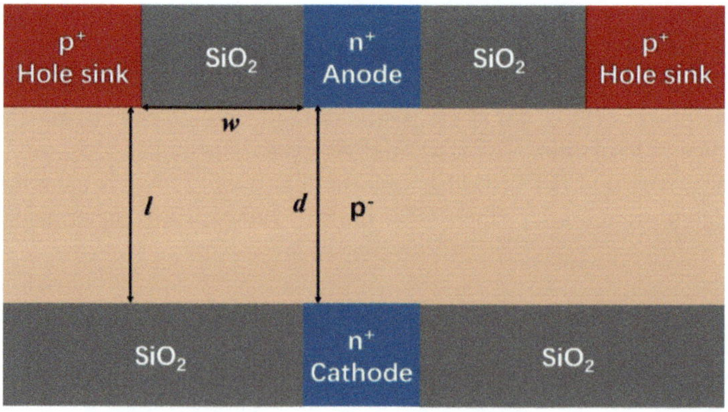

Fig. 1 Design of the SIM structure

and the length between the lower edge of hole sink to the upper edge of the cathode (l) is the same as d.

4 Results and Discussion

Different device geometric structures were simulated and compared.

When d is smaller than 8 μm, the gain decreases quickly with decreasing d as shown in Fig. 2.

In the conditions with smaller *ds*, the electric field is low between the anode and the cathode but high between the cathode and hole sinks, as shown in Fig. 3a. From the potential distribution shown in Fig. 3b, there is little potential drop between the cathode and the anode. This is caused by the fact that Si in this region is unable to support the high electric field such that the cathode and the anode are shorted. Hence, the high electric field is formed between the cathode–anode pair and hold sinks as shown in Fig. 3a. This low electric field caused the small gain of 1.7 at the voltage of 60 V when d is 1 μm.

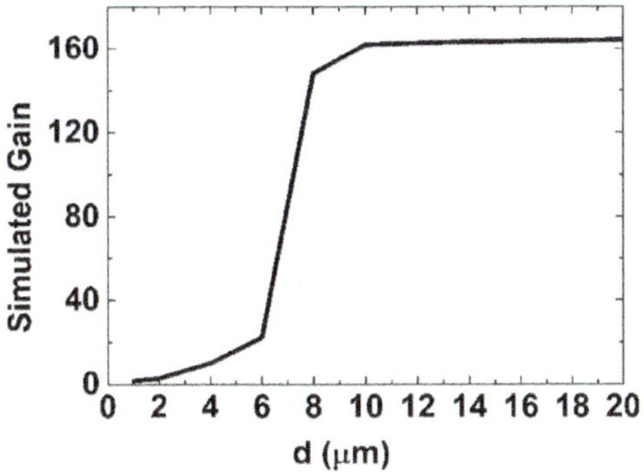

Fig. 2 Simulated gain versus d at the voltage of 60 V

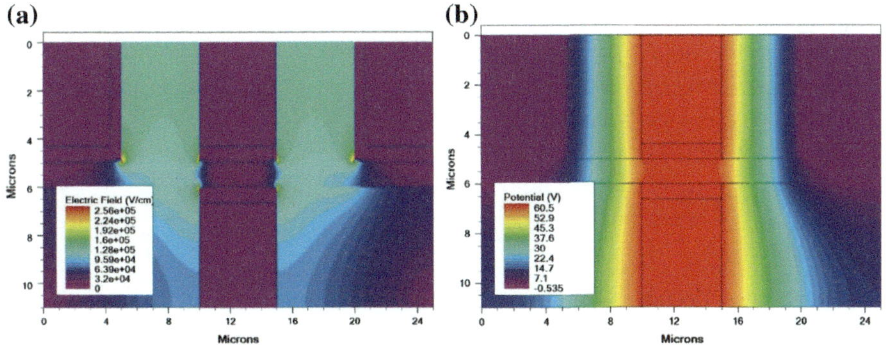

Fig. 3 Electric field and potential distribution at the voltage of 60 V with d of 1 μm. **a** Electric field distribution; **b** potential distribution

Fig. 4 Simulated gain versus *l* at the voltage of 60 V

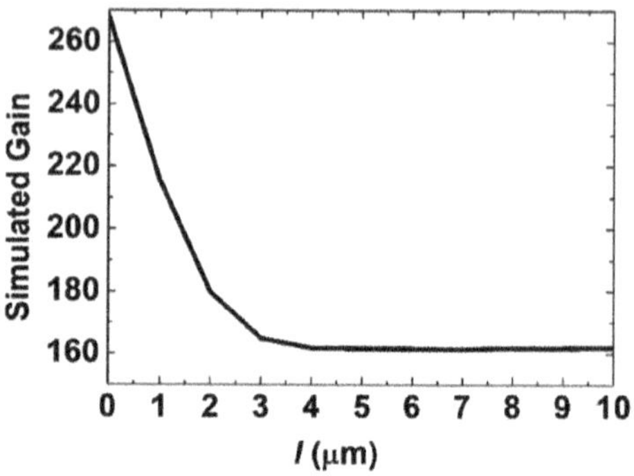

Fig. 5 Simulated gain versus *w* at the voltage of 60 V

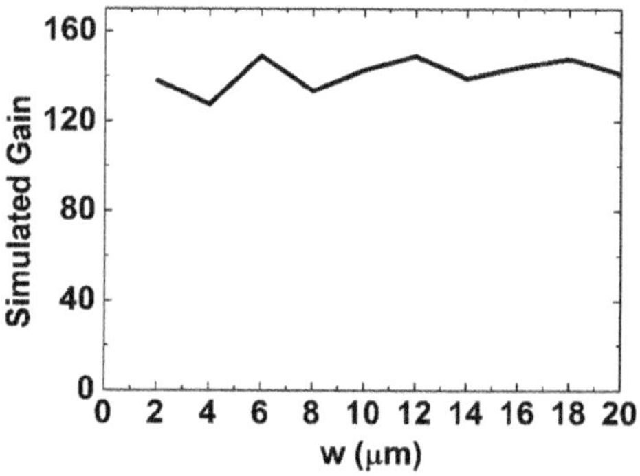

When *d* gets larger, the electric field and potential drop between the anode and the cathode are increased. This increases the gain to 160 at the voltage of 60 V. When *d* is larger than the threshold distance of 10 μm, the gain changes little because potential change stops, but the electric field drops with the increase of *d*.

With the increase of *l*, the hole sinks get farther to the cathode and the gain keeps decreasing as shown in Fig. 4.

The length of *w* is not critical as shown in Fig. 5. The simulated gain is around 140 for various *w*. The reason behind this is that the electric field distribution is similar for different *w*, which leads to the similar gains.

These simulation results show that the higher gain could be obtained by smaller *l* and larger *d* while *w* is not critical.

5 Conclusion

The SIM design has been simulated and optimized. The geometric structure has a great effect on the gain of SIM. Higher gain could be obtained by smaller *l* and larger *d*, which increase the gain from 160 to 270 at the voltage of 60 V. These results make SIMs highly potential to be used in LiDARs as an alternative to APDs. Future work will be focused on device fabrication based on this simulation work.

Acknowledgements This work was supported in part by the Shenzhen Fundamental Research fund under Grant No. JCYJ20170306100015508.

References

1. Maes, W., De Meyer, K., Van Overstraeten, R.: Impact ionization in silicon: a review and update. Solid-State Electron. **33**(6), 705–718 (1990)
2. Campbell, J.C.: Recent advances in avalanche photodiodes. J. Lightwave Technol. **34**(2), 278–285 (2016)
3. McIntyre, R.J.: Recent developments in silicon avalanche photodiodes. Measurement **3**(4), 146–152 (1985)
4. David, J.P., Tan, C.H.: Material considerations for avalanche photodiodes. IEEE J. Sel. Top. Quantum Electron. **14**(4), 998–1009 (2008)
5. Campbell, J.C., Demiguel, S., Feng, M.A., Beck, A., Guo, X., Wang, S., Zheng, X., Xiaowei, L.I., Beck, J.D., Kinch, M.A., Huntington, A., Coldren, L.A., Decobert, J., Tscherptner, N.: Recent advances in avalanche photodiodes. IEEE J. Sel. Top. Quantum Electron. **10**(4), 777–787 (2004)
6. Campbell, J.C.: Recent advances in telecommunications avalanche photodiodes. J. Lightwave Technol. **25**(1), 109–121 (2007)
7. Lee, H.W., Hawkins, A.R.: Solid-state current amplifier based on impact ionization. Appl. Phys. Lett. **87**(7), 3511 (2005)
8. Lee, H.W., Beutler, J., Hawkins, A.: Surface structure silicon based impact-ionization multiplier for optical detection. Opt. Express **13**(22), 8760–8765 (2005)

Iris Localization Based on Spiking Neural Networks

Jinqing Liu and Yin Liu

Abstract In this paper, a morphological operator is used to remove eyelash on eye image instead of median filtering. Spiking neural networks, which inherit the parallel mechanism from biological system, are used to extract the edge of iris. Since spiking neural networks are able to remove the unimportant details and retain the main outline features, the edge images, which extracted by spiking neural networks, are good for the eye positioning. Based on the edge images, the least square method is used to fit outer boundary so that the complexity of computation is reduced. Experiments show that the proposed algorithm is very efficient. Furthermore, the algorithm can be transformed into GPU platform and can speed up dramatically.

Keywords Iris location · Morphological operator · Spiking neural network · Least square method

1 Introduction

As developing of the modern society, technology of identification is used in more and more fields. The biometric technologies of iris, fingerprints, face and voice are developing to replace the traditional form of identification. Especially, the iris identification is recognized as a kind of ideal identification method because of its security, accuracy and anti-counterfeiting. Iris is located between pupil and sclera, and contains much texture information. The inner and external boundaries of the iris are rounds approximately.

J. Liu (✉)
Fuzhou University of International Studies and Trade, Fuzhou 350202, Fujian, China
e-mail: jqliu8208@fjnu.edu.cn

Y. Liu
The Department of Communication Engineering, Xiamen University, Xiamen 361005, Fujian, China

V. Jain et al. (eds.), *Recent Trends in Intelligent Computing, Communication and Devices*, Advances in Intelligent Systems and Computing 1031,
https://doi.org/10.1007/978-981-13-9406-5_47

Iris recognition can be divided into three steps: iris localization, iris feature extraction and iris recognition. As the iris localization algorithm is the foundation of the whole iris recognition technology, how to positioning iris quickly and accurately under the interference of eyelash and eyelid is one of most important issues in the whole iris recognition system.

At present, there are two kinds of iris localization algorithms, i.e., calculus localization algorithm [1] by Daugman and edge detection combined with Hough transform voting localization algorithm [2] by Wildes. Many other methods always evolved these two kinds of algorithms. In these methods, Daugman's calculus algorithm has a large amount of calculation and the positioning can be easily influenced by the light source and eyelash; Hough transform is to obtain parameters in three-dimensional spaces by vote, so it is time-consuming. Qingrong Li put forward a method combined coarse position and fine positioning [3]. This method uses the characteristics of gray projection quantity distribution to perform a rough positioning and then makes the accurate positioning with round template. It improves the accuracy and speed of positioning and solves the problems in iris image with a large face region. This paper proposes a new approach based on spiking neural networks to improve the accuracy further.

2 Eyelash Detection and Elimination

2.1 *Smooth Process of the Eye Image*

The boundary between iris and sclera is not clear because of light or other environmental influences. The information of the texture will be easily covered by noise. The light spot, which the reflection of light leaves in the eye image, will produce adverse effect in edge extraction. In order to reduce the influence of the noise, the common method is to smooth the eye image using a filter. The methods include average filtering, median filtering and Gaussian filter. This paper using wiener filtering [4], wiener filtering is an optimal estimator for stationary processes based on the minimum mean square error criterion. The mean square error between the output and the expected output of this filter is the smallest, so it is an optimal filtering system. Local mean u, the variance σ^2 and the estimated value in Wiener filtering $b(n_1, n_2)$ are as follows.

$$u = \frac{1}{MN} \sum_{n_1, n_2 \in \eta} a(n_1, n_2) \tag{1}$$

$$\sigma^2 = \frac{1}{MN} \sum_{n_1, n_2 \in \eta} a(n_1, n_2) - u^2 \tag{2}$$

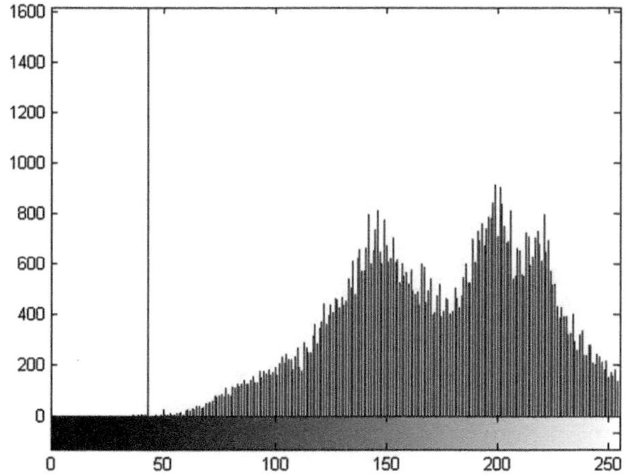

Fig. 1 Gray histogram before Wiener filter

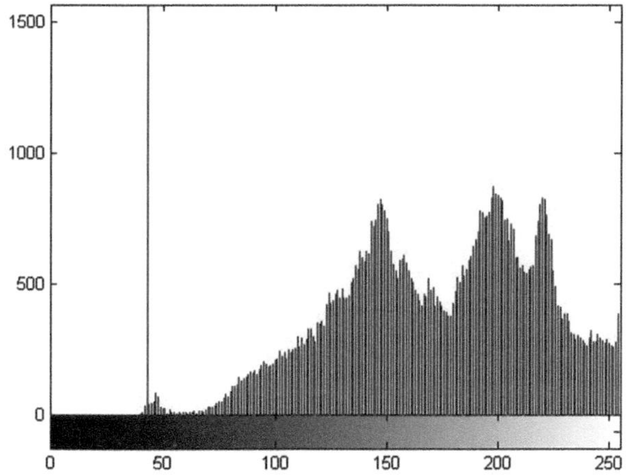

Fig. 2 Gray histogram after Wiener filter

$$b(n_1, n_2) = u^2 + \frac{\sigma^2 - v^2}{\sigma^2}(a(n_1, n_2) - u) \tag{3}$$

Wiener filter keeps the edge of image and high-frequency detail better than linear filter. And the amount of calculation of Wiener filtering is small. Comparing Figs. 1 and 2, it is obvious that the histogram becomes more smooth after using Wiener filter.

2.2 Remove Eyelash Using Morphological Operator

Eyelash hides some parts of the iris, and the pupil's gray scale is similar to the scleras, which is shown in Fig. 3. This directly affects iris boundary localization. The article [5] uses the median filter to remove the noise of iris images, so the

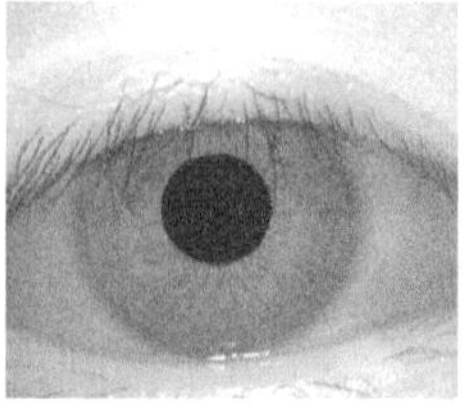

Fig. 3 Picture in library

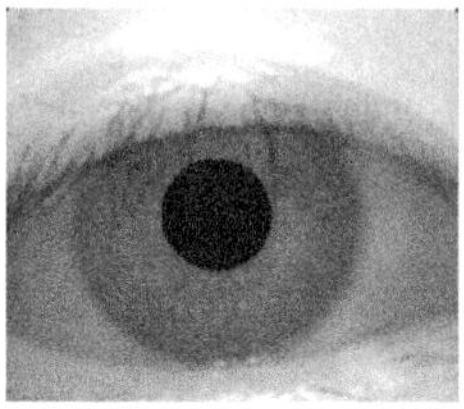

Fig. 4 Median filtering

eyelash in Fig. 4 becomes less clear but still looks obvious, and the external boundary becomes fuzzy. All of these affect the accuracy of positioning. In this paper, the eyelash is regarded as a slender, irregular black noise. Using a gray image morphological operator [6], dilation and erosion operators are combined to form a filter of excellent performance. The dilation expansion of gray image *f* with structural element *b* is expressed in Eq. (4).

$$(f \oplus b)(x,y) = \max\{f(x - x', y - y') + b(x', y') | (x', y') \in D\} \tag{4}$$

The erosion expansion of gray image *f* with structural element *b* is expressed in Eq. (5):

$$(f \Theta b)(x,y) = \min\{f(x + x', y + y') - b(x', y') | (x', y') \in D\} \tag{5}$$

D is the domain of structure element *b*. Close operation dilates the image before eroding: $f \cdot b = (f \oplus b) \Theta\, b$. Close operation can remove the regions for the size smaller than structure element *b*. Choosing the structure element as 5 * 5 square matrix of ones to process the image with close operation, the result is shown in Fig. 5.

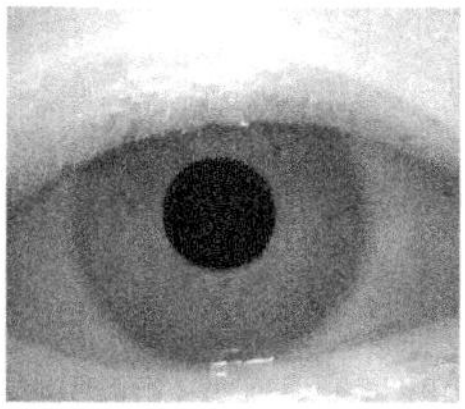

Fig. 5 Close operation

3 Edge Extraction of Spiking Neural Networks

Since the 1990s, scientists put forward a set of pulse neural network theory based on the neuron model of the Nobel Prize winner Hodgkin [7] which were more close to the biological neural network. Combining biological visual information processing mechanism and spiking neural networks to study image information processing becomes the interdisciplinary of computer vision, neuroscience and intelligent science.

In digital image processing, different pixels can be regarded as different strength stimuli to neurons; different stimuli induce neuron membrane potential up to a threshold value; if the potential gets to the threshold value, the spiking neuron will release a spike train. When the input is a square of smooth gray value in the receptive field, spiking neural network produces a low-frequency spike train; when a square of gradient gray values is input, such as the edge of an image, the spike neurons respond with high-frequency impulses. Figure 6 shows the response to edge by spiking neurons [8].

The structure of edge detection of spiking neural network [9] is shown in Fig. 7. The network is formed of three layers: The first layer is light receptor, and each pixel corresponds to a receiver, so the pixel can be changed into the spiking signal.

The middle layer composed of different types of neuron arrays which are connected to receptive fields in the first layer, where there are four kinds of neurons, respectively, can respond to right, left, up and down edges, and each neuron connects receptive fields with connection weights. 'Δ' represents inhibitory synapse, and 'X' represents excitatory synapse. The principle is as follows: for example, the up edge; if the receptive fields receive the smooth gray value, the upper inhibitory synapses will suppress membrane potential of neurons N1 while the lower excitatory synapses strengthen the membrane potential of N1 neurons. Overall the N1 membrane potential has not changed. N1 does not produce spikes (pulses). If the receptive fields receive the gradient gray value, excitatory signals are larger than inhibitory signals, and the N1 membrane potential rises fast. Therefore, N1 produces spikes. If there are enough pixels corresponding to excitatory

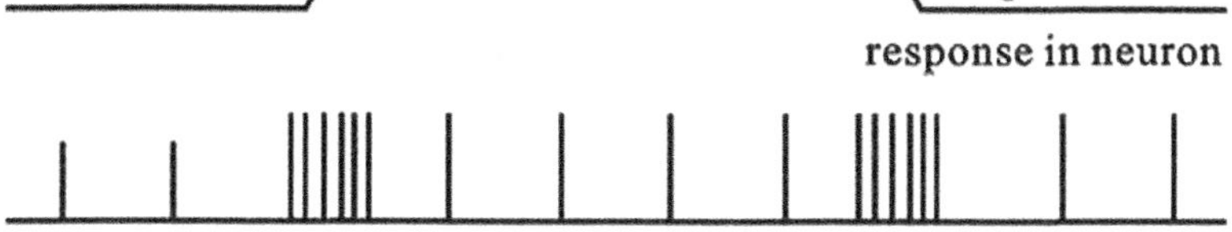

Fig. 6 Response to the edge of spiking neurons

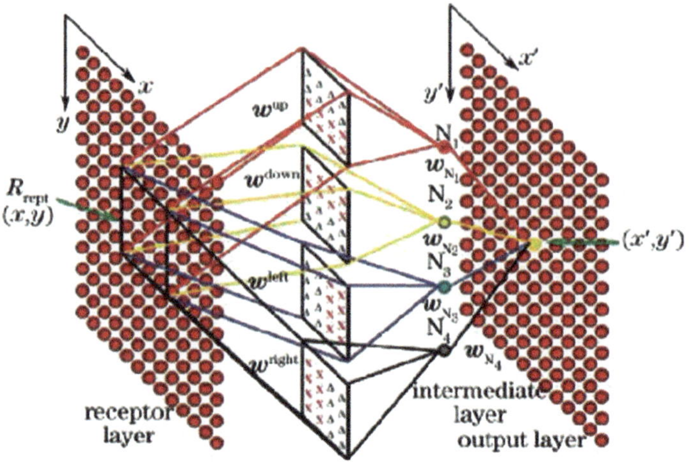

Fig. 7 Structure of edge detection of spiking neural network

synapses, spiking neurons will generate spikes at a different time to form a certain frequency spike train.

The third layer is output layer, and neuron (x', y') integrates spike trains s from different neuron arrays of the middle layer. The high spike rate will be obtained from the neurons corresponding to the edge. Figure 8 shows the result of different methods of edge extraction, it is obvious that the main outline features are clearer and consecutive, and the unimportant details become less in the spike rate image of spiking neural networks.

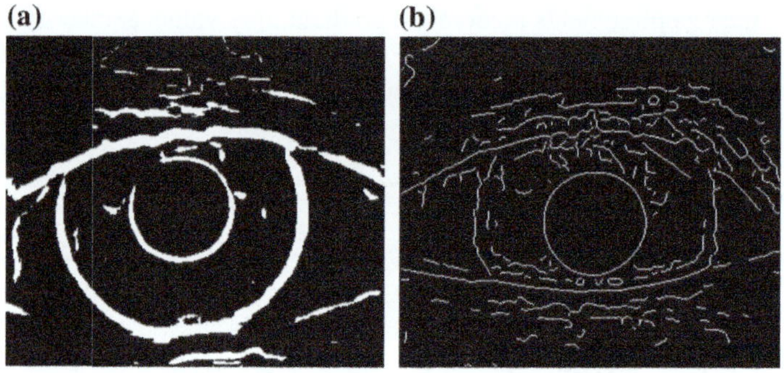

Fig. 8 Different results of different methods: **a** edge extraction of spiking neural networks and **b** edge extraction of canny operator

4 External Boundary Location of the Iris

4.1 Inner Boundary Location of the Iris

The shape of the pupil can be seen as round approximately. The gray values in pupil area are close to black. This character can be used to segment pupil. The external boundary of pupil is the iris' inner boundary. Figure 9 is gray-level histogram of eye image where eyelash is removed. The lower peak of the histogram represents the pupil area. Taking the bottom after the peak as the threshold to get binary, the result is shown in Fig. 10. Papers [10, 11] have segmented the eye image directly so that influence of eyelash has not been eliminated. In this paper, we use morphological operator to remove eyelash so that the accurate positioning of edge can be achieved.

Presented Fig. 10 to spiking neural networks, a round edge is obtained in the output. Label the connected pixel in Fig. 11, and the point coordinates (x_1, y_1), $(x_2, y_2) \cdots (x_i, y_i) \cdots (x_n, y_n)$ in the edge can be obtained if the edge image is regarded as a marked matrix. Central coordinates (x, y) and radius r can be calculated by expression (6). The inner edge location of the iris is shown in Fig. 12.

$$\begin{aligned} x &= \frac{1}{n}\sum_{i=1}^{n} x_i, \\ y &= \frac{1}{n}\sum_{i=1}^{n} y_i \\ r &= \frac{1}{n}\sum_{i=1}^{n} \sqrt{(x - x_i)^2 + (y - y_i)^2} \end{aligned} \tag{6}$$

Fig. 9 Gray-level histogram

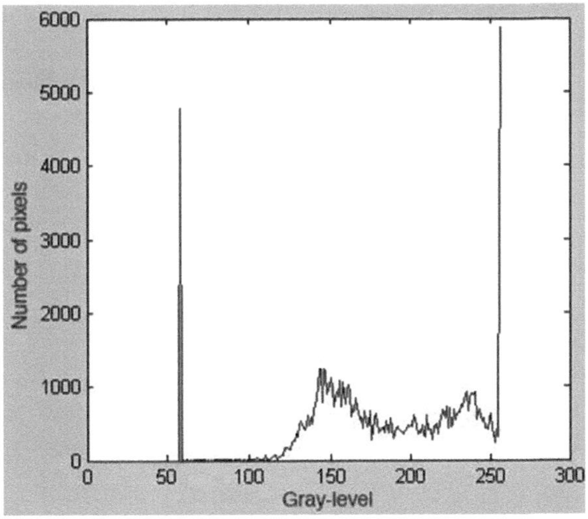

Fig. 10 Binary

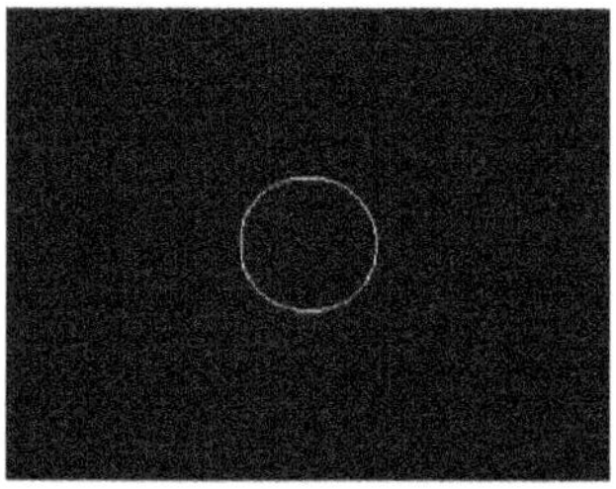

Fig. 11 Inner edge extraction

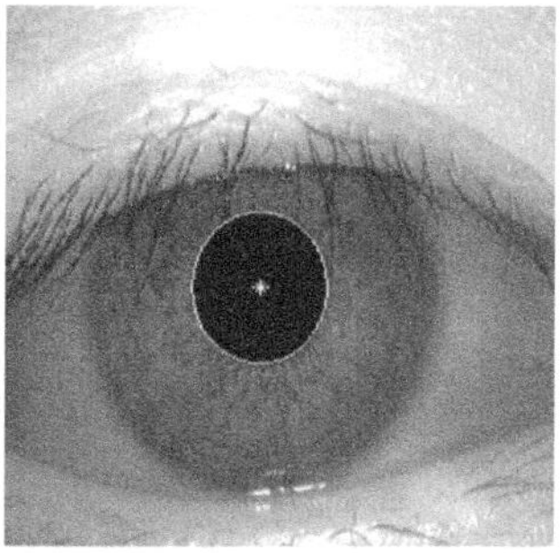

Fig. 12 Inner edge location

4.2 Circle Fitting Method Based on Hough Transform

The basic strategy of Hough transform is to calculate amount of the points which meet the curve with certain parameters. The parameters of the most points are the parameters of the boundary. Let following equation represent outside boundary of iris:

$$(x - a)^2 + (y - b)^2 = r^2 \tag{7}$$

Hough transform scanning (a, b, r) is carried out in three-dimensional parameter space, and three-dimensional accumulator array $H(a, b, r)$ is established. When point (a, b) moves in the image, if boundary point (x, y) of the boundary with radius

r fell on the circle with a certain parameter, then it is added to the accumulator array $H(a, b, r)$, according to the voting decision. If

$$H(x_0, y_0, r_0) = \mathrm{MAX}(\cup(H(x, y, r))) \tag{8}$$

then the parameters x_0, y_0, r_0 get the most votes. It indicates that the circle with a, b, r is on the circle which is most close to the boundary. As it requires prior knowledge before using Hough transform, the results are not susceptible to noise and discontinuous curve induced by vote mechanism. However, this algorithm leads to time-consuming and it is not suitable to use in real time. A new method is proposed in the next session.

4.3 External Boundary Location of the Iris Based on the Effective Arc

Eyes will turn in the process of collecting iris images, so iris in the top and bottom of the pupil often gets out of eyelid, as shown in Fig. 14a. This can affect the accuracy of iris positioning, and this paper uses the iris area in both sides of pupil. Iris' boundary and external boundary are concentric circles approximately. In order to get the effective arc in the external edge in Fig. 8a, the positioning of the external boundary is changed into solving the parameters of the circle according to the circular arc. The intercept processing is shown in Fig. 13. The pupil radius r and center (x, y) can be used to remove noise in Fig. 13b. Firstly, we remove the noise points whose distance from the center is slightly longer than radius r. Secondly, we remove the noise points outside the external boundary of iris based on the fact that the distance between points outside the iris and the center of the pupil is less than $3.5 * r$ [12], where r represents the radius of pupil.

Some researchers detect a round with Hough transform [13] or the evolution of Hough transform [14], but these methods have a large amount of calculation. This paper fit outer boundary through the least square method that can find the cure $y = p$

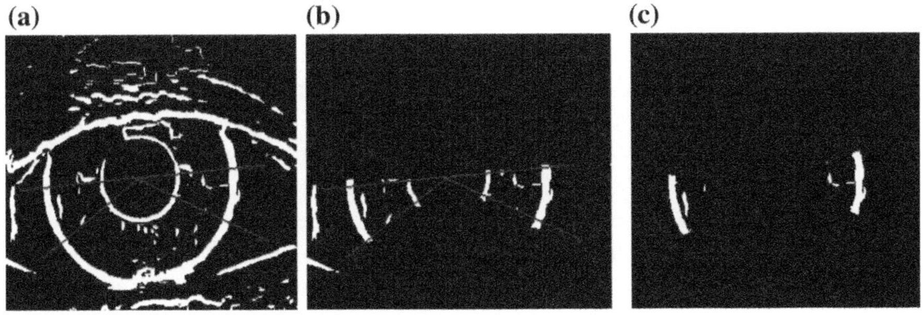

Fig. 13 Intercept processing of effective arc in external boundary

(x) which forms the least square summation of the distance between the points in $y = p(x)$ and the designated points $\{(X_i, Y_i)\}$ ($i = 0, 1, \ldots, m$).

Let a, b and c represent undetermined parameters, and the general formula of round is shown in Eq. (9)

$$x^2 + y^2 + ax + by + c = 0 \tag{9}$$

Transform it to standard equation of a round.

$$\frac{1}{4}\left(a^2 + b^2 - 4c\right) = \left(x - \left(-\frac{a}{2}\right)\right)^2 + \left(y - \left(-\frac{b}{2}\right)\right)^2 \tag{10}$$

We have central coordinates (X_c, Y_c) and radius R for external boundary.

$$X_c = -\frac{a}{2}, \quad Y_c = -\frac{b}{2}, \quad R = \frac{1}{2}\sqrt{a^2 + b^2 - 4c} \tag{11}$$

The known distance D_i from points (X_i, Y_i) ($i = 1, 2, 3, \ldots, N$) of arc to the center is as follows

$$D_i^2 = (X_i - X_c)^2 + (Y_i - Y_c)^2 \tag{12}$$

$$F(a, b, c) = \sum_{i=1}^{N} \left(D_i^2 - R^2\right)^2 = \sum_{i=1}^{N} \left(X_i^2 + Y_i^2 + aX_i + bY_i + c\right)^2 \tag{13}$$

Minimize $F(a, b, c)$ to make D_i as close as the round radius R, and then get parameters a, b and c. Firstly, partial derivatives on a, b and c are calculated. Secondly, parameters a, b and c can be obtained by making the partial derivative equal to 0, where the radius R and central (X_c, Y_c) come from Eq. (11).

5 Simulation and Analysis of Result

The algorithm has been verified using the Iris database of Chinese Academy of Sciences (CASIA1.0), which contains 108 types of iris. The Iris images are clear but noised by light, eyelid and eyelash. The experiment is done on PC which has the main frequency for 2 GHZ of dual-core CPU and memory for 2 G. The simulation environment is MatlabR2010b. Results of the positioning of three iris images are shown in Fig. 14a′–c′. It is obvious that the algorithm in this paper can locate the iris accurately even if the images are noised. Figure 14a–c shows the position results using the algorithm of canny + hough [15]. The results show that even if the iris image is disturbed by eyelid and eye rotation, the proposed algorithm can locate the iris image more accurately.

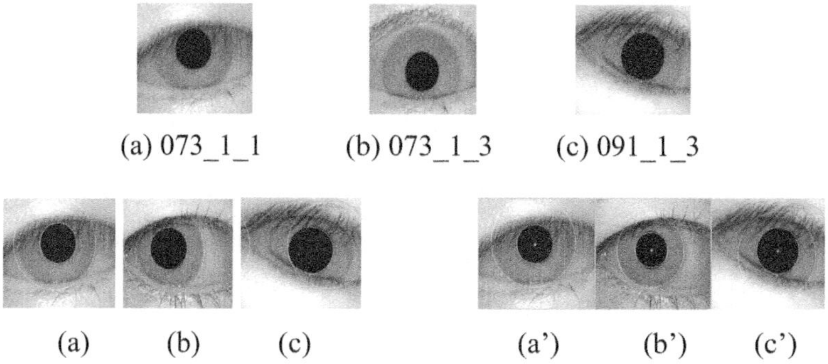

Fig. 14 Comparison of different eye image positioning methods

In this paper, we use the spiking neural networks which inherit the parallel mechanism from biological system to extract the inside and outside edge of the iris. The algorithm has been implemented using GPU platform. Experiments show that the speed of edge extraction will increase by about 30 times on GPU, about 0.5 s.

6 Conclusion

In this paper, morphological method is used to reduce the noise of eyelash on iris detection. The spiking neural networks, which inherit the parallel mechanism from biological system, are introduced to extract the inside and outside edge of the iris. The Hough transform has not been used because of its time-consuming and large amount of calculation. The effective iris' outside edge arc is extracted by the prior knowledge, fitting outer boundary through the least square method. Experiments show that the algorithm in this article can find the iris location accurately under the interference of eyelash, eyelid and eyeball movement. The speed of the algorithm can be largely increased by transplanting it to GPU platform.

Acknowledgements The authors gratefully acknowledge the support from the National Natural Science Foundation of China (61179011). The authors also acknowledge the editor and colleagues who provided technical support.

References

1. Daugman, J.: Biometric personal identification system based on iris analysis. No. 5560 (Patent) (1994)
2. Wildes, R.P.: Iris recognition: An emerging biometric technology. In: Proceedings of the IEEE 85.9, pp. 1348–1363 (1997)

3. Li, Q.: A iris location algorithm. J Univ. Electron. Sci. Technol. China **1**(31), 7–9 (2002)
4. Qiuqi, R.: Digital image processing. Publishing House of Electronics Industry (2001)
5. Le, C., Yuan, W.: A new effective method of eyelash and eyelid detection. Microelectron. Comput. **4**(28), 122–130 (2011)
6. Rafael, C., Richard, E., Steven, L.: Digital Image Processing Using MATLAB, 1st edn, pp. 253–263. Publishing House of Electronics Industry, Beijing (2009)
7. Hodgkin, A., Huxley, A.: A quantitative description of membrane current and its application to conduction and excitation in nerve. J. Physiol. **4**(3), 500–544 (1952)
8. Dayan, P., Abbott, L.: Theoretical Neuroscience: Computational and Mathematical Modeling of Neural Systems, 2nd edn, p. 254. MIT Press, Massachusetts (2001)
9. Li, X., Wu, Q., Kou, Y., et al.: Lane detection based on spiking neural network and hough transform. In: International Congress on Image and Signal Processing, pp. 626–630. IEEE (2016)
10. Guo, P.Y., Hu, X.H.: A fast iris localization algorithm based on line scanning. Microelectron. Comput. **6**(27), 105–108 (2010)
11. Luo, M.M., Wu, X.S.: Improved fast iris localization algorithm. Comput. Eng. Appl. **46**(17), 200–203 (2010)
12. Shi, C.L., Jiang, G.Y., Qi, H., et al.: An efficient and fast iris location algorithm. Appl. Mech. Mater. **263–266**, 2549–2552 (2013)
13. Radman, A., Jumari, K., Zainal, N.: Fast and reliable iris segmentation algorithm. IET Image Proc. **7**(1), 42–49 (2013)
14. Liu, N., Su, H., Guo, C.H., Zhou, J.: Improvement of eye location method based on circle examination of Hough transformation. Comput. Eng. Des. **32**(4), 1359–1362 (2011)
15. Xu, X.R., Li, Y.J.: A fast iris location method based on combined coarse- and fine-treatment. J. Lanzhou Univ. Technol. **37**(3), 104–107 (2011)

A Survey of Digital Twin Technology for PHM

Wang Xiaodong, Liu Feng, Ren Junhua and Liang Rongyu

Abstract Digital twin uses digital technologies to describe and model the characteristics, behaviors, and processes of a device or a system in real world, which has aroused wide concern in aerospace engineering, smart factory and smart building. This paper inspected the various definitions and characteristics of digital twin, and summarized the perspectives and process when using digital twin in PHM area. A digital twin–driven PHM framework of high–speed railway (Electric Multiple Unit) EMU maintenance has been proposed.

Keywords Digital twin · Prognostics and health management · EMU · Complex equipment · Condition-based maintenance

1 Introduction

Prognostics and health management (PHM) is an important technical means to achieve condition-based maintenance (CBM), improve system reliability, and reduce system maintenance costs. PHM uses sensors to monitor the states of the device in real time, uses various models and algorithms to perform fault diagnosis, fault prognostics, and remaining life prediction, and forms the optimal maintenance plan. With the rapid development of big data and artificial intelligence technology, it is feasible to collect, transmit, and analyze device data in real time, and to perform more fine-grained and personalized health management and fault diagnosis on the devices.

Digital twin is an essential enabling technology for PHM. Digital twin refers to the process and method of using digital technology to describe and model the characteristics, behavior, process, and performance of physical objects [1]. The operation and maintenance management of complex equipment is one of the reasons why digital twin comes up with from the beginning in aerospace, and it is also

W. Xiaodong · L. Feng (✉) · R. Junhua · L. Rongyu
Beijing Jiaotosng University, Beijing 100044, China
e-mail: fliu@bjtu.edu.cn

V. Jain et al. (eds.), *Recent Trends in Intelligent Computing, Communication and Devices*, Advances in Intelligent Systems and Computing 1031,
https://doi.org/10.1007/978-981-13-9406-5_48

one of the most important research directions. In the future, objects in real world will have corresponding digital twins in the virtual space, on which the health management, condition-based maintenance, and real-time control could be performed.

The emergence and development of digital twin has brought new opportunities and challenges to PHM. As a new technology originating from the aerospace industry and widely used in the field of intelligent manufacturing, the research and application of digital twin in the complex equipment PHM is still in the early stage. This paper studies the various concepts and definitions of digital twin for PHM. The main characteristics of digital twin for PHM are proposed. An example of how to use digital twin-based PHM in high-speed railway train maintenance is proposed.

2 Concepts, Functions, and Perspectives

2.1 Concepts and Evolution

There is currently no uniform definition of digital twin for PHM. This section analyzes the content related to equipment health management and prognostics in the existing literature on digital twin, as shown in Table 1.

Table 1 Various definitions of digital twin

Year	Definition	Concern
2005 [15]	a virtual, digital equivalent to a physical product	PLM, intelligent manufacturing
2012 [1]	an integrated multi-physics, multi-scale, probabilistic simulation of a vehicle or system that uses the best available physical models, sensor updates, fleet history, etc., to mirror the life of its flying twin	Simulation-based product design and maintenance
2016 [2]	A virtual substitutes of real world objects consisting of virtual representations and communication capabilities making up smart objects acting as intelligent nodes inside the internet of things and services	Simulation-based virtual verification
2016 [16]	a digital representation of a real world object	PLM data model
2017 [8]	a model capable of rendering the state and behaviour of a unique real asset in (close to) real time	Relations between model-based and data-driven digital twin
2016 [17]	a model of a physical asset, that encapsulates an information/data model, a simulation model and model-based and data driven analytics, and possibly a dependability model and possibly a visualisation model	Condition assessment of complex marine machinery systems
2018 [18]	a highly-detailed simulation model which tries to reproduce its physical behavior in virtual world	Plant's lifecycle in Industry 4.0

In 2012, the National Aeronautics and Space Administration (NASA) proposed digital twin as a key technology for "Simulation-based Systems Engineering" in its published "Modeling, Simulation, Information Technology and Processing Roadmap" [1]. Digital twin was defined as "an integrated multi-physics, multi-scale, probabilistic simulation of a vehicle or system that uses the best available physical models, sensor updates, fleet history, etc., to mirror the life of its flying twin". In the same year, NASA and the U.S. Air Force Research Laboratory jointly proposed a digital twin paradigm for future aircraft [2]. The main uses of digital twin at this time are equipment condition monitoring, remaining useful life prediction, and providing guidance on how the system will handle safety-related events.

In the study of digital twin applied to fault diagnosis, US Air Force Research Institute proposed the idea of delivering the aircraft digital twin model when delivering the aircraft [3]. The model is "a 1000 billion degree-of-freedom (DOF) hierarchical, computational structures model, and is ultrarealistic in geometric detail and material detail".

In recent years, driven by simulation technology and big data, and artificial intelligence technology, digital twin has developed rapidly in the research and application of product manufacturing and operation and maintenance. The scope of application has expanded from aircraft simulation to intelligent manufacturing, product design, and operation and maintenance.

This paper defines digital twin for PHM as "entities, processes or systems of the physical world built digitally in a virtual space, by means of product lifecycle data fusion, multi-physics simulation modeling, virtual and real interaction, machine learning, artificial intelligence and other techniques to monitor the physical entity status, and automatically make correspond according to changes in physical entities".

2.2 *Functionalities*

The main functions of digital twin within PHM area are:

(1) Full lifecycle data model Digital twin is the lifecycle data repository of physical entities, and covers (1) the data from product design stage, such as geometric appearance [4], material [5] and CAD model [6]; (2) the data from manufacturing phase of the product, such as plant data, measurement data [7], (3) product usage and product failure data during the product operation and maintenance phase [4]. Through the integration and fusion of various dimensional information, digital twin can be used as a single data source of product.

The digital twin of complex equipment is usually not a single node, but rather a group of nodes with information interaction with each other. The possibility of using digital twin as the lifecycle management of industrial IoT devices was discussed in [8]. It is considered that a real-world object is not represented by a single node of digital twin, but by a subgraph of nodes and edges. A digital twin platform for CPS [9] was proposed to solve the inconsistency of resolution, complexity, modeling language and format among different digital twins. The components of a digital twin may be heterogeneous systems, so the interoperability and information exchange between components becomes basis of other operations. In terms of digital twin information communication, Seongjin Yun et al. [9] compared the communication middleware commonly used in CPS for digital twins. In terms of data fusion, Schroeder et al. [10] proposed a data model for data exchange and communication using AutomationML.

(2) Device condition monitoring and health management It is one of the most important research and application directions of digital twin. Through the high-fidelity and ultrarealistic digital twins built in virtual space, as well as integrated lifecycle data and multi-level and multi-physical models, digital twin can further improve the performance and reliability of condition monitoring, fault diagnosis, health management and maintenance task scheduling. The current research trends are concentrated on the structural health monitoring of spacecraft. In [3], a fusion model composed of the finite element analysis model and fatigue life prediction model is proposed to address the discrete models existing in the aircraft structural health cannot work cooperatively. In [11], a method combining porous metal plasticity model (physical model) and finite element model (numerical model) of as-manufactured specimens is generated and subsequently analyzed to resolve the crack-path ambiguity. Simulation based technology and data-driven technology are the two main ways to build digital twins.

(3) Testing and verification As long as an integrated ultrarealistic digital twin of systems is established, it is feasible to perform testing and verifying on the digital twin in the virtual space instead of on the physical objects in the real world. The process is called virtual testing and virtual verification. Since the virtuality of those procedures, the need for physical equipment and the damage to physical equipment caused by destructive testing could be reduced.

In [12], a roadmap for Transglobal Integrated Tests and Analyses Network for Structures based on digital twin is proposed as a blueprint for improvements in aircraft fatigue modelling and experimentation, since it is not easy to reliably and accurately model structural fatigue by any single team at the airframe level due to the limitation of time, cost, and data. Digital twin provides a uniform view for upstream and downstream vendors to work on cooperatively.

3 Digital Twin-Driven PHM Process

In terms of digital twin models, Tao Fei [13] proposed a five-dimensional structural model of digital twin, including physical entities, virtual models, services, data, and connections, and applied to the system architecture design of digital twin workshops [14]. According to the four perspectives of digital twin, the digital twin-driven PHM process is proposed, including the following steps.

(1) Data access: As the device lifecycle data repository, digital twin provides data access services for PHM applications. The components of a PHM system may be heterogeneous—taking the various interfaces and various formats among IT systems and devices into consideration—the information exchange, data fusion, and interoperability between components are the basis for other operations. The confusion could be alleviated if different systems and different vendors work on a uniform digital twin data repository, which will provide a holistic, structured data support for PHM and other applications.

(2) Condition monitoring, fault diagnosis and prediction: These processes should be carried out synergistically with multi-physics, multi-model, and multi-level. Take bearing as example, all the possible data from bearing should be considered, such as bearing temperature, bearing vibration, bearing maintenance, and replacement history. Different models would be established based on different data format, and those models should be working collaborative, no matter it is a mechanical vibration simulation model, or a Bayesian inference model, or a deep learning model. Since states and failure occurrences may have potential correlations among subsystems and components, analysis models of different components must be arranged, communicated, and integrated.

(3) Maintenance strategy formulation: Under the condition of precisely health status prediction, the preventive maintenance decision and inventory management plan could be made considering the device operation tasks, maintenance costs, and inventory status. Unnecessary maintenance work could be reduced, and the reliability and safety risks brought by over-maintenance could be minimized.

(4) Testing and verification: Testing and verification are important measures to improve the reliability and credibility of PHM systems. A common practice is to establish a test bed to test and verify algorithms through historical data and fault injection data. Digital twin is the high-fidelity models of the devices under test, and testing and verification could be performed on digital twin instead of real devices, which can reduce the demand for physical devices and reduce the damage to the physical devices caused by the destructive test.

4 Applications of Digital Twin for PHM

The high-speed train EMU has high running speed, long running mileage, and limited maintenance time, which has high requirements for the healthy operation of the EMU. Researchers have carried out relevant studies on how applying PHM to the key parts of EMU, such as the high-voltage power supply system, traction motor, bearing, and gear box. Figure 1 presents a digital twin-based EMU PHM process. Data are collected and fused from the design, manufacturing phase, operation and maintenance phase. Then multi-physical simulation models and data-driven models are built to provide services, including condition monitoring, fault diagnosis, reliability analysis, remaining useful life prediction, and failure prognostics. The decisions of condition-based maintenance and predictive

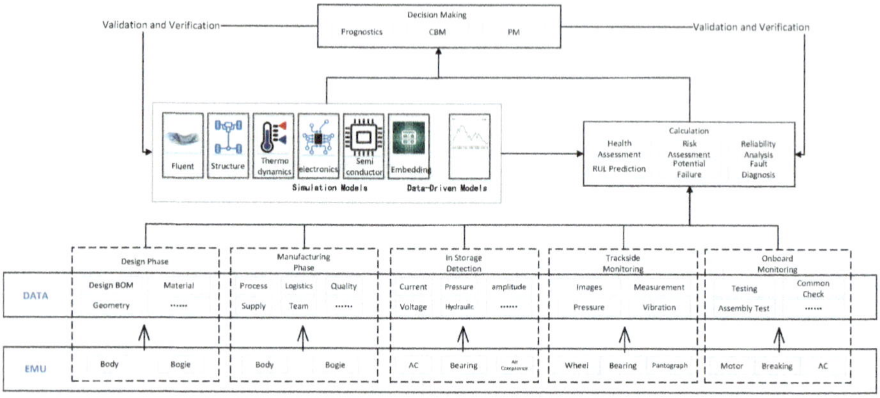

Fig. 1 Digital twin driven PHM process of EMU

maintenance could be made based on the services. Virtual verification and validation through digital twin can maximize the reliability of the service and reduce the dependence on the physical objects.

5 Conclusion

This paper examines definitions and characteristics of digital twins from different perspectives and proposes four perspectives of digital twin, summarizes the main research progress and functions of digital twin for PHM, and proposes the main process of digital twin-driven PHM. The application of the high-speed train in the field of PHM is prospected.

Acknowledgments This work was supported by Ministry of Industry and Information Technology New Model Application Project of Intelligent Manufacturing, China (2016ZNZZ01-01).

References

1. Rocessing, P.R., Kemp, C.: Modeling, Simulation, Information Technology and Processing Roadmap. Simulation, no. May 2010 (2010)
2. Glaessgen, E., Stargel, D.: The digital twin paradigm for future NASA and U.S. air force vehicles. In: 53rd AIAA/ASME/ASCE/AHS/ASC Structures, Structural Dynamics and Materials Conference 20th AIAA/ASME/AHS Adaptive Structures Conference 14th AIAA, pp. 1–14 (2012)
3. Tuegel, E.J., Ingraffea, A.R., Eason, T.G., Spottswood, S.M.: Reengineering aircraft structural life prediction using a digital twin. Int. J. Aerosp. Eng. **2011**, 1–14 (2011)
4. Schleich, B., Anwer, N., Mathieu, L., Wartzack, S.: Shaping the digital twin for design and production engineering. CIRP Ann. Manuf. Technol. **66**, 141–144 (2017)

5. Hochhalter, J.D., et al.: Coupling Damage-Sensing Particles to the Digital Twin Concept. no. April (2014)
6. Tao, F., et al.: Digital twin-driven product design framework. Int. J. Prod. Res. **7543**, 1–19 (2018)
7. Söderberg, R., Wärmefjord, K., Carlson, J.S., Lindkvist, L.: Toward a digital twin for real-time geometry assurance in individualized production. CIRP Ann—Manuf. Technol. **66**(1), 137–140 (2017)
8. Canedo, A.: Industrial IoT lifecycle via digital twins. In: Proceedings of the Eleventh IEEE/ ACM/IFIP International Conference on Hardware/Software Codesign and System Synthesis, pp. 1–1 (2016)
9. Yun, S., Park, J.H., Kim, W.T.: Data-centric middleware based digital twin platform for dependable cyber-physical systems, pp. 922–926. Int. Conf. Ubiquitous Futur. Networks, ICUFN (2017)
10. Schroeder, G.N., Steinmetz, C., Pereira, C.E., Espindola, D.B.: Digital twin data modeling with automationml and a communication methodology for data exchange. IFAC-PapersOnLine **49**(30), 12–17 (2016)
11. Cerrone, A., Hochhalter, J., Heber, G., Ingraffea, A.: On the effects of modeling as-manufactured geometry: toward digital twin. Int. J. Aerosp. Eng. **2014**, 1–10 (2014)
12. Wong, A.K.: Blueprint TITANS : A Roadmap towards the Virtual Fatigue Test through a Collaborative International Effort. no. June, pp. 7–9 (2017)
13. C. Integrated and M. Systems: Digital Twin and Its Potential Application Exploration. no. February (2018)
14. Tao, F., Zhang, M.: Digital twin shop-floor: a new shop-floor paradigm towards smart manufacturing. IEEE Access **5**(c), 20418–20427 (2017)
15. Grieves, M.: Digital Twin: Manufacturing Excellence through Virtual Factory Replication, pp. 1–7 (2014)
16. Schluse, M., Rossmann, J.: From simulation to experimentable digital twins: simulation-based development and operation of complex technical systems. In: 2016 IEEE International Symposium on Systems Engineering (ISSE) (2016
17. Erikstad, S.O.: Merging Physics, Big Data Analytics and Simulation for the Next-Generation Digital Twins. Hiper, no. September, pp. 139–149 (2017)
18. Zymaris, A.S., Alnes, Ø.Å., Knutsen, K.E., Kakalis, N.M.P.: Towards a model-based condition assessment of complex marine machinery systems using systems engineering. In: European Conference of the PHM Society (2016)

Image Inpainting of Patch Matching with Boundary and Region Constraints

Huaming Liu, Xuehui Bi, Guanming Lu, Jingjie Yan, Jian Wei and Xiuyou Wang

Abstract Image inpainting aims to fill missing regions or damaged regions. The inpainting results need to meet human visual consistency. Patch matching technique is often applied to search similar patch for fill missing parts. However, only using the distance between pixels cannot guarantee to find the best similar patch. In our paper, we proposed a novel patch matching technique which adopts region and boundary constraints. There are two advantages: (1) boundary constraint can alleviate blockiness or strong discontinuities; and (2) region constraint is more conducive to finding similar patches. By weighting the distance between two patches to constrain the matching block. Using exemplar-based technique proposed by Criminisi et al., we demonstrate the validity of our block match.

Keywords Boundary constraint · Region constraint · Image inpainting · Patch matching

1 Introduction

Digital images are increasingly exposed to our lives, such as cell phones, camera photographs. Removing or repairing some areas is often encountered in image editing. In recent years, image restoration technology has become more and more popular. Removing noise and deburring are blind inpainting. Our work focuses on non-blind inpainting. Patch matching technique is often used for non-blind inpainting [1–4]. Block matching technology initially uses distance of color for measuring the similarity between two patches. Sometimes, the patch with the smallest distance is not necessarily the most similar block.

H. Liu (✉) · X. Bi · X. Wang
Fuyang Normal University, Fuyang, China
e-mail: liuhuaming888@126.com

H. Liu · G. Lu · J. Yan · J. Wei
Nanjing University of Posts and Telecommunications, Jiangsu, China

V. Jain et al. (eds.), *Recent Trends in Intelligent Computing, Communication and Devices*, Advances in Intelligent Systems and Computing 1031,
https://doi.org/10.1007/978-981-13-9406-5_49

In addition, blockiness occurs frequently. In [5], spatial patch blending is to solve this problem. In image editing [6], copy-and-paste technique often generates inconsistence with around regions. Boundary blending can effectively alleviate blockiness. To improve patch matching technique, we look for similar blocks with boundary and region constraints. In this way, the boundary similarity problem is taken into account. By this way, blockiness can be effectively alleviated. Similarity measurement between blocks is computed jointly by region and boundary distance. To overcome limitation of error matching, a new patch matching with boundary and region constraints for image inpainting is proposed which makes the following contributions:

(1) Contrary to traditional patch matching, we pose patch matching as a combination of region matching and boundary matching. (2) Boundary constraint can alleviate blockiness, since it emphasized the importance of boundary constraint. (3) The weights of the boundary and regional distances are automatically determined, avoiding the disadvantages of manual intervention. Our patch matching method can be applied for the inpainting using patch matching method.

The paper is organized as follows. Section 2 presents patch matching technique. Patch matching with boundary and region constraints is introduced in Sect. 3. Section 4 introduces experiment results. Conclusion of our work is presented in the Sect. 5.

2 Related Work

2.1 Patch Matching Technique

Exemplar-based inpainting technique [1] uses patch matching for searching similar patches. Figure 1 shows the process of filling using exemplar-based technique. Ω denotes the target region. $\partial\Omega$ refers to the boundary of target region. We compute priority of a patch which center in the $\partial\Omega$. If the patch has the highest priority, it will be filled first in the order of filling. Assume that a patch ψ_p whose center is p has the highest priority as shown in Fig. 1b.

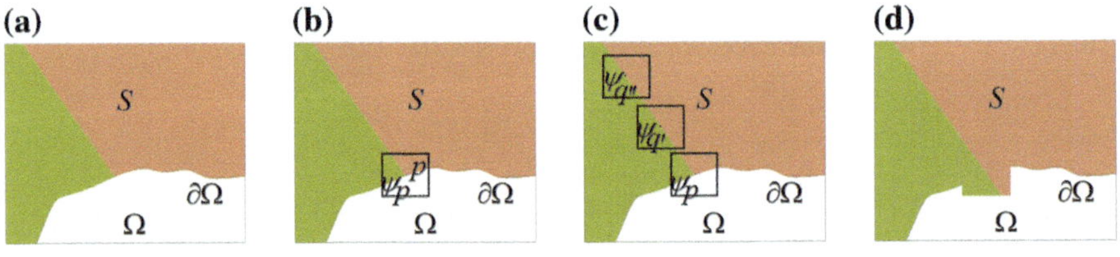

Fig. 1 Exemplar-based method. **a** Target region Ω, **b** select a patch to be filled, **c** look for similar patches, **d** fill the target patch

The target image is denoted as I. The patch ψ_p will be filled by similar patch. Similar patches can be found by patch matching. The patch matching is to compute distance of color between two patches by (1).

$$d_{ssd}(\psi_p, \psi_q) = \frac{1}{N}\sqrt{\sum_{c=1}^{m} \sum_{i=1\psi_{pi}\in\psi_p \cap S\psi_{qi}\in S}^{N} (\psi_{pi} - \psi_{qi})^2} \tag{1}$$

where c is the channel number of image, when $c = 1$, the image denotes gray-level image, and when $c = 3$, the image refers to color image. N is the number of known pixels in the ψ_p. S is sample resource ($S \cup \Omega = I$). The patch matching for searching patch is shown in Fig. 1c.

The best similar patch can be searched in the full image by (1). The best similar patch is computed by (2).

$$\psi_{\hat{q}} = \arg\ \min\{d_{ssd}(\psi_p, \psi_q)\} \tag{2}$$

The best similar patch $\psi_{\hat{q}}$ is the patch which has the smallest distance. The best similar patch is used to fill the target patch ψ_p. The result of filling is shown in Fig. 1d.

2.2 *The Insufficiency of Patch Matching*

If there are three patches as shown in Fig. 2. The target patch is shown in Fig. 2a. The patches in Fig. 2b, c are similar to target patch. But the distribution of pixel value is different. Therefore, error matching occurs sometimes.

(a)

1	2	3
4	5	6
7	8	9

(b)

1	2	3
6	5	4
7	8	9

(c)

1	2	3
8	9	4
7	6	5

Fig. 2 Insufficiency of patch matching. **a** The target patch. **b** The patch is similar to (**a**). **c** The patch is similar to (**a**)

3 Proposed Patch Matching

3.1 Boundary and Region Separation

Suppose the target patch is ψ_p as shown in Fig. 3a. The known part is composed by boundary part ψ_p^b and region part ψ_p^r. The unknown part will not be used in patch matching. The red area is the boundary part and plays an important role in the boundary matching. The region part, which guarantees the matching of regions, is labeled by yellow color as shown in Fig. 3a.

3.2 Patch Matching with Boundary and Region Constraints

We use boundary and region constraints for patch matching. The computation of distance between two patches is given by (3).

$$d(\psi_p, \psi_q) = \alpha d_{ssd}(\psi_p^r - \psi_q^r) + (1-\alpha) d_{ssd}(\psi_p^b - \psi_q^b) \tag{3}$$

where $d_{ssd}(\psi_p^r - \psi_q^r)$ and $d_{ssd}(\psi_p^b - \psi_q^b)$ are the distance of region part and the distance of boundary part between ψ_p and ψ_q, respectively. α is the weight. $d_{ssd}(\psi_p^r - \psi_q^r)$ and $d_{ssd}(\psi_p^b - \psi_q^b)$ are computed by (4) and (5).

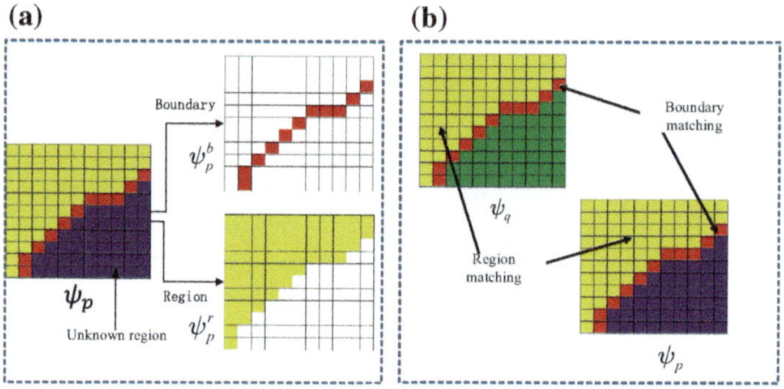

Fig. 3 Separation of target patch. **a** ψ_p is a target patch to be filled. The red region and yellow region are the known part, violet region is unknown part (missing part), the boundary is red part denoted by ψ_p^b, and region is yellow part denoted by ψ_p^r. **b** The patch matching is combination of boundary and region constraints between two patches

$$d_{ssd}(\psi^r_p - \psi^r_q) = \frac{1}{M}\sqrt{\sum_{c=1}^{3}\sum_{i=1}^{M}(\psi^r_{pi} - \psi^r_{qi})^2} \quad (4)$$

$$d_{ssd}(\psi^b_p - \psi^b_q) = \frac{1}{N}\sqrt{\sum_{c=1}^{3}\sum_{j=1}^{N}(\psi^b_{pj} - \psi^b_{qj})^2} \quad (5)$$

where M and N are the number of pixel in the region part and boundary part in the ψ_p and α is determined automatically by (6).

$$\alpha = \frac{M}{M+N} \quad (6)$$

3.3 The Proportion of Boundary and Region Parts

Table 1 shows the average proportion when using patches of different sizes. If we use patches of different sizes, the regional proportion will vary. In the experiments, the value of α is variable when using patches of the same size, because the shape of the missing region is often change. The change of α will affect the result of repairing. So we use the average of α when to fix the size of patch, and Table 1 shows the average of α when we take patches of different sizes.

Table 1 Region proportion (α) with different patch sizes

Patch size	Average of α	Average of $1 - \alpha$
7 × 7	0.756109	0.243891
9 × 9	0.801354	0.198646
11 × 11	0.8303	0.1697
13 × 13	0.853203	0.146797
15 × 15	0.871072	0.128928
17 × 17	0.874048	0.125952
19 × 19	0.891978	0.108022
21 × 21	0.905896	0.094104
23 × 23	0.913188	0.086812
25 × 25	0.921969	0.078031
27 × 27	0.928777	0.071223
29 × 29	0.937235	0.062765

4 Results

We use exemplar-based technique [1] to test our patch matching. Figures 4 and 5 give the some results of repairing using other algorithms and our method. c–h in Figs. 4 and 5 are results of inpainting by Criminisi's method [1], Deng's method [7], Jurio's method [8], Barnes's method [9], Newson's method [10], and Our method. Imperfect repair results can be seen in Figs. 4c–e and 5c–e. Figures 4f–h and 5f–h show the pleasing results.

Figure 6 shows structure repairing. Figure 6c–e is results of inpainting by Crimini's method [1], Anupam's method [11], Jurio's method [8], respectively. However, they all obtain unpleasing results. Figure 6f–h is results of inpainting by Barnes's method [9], Newson's method [10], and Our method. They give pleasing results. The edge of the stairs is fuzzy in Fig. 6g occurs blockiness, and our method can alleviate some blockiness as shown in Fig. 6h.

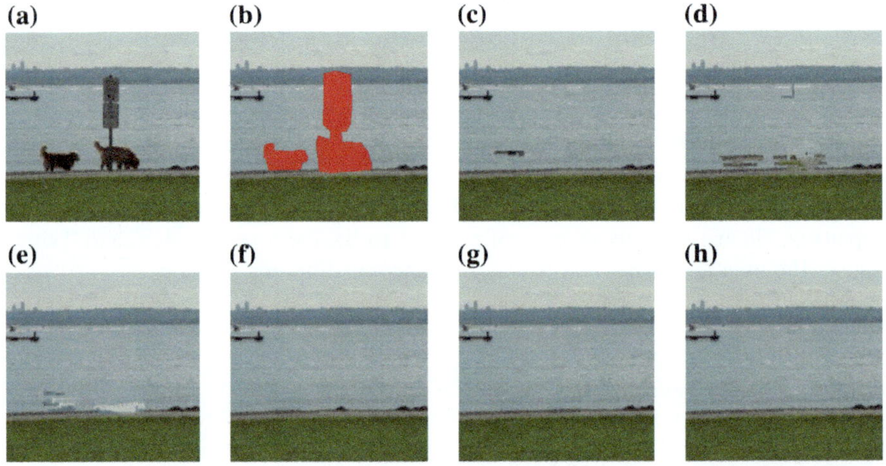

Fig. 4 Object removing. **a** Origin image, **b** mask image, **c** Criminisi's method [1], **d** Deng's method [7], **e** Jurio's method [8], **f** Barnes's method [9], **g** Newson's method [10], **h** our method

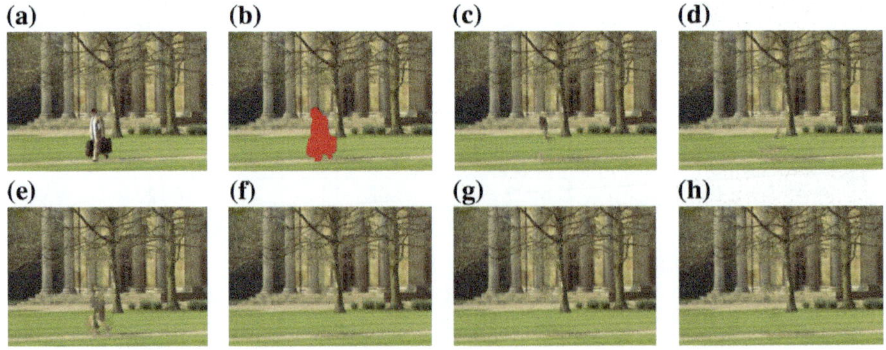

Fig. 5 People removing. **a** Origin image, **b** Mask image, **c** Criminisi's method [1], **d** Deng's method [7], **e** Jurio's method [8], **f** Barnes's method [9], **g** Newson's method [10], **h** Our method

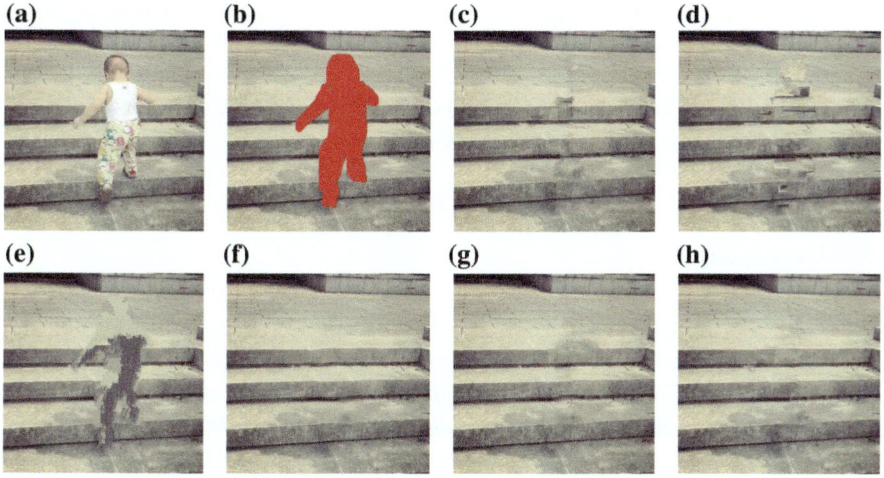

Fig. 6 Object removing. **a** Origin image, **b** Mask image, **c** Criminisi's method [1], **d** Anupam's method [11], **e** Jurio's method [8], **f** Barnes's method [9], **g** Newson's method [10], **h** Our method

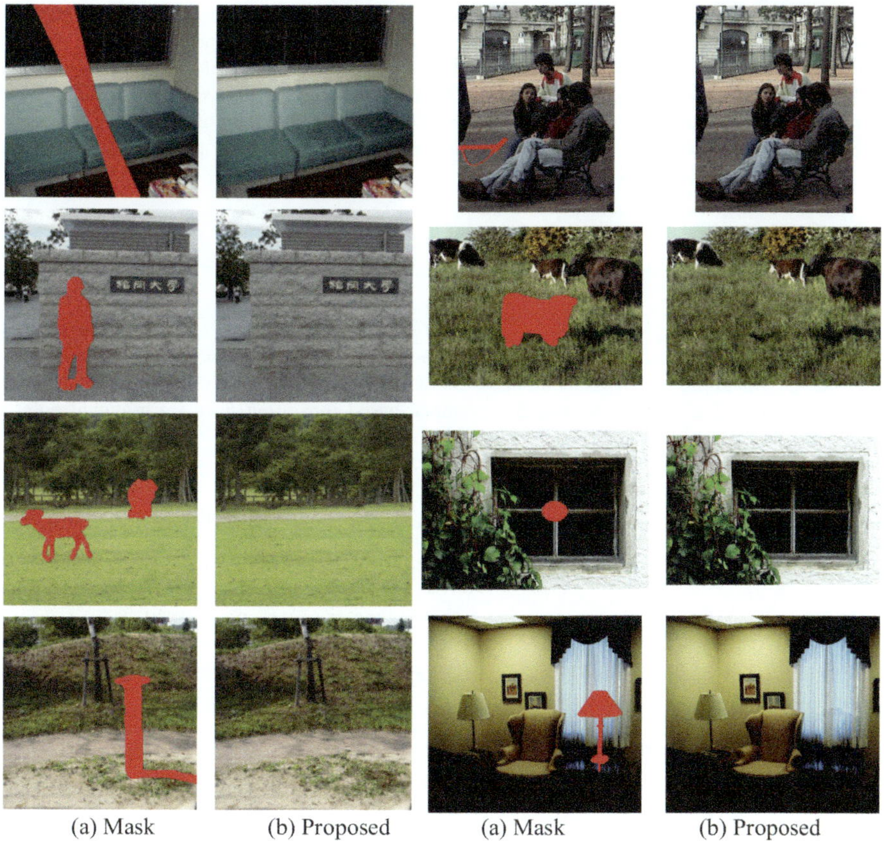

Fig. 7 More examples of inpainting by proposed method

Figure 7 gives more examples of inpainting using our patch matching algorithm. Those images contain structure and texture. Using our patch matching algorithm, we can obtain pleasing results. Some images in Fig. 7 are from http://yokoya.naist.jp/research2/inpainting/, https://wenfulee.github.io/CS-766-Computer-Vision/, or have been published in previous researches [12].

5 Conclusion

We propose a novel patch matching technique which is constrained by boundary and region. It can alleviate the blockiness. It can alleviate the blockiness and reduce error filling. Patch matching using boundary and region constrained, respectively, can improve patch matching. We give more examples to show that patch matching proposed can alleviate blockiness and easily inpaint images. On the other hand, we also find that patch size also affects the quality of repair. Next, we will optimize the selection of block size.

Acknowledgements This research was funded by the key projects of natural science research in Anhui colleges and universities (No.KJ2018A0345), the key fund projects of Young Talents in Fuyang Normal University (No.rcxm201706), the National Natural Science Foundation of China (61772430), Postgraduate Research and Practice Innovation Program of Jiangsu Province (No. KYCX18_0901), the Natural Fund Project in Fuyang Normal University(No.2017FSKJ17). This work is also supported by the Key Research and Development Program of Jiangsu Province (No. BE2016775). The 2017 horizontal cooperation project of Fuyang municipal government-Fuyang Normal College (No.XDHX201732).

References

1. Criminisi, A., Pérez, P., Toyama, K.: Region filling and object removal by exemplar-based image inpainting. IEEE Trans. Image Process. **13**(9), 1200–1212 (2004)
2. Aujol, J.F., Ladjal, S., Masnou, S.: Exemplar-based inpainting from a variational point of view. Siam J. Math. Anal. **42**(3), 1246–1285 (2010)
3. Cao, F., Gousseau, Y., Masnou, S., Perez, P.: Geometrically guided exemplar-based inpainting. Siam J. Imag. Sci. **4**(4), 1143–1179 (2011)
4. Liu, Y.Q., Caselles, V.: Exemplar-based image inpainting using multiscale graph cuts. IEEE Trans. Image Process. **22**(5), 1699–1711 (2013)
5. Daisy, M., Tschumperlé, D., Lézoray, O.: Spatial patch blending for artefact reduction in pattern-based inpainting techniques. In: Proceeding of 15th International Conference on Computer Analysis of Images and Patterns, 523–530. Springer, Berlin Heidelberg, Germany, York, UK, Aug. 27–29, 2013
6. Martino, J.M.D., Facciolo, G., Meinhardt-Llopis, E.: Poisson image editing 5. Image Process. On Line **18**, 300–325 (2016)
7. Deng, L.J., Huang, T.Z., Zhao, X.L.: Exemplar-based image inpainting using a modified priority definition. PLoS ONE **10**(10), 1–18 (2015)

8. Jurio, A., Paternain, D., Pagola, M., Marco-Detchart, C., Bustince, H.: Two-step algorithm for image inpainting. In: Proceeding of Advances in Intelligent Systems and Computing, pp. 302–313. Springer, Cham, Switzerland, Warsaw, Poland, Sep. 11–15, 2017
9. Barnes, C., Shechtman, E., Finkelstein, A., Goldman, D.B.: PatchMatch: a randomized correspondence algorithm for structural image editing. ACM Trans. Graph. **28**(3), 1–11 (2009)
10. Newson, A., Almansa, A., Gousseau, Y., Pérez, P.: Non-local patch-based image inpainting. Image Process. On Line **7**, 373–385 (2017)
11. Goyal, P., Diwakar, S.: Fast and enhanced algorithm for exemplar based image inpainting. In: Proceeding of 2010 Fourth Pacific-Rim Symposium on Image and Video Technology, pp. 325–330. Nov 14–17, 2010, IEEE, New York, NY, USA
12. Shen, J., Jin, X., Zhou, C.: Gradient based image completion by solving poisson equation. In: Proceeding of Pacific-Rim Conference on Multimedia, pp. 257–268. Jeju Island, Korea, Republic of, Nov. 13–16, 2005, Springer, Berlin, Heidelberg

Design of High-Availability E-Reading Platform

Jianming Huang and Yu Wang

Abstract In recent years, people have become more and more accepting of e-reading, and the e-reading industry has developed rapidly. With the increasing demand for e-reading platforms, traditional e-reading platforms have become increasingly unable to satisfy readers' demand. In view of the above situation, this paper studies and designs a high-availability e-reading platform architecture strategy, realizes hierarchical de-coupling through Spring + Spring MVC + MyBatis framework, improves productivity, adopts cluster load balancing technology, and uses LVS + Keepalived + Nginx as load balancing framework and design a load balancing strategy based on machine learning-decision tree algorithm to further ensure high availability of the platform.

Keywords High availability · Electronic reading · SSM framework · Load balancing

1 Introduction

With the rapid development of the mobile Internet, people's reading behavior has begun to shift from traditional paper media to new media such as computers, mobile phones, and e-book readers. As early as 2011, Amazon, the largest online bookstore in the USA, sold more e-books than traditional books. With the rapid development of the electronic reading industry, readers' demand for electronic reading platforms is also more diversified. The realization of these requirements has higher requirements on the rationality and stability of the design structure of the electronic reading

J. Huang
Intelligent Traffic Laboratory, Beijing University of Posts and Telecommunications, Beijing, China

Y. Wang (✉)
School of Electronic Engineering, Beijing University of Posts and Telecommunications, Beijing, China
e-mail: wangsy613@163.com

V. Jain et al. (eds.), *Recent Trends in Intelligent Computing, Communication and Devices*, Advances in Intelligent Systems and Computing 1031,
https://doi.org/10.1007/978-981-13-9406-5_50

platform. This article is only from the system architecture analysis which is committed to design a high-availability e-reading platform solution.

2 SSM Frame Design

The SSM framework is the abbreviation of Spring + Spring MVC + MyBatis. This is the mainstream Java EE enterprise framework after SSH, which is suitable for building various large-scale enterprise application systems [1].

2.1 Spring

Spring is an open-source design-level framework that addresses the loose coupling between layers, so it applies interface-oriented programming ideas throughout the system. Spring is a Java development framework created by Rod Johnson and emerging in 2003. The core of it is inversion of control (IOC) and aspect-oriented programming (AOP). Inversion of control (IOC) is also called dependency injection. Using the factory pattern, the spring container can generate instance objects and manage objects by configuring beans and properties. AOP can be seen as a continuation of object-oriented programming (OOP). It is a programming idea, not a certain technology. Using AOP's ideas, business logic code can be divided to achieve decoupling between the caller and the callee, thereby improving stability and usability.

2.2 Spring MVC

MVC is an abbreviation of model–view–controller, which is a design pattern that can improve code reusability [2]. The core principle is to reorganize the code in a separate way and then bring together a lot of business logic. Spring MVC implements the core concept of ready-to-use MVC, which provides a number of related functions for controllers and handlers. Spring MVC makes applications highly decoupled and can be dynamically changed with simple configuration changes.

2.3 MyBatis

MyBatis was formerly the open-source project iBatis, and iBatis moved to google code in 2010 and changed its name to MyBatis [3]. MyBatis is a Java-based persistence layer framework that supports custom SQL, stored procedures, and advanced mapping. With simple XML or annotations, MyBatis can map interfaces and Java POJOs to records in the database. Flexible use, does not affect the database, reducing coupling and reusability are the reasons for the success of MyBatis [4].

3 Cluster Load Balancing

Since the reading platform is in a low-pressure environment most of the time, the stability of the platform depends largely on a few high-concurrency and high-pressure scenarios, such as the purchase of special books and the downloading of free books. Load balancing is one of the keys to high availability of the platform. The following is a detailed analysis of the architecture and algorithms of load balancing.

3.1 Load Balancing Architecture

Load balancing is explained in the architecture. It refers to an architecture technology that uses a certain technology to achieve load sharing among hosts in a cluster. When the performance of a single machine reaches a bottleneck, it is necessary to use cluster technology to build a high-performance server cluster. Common load balancing architectures are LVS with Keepalived and Nginx with Keepalived. LVS is short for Linux virtual server, has good throughput, and can shield out faulty servers. However, the cluster software does not support regular processing. Nginx is a lightweight Web server/reverse proxy server and email proxy server that supports Rails, PHP, and HTTP proxy servers [5]. Based on the above, the load balancing architecture of this paper uses LVS + Keepalived + Nginx. The LVS is introduced to be responsible for the first layer of load, and Keepalived is used to manage and monitor the status of each service node in the LVS cluster system. The request is routed to the backend Nginx through LVS, and Nginx forwards it to one of the internal application servers for processing. The overall architecture is shown in Fig. 1.

3.2 Choice of Machine Learning Algorithms

This algorithm uses independent nodes to train the model, reducing the load on the server when the load balancer is high concurrency. The overall structure of the system is shown in Fig. 2.

The training model used in this paper is the decision tree algorithm. The decision tree is a tree structure similar to the flowchart. Each non-leaf node is represented by

Fig. 1 Load balancing overall architecture diagram

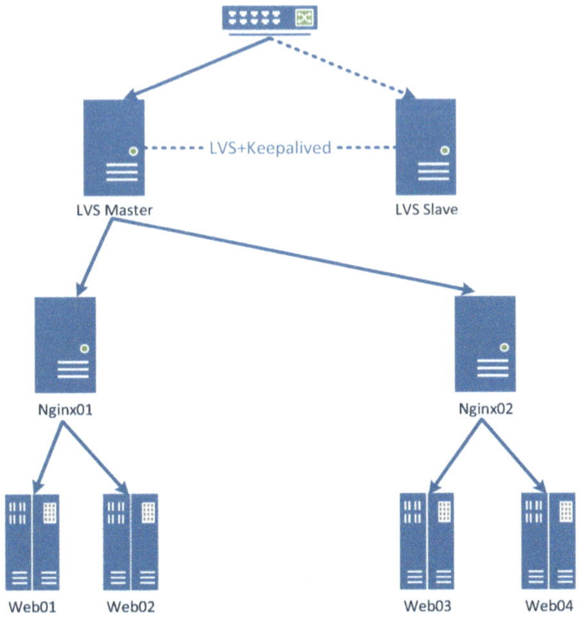

Fig. 2 System overall structure

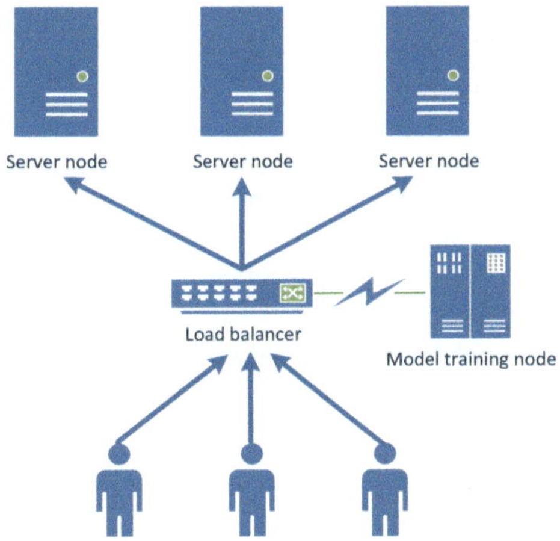

a test on a feature. The branch under the node indicates the test. The output and each leaf node represent a classification [6]. The decision tree algorithm has the following advantages: The model is simple and easy to interpret, the amount of calculation is relatively small, and a large-scale data set can be trained in a short time.

Let us call the attribute that needs to be studied as "classification attribute." Assume that the training data set D has a class m sample, and the sum of each of them in the data table is d_1, d_2, … d_m. Let $d = d_1 + d_1 + \ldots + d_m$, $p_i = d_i/d$, and then for any case, the information that determines the category it belongs to can be calculated using the following formula:

$$I(e_r) = -\sum_{i-1}^{m} p_i \log_2(p_i) \quad (1)$$

It is also assumed that the attribute A has v different values $\{v_1, v_2, \ldots v_v\}$. Attribute A divides D into v subsets $\{e_1, e_2, \ldots e_v\}$; the entropy or expectation information according to the subset divided by A is given by:

$$E(A) = \sum_{j-1}^{v} \frac{|e_r|}{|D|} \times \mathrm{I}(e_r) \quad (2)$$

From (1) and (2), the information gain of the branch on A is:

$$\mathrm{Gain}(A) = I(e_r) - E(A) \quad (3)$$

ID3 selects the largest value of gain (A), that is, the attribute A with the smallest $E(A)$ as the root node. After the model training node trains the model, it will output a “decision tree.” After the load balancer receives the new request, it only needs to use the if–then rule on the tree. The complexity of response time prediction depends on the depth of the decision tree. When the model is trained, only the depth of the decision tree is controlled within the controllable range, so that the complexity of the response time prediction is not too high. Make the calculation of the load balancer not very large.

3.3 Algorithm Flow Description

The load balancing algorithm obtains a response time prediction model by training historical data to predict the response time of the new request and allocates the request to the server node with the least response time according to the estimated response time of each server node, thereby improving the cluster. In the balance of request allocation, improve the efficiency of the cluster. The specific algorithm flowchart is shown in Fig. 3:

The load balancing module receives the request, obtains the request parameter A, and requests the type parameter A and the time prediction model to obtain the estimated response time R_p of each node. Select a server node with the least response time among each node, update the R_p value of the node, forward the request to the service server node, obtain the actual processing time of the request, and correct the prediction model with the actual processing time. The algorithm obtains the following response time prediction model through data training:

Fig. 3 Algorithm flowchart

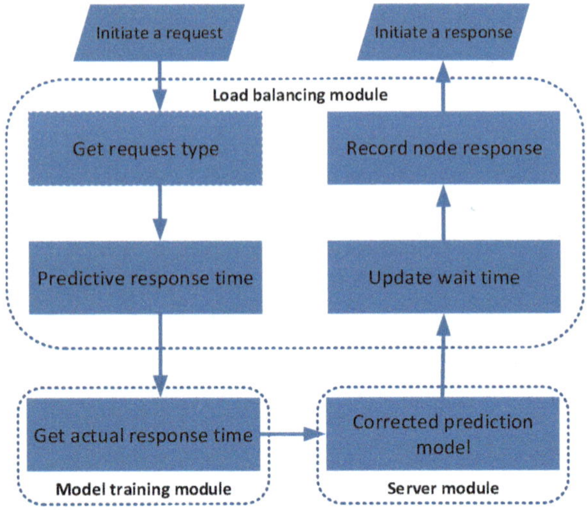

$R_p = f(A, P, T_w)$ where R_p is the request response time as the output parameter, A is the request type, P is the performance of the server node, and T_w is estimated request wait time.

4 Experimental Verification

4.1 Architecture Test

When the Keepalived service of the primary LVS is turned off, the standby LVS server enters the master role. The nginx1 service is turned off and subsequent requests continue to be forwarded to the nignx2 (103 node) server. Restart the Nginx service of nginx1 (104 nodes), and LVS automatically adds the server to the service.

4.2 Contrast Test

In order to demonstrate the advantages of machine learning-based algorithms, we use the traditional source address hash load balancing algorithm as a control group for comparative testing. The comparison is shown in Fig. 4.

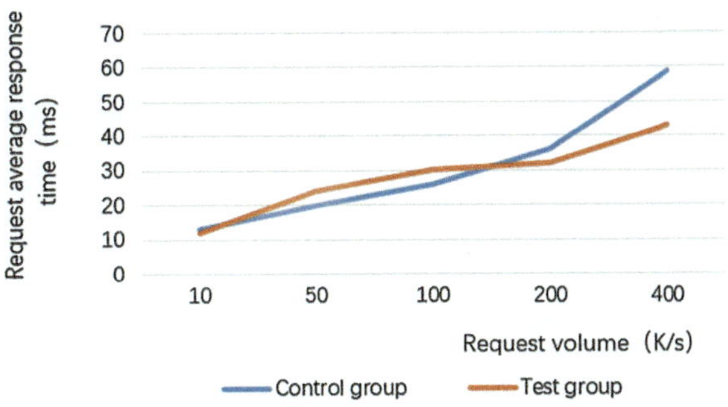

Fig. 4 Experimental comparison chart

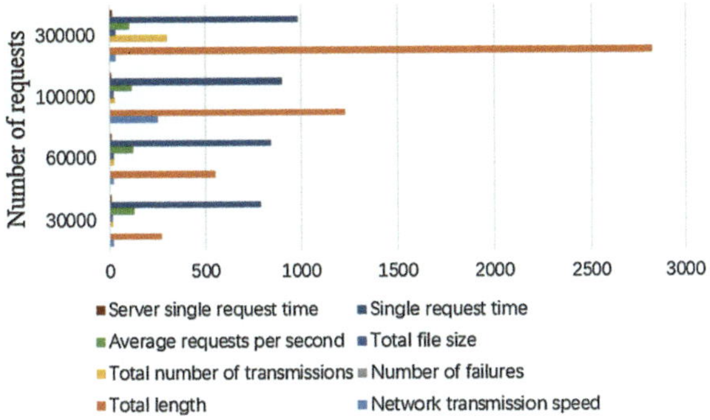

Fig. 5 Experimental analysis chart

4.3 *Pressure Test*

This time I used Apache's ab stress testing tool. Apache's ab command can simulate multi-threaded concurrent requests, test server load pressure, and test the pressure of other Web servers such as Nginx and IIS. After finishing the experimental data, the analysis is shown in Fig. 5.

5 Conclusions

This paper designs a set of highly available e-reading platform architecture for the development of e-reading platforms. The whole architecture is based on Spring + Spring MVC + Mybatis, using the load balancing technology based on LVS + Keepalived + Nginx hybrid architecture to alleviate server pressure and achieve dynamic load balancing through decision tree algorithm. The experiment proves that the research scheme proposed in this paper basically solves the high-availability problem of the new electronic reading platform and has certain practical application value.

References

1. Li, Y.: Design and implementation of SSM framework in web application development. Comput. Technol. Dev. **26**(12), 190 (2016)
2. Deinum, M., Serneels, K., Yates, C., Ladd, S., Vanfleteren, C.: Spring MVC Architecture (2012)
3. Song, L.P., Sun, R.L.: Research and implementation of interceptor based general physical pagination component of Mybatis. Appl. Mech. Mater. **513**, 1299 (2014)
4. Rong, Y.: Application research of Mybatis persistence layer framework. Inf. Secur. Technol. **12**, 86 (2015)
5. Zhang, Yan: Practical nginx: replacing Apache's high-performance web server. Publishing House of Electronics Industry, Beijing (2010)
6. Lombardo, L., Cama, M., Conoscenti, C.: Binary logistic regression versus stochastic gradient boosted decision trees in assessing landslide susceptibility for multiple-occurring landslide events: application to the 2009 storm event in Messina (Sicily, Southern Italy). Natural Hazards (2015)

Spatiotemporal Evolution Simulation of Volcanic Ash Cloud from Remote Sensing Image

Cheng Fan Li, Lan Liu, Xian Kun Sun, Jun Juan Zhao, Yuan Yuan and Jing Yuan Yin

Abstract Aiming at the dynamic monitoring and diffusion forecasting of volcanic ash cloud, this paper presents a new spatiotemporal evolution simulation method of volcanic ash cloud from the satellite remote sensing image. It couples on cellular automata (CA) and artificial neural network (ANN) (i.e., CA-ANN) in terms of cell spatiotemporal database construction, neural network structure confirmation and model training and verifying, and finally realized the simulation and forecast of volcanic ash cloud in space and time. The simulation is performed with test data and the real volcanic ash cloud spatiotemporal evolution case is discussed and verified in our empirical study. Experimental results show that the proposed method has the advantages of high accuracy and good image quality compared with the traditional methods.

Keywords Cellular automata (CA) · Artificial neural network (ANN) · Remote sensing image · Volcanic ash cloud · Spatiotemporal evolution

1 Introduction

Satellite remote sensing has the advantages of timeliness, economy and great potential for monitoring of volcanic ash cloud [1]. Compared to the traditional digital image, the satellite remote sensing image is complex, and the identification of volcanic ash cloud is also more difficult [2]. How to accurate identify and

C. F. Li (✉) · J. J. Zhao
Shanghai Institute for Advanced Communication and Data Science, School of Computer Engineering and Science, Shanghai University, Shanghai 200044, China
e-mail: lchf@shu.edu.cn

L. Liu · X. K. Sun
School of Electronic and Electrical Engineering, Shanghai University of Engineering Science, Shanghai 201620, China

Y. Yuan · J. Y. Yin
Earthquake Administration of Shanghai Municipality, Shanghai 200062, China

V. Jain et al. (eds.), *Recent Trends in Intelligent Computing, Communication and Devices*, Advances in Intelligent Systems and Computing 1031,
https://doi.org/10.1007/978-981-13-9406-5_51

forecast the spatiotemporal evolution of volcanic ash cloud from satellite remote sensing image has been critical to image processing and volcanic hazard monitoring.

The cellular automata (CA) have enormous computing power, simple framework, and open source itself and can effectively simulate the spatiotemporal dynamic evolution of complex system [3–5]. Artificial neural network (ANN) which abstracts and learns the nerve element cerebrum from the information processing is a new simple model that can simulate many kinds of engineering and scientific problems. Due to the powerful modeling capabilities of CA and learning ability of ANN, numerous studies on disaster monitoring, land use, population migration have been carried out and many achievements have also been made in the recent years [6–8]. However, this is very little involvement in the current monitoring study and spatiotemporal evolution of volcanic ash cloud because of the particularity of volcanic eruption and hazards, and the related achievement is not as so much.

The rest of the paper is constructed as follows: Sect. 2 presents brief overview of the theoretical basis and proposed method. Section 3 describes the simulation of our study, and Sect. 4 devotes the empirical study of spatiotemporal evolution. Finally, conclusion is drawn in Sect. 5.

2 Theoretical Basis

2.1 CA and ANN

The structure of CA model is shown in Fig. 1. It contains cell space, state, neighbor and evolutionary rule [9–11]. Evolutionary rule (also called state transfer function)

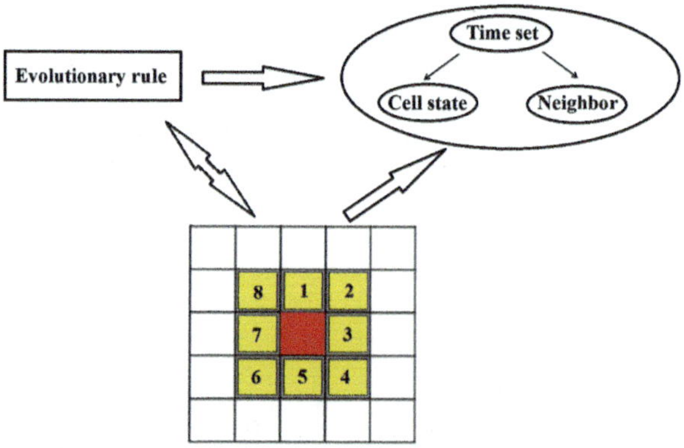

Fig. 1 Structure of CA model: the red square is the center cell; the yellow square is the neighborhood and the whole square area is the cell space

refers to the rules that determines the cell's state at the next moment in terms of the current cell and its neighborhood, and further updates by the interaction among cells:

$$S_i^{t+1} = f(S_i^t, S_{i+1}^t, \ldots, S_{i+8}^t) \tag{1}$$

where f is the evolution rule, and S_i^{t+1} is the state of i cell at time $t + 1$.

ANN is a mathematical model which handles with the distributed parallel information by simulating the behavior characteristics of biology neural network [10]. In ANN model, the cell node and arrow separately represent the artificial neuron and connection between an output of artificial neuron and the input of another. At present, there have been many ANN models such as perceptron and Hopfield network.

2.2 CA-ANN Model Construction

Construction of cell spatiotemporal database: Let the spatiotemporal evolution of volcanic ash cloud be a spatiotemporal database based on cells. Each record in the database corresponds to all the attributes of a cell. Namely, it contains cell coordinate (X, Y) at time t, cell attribute (vegetation coverage, wind direction, wind speed, altitude, cell state at time t and $t + 1$), cell neighborhood (N_1S_t, N_2S_t, N_3S_t, N_4S_t, N_5S_t, N_6S_t, N_7S_t, N_8S_t). Then, the cell spatiotemporal evolution database at time $t + 1$ is shown in Table 1.

Confirmation neural network structure: There are 6 neurons in the input layer and correspond to the cell attributes. In line with Kolmogorov theorem, there is at least 2/3 neurons numbers of input neurons' in hidden layer for nonlinear neurons, and the number of neurons in hidden layer was taken as 5. Similarly, there are 6 neurons in the output layer and corresponds to the transfer probability of different cell states. The detailed parameters contain the maximum numbers of training iteration is 10,000, the minimum mean square error (MSE) is 0.01 and the initial learning rate is 0.5.

Model training, verifying and forecasting: The training database was inputted the CA-ANN model so as to acquire the parameters, then verified the results via comparing to the actual situation. Next, the suitable threshold value was obtained in the experiment. Once the verified result is greater or equal to the threshold value, the CA-ANN model is used to simulate the cell state at the next moment (i.e., spatiotemporal evolution forecast). In addition, CA-ANN model is designed with object-oriented method and programmed by neural network toolbox in MATLAB software, and the Levenberg–Marquardt (LM) algorithm is used to optimize the CA-ANN model.

Table 1 Data structure of cell spatiotemporal evolution

Coordinate		Cell neighborhood								Cell attribute					
X	Y	N_1S_t	N_2S_t	N_3S_t	N_4S_t	N_5S_t	N_6S_t	N_7S_t	N_8S_t	V	WD	WS	A	S_t	S_{t+1}

3 Simulation Experiment

In this section, simulation experiments are performed on PC with Intel (R) Core (TM) i5@1.70 GHz CPU and 4GB memory in MATLAB R2014b environment. In Moore neighbor, the proportion of variational cell to the total number of cells was calculated and represented by *x*, *y*, and *z*, respectively. In the experiment, the LM algorithm is used to mine and optimize the evolutionary rule of CA-ANN model.

Next, the state transfer probability of test group cell was simulated and forecasted by the trained CA-ANN model, and the spatiotemporal evolution process of forecasting group cell was obtained in MATLAB platform. That is, using the evolution results of the previous moment to simulate the evolution state of the next moment. Figure 2 shows the detailed state transfer process of node data at different times.

It can be clearly seen from Fig. 2 that the state transfer probability of node data changes along with the time. Compared to the time t, the normal probability means value of node data at the time $t + 1$ is greater than the variational probability; it means the system is in a stable state. In contrast, the normal probability means value of node data; at the time, $t + 2$ is less than the variational probability; it also means the cells have occurred significantly variation and the system is unstable. And then the accuracy and MSE were used to evaluate the simulation results in case of the number of iterations is 10,000, also shown in Table 2.

In Table 2, the variation trend of the simulation results of cell state is basically consistent with the changes in Fig. 2. It also indicates the validity of the propose CA-ANN method in spatiotemporal evolution forecast.

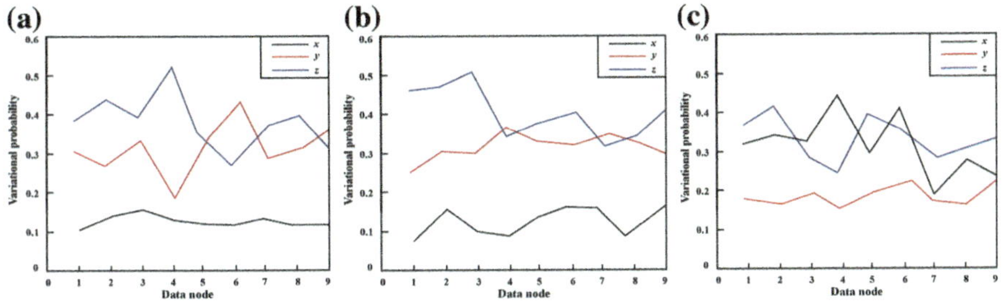

Fig. 2 State transfer process: **a** t, **b** $t + 1$, and **c** $t + 2$

Table 2 Precision evaluation

	Accuracy (%)	MSE	Number of iterations
x	78.25	0.015	10,000
y	79.68	0.012	10,000
z	70.04	0.032	10,000

4 Empirical Study

4.1 Data Source and Preprocessing

In this section, the visible and infrared radiometer (VIRR) Level-1 data with HDF5 format as a data source in our study. The data preprocessing mainly includes geometric correction, striping processing, and cropping. Geometric correction was implemented by geographic lookup table (GLT) with dataset; radiometric calibration was demarcated by the parameters published by the National Meteorological Satellite Center. The preprocessing is programmed and implemented on ENVI 4.6 software.

4.2 Simulation of Spatiotemporal Evolution

Next, the VIRR images of volcanic ash cloud on May 7, 13, and 14, 2010, were implemented by the proposed CA-ANN model in this study, and the simulation results are shown in Fig. 3.

It can be clearly seen from Fig. 3 that the volcanic ash cloud distribution on May 7, 13, and 14, 2010, were accurately simulated by the proposed CA-ANN model. And what's more, the simulated volcanic ash cloud image quality and the visual effect are good, and the broken spots are less.

4.3 Discussion

In this section, the 1000 proofreading data points were randomly generated in the experiment and the performance of volcanic ash cloud spatiotemporal simulation was evaluated by confusion matrix method. Whereinto, the reference data is the simultaneous VIRR satellite images (Fig. 4).

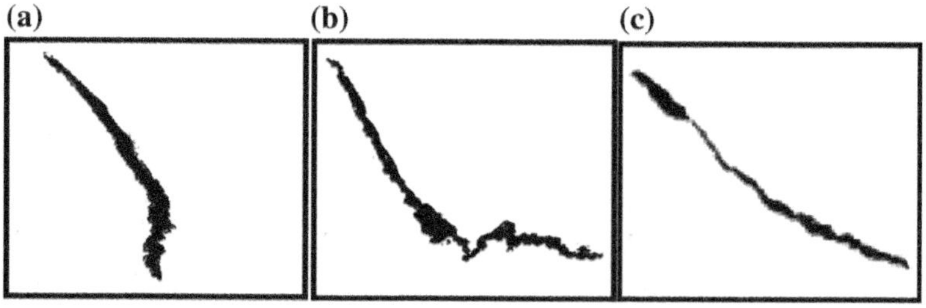

Fig. 3 Simulation results: **a** May 7, 2010, **b** May 13, 2010 and **c** May 14, 2010

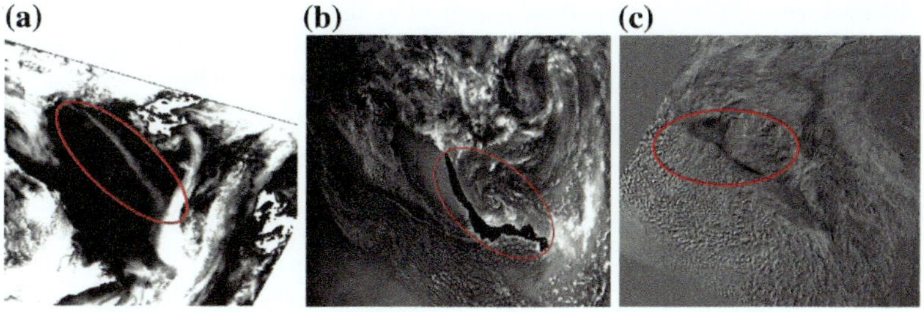

Fig. 4 VIRR images: **a** May 7, 2010, **b** May 13, 2010, and **c** May 14, 2010, the red ellipse is the distribution of volcanic ash cloud

Table 3 Accuracy evaluation

	Total precision (%)	Kappa coefficient
May 13, 2010	86.32	0.8001
May 14, 2010	87.18	0.8095

Table 3 shows the detailed accuracy evaluation. It clearly shows that the proposed CA-ANN model has achieved a good simulation effect of volcanic ash cloud spatiotemporal evolution from VIRR satellite images. The total accuracy of spatiotemporal evolution on May 13 and 14, 2010, reached 86.32 and 87.18%, respectively, and the Kappa coefficient reached 0.8001 and 0.8095, respectively. This is also consistent with the above volcanic ash cloud simulation results (Figs.3 and 4).

As can be seen from Figs. 3 and 4, and Table 3 that volcanic ash cloud spatiotemporal evolution simulated by CA-ANN method is basically consistent with the existing distribution (the red ellipse in Fig. 4) of volcanic ash cloud in VIRR images. It partially indicates that the proposed CA-ANN method is effective and feasible in spatiotemporal evolution at various stages of volcanic ash cloud.

5 Conclusion

In this study, the spatial dynamic monitoring of CA and self-learning of ANN were comprehensive utilized to simulate the spatiotemporal evolution of volcanic ash cloud from VIRR images. The result shows that the forecast diffusion is basically consistent with the monitoring of real satellite remote sensing images. To simplify model parameters and reduce computational complexity of CA-ANN, in this study the six attribute factors of volcanic ash cloud were chosen and trained in CA-ANN model. However, in fact there are multiple factors in the image, and it may have many coupling cases in accordance with the different application. So next, we will try to choose more cell attribute factors to take part in the model training and learning on the basis of satellite remote sensing image.

Acknowledgements This work has been supported by the National Natural Science Foundation of China under project No. 41404024 and the Science and Technology Development Foundation of Shanghai of China under project No. 16dz1206000 and 16142203000.

References

1. Millington, S.C., Saunders, R.W., Francis, P.N., Webster, H.N.: Simulated volcanic ash imagery: a method to compare NAME ash concentration forecasts with SEVIRI imagery for the Eyjafjallajökull eruption in 2010. J. Geophys. Res. **117**(D20), 1–14 (2012)
2. Liu, L., Li, C.F., Lei, Y.M., Yin, J.Y., Zhao, J.J.: Volcanic ash cloud detection from MODIS image based on CPIWS method. Acta Geophys. **65**(1), 151–163 (2017)
3. Yuan, Q., Liao, H.S.: A software based on cellular automata used to simulate time and space dynamic change in geography. J. Nanjing Univ. (Nat. Sci.) **41**(3), 857–861 (2005)
4. Zhang, Y.H., Li, X., Liu, X.P., Qiao, J.G., He, Z.Q.: Urban expansion simulation by coupling remote sensing observations and cellular automata. J. Rem. Sens. **17**(4), 872–886 (2015)
5. Rui, X.P., Wu, B., Li, Z.Z., Song, X.F., Li, Y.: A pollutants diffusion simulation algorithm on water table based on cellular automata. Sci. Technol. Eng. **17**(30), 239–245 (2017)
6. Meng, J.E., Zhang, Y., Wang, N., Mahardhika, P.: Attention pooling-based conventional neural network for sentence modeling. Inf. Sci. Int. J. **373**, 388–403 (2016)
7. Musa, B.: Co-combustion of peanut hull and coal blends: artificial neural networks modeling, particle swarm optimization and Monte Carlo simulation. Biores. Technol. **216**(1), 280–286 (2016)
8. Yang, J.S., Mei, T.C., Zhong, S.D.: Application of convolution neural network using region information to remote sensing image classification. Comput. Eng. Appl. **54**(7), 188–195 (2018)
9. Dong, T.T., Zuo, L.J., Zhang, Z.X.: A study of spacetime evolution of soil erosion based on ANN-CA model. J. Geo-inf. Sci. **11**(1), 132–138 (2009)
10. Jiang, D.R., Chen, G.L., Jia, Y., Xiang, H.W., Wang, P.Y.: Security assessment of cyber physical power system based on cellular automata and multilayer feedforward neural network. J. Chongqing Univ. Technol. (Nat. Sci.) **32**(3), 217–226 (2018)
11. Jing, C.Q., Zhang, Y.F., Yang, X.D.: Approach of dynamic evolution model of urban land use based on the integration of ANN and CA. Arid Zone Res. **27**(6), 854–860 (2010)

A Novel Selection Criterion Based on Diversity Preservation for Non-dominated Solutions

Ali Metiaf and Qianhong Wu

Abstract Solving the multi-objective optimization (MOO) problem with maintaining a good exploration and a uniform distribution is crucial. In this paper, we propose a novel selection criterion for maintaining efficient exploration and uniform distribution for the solution. In the proposed strategy, angular sectors of the solutions are used to preserve and maintain a good non-dominated solution in the searching space in all possible directions equally. Concerning the evaluation, we replace the crowding distance into the NSGA-II by our proposed criterion that named as angular sectors, and the resulting NSGA-II-AS algorithm was compared with NSGA-III according to the function which is widely used in the literature. The results show that NSGA-II-AS outperforms NSGA-III.

Keywords Multi-objective optimization · NSGA-II · Evolutionary algorithms · Crowding distance · Angular sectors

1 Introduction

The optimization is a crucial requirement in the real world, due to the conflicting objectives that require to be achieved simultaneously [1]. The multi-objective optimization algorithm is considered as a useful tool to resolve this problem and is applied to solve single-objective problem [2]. However, we can't avoid the multiple conflicting objectives' problem in real life, that cannot be satisfied by a single solution, where with trying to decrease one objective, we lose the others (increased the objectives) in the final result [3]. Under those circumstances, the MOO problem requires a specific solution termed non-dominated solutions or a Pareto optimal solution, and this Pareto set is mapped in the Pareto front (PF) in the objective function space [4, 5]. Now we can define a multi-objective optimization as the way to optimizing two or more contradictory objective at the same time, and we find this

---

A. Metiaf (✉) · Q. Wu
School of Electronics and Information Engineering, Beihang University, Beijing, China
e-mail: alimetiaf23@buaa.edu.cn

V. Jain et al. (eds.), *Recent Trends in Intelligent Computing, Communication and Devices*, Advances in Intelligent Systems and Computing 1031,
https://doi.org/10.1007/978-981-13-9406-5_52

problem in many fields when should be taken optimal decisions in the presence of conflicting between objectives [6].

Generally, to prove the quality of the solution in the multi-objective optimization two measures should be take it into account; the convergence of the obtained solution to the true Pareto front (TP) at first and second the diversity of the solutions [7]. In the literature, many efforts have been already made to enhance the non-dominated solution in the multi-objective optimization algorithms where these searches are focused on the second aspect by improving selection operators to maintaining solution diversity. The crowding distance method was initially proposed in NSGA-II [8], to achieve a high level of exploration and solution diversity. Over the last years, various researchers have attempted to modify the formula for crowding distance. For example, a new crowding distance was proposed in [9] termed the novel dynamic crowding distance (NDCD). It is determined according to the deviation degree of each solution to its adjacent it and incorporates a removal mechanism for solutions with the lowest crowding distance. This method results in much better exploration. Similarly, a novel procedure for computing crowding distances was proposed in [10]. In this procedure, solutions in less crowded regions being preferred, and simple 'hill-climbing' is employed to avoid the local optima. A new indicator for selecting a suitable solution was proposed in [11] named as AELMOEA/D-CD by combining two different methods: (i) the exploitation based on the crowding distance; (ii) based on selector vector to explore widely select. Also, another work to realize diversity and convergence together was proposed in [12] by combined The vector angle [13] and the shift-based density estimation [14], named AnD that use the two strategies in each iteration to eliminate the weak solutions that have minimum vector angle according to the angle-based selection, and that have reduced diversity and convergence according to the shift-based density estimation.

In this paper, we propose a novel selection criterion for sorting and selecting non-dominated solution using angular sectors that are resulting from dividing the searching space on angle θ to explore all possible directions. Recognizing this direction is essential for obtaining an equal distribution of the solution in the search region regarding its potential to provide better solutions, and we count the solutions that are located inside each sector where the sector that not select his solutions previously has a higher probability of being preserved his solutions for next generation. We replace the crowding distance into the NSGA-II [8] by our proposed criterion that named as angular sectors, and the resulting NSGA-II-AS algorithm was compared with NSGA-III according to the function which is widely used in the literature. The results show that NSGA-II-AS outperforms NSGA-III concerning the hyper-volume measures that prove the diversity achieved in the solutions.

2 Basic Concepts and Terms

In this section, we describe basic definitions of multi-objective optimization (MOO) and introduce the concept of NSGA-III. In the following, we define the MOO as a mathematical problem:

Let us assume that we have minimization problem with m objective given by functions

$$f(x) = [f_1(x), f_2(x), \ldots f_m(x)] \in R^m \text{ where } x = [x_1, x_2, \ldots x_n] \in R^n.$$

Therefore, to obtain a decision maker, that achieve the trade-offs between the conflicting objectives, where the single solution cannot satisfy all the objective functions m simultaneously a significant improvement must be made to the search process.

The given two solution vectors $x, y \in R^n$ the solution x is dominating y ($x \prec y$), if:

$$\forall i \in \{1, 2, \ldots, m\}: f_i(x) \leq f_i(y), \quad \exists i \in \{1, 2, \ldots, m\}: f_i(x) < f_i(y)$$

In addition, a solution $x^* \in R^n$ is called non-dominated solutions or Pareto optimal solution if there is no solution x dominated it that mean the values of any of the elements of x^* are at least equal to the values in x, and one element of x^* is surely lower than the corresponding value in x [15].

The basic context of NSGA-III is same with the original context of NSGA-II [8] with significant changes in its selection operators. In NSGA-III, the selection is based on a set of reference points to preserve the solutions diversity [16], unlike NSGA-II that use the crowding distance. The main procedure of NSGA-III starts which initializes the initial population *N* randomly and defines a set of reference points *H* on the hyperplane. The next steps are iterated until creating new generation *Ngen*, and the existing population is used as a parent to create the offspring by the genetic operation (crossover and mutation). Then, the two populations, parent and offspring, are combined to produce a population with size *2N*, and sorting it according to non-domination, into different non-domination ranks. After that selecting the best solution to preserve elite members for *Ngen* with size *N*, starting with the first rank until reaching at the rank where the number of solutions exceeds the population size *N* the NSGA-III algorithm selects the remaining number of solutions for *Ngen* from the last rank according to the reference points. The solutions that have the minimum perpendicular distance to the reference line are selected as shown in Fig. 1.

Fig. 1 NSGA-III concept

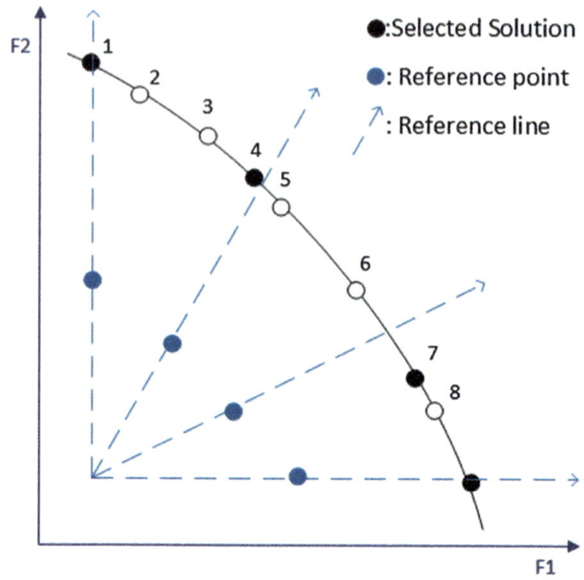

3 Proposed Strategy

Our work aims to enhance the non-dominated solution of the MOO algorithms, by improving the selection process. However, the efficient exploration of the population has a vital role to maintain a good non-dominated solution in evolutionary algorithms. The framework of the suggested strategy is similar to the NSGA-II [8] and NSGA-III [16] with significant changes in selection operators. In the proposed strategy, we maintain the best uniform distribution for the optimal solution in the Pareto front using angular sectors that are resulting from dividing the searching space on angle θ to explore all searching space. As mentioned before in NSGA-III after the combined population (parent + offspring) is sorted according to non-domination, and selecting the best solution for *Ngen* starting with the first rank until reaching the rank in which the number of solutions exceeds the required number for *Ngen*. We select the remaining number of solution for *Ngen* from the last rank in all possible direction based on the angular sectors $I_{\text{anglesector}}$ that preserves the diversity where this criterion gives priority to the sectors that is not select his solutions previously or that have lower $I_{\text{anglesector}}$. Figure 2 and Algorithm 1 illustrate the proposed strategy concept, where the solutions 1, 2, 5, 7 are preferred.

Algorithm 1 shows the pseudocode for extracting solutions from the combined population (parent plus offspring) for the following generation. This algorithm starts with selecting first rank (the next lower rank) until reaching a rank at which the number of solutions exceeds the population size. The algorithm enables the new criterion $I_{\text{anglesector}}$ to select solutions from the last rank in all directions.

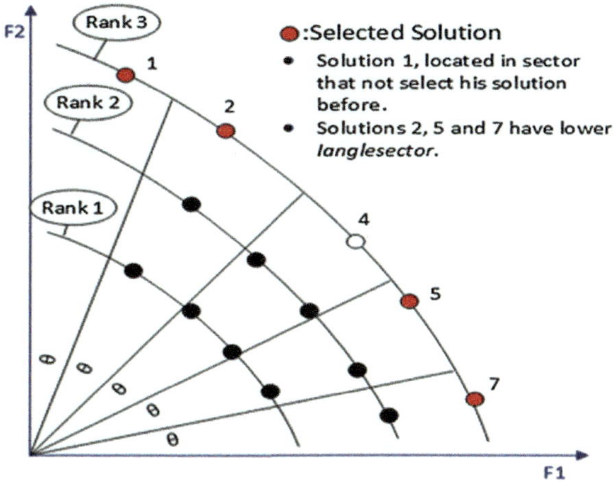

Fig. 2 NSGA-II-AS concept

Algorithm 1 Selecting solutions

Input: parent + offspring (ranked)
Output: *Ngen*.
$counter \leftarrow 0, rank \leftarrow 1, n \leftarrow 0$
NS = number of solution selected
N = number of solution in the rank
While $(N + NS < = Ngen)$

$rank \leftarrow rank + 1$
Add solution of the current rank to the *Ngen*
$NS \leftarrow N + NS$

End While
While $(counter < (Ngen - NS))$

$counter \leftarrow counter + 1$
Select a solution with minimum $I_{\mathrm{anglesector}}$
Add the selected solution to the *Ngen*

End While

4 Experiments

To validate the effectiveness of the developed strategy, a set of mathematical benchmark functions with multi-objective nature have been selected which include continuous, discontinuous, convex, and nonconvex problems are presented in

Table 1 Used parameters

Number of solutions	Number of generations	Crossover option		Mutation option			θ(theta)
		Fraction	Ratio	Fraction	Scale	Shrink	
100	500	$2/n$	1.2	$2/n$	0.1	0.5	0.0000001 for all test

Table 2. The parameters used for experiments are shown in Table 1. We replaced the crowding distance into NSGA-II by the angular sectors, the resulting algorithm named as NSGA-II-AS.

5 Results and Discussion

The widely used performance metric in evolutionary multi-objective optimization, i.e., hyper-volume (HV), were employed to compare NSGA-II-AS with NSGA-III, where this single metric can provide enough information about the achieved solution diversity. However, high values of this metric indicate more acceptable solutions.

Hyper-Volume Metric
The HV metric estimates the volume of the dominated part of the objective space affiliated to a reference point (worst solution); this region is the union of the hypercube whose diagonal is the distance between the reference point and a solution x from the Pareto set P(S) [17]. The flowing equation gives HV.

$$\mathrm{HV} = \mathrm{volume}\left(\bigcup_{x \in P_s} \mathrm{Hypercube}(x)\right) \tag{2}$$

The evaluation of the six mathematical testing functions is provided in Fig. 3. Naturally, the NSGA-II-AS algorithm has achieved superiority over NSGA-III regarding FON, KUR, POL, ZDT6 mathematical functions concerning the hyper-volume metric, and this measure is preferred to be higher. Notably, the NSGA-II-AS algorithm has also achieved higher values than NSGA-III.

6 Conclusion

The selection process in multi-objective optimization algorithm has a vital role in increasing the quality of non-dominated solutions that must be distributed equally in the searching areas in all directions of the searching space. In this paper, we present a novel selection criterion (angular sectors) in order to preserve the

Table 2 Mathematical test functions

Problem	Variable bounds	Objective functions	**Comments**
FON	[−4, 4]	$f_1(x) = 1 - \exp\left(-\sum_{i=1}^{3}\left(x_i - \frac{1}{\sqrt{3}}\right)^2\right)$; $f_2(x) = 1 - \exp\left(-\sum_{i=1}^{3}\left(x_i + \frac{1}{\sqrt{3}}\right)^2\right)$	Nonconvex
KUR	[−5, 5]	$f_1(x) = \sum_{i=1}^{n-1}\left(-10\exp\left(-0.2\sqrt{x_i^2 + x_{i+1}^2}\right)\right)$; $f_2(x) = \sum_{i=1}^{n}\left(\lvert x_i\rvert^{0.8} + 5\sin x_i^3\right)$	Nonconvex
POL	[−π, π]	$f_1(x,y) = 1 + (A_1 - B_1)^2 + (A_2 - B_2)^2$; $f_2(x,y) = (x+3)^2 + (y+1)^2$ $A_1 = 0.5\sin 1 - 2\cos 2 + 1.5\cos 2$; $A_2 = 1.5\sin 1 - \cos 1 + 2\sin 2 - 0.5\cos 2$ $B_1 = 0.5\sin x - 2\cos x + 1.5\cos y$; $B_2 = 1.5\sin x - \cos x + 2\sin y - 0.5\cos y$	Convex
SCH	[−3, 3]	$f_1(x) = x^2$; $f_2(x) = (x-5)^2$	Nonconvex
ZDT4	[0, 1]	$f_1(x) = x_1$; $f_2(x) = g(x)\left[1 - \sqrt{f_1(x)/g(x)}\right]$; $g(x) = 91 + \sum_{i=2}^{10}(x_i^2 - 10\cos(4\pi x_i))$	Convex, Disconnected
ZDT6	[0, 1]	$f_1(x) = 1 - \exp(-4x_1)\sin^6(6\pi x_1)$; $f_2(x) = g(x)\left[1 - (f_{1(x)}/g(x))^2\right]$; $g(x) = 1 + 9\left[\left(\sum_{i=2}^{n} x_i\right)/(n-1)\right]^{0.25}$	Convex, Non-uniformly spaced

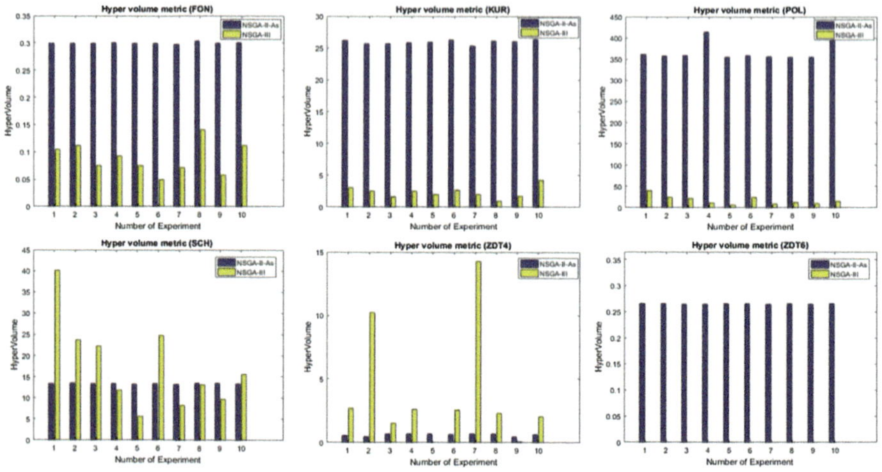

Fig. 3 Hyper-volume results of NSGA-II-AS vs NSGA-III

diversity, which replaced the crowding distance into the NSGA-II. An evaluation of NSGA-II-AS, under a set of multi-objective optimization mathematical functions, is applied and compared with NSGA-III. The results clarify that the proposed strategy has proved the diversity performance significantly.

Acknowledgements The authors thank the anonymous reviewers and the editors of this conference sincerely for their future helpful observations and detailed recommendations that will help in increasing the quality of this paper.

References

1. Metiaf, A., Elkazzaz, F., Qian Hong, W., Abozied, M.: Multi-objective optimization of supply chain problem based NSGA-II-Cuckoo Search Algorithm. In: IOP Conference Series: Materials Science and Engineering, vol. 435, p. 012030 (2018)
2. Mukhopadhyay, A., Member, S., Maulik, U., Member, S.: A survey of multiobjective evolutionary algorithms for data mining: part I. **18**(1), 4–19 (2014)
3. Zitzler, E., Laumanns, M., Bleuler, S.: A tutorial on evolutionary multiobjective optimization, vol. 535 (2004)
4. Coello, C.A.C., Pulido, G.T., Lechuga, M.S., Clerc, M.: Handling multiple objectives with particle swarm optimization. IEEE Trans. Evol. Comput. **8** (2004)
5. Li, H., Zhang, Q.: Multiobjective optimization problems with complicated Pareto sets, MOEA/D and NSGA-II. IEEE Trans. Evol. Comput. **13**(2), 284–302 (2009)
6. Xing, L.N., Chen, Y.W., Yang, K.W.: An efficient search method for multi-objective flexible job shop scheduling problems. J. Intell. Manuf. **20**(3), 283–293 (2009)
7. Cheng, R., Jin, Y., Olhofer, M., Sendhoff, B., Member, S.: A reference vector guided evolutionary algorithm for many-objective optimization **20**(5), 773–791 (2016)
8. Deb, K., Pratap, A., Agarwal, S., Meyarivan, T.: A fast and elitist multiobjective genetic algorithm: NSGA-II. IEEE Trans. Evol. Comput. **6** (2002)

9. Yang, L., Guan, Y., Sheng, W.: A novel dynamic crowding distance based diversity maintenance strategy for MOEAs. In: 2017 International Conference on Machine Learning and Cybernetics, pp. 211–216 (2017)
10. Sun, C.: An improved differential evolution and novel crowding distance metric for multi-objective optimization. In: 2010 Third International Symposium on Knowledge Acquisition and Modeling, no. x, pp. 265–268 (2010)
11. Harada, T., Kaidan, M., Thawonmas, R.: Crowding distance based promising solution selection in surrogate assisted asynchronous multi-objective evolutionary algorithm. In: Proceedings of the Genetic and Evolutionary Computation Conference Companion—GECCO'18, no. 1, pp. 253–254 (2018)
12. Liu, Z.Z., Wang, Y., Huang, P.Q.: A many-objective evolutionary algorithm with angle-based selection and shift-based density estimation. Inf. Sci. (Ny). 1–16 (2017)
13. Xiang, Y., Zhou, Y., Li, M., Chen, Z.: A Vector angle-based evolutionary algorithm for unconstrained many-objective optimization. IEEE Trans. Evol. Comput. **21**(1), 131–152 (2017)
14. Li, M., Yang, S., Liu, X.: Shift-based density estimation for Pareto-based algorithms in many-objective optimization. IEEE Trans. Evol. Comput. **18**(3), 348–365 (2014)
15. Bi, X., Wang, C.: A niche-elimination operation based NSGA-III algorithm for many-objective optimization. Appl. Intell. **48**(1), 118–141 (2017)
16. Deb, K., Jain, H.: An evolutionary many-objective optimization algorithm using reference-point based non-dominated sorting approach, part i: solving problems with box constraints. Ieeexplore.Ieee.Org **18**(c), 1–1 (2014)
17. Zitzler, E., Thiele, L.: Multiobjective evolutionary algorithms: a comparative case study and the strength Pareto approach. IEEE Trans. Evol. Comput. **3**(4), 257–271 (1999)

Simulation Study of the High Gain Effect of Reach-Through Solid-State Impact Ionization Multipliers

Yu Geng, Qin Li and Wan Qiu

Abstract A simulation study of the high gain effect of reach-through solid-state impact ionization multipliers (SIMs) is reported. This design combines the advances of Si avalanche photodiodes (APDs) used in LiDARs and SIMs. The SIMs without reach-through design have gain of 3 at the voltage of 15 V. With the help the optimized reach-through design, the reach-through SIMs obtain high gain of 5500 at the voltage of 15 V. The optimized design has great potential for future application in LiDARs for autonomous driving.

Keywords APD · Simulation · RSIM · LiDAR · Autonomous driving

1 Introduction

Avalanche photodiodes (APDs) are widely used in LiDARs for autonomous driving due to their high gain, high bandwidth, and low noise. Most LiDARs use 905 nm lasers with mature Si APDs. Si APDs have the advantage of low price and large-scale fabrication [1–3]. For autonomous driving, eye safety is an important issue. But the wavelength of 905 nm can penetrate eyes, which restricts the power of 905 nm lasers. Hence, longer wavelengths such as 1550 nm were more suitable for the sake of eye safety. However, the expensive and complicated III–V APDs must be used in this situation due to their high absorption ratio at this wavelength range. Solid-state impact ionization multipliers (SIMs) were developed as an alternative of APDs, which can be integrated with any light detector [4, 5].

The SIMs have to be operated at voltages larger than 50 V in order to obtain high gain [4]. Lower operating voltage is highly preferred which would reduce

Y. Geng (✉) · Q. Li
School of Engineering, Shenzhen Institute of Information Technology, Shenzhen, China
e-mail: gengy@sziit.edu.cn

W. Qiu
Shenzhen Zhaoyang Institute of Information Technology, Shenzhen, China

V. Jain et al. (eds.), *Recent Trends in Intelligent Computing, Communication and Devices*, Advances in Intelligent Systems and Computing 1031,
https://doi.org/10.1007/978-981-13-9406-5_53

power usage and increase safety. In this work, a reach-through SIM (RSIM) design was introduced in order to lower the voltage.

2 Design

The structure of the SIM was optimized [6]. The RSIMs were based on this optimized structure and introduced the *p* region to the SIM as shown in Fig. 1. This differs from the original SIM design in that it consists of $p^+–\pi–p–n^+$ layers as in reach-through APDs instead of only $p^+–\pi–n^+$ layers. For RSIMs, the depletion layer widens across the *p* region by reaching to the unintentionally doped π-region when the reverse bias voltage is high enough. It provides high gain with low noise and a relatively large bandwidth [7, 8]. The original SIM structure is either with Schottky contact-$\pi–n^+$ or with $n^+–\pi–n^+$ structure, which does not adopt the reach-through design. A RSIM with a $n^+–\pi–p–n^+$ structure has the potential of operating with higher gains at lower voltages.

3 Simulation

The RSIM is simulated by Silvaco. The widths of the anode and the cathode are both set to be 5 μm, while the distance between them is set as 10 μm. The width of SiO_2 insulation region (*w*) is 5 μm. Other parameters can be seen in reference 10.

The control design is *p* region with height (*r*) of 5 μm and doping concentration of 1×10^{17} cm^{-3}. The wider *p* region compared with the cathode is for the sake of future fabrication.

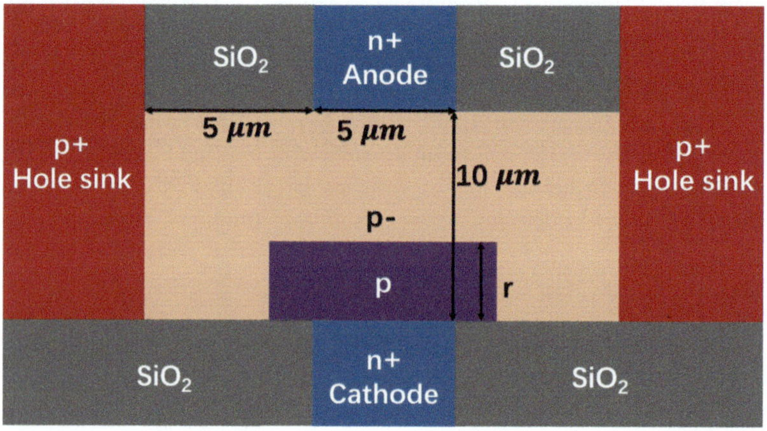

Fig. 1 Design of the SIM structure

4 Results and Discussion

First, the doping concentration of 1×10^{16} cm^{-3} and 1×10^{17} cm^{-3} were compared. The simulation shows that the gain at 15 V is 1.5 for the case of doping concentration of 1×10^{16} cm^{-3} and 5500 for the other case. This is caused by the fact that the electric field is greatly lifted at the interface of p region and the cathode when the doping concentration is 1×10^{17} cm^{-3} as shown in Fig. 2.

Then different values of r are simulated. The result in Fig. 3 shows that r is not a critical variable. Minimal change of electric field was occurred with r because the electric field drops mainly at the interface of p region and the cathode.

The gain comparison of optimized RSIM and SIM at different voltages is shown in Fig. 4. It can be seen that with reach-through design, the gain increased greatly from 3 to around 5500.

Based on the design above, a new structure was designed for future fabrication. r is designed to be 3 μm as shorter r is hard to fabricate and longer r is not necessary and may give rise to larger noise. The doping concentration is designed to be 1×10^{17} cm^{-3}.

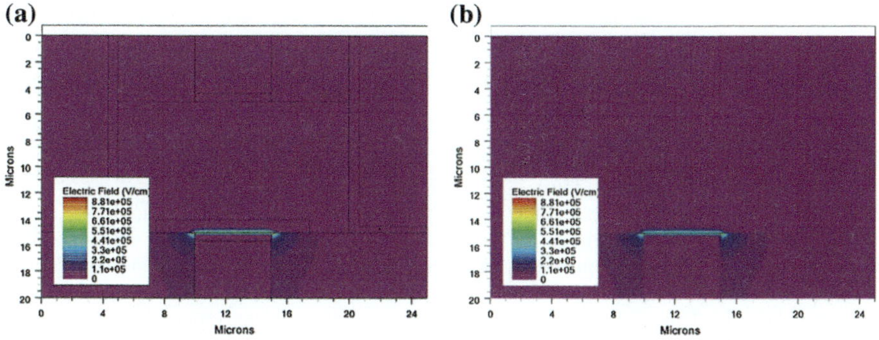

Fig. 2 Electric field distribution with different r at the voltage of 15 V. **a** $r = 1$ μm. **b** $r = 5$ μm

Fig. 3 Simulated gain versus r at the voltage of 15 V

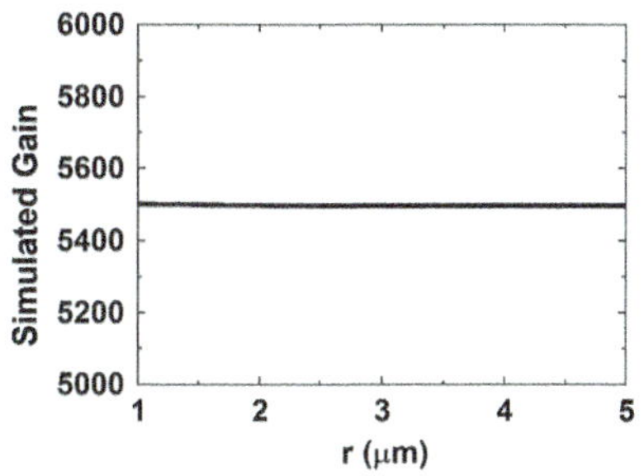

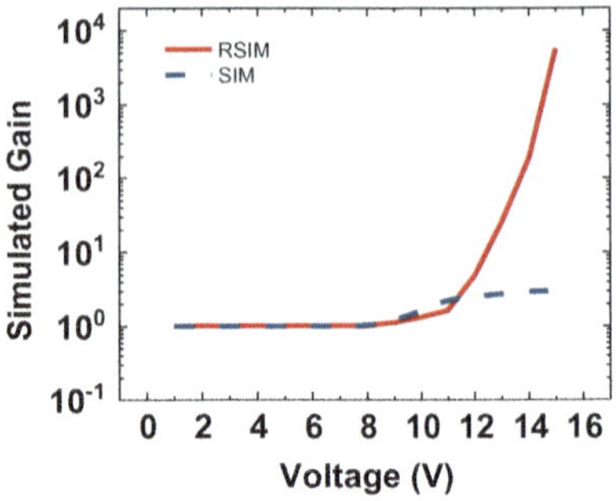

Fig. 4 Gain comparison of RSIM and SIM at different voltages

5 Conclusion

The RSIM design has been designed and simulated. The reach-through design is the key to increase the gain, with which the simulated gain can be increased to 5500 at a low voltage of 15 V. This result shows that RSIM has a great potential to be used in LiDARs integrated with any other light detector.

Acknowledgements This work was supported in part by the Shenzhen Fundamental Research fund under Grant (No. JCYJ20170306100015508).

References

1. Maes, W., De Meyer, K., Van Overstraeten, R.: Impact ionization in silicon: a review and update. Solid-State Electron. **33**(6), 705–718 (1990)
2. Campbell, J.C.: Recent advances in avalanche photodiodes. J. Lightwave Technol. **34**(2), 278–285 (2016)
3. McIntyre, R.J.: Recent developments in silicon avalanche photodiodes. Measurement **3**(4), 146–152 (1985)
4. Lee, H.W., Hawkins, A.R.: Solid-state current amplifier based on impact ionization. Appl. Phys. Lett. **87**(7), 3511 (2005)
5. Lee, H.-W., Beutler, J., Hawkins, A.: Surface structure silicon based impact-ionization multiplier for optical detection. Opt. Express **13**(22), 8760–8765 (2005)
6. Yu, G., Qin, L., Wan, Q.: Simulation of gain effect of solid-state impact ionization multipliers. In: The 4th International Conference on Intelligent Computing, Communication & Devices, Dec 2018
7. Ruegg, H.W.: An optimized avalanche photodiode. IEEE Trans. Electron Devices **14**(5), 239–251 (1967)
8. Kaneda, T., Matsumoto, H., Yamaoka, T.: A model for reach-through avalanche photodiodes (RAPDs). J. Appl. Phys. **47**(7), 3135–3139 (1976)

Big Data Analysis on Learning of Freshmen Based on Open Teaching

Tian Fang, Tan Han and Yao Juan

Abstract The open teaching platform is a teaching platform designed by Huazhong Agricultural University independently for the study of university computer basic courses. The platform has been used since 2012 and has accumulated a large amount of data. Based on big data technology, this paper analyzes the learning behavior of information course of freshmen, aiming to help the universities to explore new teaching mode based on artificial intelligence, reconstruct the teaching process, and establish a multi-dimensional comprehensive intelligent evaluation system based on big data.

Keywords Big data analysis · Open type

1 Introduction

The Ministry of Education issued Innovative Action Plan for Artificial Intelligence in Colleges and Universities, proposing "promoting the reform of school education and teaching, evolving to a smart campus on the basis of a digital campus, constructing a technically empowered teaching environment, exploring a new teaching model based on artificial intelligence, reconstructing the teaching process, using artificial intelligence to carry out teaching process monitoring, academic analysis and academic level diagnosis, establishing multi-dimensional comprehensive intelligent evaluation based on big data, accurately evaluating the performance of teaching and learning, and realizing the teaching according to the aptitude;" [1]. Therefore, the analysis of open teaching big data can help universities to track and

T. Fang · Y. Juan (✉)
Huazhong Agricultural University, Wuhan, Hubei, China
e-mail: yaojuan@mail.hzau.edu.cn

T. Han
Wuhan Vocational College of Software and Engineering, Wuhan, Hubei, China

V. Jain et al. (eds.), *Recent Trends in Intelligent Computing, Communication and Devices*, Advances in Intelligent Systems and Computing 1031,
https://doi.org/10.1007/978-981-13-9406-5_54

monitor students' learning situation and points out the direction for reconstructing teaching process and establishing intelligent evaluation system based on artificial intelligence.

2 Open Teaching Platform

The Open University Computer Basic Experiment Platform is an independent design of the Information School of Huazhong Agricultural University. It is an innovative platform aiming to cultivate the students' information technology capability and self-teaching ability as there is a big gap in the IT capability of the freshmen. The open experiment teaching system (Table 1) breaks the "Four Fixed" mode (i.e., fixed time, fixed class, fixed location, and fixed contents), and opens up the "Four Open" teaching mode, i.e. open time, open location, open contents and open interest.

The open teaching platform was designed and launched into use in 2012. It includes functions such as experiment background management, experiment appointment, experiment preparation, experiment preparation review score, experiment review score, and paperless examination system. Appointment—preparation—Experiment—Exam, this approach ensures that the whole process data can be stored. Learning time—study location—student's major—student's geography, with this, data in the software platform enable the learning behavior data of students can be stored. The vast amount of data generated by the platform provides a data foundation for analyzing the learning behavior of the freshmen when they learn information technology courses.

3 Data Set Based on the Open Teaching Platform

(This paper focuses on the analysis results of data; the method of big data storage and analysis is omitted in this paper.)

Table 1 Traditional teaching and open teaching

	Traditional teaching	Open teaching
Class	Fixed—administrative class	Open—the unit is individual
Time	Fixed	Students choose class time through the appointment system
Location	Fixed	Students choose class location through the appointment system
Contents	Fixed	Students choose learning content according to their own interests

3.1 *Data Set of Students' Characteristics*

Study time—study location—student's major—student's geography, the data set organizes the data generated in the web log, system login data, and user data. The main data are as given in Table 2.

3.2 *Teaching Process Data Set*

Appointment—preparation—Experiment—Exam. The entire process data can be stored. It contains the main data as given in Table 3.

Table 2 Data set of students' characteristics

Data	Data description	Data source
date	Date	Log
time	Time	Log
IP	Access IP	Log
url	Access website	Log
Username	Access user	Platform
U_depart	User's major	Platform
U_city	User source	Platform
U_nation	User's nationality	Platform
U_type	User type	Platform
U_sex	User's sex	Platform

Table 3 Teaching process data set

Data	Data description	Data source (2)
Date–time	Date and time of appointment	Appointment system
Book_type	Appointment method (mobile phone, in-campus address, external network address)	Log
Book_times	Number of appointments	Appointment system
Prepare_time	Preparation time	Log + platform
Prepare_project	Preparation project	Platform
Prepare_score	Preparation score	Platform
Lab_time	Lab time	Log + platform
Lab_project	Lab project	Platform
Lab_score	Lab score	Platform
Score	Score of final exam	Platform

4 Analysis of Students' Learning Behavior Based on Big Data Technology

The Academic Affairs Office of Anhui University of Technology analyzes the results of students in all grades of a university through the teaching information big data. According to the analysis, the first year of university is a key period after the students enter the school. The forming of study habits and learning style are crucial in this period [2]. The campus big data analysis system uses big data to draw "digital portraits" of students, which can mine the basic situation of students and analyze and predict them [3]. Therefore, to analyze the learning situation of freshmen in the school, and then build an intelligent learning platform to promote the process evaluation and truly realize the teaching in accordance with their aptitude through the study of university computer basic courses becomes a new topic that modern education needs to solve.

Based on the data of the entire open teaching platform, this paper uses Python program to analyze the data, which can help the development of grading strategy of the university computer basic course. It also provides the basis for reconstructing the teaching flow of artificial intelligence for recommending learning content according to interest and the solution algorithm for the new subject of constructing intelligent evaluation system.

4.1 Study Time

Figure 1 shows the distribution ratio of students' learning time points. From the data, the author finds that the students study the most from 4:00 p.m. to 7:00 p.m., followed by 9:00 a.m. to 11:00 a.m. and 7:00 p.m. to 9:00 p.m., and other time periods. This can be used for intelligent allocation of system resources.

Fig. 1 Proportional distribution of learning time point

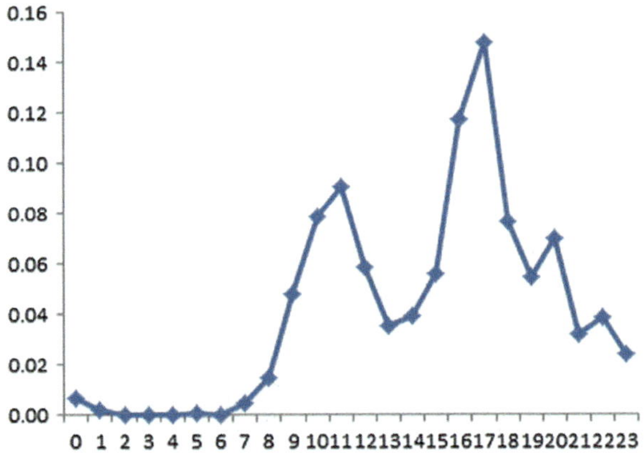

4.2 Average Preparation Interval

Figure 2 reflects the relationship between the average time interval and the average score of the students each time they prepare the experiment. The author finds that the time intervals for failing are short, and some are even completed within one period, and the students who make preparation before the experiment can reasonably plan and arrange their own learning process, and such students have excellent grades. Therefore, the preparation interval can be used as an early warning reference for students' bad learning status, and can also be used as one of the parameters of multi-dimensional comprehensive intelligent evaluation.

4.3 Effect of Geography

Figure 3 is a comparison of the average scores of various provinces. Figure 4 is a comparison of the average number of experiments performed by students in each province (the system allows students to re-do experiments at specified times). Through the analysis of the data, this paper provides the basis for the regional parameters in the intelligent recommendation algorithm of the learning content difficulty and the individualized customization scheme for students in different provinces.

4.4 Effect of Majors

Figure 5 is a comparison of the average scores of majors, and Fig. 6 is a comparison of the average number of experiments done by the students of each major. The author thinks that the major parameters are also very important in the intelligent recommendation algorithm.

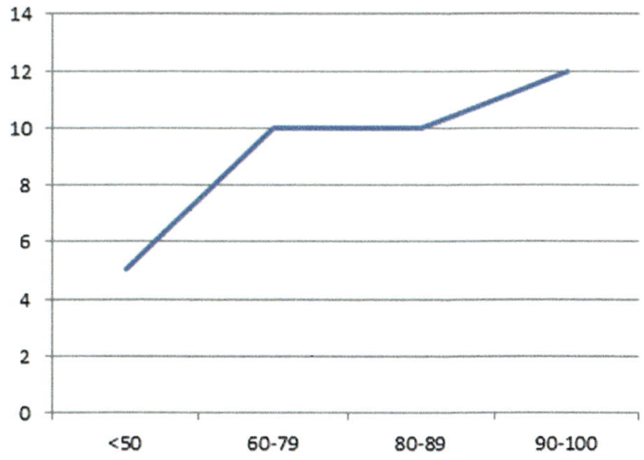

Fig. 2 Average preparation interval

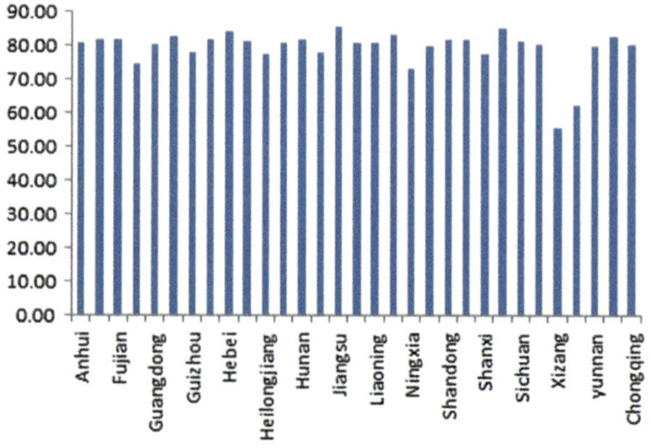

Fig. 3 Comparison of average score in different regions

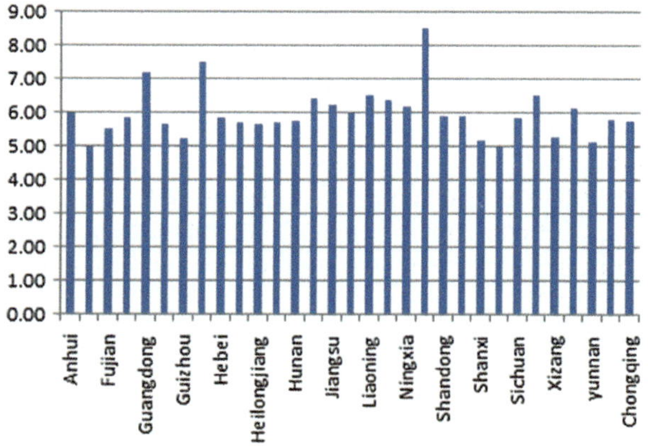

Fig. 4 Comparison of average experiment times in different regions

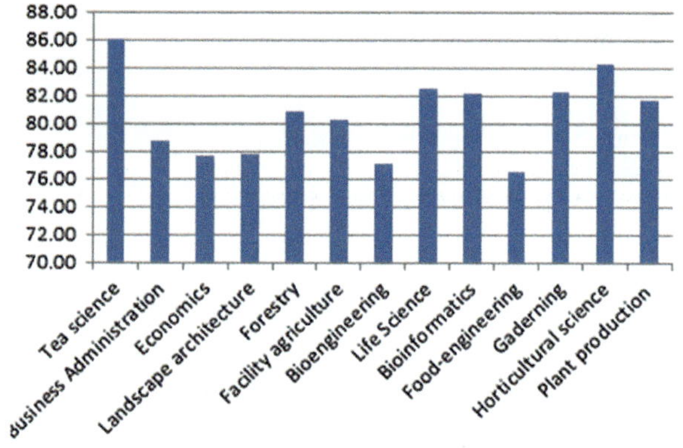

Fig. 5 Comparison of average scores of majors

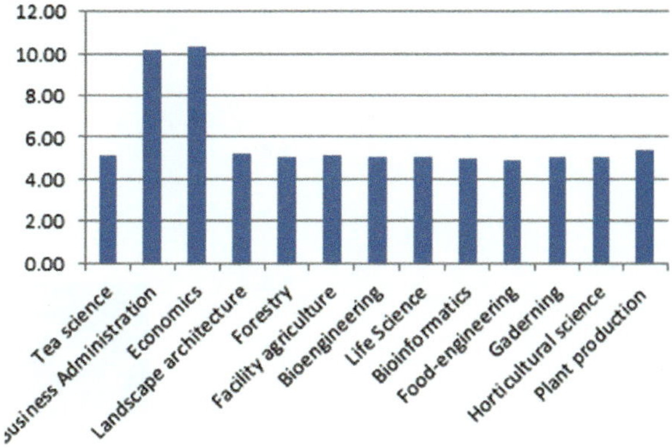

Fig. 6 Comparison of average experiment times of majors

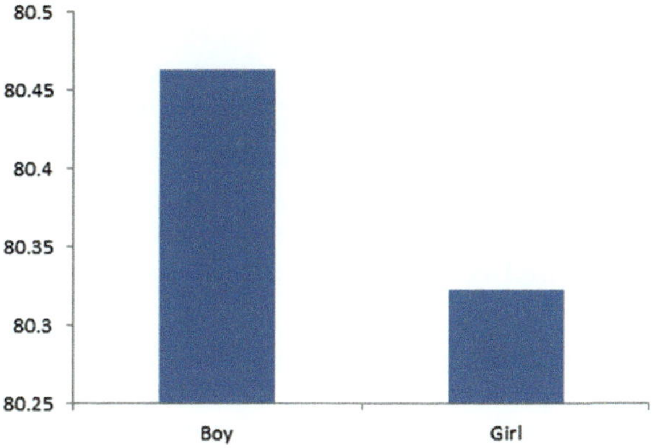

Fig. 7 Comparison of average scores of two genders

4.5 *Effect of Gender*

Figure 7 is a comparison of the average scores of two genders, and Fig. 8 is a comparison of the average number of experiments done by two genders. It is not difficult to see that boys have an advantage in the study of computer courses. Although the number of experiments is small, the results are good. Therefore, whether the learning programs should be gender-specific is also a question to be explored.

4.6 *Others*

In July 2018, big data classroom analysis based on Moso Teach was applied to teaching. In this paper, we will take part in activities, look at resources, and quantify class attendance into empirical values, so as to find out the students with learning

Fig. 8 Comparison of average experiment times of two genders

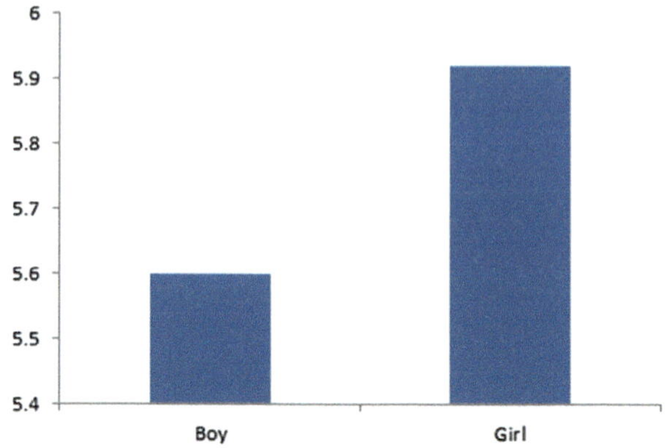

abilities. The platform can be used by this kind of students to expand learning [4]. Drawing lessons from the experience value, the author thinks that from the log, we can mine more valuable data through analysis, and form a better learning recommendation system.

5 Conclusion

Based on the big data of open teaching platform, this paper explores the impact of gender, province, major, average learning time, and average learning time interval on students' learning effect, and provides a good foundation for carrying out monitoring of teaching process and accurate evaluation of learning effect for the subsequent reconstruction of teaching process based on artificial intelligence.

References

1. Innovative Action Plan for Artificial Intelligence in Colleges and Universities. Teaching Techniques No. 3 (2018)
2. Xu, H., Tian, X., Qian, X.: Research on university academic situation based on big data. J. Anhui Univ. Technol. (2018)
3. Wang, Z.: Design and Research of Academic Analysis System Based on Big Data Technology. Regional Governance
4. Guo, J., Cao, L., Pan, X.: Probing into application of classroom big data analysis to teaching based on Moso Teach. Comput. Knowl. Technol. **14**(19) (2018)

A Depth Map Inpainting Method-Based Background and Texture for RGB-D Image

Zhang Yan, Wang Jian and Che Dong-Juan

Abstract The rendering technology-based depth map can greatly reduce amount of data in encoding, decoding, and transmission. Similarly, this method is important for synthesizing virtual viewpoints. RGB and depth (RGB-D) image can be acquainted by Microsoft Kinect camera. But hole and noise problems for depth map must be inpainted and improved. In this paper, based on background of RGB-D image and frame difference, the depth map is inpainted. The experimental results show that our method can obtain the better depth map.

Keywords Depth map · RGB-D image · The difference map

1 Introduction

In free viewpoint video (FTV) system, the rendering technology-based depth map can greatly reduce amount of data in encoding, decoding, and transmission [1]. Then the depth map is important. Virtual viewpoints are synthesized by texture image and corresponding depth map. Depth camera such as Microsoft Kinect can produce depth map and RGB-D image. While there are many problems for these depth maps such as a wide range of holes in the foreground and low reflection area, Wang et al. used *k*-means algorithm to fill the holes [2]. It effectively solves the problem of large area holes in depth maps, but details like edges are difficult to inpaint perfectly. Li et al. proposed an inpainting and error compensation method for depth image based on optimization estimation in which multiple frames averaging method is used, and however, texture information is ignored [3]. In this paper, we use the difference map between texture and depth map to get hole areas and the frame difference of depth maps to get background. The depth map is

Z. Yan (✉) · W. Jian · C. Dong-Juan
School of Computer and Remote Sensing Information Technology, North China Institute of Aerospace Engineering, Langfang 065000, China
e-mail: zhangyan_20142015@163.com

V. Jain et al. (eds.), *Recent Trends in Intelligent Computing, Communication and Devices*, Advances in Intelligent Systems and Computing 1031,
https://doi.org/10.1007/978-981-13-9406-5_55

inpainted-based background and texture image. The experimental results show that our method is effective and can obtain a better depth map.

In Sect. 2, the paper proceeds with a description of the RGB-D camera. In Sect. 3, the proposed detailed method is proposed. Experiment results are shown in Sect. 4. Section 5 is the conclusion.

2 Depth Map Acquisition Principle

As shown in Fig. 1, RGB-D camera, i.e., Kinect has three lenses, RGB camera, infrared sensor, and infrared cameras, respectively. Infrared sensor and infrared cameras are depth sensor. Lacking depth in the occlusion region and black or transparent objects are the main sources of holes. We can see that there are many holes in the depth map especially on the object edge shown in Fig. 2.

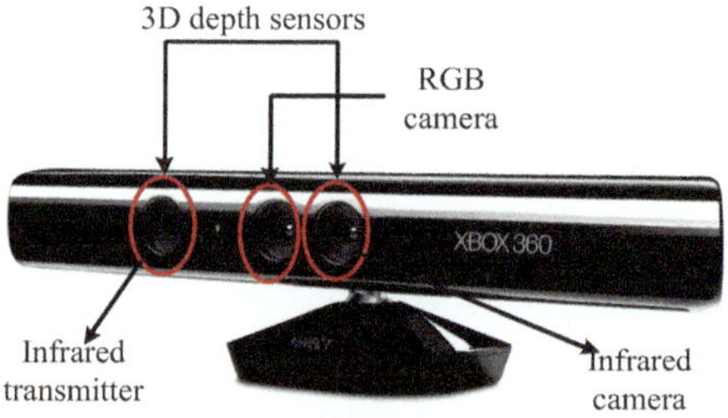

Fig. 1 Kinect camera

Fig. 2 Depth map by Kinect camera

3 Inpainting Method-Based Background and Texture

Texture images and depth maps can be acquired simultaneously for RGB-D image. In this paper, the main process is the following steps: (1) obtain the difference map between texture and depth map, (2) distinguish the background area and the foreground object area in the difference map and carry out the corresponding filling in the differentiated areas, (3) use 16 * 16 block to further analyze and fill the holes, and (4) adopt the median filtering to denoise.

Firstly, the difference map is obtained by texture map subtracting depth map after transforming texture image to gray image. Because the holes in depth maps are black, that is to say, the color value is 0, we only study the areas where the difference maps and gray texture images which have the same values. The difference map is set to avoid that other good places are mishandled.

Secondly, the background area and the foreground object area are distinguished in the difference map. Background is recognized by the front and back frames of the depth maps. The foreground object area can change when moving, while the background area cannot change. The foreground and background parts of the image are distinguished by the frame difference based on pixels between adjacent frames in depth map video as shown in Eq. (1).

$$\begin{cases} F_{k-1}(i,j) = |f_{k-1}(i,j) - f_k(i,j)| \\ F_{k+1}(i,j) = |f_{k+1}(i,j) - f_k(i,j)| \end{cases} \tag{1}$$

where $F_{k-1}(i,j)$, $F_{k+1}(i,j)$ are frame difference maps for $k-1$ and k, $k+1$ and k frames, respectively in (i, j). $f_{k-1}(i,j)$, $f_k(i,j)$, and $f_{k+1}(i,j)$ represent the pixel values at coordinates (i, j) in frames $k-1$, k, $k+1$, respectively.

For the current frame in the difference map, we fill the hole area by the brightness of the previous frame subtracting the current frame firstly $(F_{k-1}(i,j))$, and then fill the remaining area by the brightness of the latter frame subtracting the current frame $(F_{k+1}(i,j))$.

Thirdly, for fewer holes after the background of frame difference method after adding the processed difference map to the depth map, we use 16 * 16 block to analyze and fill the holes. The main idea is to fill the current pixel position with the brightness of the same most brightness which is not zero.

Finally, adopt the median filtering to denoise the depth maps' noises.

4 Experimental Results

The depth map obtained by RGB-D camera is processed according to the Sect. 3. Figure 3 is the difference map between texture and depth map. Figure 4 is the processed difference map which has the same values with gray texture images. Figures 5 and 6 are the frame difference maps. The result by the background of

Fig. 3 Difference map between texture and depth map

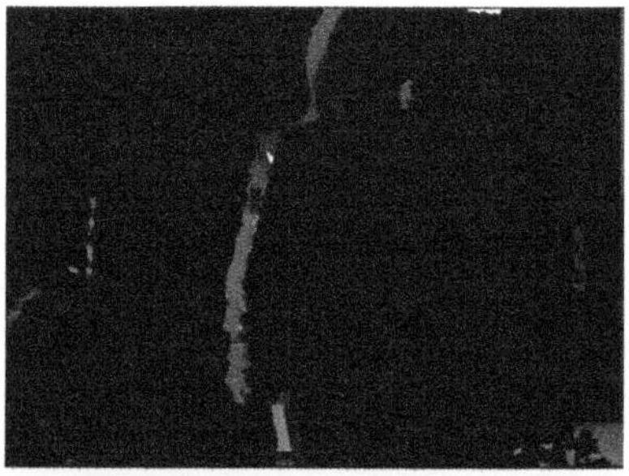

Fig. 4 Processed difference map

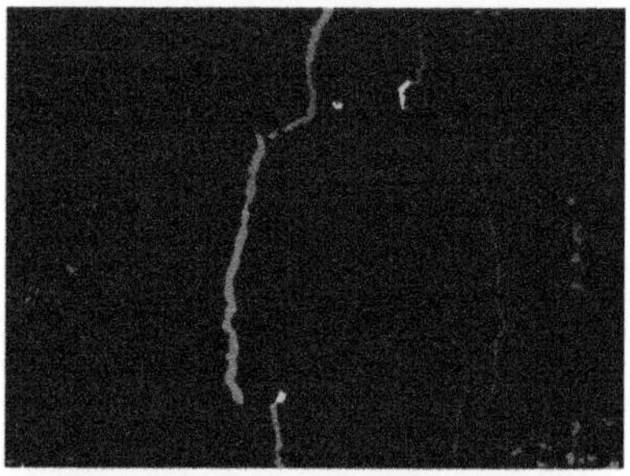

Fig. 5 Frame difference between $k - 1$ and k frames

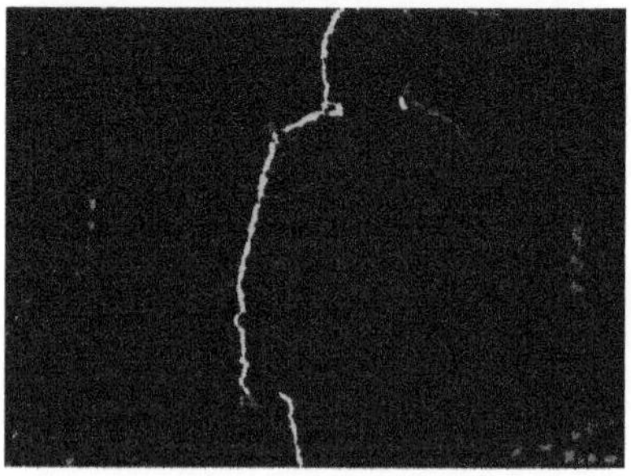

Fig. 6 Frame difference between $k + 1$ and k frames

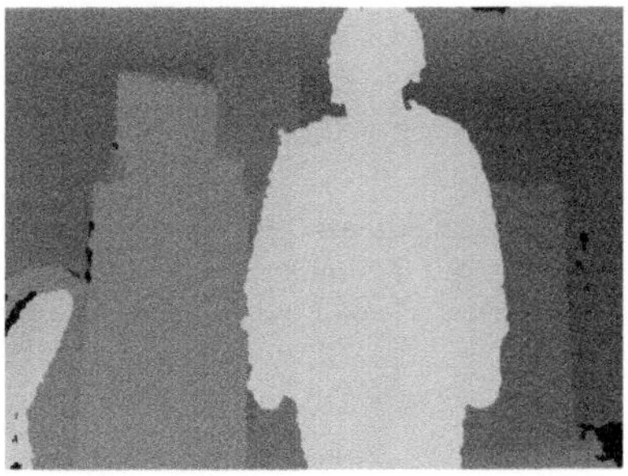

Fig. 7 Result by the background of frame difference method

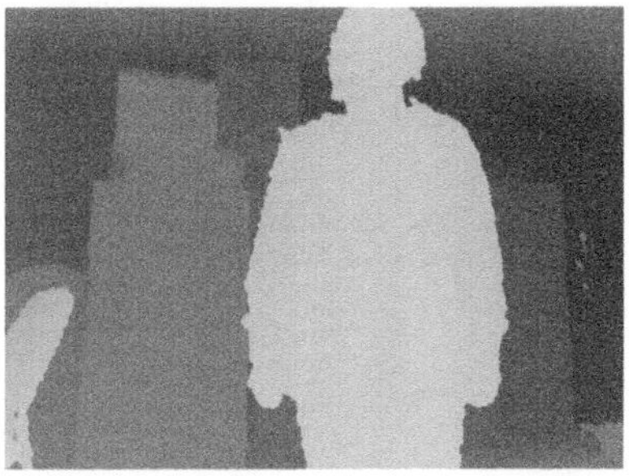

Fig. 8 Result after 16 * 16 block

Fig. 9 Processed by median filtering

frame difference method is shown in Fig. 7. Figure 8 is the proposed result after 16 * 16 block. The result by the median filtering is as shown in Fig. 9. We can see that the processed depth map is better.

5 Summary

This paper proposed an inpainting method for Kinect depth map. The method mainly inpaints and optimizes the depth map-based background and texture image. Specifically, the texture image is used to get the difference maps which find the hole position, and the depth map frame sequence is used to get the frame difference maps which obtain the background information. Finally, the median filtering is adopted. The experimental results show that the proposed method in this paper can effectively improve the depth map quality.

Acknowledgements This research was financially supported by the Youth Foundation of Hebei Educational Committee (QN2015113), the Doctoral Scientific Research Foundation (BKY-2014-04) and the Project of Langfang science and Technology Bureau (2018011002).

References

1. Ahn, I., Kim, C.: A novel depth-based virtual view synthesis method for free viewpoint video. IEEE Trans. Broadcast. **59**(4), 45–614–626 (2016)
2. Wang, Y., Jiang, A., Xu, L.: Kinect depth hole filling algorithm based on K-means. Microprocessors **4**:42–44 2015
3. Li, L., Zou, B., Zhou, G., et al.: Inpaint and error compensation method for depth image based on optimization estimation. J. Appl. Opt. **39**(1), 45–50 (2018)

Study on Vegetation Cover Change of Huang Huai Hai Plain Based on MODIS EVI

Yi Huang, Zhaodan Zhang, Xingxing Huang, Chuqiao Hong, Mingyue Wang, Rongrong Zhang, Xianmeng Zhang and Jingyu Zeng

Abstract As the natural "link" between the atmosphere, soil and water, and vegetation plays an indicative role in the process of global change (Sun et al. in J Remote Sens 2(3):204–210, [9]). Vegetation index is usually used for the study of land cover, vegetation classification, and other fields. Among them, EVI improved the disadvantages of NDVI, including atmospheric noise, to better reflect the spatial difference of vegetation cover. In this paper, MOD13A3 data were selected to study the spatial and temporal pattern and changing trend of EVI in the Huang Huai Hai plain from 2001 to 2018. The results are as follows: (1) In 2018, the maximum EVI value of the Huang Huai Hai plain was 0.94, showing a pattern of more in the west and less in the east. The year was divided into three seasons: spring, summer, and autumn with the maximum value in summer. (2) EVI fluctuated slowly in the last 18 years, with a linear growth rate of 0.28%/5a. The vegetation in more than half of the study area showed an increasing trend. Although there was partial degradation, the overall environment developed in a good direction.

Keywords MODIS · EVI · Huang Huai Hai plain

Y. Huang · Z. Zhang · X. Huang · C. Hong · M. Wang · R. Zhang · X. Zhang · J. Zeng (✉)
College of Environment and Resources, Fuzhou University, Fuzhou 350116, China
e-mail: zengjingyufzu@163.com

J. Zeng
Key Laboratory of Spatial Data Mining & Information Sharing, Ministry of Education of China, Fuzhou 350116, China

Fujian Provincial Key Laboratory of Remote Sensing of Soil Erosion and Disaster Protection, Fuzhou 350116, China

V. Jain et al. (eds.), *Recent Trends in Intelligent Computing, Communication and Devices*, Advances in Intelligent Systems and Computing 1031,
https://doi.org/10.1007/978-981-13-9406-5_56

1 Introduction

As an important part of the terrestrial ecosystem, vegetation not only connects the material circulation and energy flow of the soil circle, hydrosphere, and atmosphere, but also plays an important role in energy exchange and water circulation of the terrestrial ecosystem [2]. Dynamic monitoring of spatio-temporal evolution of vegetation cover is vital for revealing the evolution of regional environmental conditions and predicting future terrestrial ecosystems [7].

The time series of vegetation index (VI) can reflect the vegetation growth status and density well, especially the enhanced vegetation index (EVI), which improves the normalized difference vegetation index (NDVI) for its defects in atmospheric noise, saturation, soil background, and other aspects.

It has greater advantages in vegetation information extraction and ground object recognition [1, 5, 11]. Ye et al. [13] compared the spatial and temporal distribution characteristics of vegetation cover reflected by NDVI and EVI in Wanjiang River Basin and found that in the period of vigorous vegetation growth, EVI could better reflect the vegetation cover status than NDVI, and EVI could better reflect the spatial difference of vegetation cover than NDVI at the same spatial resolution. Mi et al. [8] used MODIS vegetation index to estimate the biomass and vegetation dynamics of alpine grassland, and found that EVI more stable in reflecting the above-ground vegetation growth.

In this study, we use MOD13A3 data from 2001 to 2018 to get the annual EVI maximum value synthesized by MVC method. The vegetation cover situation in Huang Huai Hai area was inverted by EVI to get the vegetation variation characteristics of the study area, which provided a basis for the follow-up work.

2 Study Area and Data

2.1 Introduction to Study Area

Huang Huai Hai plain (114°–120°E, 32°–40°N, Fig. 1) is China's second largest plain [6]. It is located in the third step of the three terraces of China. The terrain is flat with high in the west and low in the east. It is not only the political and cultural center of China but also an important agricultural and industrial base of China. With temperate continental climate, the annual precipitation varies greatly and the regional precipitation is uneven.

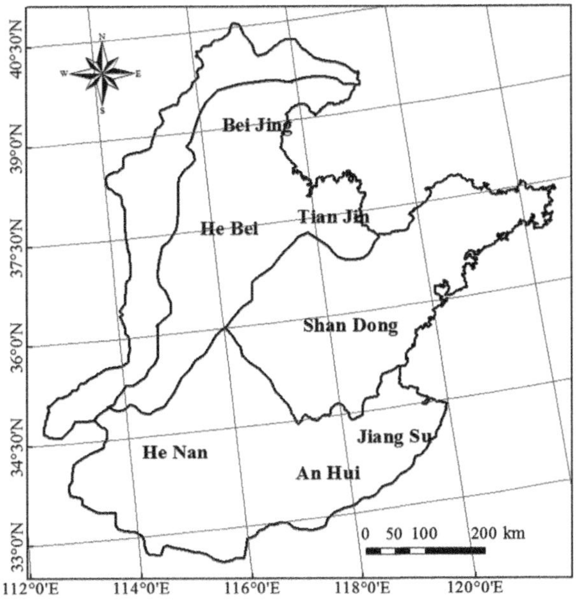

Fig. 1 Geographic location map of Huang Huai Hai plain

2.2 *Data and Processing*

In this paper, the L3 product MOD13A3 of moderate-resolution imaging spectroradiometer (MODIS) is selected, which provides EVI with a resolution of 1 km per month. Strip number: h26v04, h26v05, h27v04, h27v05, h28v05, the research year is 2001, 2005, 2010, 2015, 2018.

A series of preprocessing, such as stitching, projection transformation, and clipping, were carried out on the image by referring to the attribute information. The projection was converted into the Albers isoconic projection using neighboring pixel method, and the coordinate system was converted into WGS-84 and the image was clipped with the vector boundary of Huang Huai Hai plain.

In order to eliminate the influence of atmosphere, cloud, snow, and solar altitude angle [12], the five-year monthly EVI was synthesized by using the internationally used maximum value composite (MVC) method. Because the Huang Huai Hai Plain is mostly covered by heavy snow in winter, there are only three seasons for maximum synthesis: spring (February to April), summer (May to July), and autumn (August to October).

3 Results and Analysis

3.1 *Spatial and Temporal Distribution Characteristics of Vegetation in the Huang Huai Hai Plain*

EVI is positively correlated with vegetation coverage, so it can directly reflect the change of vegetation coverage. According to EVI index, vegetation coverage can be divided into six grades, namely bare land, sparse vegetation, less vegetation, medium vegetation, dense vegetation, and very dense vegetation [3] as shown in Table 1.

After classifying the maximum EVI of year 2018, the spatial distribution shows the pattern of more vegetation in the west and less in the east. The average EVI of classified images was 0.3890, in which bare land accounted for less than 1%, low-grade vegetation accounted for 4.03%, moderate vegetation accounted for 14.43%, and heavy vegetation accounted for 81.01%.

Figure 2 shows the EVI distribution of the Huang Huai Hai plain in three seasons of 2018. The mean EVI is 0.383 in spring, 0.536 in summer, and 0.426 in autumn. EVI increased first and then decreased with the seasons. In summer, vegetation coverage is the largest. In spring, vegetation is more sparse and showed a decreasing trend from south to north. It is more evenly distributed in summer and autumn.

Table 1 Grades of vegetation coverage

NDVI	Vegetation coverage levels
NDVI $\leq$ 0.10	Bare land
010 < NDVI $\leq$ 0.15	Sparse vegetation
0.15 < NDVI $\leq$ 0.30	Less vegetation
0.30 < NDVI $\leq$ 0.45	Moderate vegetation
0.45 < NDVI $\leq$ 0.60	Dense vegetation
NDVI > 0.60	Very dense

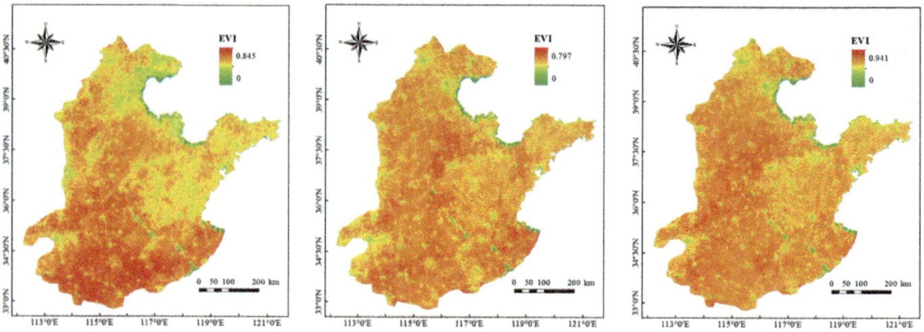

(a) Spring (b) Summer (c) Autumn

Fig. 2 EVI distribution of Huang Huai Hai plain in spring, summer, and autumn of 2018

3.2 EVI Time Series Change Characteristics

The EVI changes in Huang Huai Hai plain from 2001 to 2018 are shown in Fig. 3. During the 18 years, EVI fluctuated slowly with a linear growth rate of 0.28%/5a. The maximum of EVI appeared in 2010 and the minimum of EVI appeared in 2005. In the past 18 years, the annual average EVI was 0.260. Before 2005, the annual average EVI was mostly lower than the average, and after that, it was above the average. From 2005 to 2010, EVI increased rapidly with a growth rate of 1.8%/5a.

In order to monitor the vegetation cover changes in the long time series of the Huang Huai Hai plain, the annual (quarterly) maximum EVI and the time were analyzed by regression, and the linear regression equation with one variable was obtained. Slope was calculated to study vegetation variation trend. When Slope <0, it indicates the degradation trend of vegetation coverage, and Slope >0 indicates the improvement trend of vegetation coverage [10]. The calculation formula of Slope was as follows:

$$slope = \frac{n \times \sum_{i=1}^{n} i \times \mathrm{NDVI}_i - \left(\sum_{i=1}^{n} i\right)\left(\sum_{i=1}^{n} \mathrm{NDVI}_i\right)}{n \times \sum_{i=1}^{n} i^2 - \left(\sum_{i=1}^{n} \mathrm{NDVI}_i\right)^2} \tag{1}$$

where n represents the research year, NDVIi represents the largest NDVI of the year (season), i varies from 1 to n. Values are divided into the following seven categories [4] (Table 2):

By subtracting the images of 2001 and 2018, the overall changes of the study area over the past 18 years are obtained, as shown in Fig. 4. During the past 18 years, the area with increasing vegetation accounted for 51.99% of the study area while the degraded area accounted for 39.05%. Overall, the environment of the study area developed better and the vegetation coverage increased. Among them,

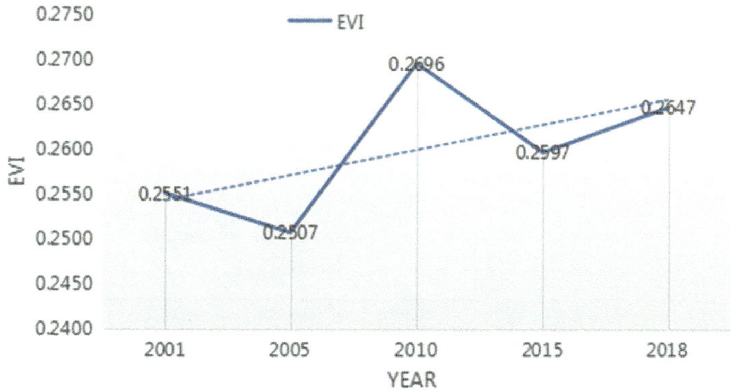

Fig. 3 Temporal variation of EVI in Huang Huai Hai plain from 2001 to 2018

Table 2 Grades of NDVI change trend

Slope	Change level	Area ratio (%)				
		2001–2018	2001–2005	2005–2010	2010–2015	2015–2018
<−0.0090	Seriously degraded	2.83	32.15	15.33	35.76	19.30
−0.0090 to −0.0045	Moderate degraded	8.97	11.13	7.48	12.73	10.58
−0.0045 to −0.0010	Slight degraded	19.15	9.08	7.17	10.27	10.21
−0.0010 to 0.0010	Unchanged	16.92	5.47	4.93	6.13	6.67
0.0010 to 0.0045	Slightly improved	29.41	8.46	8.28	8.89	11.15
0.0045 to 0.0090	Moderate improved	17.88	9.36	10.98	8.92	12.93
>0.0090	Obviously improved	4.70	24.36	45.81	17.38	29.25

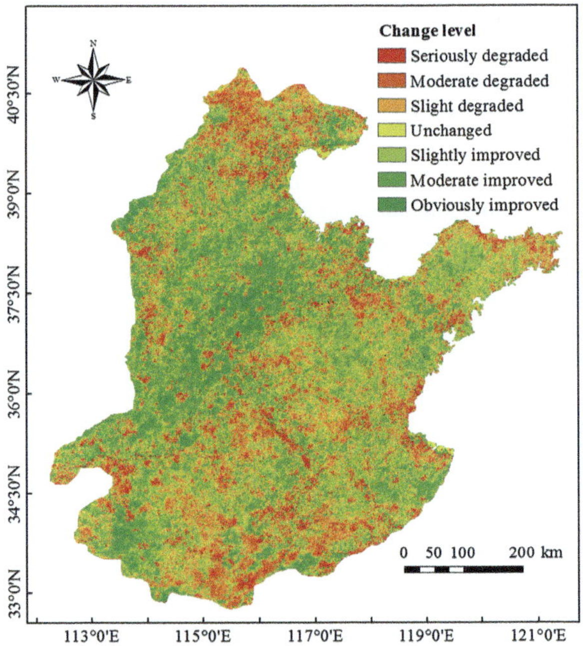

Fig. 4 Vegetation coverage change from 2001 to 2018

the largest proportion of vegetation growth areas was 65.7% in 2005–2010. From 2005 to 2010, vegetation growth regions accounted for the largest proportion, accounting for 65.7%.

4 Conclusion and Discussion

Based on the EVI data of MOD13A3 from 2001 to 2018, the characteristics of vegetation change in Huang Huai Hai plain are analyzed by trend analysis method. The main conclusions are as follows: (1) The vegetation cover in the study area is fine, the maximum value of EVI can reach 0.94. EVI distribution shows a pattern of more in the west and less in the east. EVI gets its largest value in summer. (2) During the 18 years, EVI showed a slow fluctuating upward trend, with a linear growth rate of 0.28%/5a. Vegetation coverage showed a slow overall increase and a local degradation trend. The vegetation over half of the study area showed an increasing trend.

Acknowledgements This research received financial support from the 23rd Phase of Scientific Research Training Program for Undergraduates of Fuzhou University (No. 23102) and National College Students Innovation and Entrepreneurship Training Program 2017 (No. 201710386021). We'd like to thank Qianfeng Wang for his helpful support.

References

1. Chen, P.Y., Fedosejevs, G., Tiscarelo-López, M., et al., Assessment of MODIS-EVI, MODIS-NDVI and VEGETATION-NDVI composite data using agricultural measurements: an example at corn fields in western Mexico. Environ. Monit. Assess. **119**(1/3), 69–82 (2006)
2. Chen, X.Q., Wang, H.: Spatial and temporal changes of vegetation belt and vegetation coverage in Inner Mongolia from 1982 to 2003. J. Geogr. **64**(1), 84–94 (2009)
3. Guo, N., Li, M., Ma, S.: Oasis vegetation change in Gansu Hexi area by the metrological satellite monitoring. J. Arid Meteorol. **20**(1), 033–035 (2002)
4. Han, R., Liu, P., Ma, C., et al.: NDVI 3 g trend and its response to climate change in Ordos during period from 1982 to 2012. Bull. od Soil Water Conserv. **36**(5), 028–033 (2016)
5. Jakubauskas, M.E., Legates, D.R., Kastens, J.H.: Crop identification using harmonic analysis of time-series AVHRR NDVI data[J]. Comput. Electron. Agric. **37**(1/3), 127–139 (2002)
6. Liu, Y.Q., Long, H.L.: Land use transformation and its dynamic mechanism in agricultural areas of Huang Huai Hai plain. J. Geogr. **32**(5), 1479–1485 (2012)
7. Mao, D.H., Wang, Z.M., Luo, L., et al.: Response of NDVI to climate change and increase of CO_2 volume fraction during vegetation growth season in Northeast Permafrost Region from 1982 to 2008. J. Environ. Sci. **30**(11), 2332–2343 (2010)
8. Mi, Z.R., Zhang, Y.S., Zhao, X.Q., et al.: Comparison NDVI with EVI in the herbage fresh weight estimation and vegetation dynamics for alpine grassland. Pratacultural Sci. **27**(6), 13–19 (2010)
9. Sun H., Wang C., Niu Z., et al., Analysis of the vegetation cover change and the relationship betw een NDVI and environmental factors by using NOAA time series data. J. Remote Sens. **2**(3), 204–210 (1998)
10. Song, Y., Ma, G.M.: Analysis of vegetation cover change in Northwest China based on spot vegetation data. Desert of China **27**(1), 89–93 (2007)
11. Wang, Z.X., Liu, C., Huete, A.: From AVHRR-DVI to MODISEVI: advances in vegetation index research. Acta Ecol. Sin. **23**(5), 979–987 (2003)

12. Xu, X., Li, X.B., Liang, H.W., et al., Change of vegetation coverage and its relationship with meteorological factors in temperate grassland of Inner Mongolia. J. Ecol. **30**(14), 3733–3743 (2010)
13. Ye, Q., Zhao, P., Sun, J.: Comparative analysis of vegetation coverage along the Yangtze River in Anhui Province based on MODIS/NDVI and EVI. Resour. Environ. Yangtze Basin **21**(3), 361–368 (2012)

Research on Chinese Chess Detection and Recognition Based on Convolutional Neural Network

Conghao Li and Guoliang Chen

Abstract A deep learning method based on convolutional neural network is introduced into the human–machine game of Chinese chess to detect and recognite pieces, which uses the faster R-CNN target detection architecture and the GoogLeNet convolutional neural network. Based on the shape features of pieces, an anchor mechanism is designed to detect pieces. The method of extracting eight-direction features and rotating invariant HOG features are used to train the ResNet convolutional neural network, which can improve the problem of fewer types of training-intensive chess pieces, single rotating pictures, etc. Experimental results show that the proposed convolutional neural network for chess detection and recognition can greatly enhance the detection of environmental adaptability, robustness, and improve the detection speed and the recognition accuracy.

Keywords Chess detection and recognition · Computer vision · Convolutional neural network · Recognition

1 Introduction

The robot in the human–machine Chinese chess game uses the visual system to locate, identify the pieces, and capture the board changes. In the process of human–machine interaction, the environment where the robot is located in relatively different (such as lighting conditions, shooting angle distance), and the positions of the chess pieces placed by players are different. Therefore, it is urgent to increase the environmental adaptability, robustness, and speed of the visual system. Full checkerboard scanning method is a commonly used method for positioning chess pieces. Firstly, the collected images are processed by thresholding, and the pieces are intercepted at a fixed position and then positioned and recognized. Sometimes,

C. Li · G. Chen (✉)
School of Mechanical and Electrical Engineering, Wuhan University of Technology, Wuhan 430070, China
e-mail: glchen@whut.edu.cn

V. Jain et al. (eds.), *Recent Trends in Intelligent Computing, Communication and Devices*, Advances in Intelligent Systems and Computing 1031,
https://doi.org/10.1007/978-981-13-9406-5_57

the collected images have shadows and reflections, and it is easy to distribute uneven grayscale. The background is segmented and many false goals are added, directly affecting the subsequent chess recognition and anomaly detection. The methods of chess piece identification mainly include the method of identifying pieces based on the characteristics of Fourier descriptors, the method of identifying pieces based on statistical features, the method of identifying pieces using the combination of color recognition and text features, and the method of using pieces of locally invariant SIFT and SURF [1]. Recognition method: Chinese chess recognition method using neural network and SIFT algorithm; these methods require a lot of pictures processing, a large number of repetitive calculations after each move, and the color, grayscale is more sensitive to light, noise, shadow, so these algorithms have poor adaptability to the environment, and the recognition rate is very unstable, especially for nonuniform illumination, complex shapes and objects with rotating angles [1, 2]. The recognition effect is poor and the noise immunity is poor. The main drawbacks of this type of method are poor environmental adaptability, slow detection speed, and low recognition accuracy.

With the development of machine learning and image processing technologies, there have been many methods of using deep learning to do target detection and recognition. Deep learning has achieved great success in many areas of computer vision such as image classification, target recognition, object detection, etc. [2]. The deep learning model based on convolution neural network (CNN) is the most effective model, and its network model in the ImageNet large-scale object detection and recognition contest achieved amazing results [3].

To solve the problems of low accuracy, poor environmental adaptability, and slow detection speed of the chess robot in detecting the chess pieces, the paper uses the faster R-CNN target detection architecture, GoogLeNet convolutional neural network, and ResNet convolutional neural network. Using a large number of chessboard pieces to train the network model, two deep learning models were obtained. Firstly, using the faster R-CNN target detection model to detect the chess pieces and their coordinates; then use the ResNet convolution network to classify and identify the detected piece, and effectively complete the work of checking, positioning, and recognition of the chess pieces.

2 Chinese Chess Detection [4–8]

2.1 Chess Features and RPN Networks

The purpose of the chess piece detection is to locate the chess pieces in the image and detect the type and location of the chess piece. The chess pieces in the board are circular objects, and there is no check object that is very similar to the piece on the chessboard; the chess pieces have the same size in the image, the arrangement and the combination are more regular, and generally there is no overlap, occlusion, and incompleteness, so the chess pieces are detected. Bring a certain degree of convenience; the paper uses the faster R-CNN framework as the framework for the detection of the target of the checkerboard chess pieces, and the faster R-CNN standard framework is to access the region proposal network (RPN) after the feature map of the underlying convolutional neural network, the RPN network and feature layer are then connected to the ROI pooling layer, which is connected to the softmax classification layer and the bounding box regressive layer after two layers of full connectivity.

2.2 *Anchor Mechanism*

According to the characteristics of the type and size of the chess pieces, the paper discards the anchor method to expand or reduce the length and width of the pre-selected box at a fixed ratio. Considering that the chess pieces are round in shape and that the main search target on the board is a chess piece, there are not many others. Length and width are different from those of chess pieces, so the anchor mechanism's anchor point expands or shrinks the pre-selected boxes by two types of squares and rectangles. The feature map corresponds to the highly refined and compressed original image, and the pieces' images and features are greatly compressed. The search scope and pre-selected box categories are greatly reduced; the main reason for the slow detection of chess pieces is to do a lot of repetitive work, according to the idea of quick-sort algorithm inspired and rough search and then micro-search, designed a new anchor Mechanism; RPN network divides the feature map 2 into 2^9, 2^8, 2^7, 2^6, 2^5, 2^4, 2^3, 2^2, 2^1, 2^0 pre-selected areas, then use the classification the layer determines the number of valid pre-selected regions obtained by the ten dichotomys, determines the dichotomy that yields the most effective number of pre-selected regions, and then traverses the entire image using several types of search boxes of corresponding sizes to detect valid pre-selected regions that may be missed. All pre-selected areas enter the classification layer and the regression layer.

2.3 *Faster R-CNN Model Building*

The paper uses 8×8 convolutional layer and an average pool layer of complex sparse structure in GoogLeNet V3 instead of two fully connected layers in order to reduce the model parameters and prevent overfitting. At the same time, the RPN network and the ROI pooling layer are directly connected to the Filter Concat of the 17×17 convolution units; At the same time, the convolution layer adopts a combination of ReLU + BN (Batch Normalization), which greatly reduces the number of iterations and training time through BN and greatly improves the classification accuracy.

GoogLeNet V3 splits the larger convolutional network into smaller one-dimensional networks, designs 35 × 35, 17 × 17, 8 × 8 three inception module structures and uses a simultaneous convolution and pooling of step size 2, and then a novel-pooled structure composed of aggregates is used in the pooling layer in GoogLeNet V3.

3 Chinese Chess Recognition

3.1 Analysis of Chess Features

The addition of eight-direction feature and histogram of oriented gradients (HOG) feature in Chinese character recognition can effectively improve the recognition accuracy of chess pieces and optimize system performance. The eight-direction features are a gradient feature in which image gradient features are decomposed into eight directions and is a commonly used method in pattern recognition. Gradient features of the extracted image with the Sobel operator, the Sobel operator can extract the horizontal and vertical gradients of the image pixels, and the Sobel operator's two-dimensional mask is shown in Fig. 1.

According to the two-dimensional mask, the pixel value $f(x, y)$ at any point coordinate (x, y) in the original image plane coordinate system, and the gradient $\mathrm{grad}(x, y)$ of this pixel is

$$\begin{aligned} \mathrm{grad}_x(x, y) &= f(x+1, y-1) + 2f(x+1, y) + f(x+1, y+1) \\ &\quad - f(x-1, y-1) - 2f(x-1, y) - f(x-1, y+1) \\ \mathrm{grad}_y(x, y) &= f(x-1, y-1) + 2f(x, y-1) + f(x+1, y-1) \\ &\quad - f(x-1, y+1) - 2f(x, y+1) - f(x+1, y+1) \end{aligned} \tag{1}$$

The gradient vector is decomposed in eight directions to obtain an eight-direction gradient feature vector.

(a)

(b)

1	2	1
0	0	0
-1	-2	-1

(c)

1	0	-1
2	0	-2
1	0	-1

Fig. 1 Eight-direction features of Chinese chess. **a** Two-way mask of the chess piece's eight-direction feature, **b** and **c** Sobel operator's horizontal vertical direction

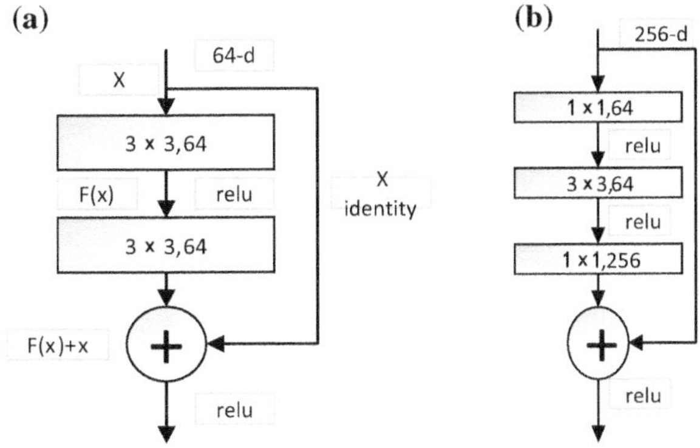

Fig. 2 Calculation process of layer-to-layer residual. **a** Residual unit building block, **b** bottleneck

3.2 Analysis of Chess Features

ResNet solves the degeneracy problem by introducing a deep residual learning framework. It uses the congruent map to directly pass the output of the previous layer to the convolution layer. The core idea of ResNet is to calculate the layer-to-layer residual, that is, $F(x)$. Assume that the input of a certain neural network is x and the expected output is $H(x)$. If the input x is directly transmitted to the output, the learning objective is $F(x) = H(x) - x$ as shown in Fig. 2.

ResNet simplifies the deep model to make the deep network feasible; it can construct the depth model of 34 layers, 54 layers, 152 layers or even 1202 layers by continuously superimposing residual blocks or bottleneck. Each ResNet layer uses a ReLU + BN (Batch Normalization) combination.

4 Experimental and Analysis

The experiment is divided into two parts: target detection (detection of the target of the pawn) and chess piece recognition. The training set image was labeled with LabelImg software. Experimental tests were performed on the target's zoom change, noise interference, ambient light change, shooting angle, distance change, and chessboard piece size change. In practical applications, the chess pieces on the board will not overlap, block, and interfere with serious noise so that the target detection achieves good results.

Experiments show that the faster R-CNN detection model has strong robustness and adaptability to the environment. Under various environments and noise interference, the detection task can be completed very well, and the average detection accuracy rate is 94.55%. In practical applications, the shooting environment is good (that is, there is no image noise, no occlusion, shooting angle is relatively fixed), the accuracy rate can reach 99.9%, and it can fully meet the needs of practical applications. The coordinate of the positioning model of the detection model is the coordinate in the picture. The coordinate transformation and image correction need to be done to get the corresponding position coordinates.

In the training process of the ResNet model, feature fusion is fully used, that is, the different levels of features in the network are input together into the final convolution layer and the average pooling layer; in order to expand the data set, the training set and the test set contain a large number of pieces images, corresponding Chinese character images, and images captured by web crawlers.

The deep learning model has more advantages than the character recognition algorithm in the recognition of the chess piece. The recognition accuracy rate is significantly improved, the adaptability to the environment is strong, and the speed of chess piece recognition is greatly improved. The manually extracted rotation-invariant HOG feature and gradient feature can also effectively improve the accuracy of the chess piece's recognition. The chess character recognition algorithm, SURF features, and ORB feature recognition algorithms are less adaptable to the environment. There are noise disturbances in the piece images, changes in the ambient light, etc., and the recognition rate will be greatly reduced, and the time required for transmission will be long. Milliseconds or even longer, the trained deep learning model requires an average of 101.67 ms to identify the target. It has fast recognition speed, strong adaptability to the environment, high accuracy and can meet real-time requirements.

5 Conclusion

The paper uses convolutional neural network object chess pieces to detect and recognize and trains the faster R-CNN frame object detection model and ResNet piece recognition model. Contrary to the traditional method of using artificial features to detect and identify the pieces, the convolutional neural network can be extremely useful, which greatly enhances the environmental adaptability and the accuarcy of detection and recognition. At the same time, for character recognition, extracting features by human analysis can help the network model to better understand the recognition objects. Extracting eight-direction features and rotation-invariant HOG features can improve the picture types and reduce the number of pictures. Rotation problems caused by the small number of rotating pictures. Deep learning model can better solve the problems of poor environmental adaptability, slow detection speed, low recognition accuracy, and instability in chess piece detection and recognition.

Acknowledgements The paper was supported by the Natural Science Foundation of China (61672396, 61373110).

References

1. Dang, H.S., Zhang, C., Pang, Y., et al.: Research on chess rapid recognition and positioning system based on ORB algorithm. Sci. Technol. Eng. **17**(7), 52–57 (2017)
2. Yang, W., Jin, L., Xie, Z., et al.: Improved deep convolutional neural network for online handwritten Chinese character recognition using domain-specific knowledge. Guangzhou **15** (6), 551–555 (2015)
3. Ren, S., He, K., Girshick, R., et al.: Faster R-CNN: towards real-time object detection with region proposal networks. In: International Conference on Neural Information Processing Systems. MIT Press, pp. 91–99, 2015
4. Sabokrou, M., Fayyaz, M., Fathy, M., et al.: Deep-anomaly: fully convolutional neural network for fast anomaly detection in crowded scenes. Comput. Vis. Image Underst. (2018)
5. Szegedy, C., Reed, S., Erhan, D., Anguelov, D.: Scalable, high-quality object detection. arXiv, 2015(2), 1412–1441 (2015)
6. Szegedy, C., Liu, W., Jia, Y., et al.: Going deeper with convolutions. In: IEEE Conference on Computer Vision and Pattern Recognition. IEEE Computer Society, pp. 1–9, 2015
7. Ioffe, S., Szegedy, C.: Batch normalization: accelerating deep network training by reducing internal covariate shift. 2015, 448–456 (2015)
8. Zhong, Z., Jin, L., Xie, Z.: High performance offline handwritten Chinese character recognition using GoogLeNet and directional feature maps. In: International Conference on Document Analysis and Recognition, pp. 846–850. IEEE, 2015

Spatial–Temporal Change Characteristics of Vegetation in Huang-Huai-Hai Plain Based on MODIS NDVI

Chuqiao Hong, Mingyue Wang, Rongrong Zhang, Xianmeng Zhang, Jingyu Zeng, Yi Huang, Zhaodan Zhang and Xingxing Huang

Abstract In this paper, MODIS normal difference vegetation index (MODIS NDVI) products of Huang-Huai-Hai Plain in 2001, 2005, 2010, 2015, and 2018 are used as data sources. With the help of MRT, ENVI, ArcGIS, and Excel software operating platforms, at the same time, using maximum value composites method, mean method, unary linear regression method, and trend analysis method to generate the time series data of monthly, seasonal, and annual NDVI, which is to study and analyze the spatial and temporal change characteristics and laws of vegetation in Huang-Huai-Hai Plain. The results show that in the past 18 years, the average annual NDVI has shown a downward trend. On the time scale, the NDVI has obvious seasonal and interannual variations with growing fastest in summer and the maximum occurring in July, while the growth rate in spring and autumn is relatively flat and the minimum occurring in December. On the spatial scale, the NDVI has fewer growth regions that are mainly distributed in the central and western regions, while the regions where NDVI declines are relatively more and mainly distributed in the northern and southern.

Keywords MODIS NDVI · Huang-Huai-Hai Plain · Spatial–temporal change · Vegetation

1 Introduction

Vegetation is the main body of terrestrial ecosystems [1]. It regulates the energy balance and material circulation in the ecosystem, and it is an important link connecting the atmosphere, hydrosphere, biosphere, and soil circle [2]. The normal difference vegetation index (NDVI) is the most commonly used indicator for characterizing vegetation conditions currently [3]. NDVI can not only reflect the status of vegetation cover but also it is possible to monitor changes in surface

C. Hong · M. Wang (✉) · R. Zhang · X. Zhang · J. Zeng · Y. Huang · Z. Zhang · X. Huang
College of Environment and Resources, Fuzhou University, Fuzhou, China
e-mail: 13023890836@163.com

V. Jain et al. (eds.), *Recent Trends in Intelligent Computing, Communication and Devices*, Advances in Intelligent Systems and Computing 1031,
https://doi.org/10.1007/978-981-13-9406-5_58

vegetation cover on a large scale and in a long time series [2, 4, 5]. In this paper, based on the MODIS NDVI data of five years in 2001, 2005, 2010, 2015, and 2018, the temporal and spatial change characteristics of vegetation in Huang-Huai-Hai Plain were studied by using the maximum value composites (MVC), mean method and trend analysis method, so as to provide constructive suggestions for the protection and restoration of ecological environment in this area.

The Huang-Huai-Hai Plain is located in the eastern part of China (32°–42°N, 113°–120°E), and it is also the largest plain area in China. The area is below 200 m above sea level. It has a large latitude span and the annual average daily temperature decreases with latitude from south to north.

2 Date and Methods

2.1 Data Sources and Processing

This paper uses MODIS NDVI synthetic products released by NASA in 2001, 2005, 2010, 2015, and 2018 for five years. The product grade is MOD13A3 and the spatial resolution is 1000 m. The data is the monthly average of vegetation index. First, MRT, ENVI, and ArcGIS software operating platforms are used to stitch, project, and tailor product data. Then, obtain the maximum seasonal and annual NDVI data sets by the means of the maximum value composites method. Finally, based on the NDVI data sets, the monthly, seasonal and annual NDVI data sets of vegetation for many years are obtained by the means of the mean method.

2.2 Methods

(1) Maximum value composites processing

On the one hand, due to the large amount of data, it is more convenient to implement maximizing processing. On the other hand, the maximum value of vegetation index can reflect the best growth of vegetation with high accuracy. So, in this paper, we use the internationally popular maximum value composites (MVC) method to maximize NDVI data by using the MVC tool plugin in ENVI software, so as to obtain the maximum seasonal and annual NDVI data, and to reduce the phenomenon of cloud and aerosol pixel doping as well as the influence of the sensor observation angle.

(2) Trend analysis based on linear regression

It can calculate the change characteristics of each grid through the linear regression equation. According to the change characteristics, the vegetation growth trend of the grid can be reversed. This method can simulate the spatial evolution of regionalized NDVI and is commonly used to analyze vegetation change trend of long time series [6–8]. In this paper, we calculate the regression slope between NDVI and time of all pixels in the data set by using the least squares method for time-dependent variable and NDVI-dependent variable data. If Slope > 0, it shows that the change trend of NDVI is increasing in the past 18 years, and vice versa, decreasing.

$$\mathrm{Slope} = \frac{n\sum_{k=1}^{n}(k\mathrm{NDVI}_k) - \sum_{k=1}^{n} k_{k=1}^{n}\mathrm{NDVI}_k}{n\sum_{k=1}^{n} k^2 - \left(\sum_{k=1}^{n} k\right)^2} \quad (1)$$

where, slope is the trend of NDVI (i.e., the slope value of the fitting curve of interannual change), n is the number of years of study, and NDVI_k is the value of NDVI in the k year. Using the raster calculation function module of ArcGIS, the trend map of NDVI during 2001–2018 can be calculated, which can directly reflect the change trend of vegetation NDVI in the 18-year time series.

3 Results and Analysis

3.1 NDVI Change During the Year

Figure 1 shows the monthly average NDVI of vegetation in Huang-Huai-Hai Plain from 2001 to 2018. It can be seen that the NDVI has a large variation during the years, and the NDVI value shows a trend of increasing first and then decreasing. Seen from the month of, the NDVI value is relatively small in December and January with the minimum value appearing in December (0.1281). In June, July, and August, the NDVI value is relatively large, with the maximum value appearing in July (0.3504). Among them, from March to June, the NDVI in the study area increases sharply, and it begins to decline rapidly from September. The NDVI is generally higher from June to August with little change. The main growth period of vegetation is from March to June, and it reaches saturation in July. After summer, the vegetation tends to wither and the NDVI value drops rapidly.

Fig. 1 Monthly average NDVI of vegetation

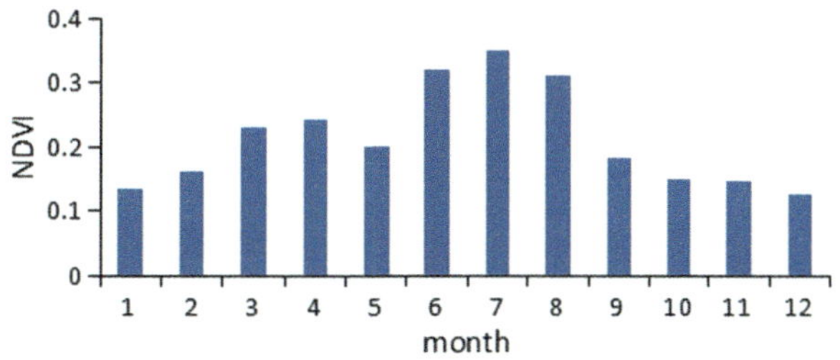

3.2 *Interannual Change of NDVI*

Figure 2 shows the interannual variation of NDVI in Huang-Huai-Hai Plain. It can be seen that the NDVI has shown an alternating ups and downs in the past 18 years. And overall, it is a downward trend with the annual degradation rate of 0.1074. The vegetation growth is relatively good in 2001 and 2005.

3.3 *NDVI Feature Changes*

The change characteristics of NDVI in each year can be seen from the average value of each season. As is shown in Fig. 3, it can be seen that the seasonal variation of the average value of NDVI during the year is unimodal, and the change characteristic is first increased and then decreased with the maximum value appearing in summer. It can also be seen that the seasonal variation has nothing to do with the year. This change has obvious seasonal characteristics, and it is related to climate precipitation.

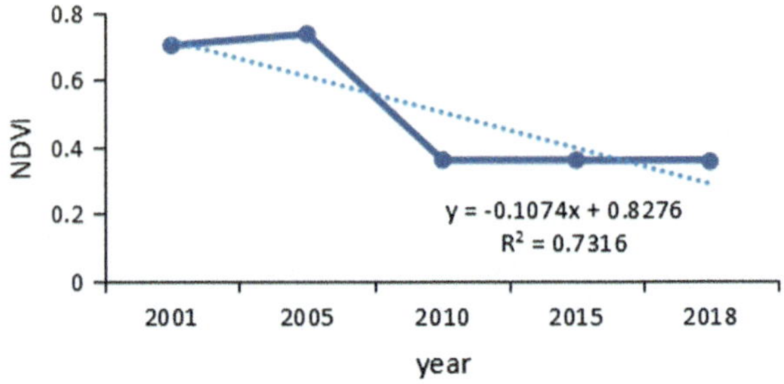

Fig. 2 Interannual changes of NDVI

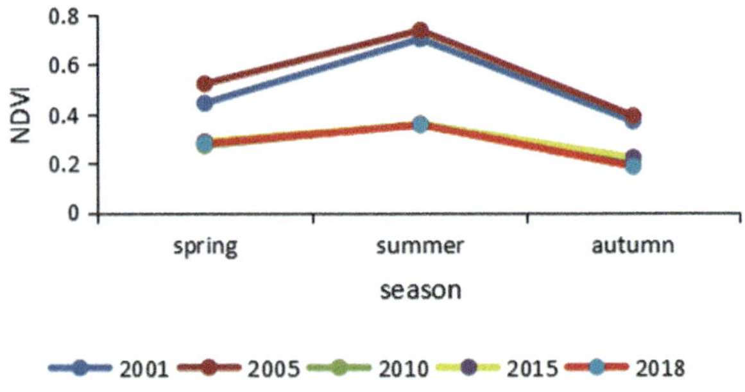

Fig. 3 Seasonal variation of NDVI

3.4 Spatial Variation of NDVI

In order to understand the trend of the cover value of Huang-Huai-Hai Plain, in this paper, we use the trend slope method to map in ArcGIS software platform to visually reflect the trend of vegetation growth with time. Figure 4 shows the overall trend of vegetation change from 2001 to 2018, the green area represents vegetation improvement, and the red area represents vegetation degradation. The darker the color, the more severe it is. It can be seen from the figure that the NDVI has generally shown a degrading trend in the past 18 years, with fewer growing regions and more degraded regions. Among them, the increase in the central and western regions is the most obvious (the green part in the picture), the value range of Slope is mainly above 0.008. But the northern and southern regions show a decreasing trend (the red part in the figure), the value range of Slope is mainly below −0.005. This also reflects the trend of decreasing from the middle to the periphery.

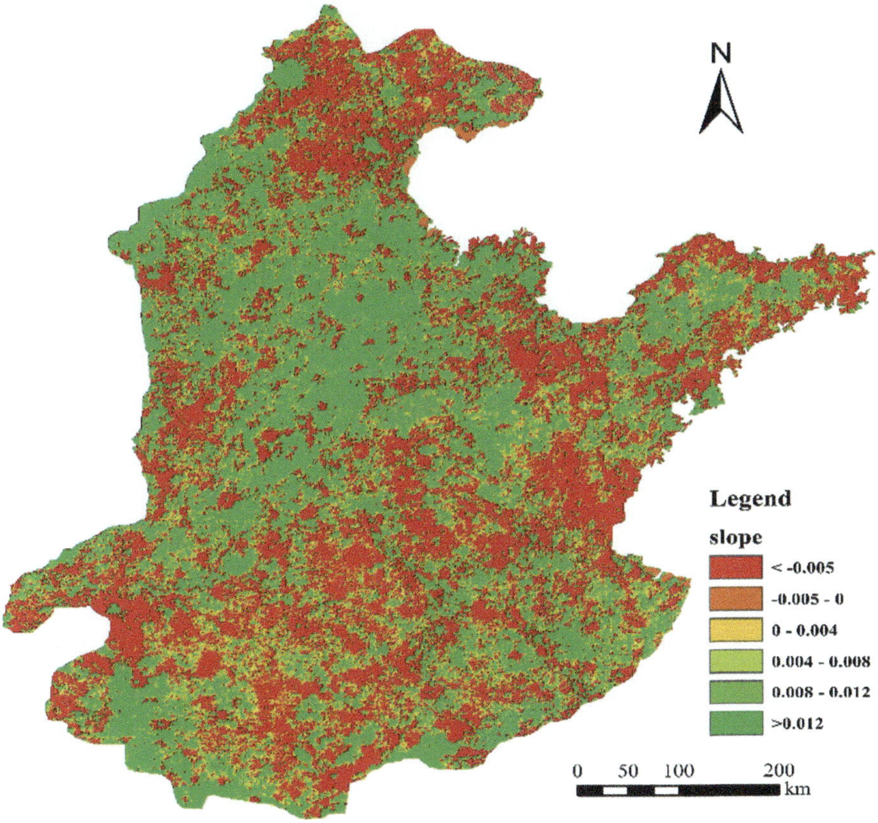

Fig. 4 Regression trend spatial distribution map

4 Conclusions

In the past 18 years, the vegetation cover in Huang-Huai-Hai Plain has shown a degraded trend, and the local area has a slight improvement trend. The degraded regions are mainly concentrated in the northern and southern, and the improved regions are mainly distributed in the central and western.

The vegetation in Huang-Huai-Hai Plain grows faster in summer and spring, with the highest vegetation coverage and the maximum NDVI occurring in July; in winter and autumn, the growth is slow with the smallest vegetation coverage and the minimum NDVI occurring in December. In the past 18 years, the vegetation change trends in different seasons have different characteristics: In spring and autumn, vegetation tends to improve; in summer, it remains stable; and in winter, it tends to degenerate. In 2001 and 2005, the vegetation grows better, but the vegetation growth situation deteriorates in the following years. In general, vegetation growth shows a stable degradation trend during the whole study period.

Acknowledgements This paper was conducted under the support of the 2017 National College Students Innovation and Entrepreneurship Training Program (201710386021) and The 23rd Phase of Scientific Research Training Program for Undergraduates of Fuzhou University (23102). For which, we are very to grateful to them.

References

1. Suo, Y., Wang, Z., Liu, C., et al.: Relationship between NDVI and precipitation and temperature in Middle Asia during 1982–2002. Resour. Sci. **38**(8), 1422–1429 (2009)
2. Sun, H., Wang, C., Niu, Z., et al.: Changes of surface vegetation coverage in China and its relationship with climate factors-based on NOAA time series data analysis. J. Remote Sens. **2** (3), 204–210 (1998)
3. Bai, J., Bai, J., Wang, L.: Spatio-temporal change of vegetation NDVI and its relations with regional climate in northern Shanxi Province in 2000–2010. Sci. Geogr. Sin. **34**(7), 882–888 (2014)
4. Tian, Q., Min, X.: Advances in study on vegetation indices. Adv. Earth Sci. **13**(4), 327–334 (1998)
5. Zhujun, G., Zeng, Z.: Overview of researches on vegetation coverage in remote sensing. Res. Soil Water Conserv. **12**(5), 18–21 (2005)
6. Pan, X., Ma, Y., Gao, W., et al.: Eco-environmental evolution in arid area of west China. J. Desert Res. **24**(6), 663–673 (2004)
7. Sun, Y., Guo, P.: Spationtemporal variation of vegetation coverage index in North China during the period from 1982 to 2006. Arid Zone Res. **29**(2), 187–193 (2012)
8. Song, Y., Ma, M.: Variation of AVHRR NDVI and its relationship with climate in Chinese arid and cold regions. J. Remote Sens. **12**(3), 499–505 (2008)

Identification of FARARX Models with Errors in Variables

D. V. Ivanov, I. L. Sandler, O. A. Katsyuba and V. N. Vlasova

Abstract Identification of real processes and systems can be implemented using various classes of models. Models based on equations with derivatives and differences of fractional order find wide application in various fields of science and technology. The paper discusses the identification of FARARX (fractional differencing autoregressive with exogenous input) models of fractional order with interference in the input and output signals. The proposed three-stage algorithm to obtain estimates of the parameters is based on the parameter estimates of the extended FARX model (fractional differencing autoregressive with exogenous input). On the first stage, the parameters of the extended FARX model are estimated. On the second stage, the parameters of the FARARX model are estimated based on extended model coefficients. On the third stage, the noise model coefficients are estimated. Simulation has shown high precession of the proposed algorithm.

Keywords ARARX model · Consistent estimates · Fractional calculus · Difference of fractional order · Error in the equation · Errors in variables · Least squares method

1 Introduction

Mathematical models based on fractional-order differential and difference equations are widely used both in natural science (heat exchange [1], polymer modeling [2, 3], acoustics [4] and in engineering (robotics [5], PID control [6], modeling of long lines [7], etc.). In economic applications, "long memory" time series models [8] and growth models [9] are used.

D. V. Ivanov (✉) · I. L. Sandler · O. A. Katsyuba
Samara State University of Transport, Samara, Russia
e-mail: dvi85@list.com

V. N. Vlasova
Moscow State University of Technologies and Management, Moscow, Russia

V. Jain et al. (eds.), *Recent Trends in Intelligent Computing, Communication and Devices*, Advances in Intelligent Systems and Computing 1031,
https://doi.org/10.1007/978-981-13-9406-5_59

Therefore, the development of methods for model identification based on difference and differential equations is a crucial task. Of special interest are methods for identification of systems with errors in variables. Nowadays, only the small number of articles, related to this topic, is available [10–12].

In Diversi et al. [13], a three-step algorithm for an estimate of ARARX (autoregressive with autoregressive exogenous input) parameters of models with errors in variables is proposed.

The results of the works [11–13] for FARARX (fractional differencing autoregressive with exogenous input) model class are generalized in the paper.

2 Problem Statement

Suppose the model is described as discrete equations with fractional-order differences (Fig. 1):

$$\Delta^{\alpha}\left(z_i+\sum_{m=1}^{r} b_0^{(m)} z_{i-m}\right)=\sum_{m=0}^{r_1} a_0^{(m)} x_{i-m}+\varphi_i,$$
$$y_i=z_i+\xi_i, w_i=x_i+\zeta_i, \varphi_i=\sum_{m=1}^{r_2} c_0^{(m)}\varphi_{i-m}+\varsigma_i, \tag{1}$$

where

$$\Delta^{\alpha} z_i=\sum_{j=0}^{i}(-1)^j\binom{\alpha}{j} z_{i-j}, \binom{\alpha}{j}=\frac{\Gamma(\alpha+1)}{\Gamma(j+1)\Gamma(\alpha-j+1)},$$
$$-1/2<\alpha<1/2, \Gamma(\alpha)=\int_0^{\infty} e^{-t}t^{\alpha-1}dt.$$

Suppose the conditions are met:

A1. The dynamic system is asymptotically stable. The system's true parameters belong to a compact set $\tilde{\mathrm{B}}$.
A2. Noise sequences $\{\xi_i\},\{\zeta_i\},\{\varsigma_i\}$ are zero mean white noise signals.

Fig. 1 FARARX model with errors in variables

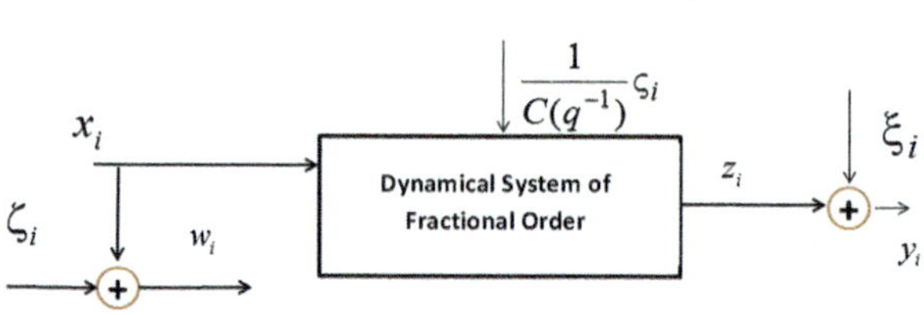

A3. Noise sequences $\{\zeta_i\}$,$\{\xi_i\}$,$\{\varsigma_i\}$ are statistically independent of $\{x_i\}$.
A4. A priori, relations $\gamma_\zeta = \sigma_\zeta^2 / \sigma_\xi^2$, $\gamma_\varsigma = \sigma_\varsigma^2 / \sigma_\xi^2$ are known.
A5. The input signal x_i is random and satisfies the condition of continuous excitation of $r_1 + r_2$ order.

It is required to find estimates of the coefficients of the system $\hat{a}^{(m)}$, $\hat{b}^{(m)}$, $\hat{c}^{(m)}$ described with Eq. (1) based on observations y_i,w_i under known orders r, r_1, r_2, α should be defined.

3 Parametric Identification Algorithm

The three-stage identification algorithm was proposed.

First stage (definition of consistent estimates of extended FARX system parameters). Equation (1) may be represented in the operator form:

$$\left(1 - q^{-1}\right)^\alpha \left(B(q^{-1})z_i\right) = A(q^{-1})x_i + \frac{1}{C(q^{-1})}\varsigma_i, y_i = z_i + \xi_i, w_i = x_i + \zeta_i,$$

where $B(q^{-1}) = 1 + \sum_{m=1}^{r} b_0^{(m)} q^{-m}$, $A(q^{-1}) = \sum_{m=0}^{r_1} a_0^{(m)} q^{-m}$, $C(q^{-1}) = 1 + \sum_{m=1}^{r_2} c_0^{(m)} q^{-m}$, is a backward shift operator $q^{-1}z_i = z_{i-1}$.

After both sides of the equation are multiplied by $C(q^{-1})$, the model will take the form:

$$\left(1 - q^{-1}\right)^\alpha \left(\bar{B}(q^{-1})z_i\right) = \bar{A}(q^{-1})x_i + \varsigma_i, y_i = z_i + \xi_i, w_i = x_i + \zeta_i,$$

where $\bar{A}(q^{-1}) = A(q^{-1})C(q^{-1})$, $\bar{B}(q^{-1}) = B(q^{-1})C(q^{-1})$.

The model will be saved in the time domain

$$\Delta^\alpha \left(z_i + \sum_{m=1}^{r+r_2} \bar{b}_0^{(m)} z_{i-m} \right) = \sum_{m=0}^{r_1+r_2} \bar{a}_0^{(m)} x_{i-m} + \varsigma_i, y_i = z_i + \xi_i, w_i = x_i + \zeta_i. \quad (2)$$

We will define the estimates of extended ARX model based on the condition of minimum of weighted least squares criterion [11]:

$$\min_{\theta \in \bar{B}} \sum_{i=1}^{N} \frac{\left(\Delta^\alpha y_i - \bar{\varphi}_i^T \bar{\theta}\right)^2}{h_\alpha(0) + \gamma_\varsigma + \bar{b}^T H_\alpha \bar{b} - 2\tilde{h}_\alpha \bar{b} + \gamma_\zeta \bar{a}^T \bar{a}} \quad (3)$$

where $\bar{\theta}=\left(\bar{b}^{(1)}\ldots\bar{b}^{(r+r_2)}\middle|\bar{a}^{(0)}\ldots\bar{a}^{(r_1+r_2)}\right)^T$, $\bar{\varphi}^{(i)}=\left(\bar{\varphi}_y^{(i)}\middle|\bar{\varphi}_x^{(i)}\right)^T$, $\varphi_x^{(i)}=(x_i,\ldots x_{i-r_1-r_2})^T$,

$\varphi_y^{(i)}=\left(\Delta^\alpha y_{i-j-1},\ldots,\Delta^\alpha y_{i-j-r-r_2}\right)^T$, $H_\alpha=\begin{pmatrix} h_\alpha(0) & \cdots & h_\alpha(r+r_2-1)\\ \vdots & \ddots & \vdots \\ h_\alpha(r+r_2-1) & \cdots & h_\alpha(0)\end{pmatrix}$,

$h_\alpha(m)=\lim_{N\to\infty}\frac{1}{N}\sum_{j=0}^{N-1}\binom{\alpha}{j+m}\binom{\alpha}{j}\frac{N-j}{N}$, $\tilde{h}_\alpha=(h_\alpha(1),\ldots,h_\alpha(r+r_2))$.

Theorem *Suppose the dynamic system is described by* Eq. (1) *with initial zero conditions and assumptions* ***A1–A5*** *are fulfilled. Then, the estimate of coefficients* $\hat{\hat{\theta}}(N)$ *that is defined by expression* (3) *exists. It is unique, and it converges to the true value of coefficients with a probability of 1, i.e.,* $\hat{\hat{\theta}}(N)\xrightarrow[N\to\infty]{a.s.}\bar{\theta}_0$.

Second stage. An estimate of model coefficients $\hat{a}^{(m)}$, $\hat{b}^{(m)}$ [14] based on coefficients of extended ARX model. For true values of coefficients, we have the equality:

$$\bar{A}(q^{-1})B(q^{-1})-A(q^{-1})\bar{B}(q^{-1})=0. \tag{4}$$

Equation (4) may be represented in a matrix form:

$$S^T\begin{pmatrix}1\\ \overline{\theta_0}\end{pmatrix}=0,$$

where $\theta_0=\left(b_0^{(1)}\ldots b_0^{(r)}\middle|a_0^{(0)}\ldots a_0^{(r_1)}\right)^T$,

S—Silvester matrix:

$$S=\left(\begin{array}{ccccccc}
\bar{a}^{(0)} & \bar{a}^{(1)} & \cdots & \bar{a}^{(r_1+r_2)} & 0 & \cdots & 0\\
0 & \bar{a}^{(0)} & \bar{a}^{(1)} & \cdots & \bar{a}^{(r_1+r_2)} & \cdots & 0\\
\vdots & & \ddots & \ddots & & \ddots & \vdots\\
0 & \cdots & 0 & \bar{a}^{(0)} & \bar{a}^{(1)} & \cdots & \bar{a}^{(r_1+r_2)}\\
\hline
-1 & -\bar{b}^{(1)} & & & 0 & \cdots & 0\\
0 & -1 & -\bar{b}^{(1)} & & -\bar{b}^{(r+r_2)} & \cdots & 0\\
\vdots & & \ddots & \ddots & & \ddots & \vdots\\
0 & \cdots & 0 & -1 & -\bar{b}^{(1)} & \cdots & -\bar{b}^{(r+r_2)}
\end{array}\right).$$

Representing the S^T matrix in the form of $S^T=\left(m\middle|M\right)$ block matrix, we can write:

$$m + M\theta_0 = 0. \tag{5}$$

Then, the evaluation of $\hat{\theta}$ parameter vector may be derived from (5) with a least squares method:

$$\hat{\theta} = \left(\hat{M}^T\hat{M}\right)^{-1}\hat{M}\hat{m}.$$

Third stage. Estimation of noise coefficients $\hat{c}^{(m)}$. In a matrix form, $\bar{A}(q^{-1})$, $\bar{B}(q^{-1})$ polynomials may be represented as:method:

$$\begin{pmatrix} 1 \\ \bar{\theta} \end{pmatrix} = G\tilde{c},$$

where $\tilde{c} = \left(1 \,\middle|\, c_0^{(1)}, \ldots, c_0^{(r_2)}\right)^T$,

$$G = \left(\begin{array}{ccccccc|cccccc} 1 & b^{(1)} & \ldots & b^{(r)} & 0 & \ldots & 0 & a^{(0)} & \ldots & a^{(r_1)} & 0 & \ldots & 0 \\ 0 & 1 & b^{(1)} & & b^{(r)} & & 0 & 0 & a^{(0)} & \ldots & a^{(r_1)} & & 0 \\ \vdots & & \ddots & \ddots & \vdots & \ddots & \vdots & \vdots & & \ddots & \vdots & \ddots & \vdots \\ 0 & \ldots & 0 & 1 & b^{(1)} & \ldots & b^{(r)} & 0 & \ldots & 0 & a^{(0)} & \ldots & a^{(r_1)} \end{array}\right).$$

Using evaluations of extended ARX model coefficients determined on the first stage as well as model coefficient evaluations $\hat{a}^{(m)}$, $\hat{b}^{(m)}$, the coefficient evaluation $\hat{c}^{(m)}$ may be determined using the least squares method:

$$\hat{\tilde{c}} = \left(\hat{G}^T\hat{G}\right)^{-1}\hat{G}\begin{pmatrix} 1 \\ \hat{\bar{\theta}} \end{pmatrix}.$$

The numerical experiments conducted have proven the high accuracy of determined estimates, as compared with the least squares method.

4 Generalization in Case of Gegenbauer Difference

The resulted algorithm may be used to identify parameters of models with the Gegenbauer difference [15]:method:

$$\begin{gathered} \nabla_v^\alpha\left(z_i + \sum_{m=1}^{r} b_0^{(m)} z_{i-m}\right) = \sum_{m=0}^{r_1} a_0^{(m)} x_{i-m} + \varphi_i, \\ y_i = z_i + \xi_i, w_i = x_i + \zeta_i, \varphi_i = \sum_{m=1}^{r_2} c_0^{(m)} \varphi_{i-m} + \varsigma_i. \end{gathered} \tag{6}$$

The Gegenbauer difference ∇_v^α is a generalization of the fractional-order difference, and it is equal to:

$$\nabla_v^\alpha = \left(1 - 2vq^{-1} + q^{-2}\right)^\alpha,$$

where $0 < \alpha$, $0 < v \leq 1$.

The difference ∇_v^α may be represented in the form

$$\nabla_v^\alpha z_i = \sum_{j=0}^{i} C_j^\alpha(v) z_{i-j},\ C_j^\alpha(v) = \sum_{k=0}^{[j/2]} (-1)^k \Gamma(\alpha + j - k) \frac{(2v)^{j-2k}}{\Gamma(\alpha)\Gamma(k+1)\Gamma(j-2k+1)}.$$

5 Simulation Results

The proposed algorithm has been realized in MATLAB and compared with the least squares (LS) method. The dynamical system is defined by the equation

$$\Delta^{0.5}(z_i - 0.5z_{i-1} - 0.2z_{i-1}) = x_i + 0.5x_{i-1},$$

The noise-free input is defined as

$$x_i + 0.5x_{i-1} = e_i - 0.2 \cdot e_{i-1} - 0.75 \cdot e_{i-2} + e_{i-4},$$

where e_i is the white noise.

Test cases were compared by the normalized root-mean-square error of parameter estimation, defined as

$$\delta\theta = \sqrt{\left\|\hat{\theta} - \theta_0\right\|^2 \Big/ \|\theta_0\|^2} \cdot 100\,\%,$$

The number of data points is N = 1000 (Table 1).

Table 1 Normalized root-mean-square error of parameter estimation

σ_ξ/σ_z	$\sigma_\varsigma/\sigma_z$	σ_ζ/σ_x	$\delta\theta$, %	$\delta\theta_{LS}$, %
0.05	0.5	0.05	3.82	9.47
0.1	0.5	0.1	9.77	21.42
0.1	0.5	0.2	13.32	43.41

6 Conclusion

The identification algorithm, as proposed in this work, may be used in simulation and forecasting of processes, described with equations with fractional-order differences. Further generalization of obtained results for the case of the model unknown structure [16] is supposed.

References

1. Jesus, I., Barbosa, R., Machado, J., Cunha J.: Strategies for the control of heat diffusion systems based on fractional calculus. In: Proceedings of the IEEE International Conference on Computational Cybernetics (ICCC '06), Budapest, Hungary (2006)
2. Stiassnie, M.: On the application of fractional calculus for the formulation of viscoelastic models. Appl. Math. Model. **3**(4), 300–302 (1979)
3. Bagley, R., Torvik, P.: Fractional calculus—a different approach to the analysis of viscoelastically damped structures. AIAA J. **21**(5), 741–748 (1983)
4. Holm, S., Näsholm, S.: A causal and fractional all-frequency wave equation for lossy media. J. Acoust. Soc. Am. **130**(4), 2195–2201 (2011)
5. Silva, M., Machado, J., Jesus, I.: Modelling and simulation of walking robots with 3 dof legs. In: Proceedings of the 25th IASTED International Conference on Modelling, Identification and Control (MIC '06), pp. 271–276, Lanzarote, Spain (2006)
6. Valério, D., Costa, J.: Tuning of fractional PID controllers with Ziegler-Nichols-type rules. Signal Process. **86**(10), 2771–2784 (2006)
7. Ivanov, D.V., Katsyuba, O.A., Dubinin, A.E.: Identification of models of long lines on the basis of adaptive filters with fractional-order differences. Russ. Electr. Eng. **88**(3), 120–122 (2017)
8. Granger, C.W., Joyeux, R.: An introduction to long-range time series models and fractional differencing. J. Time Ser. Anal. **1**, 15–30 (1980)
9. Tejado, I., Valério, D., Valério, N.: Fractional calculus in economic growth modeling. The Portuguese case. In: Proceedings of the 2014 International Conference on Fractional Differentiation and Its Applications (ICFDA 2014), Catania, Italy (2014)
10. Chetoui, M., Thomassin, M., Malti, R., Aoun, M., Najar, S., Abdelkrim, M., Oustaloup, A.: New consistent methods for order and coefficient estimation of continuous-time errors-in-variables fractional models. Comput. Math Appl. **66**(5), 860–872 (2013)
11. Ivanov, D.V.: Identification discrete fractional order linear dynamic systems with errors-in-variables. In: Proceedings of IEEE East-West Design & Test Symposium (EWDTS'13), pp. 374–377. Rostov-on-Don. Russia (2013)
12. Ivanov, D.V, Ivanov, A.V.: Identification Fractional Linear Dynamic Systems with fractional errors-in-variables. J. Phys.: Conf. Ser. **803**(1) (2017)
13. Diversi, R., Guidorzi, R., Soverini, U.: Identification of ARX and ARARX models in the presence of input and output noises. Eur. J. Control **16**(3), 242–255 (2010)
14. Stoica, P., Söderström, T.: Common factor detection and estimation. Automatica **33**(1), 985–989 (1996)
15. Guegan, D.: A prospective study of the k-factor Gegenbauer processes with heteroscedastic errors and an application to inflation rates. Finance India **17**(1), 165–197 (2003)
16. Engelgardt, V.V., Ivanov, D.V., Katsyuba, O.A.: Structural and parametric identification of linear dynamic systems of fractional order with noise on input and output. In: Proceedings of International Siberian Conference on the Control and Communications (SIBCON'17), IEEE, Astana, Kazakhstan (2017)

A Proposed Authentication Approach Based on Voice and Fuzzy Logic

Alia Karim Abdul-Hassan and Iman Hassoon Hadi

Abstract Conventional authentication methods which depend on what the user knows that is usually a password do not guarantee sufficient protection to the information system, because they do not depend on the identity of the authenticated user. The proposed intelligent authentication makes use of digital identity which relies on the concept "who is the user," and this identity consists of user biometric (voice) attributes, like Mel-frequency cepstral coefficient (MFCC). The proposed authentication approach has two options for feature matching; the first option uses dynamic time warping (DTW) as crisp matching to find the distance between two samples of voice to recognize the identity of an authenticated user (speaker) using MFCC features. The second option uses the fuzzy inner product between fuzzy feature vectors to compute the approaching values between two voice samples using the same features. The experiments are implemented using (ELSDSR) data set, in order to recognize the identity of claimed user. The accuracy of recognition using MFCC with fuzzy vectors was better than recognition using MFCC with dynamic time warping (DTW), and the accuracy of the recognition was 95.45% when using fuzzy logic in MFCC features matching technique.

Keywords Authentication · Speaker · Identity · Voice · MFCC · Fuzzy vector · DTW

1 Introduction

One of the most critical tasks of information security is developing an authentication method based on the digitization of identity that is required to access these systems to an acceptable level of trust. The identity refers to a group of well-defined properties that make an entity recognized compared to other entities [1], while digital identity is a set of features owned by an entity used by information systems

A. K. Abdul-Hassan (✉) · I. H. Hadi
Computer Sciences Department, University of Technology—Iraq, Baghdad, Iraq
e-mail: 110018@uotechnology.edu.iq

V. Jain et al. (eds.), *Recent Trends in Intelligent Computing, Communication and Devices*, Advances in Intelligent Systems and Computing 1031,
https://doi.org/10.1007/978-981-13-9406-5_60

to represent an identity (individual, organization, application, or device). Voice is one of such features that could be digitized to recognize between users.

1.1 Related Work

The authentication process must depends on unique user attributes, which discriminate the authenticated user. Biometric attributes have the uniqueness feature like voice attributes, so many researches proposed methods that depend on biometric attributes, like [2] proposed authentication method using the characteristic of the voice by extracting features called Mel Frequency Cepstral Coefficients (MFCC) and matching these features using Vector Quantization (VQ), this method modified with a predefined threshold and the true acceptance using trained data is around 86%, but this proposed system influence by the spoken word, because the difference in the spoken words in the trained sample and verification sample will reduce the system's accuracy. [3] Proposed speaker identification based on (MFCC) features and Support Vector Machine (SVM), where SVM classifies the speaker with accuracy 92%. [4] Proposed speaker recognition model using MFCC features and Dynamic Time Warping (DTW) to compare between these feature patterns. [5] Used MFCC and Shifted MFCC and Vector Quantization and fuzzy logic for feature matching, the accuracy of using MFCC using Fuzzy around 83% and around 86% of shifted MFCC using Fuzzy. [6] Used short term spectral features learnt from the Deep Belief Networks (DBN) augmented with MFCC features to perform the task of speaker recognition, they achieved a recognition accuracy of 95% when using MFCC features extraction on the ELSDSR dataset. It is clear from this review, there was no research used the fuzzy inner product as MFCC features matching method as proposed in this paper. This matching method is proposed for speaker authentication process as will be explained in the following sections.

1.2 Voice as Biometric Recognition Feature

Voice biometric feature has a uniqueness attribute that can be used efficiently to discriminate between identities in the recognition process. Another reason for using voice as a biometric feature is it is not an intrusive feature [7]. Nevertheless, the variability aspect may be due to illnesses, samples are quite repeatable, and one or at most a handful of samples would be enough for acceptable recognition [8]. To overcome the variability in voice, it is possible to create a monitoring technique to keep track of the changes in voice during the use of the authentication system.

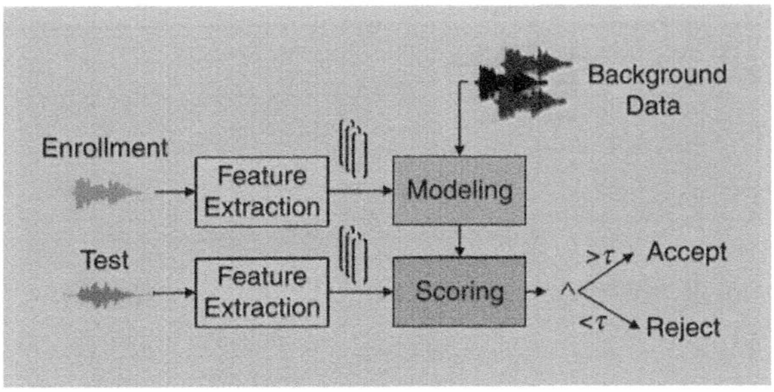

Fig. 1 General speaker recognition system

2 Proposed Speaker Recognition Methodology

Speaker recognition is a powerful tool for verifying identity in many applications. Speaker recognition may work on the user voice sample that is text dependent or text independent. This is more suitable in authentication systems—where a claimed user says specific phrase, such as a password or personal identification number, to be authenticated to access to the information system. In the proposed intelligent authentication, a claimed speaker claims an identity, and the main task is to verify if this identity is true. This is done by comparing his voice sample with a set of authenticated speaker samples and deciding if the claimed speaker is authenticated [9]. The data used in the proposed speaker recognition system are divided into two kinds: design (or training) data and test data; train samples are labeled (the speaker identification to which this sample belongs). Test data are samples of voice that belongs to authenticated speakers which are labeled to testing the overall performance of the recognition process.

Figure 1 shows the general speaker recognition system [9]. MFCC features are extracted from the voice samples that are designed to represent the characteristics of the speaker's voice. For the claimed speaker voice sample, the same features are extracted and are compared against the features of other speakers. The comparison score (a scalar value) indicates whether the two voice samples refer to the same speaker. If this score is higher (or lower) than a predefined threshold, then the system accepts (or rejects) the test speaker.

2.1 *Short-Term Features (MFCC) Extraction*

The most popular short-term acoustic features are the Mel-frequency cepstral coefficients (MFCCs), these features are extracted from short voice frames of duration within 20–25 ms [9], this extraction process mimics the human hearing system (see Fig. 2), and the following steps are used to compute MFCCs:

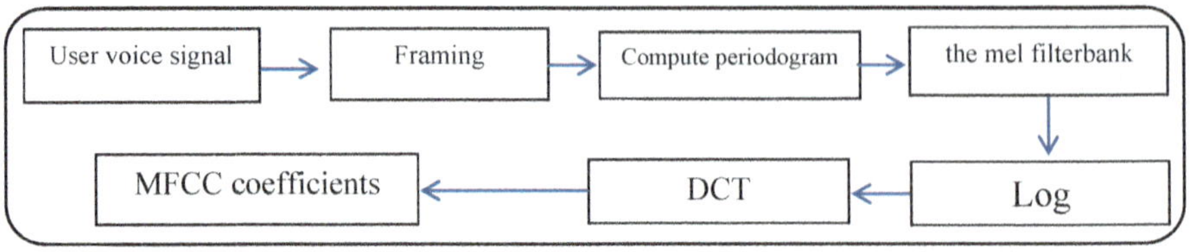

Fig. 2 MFCC extraction diagram

1. Segment the signal to small frames.
2. For each frame, compute the periodogram estimate of the power spectrum.
3. Apply the Mel filterbank to the power spectra, and sum the energy in each filter.
4. Compute logarithm of all filterbank energies.
5. Compute the DCT of log filterbank energies.
6. Keep DCT coefficients 1–13, and discard the rest.

2.2 *MFCC Features Matching*

2.2.1 Crisp Matching Using Dynamic Time Warping

Dynamic time warping is for measuring similarity between time series that vary in length, by calculating all possible distances between them and choosing the minimum distance. The warping between two series can then be used to determine the similarity between the two time series [4]. The main task of DTW is to compare two sets $X = (x1, x2, \ldots, xn)$ of length $n \in \mathrm{N}$ and $Y = (y1, y2, \ldots, ym)$ of length $n \in \mathrm{N}$. A feature space is denoted by F, and let $xn, ym \in \mathrm{F}$ for $n \in [1\!: \mathrm{N}]$ and $m \in [1\!: \mathrm{M}]$. To find the similarity of two different features $x, y \in \mathrm{F}$, one needs a local cost measure (local distance measure), which is defined to be a function [10]:

$$c : \mathrm{F} \times \mathrm{F} \rightarrow \mathrm{R} \tag{1}$$

$c(x, y)$ is small (minimum cost) if x and y are similar, and otherwise $c(x, y)$ is large (maximum cost). The matching between the claimed user MFCC feature vector and the feature vectors of authenticated users in the data set will be done using a DTW algorithm to compute distances between them. The minimum distance refers to recognition score if it is under a predefined threshold (100) which is specified due to experiments, and then it is authenticated.

2.2.2 Fuzzy Matching Using Fuzzy Vector

Fuzzy logic is based on the fuzzy set theory that is approximate rather than classical predicate logic. Fuzzy truth represents membership in defined sets, similar to the likelihood of some event or condition, and fuzzy sets are based on vague definitions of sets, not randomness [11]. The two categories of voice sample data (training and test) as generated from speaker voice feature extraction are too close values, so fuzzification of these feature vectors can enhance recognition performance.

Features can be represented as fuzzy sets, let A typical pattern features represented as fuzzy sets *Ai* on *X*(i = 1, 2, …, m) as a group of known features patterns and *B*, a new unknown feature pattern. The recognition process goal is to find which element in *Ai* the sample *B* most matches. To achieve this issue, the technique of fuzzy vectors is used.

If we have a vector *a*, where all elements *ai* satisfy the condition, 0 =< *ai* => 1 for i = 1, 2, …, n, then *a* is a fuzzy vector. Usually, this could be done by fuzzy membership function; in our work, we choose Gaussian membership function. One of the most important operations on fuzzy vectors that are used in pattern recognition is the fuzzy inner product. Let us define *a* and *b* as fuzzy vectors of length *n*, then the fuzzy inner product as follows [12]:

$$\bigwedge_{i=1}^{n} (ai \wedge bi) \tag{2}$$

And fuzzy outer product:

$$\bigwedge_{i=1}^{n} (ai \vee bi) \tag{3}$$

If two separate fuzzy vectors are identical, $a = b$, the inner product reaches a maximum while the outer product reaches a minimum. These operations are very useful when used in a metric of similarity between two vectors [12]. The inner product of two fuzzy vectors could compute using Gaussian membership function as follows:

Let $X = [-\infty,\infty]$, a one-dimensional universe on the real line, *A* and *B* are two fuzzy sets having normal, and Gaussian membership is defined mathematically as:

$$\begin{aligned} \mu A(x) &= \exp\left[-(x-a)^\wedge 2/\sigma_a^2\right] \\ \mu B(x) &= \exp\left[-(x-b)^\wedge 2/\sigma_b^2\right] \end{aligned} \tag{4}$$

where σ is standard deviation.

As shown in Fig. 3, the inner product of *A* and *B* could be computed as follows:

Fig. 3 Fuzzy inner product of A and B

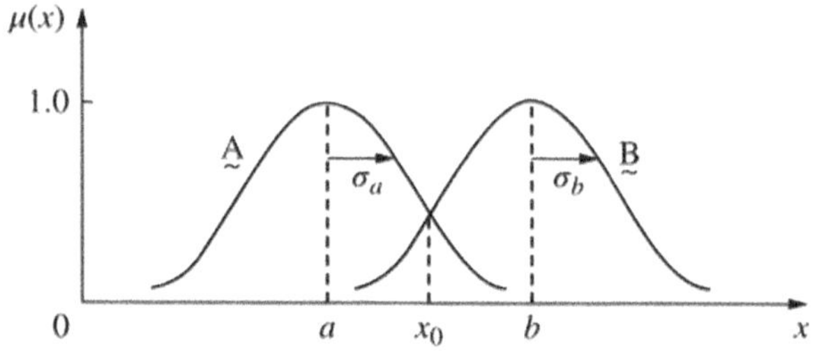

$$= \exp\left[-(a-b)^\wedge 2/(\sigma a+\sigma b)^\wedge 2\right] = \mu A(x0) = \mu B(x0) \tag{5}$$

The fuzzy inner product can be used to compare two voice features vector. This inner product is repeated between test features vector and a group of known features patterns and then select the pair with the largest approaching degree value. The recognized pattern with the maximum approaching degree value is the pattern that the data sample most closely to the unknown pattern. This concept is defined as the maximum approaching degree [12].

The proposed algorithm (Speaker Identification using Fuzzy Logic—SIFL) includes the main steps to show how to identify the identity of the speaker for authentication using MFCC feature vectors and matching process using inner product of fuzzy vectors (see Eq. 2) with Gaussian membership function (see algorithm SIFL).

Algorithm SIFL

Input: The claimed speaker (speaker1) voice signal and the authenticated speaker voice signals (speaker2), data set
Output: The speaker recognized identity (id)

Step1: Compute the MFCC array of speaker, and MFCC1(T1,13) //T1 represents length of voice signal and 13 number of MFCCs
Step2: Repeat for each speaker voice signal (speaker2) in data set of authenticated user
Step3: Compute MFCC array for speaker2, MFCC2(T2, 13)
Step4: Minimize MFCC1 and MFCC2, by computing mean value of each MFCC to get speaker1_vector(13), speaker2_vector(13)
Step5: Compute fuzzy inner product between speaker1_vector and speaker2_vector
Step6: Append the fuzzy inner product value of corresponding speaker2 to recognition_test_list
Step7: Until the last speaker in data set
Step8: Select the identity number (id) corresponding to the maximum value in recognition_test_list
Return the speaker recognized identity (id)

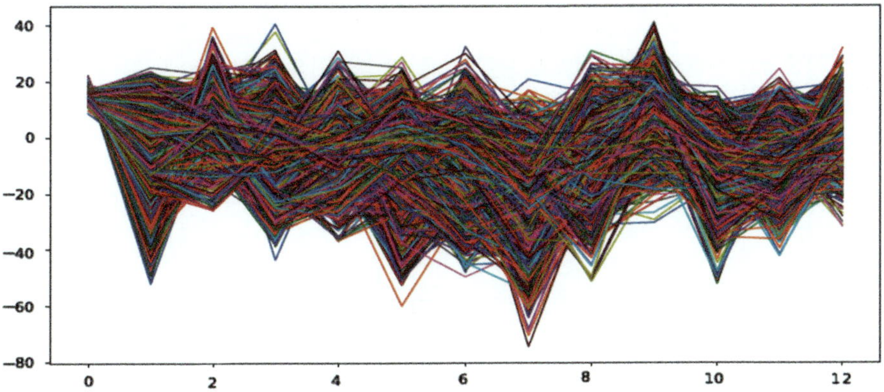

Fig. 4 MFCC features of one speaker

3 Experiments and Result

The main implementation was done using English language speech database for speaker recognition (ELSDSR) which consists of 7 audio file samples for each speaker; the total number of speakers is 22 volunteers. The text language is English [13]. The voice samples recorded into file type (.wav). The sampling frequency is chosen 16 kHz with a bit rate of 16.

The wave file of the claimed user voice is loaded using python language library. Then, the MFCC feature matrix is extracted as explained in Sect. 2.1 (see Fig. 4) where y-axis represents the value of MFCCs that result from feature extraction and x-axis represents the sequence of 13 MFCCs (from 0 to 12).

In the training phase, the MFCC feature matrices of the selected audio files that are stored in this data set (authenticated users) are extracted and normalized by calculated mean of MFCCs (see Table 1).

3.1 Feature Matching Using DTW

In the test phase, the MFCC matrix of the unknown user audio file is extracted and matched with all 22 users using DTW. If the file is identical with any audio file, the path in the cost matrix of the DTW process will be a straight line as shown in Fig. 5a, while different files (another speaker audio file or the same speaker, but with a different sentence) can be seen in Fig. 5b.

The distance values between the claimed user MFCC feature matrix and all MFCC features of audio files of the authenticated users in the data set will be stored in recognition list. The identity of the claimed user, if he is authenticated, will be the identity number which belongs to the minimum distance value that results from the

Table 1 MFCC features of 22 speakers

Id	MFCC1	MFCC2	MFCC3	MFCC4	MFCC5	MFCC6	MFCC7	MFCC8	MFCC9	MFCC10	MFCC11	MFCC12	MFCC13
1	2.22	15.98	11.18	13.86	11.31	15.57	14.34	15.51	14.39	11.16	11.01	11.81	9.50
2	3.91	15.75	12.75	11.79	12.79	12.83	13.31	12.35	12.43	9.49	11.72	10.68	8.75
3	2.97	14.60	10.69	14.41	10.44	16.01	12.28	12.15	14.27	10.66	11.21	11.08	9.44
4	3.39	16.16	13.31	10.11	10.95	12.31	12.11	13.58	14.58	11.09	11.60	13.60	11.34
5	2.24	15.63	13.55	15.90	12.16	14.32	13.58	14.72	12.26	12.30	12.29	10.81	11.48
6	2.49	14.49	13.67	11.35	14.96	12.91	15.06	18.62	12.46	10.89	11.26	11.92	11.28
7	2.27	15.20	11.23	11.54	9.98	13.29	12.62	12.12	12.72	10.54	10.94	13.03	8.70
8	3.40	15.79	11.21	10.92	11.44	13.43	12.52	15.86	12.62	11.96	10.06	11.59	8.75
9	3.58	16.05	11.08	10.49	10.37	14.83	10.86	14.46	12.37	11.45	11.74	11.57	10.01
10	2.64	14.48	11.82	13.74	12.14	12.23	14.99	15.99	14.04	14.08	10.17	12.60	13.08
11	2.00	15.29	8.95	10.54	10.93	13.05	13.43	13.13	15.92	10.35	12.00	11.33	11.38
12	2.70	15.08	11.08	13.43	9.96	14.38	11.15	16.42	15.01	11.61	10.14	10.32	8.60
13	2.52	13.31	11.65	12.99	13.08	13.07	14.56	16.90	16.55	14.30	11.81	12.76	12.88
14	2.65	15.47	12.63	13.79	10.43	16.56	13.80	14.41	15.65	11.05	11.13	12.67	12.35
15	2.54	16.35	11.48	13.53	10.76	14.20	14.49	15.89	18.43	13.89	13.38	11.22	11.09
16	2.49	16.54	10.10	10.42	11.48	12.83	12.02	12.72	13.88	12.82	9.69	11.94	9.99
17	3.40	15.05	10.72	13.39	13.88	11.86	14.15	16.23	14.32	14.41	11.72	10.75	8.92
18	3.03	14.90	9.81	11.75	11.53	10.40	14.03	16.29	14.57	12.02	10.13	14.37	10.11
19	3.83	16.19	9.12	12.64	10.36	10.94	12.99	15.82	12.86	12.11	10.71	13.12	10.00
20	3.41	14.31	12.34	16.75	11.92	15.05	14.62	16.19	13.15	13.34	10.35	12.30	10.95
21	2.19	13.95	11.34	11.70	9.63	10.33	13.57	15.34	13.77	14.51	9.80	10.12	12.55
22	5.24	16.73	11.79	6.60	8.55	12.68	7.011	5.88	10.91	5.76	10.01	6.76	5.30

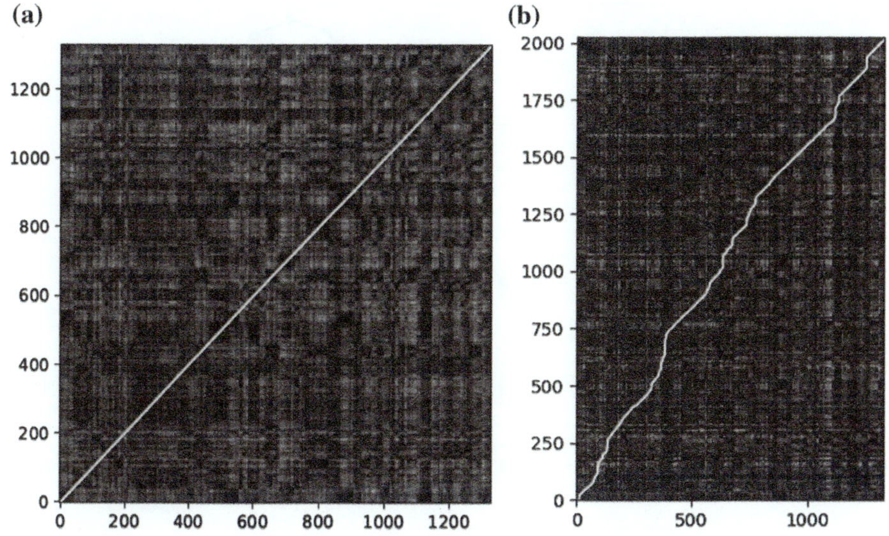

Fig. 5 **a** DTW path of identical audio files and **b** DTW path of different audio files

Table 2 DTW matching

Id	MFCC_DTW	Notes
1	0.0	Minimum matching score with speaker1 identical audio file
2	85.9	Matching score with speaker1 different audio file
3	69.0	Matching score with speaker2 different audio file
4	64.5	Matching score with speaker3 audio file
5	80.3	Matching score with speaker4 audio file

comparison process (see row 1 in Table 2); as a result, the identity value is (1) (the authenticated user is speaker 1), which is correct, and this is because the two audio files contain the same spoken words in train and test voice samples. The recognition process using DTW is not always accurate, especially if the two audio files of the same speaker have different words as shown in row 2 in Table 2 where the distance was 85.9 which is bigger than distance (64.5) in row 4 in Table 2 related to speaker 3 feature matching. So, DTW is accurate when the feature matching is done with the same spoken words in both the training and the testing voice samples, while it is less accurate with different samples.

3.2 *Fuzzy Matching*

Since the DTW matching is not always accurate, it is useful to use fuzzy matching to get the accurate recognition result. Figures 6 and 7 show the fuzzification of

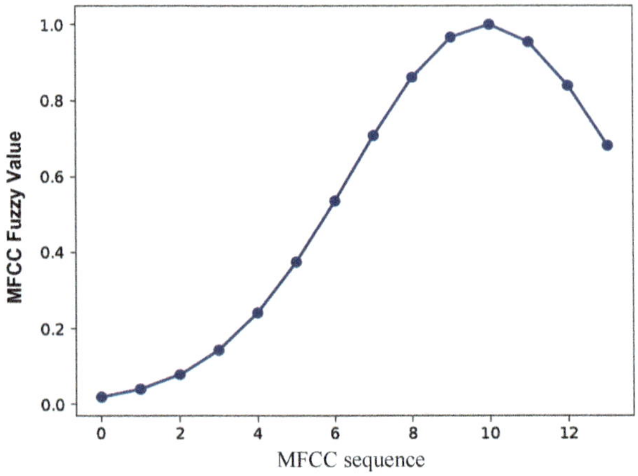

Fig. 6 Fuzzification of one speaker MFCC feature vector

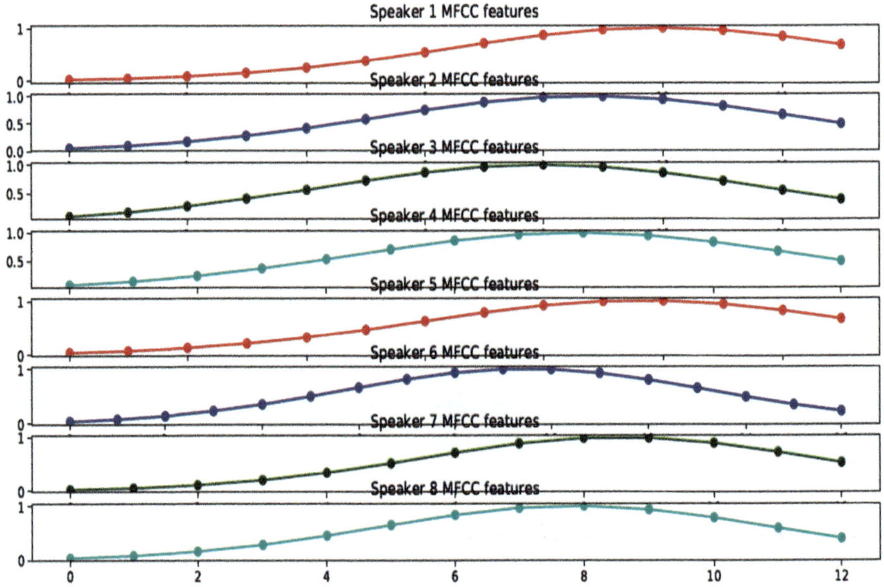

Fig. 7 Fuzzy membership functions of MFCC features of eight speakers

Table 3 MFCC feature vector fuzzy matching

Speaker Id	1	1	2	3	4	5	6	7	8	9	10	11	12	13	14	15	16	17	18	19	20	21	22
Fuzzy MFCC matching	13	13	12	12	12	2	12	12	12	12	11	12	12	12	12	12	12	12	12	12	12	11	11

Table 4 MFCC fuzzy matching result

Authenticated speaker identity	Recognized identity	Fuzzy matching
1	1	13
		12.7
2	2	13
		12.91
3	3	13
		12.86
4	4	13
		12.78
5	5	13
		12.75
6	6	13
		12.79
7	7	13
		12.82
8	8	13
		12.88
9	9	13
		12.95
10	10	13
		12.83
11	11	13
		12.76
12	12	13
		12.83
13	13	13
		12.81
14	14	13
		12.70
15	15	13
		12.83
16	16	13
		12.71
17	17	13
		12.71
18	18	13
	22	12.37
19	19	13
		12.75
20	20	13
		12.85
21	21	13
		12.79
22	22	13
		12.72

MFCC vectors where x-axis represents the sequence of MFCC sequence and y-axis represents the fuzzy value of MFCC value result from fuzzy membership function (see Eq. 4). Table 3 shows the fuzzy matching of MFCC vectors using fuzzy inner product operation, to recognize the identity of speaker 1 regardless of spoken words as shown in rows 1 and 2 in Table 3 which represent the inner products of the speaker 1 voice sample files, while the rows from 3 to 20 in Table 3 show the values of the fuzzy vector inner product between speaker 1 and other speakers' voice files, where the maximum value is the criteria for recognition of the authenticated speaker.

In Table 4, test stage is implemented to compute the performance of the speaker recognition process, where each user is recognized by matching two voice samples, as shown in the third column (fuzzy matching) in Table 4 which holds two values (two sub-rows); first sub-row is the result of matching of two identical voice samples, and the second sub-row is the result of matching of different spoken words that belong to voice samples of the same speaker. Most recognized identities (column 2 in Table 4) of authenticated speakers in data set were true identities, except speaker 18 where the matching result between his voice samples with the 22nd speaker voice sample was too close as shown in row 18 in Table 4.

Speaker recognition performance metric from Table 4 is as follows:

$$\text{TrueIdentity rate} = 21/22 * 100 = 95.45\%$$
$$\text{FalseIdentity rate} = 1/22 * 100 = 4.54\%$$

As a result, the accuracy of the proposed approach was 95.45% using MFCC as voice features and the inner product of fuzzy vectors as features matching.

4 Conclusion

The proposed authentication approach depends on voice attributes (MFCC) to recognize the identity of authenticated user using two options for feature matching; the first option was crisp matching using the DTW, and the second option was fuzzy matching using the inner product of fuzzy vectors. The matching score results using DTW of two MFCC metrics features influenced by the spoken words. The two voice samples must have the same words to get accurate recognition rate. MFCC features matching using fuzzy vector matching is more accurate than DTW matching since it did not require the identical spoken words in the matching process between the test voice sample and the labeled voice sample of the same speaker.

5 Future Work

It is important to use other types of voice features for authentication to enhance recognition process, so we planned to use long-term voice features like fundamental frequency and zero-crossing rate with another data set.

References

1. Ayed, G.B.: Architecting User-Centric Privacy-as-a-Set-of-Services: Digital Identity-Related Privacy Framework. Springer (2014)
2. Sadewa, R.A., Wirayuda, T.A.B., Sa'Adah, S.: Speaker recognition implementation for authentication using filtered MFCC—VQ and a thresholding method. In: 3rd International Conference on Information and Communication Technology, ICoICT (2015)
3. Kumar Yadav, R., Pandit, S., Singh, P., Arya, P., Prasad, R.S.: Speaker recognition using MFCC and support vector machine. Imp. J. Interdiscip. Res. (IJIR) **3** (2017)
4. Muda, L., Begam, M., Elamvazuthi, I.: Voice Recognition Algorithms using Mel Frequency Cepstral Coefficient (MFCC) and Dynamic Time Warping (DTW) Techniques (2010)
5. Bansal, P., Imam, S.A., Bharti, R.: Speaker recognition using MFCC, shifted MFCC with vector quantization and fuzzy. In: International Conference on Soft Computing Techniques and Implementations, ICSCTI 2015, pp. 41–44 (2015)
6. Banerjee, A., Dubey, A., Menon, A., Nanda, S., Chand Nandi, G.: Speaker Recognition using Deep Belief Networks to CCIS Proceedings. Robotics and Artificial Intelligence Laborotary, Indian Institute of Information Technology, Allahabad, India
7. Karray, F.O., De Silva, C.: Soft Computing and Intelligent Systems Design: Theory, Tools, and Applications (2004)
8. Beigi, H.: Fundamentals of Speaker Recognition (2011)
9. Hansen, J.H.L., T. Hasan, T.: Speaker recognition by machines and humans. IEEE Signal Process. Mag. (2015)
10. Müller, M.: Fundamentals of Music Processing : Audio, Analysis, Algorithms, Application. Springer (2015)
11. Venugopal, K.R., Srinivasa, K.G., Patnaik, L.M.: Soft Computing for Data Mining Applications. Springer (2009)
12. Ross, T.J.: Fuzzy Logic With Engineering Applications, 4th edn. (2017)
13. Feng, L.: English Language Speech Database for Speaker Recognition (ELSDSR). Department of Informatics and mathematical modelling, Technical University of Denmark (DTU) (2004)

Process and Subprocess Studies to Implement the Paraconsistent Artificial Neural Networks for Decision-Making

Luiz Antonio de Lima, Jair Minoro Abe, Angel Antonio Gonzalez Martinez, Alvaro Corrêa de Frederico, Kazumi Nakamatsu and Jonatas Santos

Abstract Since the time of Aristotle's thinking of logic to being a tool for orderly think, has maintained its importance until the present times. So soon, studies of the non-classical logical calls (Abe in 4th International Workshop on Soft Computing Applications. IEEE, pp. 11–18, 2010 [1]) have become a powerful tool as an aid in the making of decisions. The Paraconsistent logic calls attention to the clarity of containing provisions contrary to some of the basic principles of Aristotelian logic, such as the principle of contradiction. In this article, the use of technology allows proposing a structured organization in the process for use of Paraconsistent artificial neural networks. This process aims to be a facilitator in supporting the construction of the decision support (Abe in 4th International Workshop on Soft Computing Applications. IEEE, pp. 11–18, 2010 [1]) with the announcements for project recount in the function point analysis technique.

Keywords Paraconsistent artificial neural networks · Paraconsistent logic · Non-classical logics · Paraconsistency · Paracompleteness · Decision making · Software engineering · Software metrics

L. A. de Lima (✉) · J. M. Abe · A. A. G. Martinez · A. C. de Frederico · J. Santos
Graduate Program in Production Engineering, Paulista University, Rua Dr. Bacelar 1212, CEP, 04026-002 São Paulo, SP, Brazil
e-mail: luiz@wcisp.com.br

K. Nakamatsu
School of Human Science and Environment/H.S.E., University of Hyogo, Hyogo, Japan

V. Jain et al. (eds.), *Recent Trends in Intelligent Computing, Communication and Devices*, Advances in Intelligent Systems and Computing 1031,
https://doi.org/10.1007/978-981-13-9406-5_61

1 Introduction

Throughout the scientific community of software engineering [2], some kind of improvement in the accuracy in software estimation, through attempts of models based on the measurement of software size, is being investigated, such model continues to evolve from the model LOC (Lines-of-code-Metrics) [3–5], UCP created by Gustav Karner [6], NESMA (Association of Software Metrics Users of the Netherlands) [7].

The International Function Point Users Group—IFPUG, in 1986, was created to control and standardize the technique through the CPM (Counting Practices Manual), currently in version 4.3.1 [8].

In general, error-free, timely, and budget-driven software projects are not often found between demands delivered by software factory suppliers because they are primarily underestimated and even with inaccuracies in their initial estimates. This lack of effort often reflects increased costs, deadlines, and throughout the project lifecycle delays in deliverables, resulting in contractual penalties between customers and suppliers.

In face with these occurrences, there are significant losses of new business and renovation in the continuity of projects. Despite imprecise estimates, companies can gain insight into data stored on historical bases for a given period. Such knowledge produces reliable quality and productivity indicators [9] to aid in the choice of systems measurement.

The paraconsistent artificial neural networks may be able to process large amounts of data, which are fostered by social media, and with the possibility of analyzing these amounts of generated data. Artificial neurons have become a great tool capable of processing information for decision making, which is often inconsistent and uncertain, and that within a non-classical logic, paraconsistent logic can innovate and propose a series of techniques and algorithms to analyze and obtain outputs of answers that must be taken into account in the decision-making process.

The paraconsistent artificial neural networks analysis can be used in applications that focus on the use of artificial intelligence, being very efficient to assist in situations that the classic logic does not meet in its fullness, for the simple fact of operating with only two states, true or false, so paraconsistent logic may fill a gap to assist in decision making by identifying accepting and identifying data previously considered to be inconsistent, for the full real world. Managers in public or private sector rely on information generated by specialties, many of which are large or even unreliable, inconsistent, contradictory to decide something, or even this decision making in hiring new projects.

2 Methodology

The methodology applied in this article followed the paradigm design science research [10]. According to Hevner [11], this research reflects on cycles of related activities. Each cycle [12] should be matched first in the relevance of the subject addressed, along with the elicitation of requirements and criteria for evaluation of the research. In the second cycle, a process must be sought to solve the problem according to relevance, requirements, and criteria raised.

The problem addressed by this research concerns the process of recounting in function points, by a new expert in measurements counted by a group of specialists. The identification of the problem occurred from the literature and the consulting work on improving processes in the acquisition of software and services of software factories. In order to learn the state of the art about counting discrepancies that involve contracting systems and services, it was necessary to review the literature and mapping processes that resulted in the discovery of gaps in the research theme, mainly: project managers, the size discrepancies of the same software among counts made by specialists, the lack of use of tools with artificial intelligence and non-classical logic (paraconsistent logic), among others.

Considering the problem raised and identified gaps, it proposes processes and subprocesses of the paraconsistent logic that can guide the implementation of the tool as support in decision making. This proposal should address evaluations and provide benefits and efficacy to the process of deciding to hire resources for recounting in new projects that use point-of-function analysis such as systems measurement. His evaluation was based on a case study of software factories consulting in counts defended by specialists.

The contribution in the generation of knowledge is in the proper application of existing methodologies contributing to the construction of a knowledge base. In this research, some foundations of literature review, artificial intelligence with paraconsistent logic, and evaluation methods were studied. It stands out as the main advance, the new process for the use of paraconsistent annotated logic (UPPAL) as an aid in the decision making of recounting. The solution can be used by other institutions and even contribute to the creation of tools.

2.1 Results

The purpose of literary review identifies how companies engage in software factory services. Public and private sector companies use metrics such as Use Case Points (UCP) created by Gustav Karner [6] as described by Schneider and Winters [13] and Ribu [14], FPA (function point metrics) maintained by BFPUG [15] (Brazilian Function Point Users Group) in Brazil, the latter being adopted by several companies, due to its international acceptance [8], but among several solutions found, the function point metrics are most effective in choosing suppliers by the customer

According to the author, success in hiring must be directly related to the level of maturity and capacity of the suppliers to be hired.

According to Ash [16], it clarifies that managing project success depends on important aspects such as resources, term, and offer price that even in the hiring of new projects. At this point, we recommend the Guide to the Project Management Body of Knowledge [17].

The counting process (as shown in Fig. 1) used in the development projects should be characterized as new projects, and thus, to identify the project frontier to be measured about the other systems, making clear the scope of the project count.

The paraconsistent logic evidential logic $E\tau$ [1] is a class of paraconsistent logic that works with propositions of type p (μ, λ), where p is a proposition and (μ, λ) indicate the degrees of favorable evidence and contrary evidence, respectively. The pair (μ, λ) is called the annotation constant, with the values of μ and λ being limited [18–21] between 0 and 1.

One way of representing (as shown in Fig. 2) the paraconsistent logic that allows to perceive the real reach and thus extract results to support in the decision-making, is faced with the understanding of the diagram and its degrees of certainty and uncertainty, grouped in extreme states identified in the results (1–4) and

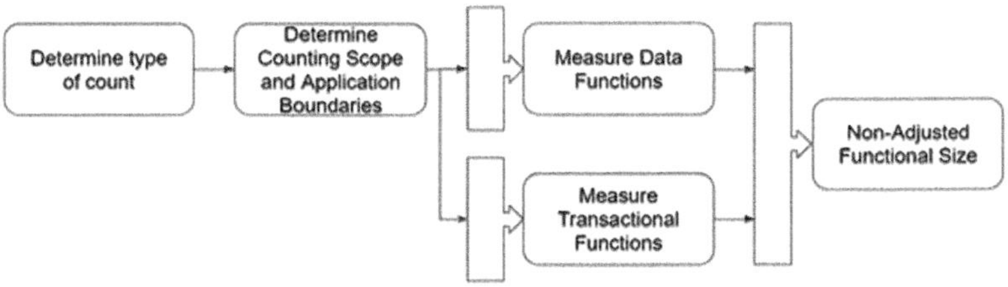

Fig. 1 Process of counting function points in development projects. *Source* Author

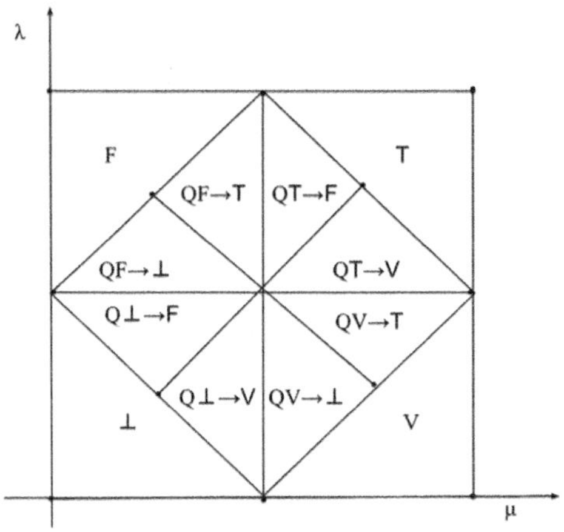

Fig. 2 Diagram with degrees of certainty and uncertainty, with adjustable values of limit control, indicated in the axes. *Source* Abe [18]

Table 1 Extreme and non-extreme states

Extreme states	Symbol	Non-extreme states	Symbol
True	V	Quasi-true tending to Inconsistent	QV → T
False	F	Quasi-true tending to Paracomplete	QV → ⊥
Inconsistent	T	Quasi-false tending to Inconsistent	QF → T
Paracomplete	⊥	Quasi-false tending to Paracomplete	QF → ⊥
		Quasi-inconsistent tending to True	QT → V
		Quasi-inconsistent tending to False	QT → F
		Quasi-paracomplete tending to True	Q⊥ → V
		Quasi-paracomplete tending to False	Q⊥ → F

non-extreme states shown in the results (5–12), with adjustable control values representing limit values: $C1 = C3 = 1/2$ and $C2 = C4 = -1/2$;

- $C1$: Vcve = maximum value of certainty control;
- $C2$: Vcfa = minimum value of certainty control;
- $C3$: Vcic = maximum value of the uncertainty control;
- $C4$: Vcpa = minimum value of the uncertainty control.

In the representation of the diagram, the following understandings with symbols [18] and their 12 possible results were used, being 1–4 extreme states and 5–20 non-extreme states.

Another way of representing the paraconsistent logic that allows to perceive the real reach and thus to extract more results to support in the decision making is faced with the understanding of the diagram and its degrees of certainty and uncertainty, grouped in states, as Table 1.

The definition of the Paraconsistent Decision Method (MPD) proposed in the studies [1] reflects the method that assists decision making through paraconsistent logic [1]. By using paraconsistent annotated logic (LPA) in support of decision-making in project recount in the function point technique, one can mitigate countless defenses between clients and suppliers that cause a longer design time and bottlenecks of new demand.

3 Proposal of the UPPAL Process Organized as an Aid in Decision Making in Recounting New Projects

The decision-making process consists in choosing one of several alternatives, then proposing the unified process of paraconsistent logic as an aid in the decision making of recounting. In this article, I had the perception of the importance of the research with regard to the organization in the implementation of the Annotated Paraconsistent Logic, which was proposed in six processes (definition, transformation, calculation, parameterization, processing, decision aid) with respective subprocesses necessary for the success of a structuring of the technological tool focused on supporting the decision making in projects with needs to meet a proposal in the recount of projects in function point analysis by specialists.

Function point specialists count systems using the CPM manual so that it is possible for project managers to seek results through the mental-cognitive process of the specialists or group of experts who have worked on the measurement in function point analysis. The decision-making process consists of choosing one of several alternatives. The unified process of paraconsistent logic is proposed as an aid in the decision making of recounting, as Table 2.

Description of Process and Subprocess

1. Definition: List information necessary to assist decision making.
   1.1. Define Proposition: Define appropriate proposition to propose assistance in the decision to recount function points in new software projects.
   1.2. Define Factors: Identify the major factors (even weight) that influence the most success [type data, type transaction, lower point function, historical higher point function counted, postage size] or failure [type data, type transaction, higher point function, lower historical point function counted, and decision-making aid.
   1.3. Define Section: Identify the sections of each factor that allow giving conditions to the factors (success or failure) that will aid in the decision making. Example 1: found sections [ILF, EIF, EI, EQ, EO]; example 2: not found sections [ILF, EIF, EI, EQ, EO]; example 3: fewer sections found [ILF, EIF, EI, EQ, EO]; and example 4: found more sections [ILF, EIF, EI, EQ, EO].
   1.4. Define Database: Data Collection: Collect data and organize it according to the sections that meet the factors.

Table 2 Process for the use of paraconsistent annotated logic

Process for use of paraconsistent annotated logic (UPPAL)		
Item	Process	Subprocess
1	Definition	Define proposition; define factors; define section; define database
2	Transformation	Generate normalization; use evidence (favorable and unfavorable)
3	Calculation	Calculate maximization; calculate minimization Calculate evidence (resultant min, resultant max) Calculate degree (Gce: certainty, Gco: contradiction) Calculate global analysis (Gce: certainty, Gco: contradiction)
4	Parameterization	Parametrize limit values
5	Processing	Process para-analyzer algorithm with paraconsistent artificial neural network
6	Decision-making support	Assists decision making

Source Author

2. Transformation: This is a way of representing input data for paraconsistent annotated logic processing.
   2.1. Normalize Data: Normalizes organized data (by sections) to represent the inputs in the paraconsistent logic. Example 1: Linear in the range [0 and 1]; Example 2: Maximum value of the elements; and Example 3: Z-Score Standard;
   2.2. Define Favorable Evidence Ef (μ): collected data reflecting expert opinions (by sections). These data, after normalized, represent the Ef (μ) inputs for the Paraconsistent annotated logic processing.
   2.3. Define Unfavorable Evidence Ed (λ): collected data that reflects expert opinions (by sections). These data, after normalized, represent the Ed entries (λ) for the paraconsistent annotated logic processing.
3. Calculation
   3.1. Calculate Maximization MaxEf (μ): In each data (by sections) collected as evidence favorable Ef (μ), use the highest value among them (by sections) to represent the maximization of favorable evidence Ef (μ).
   3.2. Calculate MinEd Minimization (λ): In each data (by sections) collected as evidence unfavorable Ed (λ), use the smallest value among them (by sections) to represent the minimization of the unfavorable evidence Ed (λ).
   3.3. Calculate Evidence Resulting MinEf (μ): The resultant must be used, when the data are grouped by specialists and need to cross between their entities (customer X suppliers). In each data (by sections) collected as evidence favorable Ef (μ), use the lowest value (by sections) between client and suppliers to represent the result of minimizing favorable evidence Ef (μ) example 1: the lower value between favorable evidence Ef (μ) customer and favorable evidence Ef (μ) suppliers.
   3.4. Calculate Resultant EvidenceMaxEd (λ): The resultant should be used when the data are grouped by specialists and need to cross between their entities (customer X suppliers). In each die (by sections) collected as evidence unfavorable Ed (λ), use the highest value (by sections) between client and suppliers to represent the result of the maximization of unfavorable evidence Ed (λ). Example 1: higher value between unfavorable evidence Ed (λ) customer and unfavorable evidence Ed (λ) providers.
   3.5. Calculate Degree of Certainty (Gce): Based on the evidence (by sections) collected, it is possible to calculate the degree of certainty, since it is enough to make the difference between the favorable evidence Ef (μ) and the unfavorable evidence Ed (λ) the degree of certainty (Gce). Example 1: GCe = Ef (μ) − Ed (λ).
   3.6. Calculate Degree of Contradiction (Gco): Based on the evidence (by sections) collected, it is possible to calculate the degree of contradiction, since it is sufficient to make the sum between the favorable evidence Ef (μ) and the unfavorable evidence Ed (λ), using result of the sum in the extraction of

a unit (1) and thus obtain the degree of contradiction (Gco). Example 1: GCe = (Ef[μ] + Ed[λ]) − 1.

3.7. Calculate Global Certainty Analysis (BGce): Based on the degrees (by sections) of certainty calculated, it is possible to calculate the global analysis as the arithmetic mean of the degrees of certainty and thus result in the global analysis of the degree of certainty (Gce). Example 1: BGCe = ΣGce/Gce Amount.

3.8. Compute Global Analysis of Degree of Contradiction (Gco): Based on the degrees (by sections) of calculated contradiction, it is possible to calculate the global analysis as the arithmetic mean of the degrees of contradiction and thus result in the global analysis of degree of contradiction (Gco). Example 1: BGCo = ΣGco/Amount of Gco.

4. Parametrization: Are the boundaries that limit the regions for analysis (values high enough to be considered) regardless of logical principles.

   4.1. Parametrize limit TLV (True limit value): These are conditions parameterized by the knowledge engineer with the objective of obtaining acceptable answers as true, under the conditions in which the value of the degree is smaller, greater than or equal to the value of the parameter.
   4.2. Parametrize limit FLV (False limit value): These parameters are parameterized by the knowledge engineer in order to obtain acceptable answers as false, in the conditions in which the value of the degree is smaller, greater than or equal to the value of the parameter.
   4.3. Parametrize limit PLV (Limit Paracompletenessn value): These parameters are parameterized by the knowledge engineer with the objective of obtaining acceptable answers as full Paracompletenessn, in the conditions in which the value of the degree is smaller, greater or equal to the value of the parameter.
   4.4. Parametrize limit ILV (Inconsistent limit value): These are conditions parameterized by the knowledge engineer with the objective of obtaining acceptable answers as inconsistent, under the conditions in which the value of the degree is smaller, greater or equal to the value of the parameter.

5. Process: In this process, the objective must be to execute the para-analyzer algorithm to obtain the parameters according to the input of the data.
6. Decision Making: In this process, the objective is to analyze the degree of contradiction, which may have value both upwards and downwards. In the existence of a high degree of contradiction (Gco), it indicates that there is no certainty to aid decision making, and therefore, one must seek new evidence. Moreover, in the existence of a low degree of contradiction (Gco), together with a high degree of certainty (Gce), indicates the possibility of a conclusive analysis on the proposition.

4 Discussion

Studies demonstrate that the metric in the point-of-function analysis is widely used as a way of hiring new software projects between the public and private sectors [22]. The decision by project managers to hire new specialists for point of function recount, as they recognize that the function point can be used to measure the size of new project demands at any stage of the project lifecycle. However, many companies (customers) have difficulty in accepting countless differences between counts made by vendor specialists. In this article, we highlight as results of literary review an approach to artificial intelligence techniques using paraconsistent logic to support project managers. However, the main contribution of this work was the proposal of a unified process for success in aid decision making in conjunction with extreme and non-extreme values proposed by ABE [18]. By this precept, the use of paraconsistent artificial neural networks in decision making brings a significant contribution to project managers.

A neural architecture found in the paraconsistent neural network family (artificial neural cell paraconsistent analytic, real, detection of equality and decision) allows gains when proposing possible solutions (as shown in Fig. 3) to problem answers.

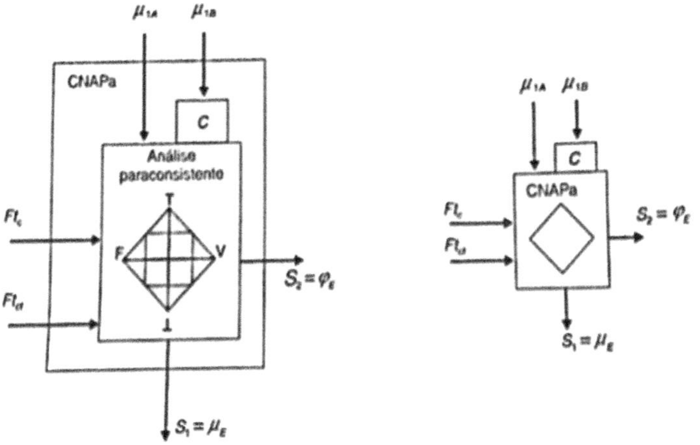

Fig. 3 Scheme of the paraconsistent artificial neural analytical cell. *Source* Abe [18]

References

1. Abe, J.M.: Paraconsistent logics and applications. In: 4th International Workshop on Soft Computing Applications, pp. 11–18. IEEE (2010)
2. Pressman, R.S.: Software engineering: a practitioner's approach, 7th edn. The McGraw-Hill Companies, Inc., New York (2011). ISBN: 0073375977/9780073375977
3. Jones, C.: Software Engineering Best Practices. McGraw-Hill (2010)
4. Jones, C.: A short history of lines of code (LOC) metrics (2008)
5. Boehm, B., Valerdi, R., Lane, J., Brown, A.: COCOMO suite methodology and evolution. Crosstalk **18**(4), 20–25 (2005)
6. ENASE (conference), Leszekmaciaszek, and Kang Zhang (2015): Evaluation of Novel Approaches to Software Engineering: 9th International Conference, ENASE 2014, Lisbon, Portugal, 28–30 Apr, revised selected papers (2014). http://rave.ohiolink.edu/ebooks/ebc/9783319272184
7. ISO/IEC 24570: 2005 (NESMA FPA ver.2.1) preview software Engineering—NESMA Functional size measurement method version 2.1—definitions and counting guidelines for the application of function point analysis (2005)
8. IFPUG: If pug 4.3.1 Unadjusted functional size measurement method—counting practices manual, international organization for standardization, ISO/IEC 20926 (2010)
9. IFPUG: If pug 4.2 Unadjusted functional size measurement method—counting practices manual, international organization for standardization, ISO/IEC 20926 (2009)
10. Hevner, A.R., March, S.T., Park, J., Ram, S.: Design science in information systems research. MIS Q. **28**, 75–105 (2004)
11. Hevner, Alan R.: A three cycle view of design science research. Scand. J. Inf. Syst. **19**, 87–92 (2007)
12. Wrigley, C.D., Dexter, A.S.: Design science in information systems research. MIS Q. **15**, 245–257 (1991)
13. Schneider, G., Winters, J.P.: Agile software development with Scrum. Prentice Hall. Applying Use Cases: A Practical Guide. Addison Wesley (1998)
14. Ribu, K.: Estimating object-oriented software projects with use Cases. Master of Science thesis, University of Oslo, Department of informatics. Schwaber, Ken, and Mike Beedle (2001)
15. BFPUG: Brazilian Function Point User Group, www.bfpug.com.br (2006)
16. Ash, C.: Convergence of IT Project Management and Change Management: A Comparative Study. Cepis Promotes **47** (2007)
17. PMI: Project Management Institute, Guide PMBOK®—6th edn. Project Management Institute Inc. (2017)
18. Abe, J.M., Akama, S., Nakamatsu, K.: Paraconsistent Intelligent-Based Systems—New Trends in the Applications of Paraconsistency, 1st edn, vol. 1. Springer International Publishing, Switzerland (2015). ISBN https://doi.org/10.1007/978-3-319-17912-4
19. Abe, J.M., Akama, S., Nakamatsu, K.: Introduction to Annotated Logics—Foundations for Paracomplete and Paraconsistent Reasoning, Intelligent Systems Reference Library, 1st edn., vol. 88, p. 190. Publisher Springer International Publishing (2015). https://doi.org/10.1007/978-3-319-17912-4, Hardcover ISBN: 978-3-319-17911-7, Series ISSN: 1868-4394
20. Abe, J.M.: Paraconsistent Intelligent Based-Systems: New Trends in the Applications of Paraconsistency. Intelligent Systems Reference Library, vol. 94, p. 306. Springer, Germany (2015). ISBN: 978-3-319-19721-0
21. Akama, S.: Towards Paraconsistent Engineering, Intelligent Systems Reference Library, vol. 110, p. 234. Publisher Springer International Publishing (2016). ISBN: 978-3-319-40417-2 (Print) 978-3-319-40418-9 (Online), Series ISSN: 1868-4394. https://doi.org/10.1007/978-3-319-40418-9
22. Görög, M.: A broader approach to organizational project management maturity assessment. Int. J. Project Manag. **34**(8), 1658–1669 (2016)

A Way to Detect the Opinion Sentences from Short Texts by the Vote-AdaBoost Combining Classify Method

Nan Liu and Wei He

Abstract Because of the sparse sentiment features of network short texts, the identify result of common classification methods which are efficacious in texts with tradition structures are unsatisfactory. This article tries to combine multiple classifiers combination and integrated learning methods to solve this problem. Vote-AdaBoost combining classify method is constructed to optimize the appropriate classifier as the voting combination in iterative learning process. Finally, an effective classification method to detect the opinion sentences of short texts is obtained, and the effectiveness of this method is verified by experiments.

Keywords Vote-AdaBoost model · Combining classify method · Opinion sentences detection · Sentiment analysis

1 Introduction

Huge number of texts was created by users from social network every day. The texts have included much information about the viewpoints and sentiments of users. Researchers try to use these texts to extract the information, and use them on many research areas, such as sentiment analysis, opinion identification, and interest recommendation. However, there still many noises in these texts. If the noises from these texts can be clean before extracting, the result of these researches may be better. The detection of opinion sentences is one way to clear noises, and many detection methods are used. Some detection methods try to identify the key words. The sentences which included key words will be asserted to opinion sentences. The

N. Liu (✉)
Department of Quartermaster and Procurement, Army Logistics University of PLA, Chongqing 401331, China
e-mail: liunans@163.com

W. He
School of Computer and Information Technology, Xinyang Normal University, Xinyang 464000, China

V. Jain et al. (eds.), *Recent Trends in Intelligent Computing, Communication and Devices*, Advances in Intelligent Systems and Computing 1031,
https://doi.org/10.1007/978-981-13-9406-5_62

accuracy of these methods is depending on the dictionary of key words. The other methods try to analyze the syntactic parsing of sentences. These methods are efficacious in texts with tradition structures, such as novels. But in new types of short texts, such as the user comment on e-shopping Web sites and the communication records on Facebook or Microblog, maybe not efficiently. It is hard to find a common way to adapt all texts. Accordingly, using combining methods to identify may be useful in complex texts.

2 The Weighting Voting Combining Classification Model

2.1 Weighting Voting Method

Voting is always used in daily life. If this way is used to build classification model, the classifiers can add classify results together as the total classify result. But because of the different performance of these classifiers, the total classify result may be not the best result. The weight of each classifier can be adjusted by its performance, then the results are added together, and the total classifies result may be better.

By the different way to choose the weight, there are many voting methods. In these methods, majority vote algorithm is used widely [1]. This method transforms the classify result of each classifiers to [0, 1]. The classify results are added together, and the Max value of them is used to identify. The weighting voting method transforms the classify result by different weight. The classify result multiplies its weight, and then they are added together as the total classify result. The value of total classify result is used to identify.

The method can be expressed in Formulae 1 and 2. For m classes $W = (w_1, w_2, \ldots, w_m)$, n classifiers are existed $T = (t_1, t_2, \ldots, t_n)$. $P(w_j|t_i)$ means the probability of the result of classifier t_i in class w_j. $P(w_j)$ means the prior probability of class w_j. $P(t_i)$ means the weight of t_i. The total classify result is Δw.

$$E_{ji} = \begin{cases} 1, & \text{if } P(w_j|t_i) = \max_k P(w_k|t_i) \\ 0, & \text{otherwise} \end{cases} \tag{1}$$

$$\Delta w = \arg\max_j \sum_i \left[E_{ji} \times P(t_i)\right] \tag{2}$$

2.2 The Training of Weight

The performance of weighting voting method is relying on the value of the weight. The accuracy of classifier to identify the sample dataset always is used as the weight. Some researchers consider that the higher accuracy of classifier, the better performance [2]. The accuracy often used as the weight when distribution of classes are equaled in sample dataset, but if is lack of balance in the prior probability of sample data, the identify result may be not good [3]. In many times, the balance of sample dataset cannot be ensuring. For example, the numbers of subjective sentences and objective sentences which are crawled from Web site may be not balanced.

2.3 The Selection of Classifier

The classifier which is selected in the voting method influences the identify result. The useful classifier has two characters, one is high accuracy, and another is varying type. The high-accuracy character makes the identify result available, and the vary type character can make the classifiers cover all attributes in sample data as much as possible.

By our attention on detecting the opinion sentences, we choose three types of common machine learning methods as the base classifiers, they are *k*-nearest neighbor (KNN), Naïve Bayesian model (NBM), and support vector machine (SVM). In the feature selection of these classifiers, we choose the sentiment features of short texts from Web sites. These sentiment features include n-GRAM, n-POS, and context relationship. Each classifier can select one type or more sentiment features, that means a big number of combinations of classifier and features are created. If we add the numbers of feature or classifier, the computation will be more complex. So we need an easy way to choose the best combination of features and classifiers.

3 The Vote-AdaBoost Combining Classify Method

3.1 Ensemble Learning Model

Ensemble learning model is another way to enhance the performance of combining classify method. Different from the voting method which combines classifiers together, ensemble learning model use iterative training on sample data to improve the accuracy of base classifier [4]. In many areas, the ensemble learning model has shown better performance than base classifier. There are two ways to construct the ensemble learning model. One way is iterative learning by data, such as bagging model and boosting model [5]. Another way is iterative learning by features, such as random subspace model [6].

3.2 AdaBoost Model

AdaBoost is a short name of adaptive boosting model [7]. The advantage of AdaBoost model is the performance of classifier can be promoted by adjusting the weight of training samples, even if the minimum accuracy rate of the classifier is not being detected. Some researchers found that the AdaBoost model often have good accuracy rates on classifiers which used unsteadiness machine learning algorithm, such as decision tree algorithm [8] and artificial neural network algorithm [9]. But sometimes, in classifiers which worked steadily, like Naïve Bayesian model algorithm, the promoting is not apparent [10]. In special research domains, like the texts samples for sentiment analysis, the sentiments of texts are tagged by people, the tags may mistake, or by someone's bias. Taken more attention on these mistakes, samples may causes the disturbance of identify result.

In the work to detect the opinion sentences, the process of AdaBoost model can be shown below. In the training sample dataset D_{train} which is mixed by opinion sentences and objective sentences, the sentence set $X = (x_1, x_2, \ldots, x_n)$ is created by n sentences. The opinion sentences are tagged to 1, and the objective sentences are tagged to −1, then a set $Y = (y_1, y_2, \ldots, y_n), y_i \in \{1, -1\}$ is created.

Step 1: The weights of sentences are set when the process is initialized. The value is normalized: $D_1(i) = 1/n$. $D_1(i)$ means the weight of sentence x_i in round 1. And in round t, $D_t = [D_t(1), D_t(2), \ldots, D_t(n)]$.

Step 2: In round t, a classifier can be trained by sample data and its weight set D_t, the classify result set $H_t = [h_t(1), h_t(2), \ldots, h_t(n)]$ is created, and $h_t(x) : x \to \{1, -1\}$. The classify result set H_t can be revised by checking the error rate on identify the sentence $h_t(i)$. The error rate of $h_t(i)$ is $\varepsilon_t = \sum_{i=1}^{n} D_t(i)[h_t(i) \neq y_i]$, and two variables α, β are set in each round. $\beta_t = \frac{\varepsilon_t}{(1-\varepsilon_t)}$, $\alpha_t = \frac{1}{2}\log\left(\frac{1}{\beta_t}\right)$.

Step 3: $D_{t+1}(i)$ is the weight of these sentences in round $t + 1$. It can be adjust by variables and $D_t(i)$ in Formula 3. In the formula, z_t is the weight of $D_t(i)$, $z_t = D_t(i)\exp(-\alpha_t y_i h_t(i))$.

$$D_{t+1}(i) = \frac{D_t(i)}{z_t}\begin{cases} e^{-\alpha_t}, & \text{if } h_t(i) = y_i \\ e^{\alpha_t}, & \text{if } h_t(i) \neq y_i \end{cases} \tag{3}$$

Step 4: The round is over when $|\varepsilon_t - 0.5| < \gamma$, then go to Step 5. Otherwise go to Step 2. The threshold of variable γ can be set by experience. In this process, we set = 0.05.

Step 5: After these rounds, a robust classifier is created, the final classify result $H = \text{sign}\left(\sum_{t=1}^{T} \alpha_t H_t\right)$.

3.3 *Vote-AdaBoost Model*

3.3.1 Idea of Vote-AdaBoost Model

In the weighting voting combining classification model, if all the base classifiers and the combination of features are added into the voting combination, it will lead to huge computation and waste of resources. We propose the Vote-AdaBoost combining classification model. In the voting process, based on the multiple random selections of training samples, the optimal classifier combination is obtained, and then use AdaBoost model to promote to weight of each classifier in the classifier combination. The boosted classifier will instead the original classifier only when the accuracy is improved. The process is shown in Fig. 1.

3.3.2 Training Method

We divide the opinion sentences used for identification into training samples dataset and testing samples dataset. The training samples dataset have been manually tagged. In the training process, random sampling method is adopted to construct small-scale training subsamples and testing subsamples from the training sample dataset. The origin classifier combination is constructed by multiple verifications of subsamples. Then use the total training sample dataset and AdaBoost model to improve the classifiers in the origin classifier combination. Only the improved classifier will be updated to the final classifier combination. The final classifier combination is used to identify the testing samples dataset.

In each round of training, due to the different subsamples, the parameters in each classifier will be re-adjusted, and the classifier combination mode with the highest accuracy suitable for this data will be retained. In the selection of classifier

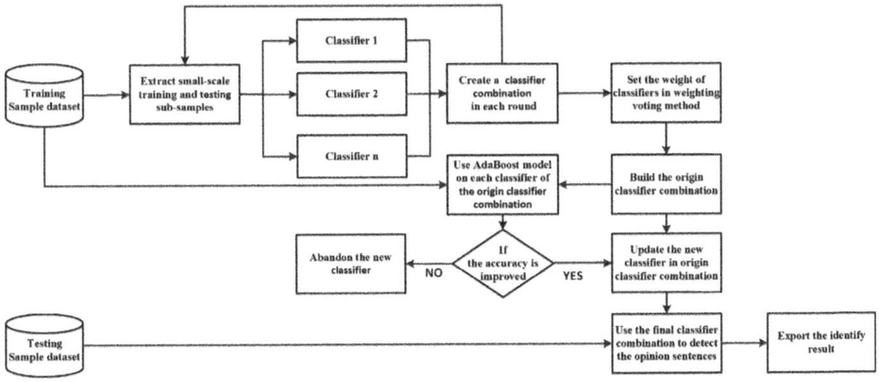

Fig. 1 Process of Vote-AdaBoost model

combination, the combination method to improve the F-score of lowest single class should be selected to improve the final accuracy.

F-score is created by precision rate and recall rate. If the classifier t_i has precision rate P_i and recall rate R_i to identify sample data, the F-score of t_i is F_i. It can be calculated by Formula 4.

$$F_i = \frac{2 \times P_i \times R_i}{P_i + R_i} \times 100\% \quad (4)$$

3.3.3 The Process of Vote-AdaBoost Model

The process of Vote-AdaBoost Model is shown below. The classifiers set $T = (t_1, t_2, \ldots, t_n)$ with different features and classification methods is used to identify m classes $W = (w_1, w_2, \ldots, w_m)$.

Step 1: Set turn $i = 0$, input the maximum number of turns T.

Step 2: Random select small-scale training sub-samples D^i_{train} and testing sub-samples D^i_{test} from the training sample dataset D_{train}.

Step 3: Use D^i_{train} to train the parameters of classifier t_j, and use D^i_{test} to verify them, then the precision rate P_{jk}, recall rate R_{jk} and F-score F_{jk} of classifier t_j to identify class k are acquired.

Step 4: Choose top three classifiers which are sorted by the F-score F_{jk}, as the combining classifiers of weighting vote U_i.

Step 5: Use the combining classifiers of identify the testing samples dataset D^i_{test}, and gain the precision rate P^i_{test}.

Step 6: Try to add a classifier w_u to initial combining classifiers from the F-score sorted classifiers list.

Step 7: If the precision rate P^i_{test} of the combining classifiers U_i is increased by adding the new classifier, then save this change, $U_i = U_i \cup W_u$, otherwise try to add the next one.

Step 8: The classifier adding work will be stopped when any classifier added cannot increase the precision rate.

Step 9: Set $i = i + 1$, if $i \leq T$ then go to step 2, otherwise go to step 10.

Step 10: The origin classifier combination U is created by summarizing the combining classifiers of each turns, $U = U_1 \cup U_2 \cup \cdots \cup U_T$. $P(w_j)$ means the prior probability of classifier, $W_j \in U$, is set by the count of w_j in the combining classifiers of each turns divide T, $P(w_j) = \frac{\sum_{i=1}^{T} \{w_j \in U_i\}}{T}$.

Step 11: Use the training sample dataset D_{train} and the AdaBoost model to train the parameters of each classifier in the origin classifier combination U. The training algorithm of classifier $w_j \in U$ is in Sect. 3.2. The boosted classifier is w'_j.

Step 12: If the F-score of U to identify D_{train} is improved by replace w_j with w_j', then save this change, otherwise abandon it.

Step 13: After the boosting work, the final classifier combination U is holded. It can used to identify the Testing sample dataset D_{test}.

4 Experiment

4.1 Experiment Design

We have used Waikato Environment for Knowledge Analysis (WEKA) 3–7–9 as the experiment environment [11]. We have used LIBSVM tool kits [12] to build the base SVM classifier. The radial basis function (RBF) kernel of this SVM classifier is set, and other parameters of it are default. We choose short texts of Chinese languages in 20 topics from microblog (Weibo) Web site as datasets for the experiment, which is collected by the opinion sentence recognition subtask in NLPCC2012 evaluation task [13]. In order to ensure the balance of corpus, the same microblog text was randomly selected for each microblog topic, with a total of 20 microblog topics and 2410 sentence examples. The sentences are preprocessed by splitting the words and removing the stop words.

4.2 Experiment 1: Comparison of AdaBoost Model and Base Classifiers with Solo Feature

In the preliminary preparation, we found that the extraction methods of 2-GRAM, 3-GRAM, and other multi-lexical features could not extract enough effective features due to the small corpus size and sparse data. Therefore, only 1-POS, 2-POS, 1-GRAM, and microblog features (including emotional words, punctuation marks and emoticons) are selected in the experiment. We also conducted experiments on whether the ensemble learning method can improve the classifier of various features. The classifiers are training by AdaBoost model, and the same corpus is verified by the method of tenfold cross-validation.

The experiment result is shown in Table 1. Among these features, 1-GRAM has a high precision rate, but its recall rate is the lowest. 1-POS has the best comprehensive effect. The microblog features also have certain accuracy. Comparing with and without AdaBoost model, we can find the classification effect is better than that before learning. For 1-GRAM with poor initial effect (equivalent to a weak classifier), the F-score increases by 10% through integrated learning. However, for the microblog features, the average F-score has decreased.

Table 1 Results of subjective and objective classification by extract different features and AdaBoost model

Features	Class	Precision	Recall	*F*-score	Features	Class	Precision	Recall	*F*-score
1-POS	Obj	0.66	0.54	0.594	1-POS +AdaBoost	Obj	0.638	0.621	0.630
	Sub	0.609	0.72	0.66		Sub	0.629	0.645	0.637
	Avg	0.634	0.63	0.627		Avg	0.633	0.633	0.633
2-POS	Obj	0.588	0.784	0.672	2-POS +AdaBoost	Obj	0.607	0.757	0.674
	Sub	0.673	0.448	0.538		Sub	0.674	0.506	0.578
	Avg	0.631	0.617	0.605		Avg	0.640	0.632	0.626
1-GRAM	Obj	0.517	0.97	0.675	1-GRAM +AdaBoost	Obj	0.538	0.847	0.658
	Sub	0.745	0.089	0.159		Sub	0.636	0.269	0.378
	Avg	0.631	0.531	0.418		Avg	0.587	0.559	0.519
Microblog features	Obj	0.615	0.535	0.572	microblog features +AdaBoost	Obj	0.617	0.512	0.559
	Sub	0.586	0.662	0.662		Sub	0.580	0.679	0.626
	Avg	0.6	0.599	0.597		Avg	0.599	0.595	0.593

Therefore, we believe that if carry out AdaBoost model for all classifiers blindly, the resource consumption may not be proportional to the final effect. It is more beneficial to choose some weak classifier than all classifiers.

4.3 Experiment 2: Comparison of Vote-AdaBoost and Weighting Voting Combining Classify Method

We combined the four classifiers in Experiment 1 by weighting voting, with the *F*-score of each classification as its weight. Vote-AdaBoost method proposed in this paper is compared with this method. That is, after the classifier combination is constructed, each classifier of the multiple classifier combinations is iteratively optimized by AdaBoost model. Table 2 shows that the accuracy of both methods

Table 2 Results of subjective and objective classification by weighting voting and Vote-AdaBoost combining classify method

Combining classify method	Class	Precision	Recall	*F*-score
Weighting voting method	Obj	0.595	0.847	0.699
	Sub	0.732	0.42	0.534
	Avg	0.664	0.634	0.617
Vote-AdaBoost method	Obj	0.992	0.704	0.823
	Sub	0.772	0.994	0.869
	Avg	0.882	0.849	0.865

are better than single classifiers in Experiment 1. Vote-AdaBoost method has significantly improved precision rate and recall rate of the classification compared with the weighted voting method, thus proving the effectiveness of the method.

5 Conclusion

Because of the sparse sentiment features of network short texts, the identification of common classification methods which are efficacious in texts with traditional structures are unsatisfactory. We try to combine multiple classifiers combination and integrated learning methods to solve this problem. Vote-AdaBoost combining classify method is constructed to optimize the appropriate classifier as the voting combination in iterative learning process. Finally, an effective classification method to detect the opinion sentences of short texts is obtained, and the effectiveness of this method is verified by experiments.

Acknowledgements This research was financially supported in part by National Natural Science Foundation of China, Grants No. 6601396, in part by the Key Scientific Research Project of Colleges and Universities in Henan Province of China, Grant No. 16A520069, and in part by the Natural Science Foundation of Chongqing, China, Grant No. cstc2018jscx-msybX0182.

References

1. Dietrich, F., List, C.: Majority voting on restricted domains: a summary. J. Econ. Theory **145**(2), 512–543 (2010)
2. Zhang, Z., Iria, J., Brewster, C.: A comparative evaluation of term recognition algorithms. In: International Conference on Language Resources & Evaluation DBLP (2008)
3. Onan, A., Korukoğlu, S., Bulut, H.: A multiobjective weighted voting ensemble classifier based on differential evolution algorithm for text sentiment classification. Expert Syst. Appl. **62** (2016)
4. Polikar, R.: Ensemble based systems in decision making. IEEE Circuits Syst. Mag. **6**(3), 21–45 (2006)
5. Zhao, J., Cao, X.: Combining semantic and prior polarity for boosting twitter sentiment analysism. In: IEEE International Conference on Data Science in Cyberspace (2016)
6. Skurichina, M., Duin, R.P.W.: Bagging, boosting and the random subspace method for linear classifiers. Pattern Anal. Appl. **5**(2), 121–135 (2002)
7. Li, X., Wang, L., Sung, E.: AdaBoost with SVM-based component classifiers. Eng. Appl. Artif. Intell. **21**(5), 785–795 (2008)
8. Dinakaran, S., Thangaiah, P.R.J.: Ensemble method of effective AdaBoost algorithm for decision tree classifiers. Int. J. Artif. Intell. Tools **26**(3) (2017)
9. Liu, H., et al.: Comparison of four AdaBoost algorithm based artificial neural networks in wind speed predictions. Energy Convers. Manag. **92**, 67–81 (2015)
10. Ting, K.M., Zheng, Z.: A study of AdaBoost with naive bayesian classifiers: weakness and improvement. Comput. Intell. **19**(2), 186–200 (2010)
11. Weka tool kits, http://www.cs.waikato.ac.nz/ml/weka/
12. LIBSVM tool kits, http://www.csie.ntu.edu.tw/cjlin/libsvm/index.html
13. NLPCC2012, http://tcci.ccf.org.cn/conference/2012/index.html

Thoughts on the Development Trend of Intelligent Transportation and the Development of Intelligent Vehicles

Yingshun Wang

Abstract Social technology is growing rapidly, and every day is a new challenge and opportunity. With the development and popularization of computer technology in various industries, various industries have found a development direction that has not been thought of in the past. For the transportation industry, the increase in road pressure has made cities overwhelmed, and the development of computer technology has brought new hope to the transportation industry. It can be said that computer technology has begun to penetrate all aspects of the transportation industry, and traffic has begun to develop in an intelligent direction [1]. On the basis of consulting a large number of documents, this paper analyzes and explains the development status of intelligent transportation, the specific application of intelligent technology in the transportation industry, and the future development of computer technology in the transportation industry. I hope to promote the use of intelligent technology in the transportation industry.

Keywords Intelligent · Transportation industry · Automobile · Computer technology

1 Introduction

No matter which country or city, car ownership is increasing year by year. However, the number and capacity of roads have not increased, so the pressure on traffic is increasing and the possibility of accidents is increasing. With the development of computer technology, transportation systems are becoming more and more dependent on computer technology, because computer technology can greatly improve the efficiency of solving traffic accidents and reduce labor consumption. It is for this reason that many experts and scholars are studying this issue.

Y. Wang (✉)
Mechanical Engineering College, Guangdong University of Science & Technology, Dongguan City 523083, Guangdong Province, China
e-mail: 279397189@qq.com

V. Jain et al. (eds.), *Recent Trends in Intelligent Computing, Communication and Devices*, Advances in Intelligent Systems and Computing 1031,
https://doi.org/10.1007/978-981-13-9406-5_63

Fig. 1 Intelligent transportation

After reading a large amount of reference materials, this paper decided to explain and think about the development trend of intelligent transportation and the development of intelligent vehicles on the basis of the results of previous studies. It is expected to help the development of intelligent transportation. Figure 1 is a schematic diagram of intelligent transportation.

2 Development Status of Intelligent Transportation and Automobile Intelligence

2.1 Development Status of Intelligent Transportation

Because the traffic problems on the roads are getting more and more serious, human beings can no longer rely on the existing manpower and material resources to solve these traffic problems, and the intelligent transportation system has thus emerged. The biggest feature of the intelligent transportation system is that it can comprehensively use the road data to supervise and manage the traffic conditions in real time, and make a reasonable distinction between the vehicle, the person and the road, and at the same time ensure the safety of the three [2]. The intelligence of intelligent transportation is not only to effectively arrange traffic order but also to solve problems in the shortest time after a traffic accident. Of course, because of the imperfect technology, the intelligent transportation development system is still undergoing continuous upgrade and maintenance, and will play a greater role in the future.

2.2 *Development Status of Automotive Intelligence*

The so-called car intelligence is the combination of automotive and computer technology, and the stable use. Nowadays, automotive electronic technology has experienced a long four-generation development. No matter network technology or sensor technology has developed relatively mature, especially the user optimization control technology and electronic technology have made breakthrough progress. Therefore, the development trend of automobile intelligentization is good, and more remarkable achievements will be made in the future.

3 The Application of Computer Technology in Intelligent Transportation

3.1 *Reduce Traffic Pressure*

Today, almost every household has one or a few cars, so the pressure on road traffic is enormous. The emergence of computer technology in the transportation system has greatly reduced the manpower and material resources of the vehicle license plate recognition and charging system in traffic management [3]. For the vehicle identification system, firstly, the computer reader recognizes the vehicle license plate, collects and matches the information after collecting the information, integrates the traffic travel information of the vehicle, and automatically responds. For charging systems, intelligent systems rely mainly on electronic radio frequency technology. When the vehicle enters the ETC charge to occupy the entrance, the computer reader first identifies the license plate number, and after the information is collected, intelligent charging is performed, which greatly speeds up the parking charge or the toll collection speed.

3.2 *Precise Vehicle Navigation*

With road reforms and rapid urban changes, the construction of urban roads has become more complex, and road updates and traffic regulations are likely to change and change. It is based on the situation that many car drivers do not follow up and understand these changes in a timely manner, and there are often some violations or even some large traffic accidents. The intelligent development of automobiles is to use computer technology to accurately navigate the vehicle's navigation system and correct the driver's possible violations in time. Unlike traditional car navigation, smart car navigation can detect obstacles on the road in time and react very quickly to avoid tragedies. In addition, intelligent vehicle navigation can monitor road traffic conditions in real time, identify the distance between the car and the car, the car and the road, and the distance between the car and the pedestrian, and ensure the driver's driving safety. With the intelligent vehicle navigation system, the driver will get more timely information during the driving process, making more accurate and effective judgments to avoid bad things happening.

3.3 Auxiliary Traffic Monitoring

The application of computer technology in traffic monitoring covers many aspects, including vehicle monitoring, accident detection, and so on. The process is to use the RF equipment to get the actual traffic situation image, and then use computer technology to process these images, so as to obtain the actual situation of traffic vehicles and vehicle tracking. Among them, vehicle monitoring means that the image processing system in the computer can use the object to carry out effective monitoring. The staff monitors the vehicle through the image differentiation method, mainly covering both static and dynamic monitoring types. The working principle is to observe the video image. Sequence, analyze whether the target vehicle is moving to achieve the purpose of monitoring. In general, this monitoring method is relatively simple, easy to operate, and the data required to be calculated is not much and can achieve fast and efficient application effects, and a large number of The use of accident detection mainly means that the intelligent transportation system can obtain the video image of the accident and can analyze the speed and actual operation of the vehicle through computer technology, thereby helping the relevant staff to analyze the real cause of the accident, thereby quickly handle traffic accidents and give corresponding tips at relevant intersections to reduce the recurrence of similar accidents.

4 Application of Computer Technology in Automobile Intelligentization

4.1 Limit Car Speed and Ensure Driving Safety

Intelligent systems can limit the speed of the car is the biggest bright spot to ensure safe driving. In today's many car driving systems, speed limit is an important aspect, and the speed of the car is also displayed on the display. Generally speaking, the intelligent speed limit systems used in the intelligent driving system are mainly divided into two types, one is reflected by the information collected after the reflection of the wireless signal emitted by the vehicle itself, and the other is the wireless broadcast. Received by the car navigator. The principle of these two intelligent speed limit systems is not exactly the same, but the effect is the same, that is, to limit the speed of the car and ensure the safety of the car.

4.2 Forming Intelligent Driving and Reducing Traffic Accidents

People have a lot of uncertainty, especially in the driving process, there are many external interference factors, and people are prone to problems. After the car is intelligently developed, infrared cameras can be installed on the front and rear bumpers and on both sides of the car, so that the situation around the car can be calculated and processed, giving reasonable driving advice [4]. Another advantage of intelligent driving is that the intelligent system can replace the human brain to issue driving instructions and carry out intelligent driving to ensure safe driving.

4.3 Increase the Parallel Warning and Increase the Safety Factor

Adding a parallel warning, the main role is to limit the driver must walk on a normal road. The parallel warning system is mainly controlled by the onboard camera. A camera captures an abnormal driving picture, and the intelligent parallel warning system warns the driver to return to the normal driving path.

4.4 Intelligent Braking System to Reduce the Chance of Collision

The biggest advantage of the intelligent brake system is that the intelligent radar monitoring system installed on the car can monitor the distance between the car and the car in real time. Once the distance is too close, emergency braking will be taken to ensure the safety of the driver. During the running of the vehicle, the driver's reaction speed is different, and the driving technology is also very different, so especially for those new drivers, the distance is the last guarantee of driving safety. With the intelligent braking system, the driver can effectively reduce the possibility of collision accidents during driving and increase the safety factor of the driver's driving. Figure 2 is a schematic diagram of the intelligent braking system. The system performs braking control through the distance of the vehicle.

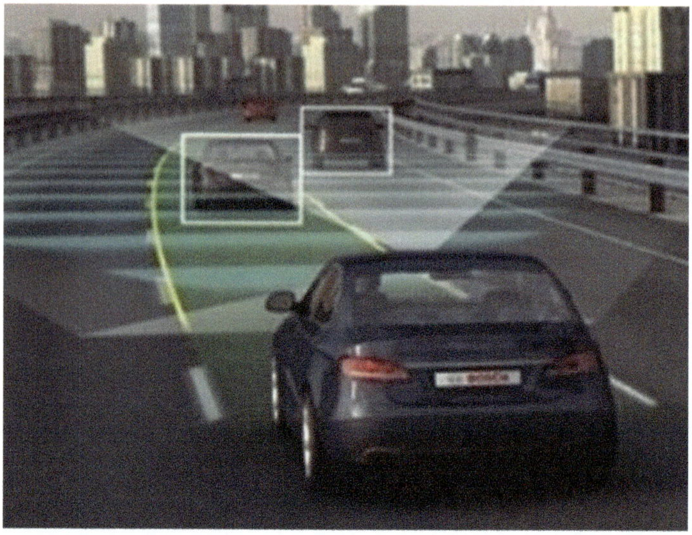

Fig. 2 Intelligent brake system

5 Future Development of Intelligent Transportation and Intelligent Car

In the past few decades and for a long time to come, computer science and technology will be the backbone of social development. Intelligent transportation and intelligent development of automobiles are inevitable trends, and even the development of smart cars may reach the situation of universal adoption of driverlessness. However, the popularization of driverless driving is certainly not a one-off event, which requires the cooperation of industry professionals and relevant government departments. For intelligent transportation, the penetration rate is already very high. However, there are still many areas for intelligent transportation that need to be improved, and the efficiency of dealing with problems can be enhanced. I believe that in the near future, the development of intelligent transportation will make our life more convenient [5]. The intelligent development of automobiles will break through our previous perceptions and our lives will be more comfortable.

In addition, "Made in China 2025" gives two goals for the development of smart cars and cars. By 2020, it is necessary to master the overall technology and key technologies of intelligent-assisted driving, and initially establish an independent research and development system and production supporting system for intelligent-networked vehicles. By 2025, it is necessary to master the overall technology of autopilot and various key technologies and establish a relatively complete independent research and development system, production supporting system and industrial cluster of intelligent networked vehicles.

6 Conclusion

The living conditions and living standards of the people are constantly rising, so it is normal for the number of cars to grow to a large extent. However, contrary to reality, the amount of land we have is constant, so the land available to everyone is getting smaller and smaller. Therefore, the pressure on urban traffic is enormous. So, as a researcher in the transportation industry, we must pay attention to this reality, strengthen our professional level as soon as possible, make rational use of the advantages of intelligent transportation, and try to solve the practical problems of transportation.

References

1. Xu, Y., Wang, R., Li, B., Li, B.: Overview of the current situation of smart vehicles in the world. Autom. Eng. (2001–05)
2. Ge, R.: Welcome to build intelligent transportation that adapts to the law of urban and traffic development in China—an interview with Quan Yongsang, director of Beijing Transportation Development Research Center. China Inf. Ind. **8**, 31–35 (2013)
3. Li, X., Zhang, W.: Research progress of intelligent vehicle navigation technology. Robot Technol. Appl. (2007–04)
4. Cheng, W., Wan, Q., Tang, X., Liu, G., Peng, G., Huang, X., Wang, Y.: Study on road recognition and control of intelligent vehicles. Servo Control (2007–06)
5. Wang, X.: Summary of research on the construction mechanism of new generation intelligent transportation system. Sci. J. **1**(6), 12–19 (2013)

Research on Key Technologies of Internet of Things and Artificial Intelligence

Min Xiao and Mei Guo

Abstract Internet of Things and artificial intelligence are the products of the current rapid development of social science and technology. Internet of Things technology is to further develop mature network technology based on the original level of Internet development. Real-time information exchange between any individual and individual, between items and items, can expand the possibility of communication and communication. Artificial intelligence is a new subject of comprehensive research on people's technology, theory, and methods, and belongs to the branch of computer science. The field of artificial intelligence involves language processing, robotics, image processing, etc., all of which are capable of simulating all aspects of human beings. This article is a scientific study and analysis after the author has made some understanding of the related research on the Internet of Things and artificial intelligence.

Keywords Internet of things · Computer · Artificial intelligence · The internet · Key technology

1 Introduction

Artificial intelligence can be simply understood as the process of symbolizing various kinds of information in human life, using computer processing programs to help humans solve difficult problems, further facilitating and accelerating human life [1].

M. Xiao · M. Guo (✉)
College of Software and Communication Engineering, Xiangnan University,
Chenzhou 423000, China
e-mail: jasmine_mm@163.com

V. Jain et al. (eds.), *Recent Trends in Intelligent Computing, Communication and Devices*, Advances in Intelligent Systems and Computing 1031,
https://doi.org/10.1007/978-981-13-9406-5_64

2 Internet of Things

2.1 Internet of Things Definition

IoT technology refers to a new type of network that realizes intelligent information exchange by connecting any external object with the network through information processing equipment such as laser scanner, global positioning system, infrared sensor, and radio frequency identification according to a certain network protocol technology. Nowadays, IoT technology can also realize the information exchange between any object and object, which is a breakthrough in the development of Internet of Things technology. Figure 1 is the design of the Internet of Things.

2.2 Key Technologies of the Internet of Things

2.2.1 RFID Tag Technology

RFID tag is a kind of radio frequency identification technology consisting of coupling components and chips. It is divided into three types: automatic, semi-active and automatic. On the technical level, RFID technology is to identify the target by RF signal. After collecting relevant information, you can use sensor technology to display what we need to know. RFID technology does not require excessive human intervention and can be used by multiple machines at the same time without causing interference.

Fig. 1 Internet of Things design

2.2.2 Embedded System Technology

Embedded system technology is one of the most influential technologies for human life. It not only allows smart terminal products to enter thousands of households but also greatly promotes the development of national defense and industrial production in our country. Embedded system is similar to embedded system technology, which is equivalent to the brain of the human body.

2.2.3 Sensor Technology

Sensor technology is also one of the key technologies in IoT technology. It is well known that almost all information that a computer can process appears in the form of digital signals. However, the signals that humans want to express or the signals they send are not all digital signals. This requires sensors to convert signals from humans or objects into digital signals that computers can recognize. Only after the conversion of the sensor can the computer process further conclusions.

2.3 Measures for IoT Technical Support

2.3.1 Importing Foreign Projects

The introduction of projects abroad is not only the quantity that must be strictly controlled but also the quality of the introduction. Especially for embedded system technology, China should solve the problem of independent research and development as soon as possible. Because the introduction of this type of high-end technology from foreign countries will not only consume a lot of funds in China, but also the repeated introduction of many different industries, resulting in waste of resources.

2.3.2 Government Increases Investment in the Internet of Things Industry

The so-called increase in government investment in the industry research of the Internet of Things means not only that the government should provide funds for industrial assistance but also that the government should take some care of the Internet of Things industry in terms of policies. Especially in the field of IC design, network development tools, etc., we must promptly give corresponding measures to accelerate the development of the Internet of Things industry [2].

2.3.3 Develop Core Applications and Promote the Development of the Internet of Things Industry

The QR code is a core application newly launched in the 12th Five-Year Development Plan of the Internet of Things. The launch of this application has helped the Internet of Things move from the previous conception to the substantive. This application is much more convenient and quicker than the previous design.

3 Artificial Intelligence

3.1 Artificial Intelligence Definition

The term artificial intelligence was first proposed at the Dartmouth Society in 1956, followed by the development of numerous research theories. Artificial intelligence is a sub-discipline of computers, mainly for the development and study of a systematic science of human intelligence. The field specifically includes language recognition, image processing, robot research, and so on. In general, artificial intelligence is not exactly the same as human intelligence, but it can transcend human intelligence and have unlimited development potential. Figure 2 is a schematic diagram of the proportion of artificial intelligence occupying the market.

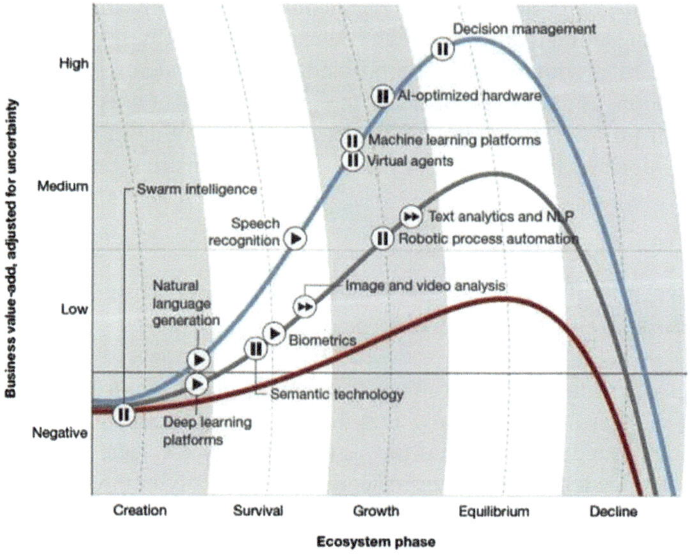

Fig. 2 Artificial intelligence design

3.2 *Key Technologies of Artificial Intelligence*

3.2.1 Machine Learning

Machine learning is actually more common in our current life, such as smart customer service in Taobao, which means that they can automatically answer some questions generated by shoppers during shopping. The key technology for this solution is machine learning. Machine learning essentially allows the machine to read a large number of similar sample data, and the system analyzes the sample data to find certain rules and form a model that conforms to the law. Usually, almost 10,000 similar sample data can be used to construct a model that fits the situation, otherwise, the probability of error will be higher. Therefore, the problem of intelligent customer service that is not within the scope of the model is unanswerable. At present, Siri of Apple's mobile phone can already reach this random question and answer without limitation. However, such a technology can only be mastered by a very small number of enterprises in a few countries, so China needs to invest a lot of time and capital costs to break through this problem.

3.2.2 Brain Science

The so-called brain science refers to letting the computer imitate the brain power of human beings to carry out a series of calculations, stresses, or responses. After the computer learns and imitates the process of brain information processing, humans can create brain-like intelligent machines or programs according to the simulation process. Brain-like science has been in advanced for more than two decades and is fundamentally different from the more connected, symbolic, and behavioralism we have heard [3]. The main core content of brain science is simulationism. It not only requires the structure to be almost identical to the human brain but also requires the level of the device to approach the neural morphology of the brain.

3.2.3 Computer Vision

Computer vision technology uses sequences of image processing operations and other techniques to decompose image analysis tasks into small, easy-to-manage tasks. For example, computer technology can assist medical devices in the diagnosis and treatment of patients' diseases, and improve the correctness of diagnosis. In addition, face recognition technology is widely used in ID card photograph recognition or network photograph analysis.

3.2.4 Bionic Storage Technology

Nature provides guidance for researchers in artificial intelligence in data storage. Some scientists say that global data usage is growing exponentially year after year. It is expected that by 2040, the demand for global data storage will exceed the supply of silicon for raw materials in flash memory devices. To this end, scientists hope to solve the above problems by using the most efficient memory cell DNA (deoxyribonucleic acid) in nature.

3.2.5 Biometrics

From a statistical point of view, human fingerprints, palms, irises, and other physiological features are unique. Therefore, these features can be used as the basis for authenticating the identity of the user. The potential for biometric applications is enormous, especially at this stage in China. This is mainly because China currently has the following realities: First, China's population base is huge, population mobility is too strong, and management is very difficult.

3.3 Measures for Artificial Intelligence Technical Support

3.3.1 The State Council Accelerates the Development and Industrialization of Intelligent Manufacturing Products

China's Ministry of Industry and Information Technology and the National Development and Reform Commission and other departments will also build artificial intelligence resources to further promote the construction and development of national smart products [4]. The state will also comprehensively strengthen the potential risk analysis and prevention of artificial intelligence development, safeguard the interests of the people and the country, and ensure that artificial intelligence is safe, reliable, and controllable. It is necessary to integrate multi-disciplinary forces, strengthen research on legal, ethical, and social issues related to artificial intelligence, and establish and improve laws, regulations, institutional systems, and ethics that ensure the healthy development of artificial intelligence. The role of artificial intelligence in human life in the future is incalculable, so state support must be the most important and most powerful.

3.3.2 Local Leaders Combine Artificial Intelligence with People's Lives

Ways and lifestyles. It is necessary to grasp the prominent contradictions and difficulties in the field of people's livelihood, and strengthen the in-depth application of artificial intelligence in the fields of education, medical and health care, sports, housing, transportation, disability and old-age care, and domestic service, and innovate intelligent service systems. In addition, local leaders can also strengthen the combination of artificial intelligence and social governance, then develop artificial intelligence systems for government services and decision-making [5].

4 Conclusion

Today, there are already many artificial intelligence research results that have entered people's daily lives. In the future, the development of artificial intelligence technology will have a greater impact on people's lives, work, and education. However, due to the limitations of the development of the times, there are still many unsolved problems in our field of Internet of Things and artificial intelligence. Therefore, we must actively explore, courageously improve our own technology, conduct research and discussion within the scope permitted by law and ethics, and do not violate ethical matters.

Acknowledgements This paper is funded by Project of:

1. School level scientific research project of Xiangnan University, Research on network security situation prediction based on data fusion (No. 2017XJ16).
2. Scientific Research Fund of Hunan Provincial Education Department, Research on Knowledge Reduction and Rule Fusion Based on Probabilistic Graph Model (No. 16C1497).

References

1. Wang, W.: The Theory and Application of Internet of Things. Higher Education Press, Beijing (2005)
2. Wu, K.: Agent technology—a new leap of artificial intelligence. The impact of science on society (1) (2000)
3. Lin, Y., Ma, S.: Introduction to Artificial Intelligence. Tsinghua University Press, Beijing (2001)
4. Sun, Y.: Talking about the development trend of artificial intelligence. IT and Network (6) (2002)
5. Wenjie, Wang: Principles and Applications of Internet of Things and Artificial Intelligence. People's Posts and Telecommunications Press, Beijing (2004)

Smart Sensor and Devices

Research on Wireless Sensor Network in Orchard Bird Repellent System Based on 6LoWPAN

Wang Pengju, Ai Zhiwei and Su Xiuzhi

Abstract Bird damage is one of the main causes of fruit damage in orchard. Aiming at the single or single-point problem in traditional bird repellent device, a networked bird repellent system model based on 6LoWPAN is proposed. Firstly, an orchard bird repellent system model and its network topology structure are designed according to the self-organization of 6LoWPAN sensor network, the uniqueness of the node address and the neighbor discovery and the networking characteristics. Secondly, the hardware design and software programming algorithm of bird repellent system are designed, and a network bird repellent algorithm is proposed. Finally, the bird repellent model and the designed bird repellent system were tested to test the effect of bird repellent algorithm. The test data and results show that the network bird repellent system based on 6LoWPAN sensor network has enhanced the effect of bird repellent ability and well protected the fruit from the bird in the orchard.

Keywords Bird damage · 6LoWPAN · Networking style · Bird repellent algorithm

1 Introduction

With the strengthened protection of environment, the habitat of birds becomes better and better, which leads to a new problem—birds' damage, especially in orchard when the fruits are around to ripe. Birds' damage makes orchard man loss a

W. Pengju (✉) · A. Zhiwei
Guilin University of Aerospace Technology, Guilin, China
e-mail: wpjxiaolu@126.com

S. Xiuzhi
Hunan Software Vocational Institute, Xiangtan, China

V. Jain et al. (eds.), *Recent Trends in Intelligent Computing, Communication and Devices*, Advances in Intelligent Systems and Computing 1031,
https://doi.org/10.1007/978-981-13-9406-5_65

lot. Even more, the damaged fruits may cause disease transmission among mankind without handling them. Bird damage has become an urgent problem. Net and drug are traditional ways to repellent birds. These ways might hurt the birds, even their lives. The development of the technology makes repellent birds possible without hurting birds [1], proposed a method which uses Doppler detecting principle to repellent birds. The application of bionic UAV in the air repellent is proposed in Ref. [2]. The proposed methods take effect in the initial stage. With time goes by, the birds immune to these methods, the effect of bird repellent reduces a lot.

There is a big address space in 6LoWPAN (IPV6 Low Power Wireless Personal Area Network), as well as a low power consumption. It supports star and mesh topology networking under Contiki stage. Wireless sensor networks based on 6LoWPAN has a good application prospect on many occasions.

The usage of wireless sensor network based on 6LoWPAN in environmental monitoring, forest fire monitoring, smart lighting, smart city, medical and farmland information monitoring is proved effective in Ref. [3–10]. This paper tries to apply the 6LoWPAN to the orchard bird control system. On the basis of the traditional single-point sound drive bird and single-point ultrasonic displacement bird, the network security model algorithm is added. Drive bird network which is made up of several signals is designed to enhance the ability of the repellent bird Ref. [11].

2 Design of Bird Repellent Net and System Structure

2.1 Model of Bird Repellent Net

6LoWPAN sensor network support tree, star and mesh topology. The mesh topology model is used in the bird repellent net system. Every spot in the system has the same structure and function. The net model is designed as follows.

(1) Self-organization: The structure of 6LoWPAN can be divided into simple 6LoWPAN, extensional 6LoWPAN and self-organization 6LoWPAN. Self-organization 6LoWPAN can be used without connecting the Internet.
(2) Node address: All nodes in 6LoWPAN share a common IPV6 address prefix in 6LoWPAN. Data can be distinguished, posted and received through an exclusive address (Fig. 1).
(3) Neighbor discovery: Neighbor discovery is a key point in 6LoWPAN. The startup and maintenance in the system is charged by the function. The node uses the neighbor discovery function to discover the other nodes on the same link to determine its link-layer address, to find the router and to maintain the information transfer between the nodes [12–15].

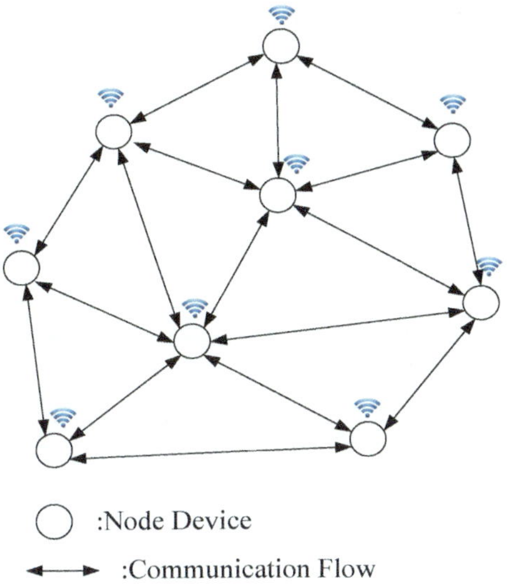

Fig. 1 Model of bird repellent net

2.2 Design of Bird Repellent Net System

The structure of the bird repellent net system is shown in Fig. 2—putting A, B, C, D or more bird repellent points in orchard.

When a bird flies to the sensing area of B node, it will be induced by the infrared sensor of B node.

Situation 1: B nodes immediately send out bird repellent sounds, and birds sense the dangerous sound signals which are emitted by B nodes and fly away from orchard areas immediately.

Situation 2: Although the bird feel the dangerous sound signals which is sent by B node, it has certain environmental adaptability and continues to stay at the B node to destroy the fruit around the B node. When it exceeds a certain time (setting the value of 5 s), the intelligent control device of the B node starts the 6LoWPAN wireless data transmission unit to the B node. The peripheral (such as A, B, C, D or other nodes in the map) signal, these nodes receive the alarm signal, immediately send out

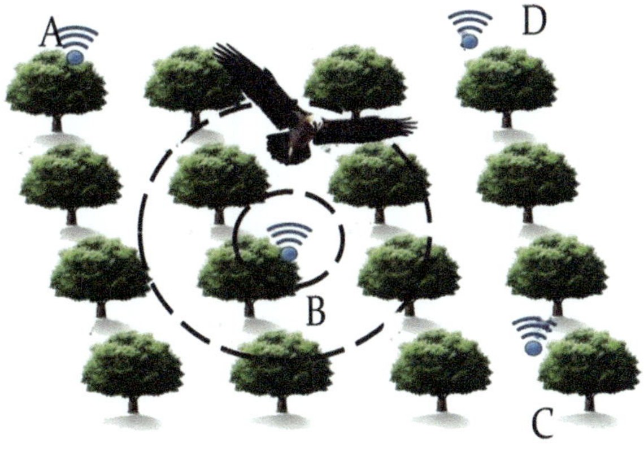

Fig. 2 Diagrammatic sketch of bird repellent

the sound of the drive bird, make the B node surrounding the drive bird sound signal stronger, birds feel the threat, frightened and fly away from the orchard.

Situation 3: Birds have adapted to the dangerous sound around them. The first two cases are not enough to drive them away, and they remain around the B nodes. The nodes continue to send out dangerous sounds of the birds because of the presence of the birds. The dangerous sounds from the A, B, C, D or other nodes are random and may simultaneously send out like wolves, the voice of cry, the voice of the machine gun, the sound of the bullets flying in the air, the screams of the birds, and so on. The birds cannot accurately determine where the real danger comes from, and the birds fly away from the orchard.

Situation 4: Birds are accustomed to sound signals around them. No voice playing around can drive them away. The birds still harming the fruits around the B node or B node. When more than a certain time (such as 10 s), the B node closes the sound signal output and sends out the frequency conversion ultrasonic signal and sends another one to the surrounding node. After receiving the signal, A, C, D nodes or other nodes closed the voice signal. At the same time, a frequency conversion ultrasonic signal is also sent. In the environment of different frequencies of ultrasonic signals, the birds will be stimulated, become restless, difficult to eat, not to the surrounding environment and fly away from the orchard.

3 Application of Bird Repellent System Based on 6LoWPAN

3.1 Design of the Nodes' Hardware

The design diagram of the hardware of bird repellent system is shown in Fig. 3.

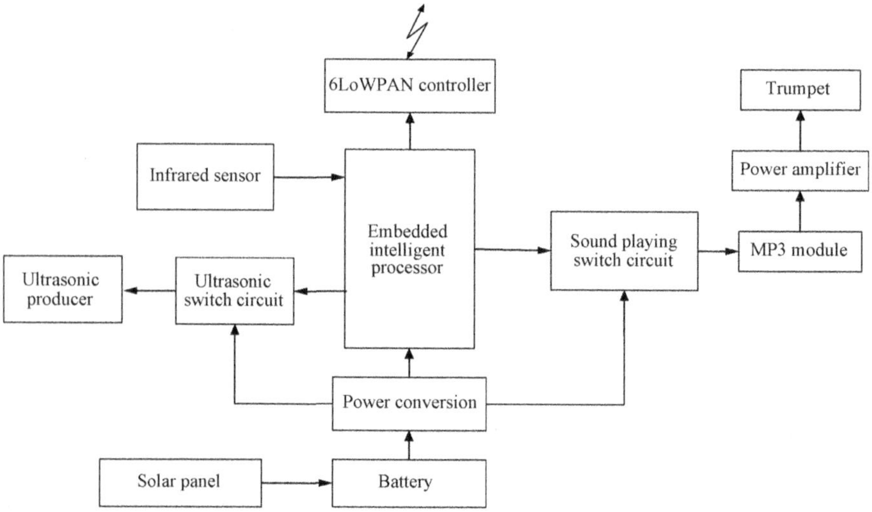

Fig. 3 Design of the nodes' hardware in bird repellent system

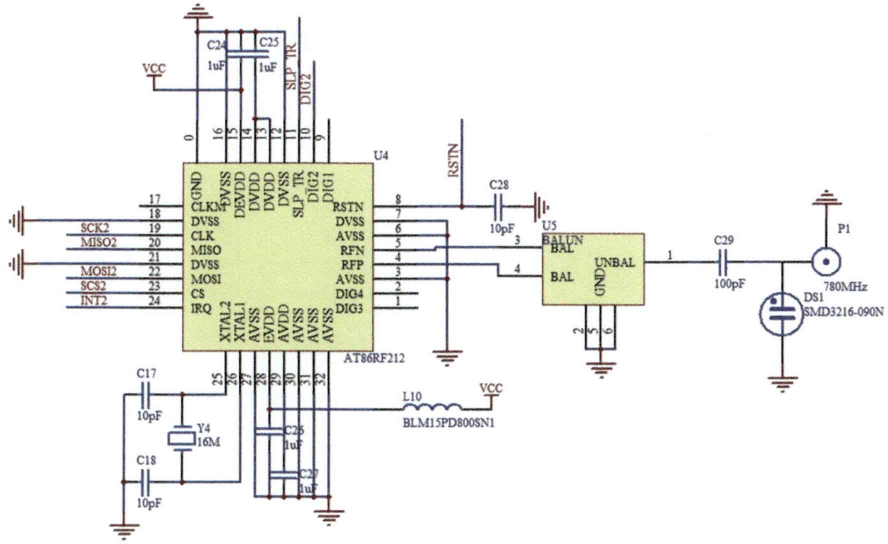

Fig. 4 Circuit of wireless data transmission chip

The node organizes the network by using the 6LoWPAN wireless module. The core chip of 6LoWPAN is AT86RF212B produced by Atmel company. The chip uses the central frequency of 780 MHz to transmit the wireless data, and its specific application circuit is shown in Fig. 4.

Using infrared sensing module to detect birds around the nodes, it can detect birds' activity within 0–5 m.

The bird repellent action uses two kinds of combination of random playing drive bird audio and random transmitting frequency conversion ultrasonic signal to reduce the energy consumption of the nodes, reduce the environmental adaptability of birds and increase the probability of the drive birds.

Using solar panels and lead–acid batteries to supply power to the entire node, nodes can be arranged arbitrarily in the orchard.

3.2 Design of Node Algorithm

The programming of software algorithm for bird driving node is one of the important contents of system design. The main program of each node is divided into 2 main tasks which are the detection of the infrared signals of the birds around this node and the detection of the wireless signals sent by other nodes around the node. In the program design process, query and interrupt are implemented in two ways, respectively. The flowchart of the program design is shown in Fig. 5.

The main program inquires the output signal of infrared sensor detection module. Once a bird or other infrared-releasing object enters the induction area, the

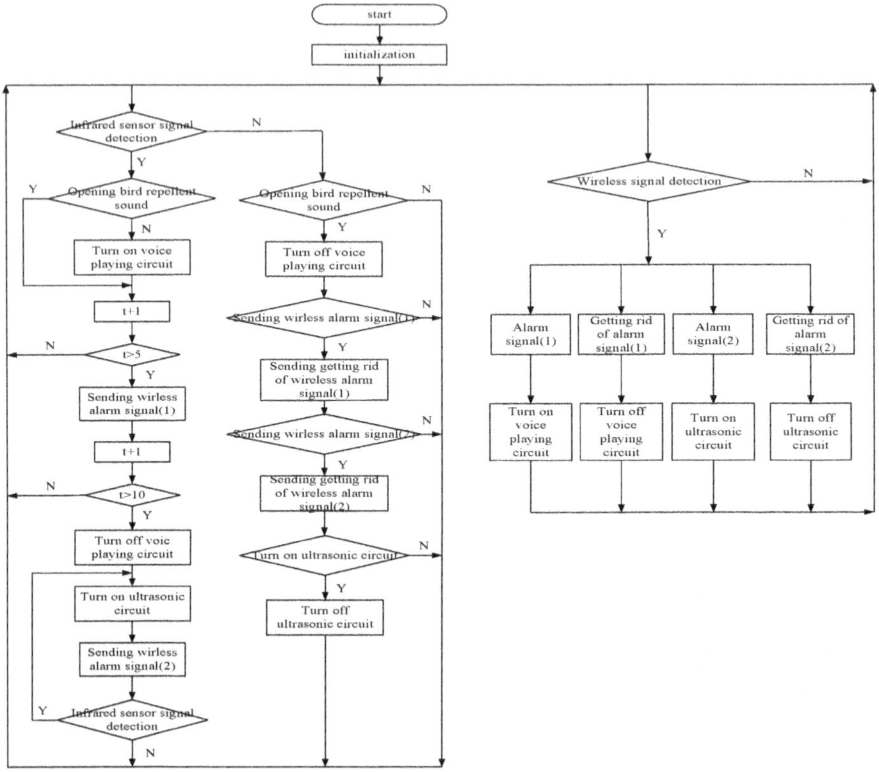

Fig. 5 Flowchart of bird repellent node algorithm

program opens the voice switch circuit immediately and opens the timer to judge the time of the bird's stay around the node. If birds fly away from the infrared sensing area of the node within 5 s, the program closes the bird repelling sound. Otherwise, an alarm signal (1) will be transmitted to other nodes through 6LoWPAN. The program closes the sound if the bird stays at a node exceeding 10 s and the ultrasonic circuit starts to work.

Through the interruption, the program detects the 6LoWPAN wireless signal around the node. After receiving the wireless signal, the node performs the action of opening voice play, closing voice play, opening ultrasonic wave, closing the ultrasonic wave according to the signal.

4 Experiment and Application

The flight status of birds in nature is not regular. The experiment and application of bird repellent system are tested by combining artificial control with the actual test data of the orchard.

4.1 Simulation of Bird Displacement Test

In an open field, four nodes are placed with a distance of about 5 meters between each other. One of the nodes is shown in Fig. 6.

Birds in the cage are used to simulate the invading birds. The cage is fixed on the remote four-wheel trolley, and a noise tester is fixed on the cage to measure the bird's audio frequency. By controlling the mobile phone, the four-wheel car is gradually close to one of the nodes. When the infrared sensor of the node senses the infrared signal of the bird in the cage, the piercing bird's voice will be displayed. The bird in the cage is frightened and flew up and down. After 5 s, the other nodes around the cage also sent out the sound of the birds. The test data of noise is shown in Fig. 7. All the sounds turn off after another 5 s. When ultrasonic turns on, the bird in the cage is frightened, the cage is impacted and the nervous system of the bird is stimulated.

Fig. 6 Photograph of the node

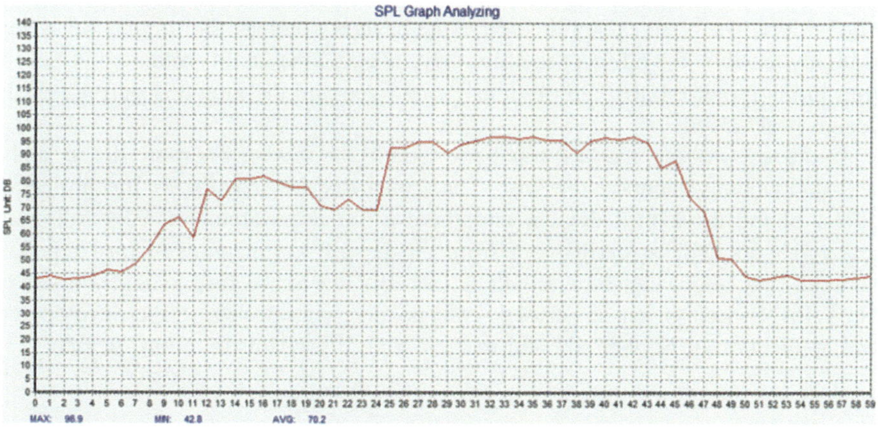

Fig. 7 Test data of the noise around the bird

Table 1 Orchard field test result

Statistical day	Number of invading birds	
	Protection zone	Observation area
1st	5	37
2nd	3	22
3rd	6	28
4th	2	43
5th	3	45

4.2 Orchard Field Test

A mature wax berry orchard is selected as the test site. A test node is placed on every fruit tree of wax berry. A total of each of the pieces of 10 adjacent wax berry trees are selected as protection objects for bird repellent system. Another 10 nearby wax berry fruit trees are selected as the observation objects. Statistics showed that the number of birds in the protected area and the observation area fell within four hours at 7:00–11:00 a.m. The results are compared which is shown in Table 1.

From the statistical data and the field test, it can be seen that with the gradual maturity of wax berry, fresh red fruits attract a large number of birds to peck, but in the protected area, only a few large birds of the bird invade and stay for a long time. The quality and quantity of wax berry are significantly higher in the protected area from the quality of the fruit picked on the spot, with an average picking more than 30% of the fruit.

5 Conclusion

An orchard bird repellent device based on 6LoWPAN is designed and implemented in the paper. By using infrared sensors, the invasion of birds around the drive bird nodes can be sensitively detected, and the bird repellent net is made up based on 6LoWPAN. The practical application in the bayberry orchard shows that the system is highly feasible, the effect of the bird drive is good, the quality and quantity of the fruit in the protected area are obviously improved and the intended purpose of the design is achieved.

In the test process, the bird repellent node in the test area is more dense, and it will undoubtedly increase the cost. Ten fruit trees are selected as the protection object of the drive bird device, and there is no large area test. The follow-up research work will be carried out from the layout optimization of bird repellent network and large area bird repellent test. But the drive bird scheme based on 6LoWPAN wireless sensor network is of positive significance to the development and application of 6LoWPAN and the research of orchard drive bird device.

Acknowledgments This work was supported by the research project of education and teaching reform in Hunan Vocational Colleges (No. ZJGB2016233); Natural Science Foundation of Hunan Province of China (No. 2017JJ5036); and Promotion of Basic Ability of Young and Middle-aged Teachers in Universities in Guangxi (No. 2018KY0660).

References

1. Lu, Y.S., Yu, L.Z.: Intelligent bird repeller based on detecting principle of doppler. Mod. Electron. Tech. **34**(24), 174–177 (2011)
2. Wang, S.Y., Liu, Z.T., Zhong, C.S.: A type of flying bionic intelligent bird repeller. Agric. Eng. **7**(6), 21–24 (2017)
3. Li, Y.M., Liu, C., Liu, Z.Y.: Environmental monitoring system based on 6LoWPAN wireless sensing technology. Instrum. Tech. Sens. **7**, 50–54 (2016)
4. Hu, Y.: Design for monitoring system of forest fire based on 6LoWPAN. Comput. Meas. Control **22**(4), 1099–1101 (2014)
5. Huang, Z.C., Yuan, F., Li, Y.: Implementation of IPV6 over low power wireless personal area network based on wireless sensor network in smart lighting. J. Comput. Appl. **34**(10), 3029–3033 (2014)
6. Feng, D.Q., Zhu, X.W.: Design and implementation of data acquisition system for 6LoWPAN smart city. Comput. Eng. **43**(11), 286–291 (2017)
7. Wang, B., Wang, W.Y., Yang, T.: Wireless monitoring system of farmland information based on 6LoWPAN. Transducer Microsyst. Technol. **30**(4), 149–152 (2011)
8. Fengi, C.W., Hu, G.Q.: Greenhouse intelligent monitoring system based on 6LoWPAN and WLAN. Microcontrol. Embed. Syst. **12**, 70–73 (2016)
9. Chen, P.: System design of farmland intelligent anti-bird system based on Internet of Things. J. Lanzhou Petrochem. Polytech. **18**(1), 24–26 (2018)
10. Gui, B., Zeng, B.: Design of medical monitoring system based on 6LoWPAN and wireless sensor networks. Autom. Inf. Eng. **36**(2), 1–7 (2015)
11. Zhang, C., Li, G.L., Jin, C.: On design and test of intelligent equipment of driving farmland birds. J. Southwest China Norm. Univ. (Nat. Sci. Ed.) **41**(5), 81–87 (2016)
12. Duan, Y., Song, H.S., Yan, J.X.: System of gas environment monitor based on 6LoWPAN technique. Comput. Meas. Control **23**(7), 2357–2362 (2015)
13. Yan, H.Y., Huang, B., Wang, X.N.: Elevator comfort monitoring system based on sensor and 6LoWPAN. Meas. Control Technol. **6**, 21–23 (2015)
14. Li, Z.Q., Song, M.X., Wang, H.F.: Design of classroom light monitoring system based on 6LoWPAN. J. Huazhong Univ. Sci. Technol. (Nat. Sci. Ed.) **44**, 210–214 (2016)
15. Wang, X.K., Wei, S.D., Wang, H.F.: Study of the wireless remote control bird repellent system. J. Chang. Univ. Sci. Technol. (Nat. Sci. Ed.) **38**(6), 163–171 (2015)

Adaptive Adversarial Example Generating Network for Object Detection

Zhiqiang Li, Junsheng Liang and Hongchen Guo

Abstract The consensus of most recent researches is that rich training data which includes more different object forms in different situations is beneficial to improve the performance of deep networks. Deep learning is very dependent on data diversity. So, we provide a novel and principle method called adaptive adversarial example generating network (AAEGN) to generate examples with occlusions for object detection. AAEGN can generate occlusion examples which are rare in current dataset and are difficult to localize and classify. Compared to the traditional adversarial network for the example generation, AAEGN employs several innovations to simplify the architecture of adversarial network and improve the example generation and training speed. What's more, AAEGN can adjust the hyper-parameters of the adversarial network in the process of training adaptively. The results of our experiments show that using AAEGN to train Fast-RCNN pipeline can get a 1.8% boost on VOC2007 dataset and 3.8% boost on VOC2012 dataset.

Keywords Deep learning · Object detection · Example generation

1 Introduction

It is the fact that the object detection task contains various complex situations, which result in low detection accuracy. For example, some important regions of an object are occluded, deformed or blurred, which are caused by background changes, conditions of illumination or the angle of shooting. Therefore, the detection models may not perform effectively due to the unobvious features extracted from such important regions. We hope that the training data can provide all possible variations

---

Z. Li · J. Liang
School of Computer Science, Beijing Institute of Technology, Beijing, China

H. Guo (✉)
Network Information Technology Center, Beijing Institute of Technology, Beijing, China
e-mail: guohongchen@bit.edu.cn

V. Jain et al. (eds.), *Recent Trends in Intelligent Computing, Communication and Devices*, Advances in Intelligent Systems and Computing 1031,
https://doi.org/10.1007/978-981-13-9406-5_66

of a visual cue and let the network to learn these invariances automatically. But it is hard to collect a dataset which includes all possible situations where some occlusions are so rare in real life. Data augmentation [1–4] has been widely applied in the training stage and achieved a great effect. But it cannot solve the above problems thoroughly.

Recent researches [5–7] try to identify discriminative regions and features of the objects for classification and detection models. The visualization results from these researches also proved it very intuitively. Hide-and-Seek [8] provides a weekly supervised method to force our models to learn other regional features except the most discriminative ones. But it is useful in object classification and weakly supervised object localization, not object detection which would handle more complex situations. A-Fast-RCNN [9] provides a new idea that using adversarial networks to identify the discriminative parts in each proposal and generates new occlusion and deformation proposals in training for object detection. To the best of our knowledge, using adversarial networks to generating examples for object detection is the first general purpose work which is mentioned in A-Fast-RCNN [9]. It uses an extra convolutional network called adversarial spatial dropout network to compute features of each proposal and generates the relevant occlusion masks. A-Fast-RCNN includes two sub-networks: adversarial spatial dropout network (ASDN) and adversarial spatial transform network (ASTN). We only refer to ASDN in this paper.

In this paper, we go deeper with it and provide a novel and principle method called adaptive adversarial example generating network (AAEGN). Instead of computing features for each proposal, we directly put the features of the whole image into an extra convolutional network and then calculate each proposal's feature map separately according to RoI pooling layer. This strategy is proposed in SPPnet [10] firstly which makes the training time reduced by 3× due to the faster proposal feature extraction. And ASDN needs to be pre-trained to initialize its parameters and then jointly optimize the network in the training of the detection model. But the prediction results generated from ASDN still cannot be too accurate to represent the most discriminative regions of each proposal. When we used randomly occlusion masks instead of masks generated from ASDN, we found that the accuracy of the final detection model did not fall a lot. What's more, the pre-trained stage of ASDN needs to cost a lot of time which is almost two to three times the whole training time of the detection model. So we remove the pre-trained process and directly use the random variables to initialize AAEGN and jointly optimize it in the training process. Instead of setting a fixed threshold to distinguish the positive and negative examples for AAEGN, we use an adaptive threshold which can get a better effect.

Our work has made some contributions based on the work of predecessors. We introduce the idea of adaptive adversarial example generating network (AAEGN) for object detection and produce state-of-the-art results on the PASCAL VOC [11] and COCO [12] dataset. Using AAEGN to train detectors can reduce the training time greatly and get a better detection accuracy compared to other relevant methods like ASDN [9] and OHEM [13]. What's more, our method AAEGN can be extended to other region-based detection frameworks very easily.

2 Related Work

In the field of object detection, great improvements have been achieved in recent years. And the accuracy and speed of the object detection are continuously improved which mainly come from the promotion of computing capacity like the operation based on GPU and the ability to learn and extract the image features by the detection model. In the early stage, the application and combination of low-level image features, such as SIFT and HOG [14, 15], are the mainstream methods in object detection tasks and have achieved great success. Instead of using manual features, convolutional neural network (CNNs) can learn image features from training data autonomously by SGD algorithm. And in 2012, Krizhevsky et al. [16] show deep CNNs can achieve substantially higher accuracy of image classification on ILSVRC [17] which make deep CNNs, the most popular framework for image classification. Object detection also has benefited from deep CNNs. For region-based detection methods, the detection task has been simplified to the task of abundant proposals' classification. R-CNN [18] is the first to do it by classifying and localizing each proposal separately. And other methods like Fast-RCNN, Faster-RCNN, RFCN, Mask-RCNN [19–22] have improved it from different aspects.

Researches in the past few years are mainly focused on the following aspects: base architecture improvement [10, 18, 20, 21], additional contextual reasoning and multi-features integration [23–25]. However, little work has focused on the data perspective. Data diversity is very important for training deep CNNs. But data collection needs much labor cost and it is hard to cover all of the cases. So, we use a lot of strategies to expand data diversity. One of the effective ways is hard exampling mining and generating. OHEM [13] algorithm provides an online hard example selection technique to train the detector which has achieved significant boost in detection performance. Hide-and-Seek [8] uses randomly occlusion image patches in the training process in order to force the detector to learn the features of the remaining regions. A-Fast-RCNN [9] provides a novel idea that we can generate hard examples using adversarial networks. It uses adversarial networks to generate occlusion examples and deformation examples. These examples are hard for our object detector to classify. It tries to generate occlusions and deformations in feature vectors instead of pixels. Especially, its adversarial spatial dropout network can generate occlusion masks for region-based features.

3 Our Method

Our method adaptive adversarial example generating network (AAEGN) has improved the performance of adversarial spatial dropout network for hard example generation in the two aspects: sharing feature computation and adaptive threshold.

Firstly, we need to create an example generation network. Instead of generating occlusion examples in pixels, we generate it in the convolutional feature space. What we need to do here is to create a generation network which would predict the most discriminative regions for each proposal. In region-based detectors, each proposal corresponds to a convolutional feature map. Through RoI pooling layer, we can generate region-based feature for each proposal. We add two convolution layers behind the last layer of the original CNNs. These two convolution layers were used to predict the most discriminative regions in the whole input image. After that, we add another RoI pooling layer to generate the most discriminative regions for each proposal. We reduce the computation of the example generation network greatly by it. And it also can make full use of the global features of the whole images which would reduce the prediction error of the network. We use the random variables to initialize it and jointly optimize it in the training process. The whole architecture has been shown in Fig. 1. The network consists of two parts: the first part is our AAEGN which is used to generate masks for blocking region-based features (on the blue background). The second part is a region-based detection network. It is used to extract features, and then, it uses these features to classify and localize objects.

The next step is to jointly optimize AAEGN in the training process. We use a read-only network to calculate each proposal's class loss before masked. It is independent of the whole network and does not have the back-propagation stage so we do not need to train it. Its parameters are mirroring the parameters of the full-connected layers in Fast-RCNN. The loss of AAEGN is equal to the class loss before masked minus the class loss after masked. The loss of AAEGN represents the impact from the occluded regions. The bigger the loss, the more important it is

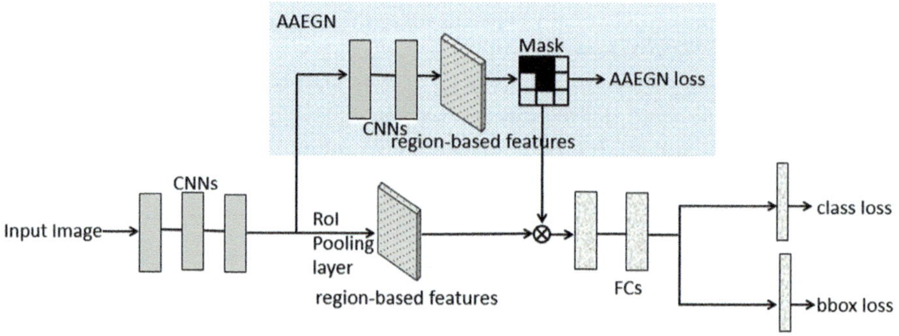

Fig. 1 AAEGN in the standard Fast-RCNN architecture

for the detector to classify the object successfully. So, according to the loss of AAEGN, we can distinguish positive and negative examples in the training stage. Here, the loss of AAEGN corresponds to each proposal. But the output generated by AAEGN is a continuous heat map. So, we need to choose the positive proposals and their corresponding masks. And then, we use these masks as positive examples to update weights in AAEGN.

Instead of using a fixed threshold to distinguish the positive and negative examples, we use an adaptive threshold which can get a better effect. At the beginning of training, we use the random variables to initialize AAEGN, so the output of AAEGN tends to have random values in the early stage. In order to make our positive examples closer to the ground truths, we firstly select the top $N\%$ positive examples and then in each mask, we randomly select $M\%$ pixels where the value was 1 before and change it to 0. With the progress of training, we will gradually increase the value of N and reduce the value of M. In the training stage, we just use these adaptive settings for back-propagation to optimize AAEGN. But in the example generation stage, we still use a fixed proportion (30%) to mask the regions in each proposal. As the output generated by AAEGN is a continuous heat map, we directly choose the top 30% pixels to mask out.

4 Experiments

We conduct most of the experiments on the PASCAL VOC2007 [11]. And we also conduct our experiments on PASCAL VOC2012 [11] and MS COCO [12]. We compare our AAEGN with ASDN [9] and OHEM [13] algorithm on the PASCAL VOC2007 and VOC2012. In the end, we present the results on MS COCO and show some category-based results and analysis.

4.1 Experimental Settings

We implemented our AAEGN based on the Fast-RCNN [19] pipeline in our experiments. And we use pre-trained parameters from ImageNet to initialize CNNs in Fast-RCNN. The standard CNN architecture we used is VGG16. For PASCAL VOC2007 dataset, we use the 'trainval' set for training and 'test' set for testing. We train our detection network for 50k iterations. The base learning rate is 0.001 and after 40k iterations, we decay the learning rate to 0.0001. Each iteration contains two batches, and we apply SGD in each iteration.

4.2 How Should We Occlude Background Examples?

In the training process, we provide foreground proposals in a batch without any occlusions and provide foreground proposals with occlusions (30%) in the other batch. And in order to avoid our detector learning the difference between foreground and background examples caused by occlusion, we also use background examples with occlusion to train our model. Background examples may contain a part of the object; when we randomly dropout the features in background examples with a high proportion, the remaining features of the background examples may be confused with the features of the foreground examples with occlusion. So we tried different combinations and chose the best one. We show the results in Table 1. When we randomly blocked the 15% regions of the background proposals only in the second batch, we got the best result, 71.2% mAP in VOC2007.

4.3 Comparisons with ASDN and OHEM

In order to show our AAEGN to get a better performance compared to other similar methods like ASDN and OHEM, we compared our AAEGN with them on the PASCAL VOC2007 and VOC2012. The results are shown in Tables 2 and 3.

Table 1 VOC2007 test detection mAP (%)

Batch 1		Batch 2		mAP (%)
Foreground (%)	Background (%)	Foreground (%)	Background (%)	
0	0	30	0	71.0
30	0	30	0	69.8
30	0	30	10	69.7
0	**0**	**30**	**15**	**71.2**
0	15	30	15	70.3
0	0	30	20	71.0
0	0	30	30	70.8

Table 2 VOC2007 test detection average precision (%). We select six class's AP to show from the 20 classes

Method	mAP	Bike	Bird	Bus	Car	Table	Train
Fast-RCNN	69.4	78.4	70.1	77.4	76.6	70.1	74.8
ASDN	70.8	80.5	70.2	80.3	79.8	69.4	78.6
OHEM	69.9	78.3	69.2	81.8	81.8	68.9	75.6
AAEGN	**71.2**	**74.3**	**69.5**	**82.2**	**82.2**	**70.0**	**77.4**

Table 3 VOC2012 test detection average precision (%). We select six class's AP to show from the 20 classes

Method	mAP	Bike	Bird	Bus	Car	Table	Train
Fast-RCNN	66.4	74.4	66.5	75.9	69.1	54.0	76.7
ASDN	68.7	75.4	69.2	76.4	70.3	58.2	77.5
OHEM	69.8	78.9	69.6	77.4	72.1	58.3	78.5
AAEGN	**70.2**	**78.3**	**70.2**	**77.9**	**72.6**	**58.6**	**77.4**

On the PASCAL VOC2007 dataset, we get the best result 71.2% mAP. And compared to ASDN, our AAEGN does not need the pre-train stage which can greatly shorten the training time.

On the PASCAL VOC2012 dataset, we also got the best result 70.2% mAP. The result indicates that when the dataset's diversity increased, ASDN's benefit obviously reduced, but our AAEGN not.

4.4 MS COCO Results

In order to prove our AAEGN can be significant on the best challenging dataset, we train the standard Fast-RCNN with our AAEGN on the MS COCO2014 dataset [12]. On the COCO2014 dataset, we use the 'train' set for training and the 'minival' set for testing. We show the results in Table 4. Our AAEGN improved the performance of the baseline Fast-RCNN from 36.7 to 40.1% mAP@0.5 on the VOC metric and from 18.8 to 20.6% AP on the standard COCO metric. It proved that our method MOEG can be effective even though datasets become larger and more challenge.

4.5 Category-Based Analysis Comparisons with ASDN and OHEM

Our experimental results show that our AAEGN can significantly improve the detection performance on some categories which can get benefit from multi-region

Table 4 MS COCO2014 minival detection average precision (%)

AP@IoU	Area	FRCN	AAEGN	Boost
[0.50: 0.95]	All	18.8	20.6	1.8
0.50	All	36.7	40.1	3.4
[0.50: 0.95]	Small	4.0	6.5	2.5
[0.50: 0.95]	Med.	19.1	21.4	2.3
[0.50: 0.95]	Large	33.4	35.1	1.7

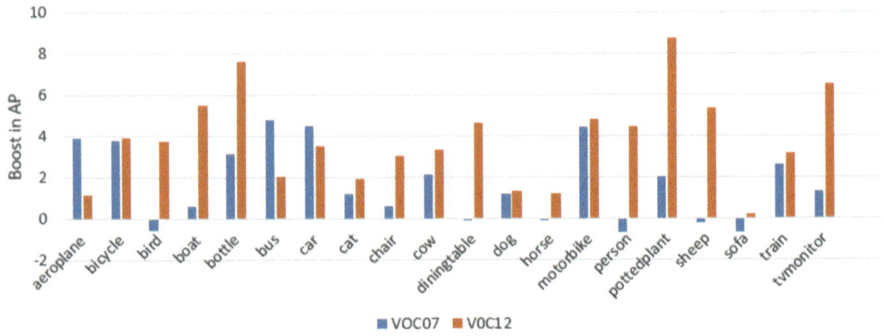

Fig. 2 Boosts of APs by using our AAEGN compared to the baseline Fast-RCNN

features. We present the boost of AP in each class by using our AAEGN and compare it with the baseline Fast-RCNN. We get the best performance on such categories like bicycle, bus, car, motorbike on PASCAL VOC datasets. And this advantage of AAEGN will be valuable to the field of autonomous vehicle (Fig. 2).

5 Conclusions

In this paper, we present an adaptive adversarial example generating network for object detection called AAEGN. It is an improvement on ASDN in the two aspects: sharing feature computation and adaptive threshold. From our experimental results, we can see that our improvement is remarkable. And compared to the state-of-the-art method like ASDN and OHEM, our AAEGN can get a higher detection accuracy and cost less training time.

References

1. Krizhevsky, A., Sutskever, I., Hinton, G.E.L.: ImageNet classification with deep convolutional neural networks. In: International Conference on Neural Information Processing Systems, pp. 1097–1105. Curran Associates Inc. (2012)
2. Simonyan, K., Zisserman, A.: Very deep convolutional networks for large-scale image recognition. Comput. Sci. (2014)
3. He, K., Zhang, X., Ren, S. et al.: Deep residual learning for image recognition. In: Proceedings of the IEEE Conference on Computer Vision and Pattern Recognition, pp. 770–778 (2015)
4. Szegedy, C., Liu, W., Jia, Y. et al.: Going deeper with convolutions. In: Proceedings of the IEEE Conference on Computer Vision and Pattern Recognition, pp. 1–9 (2014)
5. Zhou, B., Khosla, A., Lapedriza, A. et al.: Learning deep features for discriminative localization. In: IEEE Computer Vision and Pattern Recognition, pp. 2921–2929 (2016)

6. Cinbis, R.G., Verbeek, J., Schmid, C.: Weakly supervised object localization with multi-fold multiple instance learning. IEEE Computer Society (2017)
7. Oquab, M., Bottou, L., Laptev, I. et al.: Is object localization for free?—weakly-supervised learning with convolutional neural networks. In: IEEE Computer Vision and Pattern Recognition, pp. 685–694 (2015)
8. Singh, K.K., Yong, J.L.: Hide-and-seek: forcing a network to be meticulous for weakly-supervised object and action localization (2017)
9. Wang, X., Shrivastava, A., Guptam A.: A-Fast-RCNN: hard positive generation via adversary for object detection (2017)
10. He, K., Zhang, X., Ren, S., et al.: Spatial pyramid pooling in deep convolutional networks for visual recognition. IEEE Trans. Pattern Anal. Mach. Intell. **37**(9), 1904–1916 (2015)
11. Everingham, M., Gool, L.V., Williams, C.K.I., et al.: The pascal visual object classes (VOC) challenge. Int. J. Comput. Vision **88**(2), 303–338 (2010)
12. Lin, T.Y., Maire, M., Belongie, S. et al.: Microsoft COCO: common objects in context. In: European Conference on Computer Vision, vol. 8693, pp. 740–755 (2014)
13. Shrivastava, A., Gupta, A., Girshick, R.: Training region-based object detectors with online hard example mining. In: IEEE Computer Vision and Pattern Recognition, pp. 761–769 (2016)
14. Lowe, D.G.: Distinctive image features from scale-invariant keypoints. Int. J. Comput. Vision **60**(2), 91–110 (2004)
15. Dalal, N., Triggs, B.: Histograms of oriented gradients for human detection. In: IEEE Computer Society Conference on Computer Vision and Pattern Recognition, pp. 886–893 (2005)
16. Krizhevsky, A., Sutskever, I., Hinton, G.E.: ImageNet classification with deep convolutional neural networks. In: International Conference on Neural Information Processing Systems, pp. 1097–1105. Curran Associates Inc. (2012)
17. Russakovsky, O., Deng, J., Su, H., et al.: ImageNet large scale visual recognition challenge. Int. J. Comput. Vision **115**(3), 211–252 (2015)
18. Girshick, R., Donahue, J., Darrell, T. et al.: Rich feature hierarchies for accurate object detection and semantic segmentation. In: IEEE Conference on Computer Vision and Pattern Recognition. IEEE Computer Society, pp. 580–587 (2014)
19. Girshick, R.: Fast R-CNN. Comput. Sci. (2015)
20. Ren, S., He, K., Girshick, R. et al.: Faster R-CNN: towards real-time object detection with region proposal networks. In: International Conference on Neural Information Processing Systems, pp. 91–99. MIT Press (2015)
21. Dai, J., Li, Y., He, K. et al.: R-FCN: object detection via region-based fully convolutional networks (2016)
22. He, K., Gkioxari, G., Dollar, P. et al.: Mask R-CNN. IEEE Trans. Pattern Anal. Mach. Intell. (99), 1 (2017)
23. Zeng, X., Ouyang, W., Yang, B. et al.: Gated bi-directional CNN for object detection, pp. 354–369 (2016)
24. Chen, X., Gupta, A.: Spatial memory for context reasoning in object detection, pp. 4106–4116 (2017)
25. Bell, S., Zitnick, C.L., Bala, K. et al.: Inside-outside net: detecting objects in context with skip pooling and recurrent neural networks. In: IEEE Computer Vision and Pattern Recognition, pp. 2874–2883 (2016)

Laser-Radar-Based Highway Visibility Monitor

Yueqin Wang and Xiaomin Xie

Abstract Designed highway visibility real-time monitoring system. The system can monitor, analyze, and take effective early warning of the full-time real-time monitoring of atmospheric visibility along the highway. The current visibility is measured by the lidar, and the wireless communication module sends the current visibility value to the driver and display. The system is capable of preventing traffic accidents caused by low visibility, especially short-term low visibility such as random fog.

Keywords Atmospheric visibility · Lidar · Highway

1 Introduction

Research on visibility monitors began as early as the 1930s and has undergone three stages of development: transmissive, forward scattering, and backward scattering. The forward-scattering visibility meter only measures the amount of scattering of a small volume of air sample, so the climate characteristics with low performance in some areas cannot be well monitored and monitored, especially for the fog, which has low visibility and comparison of spheres of influence. Small local microclimate environments are difficult to monitor in real time. The visibility monitoring range of the highway visibility monitor must reach the standard of 10–5000 m. The visibility measurement instrument based on forward scattering is difficult to implement in the highway monitoring system.

The lidar visibility meter can detect the visibility range from 50 to 5000 m. It is suitable for highway, airport, environmental monitoring, and other aspects of visibility detection. It can change the elevation angle of the measurement arbitrarily

---

Y. Wang (✉) · X. Xie
Anhui Xinhua University, Hefei 230088, China
e-mail: wyqme@163.com

V. Jain et al. (eds.), *Recent Trends in Intelligent Computing, Communication and Devices*, Advances in Intelligent Systems and Computing 1031,
https://doi.org/10.1007/978-981-13-9406-5_67

and give the integral visibility value in any elevation direction. The stable iterative method also makes measurement results more reliable and provides reliable weather information for meteorological, transportation, and environmental protection departments.

2 Overall Design of Highway Visibility Monitor

The highway visibility monitor mainly includes a laser emission system, a receiving optical system, and a signal acquisition and control system. The overall block diagram of the system is shown in Fig. 1.

3 Principle of Highway Visibility Monitor

The laser transmitter transmits a laser signal in the atmospheric channel, and the optical signal received by the receiver is attenuated due to the scattering and absorption of the beam by molecules and particles in the atmosphere. The basic equation for visibility measurement is the Bougner–Lambert law.

$$F = F_0 \cdot \exp(-\sigma x) \tag{1}$$

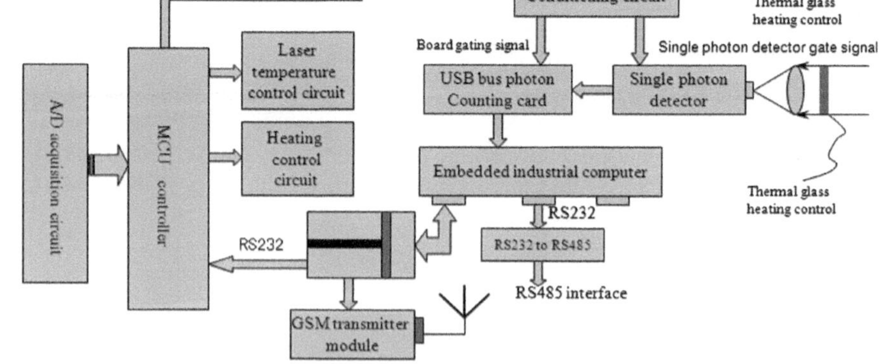

Fig. 1 Overall design block diagram

F is the light intensity of the path of light at x in the atmosphere; F_0 is the initial light intensity of the laser; σ is the extinction coefficient;

Derivation of Eq. (1):

$$\sigma = \frac{-\mathrm{d}F}{F} \cdot \frac{1}{\mathrm{d}x} \tag{2}$$

The transmission factor:

$$\mathrm{T} = \frac{F}{F_0} = \exp(-\sigma x) \tag{3}$$

According to some observers' observations of representative test targets, the contrast threshold of the human eye recognition target is 0.05, then the visibility distance can be derived from Eqs. (2) and (3) [1]:

$$V_{\mathrm{m}} = \frac{1}{\sigma} \cdot \ln^{\frac{1}{0.05}} \approx \frac{2.996}{\sigma} \tag{4}$$

This equation is for the light wave (white light) perceived by the human eye. As an approximation, the attenuation coefficient of white light can be replaced by the attenuation coefficient of the most sensitive green light (λ = 532 nm) of the human eye. For other wavelengths, the following correction should be made.

$$V_{\mathrm{m}} = \frac{2.996}{\mu} \left(\frac{\lambda}{550} \right)^{q} \tag{5}$$

μ is the extinction coefficient when detecting the atmosphere using a 532 nm laser; q values may vary in different regions and times, the most commonly used values are given by Kruse etc. [2]:

$$q = \begin{cases} 0.585 V_{\mathrm{m}}^{1/3} & V_{\mathrm{m}} < 6\,\mathrm{km} \\ 1.3 & 33\,\mathrm{km} > V_{\mathrm{m}} > 6\,\mathrm{km} \\ 1.6 & V_{\mathrm{m}} > 30\,\mathrm{km} \end{cases} \tag{6}$$

4 Configuration of Highway Visibility Monitor

The semiconductor laser radar visibility meter adopts the transceiving and split off-axis structure. Figure 2 shows the system structure. The system consists of three parts: laser emission system, receiving optical system, and signal acquisition and control system. The system is light, compact, fully solidified and modular in

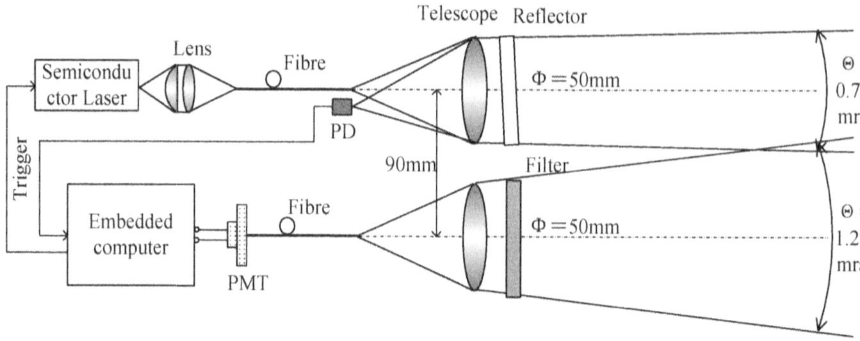

Fig. 2 Configuration of highway visibility monitor

design, compact, lightweight and stable. The lidar visibility meter is designed to detect visibility from 50 to 2000 m and output a set of visibility values every 1–10 min [3].

5 System Control and Data Acquisition Method

Automatic measurement and control of the lidar system can be achieved with an industrial control computer. The main functions of the computer are divided into two parts: system control and data processing. First, after entering the system initialization parameters, the computer manages the multi-channel digital capture card through the PCI bus. The trigger signal is given by the laser sync signal PD. Secondly, the computer receives the output signal of the photon counting detector through the digital acquisition card and processes the signal to display the visibility value. The flow of computer control instructions and data processing is shown in Fig. 3.

6 System Timing Control

Figure 4 is a timing control diagram of a portable lidar visibility meter. The timing control of the system includes laser triggering, PD triggering, detector PMT triggering, and the AD capture card triggering several parts [4].

The laser output pulse frequency is 2000 Hz, the pulse interval is 500 μs, and the pulse width is 30–100 ns. The laser emits a pulse through the emission window, and the reflected light passes through the photodiode PD as a trigger signal to control the detector and the data acquisition card to start working. The sampling frequency of the data acquisition card is 10 MHz, so the system distance resolution is 15 m.

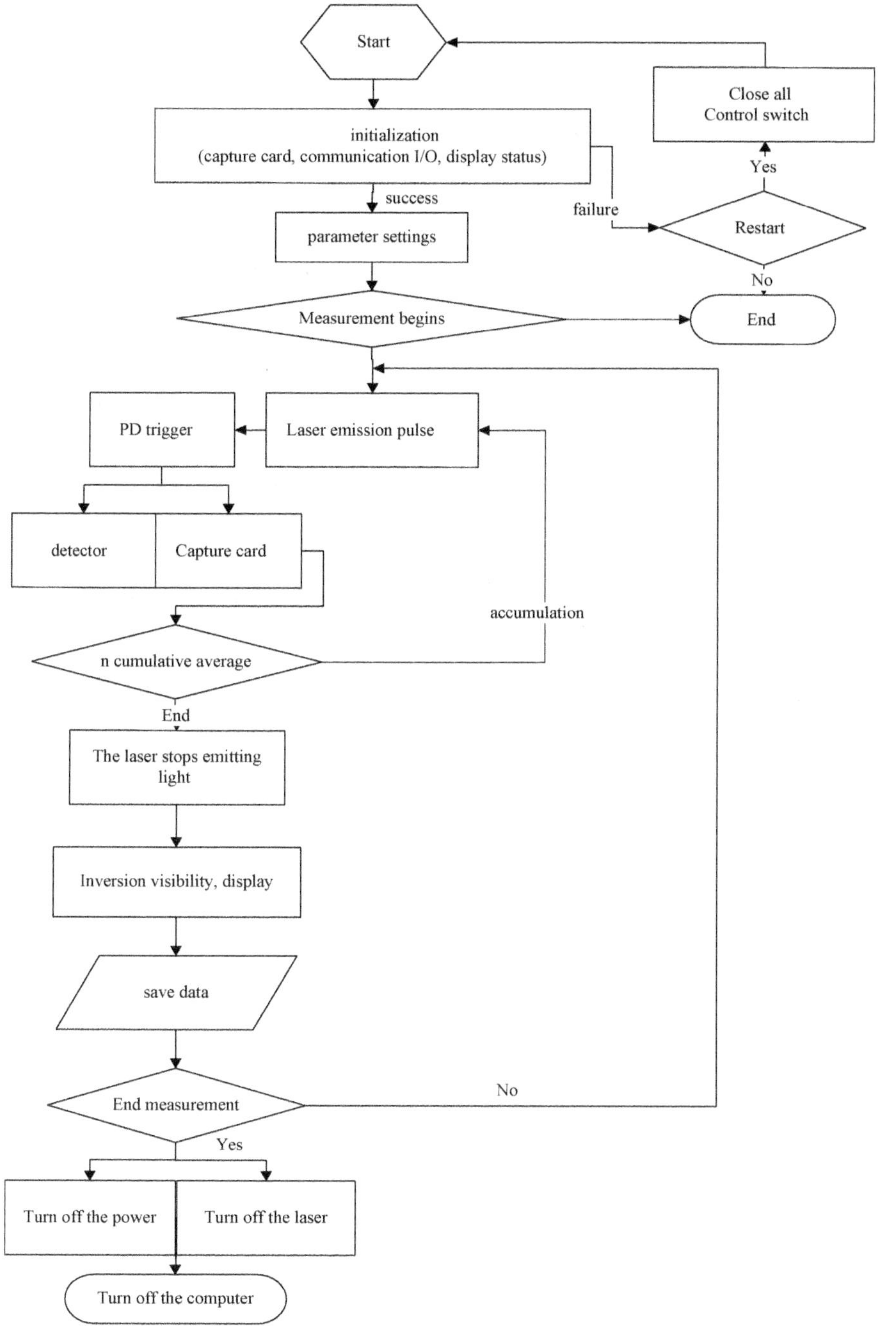

Fig. 3 System control instruction process diagram

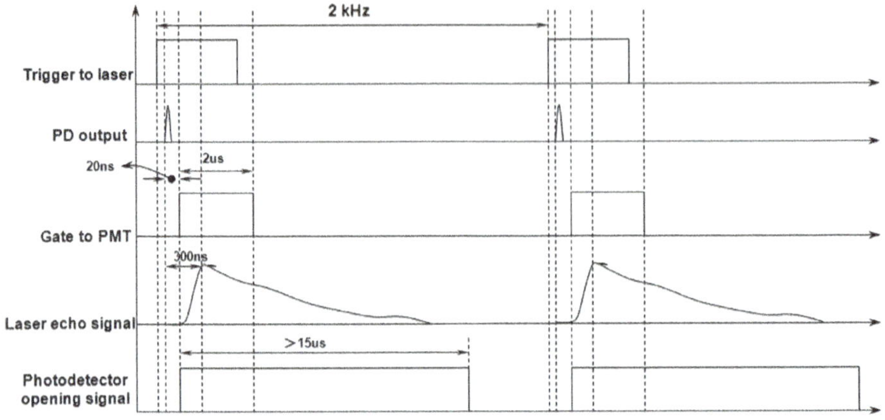

Fig. 4 System timing control highway visibility monitor

7 Inversion Algorithm of Highway Visibility Monitor

7.1 Geometric Overlap Factor Measurement

In the full-range r of the lidar, the horizontal atmospheric extinction coefficient σ_H is obtained by the slope method under the weather with uniform visibility in the clear, and the geometric overlap factor can be obtained by substituting into the lidar equation.

$$Y(r) = \frac{P(r)r^2}{\mathrm{CE}\beta \mathrm{e}^{-2r\sigma_H}} \tag{7}$$

In the observation experiment, the weather was fine and the atmosphere was evenly distributed to perform geometric overlap factor calibration. Below are several sets of data from consecutive observations for geometric overlap factor measurements. The measured geometric overlap factor is shown in Fig. 5.

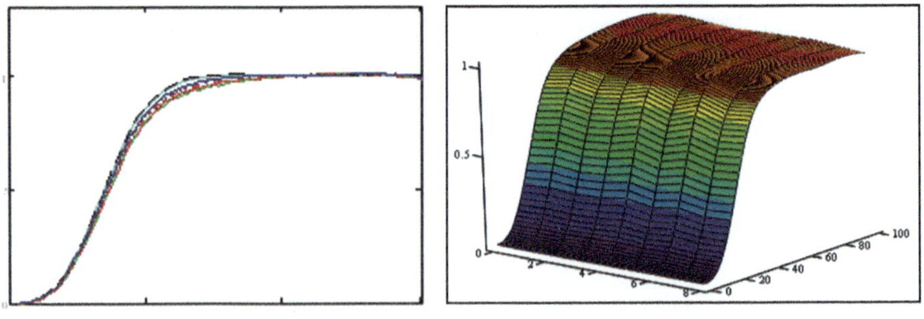

Fig. 5 Geometric overlap factor measurement curve

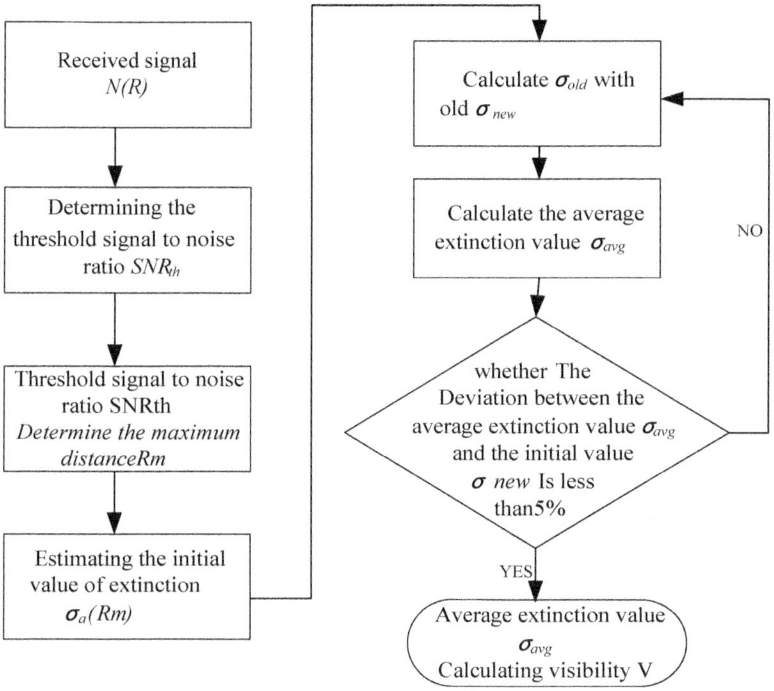

Fig. 6 Visibility calculation process

7.2 *Visibility Calculation*

Firstly, the maximum detection distance is determined according to the signal-to-noise ratio of the backscattered signal. Secondly, the initial value of the atmospheric extinction coefficient is estimated. The initial value selection has little effect on the final inversion result. Finally, the stable iterative algorithm is used to invert the atmosphere. Visibility value. The specific calculation process is shown in Fig. 6.

8 System Test

In order to test the performance of the system, the lidar visibility meter was tested separately during the day and night. Figure 7 shows the original echo signal of the visibility meter under strong daylight conditions. Figure 8 shows the echo signal of the visibility meter under weak background light at night. It can be seen that the system filter fails to effectively filter the background light, and there is strong noise

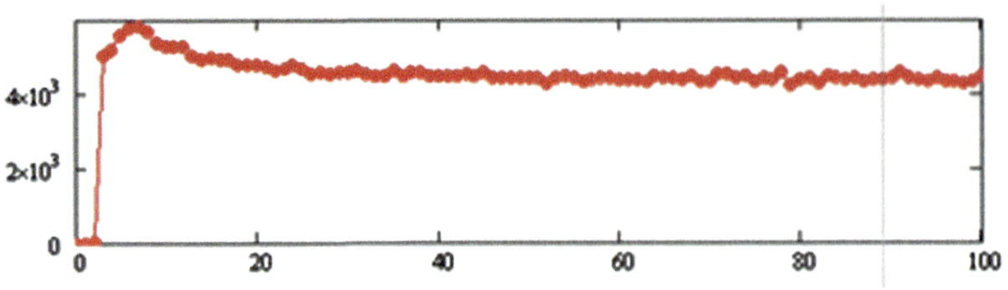

Fig. 7 Day-level return signal

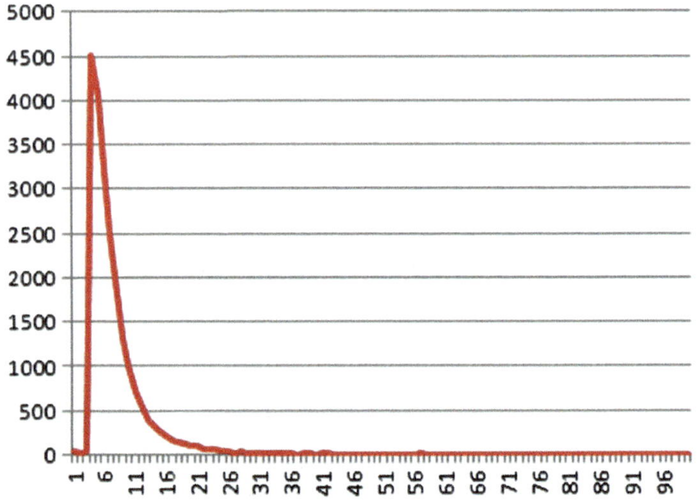

Fig. 8 Night-level return signal

in the echo signal. The real-time background compensation method can better filter out the noise signal.

Acknowledgements This work was supported by Research on the Construction of Electronic Information Major Course Group Based on " "Integration of theory and practice" (2017jyxm0521), Research on Low Conductivity Fluid Flow Measurement Technology (KJ2018A0596), Research on WLAN Intelligent Network Optimization Method Based on Mobile Internet (2015zr011), Analog circuit (2016gxkcx01).

References

1. Zibo, Zhuang, Wei, Huang, et al.: Protable visibility detector based on backward scattering lidar. Laser Technol. **39**(1), 119–123 (2015). (in Chinese)
2. Huang, Z., Cai M., Huang H., et al., Design of portable visibility meter. J. Lingnan Norm. Univ. **37**(6), 51–56 (2016) (in Chinese)

3. Bin, Yue, Dong, J., Sun, D., et al.: Measuring method of atmospheric slant visibility with semiconductor lidar. Infrared Laser Eng. **38**(1), 135–139 (2009). (in Chinese)
4. Jiang, L., Huang, W., Feng, S.: Design of laser visibility meter based on real-time background compensation. Laser & Infrared **44**(4), 399–404 (2014). (in Chinese)

A Stress Testing Method of Large-Scale Software Systems Based on Asynchronous Request

Yan Gong, Lin Huang, Hongman Wang, Lin Qie and Shangguang Wang

Abstract With a large number of concurrent users, many large-scale software systems will face a heavy computational load. In this case, a stress test to verify the performance of large-scale software systems becomes especially important. At present, most of the stress test methods adopt synchronous request and they initiate a new request only after the last request ends. However, the synchronous request-based methods do not provide large concurrency stress, nor can they faithfully reflect the maximum performance load of the system under test. This paper proposes a new stress test method adopting the asynchronous concurrent mode. With the amount of new requests sent to the system under test independent of its historical processing, it effectively ensures the tested system can run under full load. The experimental results show that our method simulation has large concurrency and high execution efficiency.

Keywords Stress test · Asynchronous request · Large-scale software systems

1 Introduction

At present, large-scale software systems in the network all need to support a large number of users, and with the development of technology, the software systems in the network architecture are more and more complicated. In the face of concurrent access by a large number of users, the system needs to find and solve a series of problems—such as slow request response and system collapse—that may occur [1]. The purpose of stress testing is to simulate concurrent stress requests to measure the performance of a system, thereby discovering its deficiencies and urging it to make

Y. Gong · L. Huang (✉)
Academy of Military Sciences PLA, Beijing, China
e-mail: huanglinwxs@163.com

H. Wang · L. Qie · S. Wang
State Key Laboratory of Networking and Switching Technology, Beijing University of Posts and Telecommunications, Beijing, China

V. Jain et al. (eds.), *Recent Trends in Intelligent Computing, Communication and Devices*, Advances in Intelligent Systems and Computing 1031,
https://doi.org/10.1007/978-981-13-9406-5_68

targeted improvements. Stress testing can also be understood as the limit test of resources. The test focuses on whether a system can operate normally when the resources are saturated or overloaded. It is a stability test under extreme load [2].

Transaction per second (TPS) refers to the number of things or transactions that a system can process per second. It is a very important indicator to measure the processing power of a system [3]. Resource utilization refers to the occupancy of system resources under certain load conditions, such as server CPU utilization, memory usage, disk utilization, and network utilization. Resource utilization is usually proportional to the number of user loads. When the utilization value of a resource is constantly kept at 100%, it indicates that the resource has become a bottleneck of the system. The number of concurrent users refers to the number of users who open a session to a same application or module at a given time. As the number of concurrent users increases, the resource utilization of a system will also increase.

Common stress testing tools include LoadRunner [4], JMeter [5], WebLOAD [6], etc. These test tools all use the test method of synchronous request, which simulate multiple users to continuously make requests to a system following the rule that the user will not initiate a new request until his/her last request ends. This test method does not provide significant concurrency and therefore does not truly reflect the performance of the system under maximum load. This paper proposes a new stress testing method, which adopts the asynchronous request mechanism.

The new demand on a system is independent of its historical processing, ensuring that the newly generated stress is constant and that the system under test can run under full load.

The remaining chapters of this paper are organized as follows: Sect. 2 introduces stress testing-related work, Sect. 3 describes our proposed asynchronous request-based software system stress testing method, and Sect. 4 is experimental construction and experimental results. Section 5 is a summary of this paper.

2 Related Works

LoadRunner tests the processing performance of the entire system by simulating thousands of user requests [7] and thereby detects the system's bottleneck. However, the following major problems exist:

(1) LoadRunner adopts synchronous concurrent method, by which a new request will not be made when the proposed request is not responded. This will cause that the new generated stress becomes increasingly smaller and the tested system does not run at the capacity limit.
(2) The management terminal of LoadRunner cannot be deployed on the Linux operating system so that the Linux-based operating system cannot install or use LoadRunner.

Similarly, test tools such as JMeter and WebLOAD all use the test method of synchronous request. The number of new requests is inversely proportional to the number of failed requests. The more requests that the system under test does not process, the fewer new requests will be generated, resulting in a continuous reduction in stress.

3 A Stress Testing Method of Large-Scale Software Systems Based on Asynchronous Request

3.1 Asynchronous Request Mechanism

This paper proposes to provide virtual stress by means of asynchronous request. Unlike the synchronous request mechanism, the testing tool applying asynchronous request mechanism constantly sends a preset quota of requests in every second regardless of whether it receives the response of the previous request from the system under test. Figure 1 depicts the difference between synchronous request and asynchronous request when sending a message.

As can be seen from Fig. 1, under the mode of synchronous request, the testing tool sends the new request after receiving the response of the system. During the second seconds, the testing tool does not impose any stress on the tested system, which gives the system a brief respite. However, under asynchronous request, the testing tool initiates request in every second regardless of whether a response to a

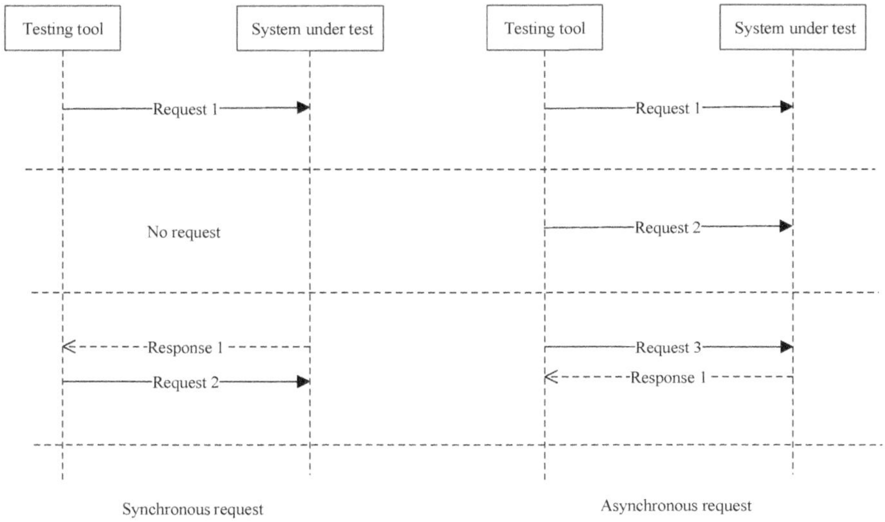

Fig. 1 Difference between synchronous request and asynchronous request when sending a message

previous request is received, ensuring continuous stress is applied to the system. Compared with synchronous request mechanism, it can better reflect the actual processing power of the system under test.

It does not imply that the asynchronous request mechanism completely ignores the response the system gives back after a request is sent. As the testing tool needs to perform corresponding statistics on the processing performance of the system under load, the asynchronous request mechanism shall be devised to be sensitive to throughput performance. When a new request needs to be generated, the previous request can continue for a period of time if it does not receive a response, and the new request will be irrelevant to previous request, thereby achieving the purpose of generating asynchronous stress.

3.2 Design of Software System Stress Test Tool Based on Asynchronous Request

The software system stress testing tool based on asynchronous request mainly includes four parts: stress generation and control subsystem, test management subsystem, test script library subsystem, and test execution subsystem. The test script library subsystem and the test management subsystem run on the test execution subsystem. The stress generation and control subsystem uses the scheduling capabilities of the automaton to generate asynchronous stress requests.

The overall architecture of the stress test tool is shown in Fig. 2. The modules and functions included in the system are described below.

(1) Stress generation and control subsystem. The stress generation and control subsystem is a core functional component of the stress test tool. The stress generation and control subsystem mainly provides control of the stress model and the concurrent model and provides the scheduling function of the test script and the statistical function of the test information.
(2) Test management subsystem. The test management subsystem is for testers. The main function is to configure and control the stress generation and control subsystems. The test management subsystem accepts the HTTP message and triggers the corresponding service according to the HTTP address information. After the service processing, the HTTP message with the HTML format is returned and returned to the browser, and the returned message is the standard HTML format content.
(3) Test script library subsystem. The test script library subsystem consists of the written test scripts, including the INSERT, SELECT, UPDATE, and DELETE operations of the database. During the test, the stress generation and control subsystem selects the response test script execution from the script library according to the configuration.
(4) Test execution subsystem. The test execution subsystem provides a test interface for use by the test script library subsystem. The test execution subsystem

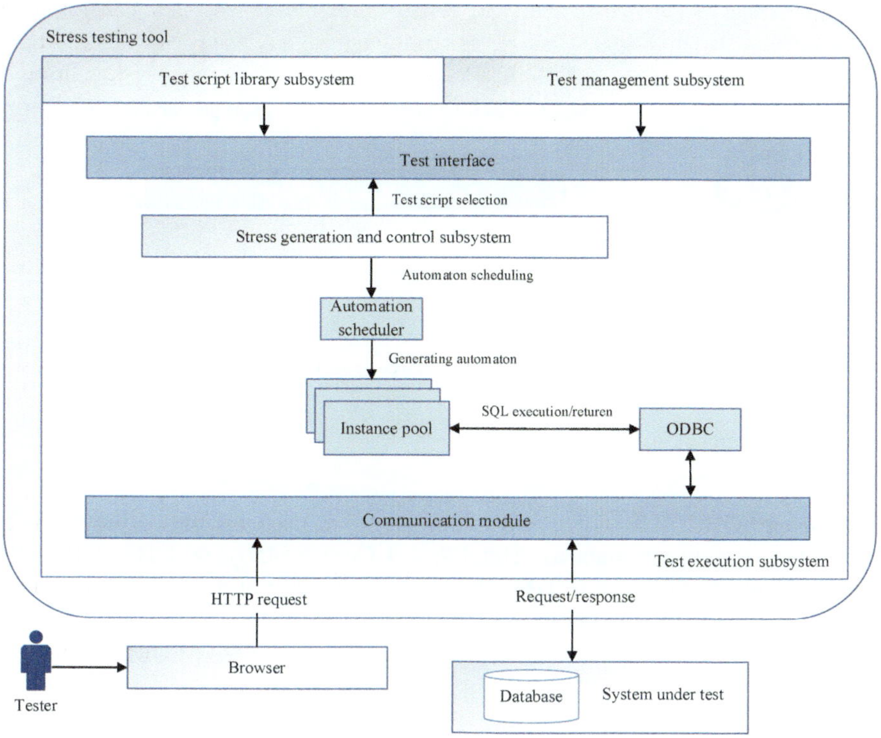

Fig. 2 Difference between synchronous request and asynchronous request when sending a message

uses the automaton scheduler to schedule automata in the automaton pool to control the creation, execution, and deletion of the automaton; the test execution subsystem uses the communication module to interact with other systems and can take protocols based on the transmitted data; the test execution subsystem uses the ODBC module to provide an access interface to the ODBC, connect to the database, and perform database operations.

4 Experimental Validation

4.1 Test Environment

The stress test tool needs to initiate a request to the system under test via a network connection. The test environment is shown in Fig. 3. User operation is performed on one PC (Intel CoreTM2 P8400 2.26 GHz, 3GBDDR), and two servers are used to deploy the stress generation node of the stress test tool (Intel 2 × 4 Core

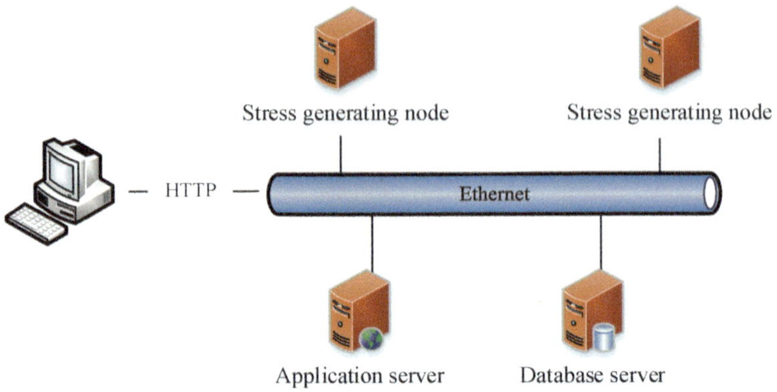

Fig. 3 Test environment

X5450 3.00 GHz, 4GBDDR), and two servers are used for the deployment of application system and database (Intel 2 × 4 Core X5450 3.00 GHz; 4GBDDR, RedHat Enterprise Linux 6.0, DM7.0). The total number of database table records exceeds 100,000, and the number of fields per record is 45.

4.2 *Software System Concurrent Stress Test*

Using the asynchronous system-based software system stress test tool to generate 10 database connections, a certain number of SELECT operations are initiated in parallel every second, and the SELECT condition matches the index. Thereafter, the number of database connections is continuously increased, the software system is forced to run beyond the rated capacity requirement state, the software system is checked for operation under the increased number of database connections, and the database query success rate is recorded. Adjust the number of database connections by the stress test tool, increase the number of database connections by 10 each time, and check the running status of the software system and the delay of the database operation.

The test results are shown in Table 1. When the number of connections is 60, the system response rate is slow, the average delay is 99 ms, and the success rate is 71%.

Table 1 System concurrency stress test results

Number of connections	Total number of requests	Test duration (s)	TPS	Average delay (s)	Success rate (%)	CPU (%)	Memory (MB)
10	375,660	600	625	0.015896	100	70.00	770
20	366,972	600	610	0.028763	100	71.90	794
30	360,063	600	600	0.049585	92	84.03	867
40	361,656	600	601	0.066602	91	73.90	794
50	360,648	600	599	0.082686	82	74.50	802
60	363,010	600	603	0.098632	71	81.00	802

5 Conclusions

The software system stress test method based on asynchronous request proposed in this paper is different from the synchronous request method. The new stress request does not depend on whether the old request is returned. Therefore, the system under test is continuously applied with constant stress, which will give more realistic data about the maximum processing power of the system. The experimental results show that the software system stress test method based on asynchronous request simulates large amount of concurrency and high execution efficiency, which can meet the stress test requirements of the system under test with increasing performance.

References

1. Weyuker, E.J., Vokolos, F.I.: Experience with performance testing of software systems: issues, an approach, and case study. IEEE Trans. Software Eng. **26**(12), 1147–1156 (2000)
2. Bainbridge D., Witten I.H., Boddie S., Thompson J.: Stress-testing general purpose digital library software. In: Research and Advanced Technology for Digital Libraries, ECDL 2009, vol. 5714, pp. 203–214. LNCS (2009)
3. Zhang, M,. Li H.H., Xiao, J.: The estimation method of common testing parameters in software stress testing. In: Proceedings of IEEE International Conference on Computer Science and Network Technology, pp. 1468–1472 (2011)
4. HP LoadRunner software, https://software.microfocus.com/zh-cn/products/loadrunner-load-testing/overview/ (2018)
5. Apache JMeter, http://jakarta.apache.org/jmeter/ (2018)
6. WebLOAD product overview, http://www.radview.com/ (2018)
7. Han, X.Y., Zhang, N., He, W., et al.: Automated warship software testing system based on LoadRunner automation API. In: Proceedings IEEE International Conference on Software Quality, Reliability and Security, pp. 51–55 (2018)

Lightweight and Fast Coordinated Update Algorithm for Hybrid SDN Networks

Changhe Yu, Julong Lan and Yuxiang Hu

Abstract Hybrid software-defined networking (SDN) networks are intermediates between traditional networks and SDN networks. However, since the hybrid control planes are structurally different from the single control plane, updating the hybrid SDN networks lacks effective techniques. In this paper, by utilizing the segment routing (SR) mechanism to update hybrid SDN networks and formalize the application of SR during updates, we propose a cooperative consistency update algorithm that can coordinate SR, two-phase commit techniques. The algorithm is capable of preserving consistency with low overhead and excellent applicability. When a network requires updating, the algorithm leverages the SR mechanism to stitch a new path using the available path segments; thus, flows can use the new path immediately. If the update cannot be completed with the SR mechanism alone, the algorithm can utilize the two-phase commit technique to update the nodes through duplication. Notably, our experiments based on real topologies indicate that, compared to prior techniques, the proposed algorithm can greatly reduce the TCAM overhead.

Keywords Network updates · Hybrid SDN · Consistency

1 Introduction

Software-defined networking (SDN) can greatly simplify the network management operation and maintenance process, bringing flexibility and fine-grained control for operators. However, due to realistic and economic considerations, the deployment of SDN networks is an incremental process. Thus, operators must maintain and manage hybrid SDN networks during the transition period. A hybrid SDN network refers to the network in which the SDN controller and traditional routing protocol (e.g., OSPF, IS-IS) exist simultaneously.

C. Yu (✉) · J. Lan · Y. Hu
National Digital Switching System Engineering & Technological R&D Center, Zheng Zhou 450002, China
e-mail: yu_changhe@hotmail.com

V. Jain et al. (eds.), *Recent Trends in Intelligent Computing, Communication and Devices*, Advances in Intelligent Systems and Computing 1031,
https://doi.org/10.1007/978-981-13-9406-5_69

Because of the dynamic characteristics of networks, network updates have become a common phenomenon. Network managers often modify the routing table, change the path weights, and access privileges to complete maintenance, traffic scheduling, and patch security vulnerabilities through updates. Consistency is a sufficient security guarantee; however, maintaining consistency is costly. Currently, hybrid networks have many use cases, such as the incremental deployment of SDN nodes and dynamic traffic engineering. In such cases, updates are necessary when flows must be scheduled or when the service requires change. However, it is more difficult to implement consistent updates in hybrid SDN networks than single control plane networks due to structural differences.

The essence of network updating is to modify the forwarding information base (FIB) of network nodes, which is a process that replaces the initial FIB with a new one. For the sake of reliability, a network update is a gradual incremental process that involves a series of intermediate states. To ensure network security and normal service, consistent maintenance is essential and fundamental. Consistency means that each flow must use the initial path or adopt the final path with no other choices.

To achieve consistent updates, various techniques have been presented in prior researches. In general, prior techniques can be divided into three categories. The first class, called two-phase commit ([1, 2]), can ensure consistency by duplicating FIBs to maintain initial and final entries simultaneously on all nodes. However, this process consumes an excessive amount of additional ternary content-addressable memory (TCAM) storage. Because TCAM storage is expensive and requires considerable power, the two-phase commit mechanism is not always suitable. The second class, based on the ordered scheduling mechanism (e.g., [3–6]), does not require additional network resources because it modifies the FIB using a calculated sequence. These techniques generally guarantee weak consistency and include policy-preserving [3], loop-free, blackhole-free [4], and congestion-free approaches [5]. Strong consistency can be ensured only by minority techniques (e.g., [6]). Nevertheless, this technique still suffers from intrinsic drawbacks because an update order that supports strong consistency may not be available. The third class (e.g., [7]) combines several mechanisms to complete network updates. The algorithm proposed in [7] combines the two-phase commit techniques and ordered scheduling to reduce overhead and achieve consistency during the update process. This technology can be used in all network types. However, it causes considerable overhead when the ordered scheduling technique is not effective. Thus, prior techniques have two shortcomings: First, they impose too much overhead in the network such as additional TCAM consumption. Second, they have limited applicability because certain techniques are merely feasible for specific topologies or protocols.

In this paper, based on the analysis of the routing characteristics of hybrid SDN networks, we explore the update mechanism of hybrid SDN networks which contain distance-vector routing and propose the LFCA (lightweight and fast coordinated update algorithm) to achieve consistent update in hybrid SDN network. The algorithm attempts to stitch the final path using the segment routing [8] and install the routing information into the packet header first. Nevertheless, for those flows which are not segmentable, two-phase commit mechanism is leveraged to update those flows. Our experiments verify that this algorithm can hugely save rule overhead than prior techniques.

2 Model and Characteristic

In this section, we abstract a generic model from the implementation of protocols and controllers for hybrid SDN networks, describing the general characteristics of the networks. Subsequently, we primarily analyze the challenges posed by the characteristics of the hybrid SDN networks. Moreover, by combining prior research, we illustrate the influence of this model on those update techniques. Finally, we detail how to handle these challenges using an SR and formalize SR application in the network updates.

2.1 Hybrid SDN Network Characteristics

Prior research has demonstrated that the SDN controller can install rules into all nodes, including those that are supported only by IGP protocols, via the installation of static routes or inject information into the IGP protocol [9] and it allows operators pervasive traditional services, such as MPLS and VPNs, as well as the advantages of SDNs without a complete deployment of SDN nodes [10]. Moreover, in this paper, pure SDN networks and IGP-only networks are viewed as special cases of hybrid SDN networks.

Our research is based on these conclusions and abstracts a generic hybrid SDN network, in which we cannot make any assumptions concerning the specific protocols that are running or how they modify the node FIBs. This control plane arbitrarily combines uncoordinated control planes. We can abstract a control plane as an operation course that acquires network information, calculates node forwarding rules, and installs them on nodes. In particular, we adopt the classification proposed in [11], in which the control planes are divided into two parts: FA (FIB aware) control planes and FU (FIB unaware) control planes. FA control planes use the FIB as input to propagate a route and react to changes in the FIB.

However, the disseminate route in the FU control plane is unrelated to FIB entries; hence, the FU control planes do not make reaction to modification of the FIB entries. Moreover, distance-vector protocols are FA control planes, whereas link-state protocols and general OpenFlow controllers are FU control planes. In particular, OpenFlow offers convenient alternatives to build controllers that react to FIB changes. In later sections, we call an update an FA-involving update if an arbitrary FA control planes modifies the FIB at any time during the update process; by contrast, updates in which all FIB entries are installed by FU control planes are called FU-only updates. The properties of FA and FU control planes are as follows.

Property 1 FIB entries installed by a FU control plane are independent of other control planes that coexist with the FU control plane.

Property 2 If the running control plane is an FA control plane, the FIB entries of nodes in the same path should be installed by the same control plane.

Interested readers are referred to [11] for a detailed description and discussion of these properties.

2.2 Limitations of Previous Technologies

Most previous techniques cannot be directly used in hybrid SDN networks because different controllers and routing protocols have different reaction mechanisms for FIB updates [11]. FU control planes, such as link-state protocols and general OpenFlow controllers, do not make reaction to FIB modifications, whereas FIB modifications in networks running FA control planes, such as EIGRP, will cause not only local effects but also remote effects. For example, if a node's route to destination d is installed by another control plane, the initial control plane will withdraw all routes traversing the node to d. As shown in Fig. 1, in the EIGRP network, the initial path of flow $f_{u,d}$ is $\langle u\, v\, w\, d\rangle$. As soon as another control plane updates the route of node v to d from $\langle v\, w\, d\rangle$ to $\langle v\, z\, d\rangle$ node v will stop propagating the routes given by the EIGRP to its neighbor nodes, and node u will withdraw its routes traveling v to d. Consequently, u will recalculate its FIB to d, which is risky and may cause inconsistencies. However, this phenomenon does not occur in FU control planes. That is, FU-only updates will cause only local effects and not the remote effects. Indeed, most previous technologies assume that control planes are indifferent with respect to FIB changes. Hence, most prior researches are unsuitable for hybrid SDN networks.

Specifically, different types of control planes have different effects on prior techniques. However, it has been proposed that prior techniques are applicable to all

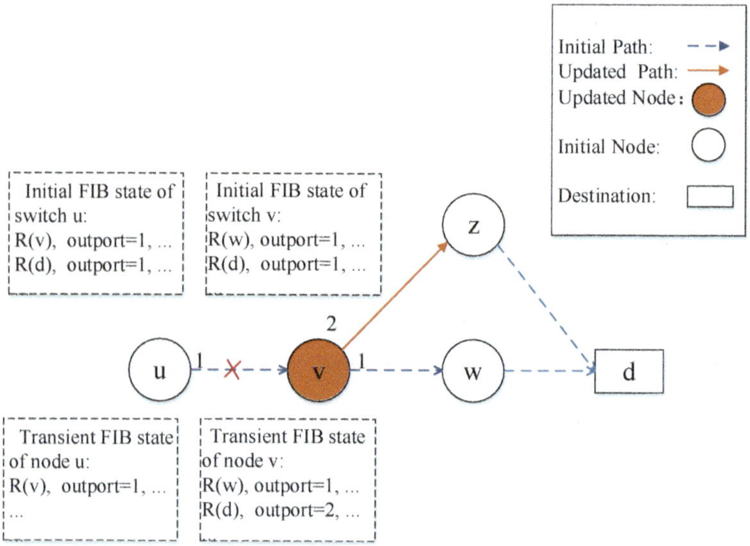

Fig. 1 An example of the remote effect of the FA control plane

types of FU-only updates interchangeably [7]. In contrast, these techniques cannot be directly applied to networks that run FA control planes. FA control planes are structurally different from FU control planes, and update algorithms for FA control planes require certain essential improvements compared with the FU update algorithm. This characteristic can be summarized as Property 3.

Property 3 Previous techniques can be used interchangeably among SDN controllers, traditional IGP protocols, and hybrid SDN networks if the control plane is FU, whereas the FA control plane does not directly support previous technologies.

In particular, prior technology is too complicated for the FU-only update. As demonstrated in [7], the greedy algorithm is sufficient to obtain the longest consistent update sequence without causing large TCAM overhead if a consistent update sequence exists.

2.3 Limitations of Previous Technologies

Using SR to stitch a specific path is an effective approach to traffic engineering. SR can instruct the forwarding of packets by adding a segment identifier (SID) to the packet header. Therefore, if packets are instructed to move forward along the final path joined by the old rules to reach designated destinations, we can also obtain the expected update effect. For example, as shown in Fig. 2, the initial path is $\langle a\,b\,d\rangle$, and the final path is $\langle a\,b\,c\,d\rangle$. If the update occurs in pure SDN networks, since path

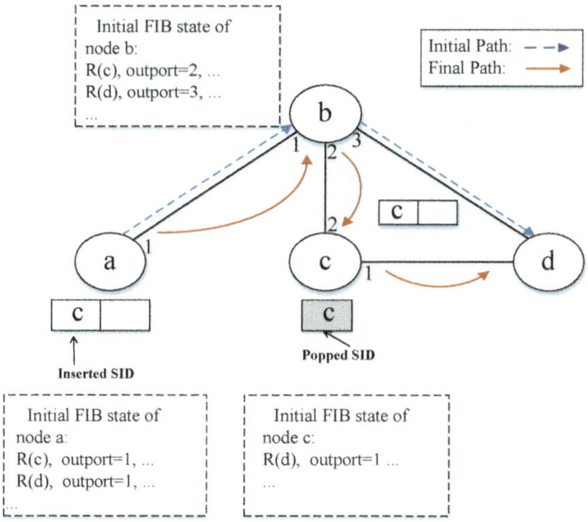

Fig. 2 Example of SR application in network update

$\langle a\,b\,c\rangle$ exists, $P_{a,c}$ can be chosen as a segment and a final path segmentation of flow $f_{a,d}$ can be $\langle P_{a,c}\,P_{c,d}\rangle$. Therefore, a single SID $\langle c\rangle$ can instruct packets forward along the final path.

However, as generic models of hybrid SDN networks and their properties have been discussed, we know that hybrid SDN updates present certain differences with respect to pure SDN updates. The simple mechanism used in pure SDN networks is not suitable for hybrid SDN updates because modifying the FIB may cause nodes to withdraw routes, which can result in blackholes. For example, if the segment containing path $\langle u\,v\,w\,d\rangle$ in Fig. 1 is adopted to instruct packet forwarding, a blackhole will occur because of the withdrawal of routes to d by node u as long as node v is updated.

If we use the segments that contain several hops to splice a new path, then blackholes will occur when the segments are updated during the update process. However, for arbitrary node a in the networks, the direct route of the previous hop node to a will never be revoked due to modifications to the FIB of node a. To address such cases, we utilize the strict source route mechanism and construct the SID list hop by hop. As shown in Fig. 2, the final path is $\langle a\,b\,c\,d\rangle$ and two feasible segmentations can splice the path are $\langle P_{a,b}\,P_{b,c}\,P_{c,d}\rangle$ and $\langle P_{a,c}\,P_{c,d}\rangle$. If the segment $P_{a,c}$ is to be updated in the update process, the only safe segmentation is $\langle P_{a,b}\,P_{b,c}\,P_{c,d}\rangle$ and the SID list is $\langle b\,c\rangle$. Essentially, we use the direct route to construct the final path. Naturally, segmentation $\langle P_{a,c}\,P_{c,d}\rangle$ can be a better choice when the segment $P_{a,c}$ does not need an update.

In particular, we extract old paths as segments and stitch them together to create a new path. Those paths have relevant rules in the FIB, and we can use SIDs to match the rules directly. Hence, the SR mechanism here requires only additional TCAM space of ingress nodes for the SR-enable rules to install SID to the corresponding packets, which dramatically reduce the TCAM overhead. Subsequently, the packets can traverse the new path immediately as nodes gradually pop the SIDs.

3 LFCA: Updating Networks via the SR Mechanism

Although SR mechanism has considerable advantages for network updates (e.g., excellent applicability, low overhead, and fast, consistent updates) due to the unique attributes of the SR mechanism, its superiority is largely dependent on the amount of segmentable flows. Therefore, the core of LFCA is identifying all segmentable flows and allocating appropriate SIDs to these flows. We introduce the details of LFCA in later sections.

Notably, the SR-enabled packets impose more bandwidth consumption and reduce the efficiency of packet processing. Therefore, we aim to find the shortest SID sequence from several feasible segmentations.

In the LFCA, we transform the segment allocation into a 0–1 ILP problem. Notably, selection of the optimal segmentation of a flow is independent from that of the other flows. The 'divide-and-conquer' technique is an effective method for dividing the initial problem into $|N|$ subproblems. Algorithm 1 illustrates the core of the LFCA in detail. The flows that must be updated can be identified through the initial and final path sets (lines 3–4). For these flows, the algorithm leverages the *solve* function to determine whether the flow is segmentable and then allocates segments to it if necessary. If the returned value of the *solve* function is not false, the flow $f_{s,d}$ is segmentable, the LFCA will calculate the SID according to the update type and then update the flow (lines 3–22). The mechanism of the SID calculation is illustrated in Sect. 2.3. The *solve* function first extracts feasible segments from the existing path for a flow $f_{s,d}$ (line 27). Subsequently, the algorithm iterates the *solve* function using the incremental variable h until optimal segment combinations are found or until variable h exceeds the maximum number (h ranges from 2 to L, where L is the number of hops of the final path of the flow). The LFCA can always find the shortest SID list as long as the corresponding flow is segmentable. Otherwise, the LFCA will record that the flow is not segmentable, and the two-phase commit algorithm will be used to complete the update.

The *solveILP* function abstracts an ILP model from the segment allocation problem. Assuming that the segment candidate set C and the supremum of segment number h are given, the algorithm can obtain all segmentable flows and corresponding SID lists by solving the ILP model, which is subject to Eqs. 1–5.

Maximize:

$$\mathrm{aim}_{s,d} \tag{1}$$

Subject to:

$$\forall S_{x,y} \in C : w_{s,d}^{x,y} \in \{0, 1\},\ \mathrm{aim}_{s,d} \in \{0, 1\} \tag{2}$$

$$\forall e \in P_{s,d}^{f} : \sum_{S_{xy} \in C_{s,d}} a_e^{x,y} \times w_{s,d}^{x,y} \leq b_e^{s,d} \tag{3}$$

$$L_{s,d} \times \mathrm{aim}_{s,d} = L_{s,d}^{*} \tag{4}$$

$$\sum_{S_{xy} \in C_{s,d}} w_{s,d}^{x,y} \leq h \tag{5}$$

Algorithm 1 LFCA: find all segmentable flows

Input: all nodes N, initial path P_i, final path P_f, all destinations D, update type *key*.
Output: SIDs of segmentable flow SL, Current path P_c, Set of unsafe pairs (node, destination) *unsafe*.

```
function LFCA (N, P_i, P_f, D, key)
SL={}   #SL store SID
changing_node=compute_changing_node (P_i, P_f, D)
for (s, d) in changing_node:
   segmentation =solve (s, d, P_i, P_f)
   if segmentation ≠∅ and key=FU-only
     SL_s,d. fetch (segments) # extract the last hop as SID
     P_c=update_network (s, d, P_i, P_f)
   else if segments ≠∅:
      for seg in segmentation:
          if seg in (P_i (a, b) for (a, b) in changing_node) #  segments need update
           SL_s,d =split_SIDs_FA(segments, seg) # extract SID hop by hop
           P_c=update_network(s, d, P_i, P_f)
        else:
           SL_s,d. fetch (SL_s,d)
           P_c=update_network (s, d, P_i, P_f)
   else:
     unsafe. add (s, d)
   SL. add(SL_s,d)
   end if
end for
return (P_c, SL, unsafe)
function solve(s, d, P_i, P_f)
segmentation={}  # segmentation set
f_s,d=P_f (s,d)          # the final path of flow f_s,d
h=2                      # the minimum number of segments
C=extract(f_s,d, P_i)
L=length(f_s,d)         # the number of hops of f_s,d
while segmentation =∅ and h<L:
  segmentation=solveILP(f_s,d, C, h) # build and solve an ILP
  h=h+1
  if segmentation≠∅:
    return segmentation
  else:
    return False
  end if
```

The optimization object maximizes the Boolean variable $\text{aim}_{s,d}$, which represents whether flow $f_{s,d}$ is segmentable. In the constraints, Boolean variable $w_{s,d}^{x,y}$ denotes whether the old path $S_{x,y}$ is chosen as a segment for flow $f_{s,d}$. In addition, variables $a_e^{x,y}$ and $b_e^{s,d}$ represent whether link e is in the initial path $S_{x,y}$ and the new path $P_{s,d}^f$, respectively. Notably, Eq. (3) contains the constraint that the spliced path must have all links in the new path. Moreover, $L_{s,d}$ and $L_{s,d}^*$ describe the length of the new and spliced paths of the flow, respectively. Obviously, the length of the new path should be equal to that of the spliced path. Finally, Eq. (5) adds the segment number constraint to the model because the number of segments cannot be greater than the number of new path links.

4 Evaluation

In this section, we evaluate the performance of our algorithm in terms of TCAM resources, updating effectiveness, and the time cost. In the following section, we introduce our experimental method and report the results and perform comparisons with both mainstream techniques and the latest techniques.

We used four publicly available Rocketfuel topologies [12] {AS 1221, AS 1755, AS 3257, AS 3967} as our dataset. Detailed information on our experimental topologies is presented in Table 1. For universality, we designed two scenarios for multiple experiments: a single-destination update and a multi-destination update. For the single-destination update, we randomly chose a node as the only destination and changed the forwarding path for it. For the multi-destination update, we randomly chose several nodes as destinations. Indeed, we mimicked the case in which flows to be updated can be any flows between any two nodes. In simulations, we calculated the shortest path of the original topology as the initial path and randomly changed the weights of 5, 20, and 50% of the links. Subsequently, the shortest path algorithm was reused in the changed topology to obtain the shortest path. We treated changed flows as the flows to be updated and the new path as the final configuration.

To verify the effectiveness of the LFCA, we compared our algorithm with recent consistent algorithms GPIA+FED [7] and the mature two-phase commit mechanism. presented in [1] in FA-involving update. Moreover, we repeated every experiment 200 times to ensure statistical significance. To evaluate the ability of our algorithm to reduce the TCAM overhead, we considered the two-phase commit technique as the baseline and computed the TCAM overhead of the other algorithms, comparing their performances with that of the two-phase commit alone. We use N_{TPC} to represent the number of additional rules imposed by the two-phase commit mechanism. Similarly, N_{LFCA} and $N_{\mathrm{GPIA+FED}}$ represent the numbers of additional rules imposed to perform updating using the LFCA and GPIA + FED, respectively, under the same conditions. Notably, the additional rules imposed by the LFCA consist of the rules installed in the ingress nodes that govern installation of the corresponding SIDs for segmentable flows and the overhead for those flows,

Table 1 Experimental topologies

Topology	AS 1221	AS 3967	AS 1755	AS 3257
Nodes	104	79	87	161
Links	302	297	322	656

which are updated by the two-phase commit. We calculated the percentage of the overhead reduction achieved by each algorithm using the formula $(N_{TPC}-N_x)/N_{TPC}$, where N_x represents N_{CCUA} or $N_{GPIA+FED}$, depending on the algorithm being tested. Our experimental outcomes are illustrated in Fig. 3, which depicts the complementary CDF (CCDF) of the overhead reduction percentage of each algorithm.

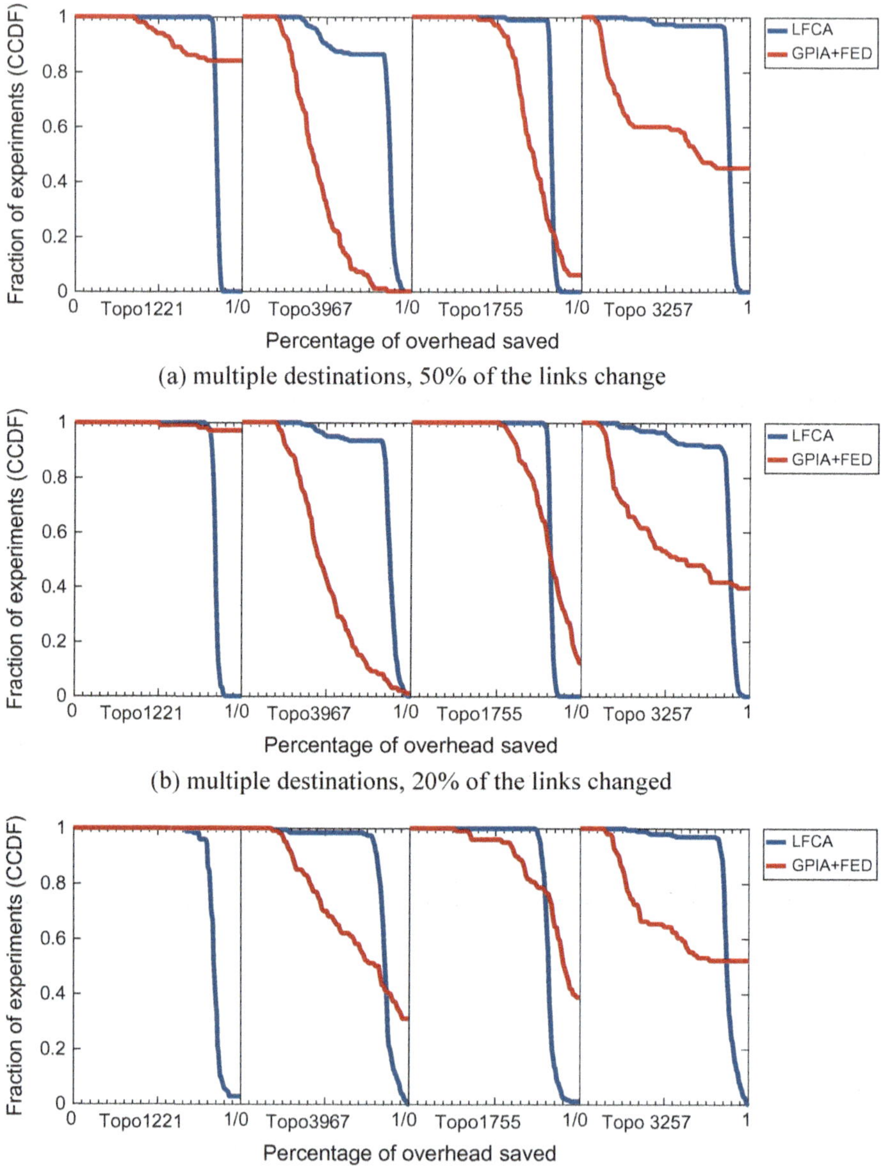

(a) multiple destinations, 50% of the links change

(b) multiple destinations, 20% of the links changed

(c) multiple destinations, 5% of the links changed

Fig. 3 CCDF of the TCAM overhead reduction

A given point in this figure represents that the rules overhead is reduced by a percentage of at least x compared with the two-phase commit algorithm in $y * 100\%$ of the experiments.

Overall, the LFCA achieves greater reductions than GPIA+FED in most experiments, particularly regarding the minimum reduction percentage. For example, in Fig. 3a Topo1221, the LFCA can reduce the number of additional rules by at least 83% and improve 47% in terms of GPIA+FED. In particular, it is difficult for the LFCA to complete the update without overhead due to the additional rule requirements of the SR, which needs to install additional rules on the ingress nodes. As illustrated in Fig. 3b, c, when the flows to be updated are relatively small-scale, GPIA+FED can achieve the updates without overhead in some experiments, whereas in the remaining experiments, GPIA+FED still suffers from large overhead. Apparently, this 'zero-overhead update' is not reliable enough.

5 Conclusion

In this paper, we deeply analyze the routing properties of hybrid SDN networks and formalize the application of SR in hybrid SDN network updates. Besides, combining SR and two-phase commit, we propose the LFCA, which can achieve a fast and consistent update in hybrid SDN networks. Algorithm performance has been proved by a series of experiments based on real topologies. Experiment results indicate that our algorithm can reduce TCAM overhead obviously.

References

1. Reitblatt, M., Foster, N., Rexford, J., et al.: Abstractions for network update. ACM SIGCOMM Comput. Commun. Rev. **42**(4), 323–334 (2012)
2. Katta, N.P., Rexford, J., Walker, D.: Incremental consistent updates. In: Proceedings of ACM SIGCOMM Workshop on Hot Topics in Software Defined NETWORKING, pp. 49–54, 12–16 Aug 2013
3. Vissicchio, S., Cittadini, L.: Safe, efficient, and robust SDN updates by combining rule replacements and additions. IEEE/ACM Trans. Netw. **25**(5), 3102–3115 (2017)
4. Förster, K.T., Mahajan, R., Wattenhofer, R.: Consistent updates in software defined networks: On dependencies, loop freedom, and blackholes. In: Proceedings of IEEE IFIP Networking Conference (IFIP Networking) and Workshops, pp. 1–9, 17–19 May 2016
5. Wang, W., He, W., Su, J., et al.: Cupid: congestion-free consistent data plane update in software defined networks. In: Proceedings of IEEE International Conference on Computer Communications, pp. 1–9, 10–15 April 2016
6. Mcclurg, J., Hojjat, H., Foster, N.: Efficient synthesis of network updates. ACM SIGPLAN Not. **50**(6), 196–207 (2015)
7. Vissicchio, S., Vanbever, L., Cittadini, L., et al.: Safe update of hybrid SDN networks. IEEE/ACM Trans. Netw. **25**(3), 1649–1662 (2017)
8. Filsfils, C., Nainar, N.K., Pignataro, C., et al.: The segment routing architecture. In: Proceedings of IEEE Global Communications Conference, pp. 1–6, 6–10 Dec 2015

9. Vissicchio, S., Tilmans, O., Vanbever, L., et al.: Central control over distributed routing. In: Proceedings of ACM Conference on Special Interest Group on Data Communication, pp. 43–56, 17–21 Aug 2015
10. Vissicchio, S., Vanbever, L., Bonaventure, O.: Opportunities and research challenges of hybrid software defined networks. ACM SIGCOMM Comput. Commun. Rev. **44**(2), 70–75 (2014)
11. Vissicchio, S., Cittadini, L., Bonaventure, O., et al.: On the co-existence of distributed and centralized routing control-planes. In: Proceedings of IEEE International Conference on Computer Communications, pp. 469–477, April 26–May 1, 2015
12. Spring, N., Mahajan, R., Wetherall, D.: Measuring ISP topologies with rocketfuel. IEEE/ACM Trans. Netw. **12**(1), 2–16 (2004)

Research on the Application of RFID in Equipment Management in Universities

Hao Zhu and Mengshu Hou

Abstract In most universities, equipment is managed through barcode tags. This method is low in automation and inefficient and cannot improve the level of equipment management in universities. RFID tags have small volume, large storage capacity, strong resistance to breakage, fast scanning speed, strong penetrability, high reliability, and reusability. Through RFID tags, universities can achieve rapid inventory of equipment, real-time monitoring equipment, and big data analysis of equipment data.

Keywords RFID · Equipment management · Rapid inventory · Real-time monitoring · Big data analysis

1 Introduction

In recent years, the enrollment scale of colleges and universities has grown rapidly in China. The types and quantities of teaching equipment and experimental equipment have also increased significantly. The workload of equipment management is also growing. In the past, the equipment management relied entirely on manual work to manually record the equipment brands and equipment numbers. This method is inefficient and error-prone, and it is difficult to find the equipment later. With the development of technology, many universities use barcodes to mark the equipment and to distinguish the information such as the name, model, date of manufacture, and recipients. The equipment acceptance, account management, sharing, and scrapping are basically done by hand, and the workload is huge and error-prone. With the development of information technology, some universities have replaced the barcodes with RFID tags for equipment management. The RFID support system performs a fast scan to simultaneously identify and read multiple

H. Zhu (✉) · M. Hou
Information Center, University of Electronic Science and Technology of China, Chengdu 611731, China
e-mail: zhuhao@uestc.edu.cn

V. Jain et al. (eds.), *Recent Trends in Intelligent Computing, Communication and Devices*, Advances in Intelligent Systems and Computing 1031,
https://doi.org/10.1007/978-981-13-9406-5_70

RFID tags. RFID tags are very resistant to damage and are easy to store. RFID tags have large storage capacity to meet the growing demand for equipment information storage. RFID tags can penetrate ordinary objects for identification and are highly reliable. Therefore, the use of RFID tags for equipment management can greatly improve work efficiency.

2 RFID Tags

At present, most universities have backward equipment management methods and there are many problems. RFID tags can effectively solve current problems and improve the efficiency of equipment management.

2.1 Problems of Equipment Management in Universities

The daily equipment management is mostly done manually by barcode tags, and the management staff has a large workload and low work efficiency. Although the digital equipment management is realized by the barcode tags, the equipment data is basically static data, and it is difficult to update in real time according to the change of the equipment. When performing equipment inventory, the management staff exports the equipment list from the database firstly, and then arranges a large number of staff to scan the equipment one by one. This method is prone to errors and omissions.

In the universities, equipment sharing is a trend, so the equipment is often adjusted between different departments. In the traditional management mode, the changes of equipment rely on manual recording, and then, the data is entered into the system; there is the possibility of errors and omissions. At the same time, the system cannot monitor the changes of the equipment in real time, there are loop-holes in the equipment management, and the equipment may be lost.

2.2 How RFID Works

RFID is an abbreviation for radio-frequency identification [1]. It is a wireless communication technology. We can identify specific targets by radio signals. It can be read and written to relevant data without establishing mechanical or optical contact between the RFID system and the identified target. RFID technology originated in the UK and was used in the Second World War to identify the enemy aircraft. It was commercialized in the 1960s.

When the RFID tag enters the magnetic field, it receives an RF signal from the reader and the passive tag will gain energy from the induced current. This energy is

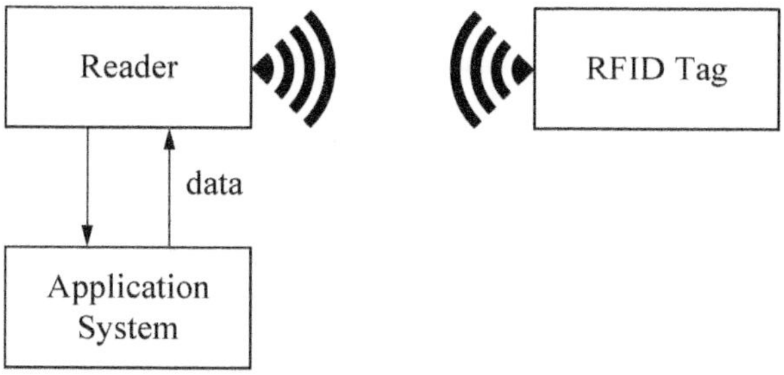

Fig. 1 Architecture of the RFID system

sufficient to transfer product information stored in the chip. The active tags can actively transmit signals without RF signals. After the reader reads the information and decodes it, it sends it to the application system for data processing. The tag contains electronically stored information that can be identified from a distance. Unlike barcodes, RFID tags do not need to be placed in the field of view of the identifier, but can also be embedded in specific objects. In the current market, there are three main types of RFID products: passive RFID products, active RFID products, and semi-active RFID products.

The RFID system consists of a RFID tag, a reader, and an application system. The system structure is shown in Fig. 1.

The RFID tag consists of an antenna, a coupling element, and a chip. Each tag has a unique electronic code. This electronic code will be attached to the object to identify the target object.

The reader consists of an antenna, a coupling element, a chip, and a device that reads (sometimes writes) tag information. Some readers are stationary readers. Some readers can be handheld readers.

The application system is at the application layer of the RFID system and is mainly used to process the collected data.

3 Advantages of RFID Tags

RFID tags can be easily read by the reader. The reader does not need a light source to read the data, it can even be read out by outsourcing, and there is a large effective recognition distance. When the active tag with the battery is used, the effective recognition distance can be more than 30 m.

RFID tags can be read quickly. Once the tag enters the magnetic field, the information can be read by the reader in real time. And multiple tags can be processed simultaneously for batch identification.

RFID tags are small in size and diverse in shape. RFID is not limited in size and shape on reading, and it does not need to match the fixed size and print quality of the paper for reading accuracy.

RFID tags have strong anti-pollution capabilities and durability. The carrier of traditional barcodes is paper, which is susceptible to contamination and damage. But RFID tags are highly resistant to substances such as water, oil, and chemicals. Since the RFID tag stores data in the chip, it can be protected from fouling.

RFID tags can be reused. The barcode cannot be changed after it has been printed. RFID tags can repeatedly add, modify, and delete stored data to facilitate information updates.

RFID tags have a large data memory capacity. The capacity of a one-dimensional barcode tag is several tens of bytes, while the maximum capacity of an RFID tag has a number of megabytes.

RFID tags have better security [2]. Since RFID tags carry electronic information, their data content can be protected by passwords, making their contents difficult to forge and change.

4 Application of RFID in Equipment Management in Universities

In reality, RFID enables efficient equipment management in universities by integrating the Internet and databases.

4.1 Realize Dynamic Management of Equipment Information

After the equipment is delivered to the school, the management staff can enter the model number, specification, price, serial number, unit of use, location of use, manufacturer, and other information in the system to generate the unique identification code in the system. The code will be stored in the chip of the RFID tag, which will be attached to the equipment. When the equipment information changes, the management staff only needs to modify the information stored in the RFID tag through the system to synchronize the information in the system database and the RFID tag. It greatly improves the efficiency and accuracy of equipment information management.

4.2 Realize Quick Inventory of Equipment

When the equipment needs to be inventoried, the equipment staff can remotely collect data from multiple equipment simultaneously through a handheld RFID reader. The collected data will then be transmitted to the application system through the network, and the inventory of the equipment will be completed in batches. RFID can improve the speed and accuracy of equipment inventory.

4.3 Realizing Big Data Analysis of Equipment Information

The information of equipment can be completely stored in the system database by RFID. We can use this data for big data analysis to provide a basis for equipment procurement, equipment allocation, equipment maintenance, equipment scrapping, and achieve accurate management of equipment by the university.

5 Conclusions

The application of RFID in equipment management in universities has changed the backward mode of manual equipment management and improved the accuracy and integrity of equipment management. It improves the efficiency of equipment management. At the same time, it also provides data support for the equipment management and makes decision-making more scientific.

References

1. Guo, P., Zhang, H., Liu, Z., Li, Y.: Design of experimental equipment management system based on RFID. Exper. Technol. Manag. **32**(3), 255–258 (2015)
2. Changjun, Zhou, Jun, Song.: Analyzing advantages of RFID technology in instruments and equipment management of colleges and universities. Exper. Technol. Manag. **33**(3), 267–271 (2016)

Extraction Technique of Cell Targets from Marine *Coscinodiscus* Microscopic Images

Kun Yu, Xinping Mo and Chunfeng Guo

Abstract Cell extraction is the most important link to construct microscopic image identification system of phytoplankton, and the extraction quality directly determines the accuracy of recognition. In this paper, a new technique that combines high-precision threshold segmentation with proper morphological operation for automatically extracting *Coscinodiscus* cells from microscopic images is put forward. In particular, firstly, the input microscopic image to grayscale image is converted and then the gray value of the grayscale image linearly to enhance its contrast is mapped. Next, threshold segmentation using the between-class maximum discrete measure matrix trace method is performed to get the binary image, whose noise is removed by successive morphological close operation and hole-filling operation. Owing to the possibility of multiple cells in the image, those larger connected components are further extracted. The obtained image and the original grayscale image are merged into a single image by using logical AND operation. The final extraction result is an image with black background and grayscale cell foreground. The extensive qualitative and quantitative experimental results demonstrate that the proposed strategy can effectively remove various types of noise and accurately extract *Coscinodiscus* cells with clear edge and complete shape.

Keywords *Coscinodiscus* · Diatom · Phytoplankton · Target extraction · Microscopic image segmentation

1 Introduction

Coscinodiscus is a very large category of diatom in phytoplankton [1], which belongs to *Coscinodiscaceae*, *Discoidales*, *Centricae*, and *Bacillariophyta*. Throughout the world, this genus contains about 450 species in total, including

K. Yu (✉) · X. Mo · C. Guo
Information Management Department, Shandong Foreign Trade Vocational College, Qingdao 266100, China
e-mail: yukunemail@163.com

V. Jain et al. (eds.), *Recent Trends in Intelligent Computing, Communication and Devices*, Advances in Intelligent Systems and Computing 1031,
https://doi.org/10.1007/978-981-13-9406-5_71

fossils, freshwater and seawater species, most of which are marine planktonic algae. China has more than 51 species.

At present, the specific species verification and identification of *Coscinodiscus* still rely mainly on an experienced algologist to observe the morphological characteristics artificially under the optical microscope. It is obvious that the microscopic observation method requires high professional knowledge and classification experience of the classified personnel, needs long-term and meticulous work, and is time-consuming and laborious. Therefore, this method is far from being able to realize real-time and rapid analysis of a large number of samples in situ [2].

The recognition method of phytoplankton microscopic image based on image processing technology that began in 1980s can accurately and quickly classify planktonic algae into species [3, 4], so it has received increasing attention and become the main development direction of identification technology in the future. The powerful image information processing ability of the computer combined with the morphological structure characteristics of phytoplankton lays a solid foundation for the analysis and pattern recognition of microscopic images, which is an effective means to automate traditional classification methods.

Cell target extraction from phytoplankton microscopic images is very important for image analysis and essential for building microscopic image identification system. In recent years, domestic and foreign research have also been carried out in this field. In the previous study, we classified the common marine red tide algae into non-setae algae species and *Chaetoceros* depending on the existence of setae and designed respective extraction strategy for each category [5]. Regarding non-setae algae, we proposed a multiscale detection algorithm for algal cell regions based on border correlation and further combined this with an improved GrabCut algorithm to segment cell targets. In respect of *Chaetoceros*, we presented a cell extraction method based on a gray surface orientation angle model to preserve the orientation information and tiny features of the setae. Jalba et al. put forward an improved watershed approach controlled by connected operator marker to extract diatom cells from microscopic images [6]. This algorithm obviously improved the initial extracted result of the automatic diatom identification and classification project [7] and has been successfully applied to the image classification field of diatom [8]. Blaschko et al. employed a snake-based technique to generate contours and an intensity-based method to determine the details of plankton images that are captured by a flow imaging device [9]. Rodenacker et al. describe a threshold method to extract cells of plankton from the microscopic images [10], whereas Sosik et al. integrated threshold-based edge detection with phase congruency to extract the cellular features of phytoplankton [11]. Verikas et al. extracted the circular cells of Prorocentrum minimum in microscopic images based on fuzzy c-means clustering algorithm [12].

Most of the studies mentioned above-adopted region-based or edge-based methods to segment several particular species of phytoplankton, but are not aimed at *Coscinodiscus*, so there are few suitable extraction methods for *Coscinodiscus* combining its microscopic image characteristics.

In this paper, we choose common *Coscinodiscus* in the China Sea as the research objects to explore extraction technique according to their biomorphological characteristics in microscopic images.

2 Cell Target Extraction of *Coscinodiscus*

The biggest difficulty of *Coscinodiscus* microscopic image segmentation is that there exist a lot of noise pollution around the cells which is caused by the impurities. These noises are not only numerous in number but also scattered in position and are very close to the cells' own gray levels.

In view of this situation, we propose a novel multicell target extraction strategy based on the between-class maximum discrete measure matrix trace method. The detailed extraction process is shown in Fig. 1.

In particular, the first step is to convert the input color microscopic image to a grayscale image and then stretch the gray scale of the grayscale image linearly to enhance its contrast.

Next, grasping the characteristic of large gray-level difference between the cell and background region in the image, we put forward a global threshold algorithm based on the regions: between-class maximum discrete measure matrix trace method, which constructs a two-dimensional (2D) grayscale histogram and uses the histogram to segment the grayscale image of *Coscinodiscus*. We use this algorithm to obtain the binary image.

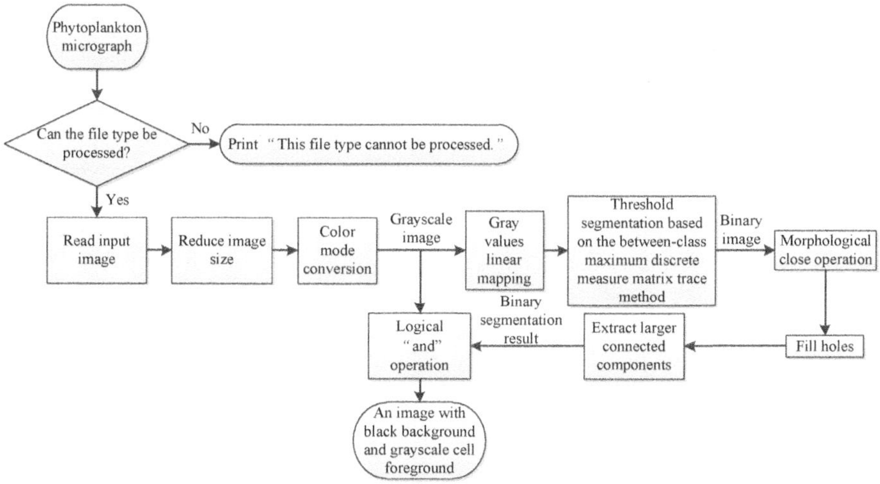

Fig. 1 Multicellular target extraction flowchart based on the between-class maximum discrete measure matrix trace method

After threshold segmentation, the image contains a lot of scattered, fragmented regions. In these regions, there exist many small holes and narrow cracks, and the edges of the regions are mostly fractured. Therefore, the current image cannot be used as the final binary segmentation result. It is necessary to adopt morphological close operation and hole-filling processing to remove all the noise.

Due to the possibility of multiple cells existing in the image, the connected components with a large area are further extracted. The area here refers to the number of pixels contained in the contour. Because there may be many cells in the image and the largest connected component must be the cell region, we extract such connected components whose area are greater than or equal to 30% of the area of the maximum connected component according to the experience and experimental verification.

Finally, the obtained image and the original grayscale image are merged into a single image by using the logical AND operation. The result is an image with black background and grayscale cell foreground.

3 Experimental Results and Discussion

3.1 Qualitative Results and Analysis

In order to test the effect of the between-class maximum discrete measure matrix trace method (referred to as BM method) on cell extraction, we compared BM method with five traditional threshold segmentation methods, which are, respectively, histogram bimodal method [13], maximum entropy method [14], expectation method (solving the average gray threshold), iterative method [15], and Otsu method [16].

Figure 2 demonstrates the binary segmentation results, where Fig. 2h is ground truth and used as the benchmark segmentation images for reference.

From the experimental results, it can be clearly found that both Otsu method and BM method have good segmentation effect and can obtain complete and accurate cell morphology for various images with different noise types and scene complexity. In contrast, maximum entropy method, expectation method, and iterative method present serious over-segmentation phenomenon when encountering the images with low contrast between cells and background, such as *Coscinodiscus apiculatus*, *Coscinodiscus nodulifer*, and *Coscinodiscus argus*. Part of the background or even the whole image is treated as the target. Histogram bimodal method has different degrees of under-segmentation, resulting in cellular edge disruption and internal structure loss.

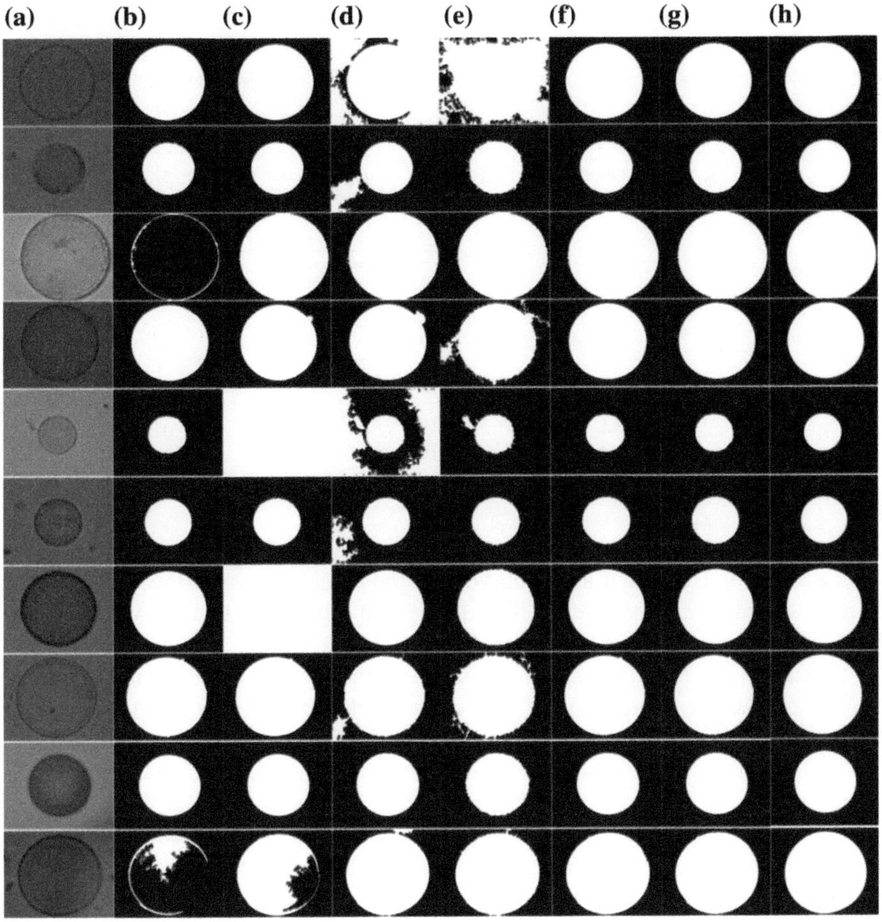

Fig. 2 Cellular binary segmentation results based on different threshold methods: **a** original microphotographs; **b** histogram bimodal method; **c** maximum entropy method; **d** expectation method; **e** iterative method; **f** Otsu method; **g** BM method; **h** ground truth. From top to bottom: *Coscinodiscus apiculatus*, *Coscinodiscus radiatus*, *Coscinodiscus curvatulus*, *Coscinodiscus oculus-iridis*, *Coscinodiscus nodulifer*, *Coscinodiscus excentricus*, *Coscinodiscus argus*, *Coscinodiscus wailesii*, *Coscinodiscus subtilis*, *Coscinodiscus asteromphalus*

3.2 *Quantitative Results and Evaluation*

In order to quantitatively analyze the experimental results of several threshold methods, PR curve is drawn in the form of discrete points according to the ground truth in Fig. 2h, as shown in Fig. 3.

In the PR curve of Fig. 3, each marker point intuitively represents a binary segmentation result. In order to distinguish, the results of each threshold method are marked with different symbols. For the convenience of observation, a reference line

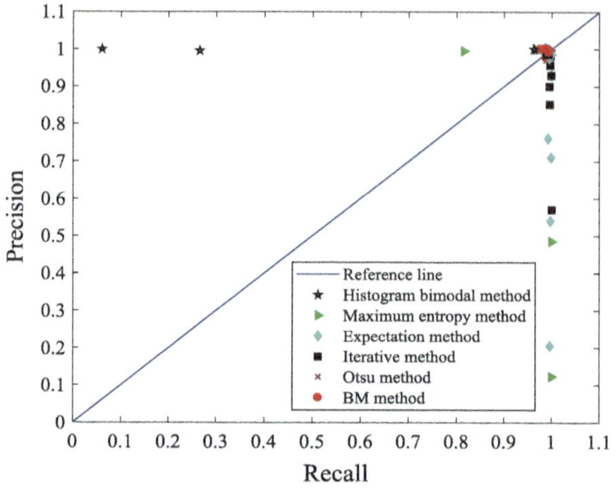

Fig. 3 PR curve in the form of discrete points

with a slope of 1 is added to the curve, and the display range of the horizontal and vertical coordinate axes is set to [0, 1.1]. It should be noted that the values of both axes cannot actually be greater than 1.

By careful observation and comparison with the location of all the marker points, it is found that the distribution of "red cross" and "red circle" is more concentrated and closer to the coordinate point (1, 1) than that of other marker points. This also illustrates that the segmentation effect of Otsu method and BM method is better.

Simply using precision and recall to evaluate the segmentation effect is not entirely convincing, because there is a mutual inhibition relationship between the two, which means that increasing one side may reduce the other side at the same time. In addition, PR curve cannot show the segmentation accuracy of specific algal cell target, so it cannot reflect the true segmentation quality scientifically and rigorously. Taking these factors into account, we also adopt comprehensive evaluation criteria F_1-measure:

$$F_1 = \frac{2P * R}{P + R} \tag{1}$$

The higher the value of F_1 is, the more effective the segmentation is. This evaluation index can be used to more objectively and comprehensively compare the segmentation quality of different algae species by various threshold methods, as shown in Table 1.

The data in Table 1 show that each threshold method has its own advantages and has the expertise to segment certain particular images. In other words, each method has the ability to obtain higher F_1-measure value. However, except for Otsu method and BM method, the F_1-measure values of the other methods are not stable enough

Table 1 F_1-measure values of various threshold methods

Algae	Methods					
	Histogram bimodal method	Maximum entropy method	Expectation method	Iterative method	Otsu method	BM method
Coscinodiscus apiculatus	0.9927	0.9931	0.7003	0.7264	0.9945	0.9947
Coscinodiscus radiatus	0.9913	0.9917	0.8612	0.9768	0.9915	0.9926
Coscinodiscus curvatulus	0.1137	0.9829	0.9923	0.9845	0.9874	0.9874
Coscinodiscus oculus-iridis	0.9949	0.9929	0.9870	0.9453	0.9952	0.9953
Coscinodiscus nodulifer	0.9819	0.2193	0.3395	0.9186	0.9821	0.9823
Coscinodiscus excentricus	0.9921	0.9913	0.8299	0.9897	0.9920	0.9924
Coscinodiscus argus	0.9938	0.6534	0.9954	0.9890	0.9955	0.9958
Coscinodiscus wailesii	0.9945	0.9950	0.9636	0.9634	0.9951	0.9951
Coscinodiscus subtilis	0.9812	0.9922	0.9919	0.9865	0.9904	0.9926
Coscinodiscus asteromphalus	0.4192	0.8961	0.9858	0.9895	0.9923	0.9927

and fluctuate in varying degrees: For example, histogram bimodal method obtains a high value 0.9913, when segmenting *Coscinodiscus radiatus*, and drops to a very low value 0.1137, when it encounters *Coscinodiscus curvatulus*. This shows that they are not adaptive to the input images and not suitable for the segmentation of a large number of various algal microscopic images. Accordingly, the segmentation quality presents a certain contingency. On the contrary, the F_1-measure values of Otsu method and BM method are always high and obviously higher than those of other threshold methods on the whole.

Although both Otsu method and BM method have achieved good results in the experiments, the latter actually outperforms the former in terms of segmentation accuracy: The F_1-measure values of BM method are mostly higher than those of Otsu method, and the equivalent case only occurs when segmenting *Coscinodiscus curvatulus* and *Coscinodiscus wailesii*.

This indicates that BM method has better segmentation quality, especially more detailed processing for morphological details (such as cell edges). On the other hand, its segmentation performance is more stable and robust, which means that it can successfully complete the segmentation task of low contrast image and resist the interference of different types of image noise.

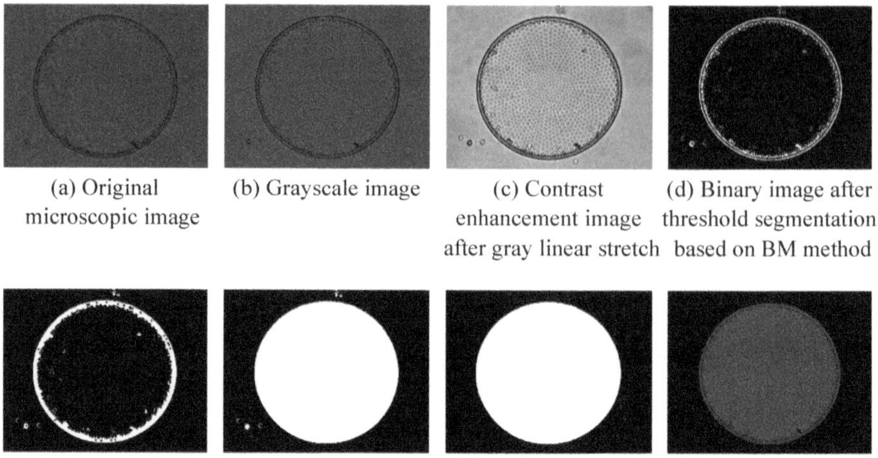

(a) Original microscopic image (b) Grayscale image (c) Contrast enhancement image after gray linear stretch (d) Binary image after threshold segmentation based on BM method

(e) Morphological close operation (f) Fill holes (g) Extract large connected components (h) Extraction result of grayscale cell target

Fig. 4 Extraction pipeline of *Coscinodiscus apiculatus* cell target

3.3 Final Cell Extraction Results

The binary segmentation results based on BM method and the original grayscale image are merged into a single image with black background and gray-level foreground by performing logical AND operation, which is taken as the final extraction result of cell target. The whole process of cell extraction of *Coscinodiscus apiculatus* is shown in Fig. 4.

4 Conclusion

This paper takes common *Coscinodiscus* in the China Sea as the research objects, to explore and study the extraction of *Coscinodiscus* cells according to their biomorphological characteristics in microscopic images.

In view of the high contrast between cellular region and background, we determined threshold segmentation as the basic idea and further designed a novel automatic extraction strategy of *Coscinodiscus* cells based on the between-class maximum discrete measure matrix trace method for both unicellular and multi-cellular targets. In addition, considering the possible serious noise pollution existing in the microscopic image, we adopted some appropriate morphological processing.

The extensive qualitative and quantitative experimental results demonstrate that the proposed strategy owns obvious advantages over the other traditional version on segmentation accuracy and stability. It can fully extract intact *Coscinodiscus* cells, maintain the natural external shape of the cells to the maximum extent, and have stronger robustness to resist noise influence when encountering a lot of images with different appearances.

Overall, our strategy realizes the automatic extraction of *Coscinodiscus* cells, thus providing the possibility of real-time and in situ processing for massive microscopic images of phytoplankton samples. It has a very important scientific significance in marine environmental monitoring.

Acknowledgements This work was supported by the Project of Shandong Province Higher Educational Science and Technology Program (No. J18KA350).

References

1. Sardet, C.: Plankton: Wonders of the Drifting World. University of Chicago Press, Chicago, USA (2015)
2. Erickson, J.S., Hashemi, N., Sullivan, J.M., et al.: In situ phytoplankton analysis: there's plenty of room at the bottom. Anal. Chem. **84**(2), 839–850 (2012)
3. Verikas, A., Gelzinis, A., Bacauskiene, M., et al.: An integrated approach to analysis of phytoplankton images. IEEE J. Oceanic Eng. **40**(2), 315–326 (2015)
4. Santhi, N., Pradeepa, C., Subashini, P., et al.: Automatic identification of algal community from microscopic images. Bioinform. Biol. Insights **7**, 327–334 (2013)
5. Yu, K., Ji, G.R., Zheng, H.Y.: Automatic cell object extraction of red tide algae in microscopic images. Chin. J. Oceanol. Limnol. **35**(2), 275–293 (2017)
6. Jalba, A.C., Wilkinson, M.H.F., Roerdink, J.B.T.M.: Automatic segmentation of diatom images for classification. Microsc. Res. Tech. **65**(1–2), 72–85 (2004)
7. Fischer, S., Gilomen, K., Bunke, H.: Identification of diatoms by grid graph matching. Lect. Notes Comput. Sci. **2396**, 94–103 (2002)
8. Dimitrovski, I., Kocev, D., Loskovska, S., et al.: Hierarchical classification of diatom images using ensembles of predictive clustering trees. Ecol. Inform. **7**(1), 19–29 (2012)
9. Blaschko, M.B, Holness, G, Mattar, M.A., et al.: Automatic in situ identification of plankton. In: IEEE Workshops on Application of Computer Vision, Breckenridge, USA, IEEE Computer Society, pp. 79–86 (2005)
10. Rodenacker, K., Hense, B., Jütting, U., et al.: Automatic analysis of aqueous specimens for phytoplankton structure recognition and population estimation. Microsc. Res. Tech. **69**(9), 708–720 (2010)
11. Sosik, H.M., Olson, R.J.: Automated taxonomic classification of phytoplankton sampled with imaging-in-flow cytometry. Limnol. Oceanogr. Methods **5**(2), 204–216 (2007)
12. Verikas, A., Gelzinis, A., Bacauskiene, M., et al.: Phase congruency-based detection of circular objects applied to analysis of phytoplankton images. Pattern Recogn. **45**(4), 1659–1670 (2012)
13. Qian, R.J., Huang, T.S.: Optimal edge detection in two-dimensional images. IEEE Trans. Image Process. **5**(7), 1215–1220 (1996)
14. Kapur, J.N., Sahoo, P.K.: A new method for gray-level picture thresholding using the entropy of the histogram. Comput. Vis. Graph. Image Process. **29**(2), 273–285 (1985)
15. Ridler, T.W., Calvard, S.: Picture thresholding using an iterative selection method. IEEE Trans. Syst. Man Cybern. **8**(8), 630–632 (1978)
16. Otsu, N.: A threshold selection method from gray-level histograms. IEEE Trans. Syst. Man Cybern. **9**(1), 62–66 (1979)

Design and Research of Wavelength Tunable Optical Receiver Module

Qian Zhu and Liang Liang

Abstract Based on TWDM-PON technology, this paper analyzes various schemes of wavelength tunable optical receivers. Aiming at the existing problems, a wavelength tunable optical receiver module is designed and implemented, which proves its high reliability, wide bandwidth, wide tunable range, and high integration.

Keywords Wavelength tunable optical · Reflective Bragg gratings · Modular

1 Introduction

Optical fiber communication is a kind of communication mode which takes light wave as an information carrier and optical fiber as transmission medium [1]. With the development of global optical communication, optical communication technology and optical communication devices are also developing in the direction of high speed, modularization, integration, flexibility, multi-function, and low cost. Based on the wavelength tunable optical technology, this paper designs and provides a wavelength tunable optical receiving module, which has the characteristics of high reliability, wide bandwidth, wide adjustable range, and high integration degree.

Q. Zhu (✉)
City College of Wuhan University of Science and Technology, Wuhan 430083, China
e-mail: zhuqian722@163.com

L. Liang
Accelink Technologies Co. Ltd., Wuhan 430205, China

V. Jain et al. (eds.), *Recent Trends in Intelligent Computing, Communication and Devices*, Advances in Intelligent Systems and Computing 1031,
https://doi.org/10.1007/978-981-13-9406-5_72

2 Research Background

The time-division and wavelength-division-multiplexing passive optical network (TWDM-PON) is the mainstream technology of the next-generation optical access network (NG-PON2) [2]. FSAN has made it clear that TWDM-PON is the technology choice of NG-PON2. Compared with DWDM, TWDM-PON has better manageability and compatibility. Wavelength tunable receiver (Wavelength Tunable ROSA) is one of the core optoelectronic devices of TWDM-PON system [3]. At present, there are many schemes to realize the wavelength tunable receiver [4], such as the thermal tunable optical receiving method of multilayer dielectric film, the rotation detection method of filter driven by a small motor, and the reception of different wavelengths by using MEMS reflector and filter.

The thermally tuned optical receiver of the multilayer dielectric film is based on the principle of the f-p cavity of the multilayer dielectric film [5]. Its passband spectrum is very narrow, the wavelength accuracy of the input light is required to be very high, and the isolation degree between channels is low. The method of small motor driving filter rotating detection has the disadvantages of large size and high power consumption. MEMS reflectors have poor stability and long-term reliability risk [6].

3 Research Contents

Aiming at the problems existing in existing technologies, the embodiment of this study provides a wavelength tunable optical receiving module, including:

1. Substrate;
2. An optical input component mounted on a substrate;
3. A group or groups of wavelength tunable components arranged on a substrate and located at the output end of an optical input component;
4. A group or groups of optical receiving units arranged on a substrate and one-to-one corresponding to the reverse output end of one or more sets of wavelength tunable components.

The optical input component is used to output the optical signal to the wavelength tunable component; the wavelength tunable component is configured to perform wavelength tuning on the incident optical signal such that the optical signal in the preset wavelength range is reflected, and the remaining optical signal continues to be transmitted; and the optical receiving component is used to receive the optical signal in the preset wavelength range reflected.

4 Concrete Implement

In order to make the purpose, technical scheme and advantages of the embodiment of this study clearer, the technical scheme in the embodiment of this study will be clearly described in combination with the attached drawings in the embodiment of this study.

4.1 *Overall Concept Design*

A schematic diagram of the overall frame structure of a wavelength tunable optical receiver assembly in an embodiment of this study is shown in Fig. 1.

The wavelength tunable optical receiving assembly in Fig. 1 includes:

1. Substrate 100;
2. An optical input component 200 is arranged on the substrate 100;
3. A group or groups of wavelength tunable components 300 are arranged on the substrate 100 and located at the output end of the optical input component 200;
4. A group or groups of optical receiving units 400 is arranged on the substrate 100 and one-to-one corresponding to the reverse output end of one or more groups of wavelength tunable components 300.

Assuming that the right side of the optical input component 200 is the output end, the wavelength tunable component 300 is located on the right side of the optical input component 200; assuming that the left lower side of the wavelength tunable component 300 is the reverse output end, the optical receiving component 400 is located on the left lower side of the wavelength tunable component 300.

Specifically, an optical receiving component 400 corresponds to a wavelength tunable component 300, and an optical receiving component 400 corresponds to a wavelength tunable component 300. If only one wavelength tunable component 300 is installed on the substrate 100, the optical input component 200 is the input

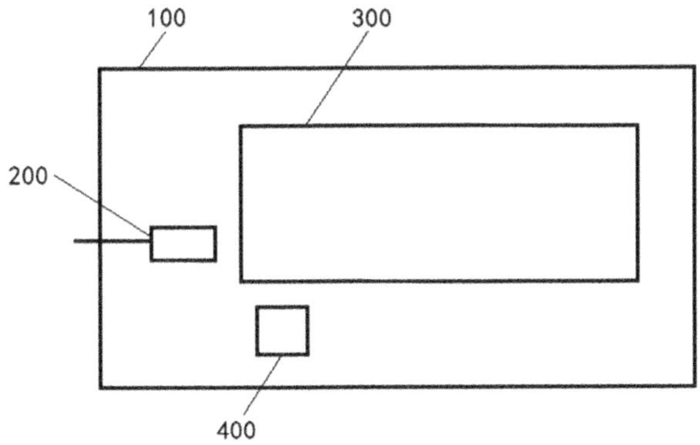

Fig. 1 Overall frame structure

collimator, and the input collimator outputs the optical signal to the wavelength tunable component 300. After the wavelength tuning of the wavelength tunable component 300, a part of the optical signal is reflected and part of the optical signal is transmitted through the medium. The reflected optical signal is an optical signal within a predetermined wavelength range after wavelength tuning by the wavelength tunable component 300 and is output to the light receiving component 400 through the inverted output end of the wavelength tunable component 300.

The wavelength tunable optical receiving module of the present embodiment has compact structure, small package size, and easy modularization integration. If only a plurality of wavelength tunable components 300 is disposed on the substrate 100, the plurality of wavelength tunable components 300 realizes one tunable filter array. Each wavelength tunable component 300 can achieve the optical signal tuning of one channel, and multiple wavelength tunable component 300 can achieve the optical signal tuning of multiple channels. At this time, the optical input component 200 is an optical fiber array, and the optical fiber array has a plurality of output optical fibers. One output optical fiber corresponds to one wavelength tunable component 300 to provide an optical signal for one wavelength tunable component 300; multiple output fibers can provide optical signals for multiple wavelength tunable components 300, respectively. The wavelength tunable optical receiving assembly of this embodiment has a larger volume than the structure of integrating a single-wavelength tunable component due to the integration of multiple wavelength tunable components, but still has a compact structure and a high degree of integration.

4.2 Detailed Implementation Plan

The present study embodies a specific embodiment of a wavelength tunable optical receiving module, as shown in Fig. 2.

Figure 2 is based on the above embodiments, and the wavelength tunable components include:

1. The micro-optical circulator 310 disposed on the substrate 100 and located at an output end of the light input member 200, and the input end of the micro-optical circulator 310 being aligned with an output end of the light input member 200;
2. The semiconductor refrigerator 320 is arranged on the substrate 100 and is located at the forward output end of the micro-optical annular device 310, which is used for temperature control;
3. The reflective Bragg grating chip 330 is arranged on the semiconductor refrigerator 320;
4. The first focusing lens 340 is coupled to the input end of the reflective Bragg grating chip 330.

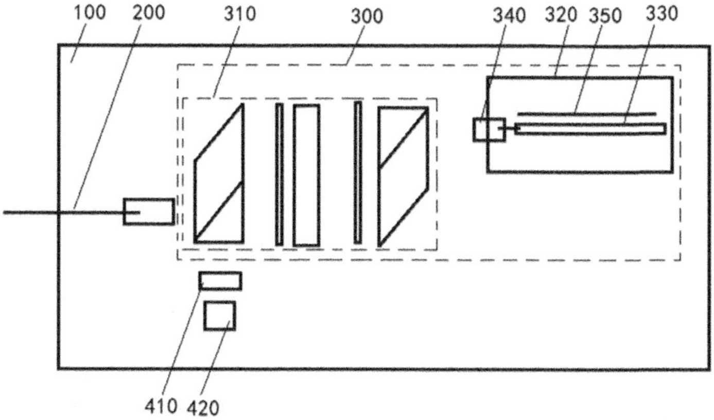

Fig. 2 Schematic diagram of embodiment structure

In the embodiment of this study, the grating in the reflective Bragg grating chip 330 is a uniform periodic microstructure, which is fabricated by photolithography, etching, and other semiconductor processes. As the optimal choice of filter bandwidth, the reflective Bragg grating is the fifth-order Bragg grating. The semiconductor refrigerator 320 is used for temperature control of reflective Bragg grating 330. The temperature of the reflective Bragg grating chip is controlled by a semiconductor cooler (TEC) to stabilize the reflective Bragg grating chip at a fixed temperature state, maintaining optimum operating characteristics and better long stability.

Preferably, the first focusing lens 340 is a C lens, and a coupling end face thereof and the reflective Bragg grating chip are directly coupled and bonded. The reflective Bragg grating chip 330 core layer can be germanium-doped silicon dioxide, SiN, polymer, or silicon. Bragg grating is made by photolithography, etching, and other semiconductor technology.

The reflective Bragg grating is a narrowband optical filter. The spectral light signal is injected into the grating, and only a very narrow spectral light signal (centered at the Bragg wavelength) is reflected within the grating. The remaining light waves will continue to be transmitted through the medium to the next grating without any loss.

Based on the above embodiments, the surface of the reflective Bragg grating chip 330 is provided with a heating electrode 350, which is used to tune the central wavelength of the reflective Bragg grating chip 330 through the thermal effect to obtain the reflected light in the preset wavelength range. Preferably, the heating electrode 350 can be disposed at a certain offset from the center of the reflective grating chip, so that the adjustment precision can be appropriately reduced to achieve a wider range of wavelength tuning.

Based on the above embodiments, the optical receiving component 400 includes a second focusing lens 410 and a detector 420; and a second focusing lens 410, which is arranged on the substrate 100 and aligned with the reverse output terminal of the wavelength tunable component 300. Among them, the reverse output of the wavelength tunable component 300 is the reverse output of the micro-optical ring 310, and the detector 420 is arranged on the substrate 100 and outside the second focusing lens 410.

The structure of the embodiment of the wavelength tunable light receiving component of the research embodiment shown in Fig. 2 can be installed and manufactured as follows:

1. The semiconductor refrigerator 320 is bonded on the substrate 100, and the reflective Bragg grating chip 330 is fixed on the semiconductor refrigerator 320.
2. The optical input component 200 outputs the optical signal and the end of the reflective Bragg grating chip 330 monitors the optical power through a single-core FA. Through the fixture clamping, the first focusing lens 340, regulating the position of the first focusing lens to maximize the coupling optical power. The first focusing lens is coupled and fixed to the end surface of the reflective Bragg grating chip by the ultraviolet glue.
3. A miniaturized micro-optical circulator 310 is placed at the input end of the first focus lens 340 to adjust the position and angle of the light input member 200 to maximize the monitored optical power.
4. Place the second focusing lens 410 and detector 420 at the reverse output of miniaturized micro-optical ring 310, adjust the position of the second focusing lens 410, and detect the DC photocurrent through detector 420, so as to maximize the photocurrent. After the position of the second focusing lens 410 and detector 420 is fixed, the optical input component 200 (such as an input collimator) is fixed on the substrate 100 through a wedge angle.

Preferably, after the whole wavelength tunable optical receiving component has been fabricated, the absorbent material can be coated at the output end of the reflective Bragg grating chip 330 to prevent the transmission spectrum from refracting in the environment to form optical crosstalk. The detector 420 is a side-entry PD with a working rate of 10 Gb/s. According to the different working rates of PD, the size of its photosensitive surface is also different, and the design parameters of the second focusing lens need to be optimized according to different PD. When the optical input component is an input collimator, the input collimator is fixed on the substrate, and the vertical center of the input collimator is consistent with the height center of the first focusing lens and the second focusing lens.

5 Conclusion

In summary, the wavelength tunable optical receiving module of the present study can achieve fast and fine wavelength switching by utilizing the thermo-optical effect of optical waveguide, which can achieve the switching speed of less than 10 ms and the wavelength accuracy of less than 0.1 nm. Compared with the external cavity and the MEMS scheme, its tunable range is wider and can cover the whole C-band. The temperature control of Bragg grating chip by TEC and the wavelength stability is better. The integrated Bragg grating chip and the miniaturized micro-optical annulus realize the compact structure, small package volume, and easy modular integration of the tunable light receiving module.

Acknowledgements This paper is supported by Hubei Provincial Education Department of Scientific Research Project of B2017420; Hubei Provincial Education Department of Humanities and social science research project of 16G251.

References

1. What is optical fiber communication. Guangming daily (2003)
2. Huang, Z.: Optoelectronic devices and components of optical fiber communication credit. Beijing: Beijing university of posts and telecommunications press, pp. 1–2 (2001)
3. Zhang, R.: High speed optical transceiver module and its progress. Integr. Circ. Commun. **4**, 35–40 (2005)
4. Su, Y.: Research on key technologies of intelligent fiber optic optoelectronics. Beijing University of Posts and Telecommunications, pp. 27–29 (2012)
5. Han, L.: Design and manufacturing technology of wavelength tunable semiconductor laser. Zhejiang University 9–11 (2008)
6. Zhao, J.: Research on wavelength tunable semiconductor laser based on novel micro-nano structure. Huazhong University of Science and Technology 16–20 (2015)

Optimizing of Spatial Activities Monitoring Using the Raspberry Pi and RFID System

Zoltán Balogh and Ivan Baláž

Abstract The article deals with the proposal and optimization of a system for monitoring small rodent spatial activity using microcomputer Raspberry Pi and RFID. The monitoring system is comprised of autonomous stations and a server. Autonomous stations constantly collect information on rodent movement using a pressure sensor and RFID chips. Captured data is stored in the local relational database and sent to the server. Each station is powered by a battery and could work without Internet access. The server secures a Web application that allows display and manipulation of the captured data. The article contains a proposal for optimal autonomous solution and installation of system monitoring small rodent spatial activity powered by photovoltaic cells.

Keywords Raspberry Pi · RFID · Monitoring · Optimization of energy consumption

1 Introduction

Nowadays, monitoring of spatial activity of small animals puts high demands on accuracy. Outdated methods of physical monitoring bring a high error rate into the observation that might affect the results of the scientific work. At present, more than 80% of all tested mammals are mice. New technologies allow the automation of the monitoring process and reduce measurement inaccuracy. Technologies like GPS or Argos offer new possibilities in the field of animal movement monitoring. These data can be captured continuously and with a minimum deviation rate. However,

Z. Balogh (✉)
Department of Informatics, Faculty of Natural Sciences, Constantine the Philosopher University in Nitra, Tr. A. Hlinku 1, 949 74 Nitra, Slovakia
e-mail: zbalogh@ukf.sk

I. Baláž
Department of Ecology and Environmental Sciences, Constantine the Philosopher University in Nitra, Tr. A. Hlinku 1, 949 74 Nitra, Slovakia

V. Jain et al. (eds.), *Recent Trends in Intelligent Computing, Communication and Devices*, Advances in Intelligent Systems and Computing 1031,
https://doi.org/10.1007/978-981-13-9406-5_73

high implementation costs may become a disadvantage of these technologies. Furthermore, when using these technologies for monitoring rodent movements, we face the dimension problem of the monitoring device.

Complex monitoring technologies such as GPS and ARGOS [1] are not suitable for monitoring small rodents. Their detection system is not accurate enough to determine the position of small rodents, and their transponders are too large for practical rodents monitoring. The most suitable technology for monitoring rodent spatial or semi-spatial activities is RFID technology. RFID transponders have dimensions small enough for practical implantation into rodent body and their low costs allow bulk testing.

2 Materials and Methods

Proposal for an optimal rodent monitoring system requires analyzing the methods used in practice. Technological developments in recent years have brought many new technologies to monitor wildlife. The main criteria of modern monitoring methods include autonomy and the ability to work without human intervention, battery consumption and battery life, distance range, and method of transferring captured data to scientists for further processing.

2.1 Animal Identification Protocols of RFID Chip

HDX and FDX-B are protocols formally used in transponders of applications requiring animal identification. Both protocols are described in ISO standards 11784 and 11785. The main difference between these two protocols is the method of their communication. The FDX-B transponder (Full Duplex) sends its data at the same time as RFID module transmits the magnetic field from its antenna. This method of communication allows FDX-B transponder to be kept underpowering in the presence of the magnetic field of the antenna. The disadvantage of this method is that a strong magnetic field drowns the transponder signal, and this leads to impaired readability [2].

FDX-B uses a 134.2 kHz frequency with biphase encoding for its communication. The biphase encoding scheme modulates the RF field and creates an explicit transition at the beginning of each bit. The logic zero has a transition in the middle of the bit period. In contrast to zeros, the logic unit does not have any transition in the bit period. For FDX-B, bit length corresponds to 32 cycles of the reader active field (Fig. 1).

FDX-B protocol takes 128 data bits in the following structure:

- 11-bit header (10000000000)
- 64 identification bits with 8 control bits

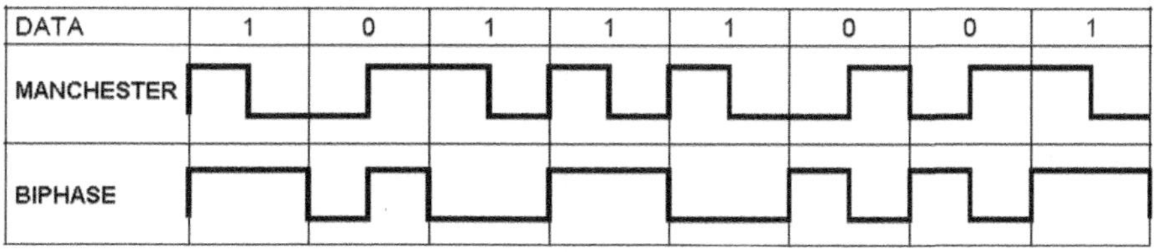

Fig. 1 Manchester a Biphase encoding scheme of the FDX-B protocol

- 16 CRC bits with two control bits
- 24 extended data bits with 3 control bits

HDX transponder (Half-duplex) is similar to FDX-B being powered by a magnetic field of the antenna but instead of the immediate communication, HDX transponder waits until the RFID module stops communicating. This means that HDX transponders can communicate with RFID module at a greater distance.

HDX works at the same frequency as FDX-B transponder with applying the essential phase shift when sending its data. For a logic unit, it sends a 16-cycle signal at 124 kHz and sends a 16-cycle signal at 134 kHz [3] for logic zero.

HDX protocol takes 112 data bits in the following structure:

- 8 header bits (01111110)
- 64 identification bits
- 16 CRC bits
- 24 extended data bits

The activation field generated by the reader is usually 50 ms in size. During this time, the reader will search for the presence of transponder with FDX-B protocol. If it cannot be found by the reader, it turns off its field and allows transponder with HDX protocol to send its data. If the RFID module does not capture the response within 3 ms, it will attempt to reconnect with the FDX-B transponder.

2.2 *Small Animals Monitoring Using RFID Tags*

The data captured on movement and behavior of rodents has been performed by manual physical observation for many years. During data capturing, the observer has gained experience to recognize unique patterns in behavior and movement of rodents. Individual rodents were tagged at the same time using color patterns on their body or amputation of parts of the limbs. The disadvantage of such data capturing was the observer's unreliability to recognize and record data. This error was induced by the mood and fatigue of the observer. Also, each new observer brings his own preoccupation into the process [4]. Since mice are night animals, capturing data on their movements has also deteriorated due to a poor visibility.

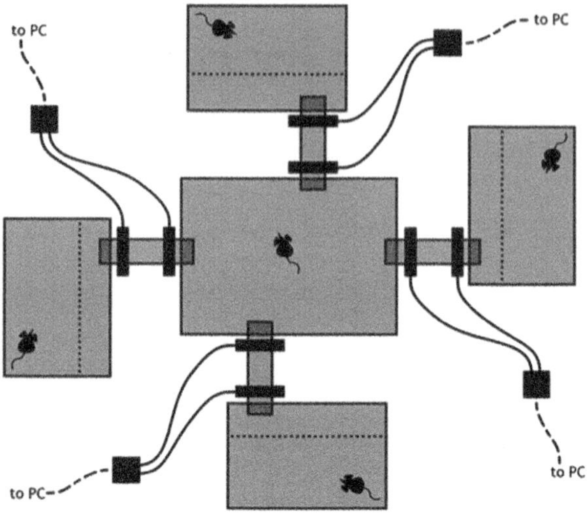

Fig. 2 Schematic layout of the RFID components of the monitoring system

The main disadvantage of commercially available rodent monitoring systems using RFID tags is their high cost and inflexibility.

An example of such a system is the system used by Miriam Linnenbrink in 2007 to test the criteria for selecting the rodent partners. Applied system comprised of five cages, while four side cages were attached to the central cage by tunnels. Each tunnel contained two RFID readers to record the movement of the mice as shown in Fig. 2. Each side cage was also divided by a grid to prevent mice mating [5].

RFID transponders are implanted into the rodent body by a syringe or by cutting a small hole into the body by scalpel under the anesthesia. To optimize the transponder reading, implanted transponders are usually inserted into the rodent body by length. Under laboratory conditions, rodents are also color-tagged, so observer could simply recognize them without the need of further manipulation [6].

3 Research Methodology

The most advantageous technology for capturing data about the small rodent movement is RFID technology. Passive RFID transponders have small dimensions enough to be implanted into small rodent body and are not costly even for massive monitoring of movement and migration. The unique benefit of passive RFID tags is that they do not need to be powered by a battery. Passive RFID tags acquire electrical power from the induction field of the reading module, which implies that for monitoring spatial activity, only the monitoring station with an RFID antenna is required [7].

The aim of the article is to describe the proposal and optimization of the system for monitoring the spatial activity of small rodents. The proposed system is based on a series comprised of Raspberry Pi microcomputers and RFID technology for passive animal tags according to ISO standard 11784/11785. Our aim was to design an optimal solution for saving energy and to increase efficiency in long-term and autonomous data capture. The proposed monitoring system powered by a battery is capable to work for at least 24 h and if the photovoltaic cell is connected, to operate continuously. We assume the proposed monitoring system will be more widely used than commercially available solutions for the spatial activity of small rodent monitoring and its cost price will be lower than the cost price of these commercially available solutions.

3.1 Proposed Solution

The proposed spatial activity data capturing system consists of two parts based on the Raspberry Pi microcomputer.

The first part of the solution is a monitoring station designed to capture data on the movement of the rodents. These data are captured using RFIDRW-E-USB RFID module, which records data about the tags implanted into rodent body in its vicinity. If the rodent does not have RFID tag implanted into its body, the station is still able to detect the presence of the rodent by a pressure switch. These data are subsequently recorded and sent for processing together with current time and identification of the station. In case, these data are not sent for any reason, the station stores them and attempts to send again. The station uses a WiFi antenna for communication and could be powered by a battery [8] (Fig. 3).

If the pressure switch is connected, the RFID module waits for a command to send the request for connecting with RFID transponder. Once the pressure switch is on, the program sends a request for at least 16-character string to the RFID module, and a 15-second timer will start. If 15 s elapse or the program gets required string, the program opens MySQL tunnel and writes the record with captured RFID tag, the identification number of station and time stamp into the database. Once the record has been stored, the program empties the front of RFID module and waits 2 s for a rodent to leave. This idle time reduces the number of duplicate and incomplete records. After the idle time is over, the station will restart to monitor the pressure switch.

If the station is configured in a mode without a pressure switch, once the connection is established, the module will automatically start searching for RFID transponders. If the RFID module captures the response from RFID transponder, then writes it to the database and waits for 1 s to restart searching.

The second part of the monitoring system is the server. The server is able to receive and process data on the movement of the rodents sent by monitoring stations. The server then displays these data on a Web page that is hosting. This Web site can also record the current status of stations and allow the export of captured data.

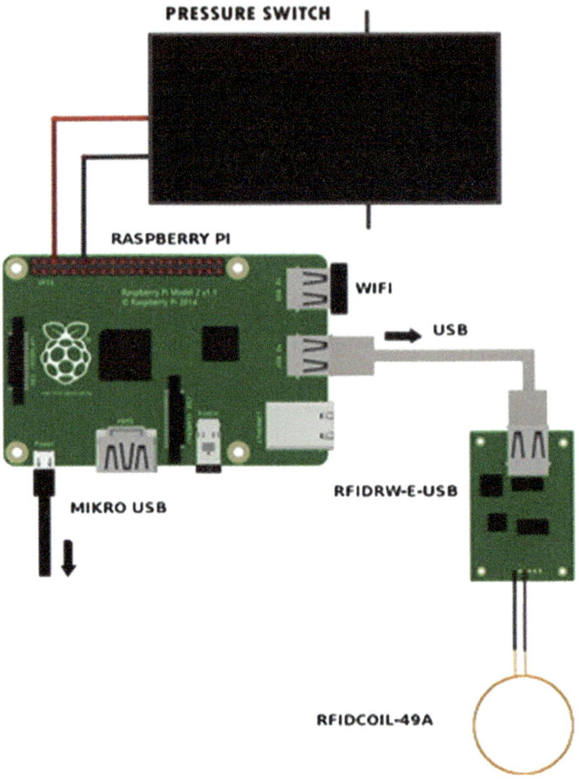

Fig. 3 Wiring diagram of components for the monitoring station

The selected components of the monitoring system are Raspberry Pi microcomputer and RFIDRW-E-USB RFID module. Raspberry Pi microcomputer matches the software and network requirements of both the server and the station and also ensures easy and fast installation. RFIDRW-E-USB RFID module was selected for its ability to work with HDX and FDX-B animal transponders and the guaranteed availability.

3.2 *Implementation and Testings*

The proposed monitoring system works on the principle of the sensor network, where the individual stations send captured data to the server without the need of feedback.

The data on rodent activities are recorded by RFID transponders and pressure switches. These data are stored in the local relational database of the station. With local data recording, individual stations are able to continue to capture the data even in case of the accidental or deliberate disconnection from the network.

Reliability of the Raspberry Pi microcomputer as an autonomous station requires a detailed analysis of the power consumption.

The approach to minimizing the power consumption of the proposed station included analyzing and testing various hardware and software configurations. Power consumption was measured externally by a digital multimeter and MTK-PE trigger USB tester. As a digital multimeter was used Vorel TO-81784, which can measure values between 0 and 600 V and 0–600 A. Raspberry Pi microcomputer is powered by a micro USB port ranging from 4.8 to 5.2 V. Monitoring the consumption of the electric current was performed by creating a crossing on stretched 5-V USB cable [9].

During optimization, the power consumption of the station was reduced from the original 1.543–1.431 W when WiFi adapter and RFID module were connected. Testing has shown that the timing of the system has a minimal impact on power consumption at high power losses. Testing has also clearly evaluated that the main source of power consumption is the USB chip and LAN9512 Ethernet connector. After connecting all peripheral devices and optimizations, Raspberry Pi microcomputer consumes 0.931 W of which 0.527 Watts requires LAN9512 chip itself.

This suggests that the last option of reducing the current consumption is to manually remove LAN9512 chip and to connect the USB connector directly to the Broadcom BCM6836 SOC processor [10]. As for Raspberry Pi B microcomputer, such connection is possible between R36 and R37 connectors. This connection will allow the functionality for one only, which means that the RFID module will have to be connected by GPIO. If the chip is removed, the estimated power consumption of the station with WiFi adapter and RFID module is reduced to $\sim$0.8 W.

An alternative option of minimizing current consumption is to replace selected components, mostly those installed in Raspberry Pi 2 microcomputer. Raspberry Pi 2 microcomputer provides high performance that is not fully used by measuring station since the measurement itself doesn't demand difficult computing. When changing the microcomputer model to Raspberry Pi B, the station consumption can be reduced to 1.283 W. When using Raspberry Pi A or Raspberry Pi Zero microcomputer, the expected reduction in consumption drops below 1 W [11].

4 Conclusion

The article describes the proposal and optimization for monitoring the spatial activity of the rodent. The system can record the movement of rodents using implanted RFID transponders and pressure switches. The monitoring system includes a server that allows manipulation and export of captured data and the series of autonomous stations. Each station requires a minimum of 37.44 W-h per day, what is almost equivalent to one 100,00 mAh battery. For 100% reliability, 10,000 mAh battery was replaced by 120,000 mAh battery in the final solution. Using a combination of 44 W battery with the 17 W solar panel and optimal sun conditions, the proposed monitoring system is able to work continuously in ideal conditions. The proposed solution is non-scalable for bigger animals due to the short range of the RFID module. RFIDRW-E-USB RFID module shown in the above schema has a maximum antenna range of 5 cm when communicating with HDX and FDX-B transponders. Practical testing on an optimized Raspberry Pi 2, however, showed that the maximum communication range is about 3 cm.

Acknowledgements This study was created with the financial support of the Scientific Grant Agency VEGA, grant no. 1/0608/16.

References

1. Argos: Argos satellites. http://www.argos-system.org/argos/how-argos-works/ (2016). 2018
2. DESIGN, P.: FDX-B Animal Identification Protocol description http://www.priority1design.com.au/fdx-b_animal_identification_protocol.html (2007). Accessed 25.11.2018
3. Vuza, D.T., Frosch, R.: RFID readers for the HDX protocol—design, simulation and testing. In: 2010 IEEE 16th International Symposium for Design and Technology in Electronic Packaging (SIITME), 23–26 Sept 2010, pp. 47–52
4. Kritzler, M., Lewejohann, L., Krüger, A.: Analysing movement and behavioural patterns of laboratory mice in a semi natural environment based on data collected via RFID-technology. In: CEUR Workshop Proceedings 2007, pp. 17–28
5. Linnenbrink, M., von Merten, S.: No speed dating please! Patterns of social preference in male and female house mice. Frontiers in Zoology 14(1) (2017). https://doi.org/10.1186/s12983-017-0224-y
6. Howerton, C.L., Garner, J.P., Mench, J.A.: A system utilizing radio frequency identification (RFID) technology to monitor individual rodent behavior in complex social settings. J. Neurosci. Methods **209**(1), 74–78 (2012). https://doi.org/10.1016/j.jneumeth.2012.06.001
7. Balogh, Z., Bízik, R., Turčáni, M., Koprda, Š.: Proposal for spatial monitoring activities using the Raspberry Pi and LF RFID technology. In: Zeng, Q.-A. (ed.) Wireless Communications, Networking and Applications, vol. 348. Lecture Notes in Electrical Engineering, pp. 641–651. Springer, India (2016)
8. Balogh, Z., Turčáni, M.: Complex design of monitoring system for small animals by the use of micro PC and RFID technology. In: Oualkadi, A.E., Moussati, A.E., Choubani, F. (eds.) Mediterranean Conference on Information and Communication Technologies, MedCT 2015, vol. 380. pp. 55–63. Springer, Verlag (2016)
9. Kaup, F., Gottschling, P., Hausheer, D.: PowerPi: measuring and modeling the power consumption of the Raspberry Pi. In: 39th Annual IEEE Conference on Local Computer Networks, 8–11 Sept 2014, pp. 236–243
10. Blecky: First Raspberry Pi Model A 512 MB off the (Re-)Production Line—part 1. http://freneticrapport.blogspot.sk/2013/01/first-raspberry-pi-model-512mb-off-re_24.html (2013). Accessed 2018-3-16 2018
11. GEERLING, J.: Power Consumption Benchmarks. https://www.pidramble.com/wiki/benchmarks/ (2018). Accessed 2018-3-16 2018

Multiple Star Node Discovery Algorithm in Social Network Based on Six Degrees of Separation and Greedy Strategy

Jinbo Bai, Hongbo Li and Jianping Chen

Abstract Because of the rocketing of information technology of this time, the extraction of large-scale social network is very common. However, mining star nodes in a large-scale network is a conundrum and has aroused the attention of many scholars. At present, there are two main problems in the existing algorithm of star node mining: On the one hand, the storage cost is comparatively high, and on the other hand, the time cost is high, too. Aiming at the deficiencies, based on six degrees of separation stemming from sociology and greedy strategy being used in numerous algorithms, we propose a multiple star node discovery algorithm. The efficiency and validity of the algorithm are proved by the experiments.

Keywords Star node discovery · Social network · Six degrees of separation · Greedy strategy

1 Introduction

Many researches demonstrate that in social networks the nodes with more contacts are more important than the average nodes and have more influence on other nodes [1–4]. However, with the development of data collection, on the one hand, large-scale social networks are very common; on the other hand, mining or discovering nodes with bigger degrees, namely star nodes, is proven a conundrum because the storage cost and time cost of the mining process are very high. Aiming at this problem, we propose a comparatively efficient algorithm in this paper. The algorithm adopts a new networking method and is implemented based on six degrees of separation and greedy strategy.

J. Bai
Economics and Management College, Zhaoqing University, Zhaoqing, China

H. Li (✉) · J. Chen
School of Computer Science and Software, Zhaoqing University, Zhaoqing, China
e-mail: islhb@126.com

V. Jain et al. (eds.), *Recent Trends in Intelligent Computing, Communication and Devices*, Advances in Intelligent Systems and Computing 1031,
https://doi.org/10.1007/978-981-13-9406-5_74

In general, for fully taking advantage of mathematical properties, in many methods, the network is built by defining a $n \times n$ matrix, where n is the number of a network [5, 6]. However, there are two main drawbacks in these methods as previously mentioned: one is comparatively high time cost, and the other is space cost. It is self-evident that if we are going to traverse or process, such as counting the degrees of the elements of a network matrix, the operational time complexity will be at least $O(n^2)$, and the size of network storage is $n \times n \times s$ bytes, where s is byte number of the data type being used to store the node's data, such as the node's serial number. The scale of a network is larger, the time and space cost to deal with it will be higher. In this paper, we will propose a new method to build a social network, which is capable of reducing the space and time costs by a large margin in the vast majority of cases.

Six degrees of separation believes that the distance between any two persons in the world, for example, Clark and Linda, is no more than six steps, where "step" is a measuring unit. If Clark and Linda are friends, then the distance of them is one step; if they are not friends, but all are Susie's friends, then the distance of them is two steps. In other words, there is a chain of fellowship which is no more than six steps between any Clark and any Linda. This hypothesis has been accepted by many people from a wild variety of domains and been applied to many researches of computer science [7–9]. Six degrees of separation suggests that in an interconnected social network any node, being taken as starting point and following a chain of 1-hop neighbor, could reach any other node in a several steps. For the sake of efficiency, the algorithm in this paper will take six degrees of separation as one of the theoretical basis.

Greedy strategy is adopted in numerous algorithms making use of heuristic information and chases a locally optimal choice at every stage with the hope of finding a global optimum solution [10]. In general, though a greedy algorithm maybe can't find a global optimal solution, it is often capable of finding the locally optimum solution. In a social network, the locally optimum solution, the node with the biggest degree in its adjacent area is as the same importance as the node with the biggest degree in the whole network, for they all can be taking as an opinion leader or objective node. Aiming at them and carrying on a marketing campaign to them, the enterprises can save a lot of advertising money. Therefore, this feature of the greedy strategy is very suitable for discovering star nodes, and it is adopted as another basis of our algorithm.

2 Related Works

In social network, mining vital nodes is an important theme in academia and a lot of wonderful research results have been achieved. "Vital nodes" are sometimes called "important nodes" or "key nodes" by some researchers and these definitions are basically different from the definition of "star nodes" in this paper. The definition of "star nodes" in this paper is similar to the definition of the degree centrality research

[11–14] and is same to the definition of star nodes discovery based on acquaintance immunization research [15].

The main deficiency of the degree centrality is its computation takes $O(n^2)$ in most of the cases. Nowadays, with the progress of data collection, the network scales have an amazing cumulative increase, getting larger and larger, so the comparatively high time cost of the computation of degree centrality seriously baffles its application.

Star nodes discovery based on acquaintance immunization is an efficient algorithm previously proposed by us, and its time complexity is no more than $O(n^2)$ [15]. This algorithm has two main deficiencies. Firstly, because of the being repeatedly selected of getting started nodes are not restricted, leading to the number of mined nodes is not enough, the scope of influence of the mined nodes is comparatively small; secondly even if the repeated selection is forbidden, the mined nodes may be repeated in a comparatively high probability, so the scope of influence of the mined nodes becomes smaller, especially in the large-scale social networks.

The main contributions of the algorithm in this paper are as follows. First, it provides a novel networking method merely making use of local information. The method can be capable of saving storing space and further improves the efficiency of the algorithm. Second, it reduces the mining time complexity to a reasonable level (less than $O(n \times \bar{d})$), where $\bar{d}$ is the average value of node degrees in a network, and is far less than the n), and is more appropriate to mining star nodes in the social network. Third, because the algorithm is on the basis of six degrees of separation and greedy strategy, mining of network is more sufficient. Fourth, the algorithm has a potential concurrency and has a great room to improve from the efficient aspect.

3 HLL: A New Networking Method

Different from the traditional networking by an adjacent matrix, our algorithm adopts heterogeneous linked list (HLL) to build network. The concrete data structures are as follows.

```
struct neighbor
{ int SN;                   // the node's serial number
  struct neighbor *next;    // neighbor node pointer        };
struct node
{ int SN;                   // the node's serial number
  int degree;               // storing a node's degree
  struct neighbor *next;    // neighbor nodepointer
  bool isStartingNode;      // starting node flag           };
```

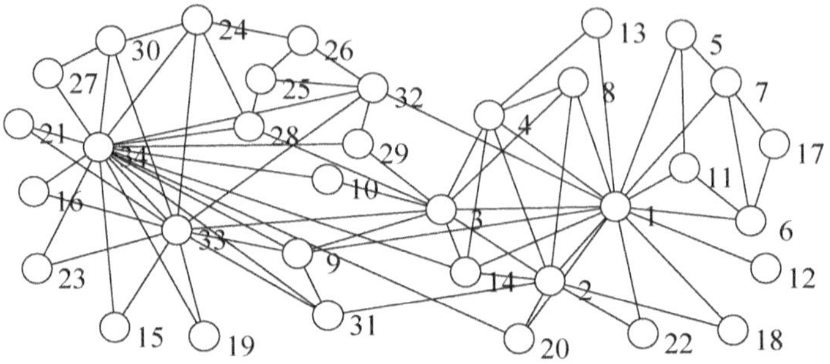

Fig. 1 Karate club network

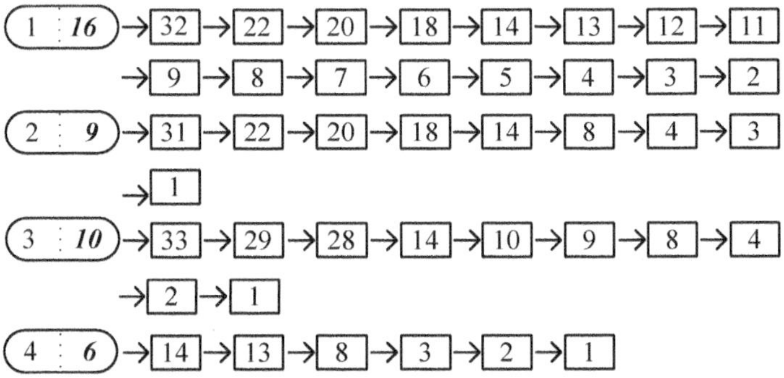

Fig. 2 Applying networking method to Karate club network (partial result)

Before building a network, a struct node array is defined, every element of the array is the head node of a heterogeneous linked list, and it will point to one of its 1-hop neighbor nodes storing by struct neighbor; the 1-hop neighbor will point to another neighbor node of the head node, if any; and so forth.

Karate club network is a classic social network, as shown in Fig. 1, and there are 34 nodes and 78 edges (meaning the total degree is 156) [16]. Applying the networking method to Karate club network, the part result is shown in Fig. 2. In Fig. 2, rectangles show the 1-hop neighbor nodes, rounded rectangles show the head nodes, and the bold italic numbers show the head nodes' degrees.

The storing size of HLL method can be computed by the following formula:

$$n \times (2 \times sInt + sBool + sPt) + (sInt + sPt) \times \sum_{i=1}^{n} d_i \tag{1}$$

where n is the number of network, d_i is the degree of a node whose serial number is i, sPt, $sBool$, and $sInt$ are pointer type, bool type, and int type, respectively.

In many traditional methods, the storing size of the matrix method can be computed by the following formula:

$$n \times n \times sInt \tag{2}$$

where n is the number of network and $sInt$ is the size of int type.

Making the assumption that the pointer type and int type all occupy 4 bytes and the bool type occupies 1 byte, according to formula (1), we can get that the storing size of Karate club network of HLL method is 1690 bytes. Compared with the size 4624 bytes (computed by formula (2)) of matrix storing, about 63.45% storing space is saved by HLL method.

After comparing the matrix method with HLL method in some networks, all the comparable data shown in Table 1, we find that in general the larger scale of network becomes, the more storing space HLL method saves, and when the node number is not less than 1000 the saving ratio is not less than 99%. For another example, the Friendster social network has 65,608,366 nodes and 1,806,067,135 edges (the node SNs range from 101 to 124,836,179) [17]. Compared with 62,336,186,480,376,964 bytes (58,055,097.68 GB) of the matrix storing, the new networking method only takes 30,519,943,187 bytes (28.42 GB), and more than 99.99% storing space is saved. In fact, the HLL method brings not only storing efficiency but also mining efficiency, which we can know hereafter.

Table 1 Storing space comparison between the matrix method and HLL method (all the data set can be found in the webs referred by [16, 17])

No.	Network name	Node number	SN Range	Matrix (MB)	HLL (MB)	Saving ratio
1	LesMiser	77	0–76	0.0226	0.0048	0.7876
2	WordAdj	112	0–111	0.0479	0.0079	0.8351
3	CollegeFoot	115	0–114	0.0504	0.0108	0.7857
4	NeuralNet	297	0–296	0.3365	0.0397	0.8820
5	Coauthor	1461	0–1588	9.6318	0.0615	0.9936
6	PowerGrid	4941	0–4940	93.1300	0.1619	0.9983
7	HETheory	7610	0–8360	266.6715	0.3440	0.9989
8	CondM99	16,264	0–16,725	1067.1962	0.9336	0.9991
9	CondM03	30,460	0–31,162	3704.5768	2.2178	0.9994
10	CondM05	39,575	0–40,418	6232.0540	3.1819	0.9995
11	ComDBLP	317,080	0–425,956	692,136.2528	21.3006	0.9999
12	ComYoutube	1,134,890	1–1,157,827	5,113,843.3911	59.9420	0.9999

4 Discovery Algorithm

4.1 Algorithm Description

The steps of the discovery algorithm are as follows.

Input: a network G being built by HLL, and a parameter $p \in (0, 1]$ controlling the starting node number.
Output: star nodes.
Step 1. Randomly selects $m = n \times p$ nodes v_1, v_2, ... v_m, and set each node's field is StartingNode = 1, where n is the node number of G.
Step 2. Visits every node of G, and if the node is a starting node begins a greedy search (GS). The search step number of GS is 6. Every step of GS finds the maximum degree neighbor node v, takes it as a star node, and begins the next step search from v.

4.2 Experiments and Analysis

How to select the controlling parameter p is a very difficult thing. If the value of p is bigger, we will have more starting nodes, and the efficiency of the algorithm decreases. Otherwise, the star nodes' influence may be impaired. Aiming at this problem, we do experiments on some networks and each network 100 times. The experiment results are shown in Table 2, and the data is the mined star nodes' average degrees. We find that when the p changes from 0.1 to 0.2 the star nodes' average degrees have a dramatically increase, but when the p changes from 0.2 to 0.3, or 0.3 to 0.4, and so on, though the average degrees have an increase, it is very slow. This situation fully conforms to the Pareto law and is illustrated in Fig. 3. Note, though there are only three networks' lines the other networks have similar lines. According to the experiment results, we strongly recommend that the parameter p should be set 0.2.

Because the bigger degree means more contacts, we suggest that the total degree of star nodes in a network be an essential measure of their influence. As shown in Table 3, we compute the fractions of coverage of star nodes' average degrees and find the influence of mined star nodes is encouraging: the minimum is 0.6849, the maximum is 0.9564, and the average is 0.8373, the median is 0.8457.

The time complexity of this algorithm is $O(6 \times \bar{d} \times n)$. In general, the average degree of a network is comparatively small, just like as shown in Table 4. Therefore, the algorithm has a linear time complexity and is very efficient.

In addition, it deserves noting that the discovery algorithm only uses the local information. The algorithm's locality means a big possibility of potential concurrency. In the future, we can take advantage of this point and design a concurrent algorithm to further improve mining efficiency.

Table 2 Parameter p testing experiments

No.	Network name	$p = 0.1$	$p = 0.2$	$p = 0.3$	$p = 0.4$	$p = 0.5$	$p = 0.6$	$p = 0.7$	$p = 0.8$	$p = 0.9$	$p = 1.0$
1	LesMiser	342	421	445	466	472	476	480	483	485	488
2	WordAdj	637	746	779	794	807	816	825	829	832	838
3	CollegeFoot	791	1128	1197	1218	1223	1224	1226	1226	1226	1226
4	NeuralNet	3520	4214	4437	4542	4593	4626	4654	4662	4678	4685
5	Coauthor	2608	3756	4425	4787	5013	5170	5256	5316	5359	5380
6	PowerGrid	7025	9182	10,328	11,047	11,502	11,852	12,102	12,297	12,457	12,583
7	HETheory	18,436	23,322	25,760	27,283	28,291	29,023	29,525	29,897	30,176	30,363
8	CondM99	59,437	73,699	73,929	85,469	88,108	89,813	90,917	91,747	92,284	92,659
9	CondM03	164,007	195,325	210,834	219,879	225,157	228,771	231,069	232,669	233,792	234,622
10	CondM05	249,127	289,753	309,415	320,340	327,253	332,151	335,512	338,143	342,579	343,782
11	ComDBLP	1,394,079	1,656,953	1,790,071	1,868,469	1,919,566	1,953,645	1,976,898	1,993,970	2,006,185	2,015,446
12	ComYoutube	4,122,248	4,482,832	4,686,972	4,823,727	4,922,332	4,996,121	5,053,494	5,099,465	5,136,797	5,167,996

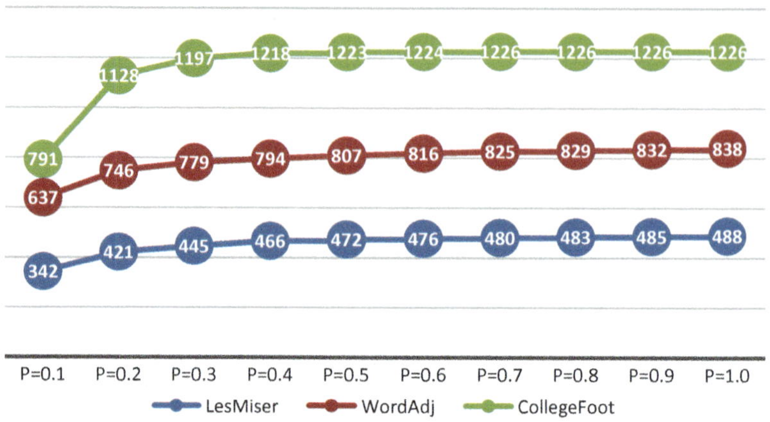

Fig. 3 Line charts of three networks

Table 3 Fraction of coverage of star nodes' average degrees

No.	Network name	$p = 0.2$	Total degree	Fraction of coverage
1	LesMiser	421	508	0.8287
2	WordAdj	746	850	0.8776
3	CollegeFoot	1128	1226	0.9201
4	NeuralNet	4214	4718	0.8932
5	Coauthor	3756	5484	0.6849
6	PowerGrid	9182	13,188	0.6962
7	HETheory	23,322	31,502	0.7403
8	CondM99	73,699	95,188	0.7742
9	CondM03	195,325	240,058	0.8137
10	CondM05	289,753	309,415	0.9365
11	ComDBLP	1,656,953	1,790,071	0.9256
12	ComYoutube	4,482,832	4,686,972	0.9564

Table 4 Average degree of some networks

No.	Network name	Node number	Total degree	$\bar{d}$
1	LesMiser	77	508	6.60
2	WordAdj	112	850	7.59
3	CollegeFoot	115	1226	10.66
4	NeuralNet	297	4718	15.89
5	Coauthor	1461	5484	3.75
6	PowerGrid	4941	13,188	2.67
7	HETheory	7610	31,502	4.14
8	CondM99	16,264	95,188	5.85
9	CondM03	30,460	240,058	7.88
10	CondM05	39,575	351,382	8.88
11	ComDBLP	317,080	2,099,732	7.45
12	ComYoutube	1,134,890	5,975,248	5.27

5 Conclusions

With the development of information technology, the scale of the social network is getting larger and larger. How to mine the important or key nodes, such as star nodes, in a large-scale social network is a hot research issue. Aiming at the deficiencies of existing algorithms, such as high storage cost and high time cost of their mining processes, based on six degrees of separation and greedy strategy, we put forward a multiple star node discovery algorithm. The experiments demonstrate the algorithm's efficiency and validity.

Acknowledgements The research is supported by Zhaoqing University Scientific Research Fund Project (201845), Guangdong natural science foundation (2016A030313015), Zhaoqing Science and technology Innovation Guiding Category Project (201704030601), Innovation and School Developing Special Fund (504-20160171).

References

1. Kapoor, K., Sharma, D., Srivastava, J.: Weighted node degree centrality for hypergraphs. In: 2013 IEEE 2nd Network Science Workshop (NSW), pp. 152–155. IEEE (2013)
2. Saxena, A., Malik, V., Iyengar, S.R.S.: Estimating the degree centrality ranking. In: 2016 8th International Conference on Communication Systems and Networks (COMSNETS), pp. 1–2. IEEE (2016)
3. Badar, K., Frantz, T.L., Jabeen, M.: Research performance and degree centrality in co-authorship networks: the moderating role of homophily. Aslib J. Inf. Manage. **68**(6), 756–771 (2016)
4. Howlader, P., Sudeep, K.S.: Degree centrality, eigenvector centrality and the relation between them in Twitter. In: IEEE International Conference on Recent Trends in Electronics, Information & Communication Technology (RTEICT), pp. 678–682. IEEE (2016)
5. Girvan, M., Newman, M.E.J.: Community structure in social and biological networks. Proc. Natl. Acad. Sci. **99**(12), 7821–7826 (2002)
6. Newman, M.E.J., Girvan, M.: Finding and evaluating community structure in networks. Phys. Rev. E **69**(2), 026113 (2004)
7. Dave, S.H., Keller, B.D., Golmer, K., et al.: Six degrees of separation: connecting research with users and cost analysis. Joule **1**(3), 410–415 (2017)
8. Lunze, J.: Six degrees of separation in multi-agent systems. In: 2016 IEEE 55th Conference on Decision and Control (CDC), pp. 6838–6844. IEEE (2016)
9. Borassi, M., Crescenzi, P., Habib, M., et al.: Fast diameter and radius BFS-based computation in (weakly connected) real-world graphs: with an application to the six degrees of separation games. Theoret. Comput. Sci. **586**, 59–80 (2015)
10. https://en.wikipedia.org/wiki/Greedy_algorithm#cite_note-NISTg-1
11. Brandes, U., Borgatti, S.P., Freeman, L.C.: Maintaining the duality of closeness and betweenness centrality. Soc. Netw. **44**, 153–159 (2016)
12. Tan, Y.J., Wu, J., Deng, H.: Evaluation method for node importance based on node contraction in complex networks. Syst. Eng. Theory Pract. **11**(11), 79–83 (2006)
13. Zhao, Y.H., Wang, Z.L., Zheng, J., Guo, X.J.: Finding most vital node by node importance contribution matrix in communication netwoks. J. Beijing Univ. Aeronaut. Astronaut. (9), 009 (2009)

14. Ye, Z., Zhan, X., Zhou, Y., Liu, C., Zhang, Z.K.: Identifying vital nodes on temporal networks: an edge-based K-shell decomposition. In: Control Conference (CCC), 2017 36th Chinese, July 26, pp. 1402–1407 (2017)
15. Li, H., Zhang, J., Yang, J., Bai, J., Zhao, Y.: Efficient star nodes discovery algorithm in social networks based on acquaintance immunization. In: ICIVC, pp. 937–940 (2017)
16. http://www-personal.umich.edu/~mejn/netdata/
17. http://snap.stanford.edu/data/#socnets

Hardware Implementation of the Sub-pixel Interpolation in the HDTV Video Codec

Mengmeng Wu

Abstract In this paper, a hardware-implemented architecture for the efficient sub-pixel interpolation in the HDTV encoder and decoder is proposed. As for the luma component, a balanced interpolator structure is adopted. The luma module consists of seven half-pixel interpolation components and a quarter-pixel interpolation component. This module can accelerate the operation in luma interpolation and reduce the possible glitches. In order to implement the eighth-pixel interpolation in the chroma component, we propose an interpolation architecture that only includes shifters and adders. The hardware proposed in this paper can compute the interpolation for different pixels with less gates and faster speed. The FPGA experiment shows that this structure can work at 65.58 MHz and can meet the requirement of video codec.

Keywords Sub-pixel interpolation · HDTV · Hardware implementation · FPGA

1 Introduction

Almost all of the standardized video coding techniques like H.264/AVC, H.265/HEVC, MPEG-4, and AVS are based on the motion-compensated prediction. In digital video technology, in order to improve the accuracy of the motion vector, sub-pixel interpolation is used by the video coding techniques, such as H.264/AVC, H.265/HEVC, and AVS. In H.264/AVC standard, half-pixel interpolation is implemented by a six-tap filter [1] and quarter-pixel interpolation is implemented by a bilinear filter [2]. In H.265/HEVC standard, three different eight-tap filters are adopted for both half-pixel and quarter-pixel interpolations [3]. In AVS standard, two different four-tap filters are used separately for half-pixel and quarter-pixel interpolations [4, 5]. Sub-pixel interpolation algorithm is complicated and is computationally intensive. Especially in the decoder, sub-pixel interpolation is the

M. Wu (✉)
School of Physics and Electronic Engineering, Linyi University, Linyi 276000, China
e-mail: wumengmeng@lyu.edu.cn

V. Jain et al. (eds.), *Recent Trends in Intelligent Computing, Communication and Devices*, Advances in Intelligent Systems and Computing 1031,
https://doi.org/10.1007/978-981-13-9406-5_75

most complex part, taking up about one-third of the computation in software implementation. Hence, some VLSI architectures of sub-pixel interpolation have been presented and accelerated the heave computing step. For example, Wang presented a parallel and pipeline architecture for the sub-pixel interpolation filter in H.264/AVC conformed HDTV decoder [6].

This paper proposed a kind of hardware-implemented architecture for the efficient sub-pixel interpolation in AVS HDTV encoder and decoder. The remainder of this paper is organized as follows. Section 2 introduces the design of the luma interpolation module. Then, Sect. 3 presents the design of the chroma interpolation module. In Sect. 4, the hardware implementation and the experimental results of the sub-pixel interpolation are given. At last, the conclusion will be shown in Sect. 5.

2 Luma Interpolation Module

As shown in Fig. 1, the luma interpolation module primarily comprises the control unit lpe_ctrl, the luma sub-pixel interpolation unit, and the memory. The memory, which is a custom 12 × 12 register array, is used to store reference pixels for the 8 × 8 macro-block. The control unit provides the address signals to the memory, and the corresponding data are presented to the interpolation unit. NUM represents the number of the 8 × 8 macro-block that needs to be interpolated. The value is decoded by the upper layer module. *X* and *Y* provide the coordinates of the pixel to be performed the interpolation algorithm.

2.1 Luma Sub-pixel Interpolation Algorithm

In the luma component of the reference picture, there are the full-pixel and sub-pixel shown in Fig. 2. The full-pixel sample position is marked by the shaded blocks with uppercase letters, and the blocks with lowercase letters represent the

Fig. 1 Details of the proposed luma interpretation module

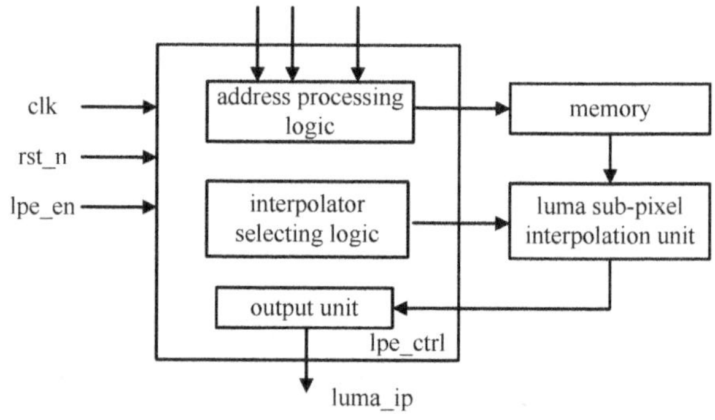

Fig. 2 Sample position of the full-pixel samples and the sub-pixel samples

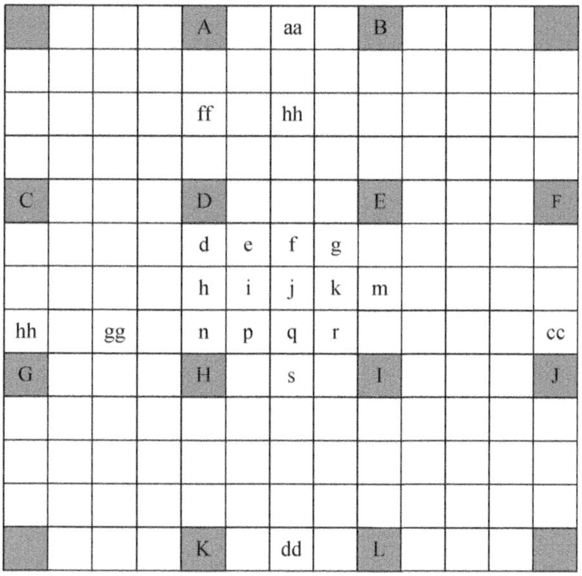

half-pixel position. Half-pixel samples are obtained by a four-tap filter F1 [−1, 5, 5, −1], and quarter-pixel samples are obtained by a four-tap filter F2 [1, 7]. The calculating process can refer to [7, 8].

2.2 *Luma Sub-pixel Interpolation Unit*

The intermediate value of the corresponding half-pixel sample needs to be calculated by the half-pixel four-tap filter F1(−1, 5, 5, −1), and then the result is fixed between 0 and 255 by the limiter. The half-pixel filter is implemented by a fourth-order interpolation filter with one unsigned operation. The data path is shown in Fig. 3.

For the quarter-pixel interpolation calculation of luma components, there are two types of algorithms. For the quarter-pixel samples, such as a, d, i, and f, the median values are calculated by four-tap filter F2(1, 7, 7, 1). And for the quarter-pixel samples, such as e, g, p, and r, the values are obtained by a linear interpolation algorithm. Then, the final predicted value is clipped under the value of 255.

In order to reduce the area of the interpolation unit, the linear interpolation algorithm also is implemented by the filter F2. The input components of weight 7 are set to 0. The architecture of the quarter-pixel interpolation is shown in Fig. 4.

The shifting and clipping operation can be divided into two types. One is adding 64 and shifting right 7 bits, and the other one is adding 512 and shifting right 10 bits. In order to make the quarter-pixel interpolator achieving the purpose of multiplexing to the maximum extent, the operation is unified to the latter. That is,

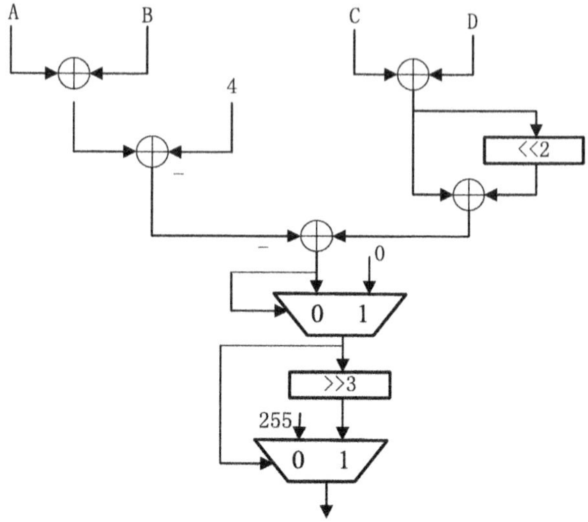

Fig. 3 Data path of the half-pixel interpolation

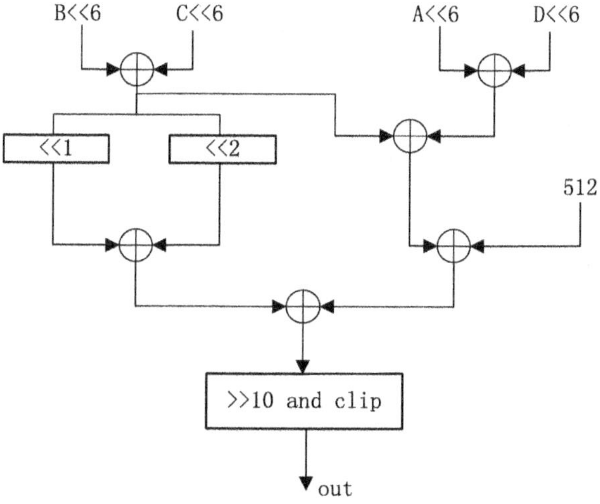

Fig. 4 Data path of the quarter-pixel interpolation

the pixel data which need to be performed adding 64 and shifting right by 7 bits are shifted to the left by 3 bits. After that, the results are sent to the unit to perform the operation of adding 512 and shifting right by 10 bits.

3 Chroma Interpolation Module

Compared to the luma interpolation module, the chroma interpolation algorithm is much simpler. In general, it is a weighted average operation to chrominance components of the reference pixels which are closest to the target pixel. The module is mainly composed of control unit, chroma interpolation calculation unit, and storage unit.

3.1 Chroma Sub-pixel Interpolation Algorithm

In the 4:2:0 sampling format, eighth-pixel resolution motion vectors in the chroma components are required. In each chroma component, the interpolated samples are generated at eighth-pixel intervals between full-pixel samples [6]. In our work, two-dimensional linear interpolation is used to generate each eighth-pixel chroma sample [6], as is shown in Fig. 5. The figure shows a 4 × 4 data block where *A*, *B*, *C*, and *D* are full-pixel chroma samples. And b, c, d, e, and f represent luma half-pixel interpolation. Thus, the pixel marked "a" represents the result of an operation of the eighth-pixel interpolation. The final prediction value can be calculated using Eq. 1.

$$\begin{aligned} PredMatrix[x,y] = Clip(((8-x)*(8-y)*A + x*(8-y)*B \\ + (8-x)*y*C + x*y*D + 32) \gg 6 \end{aligned} \quad (1)$$

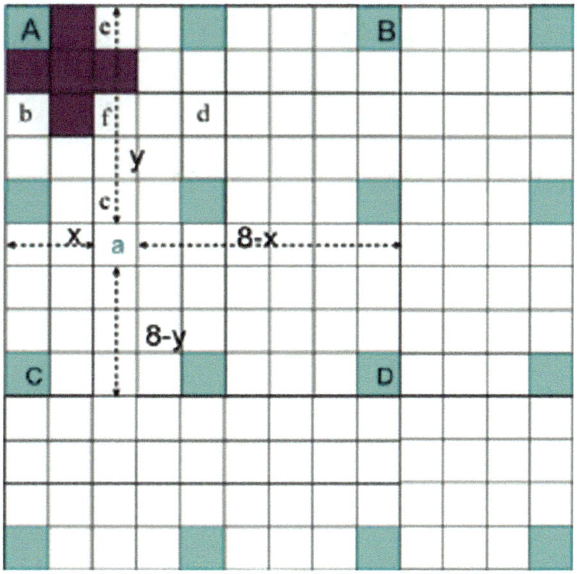

Fig. 5 Eighth-pixel chroma sample

3.2 Chroma Sub-pixel Interpolation Unit

There are two solutions to the chroma interpolation calculation. That is, multiplication operation can be performed by the hardware-implemented multiplier. It also can be achieved by the hardware shift operation. Since the high complexity and large operation delay of the multiplier, the use of the multipliers will slow down the system. What's more, the shifter can achieve less hardware overhead than the hardware multipliers. Thus, the paper adopts the shift operation to realize multiplication. Figure 6 is the shift operation data path for the chroma sub-pixel interpolation.

4 Hardware Implementation and FPGA Verification

In this work, we implement the proposed sub-pixel interpolation hardware in Verilog HDL. The whole hardware structure is shown in Fig. 7. Table 1 gives the inputs and output ports of the module and the corresponding description. Inside the interpolator, the control module is responsible for sending the clock signal, the reset signal, and the corresponding interpolation strobe signal to the corresponding luma interpolation module and the chroma interpolation module. The position decoding module converts the position signal into a signal represented by the block number. The decoded position signals input to the luma interpolation module and the chroma interpolation module. After the interpolation is completed, the results are output through the output control module.

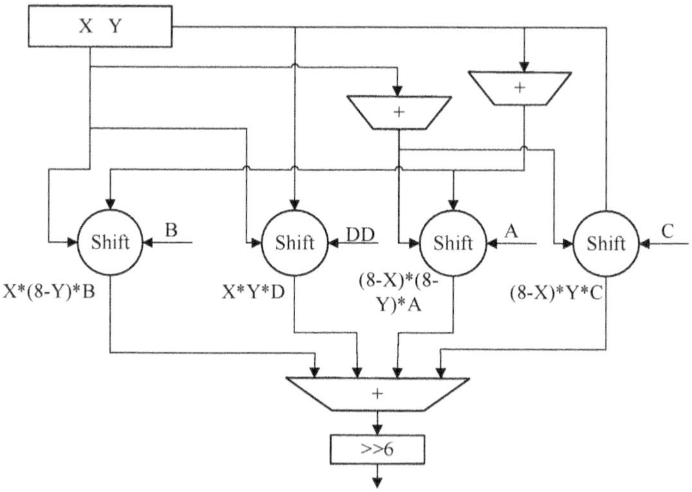

Fig. 6 Shift operation data path for the chroma sub-pixel interpolation

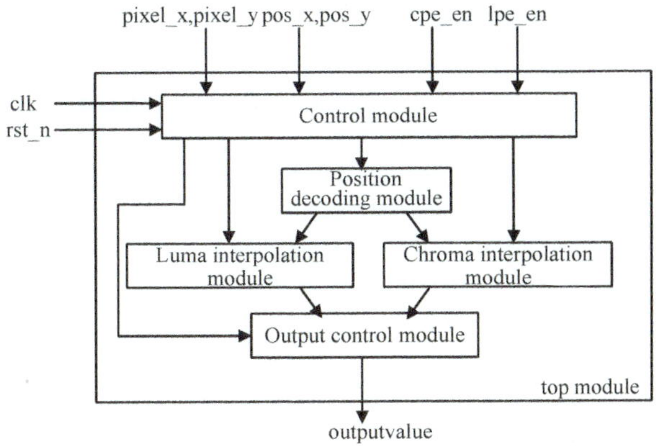

Fig. 7 Hardware structure of the proposed scheme

Table 1 Ports of the top module

Ports	Width	Description
clk, rst_n	1, 1	System signals
cpe_en, lpe_en	1, 1	Enable signals for chroma/luma interpolation
pixel_x, pixel_y	3	The position of the integer pixel
pos_x, pos_y	3	The target position of the sub-pixel
Output value	8	The output value of the interpolation module

Figure 8 shows a part of the simulation waveform for the sub-pixel interpolation hardware. The FPGA experiment shows that this structure can work at the rate of 65.58 MHz and can meet the requirement of video codec. The hardware resources of DE2 occupied by the designed sub-pixel interpolation module are about 7%. The algorithm complexity is small, and it is suitable for practical applications.

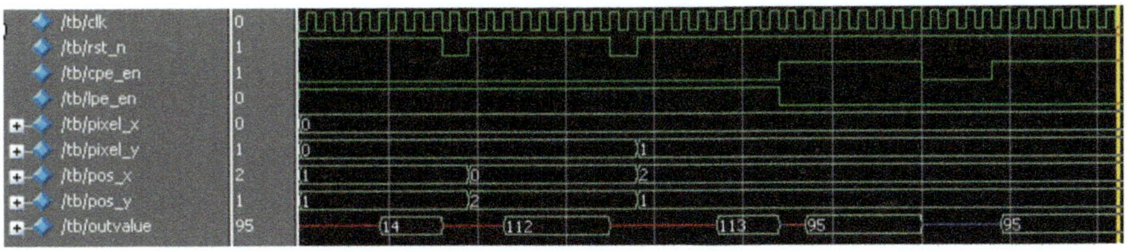

Fig. 8 Sub-pixel interpolation hardware simulation waveform

5 Conclusions

This paper presents a hardware-implemented module for the sub-pixel interpolation in the HDTV video codec. This module can be used to calculate the prediction values of the half-pixel and quarter-pixel for the luma and chroma, respectively. In order to reduce the hardware overhead and performance overhead, some shifters-based multipliers are used in this work. The experiment shows that the proposed architecture can achieve 65.58 MHz on the DE2 FPGA board with 7% resource consumption, which can meet the requirement of the target video codec.

References

1. Yalcin, S., Hamzaoglu, I.: A high performance hardware architecture for half-pixel accurate H.264 motion estimation. In: 14th IFIP International Conference on VLSI-SoC, pp. 63–67 (2006)
2. Zhao, M.: Research and analysis of a bilinear adaptive filter. In: IEEE International Nanoelectronics Conference (INEC), pp. 1–2 (2016)
3. Kalali, E., Adibelli, Y., Hamzaoglu, I.: A reconfigurable HEVC sub-pixel interpolation hardware. In: 3rd International Conference on ICCE, Berlin, pp. 125–128 (2013)
4. Audio Video Coding Standard Workgroup of China (AVS): Video Coding Standard FCDI.0, Nov 2003
5. Audio Video Coding Standard Workgroup of China (AVS): Final draft of information technology - advanced coding of audio and video - part 2: video. AVS workgroup Doc. N1214, Shanghai, China, Sep 2005
6. Wang, R., Li, M., Li, J., Zhang, Y.: High throughput and low memory access sub-pixel interpolation architecture for H.264/AVC HDTV decoder. IEEE Trans. Consumer Electron. **51** (3), 1006–1013 (2005)
7. Hu, S., Zhang, X., Yang, Z.: Efficient implementation of interpolation for AVS. Congress on Image and Signal Processing, pp. 133–138 (2008)
8. Wang, F., Li, Y., Jia, H., et al.: An efficient fractional motion estimation architecture for AVS real-time full HD video encoder. In: IEEE International Conference on Imaging Systems and Techniques Proceedings, pp. 279–284 (2012)

The Dynamic Correlation Between Civil Aviation Passenger Traffic Volume and Its Influential Factors Based on DCC-GARCH Model

Junling Cai and Ning Zhang

Abstract In this paper, the DCC-GARCH model is introduced to analyze the dynamic correlation between civil aviation passenger traffic volume and its influential factors. Results from the empirical studies using the statistical data in Beijing show that the GDP and average annual temperature directly affect the development of civil aviation passenger traffic volume, and the degree of correlation between them varies in different time backgrounds. Besides, the correlation between the civil aviation passenger traffic volume and the number of inbound tourists and the consumption level of residents is also getting closer and closer with time.

Keywords Dynamic correlation · DCC-GARCH model · Civil aviation passenger traffic volume · Influential factors

1 Introduction

As an emerging science and technology industry, the development of civil aviation industry directly affects the economic development level of the country. Attaching importance to the development of the field of civil aviation transportation is an extremely important aspect to enhance regional economic and comprehensive national strength [1]. Presently, there are many factors influencing the development of civil aviation passenger traffic volumes, such as the gross national product (GDP), number of annual inbound tourists, residents' consumption levels, and number of public transport operating lines, and the annual average temperature. It will be of great significance to the future development of the air transport industry by analyzing the dependence between each factor and passenger traffic volume.

The studies on the civil aviation transportation in China mainly focus on the revenue management [2], passenger traffic volume forecasting [3], and passenger

J. Cai (✉) · N. Zhang
School of Economics and Management, Beihang University, Beijing 100191, China
e-mail: cjunling@buaa.edu.cn

V. Jain et al. (eds.), *Recent Trends in Intelligent Computing, Communication and Devices*, Advances in Intelligent Systems and Computing 1031,
https://doi.org/10.1007/978-981-13-9406-5_76

price pricing [4], among which the dependence between various factors affecting the civil aviation passenger traffic volume is studied by China's scholars using different methods. For example, Deng [5] adopted the cluster analysis to explore the factors affecting the civil aviation passenger traffic volume according to the dependence between social and economic development and the demand of civil aviation passenger. Li [6] qualitatively and quantitatively studied the influential factors by utilizing the gray correlation theory, and the influential degree is also effectively ranked. Ji [7] analyzed the dependence between civil aviation passenger traffic volume and related factors by modeling the multiple linear regressions. Zhu [8] proved that the gross national income and per capita disposable income of residents have a significant impact on the civil aviation passenger volume.

The correlation between the civil aviation passenger volume and its influential factors is not static but changes with time, which has certain dynamic characteristics between them. The DCC-GARCH (Dynamic Conditional correlation GARCH) model is used for describing the time-varying dynamic dependence between time series [9], and it has been widely used in various fields. Xiao et al. [10] studied the dependence between onshore and offshore RMB bond markets. Zhu [11] proved that there is a dynamic correlation between stock markets in the BRIC countries based on DCC-GARCH model. Zhang [12] analyzed the dynamic relationship between international crude oil price and RMB and found that there has spillover effect of market fluctuations between them. Zinecker [13] adopted DCC-GARCH model to study the dependence of Poland, Czech, and Germany capital markets and proved that there is a significant correlation between them. Gao [14] modeled the DCC-GARCH to explore the dynamic correlation between the returns of China's carbon and fossil energy markets. However, few researches have been done using the DCC-GARCH model in the civil aviation transportation to analyze the dynamic dependence between the civil aviation passenger volume and its influential factors.

Thus, five major factors influencing the civil aviation passenger volume are selected as the object of study in this paper, including the GDP, the number of annual inbound tourists, residents' consumption levels, number of public transport operating lines, annual average temperature, to explore the dynamic correlation between them and the civil aviation passenger volume based on DCC-GARCH model. The main contributions of this paper are as follows: firstly, the DCC-GARCH model is the first applied to the civil aviation passenger transport system. Secondly, the correlation research between civil aviation passenger traffic volume and its influential factors in China is mainly based on static one, and there are relatively few literatures on the dynamic characteristics of the correlation.

The rest of the paper is organized as follows: In Sect. 2, a general description for the DCC-GARCH model is presented. Section 3 reports some empirical studies as well as the main results. Section 4 summarizes the conclusion and future work.

2 DCC-GARCH Model

The DCC-GARCH model proposed by Engle [9] assumes that the time series Y_t are composed of k series and is a multivariate normal distribution with mean 0 and the covariance matrix H_t, namely $Y_t|\Omega_{t-1} \sim N(0, H_t)$, where Ω_t indicates the information set at time t. The binary DCC-GARCH model is applied in this paper to analyze the dynamic correlation between civil aviation passenger traffic volume and its five influencing factors, respectively.

The DCC-GARCH model is defined as follows:

$$\begin{aligned} Y_t &= E(Y_t|\Omega_t) + \varepsilon_t.\varepsilon_t|\Omega_t \sim N(0, H_t) \\ H_t &= D_t R_t D_t, D_t = \mathrm{diag}\left(\sqrt{h_{ii,t}}\right) \\ R_t &= \mathrm{diag}\left(\frac{1}{\sqrt{q_{ii,t}}}\right) Q_t \mathrm{diag}\left(\frac{1}{\sqrt{q_{ii,t}}}\right) \end{aligned} \tag{1}$$

where R_t is the dynamic time-varying correlation coefficient, and $h_{ii,t}$ is conditional variance defined in all univariate GARCH models. D_t represents the diagonal matrix with conditional standard deviation as the diagonal elements. Q_t is positive definite matrix with $k \times k \times t$-dimensional, where k and t are the number of series and period number of observations, respectively.

$$\begin{aligned} Q_t &= \left(1 - \sum_p^P \alpha_p - \sum_q^Q \beta_q\right)\overline{Q} + \sum_p^P \alpha_p \varepsilon_{t-p} \varepsilon'_{t-p} + \sum_q^Q \beta_q Q_{t-q} \\ q_{ij,t} &= (1 - \alpha - \beta)\overline{q}_{ij} + \alpha e_{i,t-1} e_{j,t-1} + \beta q_{ij,t-1} \\ h_{ii,t} &= w_i + \sum_{p=1}^P \alpha_{ip} \varepsilon^2_{i,t-p} + \sum_{q=1}^Q \beta_{iq} h_{ii,t-q} \end{aligned} \tag{2}$$

where $\overline{Q} = \frac{1}{T}\sum_{t=1}^T e_t e'_t \left(e_{it} = \frac{\varepsilon_{it}}{\sqrt{h_{ii,t}}},\ i = 1, 2, \ldots, N\right)$ represents the non-conditional variance–covariance matrix with $N \times N$-dimensional. α_p and β_q indicate the parameters to be estimated for the DCC-GARCH model and should satisfy $\alpha > 0$, $\beta > 0$ and $\alpha + \beta < 1$.

3 Empirical Studies

In this study, the civil aviation passenger traffic volume, GDP, number of annual inbound tourists, residents' consumption levels, number of public transport operating lines, and the annual average temperature in Beijing are collected as the sample data, from the Beijing statistical yearbook. All data span from 1978 to 2017. In order to make the data smooth, the time series are transformed by the first-order difference method. Besides, CAPC, GDP, ITN, RCL, PTN, and YAT are used to represent each transformed series for the convenience of follow-up discussion. The DCC-GARCH model is implemented using R2012a MATLAB.

Table 1 gives the descriptive statistics of the six series. The kurtosis of six series is greater than 3, and the skewness is basically less than 0, showing that each series has a higher peak and fat tail. In addition, the J-B statistics indicate that all series reject almost the null hypothesis of normal distribution. Q(20) statistics of CAPC, GDP, and ITN are significant at the 5% confidence level.

The parameter for the DCC-GARCH model in the estimation includes mainly two steps. Firstly, the univariate GARCH models are established for each series to estimate their residuals. The estimated results are shown in Table 2, where α and β

Table 1 The estimation results for all VaR model

	CAPC	GDP	ITN	RCL	PTN	YAT
Mean	−0.138	−0.147	−0.086	−0.130	−0.056	−0.005
Median	−0.128	−0.149	−0.097	−0.108	−0.041	−0.011
Max	0.291	−0.001	0.624	−0.025	0.206	2.346
Min	−0.842	−0.275	−0.533	−0.341	−0.513	−2.293
Standard deviation	0.168	0.052	0.217	0.0748	0.107	0.556
Skewness	−1.601	−0.054	1.165	−1.263	−2.044	0.173
Kurtosis	10.337	4.086	6.097	4.002	11.556	17.807
Jarque-Bera	96.129 (0.001)	1.788 (0.229)	22.538 (0.002)	11.091 (0.011)	134.902 (0.001)	329.078 (0.001)
Q(20)	79.648 (0.000)	38.383 (0.007)	54.681 (0.000)	29.913 (0.071)	25.587 (0.179)	8.940 (0.983)
ARCH(2)	0.258 (0.836)	7.739 (0.021)	5.893 (0.052)	6.877 (0.032)	1.891 (0.388)	15.569 (0.000)

Table 2 The estimated results of parameters for each univariate GARCH(1,1) model

	ω	α	β	$\alpha + \beta$
CAPC	0.0001	0.0000	0.9996	0.9996
GDP	0.0000002	0.0000	0.9838	0.9838
ITN	0.0127	0.0000	0.7155	0.7155
RCL	0.0000059	0.1158	0.8541	0.9699
PTN	0.000804	1.0000	0.0000	1.0000
YAT	0.001031	0.9990	0.0009	0.9999

are the estimated values of ARCH(1) and GARCH(1) coefficients, respectively. Obviously, α and β are extremely significant. The greater α is, the smaller the data oscillation is. Contrarily, the greater β is, the greater the data oscillation is. In addition, $\alpha + \beta$ is close to 1, indicating that the volatility of each series is very significant.

Secondly, all residuals in the first step are standardized to estimate the dynamic correlation coefficients. The results of parameters in DCC model are given in Table 3. The larger β is and the closer it is to 1, the greater the volatility of the dynamic correlation coefficients will be. Besides, the greater and the value of $\alpha + \beta$ is and the closer it is to 1, the more obvious the characteristic of the dynamic correlation is.

Table 4 reports the statistical characteristics of the dynamic conditional correlation coefficients. Obviously, the mean and median of the coefficients of CAPC-GDP, CAPC-ITN, CAPC-RCL and CAPC-PTN are greater than 0.1, respectively, showing that there has obvious dynamic correlation between CAPC and GDP, ITN, RCL, PTN. However, the dynamic correlation between CAPC and YAT is not very significant.

According to the dynamic correlation coefficients in DCC-GARCH model, the dynamic correlation coefficient figures can be obtained, as shown in Fig. 1. The changes in the fluctuation of the dynamic correlation between CAPC and GDP, YAT are relatively large, while CAPC and ITN, RCL, PTN have less variation and are steady trend, indicating that GDP and average annual temperature directly affect the development of civil aviation passenger traffic volume in Beijing, and the degree of the correlation between them does not change with time. This can be explained by some events in reality, such as the Beijing Asian Games in 1990, SARS in 2003, Beijing Olympic Games in 2008, financial crisis in 2009, and APEC conference in 2014, which made the GDP in Beijing up and down. To some extent, this has a direct impact on the development of passenger volume.

Table 3 The estimated results of parameters in DCC model

	CAPC-GDP	CAPC-ITN	CAPC-RCL	CAPC-PTN	CAPC-YAT
α	0.2393	0.00000004	0.00000014	0.0000002	0.1034
β	0.6090	0.8239	0.9094	0.9002	0.0000023
$\alpha + \beta$	0.8483	0.8239	0.9094	0.9002	0.1034

Table 4 The statistical characteristics of the dynamic conditional correlation coefficient

	CAPC-GDP	CAPC-ITN	CAPC-RCL	CAPC-PTN	CAPC-YAT
Mean	0.3980	0.6319	0.3686	0.1145	−0.0638
Max	0.6893	0.6853	0.3920	0.2327	0.1308
Min	−0.0113	0.3053	0.3078	0.0819	−0.2609
Median	0.4839	0.6754	0.3778	0.0972	−0.0659
Standard-deviation	0.2188	0.0913	0.0241	0.0398	0.0752

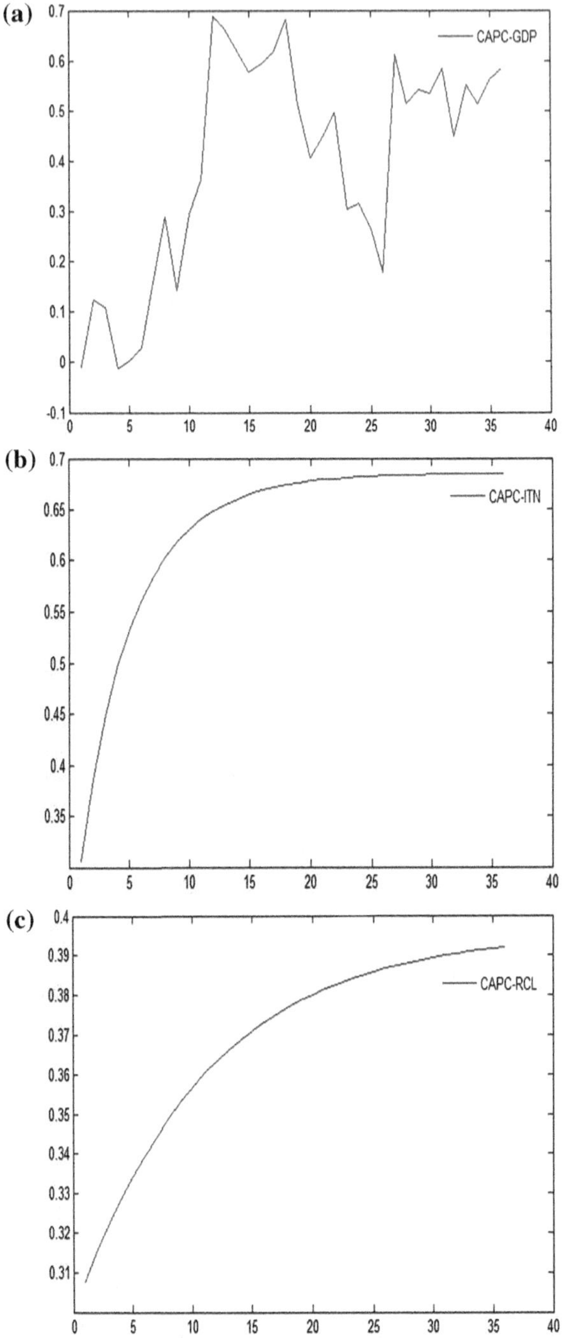

Fig. 1 The charts of the dynamic correlation coefficient: **a** CAPC-GDP; **b** CAPC-ITN; **c** CAPC-RCL; **d** CAPC-PTN; **e** CAPC-YAT

(d)
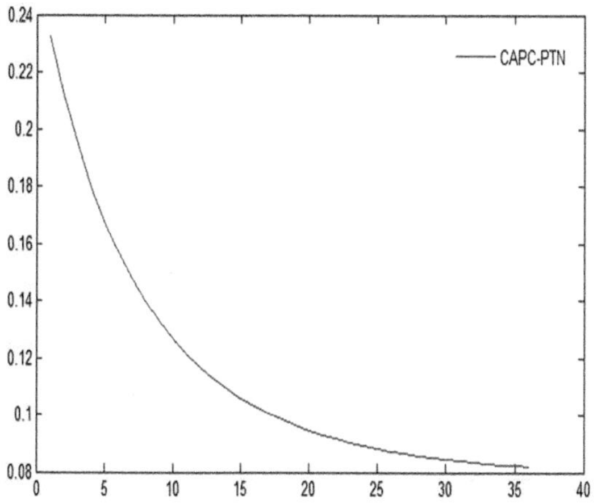

(e)
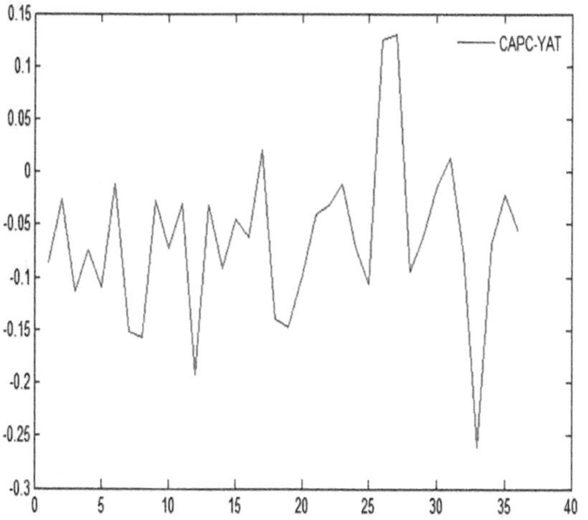

Fig. 1 (continued)

Additionally, the degree of dynamic correlation between CAPC and ITN, RCL changes gradually with time, and the coefficients between them are more than 0.3, respectively, while CAPC and PTN is the opposite, showing that in recent years, the degree of correlation between the civil aviation passenger traffic volume in Beijing and the number of inbound tourists, residents' consumption level has become increasingly close with the change of time. On the other hand, as the gradual improvement of residents' living standards and consumption level, the choice of aircraft as a long-distance transportation has become increasingly common. In addition, China is rich in tourism resources, attracting a large number of foreign tourists.

4 Conclusion

The dynamic correlation between the civil aviation passenger traffic volume and its influential factors is analyzed in this paper based on the DCC-GARCH model. By taking Beijing area as an example, it is found that GDP and the annual average temperature will have a direct impact on the development of the civil aviation passenger traffic volume, and the correlation degree is various in different time backgrounds. Additionally, the correlation between civil aviation passenger traffic volume and number of annual inbound tourists, residents' consumption level is becoming increasingly close with time. This indicates that with the gradual development of economy, the number of annual inbound tourists and residents' consumption level have become the main factors influencing the passenger volume of civil aviation, which will provide a certain basis for forecasting the civil aviation passenger traffic volume.

References

1. Wen, J., Zhang, F.R., Liu, Y., Huang, X.Q.: Analyzing the influential factors on Chinese civil aviation passenger traffic volume. Sci. Technol. Ind. **16**, 65–69 (2018)
2. Gao, Q., Zhu, J.F., Chen, K.J.: Multi-leg seat inventory control model for airline revenue management. J. Traffic Transp. Eng. **5**, 82–85 (2004)
3. Pan, W.J., Pan, Y.X., Lu, G.P., Zeng, C.: Algorithm of Hilbert-Huang transform in forecast of passenger and freight volume. Comput. Modernization 46–49 (2015)
4. Min, Z.T., Yang, X.Y.: The analysis on the pricing methods and its composition factors of China's civil aviation passenger transport price. Price Theory Pract. 9–11 (2014)
5. Deng, J.J., Luo, L.: Demand forecasting and analyzing of civil aviation passenger transport markets based on GMDH. Soft Sci. **20**, 35–38 (2006)
6. Li, C.P.: Analyzing the influential factors in China's civil aviation passenger. Sci. Technol. Ind. **11**, 59–61 (2011)
7. Ji, Y.Z., Deng, B., Qin, X.W.: The analysis of passenger civil aviation and its related factor. Math. Pract. Theory **42**, 175–183 (2012)
8. Zhu, W.W.: The analysis about influential factors in China's civil aviation passenger volume based the partial least squares regression. China Market 110–112 (2010)
9. Engle, R.F.: Dynamic conditional correlation—A simple class of multivariate GARCH models. SSRN Electron. J. **20**, 339–350 (2000)
10. Xiao, M., Wang, X.F.: The correlation analysis between onshore and offshore RMB bond markets: a research based on DCC-GARCH model. J. Central Univ. Finan. Econ. (2014)
11. Zhu, S., Zhao, H.: The dynamic correlation study between stock market in the BRIC based DCC-GARCH model. Stat. Decis. 165–167 (2015)
12. Zhang, T.: A research of the dynamic correlation between international crude oil price and dollar based on DCC-GARCH model. J. Guizhou Commercial Coll. (2015)
13. Zinecker, M., Baldzinski, M., Meluzin, T.: Application of DCC-GARCH model for analysis of Interrelations among Capital Markets of Poland, Czech Republic and Germany, Michal Pietrzak, vol. 1 (2016)
14. Gao, Q.X., Li, F.: The dynamic correlation study between carbon emission trading market and fossil energy market in China-base on DCC-(BV) GARCH model test. Environ. Sustain. Dev. **41**, 25–29 (2016)

Design of Smart Home System Based on Raspberry Pi

Lingling Zhong, Teng Lv, Changkai Li and Zhonghao Wang

Abstract For the information island problems of traditional home furnishing, such as the inability of interconnection and intercommunication between different devices, the inability of all devices to work together, and the inability of users to remotely monitor and control the devices, a smart home system based on Raspberry Pi is designed. The system presents a smart home system developed by hybrid development mode. The system is mainly composed of server management system and Raspberry Pi hardware response system. It can remotely take control of the state of home appliances through the Internet whenever and wherever it is. The system has the characteristics of low development cost, high compatibility, and stable operation, etc.

Keywords Raspberry Pi · Smart home system · Hybrid development mode

1 Introduction

Developed and advanced science and technology lead to people's higher pursuit of life, hence, the smart home industry is also emerging. At present, the information island problems of traditional home are furnishing, such as the inabilities of interconnection and intercommunication between devices, the inability of cooperation between equipments, and the inability of remote monitor and control for users, which brings inconvenience to users [1]. Therefore, a smart home furnishing system based on Raspberry Pi is proposed to realize the interaction of information with the external environment, so that people's time can be arranged more reasonably and their quality of life can be improved. The scalability of the system is improved by

L. Zhong (✉) · C. Li · Z. Wang
School of Electronic and Communication Engineering, Anhui Xinhua University, Hefei 230088, China
e-mail: zll992dz@sohu.com

T. Lv
School of Information Engineering, Anhui Xinhua University, Hefei 230088, China

V. Jain et al. (eds.), *Recent Trends in Intelligent Computing, Communication and Devices*, Advances in Intelligent Systems and Computing 1031,
https://doi.org/10.1007/978-981-13-9406-5_77

the external control unit, and its functions are practical and no longer single. The system can lower the threshold of smart home, allow more users to improve the quality of life, and facilitate users' lives.

2 Technology and Demand Analysis

2.1 Technology Introduction

The Raspberry Pi is powered by a Linux system. The Linux system is a free-to-use and freely distributed Unix-like operating system. It is a multi-user, multi-tasking, multi-threaded, and multi-CPU operating system based on POSIX and UNIX [2]. It runs major UNIX tools, applications, and network protocols. It supports both 32-bit and 64-bit hardware. Linux inherits Unix's network-centric design philosophy and is a stable multi-user network operating system. The front end uses the BootStrap framework for responsive page development. The background supporter is written in PHP [3], and the front end gets the required data through AJAX by invoking the PHP interface. The Raspberry Pi uses the shell and Python languages to get remotely sent control commands, control the state of the hardware, and finally send a reminder message to the user. The low price, small size, and powerful features of the Raspberry Pi make it the perfect choice for smart home control systems.

2.2 Requirement Analysis and Design of the Smart System

The Raspberry Pi smart home system is mainly composed of three modules, namely a login module, a remote control module, and a response module. The login module is mainly that the user can log into the remote control interface according to the information registered on the Web site, the status of the current home hardware facilities can be seen after entering in the remote control interface, and the corresponding state can be changed according to users' will. The response module drives the hardware after detecting the remotely transmitted request, causing the controlled hardware to alter according to the display state of the remote control interface. The workflow of the system is shown in Fig. 1.

2.3 Requirement Analysis of System Backend

In conjunction with the front-end remote control interface, it is divided into a login module and a save instruction module. The login module is that the user logs into the remote control interface through the account password registered by himself and

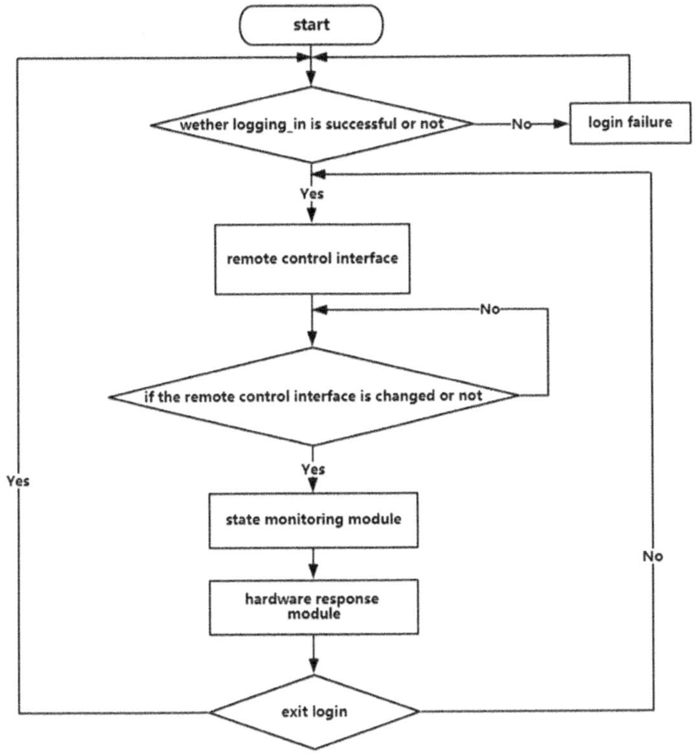

Fig. 1 Workflow of system

allows the user to operate in the remote control interface. The save instruction module is used to save the operation instructions generated from the changes of the remote control interface and store the instructions in the MySQL database.

3 The Design and Implementation of Web Front-End Module Functions

3.1 Login Module

As shown in Fig. 2, the blog site is written in HTML language, and the registration and login interface are built on this interface.

As shown in Fig. 3, the login interface uses the BootStrap grid layout. BootStrap is a framework written by JavaScript which is compatible with various types of screen sizes. This page has an input box for entering a username, password, and

Fig. 2 Login screen

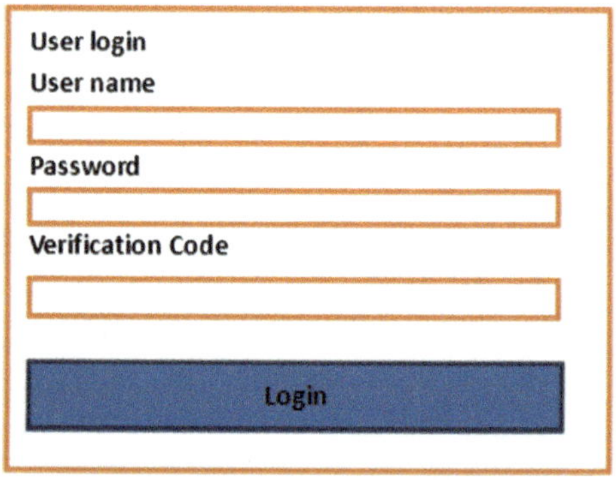

Fig. 3 Login interface

verification code. The user enters the home page according to the account that he or she has registered.

3.2 Remote Control Module

After the user logs into the management interface through the account, by clicking the hardware icon [4] that needs to be turned on or off, the system will record the status of the current interface and store it in the binary number in the database; when the monitor of Raspberry Pi notices the database contents are changed, the hardware response program can be called to remotely control the function of the corresponding electrical state in the home. The main functions are shown in Fig. 4.

3.3 PHP Background Programming

The background login code is shown in Fig. 5a, and the interface is written in PHP. When the user logs in, the front desk passes the login information to the background through AJAX technology. The background determines whether the

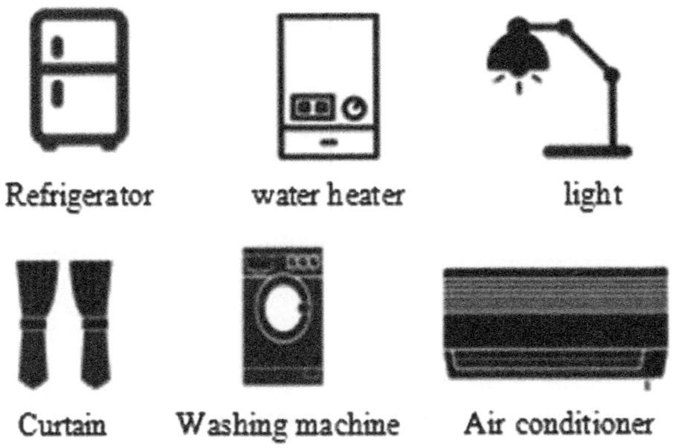

Fig. 4 Main function

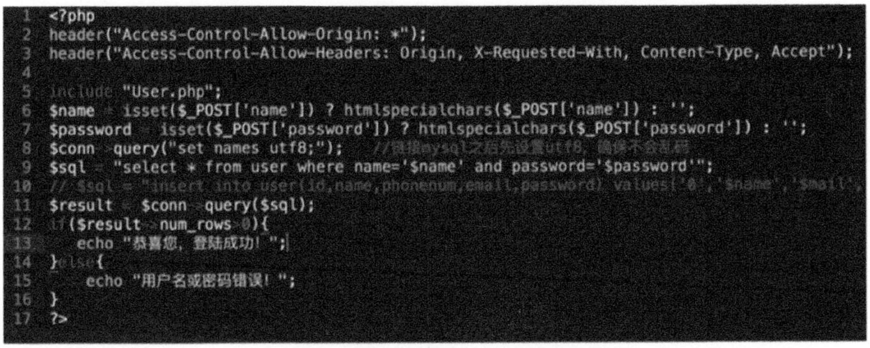

```php
<?php
header("Access-Control-Allow-Origin: *");
header("Access-Control-Allow-Headers: Origin, X-Requested-With, Content-Type, Accept");

include "User.php";
$name = isset($_POST['name']) ? htmlspecialchars($_POST['name']) : '';
$password = isset($_POST['password']) ? htmlspecialchars($_POST['password']) : '';
$conn->query("set names utf8;");    //链接mysql之后先设置utf8，确保不会乱码
$sql = "select * from user where name='$name' and password='$password'";
// $sql = "insert into user(id,name,phonenum,email,password) values('0','$name','$mail',
$result = $conn->query($sql);
if($result->num_rows>0){
   echo "恭喜您，登陆成功！";
}else{
    echo "用户名或密码错误！";
}
?>
```

(a) Login interface PHP code

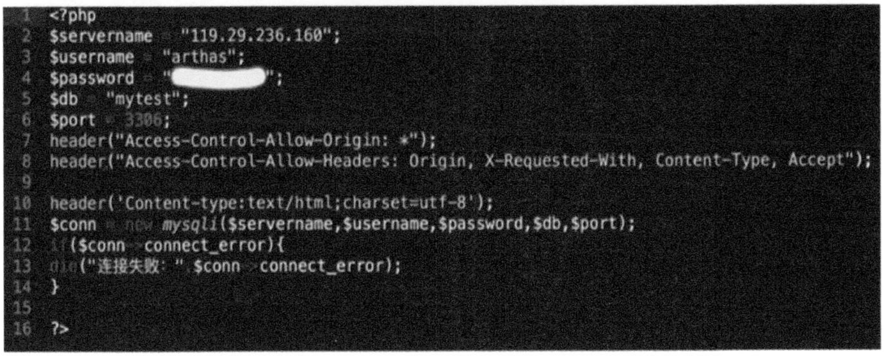

```php
<?php
$servername = "119.29.236.160";
$username = "arthas";
$password = "[redacted]";
$db = "mytest";
$port = 3306;
header("Access-Control-Allow-Origin: *");
header("Access-Control-Allow-Headers: Origin, X-Requested-With, Content-Type, Accept");

header('Content-type:text/html;charset=utf-8');
$conn = new mysqli($servername,$username,$password,$db,$port);
if($conn->connect_error){
die("连接失败: ".$conn->connect_error);
}

?>
```

(b) PHP code for database login information

Fig. 5 PHP login module diagram

user is a legitimate login by checking the database for the existence of the user and whether the password is correct.

The database login information PHP code is shown in Fig. 5b. By writing the database configuration information in a separate PHP file, you can reduce the time to repeatedly configure the database connection information when the database is read. When the PHP program needs to call the database file, directly reference the PHP file which contains the database configuration information in the program.

4 The Design and Implementation of Raspberry Pi Hardware Response Module Functions

4.1 Listening Instruction Module

The listener instruction module is mainly divided into two parts: the acquisition of instruction and the judgment of instruction. The flowchart of the listener instruction module is shown in Fig. 6. The Python language is used to connect to the remote database and obtain the instruction data. After the instruction data is obtained, it is judged whether the change field is true or not. When the change field value is true, the parsing starts. The instruction is represented by six digits of 0 and 1, which is separated by commas. Each bit corresponds to a GPIO port of the Raspberry Pi. It controls the external circuit according to outputs of GPIO port which is controlled

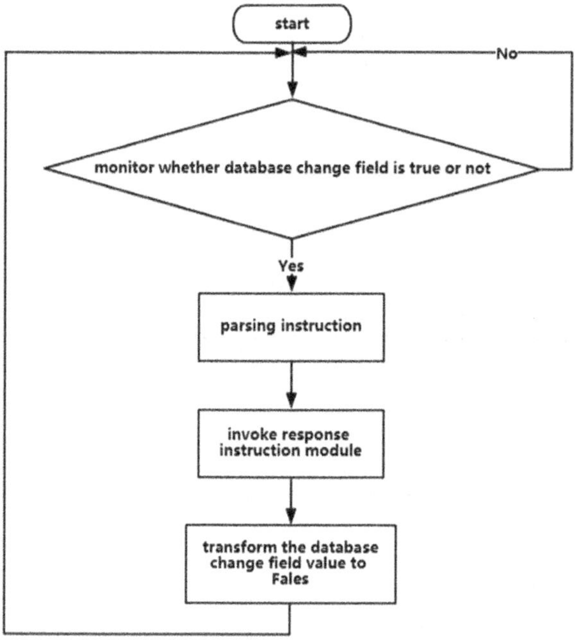

Fig. 6 Flowchart of the monitor instruction module

by 1 or 0, which corresponds to high level or low level correspondingly. If the change field value is false, skip it directly and continue to listen to the database. The whole process is implemented by looping through a Python program in a shell script.

4.2 Response Command Module

As shown in Fig. 7a, after the remote control instruction is parsed by intercepted instruction module, the control response instruction module will output a high or low electrical level to the corresponding GPIO port of the Raspberry Pi, thereby controlling the open or close of the relay connected to the GPIO port to realize the function of remotely controlling the appliance state. The Raspberry Pi GPIO port-level output code controlled by Python is shown in Fig. 7b. The Raspberry Pi is connected to the relay through the connected expansion board [5]. When the electrical level of the Raspberry Pi GPIO port changes and affects the state of the relay, the purpose of controlling the large-sized home appliance is achieved. The hardware connection diagram is shown in Fig. 7c.

4.3 E-Mail Notification Module Design

After the response instruction module completes the work, the mail program automatically sends an e-mail to the user-bound mailbox with the collected current home appliance status, reminds the user whether the operation is completed, and displays the status of the user's current home appliances, thereby allowing The user can grasp the situation of the home appliance in real time to avoid accidents. The mail program is written in Python [6], and the mail is sent through the STMP protocol. The content of the e-mail notification is as shown in Fig. 8.

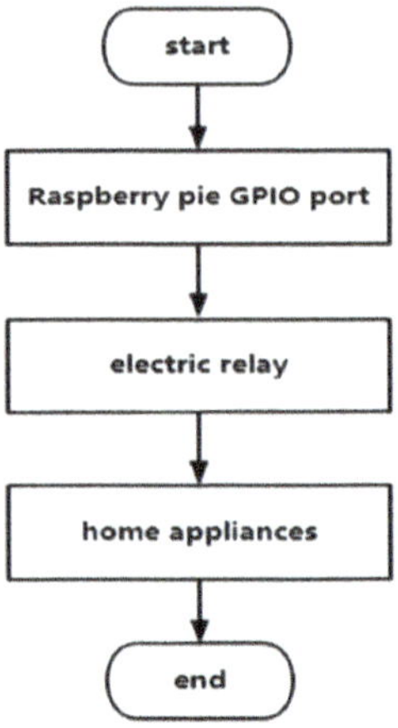

(a) Flow chart of response instruction module

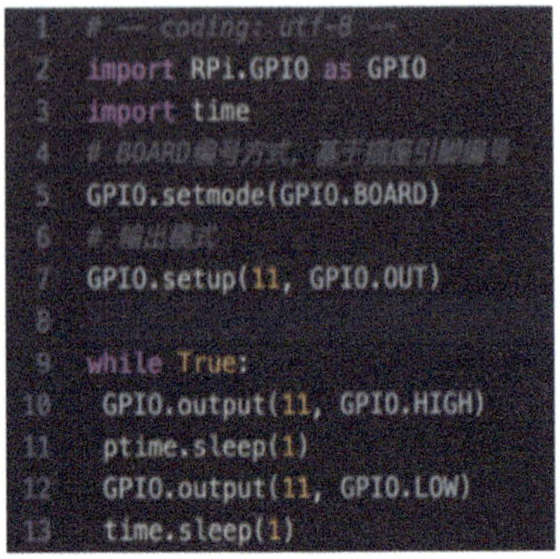

(b) Control Raspberry Pack GPIO Port Level Output Code

(c) Hardware connection diagram

Fig. 7 Response instruction module

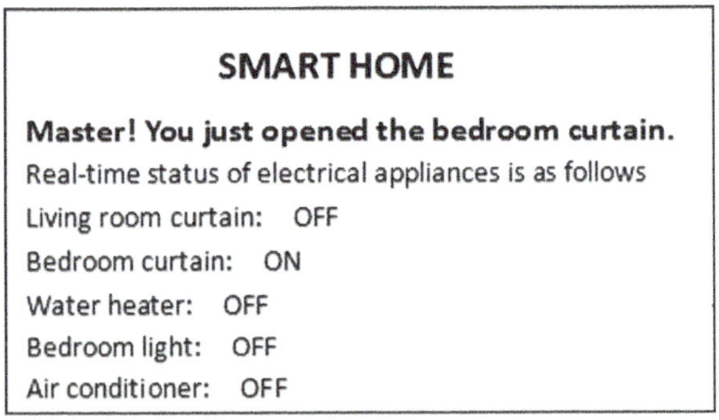

Fig. 8 E-mail notification module

5 Conclusions

This paper designs a smart home system based on the Raspberry Pi. We mainly demonstrate the design and implementation of the Web front-end development and the Raspberry Pi hardware response of the system. The test results show that the whole system can realize the basic functions of the design; i.e., the data is collected correctly, the system is stable, and the exception can be processed properly. So it has a certain marketing and reference value.

Acknowledgements The work is supported by Talent of Discipline and Specialty in Colleges and Universities of Anhui Province (No. gxbjZD54), Academic Leader Foundation (No. 2014XXK06), and Introduction of Talents Foundation of Anhui Xinhua University (No. 2015kyqd002), Reform Comprehensive Experimental Unit of Electronic Information Engineering Major (No. 2015zy073).

References

1. Zhang, Y.: Intelligent home system based on Raspberry Pie. Electron. World **15**, 53 (2016)
2. Zhang, D.: Design of robot system based on Raspberry Pi smart home sweeper. Electron. World **20**, 149–150 (2017)
3. Zhang, J.: Software System Analysis and Design Training Course. Tsinghua University Press (2016)
4. Xing, X., Gu, Y., Ma, J., et al.: Intelligent home system design scheme. Electron. Des. Eng. **25** (21), 11–13 (2017)
5. Song, K., Yao, J., Li, J.: Research on smart home control switch based on Raspberry Pi. Electron. Technol. Softw. Eng. (21), 140–141 (2015)
6. Amri, Y., Setiawan, M.A.: Improving smart home concept with the internet of things concept using Raspberry Pi and node MCU. IOP Conf. Ser. Mater. Sci. Eng. **325**, 012021 (2018)

Diagnostic of Line Loss Abnormal Causes in Transformer District Based on Big Data

Xueting Cheng, Tan Wang, Mengzan Li, Weiru Wang and Xinyuan Liu

Abstract Line loss rate is irreplaceable in evaluating the economic operation of power systems. Line loss management is one of the key management contents of power companies. The line loss in transformer district accounts for about 20% of the total loss of the power grid. It belongs to the heavy damage layer. Analyzing the important factors of affecting the line loss of the transformer district and clarifying the cause of the line loss abnormality, which is of great significance for improving the economic operation level of the distribution network. The line loss feature library of the transformer district is constructed. The improved K-means algorithm and the support vector machine classification method on the basis of PSO are used to diagnose the line loss anomaly. Taking the actual data of the actual power grid as an example, the calculation results verify the feasibility of the proposed method.

Keywords Power grid · Line loss · Transformer district

1 Introduction

Line loss rate is irreplaceable in evaluating the economic operation of power systems. Line loss management is one of the key management contents of power companies. At present, State Grid Corporation vigorously promotes the synchronous line loss system and comprehensively manages the line loss according to the four-point principle of "dividing, zoning, branching, and subdivision" and has achieved significant results. However, in the process of further deepening the application, the following problems need to be solved: The result of the quadrant loss is derived from multiple system data, also depends on various model constraints. The correctness and accuracy are affected by each. The influences of uncertain factors: some low-voltage line loss rate is obviously abnormal (such as the line loss rate is negative), and the manager lacks an effective method to locate

X. Cheng (✉) · T. Wang · M. Li · W. Wang · X. Liu
State Grid Shanxi Electric Power Research Institute, Taiyuan, China
e-mail: cheng_xueting@163.com

V. Jain et al. (eds.), *Recent Trends in Intelligent Computing, Communication and Devices*, Advances in Intelligent Systems and Computing 1031,
https://doi.org/10.1007/978-981-13-9406-5_78

the abnormality, find the cause of the abnormal line loss rate, and correct it in time. At the same time, it can be concluded from the line loss data of Shanxi Electric Power Company that the line loss of 380 V transformer region accounts for about 20% of the total loss of the power grid, which belongs to the heavy loss layer. Therefore, it is important to analyze the important factors affecting the line loss of the transformer region and to clarify the cause of the line loss anomaly in the transformer region, which is of great significance for improving the economic operation level of the distribution network.

Paper [1] proposed a method for quickly getting the line loss rate of low-voltage transformer region relied on improved K-means clustering algorithm and LM algorithm optimized BP neural network model. A reasonable random line loss rate estimation method based on the random forest algorithm for multi-source data is proposed in paper [2]. Paper [3] proposed a theoretical line loss calculation method relied on support vector regression machine (SVR-PSO) and particle swarm optimization algorithm.

1.1 Modeling Method

The research [4–7] on the cause of the abnormal line loss in the transformer region is mainly divided into two parts: the line loss characteristic library of the transformer region and the model establishment. The formation of the line loss feature library in the transformer region is the basis. The main purpose of the model building part is to establish the classification model of the transformer region and locate the line loss of the transformer region.

In this paper, the line loss feature library of the transformer district is constructed. The improved K-means algorithm and the support vector machine classification method on the basis of PSO are used to diagnose the line loss anomaly. Taking the actual data of Shanxi Electric Power Company as an example, the calculation results verify the feasibility of the proposed method.

1.2 Transformer Region Line Loss Feature Library

The input of the formation part of the characteristic database of the Taiwan area is the line loss data, the measurement data, the equipment account data, and the user file data related to the transformer region, and the output is the line loss characteristic data of the transformer region. The input data mainly comes from the grid company's synchronous line loss management system, PMS system, GIS system, and electricity information collection system.

In order to ensure the completeness of the line loss feature extraction in the transformer region, the actual grid area data is collected, with the attributes of the transformer region, the power supply radius, the low-voltage line length, the

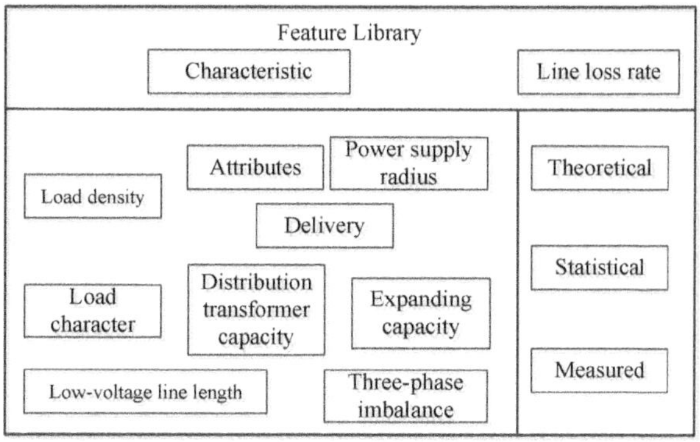

Fig. 1 Transformer region line loss feature library

distribution capacity, the total number of households, the number of power households, the power supply, and the three-phase unbalance. The theoretical line loss rate and the statistical line loss rate make up a sample set for integration analysis, and the initial transformer region line loss feature library is constructed, as shown in Fig. 1.

Load density (kWh): Consider the average capacity of the household, that is, the ratio of the amount of electricity supplied to the total number of households in the district.

Load shape factor (%): It is the ratio of the RMS current of the transformer region load to the average current, which can reflect the load characteristics of the transformer region and the three-phase unbalance of the load.

Low-voltage line length (m): the sum of the lengths of all low-voltage lines in the transformer region.

The properties of the transformer region are divided into four categories: A, B, C, and D. Power supply radius of the transformer region: less than 150 m, 150–250 m (including 150 m), 250–400 m (including 250 m), and greater than 400 m (including 400 m).

1.3 Diagnosis of the Cause of Abnormal Line Loss in Transformer Region

The diagnosis process of the line loss anomaly in Taichung is mainly divided into the following four steps. Firstly, the basic data model is constructed. Then, the improved K-means algorithm is used to cluster the causes of line loss anomalies, and the causes are identified. Finally, the clustering results are used. As a training sample, the classification model is trained to achieve a diagnosis of line loss anomalies. The detailed process is as shown in Fig. 2.

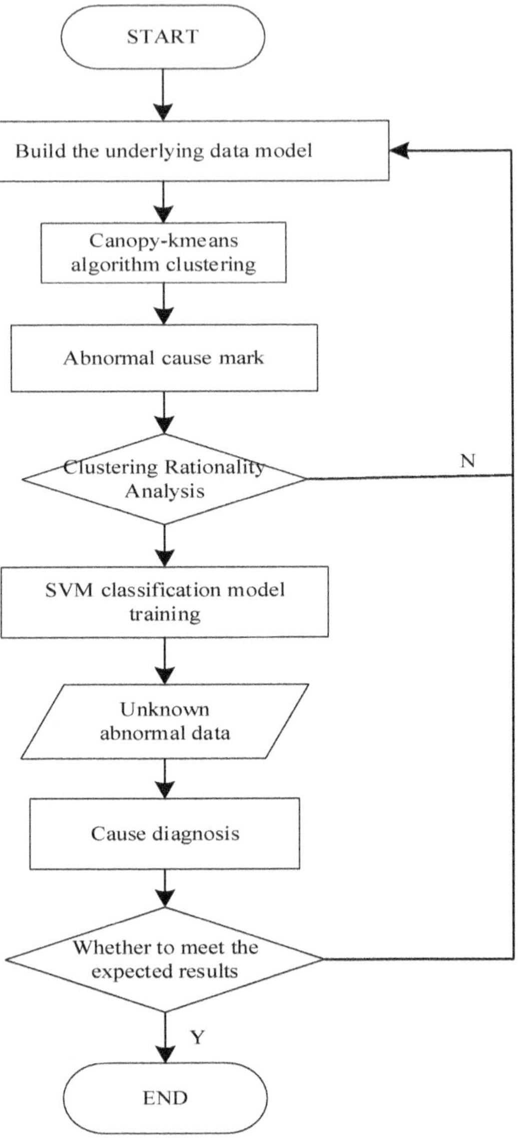

Fig. 2 Diagnostic flowchart for the cause of line loss abnormality in transformer region

In this article, the improved K-means algorithm is used to construct the clustering model, and the support vector machine classification model combined with PSO is used to classify. Finally, the feasibility of the results is proved by simulation analysis.

2 Improved K-Means Clustering Algorithm

The K-means algorithm is one of the extensively used clustering algorithms based on partitioning. The algorithm is simple in idea, fast in convergence, and easy to implement. However, the initial clustering center and the number of clusters need to be determined in advance, which are subject to subjective factors and cause the clustering results to fall into local optimum. The canopy algorithm does not need to determine the initial cluster center and the number of clusters before clustering. The clustering algorithm calculates the similarity between objects to obtain the final cluster. This paper uses canopy to improve the K-means algorithm and apply it to the transformer region line loss clustering.

The algorithm idea of improved K-means algorithm is as follows: (1) for a given set of data, set an initial center point, and set the region radius; (2) divide the set into several overlapping subsets, making all the objects fall within the scope of canopy coverage; for objects falling within the same area, the new center point is recalculated and the object belongs to the area according to the distance between the object and the new center point: Loop execution "canopy, calculation the process of the center point until the position of the k-center point no longer changes, that is, a stable classification state is reached.

Firstly, all sets are divided into K-categories on account of certain characteristics, and the characteristics of each type of data members are the same. Secondly, the canopy clustering technique is used to perform clustering in two stages: In the first place, partition the data into subsets, which is called canopy. In the second place, use the exact calculations for each point within canopy. The method is clustered again. It uses different distance metrics in two stages to form overlapping canopy. When the canopy is created, use the K-means algorithm for clustering within canopy.

3 Support Vector Machine Classification Model Combined with PSO

Given a training sample $\{x_i, y_i\}$, where $i = 1, 2, \ldots, n$ is the total number of samples and $x_i \in R^d$, $y_i \in R$ are the dimensions of the R^d space. By the nonlinear mapping $\varphi(x)$, the sample can be mapped from the original space to the high-dimensional (k-dimensional, $k > d$) feature space, then construct the optimal linear regression function as follows:

$$f(x) = w^T \cdot \varphi(x) + b$$

where x is the sample vector, w is the weight vector, and b is the threshold of the classification. When the support vector machine uses the structural risk principle, the error ξ_i (the variable allowing the misclassification) is used as the loss function in the optimization target, and the optimization problem is transformed as:

$$\min\phi(w) = \frac{1}{2}\|w\|^2 + c\sum_{i=1}^{n}\xi_i$$

The constraint is:

$$y_i\left[w^T\varphi(x_i) + b\right] \geq 1 - \xi_i, \xi_i \geq 0, \quad i = 1, 2, \ldots, n$$

where c is used to control model complexity and achieve a compromise between empirical risk and confidence range as a penalty factor; ξ_i is the relaxation factor. Solving the above optimization problem by the Lagrangian method, there is:

$$L(w, b, \xi, \alpha) = \frac{1}{2}\|w\|^2 + c\sum_{i=1}^{n}\xi_i - \sum_{i=1}^{n}\alpha_i\left(y_i\left(w^T\varphi(x_i) + b\right) - 1 + \xi_i\right)$$

where $\alpha_i(i = 1, 2, \ldots, n)$ is the Lagrangian multiplier, and the solved problem can be described as quadratic programming:

$$\max W(\alpha) = -\frac{1}{2}\sum_{i,j=1}^{n}\alpha_i\alpha_j y_i y_j K\left(x_i, x_j\right) + \sum_{i=1}^{n}\alpha_i$$

where,

$\sum_{i=1}^{n}\alpha_i y_i = 0, \quad 0 \leq \alpha_i \leq c, \quad i = 1, 2, \ldots, n; K(x, y) = \exp\left(-\frac{(\|x-y\|^2)}{2\sigma^2}\right).$

Aim to the mean square error (MSE), and construct the fitness function of the PSO algorithm as follows:

$$F = \sqrt{\sum_{i=1}^{N}\left(\frac{(y' - y_i)^2}{N}\right)}$$

where y' is the predicted value, y_i is the true value, and N is the number of sample. By the exchange and iteration of the global information among the same or different populations, the particle swarm optimization algorithm can search the optimal combination of parameters, which is a key parameter of the SVM prediction model. The entire flow is illustrated in Fig. 3.

After the clustering result is processed as the training set input, the particle swarm optimization vector machine model is used to train the station line loss classification model and the unknown station line loss anomaly generative data is used as the test set input, and the unknown line is finally obtained. The category of the abnormal data is diagnosed, thereby diagnosing the cause of the abnormal line loss.

Fig. 3 Flowchart of SVM classification based on the particle swarm

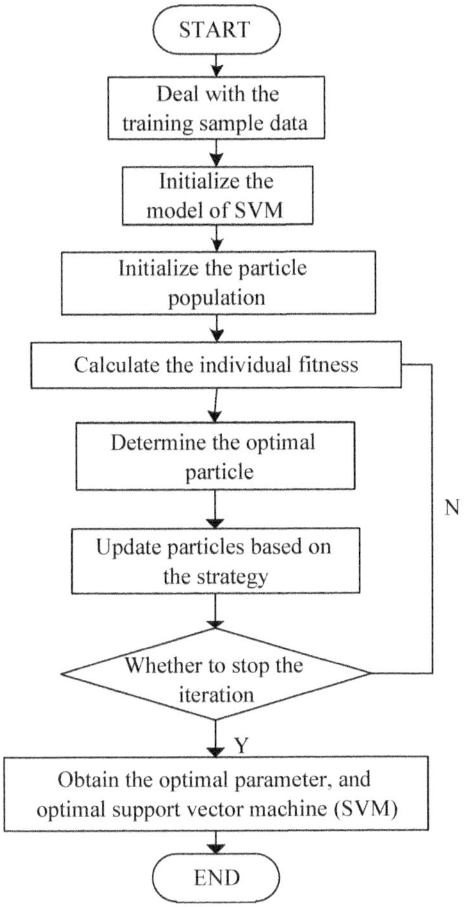

4 Numerical Simulation

Take the actual data of distribution transformer area as an example to simulate. For the sake of observation, we intercept the data of July 26, 2018, as the source of the line area anomaly analysis, on account of the attributes of the station area and actual operational data. We received 1857 line loss data for the station on the same day. There are 14 classes of station line loss anomalies based on the improved K-means algorithm. The support vector machine classification combined with the PSO algorithm is used to classify the unknown line loss anomaly data. Comparing the predicted classification result with the actual result, the SVM combined with PSO has higher accuracy and better precision. The accuracy percentage was as high as 90%.

5 Conclusion

Based on the actual situation of the line loss anomaly in the transformer region of grid, this paper proposes a relatively complete diagnosis model, including clustering and classification. First of all, in the clustering process, the cluster number, and the initial cluster core are confirmed, and then, the iterative operation is performed by the K-means algorithm, in order to obtain the satisfactory clustering result. In the second place, in the classification process, by the SVM algorithm combined with PSO algorithm, the trained SVM model is constructed and used to make abnormal causes. Through the massive actual power grid tests, the proposed method can realize the cluster and classification more and more fast and accurate, which could satisfy the actual needs of abnormal diagnosis of line loss in the transformer district.

References

1. Li, Y., Liu, L., et al.: Calculation of line loss rate in transformer district based on improved K-means clustering algorithm and BP neural network. Proc. CSEE **36**(17), 4543–4551 (2016). (in Chinese)
2. Wang, S., Zhou, K., et al.: Line loss rate estimation method of transformer district based on random forest algorithm. Electr. Power Autom. Equip. **37**(11), 39–45 (2017). (in Chinese)
3. Xu, R., Wang, Y.: Theoretical line loss calculation based on SVR and PSO for distribution system. Electr. Power Autom. Equip. **32**(5), 86–89 (2012). (in Chinese)
4. Jiang, H., An, M., Liu, X. et al.: The calculation of energy losses in distribution systems based on RBF network with dynamic clustering algorithm. Proc. CSEE **25**(10), 35–39 (2005). (in Chinese)
5. Ouyang, S., Feng, T., An, X.: Line-loss rate calculation model considering feeder clustering features for medium-voltage distribution network. Electr. Power Autom. Equip. **36**(9), 33–39 (2016). (in Chinese)
6. Zou, Y., Mei, F., Li, Y. et al.: Prediction model research of reasonable line loss for transformer district based on data mining technology. Demand Side Manag. **4**(17), 25–29 (2015). (in Chinese)
7. Wen, F., Han, Z.: The calculation of energy losses in distribution systems based upon a clustering algorithm and an artificial neutral network model. Proc CSEE **13**(3), 41–50 (1993). (in Chinese)

Research on Theoretical Structure Design of Smart Community Large Data Platform

Nan Jia, Xiao An, Jing Qian, Yongqiang Chen and Yi Liu

Abstract Smart community is the basic unit of smart city, and a community large data platform is an important guarantee to realize the safety of an intelligent community. It is pointed out in this paper that monitoring and control, forecasting and alarm, and intelligent prevention are the key techniques of risk prevention of smart community. Thus, large data platform is the foundation of risk prevention in the smart community. Finally, the structure and the construction process to carry out the theoretical structure design of intelligent community risk prevention of large data platform. The research provides theoretical and technical support for the next step of big data platform construction and risk prevention.

Keywords Smart community · Risk prevention · Key technology · Large data platform · Theoretical structure design

1 Introduction

Community is the basic unit [1, 2] of public security governance, and it is the foundation of realizing a smart city. In recent years, community risk incidents have been frequent, and community security problems have been highlighted [3, 4], such as the Shanghai community fire accident in 2010, the Beijing Hepingli community gas explosion in 2011, and the Hangzhou nanny community arson in 2018. As a carrier of disaster, the community presents vulnerability in the face of various emergencies. At the same time, the community risk factors are highly converged and interrelated with each other so that the community risk has the characteristics of

---

N. Jia · X. An · J. Qian · Y. Liu (✉)
Department of Engineering Physics, Institute of Public Safety Research, Tsinghua University, Beijing 100084, China
e-mail: liuyi@tsinghua.edu.cn

Y. Chen
Department of Mechanics and Engineering Science, College of Engineering, Peking University, Beijing 100871, China

V. Jain et al. (eds.), *Recent Trends in Intelligent Computing, Communication and Devices*, Advances in Intelligent Systems and Computing 1031,
https://doi.org/10.1007/978-981-13-9406-5_79

strong abruptness, dynamic, high uncertainty, and human–machine and material mixing. Community risk prevention is the key to ensure residents' safety and achieve stable and harmonious development of the country. Therefore, it is urgent to carry out community risk monitoring and prevention research.

Smart community risk prevention is an important way to realize community safety. It is pointed out in this paper that monitoring and control, forecasting and alarm, and intelligent prevention are the key technologies of intelligent community risk prevention, and based on the analysis of community risk data types, it is pointed out that large data platform is the foundation of community risk prevention. Finally, the theoretical structure design for the community risk prevention large data platform is carried out from the three aspects of function, structure, and construction process, which laid the foundation for the next step of platform construction and the realization of risk prevention.

2 Key Technologies of Risk Prevention of Smart Community

The monitoring and control, forecasting and alarm, and intelligent prevention of community risk are the key technologies in realizing intelligent community risk prevention and the important support of community safety guarantee.

(1) Monitoring and control, the main object of monitoring and control of community risk is human, machine, objects, as well as events, space and time, organization, etc. related to people, machines, and objects. Through the community personnel positioning [5–7] and identity identification [8], as well as community equipment operation monitoring dynamic prevention and control [9], to carry out the total factor monitoring and cycle risk control of community risk.
(2) Forecasting and alarm, in view of the multiple and interactive characteristics of community risk factors, according to the large data of community risk monitoring and control, through large-scale and rapid numerical calculation and simulation of complex mass scenarios, the key to prevention and control of accidents is effective forecasting of community risk, risk quantitative forecasting, risk evolution analysis, and timely warning.
(3) Intelligent prevention, based on the monitoring and control and forecasting and alarm, integrating the comprehensive community history information data, monitoring real-time data and forecasting and alarm calculation and analysis data, to carry out the smart community security management. To realize the comprehensive prevention of comprehensive monitoring, linked alarm and collaborative disposal of community risk.

3 Smart Community Risk Prevention Foundation Support-Large Data Platform

Large data platform is the basic technical support for the implementation of the key technology of community risk control.

(1) Monitoring and Control

Based on the analysis of community risk data, it can be found that the risk data mainly comes from the real-time monitoring and monitoring of community risk, including the real-time monitor information of the community mobile personnel and residents, and the real-time running status information of the equipment facilities, and so on. Data integration and storage are required before start risk alarm and prevention for the large data collected by the measuring and monitoring. Large data platform is the base platform of integrated storage for community risk monitoring and control data collection [10, 11].

(2) Forecasting and Alarm

Community risk forecasting and alarm technology needs to take the existing historical data, real-time monitoring data as the support to proceed risk calculation analysis. At the same time, forecasting data generated by forecasting analysis is fed back into a large data platform. Therefore, the large data platform provides the basis of computing data for community risk forecasting and provides a storage platform for forecasting analysis data.

(3) Intelligent Prevention

Intelligent prevention is through the integration and collaboration of community business, task, data, model, service to realize all-round, integrated information sharing of community risk information, this cannot be separated from the large data platform data integration, management, sharing, and service functions [12], and large data platform is the necessary support platform of the community risk intelligent prevention and implementation.

Through Fig. 1, we can find that the data interaction network based on the big data platform constitutes the key technology of community risk prevention. Through the construction of a big data platform, it can provide basic support for the realization of community risk prevention.

4 Theoretical Structure Design of Large Data Platform

In view of the multi-source heterogeneous complex characteristics of community risk data, community risk prevention large data platform needs to realize the collection, inherited storage, integration analysis, feedback update, and management distribution and share service function of multi-source heterogeneous risk data.

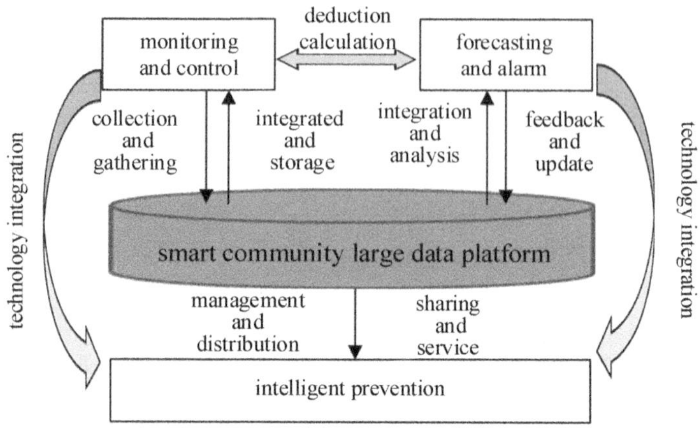

Fig. 1 Data interaction between big data platform and risk prevention key technologies

4.1 Structure Design

The large data platform for community risk prevention and control structure is designed from four aspects. Figure 2 is the structural design framework of community risk large data platform.

First, the aspect of community risk data physical integration.

The collection of risk data is the basic function of the data platform. Through the high-precision data collection, the database updating based on the basic state and the integration management of multi-source heterogeneous data, the large data physical integration oriented to community risk prevention is realized.

Key technologies include: high-precision data collection, database updating based on basic state, and multi-source heterogeneous data integration.

Second, the community risk data semantic hierarchy integration.

The data is collected and managed from the semantic point of view. Through the data association of the knowledge map, the semantic relation, logic relation, and causal relation between different data are established by the way of ontology, and the semantic interoperation and semantic integration of different data are realized.

Key technologies include: knowledge map, community risk ontology, and semantic integration.

Third, application-oriented data management integration.

Data collection is the foundation, and the integration of multi-source heterogeneous data is the necessary condition for the integrated analysis of risk. By means of unified space-time coordinate, the data of different sources, the data of different time and space are integrated, and the data hidden relationship analysis and forecasting of community risk evolution are realized.

Key technologies include: unified space-time coordinate, community risk evolution, and data hidden relationship analysis.

Forth, target-oriented data service.

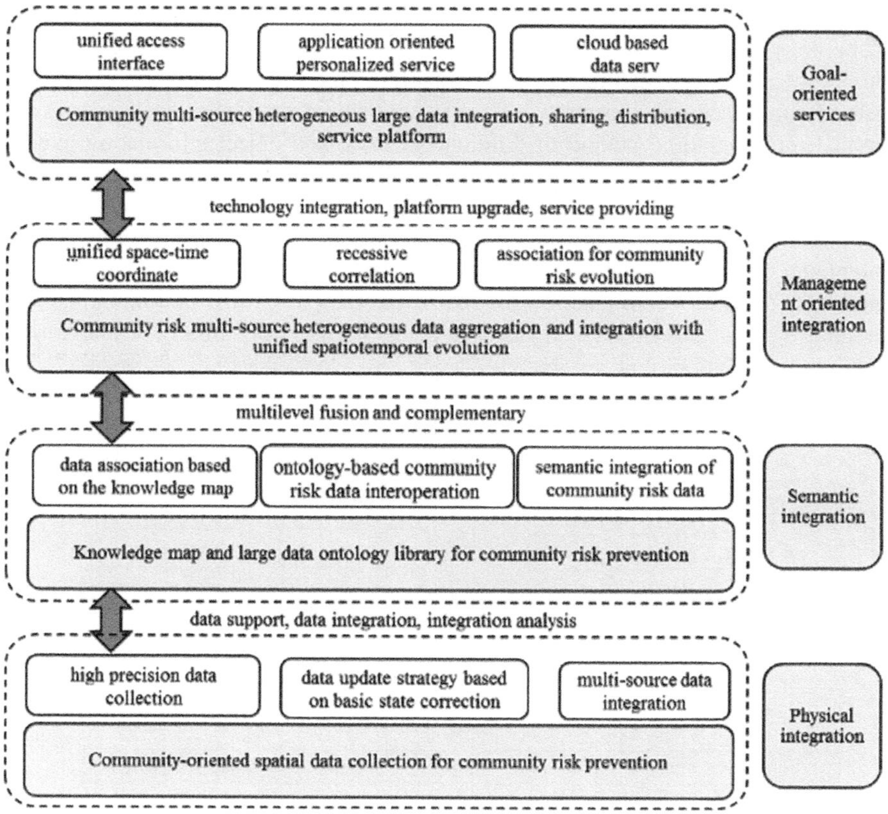

Fig. 2 Structural diagram of smart community large data platform

There are multiple management service agencies and systems in the community, and there are both separate and overlapping data requirements and applications between different agencies and systems. In the large community risk data platform, by fully consider the characteristics of the community, to achieve a unified access interface, application-oriented personalized services, and cloud-based integrated data services, to support community risk prevention cross-area integration, in-depth sharing, and multi-agent collaboration.

Key technologies include: unified access interface, application-oriented, personalized service, and cloud-based data service.

4.2 Construction Process

Based on the design of community risk large data platform, the construction process of community risk prevention large data platform is put forward, as shown in

Fig. 3. By using the high spatial resolution community spatial data collection technology for community risk monitoring, the rapid collection and three-dimensional modeling of community building and infrastructure data were carried out. Based on community security entity data, by the label classification and unified representation method of different attribute and object information, we set up a whole dimension data association relationship for community risk prevention, draw community risk prevention knowledge map and ontology model based on community's people, things, places, objects, and organizations, and carry out the integrated analysis of community risk multi-source heterogeneous data based on unified space-time coordinate; on this basis, realize platform with community risk monitoring and prevention with unified data interface for large data integration, sharing, distribution, and service.

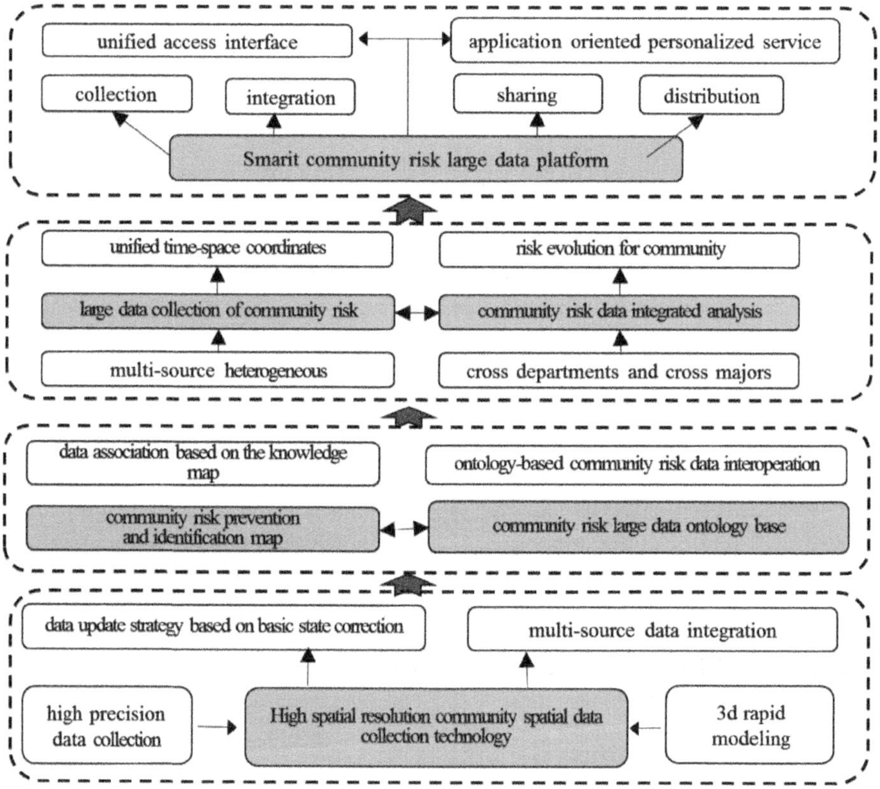

Fig. 3 Construction process of smart community large data platform

5 Conclusion

In view of the important need of intelligent community risk prevention, we put forward the key technologies of monitoring, forecasting, and intelligent prevention of smart community risk prevention, and then, by analyzing the community risk data, it is pointed out that the large data platform is the foundation support of the intelligent community. Finally, the function, structure design, and construction process of the large data platform for community risk prevention were designed, respectively.

Community risk data collection, integration analysis, and integrated sharing are the functional needs of community risk prevention data platform. Aiming at the multi-source heterogeneity of community risk data and according to the theory structure design of large data platform, the next step is to carry out research on community risk data knowledge map and large data ontology modeling and unified space-time evolution of community multi-source heterogeneous high dimension and high extensibility data collection and integrated analysis.

Acknowledgements This study is supported by National Key R&D Program of China (No. 2018YFC0809700); National Natural Science Foundation of China (No. 71904095); Beijing Natural Science Foundation (No. 9194027) and China Postdoctoral Science Foundation (No. 2019M650750).

References

1. Webster, C.J., Glasze, G., Frantz, K.: The global spread of gated communities. Environ. Plan B: Plan. Des. **29**(3), 315–320 (2002)
2. Blandy, S.: Fear of crime and of anti-social behavior and their relation to the spread of residential gated communities in England. Deviance Et Soc. **33**(4), 557–572 (2009)
3. Patterson, E.B.: Poverty, income inequality, and community crime rates. Criminology **29**(4), 755–776 (2010)
4. Pitts, J.: Crime Prevention and community safety: new directions. Safer Communities **1**(2), 46–48 (2012)
5. Wang, Y.: The development of wireless personnel positioning in Internet of Things based on ZigBee and sensors. Int. J. Dig. Content Technol. Its Appl. **6**(12), 47–54 (2012)
6. Zhang, W.Y., Yu, H., Zhang, F., et al.: Personnel positioning system of underground coal mines based on the ZigBee technology. J. Hefei Univ. Technol. **30**(9), 1087–1090 (2007)
7. Kovalenko, V., Yarovoy, A., Ligthart, L.P.: Waveform based detection of anti-personnel mines with an UWB radar [C]. In: IEEE International Conference on Ultra-wideband. IEEE (2005)
8. Bouchrika, I., Carter, J.N., Nixon, M.S.: Towards automated visual surveillance using gait for identity recognition and tracking across multiple non-intersecting cameras. Multimedia Tools Appl. **75**(2), 1201–1221 (2016)
9. Mao, G., Fidan, B., Anderson, B.D.O.: Wireless sensor network localization techniques. Comput. Netw. **51**(10), 2529–2553 (2007)
10. Zhang, F., Zhou, Z., Xu, W., et al.: Cloud manufacturing resource service platform based on intelligent perception network using fiber optic sensing. Adv. Inf. Sci. Serv. Sci. **4**(4), 366–372 (2012)

11. Alexandrov, A., Bergmann, R., Ewen, S., et al.: The stratosphere platform for big data analytics. Int J Very Large Data Bases **23**(6), 939–964 (2014)
12. Guo, D., Du, Y.: A visualization platform for spatio-temporal data: a data intensive computation framework. In: International Conference on Geoinformatics. IEEE, 1–6 (2016)

Research on Multi-sensor Data Fusion at Boar Station Based on Chaotic Thoughts

Tian Fang, Tan Han and Juan Yao

Abstract Multi-sensor information fusion refers to the handling for the uniform form of expression obtained by calibration, alliance, correlation and merging towards the data from multiple sensors. Compared with a single sensor, for the solution of the target states or the judgment of target characteristics, multi-sensor information fusion technology can enhance the reliability of the data, improve the accuracy and increase instantaneity of the systems and information utilization, etc. Multi-sensor data can supplement each other, but it may also have lower confidence coefficient and become redundancy. Therefore, the adjustment process for the measured data by the means of fusion algorithm is a key to obtain the accurate data. Based on the algorithm of chaotic thoughts, an algorithm which fused the data of temperature and humidity at a boar station was put forward to carry out an experiment in this paper. The experiment proved that this algorithm was effective to extract the useful information from different fusion data and it obtained a good fusion effect.

Keywords Chaos · Multiple sensors · Data fusion

1 Introduction

Data fusion involves various theories and technologies. There are not completely uniform algorithms adapted to all scenarios, so the relevant algorithms should be selected in the application according to different application backgrounds. Fusion methods mainly contain the weighted mean method, election decision-making method, Kalman filtering method, mathematical statistics, etc. There are some new or improved algorithms, such as weighting algorithm of improved support function

T. Fang · J. Yao (✉)
Huazhong Agricultural University, Wuhan, Hubei, China
e-mail: yaojuan@mail.hzau.edu.cn

T. Han
Wuhan Vocational College of Software and Engineering, Hubei, China

V. Jain et al. (eds.), *Recent Trends in Intelligent Computing, Communication and Devices*, Advances in Intelligent Systems and Computing 1031,
https://doi.org/10.1007/978-981-13-9406-5_80

based on Cuckoo algorithm. Compared to two kinds of support functions containing Gaussian support function and new-type support function, the weighting algorithm based on improved support function has the minimum fusion variance—0.1925 which improved the accuracy of data fusion effectively. Sun et al. [1] in High Technology Institute of Academy of Sciences in Heilongjiang Province put forward the method of multi-sensor data fusion based on Cuckoo Search Algorithm. Derived from the optimal characteristics of Cuckoo Search Algorithm such as efficient concurrency, fast convergence and the difficulty to get stuck in the local, this method has the stronger optimizing ability. Therefore, it can settle optimization problems of data fusion for the multi-sensor weighting factors better, which makes the fused data have higher accuracy and approach the targeted actual value more closely [2].

Aiming at the temperature data of multiple sensors collected by the boar station, this paper estimated the weighting coefficients in data fusion with data fusion method of the chaotic thoughts. The results showed that this algorithm could extract effectively useful information from different fused data and obtain a very good fusion effect.

2 Chaotic Mapping

2.1 Chaotic Theory

Chaotic theory was put forward by French mathematician Poincare, which was a method with isotropy thought and quantitative analysis. It was used to discuss behaviour in a dynamic system that must be explained by the complete and continuous data relationship, instead of a single data relationship. The chaos had features including regularity, sensibility, fractal dimension feature, ergodicity and randomness, etc. Therefore, chaotic mapping was introduced to data fusion algorithm to carry out optimum performance of the algorithm.

2.2 Logistic Mapping

Logistic mapping is a typical chaotic mapping system. Its equation is shown below:

$$y_{n+1} = \mu y_n(1 - y_n) \tag{1}$$

where μ is a adjustable parameter, $\mu \in (0,4]$, $n = 1, 2, \ldots, n$, yn is chaotic variable, $yn \in [0,1]$.

The experiment proved that the logistic system reached the optimal status when μ was equal to 4.

This paper selected $\mu = 4$, i.e.

$$y_{n+1} = 4y_n(1 - y_n) \tag{2}$$

3 Result and Analysis for the Experiment

3.1 *Based on Chaotic Data Fusion Algorithm*

This paper selected the data of the temperature sensors at the boar station and expressed it as below:

$$x_1, x_2\ldots, x_n \tag{3}$$

Calculate its average value

$$x_{\text{aver}} = \frac{1}{n}\sum_{i=1}^{n} x_i \tag{4}$$

The basic idea based on chaotic data fusion algorithm is to iterate its corresponding weight and make the fused data become optimal with chaotic mapping methods of different measured data under the condition of minimum total mean square error.

Select the error of x_i (average value of x_i) and x_{aver} as the initial weight, i.e.

$$w_i = |x_i - x_{\text{aver}}| \tag{5}$$

Calculate the proportion of each weight in total weight as below:

$$p_i = w_i \Big/ \sum_{i=1}^{n} w_i \tag{6}$$

Select the chaotic mapping system (8) and set its initial value as (7)

$$y_1 = \frac{x_{\text{aver}} - \min(x_i)}{\max(x_i) - \min(x_i)} \tag{7}$$

$$y_{n+1} = 4y_n(1 - y_n) \tag{8}$$

where min is the minimum of $x_1, x_2\ldots, x_n$, max is the maximum of $x_1, x_2\ldots, x_n$.

Substitute chaotic mapping value $y_1, y_2\ldots, y_n$ into the current optimal fusion value, replace the maximum x_i of pi with it to form a new sequence, then continue using the above-mentioned methods for iteration until the change of the fusion total

Table 1 Temperature data measured by different sensors

1	2	3	4	5	6	7	8
20.28	20.67	21.78	21.23	22.09	22.13	22.67	22.34
9	10	11	12	13	14	15	
21.99	21.83	22.34	22.56	23.24	23.12	22.34	

mean square error has less and the minimum fluctuation. The experiment proved that the fusion had the strong optimizing ability and good effect.

3.2 Experimental Data

A fusion experiment was carried out with the data of temperature sensors at boar station in this paper. For the data, please see Table 1. There are total of 15 sets of data. Iterations were set as 100 in the simulation experiment.

3.3 Analysis for Experimental Result

Compile Python program and calculate the fusion value of chaotic temperature as 22.34. Select 100 chaotic iterations. The experimental result is shown in Figs. 1 and 2.

The analysis showed that the data fluctuation was obvious at the initial iteration in the experimental figure. When the iterations reached 45–50, they were optimum and temperature fusion value was 22.3359 at this time.

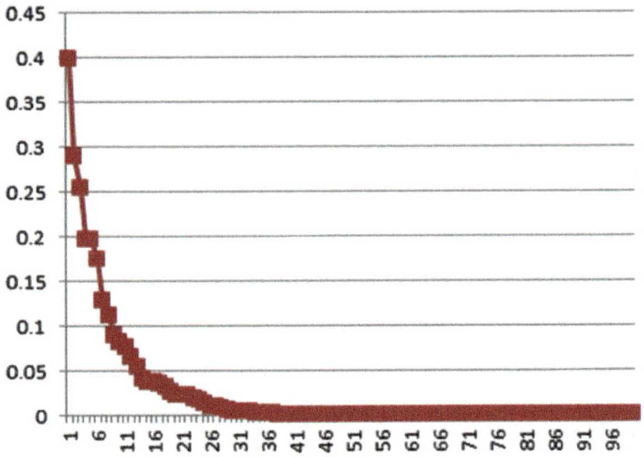

Fig. 1 Variation diagram for mean square values after 100 iterations

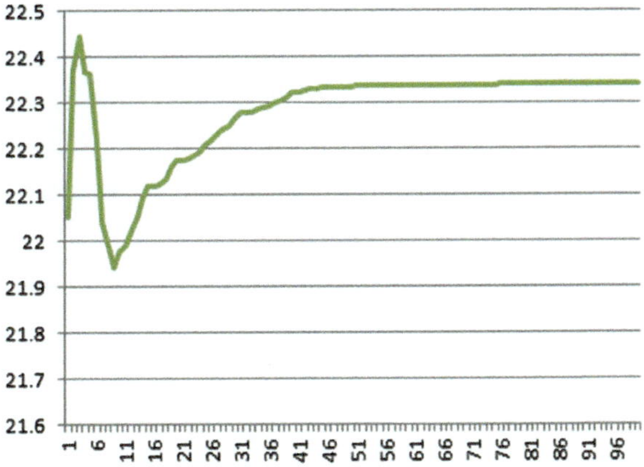

Fig. 2 Variation diagram for fusion temperature values after 100 iterations

4 Conclusions

This paper introduced a data fusion algorithm based on chaos at first. Then, this algorithm was applied to data fusion method of multi-sensor weighing factors through the transmission of the sensor data at boar station. The result showed that data fusion algorithm based on chaos had the strong optimizing ability and it could settle the problems of data fusion optimization for multi-sensor weighting factors better.

References

1. Duan, Q., Xiao, X., Liu, Y., Lu, Z., Wang, K.: Data fusion method in livestock breeding Internet of Things based on improved support functions. Agric. Eng. **33**, 239–245 (2017)
2. Sun, K., Hu, X., Wang, G.: Multi-sensor data fusion based on cuckoo search algorithm. Tech. Autom. Appl. **37**, 7 (2018)
3. Yang X.S., Deb, S.: Cuckoo search via levy flights. In: Proceedings of World Congress on Nature & Biologically Inspired Computing, pp. 210–214 (2010)

Convolutional Neural Network Applied to Remote Technical Diagnosis by Thermograms

S. P. Orlov and R. V. Girin

Abstract The article is devoted to the problem of remote determination of the technical objects operability using a thermal imager. The structure of the information-measuring system is presented. It is proposed to use a hybrid model including a convolutional neural network and a fully connected neural network to identify defects exploring the surface temperature field of the test objects. The principles of constructing such a neural network are considered, and its parameters are investigated. Experimental studies of the developed system in the technical diagnosis of electronic devices were carried out.

Keywords Artificial neural network · Machine learning · Technical diagnosis · Thermograms

1 Introduction

Technical diagnostics of complex objects develops in the direction of operational control of technical states using remote measuring instruments. This allows detecting failures not only during testing of equipment, but also during operation. In many cases, controlled technical objects have a set of measurable parameters that can identify their technical state. However, it is not always possible to build measurement channels for continuous monitoring of these parameters. Often, access to measurement is impossible due to closed design and restrictions on the weight and volume of the object to be monitored. Furthermore, the measuring procedure may contributions distortion in the object operation process. In this connection, the methods of contactless remote monitoring with the help of infrared thermography are promising [1].

S. P. Orlov (✉) · R. V. Girin
Institute of Automatic and Information Technologies, Samara State Technical University, Samara 443100, Russia
e-mail: orlovsp1946@gmail.com

V. Jain et al. (eds.), *Recent Trends in Intelligent Computing, Communication and Devices*, Advances in Intelligent Systems and Computing 1031,
https://doi.org/10.1007/978-981-13-9406-5_81

In this case, a channel for remote measurement of the temperature field of the object should be organized, including a thermal imager, a thermogram processing unit and a temperature field analysis unit, in which a decision is made on the operability of the object, and the failure facts are differentiated.

The problem of technical diagnostics using thermograms is as follows:

- The set of obtained thermograms of the object and the set of inoperative states do not have a one-to-one correspondence. Different faults and inoperative states can correspond to the same temperature distribution on the object surface. Essentially, we are incorrect inverse heat conduction problem, which is necessary to find the location and intensity of the internal heat sources in the test object.
- A large number of thermograms that are to be analyzed lead to complex algorithms for selecting the desired thermal images of the object.

2 The Information-Measuring System for Remote Technical Diagnostics

In [2], an information-measuring system (IMS) for remote monitoring of radio-electronic devices in ground tests has been described. System consists of measuring channels: (a) determining environmental parameters via thermohigrometr; (b) measuring of the temperature field of the surface based on the thermal imaging; (c) measuring of the device electrical parameters using a digital oscilloscope GDS-2104. Control of test modes of devices is carried out using a computer and FPGA XC3S500E units.

The main idea is based on a set of mathematical models of the thermal state of the device corresponding to the failures in it. Classes, corresponding to different defects in the electronic device, differentiate these models. In this case, the decision on the operability of the electronic device was made by comparing the measured thermogram with the set of calculated thermograms obtained with the help of mathematical models.

3 Hybrid Intelligent Model

It is known that the artificial neural network ANN effectively recognize complex images [3, 4]. There are a large number of ANN applications for analyzing information under uncertainty [5, 6]. In this report, we propose an approach to the regularization of incorrect inverse problem of recognition of diagnostic images, which is based on the combined use of two neural networks.

For each type of technical object, mathematical models of the thermal state of its surface are developed depending on the behaviour of internal heat sources. The developed mathematical models of the thermal state and history of previous tests of objects are used to form the fact base and the rule base that are components of the knowledge base. They are used in the expert system with a forward chaining on the production rules. Thermal states of technical objects are characterized by thermograms obtained with the thermal imager T_j, which are stored in the knowledge base of IMS. Standing heat sources Q_i inside the object are used to determine the object's performance. We define the set of states D_n in which a technical object can be located (one of failures or a working state). In this case, we assume that to all serviceable states there will correspond to one state D_0. Figure 1 shows a graph model of the technical object states.

In general, we have incorrect inverse problem, since the same surface thermograms may correspond to multiple heat sources states. Consequently, the one-to-one mapping of thermograms and inoperative states of the object is violated. We use an additional vector of measured process parameters, where V_m are the parameters monitored by measuring means. This may be, for example, input and output voltages of the electrical signals, supply voltages, frequency, and phase of the signals, and others. We carry out a regularization of the problem using more information on the connection of electrical parameters with thermal conditions. Thus, the inverse problem becomes correct by narrowing the infinite set of solutions to finite compact sets corresponding to the chosen defects.

We require the following condition: for any pair of classes T^l and T^k, $l, k \in \{1, 2, \ldots, N\}, l \neq k$, there is at least one pair of elements $T_j^l \neq T_i^k$, $j \in J_l$, $i \in J_k$ or $V_m^l \neq V_r^k$, $m \in M_l$, $r \in M_k$.

Consequently, the separation of thermograms with the help of the vector V into subsets $T^n(x, y, V) \rightarrow D_n$ puts them in a one-to-one correspondence with the inoperative states of the controlled object. Thus, the set of classes is formed:

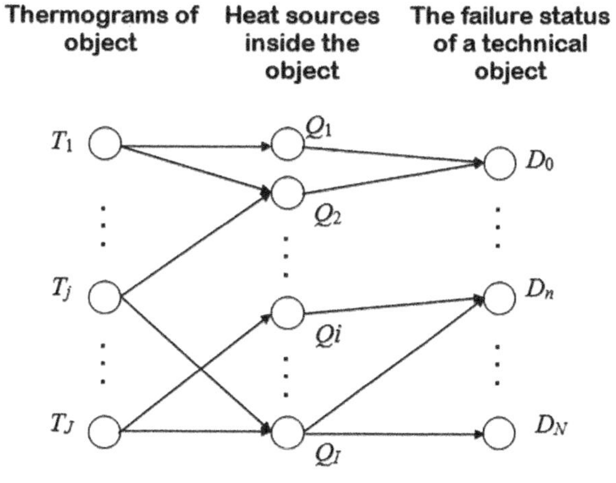

Fig. 1 Graph model of the technical object states

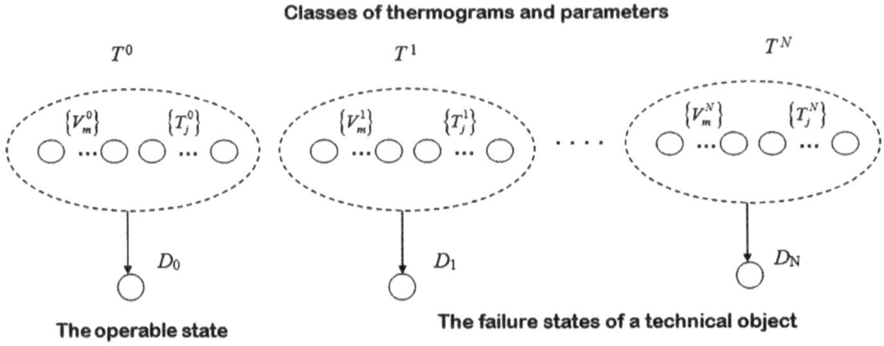

Fig. 2 Separation signs of inoperability classes

$T_C = \bigcup_n(\{T_j^n\}, \{V_m^n\})$, $j \in J_n$; $m \in M_n$, $n = \overline{0, N}$, where J_n and M_n—index sets of thermograms and vectors of the object parameters included in the nth class (Fig. 2).

4 A Measuring Channel with a Convolutional Neural Network

The thermal imager and the convolutional neural network form the main measuring channel. The organization of the convolutional network is based on the LeCun approach [3] and also combines some features which were introduced in [7, 8] and is presented in Fig. 3. The network uses batch normalization, which was first introduced in [9].

The proposed artificial neural network consists of two branches. The main branch is well-known convolutional neural network. It consists of several convolutional layers united in a feed-forward neural network. Thermograms are used as input for this branch.

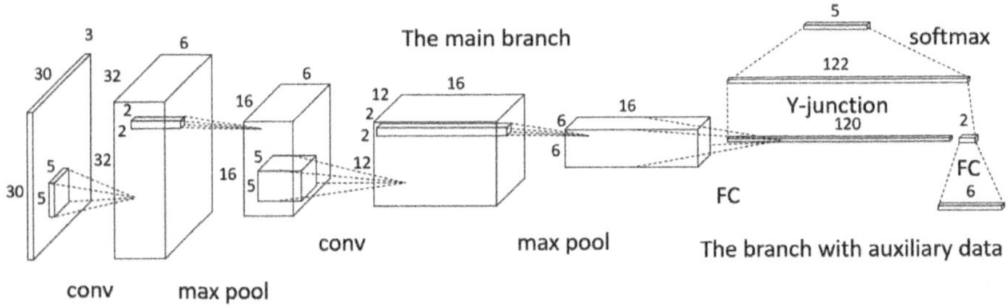

Fig. 3 Neural network with two branches used in the experiments

Thermograms as any other image can be presented as a 3D array in which every value represents colour of the corresponding pixel. The width and height of the array are the same as width and height of the image (in pixels), and the depth of the array is 3, one for each of three (red, green, blue) channels. The size of thermograms used in our experiments was 30 × 30 pixels.

Parameters of the convolutional layers, such as receptive field, stride and padding, were chosen so that with application to the given size of thermograms, the output of the last convolution layer is a vector.

In order to take into account signals from the controlled unit's built-in sensors, an auxiliary branch of ANN was introduced. This branch consists of fully connected neurons layer. Input for the branch is a vector with normalized data from the sensors which length is equal to 6. The output of the layer is also a vector of length 2. The vector is merged with output vector of the main branch, and result vector gotten after the merger is passed in as input for neurons layer that performs softmax or two-staged linear [10] normalization. This layer is the output layer of the whole ANN and performs categorization of failure in a controlled unit.

ANN was used for categorization of four major and critical failures in controlled device. Therefore, output layer contains five neurons (one per each failure category and one corresponds to the normal condition of the unit).

Some recent papers [11] consider in detail the application of some architecture that allows emulating sparsity in connections in neural networks. In our experiments, we did not use some of those techniques and used sparsity explicitly in a similar manner as it was used in [3], retrieving different subsets of feature maps for input of the third convolution layer. And as for neurons are in convolutional layer, each of them is connected to the input within its receptive field.

Considering the fact that the main branch of our neural network is widely used convolutional network, the weights for the layers can be initialized with weights of some pretrained convolutional network which parameters compatible with our network. Although we didn't use this approach and initialized all the parameters with random value from range (−2, 2) and trained, the network from scratch using pretrained weights can reduce the training time in cases when it is important.

In our experiments, we trained our network using back-propagation technique [12] with learning rate 0.001 using 100 epochs. Achieved precision of thee network is 99%. For training, we used model thermograms generated programmatically for each of classified failure and for the case when controlled unit is operating normally.

Although traditionally neural networks are trained on dataset that comprises subset of samples which network will process during its exploitation, we trained the network on model thermograms, not on thermograms taken from some controlled unit. In our case, we considered the problem from metrological point of view. Similar to any other metrological instrument, it first passes its calibration on standard data in laboratory, and then, it is used in the field. In similar manner, we collected network's training dataset from thermograms that were generated programmatically based on mathematical model of surface temperature field of our control device. In practice, even thermograms of two unit of the same model can

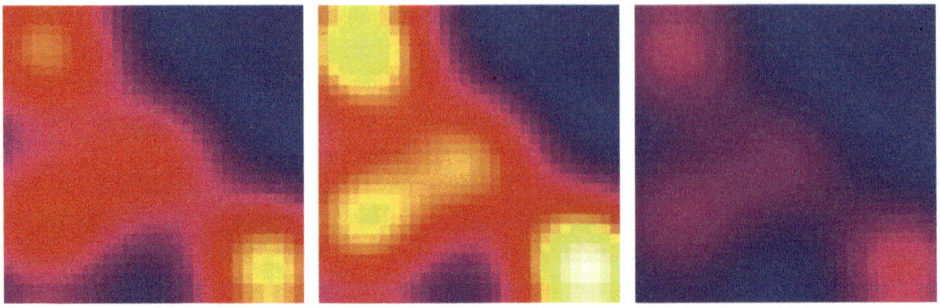

Fig. 4 Some of the thermograms used in the experiments (best viewed in electronic version)

differ slightly. By mean of using model thermograms, we reduce the influence of such variations.

Some sample thermograms used in network's training are shown in Fig. 4.

References

1. Uvaysov, S.U., Yurkov, N.K.: Method for ensuring thermal control of radio engineering devices at the design stage. Bull. Samara Univ. Aerosp. Eng. Technol. Eng (7), 16–22 (2012)
2. Orlov, S.P., Vasilchenko, A.N.: Intelligent measuring system for testing and failure analysis of electronic devices. In: Proceedings of 19th IEEE International Conference on Soft Computing and Measurements (SCM'2016), vol. 1, pp. 401–403 (2016)
3. LeCun, Y., Bottou, L., Bengio, Y., Haffner, P.: Gradient-based learning applied to document recognition. Proc. IEEE **86**(11), 2278–2324 (1998)
4. Haykin, S.: Neural networks and learning machines, 3rd edn, p. 905. Pearson Prentice Hall (2009)
5. Norvig, P., Rassell, S.: Artificial intelligence: a modern approach, 3rd edn, p. 1109. Pearson Prentice Hall (2010)
6. Nielsen, M.: Neural networks and deep learning. Free online book. http://neuralnetworksanddeeplearning.com (2017)
7. Krizhevsky, A., Sutskever, I., Hinton, G.E.: ImageNet classification with deep convolutional neural networks. In: International Conference on Neural Information Processing Systems, pp. 1097–1105. Curran Associated Inc. (2012)
8. Nair, V., Hinton, G.: Rectified linear units improve restricted Boltzmann machines. Proc. ICML **27**, 807–814 (2010)
9. Ioffe, S., Szegedy, C.: Batch normalization: accelerating deep network training reducing internal covariate shift. Cornell University Library. https://arxiv.org/abs/1502.03167v3 (2015)
10. Girin, R.V., Orlov, S.P.: Two-stage normalization of output signals of artificial neural networks. Bull. Samara State Tech. Univ. Techn. Sci. **56**(4), 7–16 (2017)
11. Szegedy, C., Liu, W., Jia, Y., et al.: Going deeper with convolutions. In: Proceedings of IEEE Conference on Computer Vision and Pattern Recognition (CVPR), pp. 1–9 (2015)
12. Rumelhart, D.E., Hinton, G.E., Williams, R.J.: Learning representation by back-propagating errors. Lett. Nature **323**, 533–536 (1986)

Verifying Control Programs Instrument for Unmanned Vehicles

A. A. Tyugashev, D. N. Frantasov, P. A. Melnikov and A. S. Klimas

Abstract This project assists developers in writing unmanned vehicles software by creating universal and public available tools. It makes possible to determine the task combination that allows the device to operate smoothly and disrupts it at any stage. All possible program solutions can be tested and discovered failure or an error occurred. Also, the program can offer developers options to get rid of conflict situations. The complete algorithm for control program branches has already been developed. The database will store the result of this algorithm. A variant of the program analysis algorithm optimization is suggested reducing the time spent on the full analysis of all branches. It is suggested that the variant of creating an expert system to assist in the analysis and errors correction. More content will be created by the developer community. Knowledge gained from Samara State Transport University basis will help students and engineers in the CP development to communicate on our platform and improve their skills.

Keywords Unmanned vehicles · Algorithm · Database · Optimization

1 Introduction

Nowadays, unmanned vehicles (UV) are one of the ways of transport efficiency improving. The requirements for the motion control non-deterministic disturbances conditions, navigational measurements errors, and control implementations increase. One of the ways of the UV control accuracy increasing is a management program (MP) optimization. Unfortunately, we face some optimization problems [1]: the optimization of power system operation modes, the subprogram composition and start time, parameters of joint work system and applied tasks.

A. A. Tyugashev · D. N. Frantasov (✉) · P. A. Melnikov · A. S. Klimas
Samara State Transport University, Samara, Russia
e-mail: frantasov@mail.ru

V. Jain et al. (eds.), *Recent Trends in Intelligent Computing, Communication and Devices*, Advances in Intelligent Systems and Computing 1031,
https://doi.org/10.1007/978-981-13-9406-5_82

This fully autonomous vehicles development will solve some problems like:

- safety of traffic;
- efficient use of energy;
- successful sharing.

The UV development has long been engaged. Eventually taking into account the technology development, most everybody could have the opportunity to create the new ones. However, during the development process, engineers face a lot of problems and one of them is the management program writing considering all possible external factors for uninterrupted movement.

At the moment, there have been already several types of unmanned vehicles that operated without human intervention programs. Each developer creates tools that are tightly linked to his area and a bit universal for the unmanned vehicles software creation [2]. It greatly hinders the development in this area, since it is very difficult for free developers to write the whole program from scratch for every new vehicle and, if necessary, even create special tools for specific cases.

Since there has been currently no publicly available program that could perform such functions as a design program for the autonomous operation of an unmanned device, it is suggested to begin with the establishment a program for the UV-uninterrupted operation. It should help engineers to design an MP. It allows to pre-calculate kind of command combinations that will make the device function smoothly and could lead to a malfunction at any stage. Thus, it is possible to test all variants of the program execution and find out on the device an error or a failure occurs, so as to make it possible to optimize the program. Also the program can offer developers conflict situations avoiding options.

2 Solution Algorithm Description

2.1 The Algorithm Analyzing

The parameters and the UV power system composition are considered as one of the main aspects, because they affect not only on the performance, but also on the weight dimensions and on the final cost. This tool computes the maximum energy consumption and its critical level in the specifying indicators. It helps to calculate in advance all possible energy costs during the UV operation and to minimize the UV power consumption in the process of executing a certain program. After the designing correct program, the software should automatically generate all the necessary documents for it and provide it to the developer in the required form.

Further, it is planned to create a developers community around the platform basing on the Web platform. It allows developing the program faster and can also give impetus to the development of the UV in many spheres of human activity.

After the "community" formation, the Web platform can become one of the centers for generating ideas and developing technologies for unmanned vehicles.

Cyclograms are compiled for the uninterrupted motion programs construction, which clearly show all the commands that make up the program. It will assess the quality and analytically identify the execution options that can lead to error in the vehicle. Currently, a tool has been developed for the designing program cyclogram control, which is free of the hardware platform and able to work with conditional control commands.

There are some program branches that arise in the event of unforeseen situations or planned systemic actions. They should be considered, since they impact on the whole program performance. That is why the algorithm has been developed for checking all possible variants of the control program branches [3] (Fig. 1).

The following database scheme is suggested for the above algorithm to record the results.

This algorithm works as follows:

1. Firstly, the program determines the total number of instructions in the analyzed program.
2. After that, the first branch is created. A branch is a sequence of commands from the very beginning of the program to some command on which the branching option is found.
3. Then, the algorithm finds the next command where branching is detected. If a new branch is detected, a new branch is created.

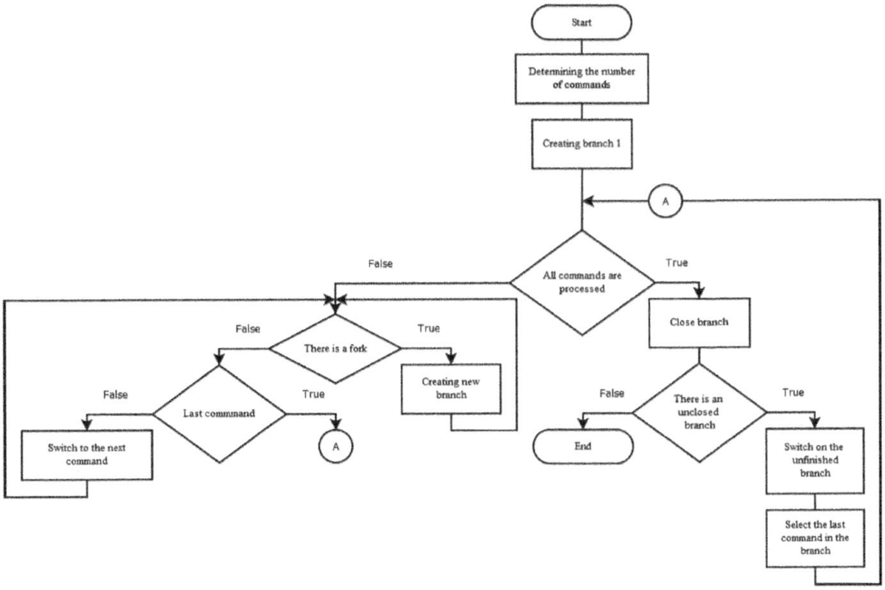

Fig. 1 Program algorithm for the determining traces

4. After creating all the possible branches, each of them is reevaluated for branches.
5. After closing all the branches, we get an information about every possible branch of the analyzed program.

Thus, after completing the algorithm, all possible variants allow to design work "tracks" (the track is the complete command sequence, which is formed from branches) and verify each line through an analyzer program for determining possible program issues.

2.2 *Development of Data Storage Structure*

As aresult of the algorithm is a data array, its dimension depends on the number of tasks in the control program and the possible branch conditions, and the array size becomes significant. Therefore, it is impossible to use the RAM for intermediate storage. It is also difficult to use files for storing the life-track data array because of the complexity. It is suggested to use databases for a long-term storage. There is the ER model of the relational database in Fig. 2; it allows storing an unlimited number of control programs and determines all possible tracks [4].

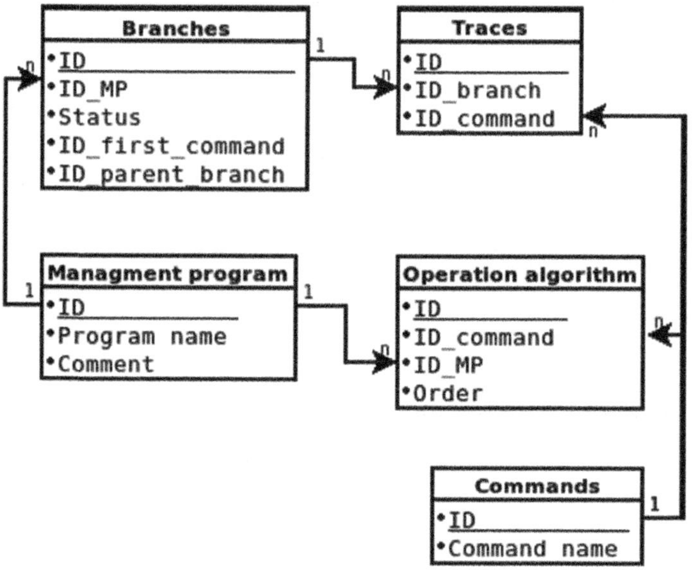

Fig. 2 Database scheme

This database works as follows:

The operation algorithm schedule records information about control programs or commands that are started during the operation of the unmanned vehicle traffic program.

Then, as a result of the algorithm operation, the branches formation commands are fixed up in the branch schedule and the branch identifiers are in the ID_Branch_progenitor that precede data, as available.

When all the branches are closed and the status field becomes equal to 1, the full-fledged variations control programs are generated in the track schedule, where a cyclogram might be designed, or transferred into an existing analyzer.

Consequently, the proposed database scheme allows analyzing and using information is obtained by the algorithm to the software developer.

This database scheme does not take into account the specific DBMS features and technologies that can be used to design the database. Therefore, it requires further refinement and can be considered only as a concept.

3 The Algorithm Optimization

It is necessary to take into account and monitor a number of criteria that lead to the UV operation failure in calculating the energy consumed.

If we take:

$P_{inst.}$ the maximum power value maintaining by the UV energy system;
$W_{inst.}$ the maximum amount of energy delivering by the UV source;
$W_{curr.}$ the current energy value providing by the UV;
P_i the *i*th active user power as the UV part;
W_i the energy consumption of the *i*th active user as the UV part.

When the power value calculates in the UV at time t, the conditions should be monitored:

$$\sum_{1}^{i} P_i(t) \leq P_y(t) \tag{1}$$

If the condition (1) is not fulfilled, it leads to the energy system overload and, probably, to the UV failure.

When all the source energy is wasted without the recharge possibility, it is always necessary to monitor the condition for the situation control:

$$W_{inst.} \leq W_{cur.}(t) \leq W_{in.}(t) - \sum_{1}^{i} W_i(t) \tag{2}$$

where $W_{in.}(t)$—the obtained energy on the UV board, if it is possible. Otherwise, $W_{in.}(t) = 0$.

Tracking the situation, when $W_{in.} < 0$, we can come to the conclusion that the energy source was completely discharged, which did not allow the UV to continue its work.

For all possible the MP tracks, it is necessary to maintain common compliance counters (1) and (2). If after the analysis, the counter value is greater than 0; the control program is declared unfit and can lead to failure or loss of communication with the UV. The $P_{inst.}$—value for MP calculation—is determined from MP section where conditions (1) or (2) are non-compliance. However, it may increase the cost or weight dimensions, because of the element base changes.

Another problem highlighted in [2, 5] is the implementation model of storing data about the MP tracks, which undoubtedly will lead to significant time delays in calculations $\sum_1^i P_i(t) \le P_y(t)$. At the stages of debugging the MP, it is suggested to resort to $\sum_1^i P_i(t)$ simplification.

Any methodology for simplification is a compromise between the program reliability and the obtaining speed of the analysis results. That's why, if you want to maintain high requirements, you need to search for other increasing productivity methods. During this process, you should identify the problem areas of the whole process (chokepoints), where the main delays occur. When PC works with a similar class of tasks, a good performance increasing is provided by document-oriented DBMSs. If the main problems arise during the special complex Ldebug, it is necessary to understand the operation principles and to use parallel processing or preprocessing of data (high-performance clusters).

While using simple analysis methods, it is suggested to consider the following simplifications:

1. When the program is formed, the priority is exposed (or use pre-prepared ranking) for tasks or subprograms. The actions that do not affect or affect insignificantly the device health are discarded based on priorities.
2. Before analysis, it should be determined the program proportion commands toward other commands using in the control program. Based on the Pareto law provisions (Pareto principle, 20/80 principle), it can be discarded commonly or rarely used commands.
3. If there are failure statistics (or the tool development that allows you to accumulate these statistics), the weak and strong programs/programmers places can be determined. Later, problem areas will have been focused on, and the points will be ignored, where problems usually do not arise.
4. Based on the statistics (average analysis time, analysis time), the desired time can be specified. Using one of the methods described above, the program in random order will discard all checks that are not comply with the time limit. The method can be used in the early development stages.

5. Analyzing the traces, if it is determined that conditions (1) or (2) are not satisfied, you should complete the analysis of this trace and put notes about its complete inapplicability and all traces unsuitability resulting from it.

The performance problems are quite obvious, when the information is searched in the DBMS with a large amount of data. In the initial stages, the problem can be reduced by the structure optimization and the usable data types, the index in use and the stored procedures on the database side, but further complications may require a transition from relational the DBMS to other types.

Acknowledgements We would like to thank, Sygurov Yury Mikhailovich, Head of 1401 sector, 1401 Department JSC SRC Progress for his great contribution to the development of some project aspects.

References

1. Frantasov, D.N., Melnikov, P.A., Klimas, A.S.: Tools for unmanned aerial vehicle control programs verification. Sci. Probl. **2**(26), 12–14 (2018)
2. Frantasov, D.N., Melnikov, P.A., Klimas, A.S.: The development and implementation algorithm for determining possible program ways of unmanned aerial vehicles control execution. In: International Scientific and Technical Conference "Advance Information Technology"—2018, vol. 1, pp. 899–902 (2018)
3. Frantasov, D.N., Melnikov, P.A., Klimas, A.S.: The development and implementation algorithm for determining possible program ways of unmanned aerial vehicles control execution. In: The International Scientific Conference of Disciples, Candidates, Students and Seniors Pupils. "The 21th Century International Science: Traditions, Innovations, Vectors of Development", vol. 1, pp. 82–83 (2018)
4. Frantasov, D.N., Melnikov, P.A., Klimas, A.S.: Optimization algorithm condition for determining possible program ways of unmanned aerial control execution. Sci. Probl. **5**(29), 22–23 (2018)
5. Filatov, A.V., Tkachenko, I.S., Tyugashev, A.A., Sopchenko, E.V.: The control system software algorithms and structure of small space vehicles. In: CEUR Seminar Proceeding, vol. 1490, pp. 246–251 (2015)

Kinect Sensor-Based Trajectory Planning Method of Collision Avoidance for Industrial Manipulator with an Dexterous Hand

Xingchen Chen, Nanfeng Xiao and Ya Chao

Abstract This paper presents Kinect sensor-based trajectory planning method of collision avoidance for an industrial manipulator with multi-fingered dexterous hand, which is used for grasping objects by using a data glove in dynamically changing environments. Each finger's joint angles of the dexterous hand are synchronously collected by the data glove and forwarded to the trajectory planning control system that navigates the dexterous hand to the grasping poses. In order to effectively calculate the solutions to the desired configuration of the manipulator with the dexterous hand, the LazyPRM algorithm is also applied to maintain a road map for answering multiple planning queries. To maintain the dynamically changing environments, the Kinect sensor is used to detect and track the moving objects that appear in the workspace and to update the environment representation to avoid the collisions. The distances between the obstacles and the camera are obtained by the depth camera of the Kinect sensor. The experimental results have verified the effectiveness of the methods.

Keywords Trajectory planning · Dexterous hand · Industrial manipulator · Object grasping · Obstacle avoidance

1 Introduction

At present, the application fields of the multi-fingered dexterous hands have been increasingly expanding by virtue of the availability and dexterity in performing a large amount of repetitive, tedious, and even dangerous tasks, which require high precision in comparison with the conventional industrial manipulator equipped with a gripper that achieves object grasping in an inflexible way. The object grasping is one of the hot spots in the field of robotics. However, the computation-intensive

X. Chen (✉) · N. Xiao · Y. Chao
School of Computer Science and Technology, South China University of Technology, Guangzhou, China
e-mail: 393171302@qq.com

V. Jain et al. (eds.), *Recent Trends in Intelligent Computing, Communication and Devices*, Advances in Intelligent Systems and Computing 1031,
https://doi.org/10.1007/978-981-13-9406-5_83

grasp planning methods which exhaustively search the entire high-dimensional state spaces limit the application in the dynamically changing environments.

Path planning in robotics seeks to find a solution to the problem of breaking down the desired movement task into discrete motions which satisfies all of the robot's constraints, the methods of path planning can be divided into many categories [1]. It has been shown that navigating a robot with a large number of degrees of freedom from its starting configuration to the goal configuration while avoiding all the obstacles is a P-SPACE-complete problem [2]. In contrast with the traditional state-of-the-art probabilistic complete approaches [3–5] due to its poor performance in reasoning over the entire high-dimensional state spaces, the sampling-based probabilistic road map methods [6, 7] which employ sampling of workspace of the robot are able to answer planning queries efficiently in a shorter time. It arises out of the need to construct a road map that connects the start and goal configuration [8] with a feasible and collision-free path based on the complicated and large state spaces. Except for the static stationary obstacles that already exist in the environment, the dynamic obstacles that appear in the workspace when performing the task can be detected and tracked using the Kinect sensors. There are many works on the vision-based collision avoidance of industrial manipulator which has been done in [9, 10] and the obstacle detection in [11, 12] and the frame difference method are used in the system due to the high efficiency and stability, and there are a lot of works which have been done in [13–15]. However, there are few works on the collision avoidance of the manipulator with multi-fingered dexterous hand.

In the present study, we developed a trajectory planning method to perform the object grasping tasks of the 6-DOF industrial manipulators with multi-fingered dexterous hands in a dynamically changing environment. The system mainly consists of three different modules: ① the kinematics solver; ② the planner; and ③ the object grasping module. The kinematics solver simply calculates the degrees of joints of the manipulator according to the pose matrix of the end effector which is the origin of dexterous hand's base reference frame. The path planning module answer queries from start configuration to the grasping configuration calculated beforehand by kinematics solver, it tries to find the solutions using multi-query sampling-based path planner, and the dynamic obstacles are captured by the Kinect sensor. When the collision-free path is found, the trajectory will be obtained using the quantic polynomial method which guarantees a smooth trajectory. The object grasping module is a master–slave system in which the grasping is achieved by a dexterous hand operated by a human operator via the data glove. The resulting experiment shows that the proposed method satisfies real-time requirements and exhibits robustness and satisfactory performance.

2 System Overview

2.1 System Architecture

The method we developed which is shown in Fig. 1 mainly consists of three parts. The kinematics solver receives the grasping pose of the manipulator and calculates the inverse solutions. There are eight close-form solutions we can be obtained from the initial configuration to the goal configuration, and it will be then sent to the planner which will perform validity checking on each candidate path and discard the disqualified ones, and select the best one with the least movement costs. The planner answers path planning queries according to the optimal inverse solution from the kinematics solver. In order to efficiently answer the multiple planning queries, the planner will first construct a road map upon the state space of the manipulator, and the validity check functions as the collision detector when sampling over the state spaces. The moving obstacle tracker captures the dynamic obstacle that appears in the workspace, then the 3D position will be obtained through a depth camera of the Kinect sensor, and the environment representation will be updated. Lastly, when the manipulator reaches the goal state, the grasping tasks can be performed. The object grasping module is a master–slave system in which the human operator is in charged to control the multi-fingered dexterous hand through the interface of the data glove.

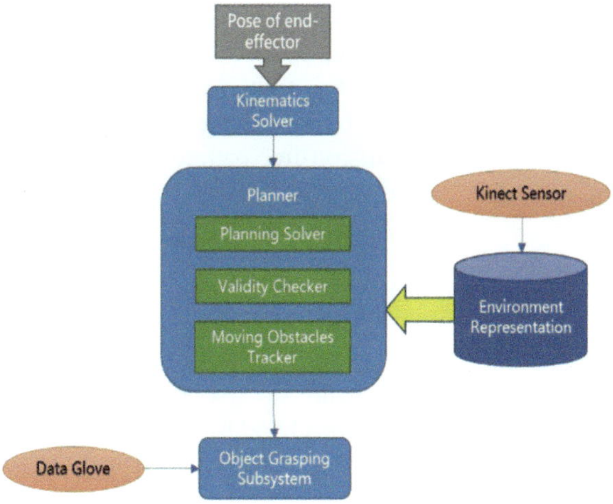

Fig. 1 System architecture

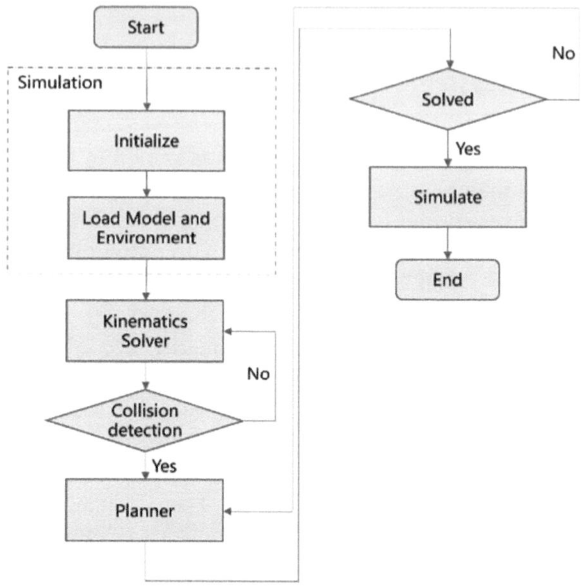

Fig. 2 System flow of the planner

2.2 *System Flow*

Figure 2 shows the system flow of the path planner in the simulation. The simulation system is developed using C++ and OpenGL graphics engine. The model of the manipulator and the dexterous hand is built in the 3Ds Max. Firstly, when the simulation system is initialized, the model of the manipulator and the dexterous hand will be imported and loaded. Secondly, the kinematics solver then receives the goal grasping pose in the form of the homogeneous matrix and calculates the inverse solutions and check if the goal pose is collision-free. Finally, the planner will search for the collision-free path to reach the goal pose if the goal pose is valid.

3 Object Grasping

In this section, we discuss the object grasping of the dexterous hand using the data glove. The dexterous hand has five fingers which are the same as the dexterous hand, to drive the dexterous hand by the data glove, the effective mapping approach of the master–slave motion should be implemented, which means that every joint of the data glove will be mapped correspondingly to the joint of the dexterous hand. As shown in Fig. 3a. The thumb has four joint, while other fingers have three joints, and each angle of the joint can be measured by the sensor of the data glove. For the accuracy of the angle mapping, the data glove should be calibrated in the first place, the 5DT SDKs provide the APIs to help the users to easily finish the calibration by

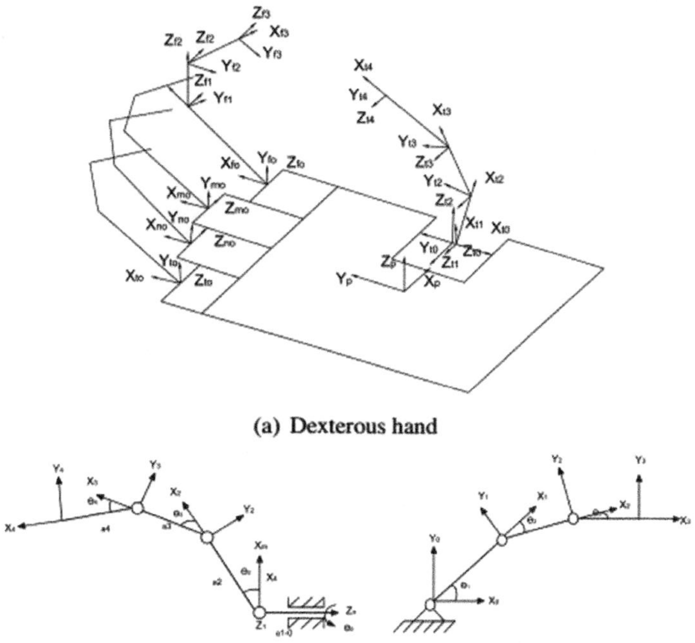

(a) Dexterous hand

(b) Thumb (c) Other Fingers

Fig. 3 Structure of the dexterous hand

fisting and flattening which diminishes the difference between the maximum raw value $\mathrm{raw}_{\max}$ and the minimum raw value $\mathrm{raw}_{\min}$.

When the calibration is done, the calibrated value can be calculated by the linear interpolation:

$$\mathrm{output} = \mathrm{maxval}\left(\frac{\mathrm{raw}_{\mathrm{cur}} - \mathrm{raw}_{\min}}{\mathrm{raw}_{\max} - \mathrm{raw}_{\min}}\right) \tag{1}$$

where the maxval denotes the maximum value that can be obtained by the sensor, and $\mathrm{raw}_{\mathrm{cur}}$ the denotes the current raw value measured by the sensor. Therefore, when the manipulator reaches the grasping pose navigated by the planner, the grasping tasks can be performed by the human operators. The object grasping module is verified in the simulation system developed using C++ and OpenGL graphics engine, which can be seen in Fig. 4, and the dexterous hand in the simulation system is controlled by the human operator.

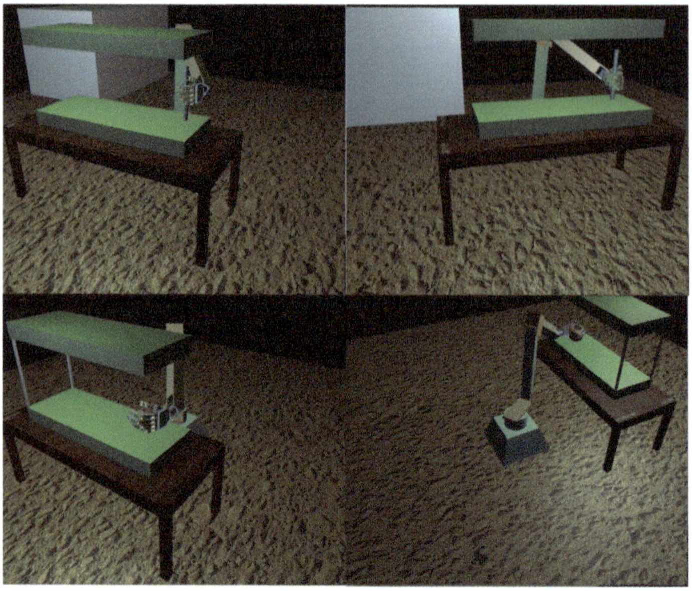

Fig. 4 Object grasping using the data glove

4 Trajectory Planner

4.1 Validity Checker

The validity checker performs the collision detection on the given configuration of the industrial manipulator. If the collision or the intersection of the bounding boxes of any pair of the object occurs, the validity check fails and the configuration will be considered as invalid. In order to efficiently performing the intersection test and reducing the computation on the unnecessary testing pairs, an octree is implemented in the space partition which performs the coarse-grained test in the first place and the fine-grained ones to determine which pairs are exactly colliding. In the case of the dynamic environment, the space partitioned by the octree should be encoded for the dynamic objects, and the encoding method can be seen in [16]. The collision check method is based on the AABB bounding box since it is faster than the intersection test per triangle mesh and still keeps an acceptable performance.

4.2 Planner

The state spaces of the industrial manipulator that the planner constructs are six-dimensional representing each arm joint, while the remaining 15 joints of the dexterous hand are not involved since the finger motion range is relatively small, and thus, it has much few chances to collide with the environment comparing to the arm, most importantly, it will result in huge state spaces and the computation costs.

Before the trajectory planner is able to properly work, the state spaces must be specified and correctly configured. As a matter of fact, state space construction is achieved offline, while the solution calculations are achieved online. We have implemented the LazyPRM algorithm which builds a road map based on the state spaces in the offline computation procedure and reduces the size of the state spaces that must be searched entirely. In order to accelerate the online computation and reduce the time cost upon the unnecessary intersection check, it postpones the road map updating when the trajectory is found invalid.

4.3 Dynamic Obstacle Detector

In order to detect the dynamic obstacles in the environments, the frame difference method is used to capture the changed pixels between the frames of the camera, and the Kinect depth camera provides the ability to obtain the depth information of each pixel, which can be transformed into world space to get the 3D point clouds of the dynamic obstacle. In order to get the different pixel in each frame, we should first convert RGB foreground and the background frames into the gray frames and calculate the subtraction

$$\delta(i,j) = |q(i,j) - b(i,j)| \tag{2}$$

where $q(i,j)$ denotes the pixel in the foreground frame, while $b(i,j)$ denotes the pixel in the background frame, and therefore, the $\partial(i,j)$ is the absolute value of the subtraction. Next, we calculate the sum of all the different pixels

$$\text{sum} = \sum_{i=1}^{\text{width}} \sum_{j=1}^{\text{height}} \delta(i,j) \tag{3}$$

If the sum $\geq$ threshold, the pixel can be considered as part of the dynamic obstacle. When performing the path planning task, the manipulator also will be considered as dynamically moving objects, and therefore, the pixel of manipulator must be removed. The current 3D points of the manipulator model and the dexterous hand can be obtained by calculating the transformation from the initial configuration to the current configuration.

The dynamic obstacle is detected and its point clouds are obtained via the depth camera of the Kinect sensor which can be seen in Fig. 5. Therefore, the representation of the environment is then updated. The manipulator is thus able to receive the newest information about the environment and is able to find the collision-free path to the desired configuration. Table 1 shows the planner of the manipulator answer planning query, the collision-free path solved here contains

Fig. 5 Detection of the dynamic obstacle

Table 1 Trajectory states

θ No.	θ_1	θ_2	θ_3	θ_4	θ_5	θ_6
0	0	0	0	0	0	0
1	107.5	−69.2	109.4	86.3	−70.9	80.9
2	113.7	−12.3	133.6	56.7	−32.3	90.3
3	87.4	−138.6	116.1	0	−86.4	180

four states, the state 0 and the state 4 are the initial state and the goal state, respectively, and the state 2 and the state 3 are the intermediate state that the planner found. Therefore, the smooth trajectory can be obtained using the quantic polynomial method which can be seen in Fig. 6.

5 Conclusions

The proposed method of the industrial manipulator with the multi-fingered dexterous hand proves to be effective and feasible, and it is a promising method for the manipulators with the dexterous hands to perform various object grasping in the dynamic environments. The LazyPRM algorithm is an efficient planning method used in the multiple query tasks which have a good performance in the real-time scenarios. The dynamic obstacles detection works quite well, and it is able to detect most of the opaque objects even in the drastically changing environment whose representation can be constantly updated to gain the current information of the

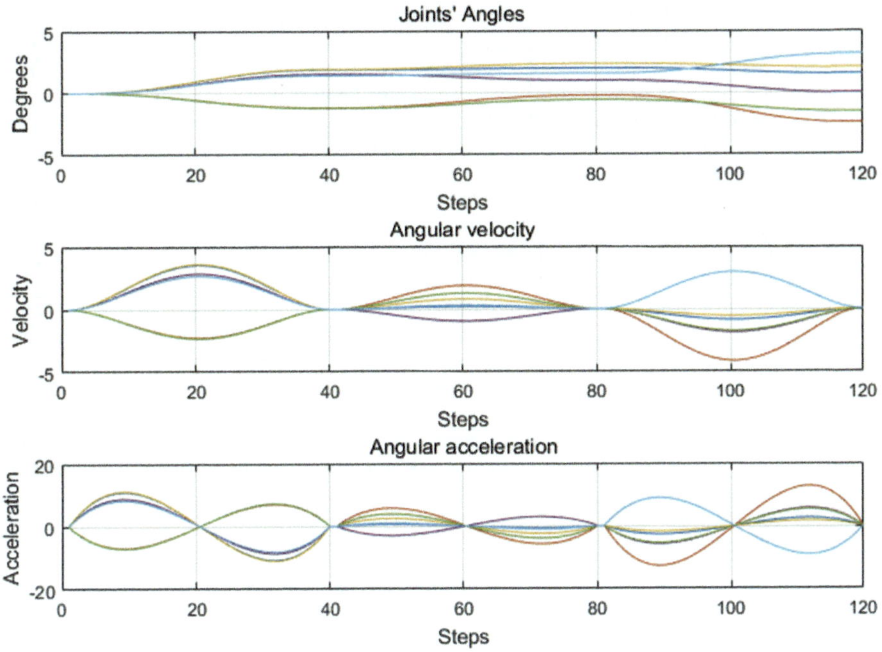

Fig. 6 Trajectory of the manipulator

surroundings. The two 5DT data gloves used in the object grasping tasks exhibit good effectiveness, and the angle mapping of the data gloves have little errors.

Acknowledgements This research is funded by the national natural science foundation (project No. 61573145), the public research and capacity building of Guangdong province (project No. 2014B010104001) and the basic and applied basic research of Guangdong province (project No. 2015A030308018), and the authors are greatly thanks to these grants.

References

1. Kavraki, L., Svestka, P., Latombe, J., et al.: Probabilistic road maps for path planning in high-dimensional configuration spaces. IEEE Trans. Robot. Autom. **12**(4), 566–580 (1994)
2. Geraerts, R., Overmars, M.H.: A comparative study of probabilistic road map planners. Springer Tracts Adv. Robot. **7**, 43–57 (2002)
3. Lozano-Pérez, T.: Spatial planning: a configuration space approach. In: Autonomous Robot Vehicles, pp. 108–120. Springer-Verlag New York, Inc. (1990)
4. Sukmanee, W., Ruchanurucks, M., Rakprayoon, P.: Obstacle modeling for manipulator using iterative least square (ILS) and iterative closest point (ICP) base on Kinect. In: IEEE International Conference on Robotics and Biomimetics, pp. 672–676. IEEE (2013)

5. Rakprayoon, P., Ruchanurucks, M., Coundoul, A.: Kinect-based obstacle detection for manipulator. In: IEEE/SICE International Symposium on System Integration, pp. 68–73. IEEE (2011)
6. Khan, S., Shah, M.: Object based segmentation of video using color, motion and spatial information. In: Proceedings of the 2001 IEEE Computer Society Conference on Computer Vision and Pattern Recognition, 2001. CVPR 2001, vol. 2, pp. II-746–II-751. IEEE (2001)
7. Rusu, R.B., Holzbach, A., Beetz, M., et al.: Detecting and segmenting objects for mobile manipulation. In: IEEE, International Conference on Computer Vision Workshops, pp. 47–54. IEEE (2009)
8. Zhan, C., Duan, X., Xu, S., et al.: An improved moving object detection algorithm based on frame difference and edge detection. In: International Conference on Image and Graphics, pp. 519–523. IEEE Computer Society (2007)
9. Ying, W., Yi, H., Han, L.U.: The methods for moving object detection. Comput. Simul. (2006)
10. Qiu, S.B., Gang, L.I., Lin, L.: New method of motion detection and background updating. J. Liaoning Inst. Technol. (2005)
11. Dai, K.X., Gou-Hui, L.I., Dan, T.U., et al.: Prospects and current studies on background subtraction techniques for moving objects detection from surveillance video. J. Image Graph. **11**(7), 919–927 (2006)
12. Kitamura, Y., Kishino, F., Tanaka, T., et al.: 3-D path planning in a dynamical environment using an octree and an artificial potential field. In: International Conference on Intelligent Robots and Systems, p. 2474. IEEE Computer Society (1995)
13. Bohlin, R., Kavraki, L.E.: Path planning using lazy PRM. In: IEEE International Conference on Robotics and Automation, vol. 1, pp. 521–528
14. Zhao, Z., Liu, Y., Zhang, Z.: Camera calibration with three noncollinear points under special motions. IEEE Trans. Image Process. **17**(12), 2393 (2008)
15. Sucan, I.A., Moll, M., Kavraki, L.E.: The open motion planning library. IEEE Robot. Autom. Mag. **19**(4), 72–82 (2012)
16. Meagher, D.: Geometric modeling using octree encoding. Comput. Graph. Image Process. **19** (2), 129–147 (1982)

NextGeneration Comm. and Networking

Application Status of Right-/Left-Handed Transmission Line in Microwave Phase Shifter

Hui-yong Zeng, Bin-feng Zong, Yan Zhao, Tong Xu and Jian An

Abstract Researches on theory and applications of RLH TL have been widely applied in microwave technique field, especially in antennas and feed network systems. With the property of wideband phase shifting, the latest researches at home and abroad are summarized of phase shifters based on right-/left-handed TL (transmission line). Two kinds of research statuses are deeply analyzed. The advantages and disadvantages of different methods are compared. The trend of phase shifter based on right/left-handed transmission line is given at the end of the paper.

Keywords Right-/left-handed TL · Phase shifter · Property of wideband phase shifting

1 Introduction

Phase shifter is an important microwave device and has a wide range of applications in the beamforming network, phase modulator, phased array antenna, and other wireless communication systems. Investigation on broadband planar phase shifter is a very meaningful work. The traditional differential phase shifter is dependent on the length difference between the two transmission lines to achieve phase shift or in

---

H. Zeng · T. Xu
Air and Missile-Defence College, Air Force Engineering University, Xi'an 710051, China

B. Zong
Unit 94710 of the PLA, Wuxi 214000, China

Y. Zhao
Air Force's Military Representative Office in Jilin, Jilin, China

J. An (✉)
Beijing Space Information Relay Transmission Technology Research Center, Beijing 102300, China
e-mail: hyzeng.1023@163.com

V. Jain et al. (eds.), *Recent Trends in Intelligent Computing, Communication and Devices*, Advances in Intelligent Systems and Computing 1031,
https://doi.org/10.1007/978-981-13-9406-5_84

the case of the same length by changing the transmission line propagation constant to achieve phase shift.

Because the right-/left-handed TLs (transmission lines) have the characteristics of broadband phase shifting, many researchers have applied it to design broadband phase shifters [1–10], many of which combine broadband phase shift line and power splitter to design broadband orthogonal power splitter and Wilkinson balun. Broadband phase shifter based on right-/left-handed TLs (transmission lines) can be divided into lumped components [1– 4] and distributed parameter mode [5 – 10].

2 Lumped Components Mode

Broadband 180° phase shifter based on right-/left-handed TLs using lumped components is designed [1], as shown in Fig. 1. The upper part of the right-/left-handed TL is a phase shift line of +90°, and the following section of the right-/left-handed TL is a phase shift line of −90°. In the frequency range 1.17–2.33 GHz (relative bandwidth of 77%), the phase difference is 180° ± 10°; 180° switch line phase shifter using coplanar waveguide multilayer structure is designed [2], as shown in Fig. 2. In the frequency range 2–3.6 GHz, the phase difference is 180° ± 7°. Broadband phase shifter using right-/left-handed TL and traditional TL is designed [3], as shown in Fig. 3. In the frequency range 1.24–3.58 GHz, the phase difference is 180° ± 10°. In [4], a wideband phase shifter using right-/left-handed TL with lumped element was designed, as shown in Fig. 4. In the frequency range 0.6 – 1.2 GHz, the phase difference is 180° ± 9.5°.

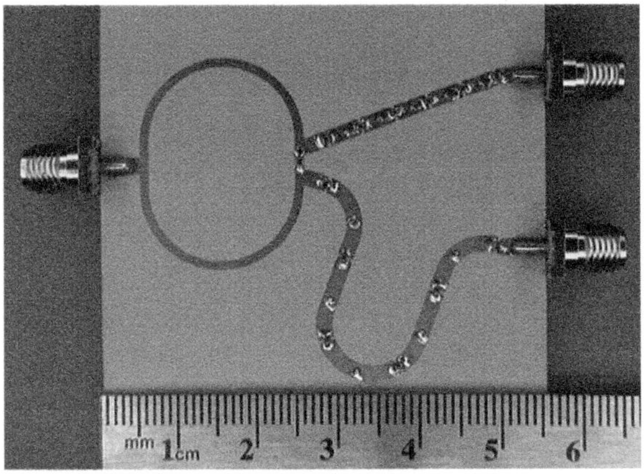

Fig. 1 Wideband phase shifter reported in the literature [1]

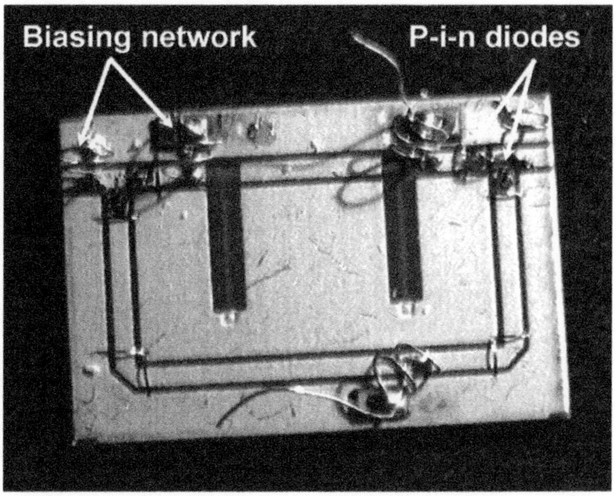

Fig. 2 Wideband phase shifter reported in the literature [2]

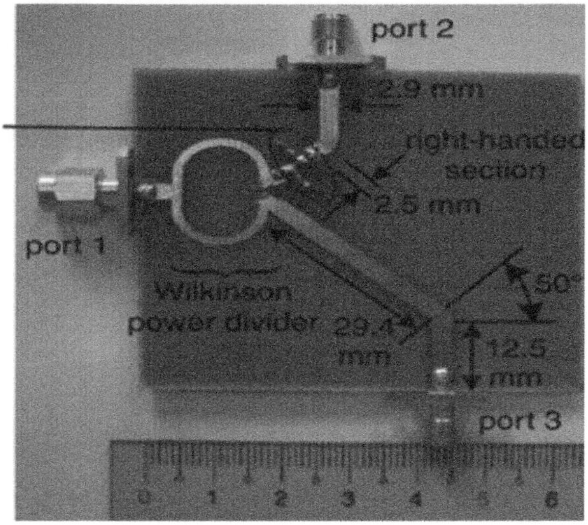

Fig. 3 Wideband phase shifter reported in the literature [3]

3 Distributed Parameter Mode

In the literature [5], broadband phase shifter using distributed right-/left-handed TL and coplanar waveguide technology is designed, as shown in Fig. 5. The phase difference is 180° ± 10° in the frequency range of 2.4–5.22 GHz. In the literature [6], broadband 90° phase shifter using right-/left-handed TL consisting of micro-strip interdigital gaps and grounded via holes is proposed. The phase difference is 90° ± 10° in the frequency range of 2.3–3.8 GHz. Broadband phase shifter using right/-left-handed TL based on complementary split-ring resonators (CSRRs) is designed [7], as shown in Fig. 6. The phase difference is 90° ± 5° in the frequency

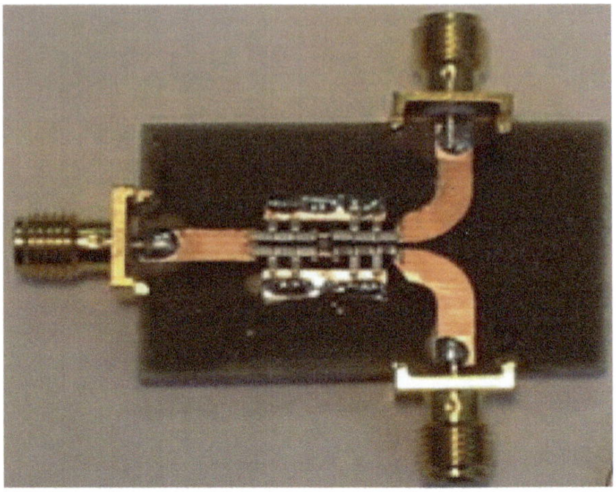

Fig. 4 Wideband phase shifter reported in the literature [4]

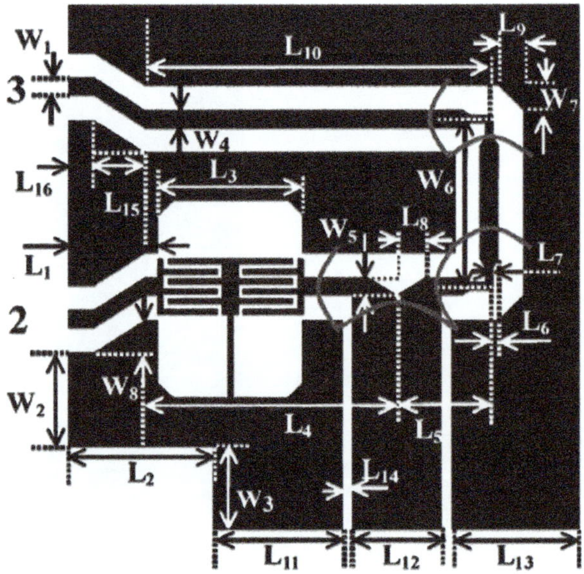

Fig. 5 Wideband phase shifter reported in the literature [5]

range of 1.4 ~ 2.1 GHz. In the literature [8], broadband 180° phase shifter using right-/left-handed TL consisting of microstrip interdigital gaps and grounded via holes is proposed, as shown in Fig. 7. The phase difference is 180° ± 3° in the frequency range of 1.3–3.5 GHz. In the literature [9], broadband 180° phase shifter is designed with odd- and even-mode method, which uses a pair of the same right-/left-handed TL consisting of interdigital gaps and virtual ground, as shown in Fig. 8. The phase difference is 180° ± 3° in the frequency range of 1.6 –3.6 GHz.

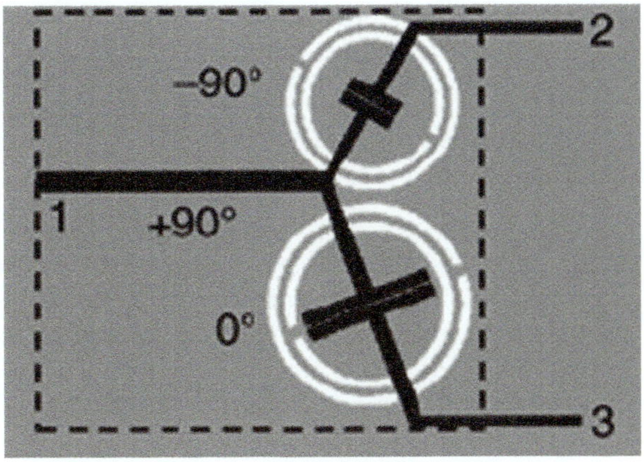

Fig. 6 Wideband phase shifter reported in the literature [7]

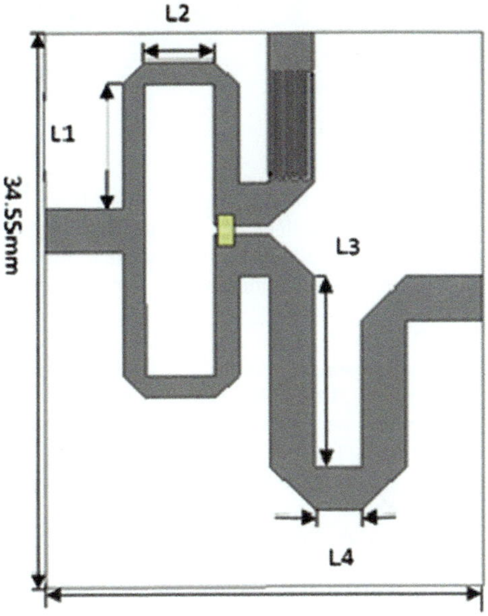

Fig. 7 Wideband phase shifter reported in the literature [8]

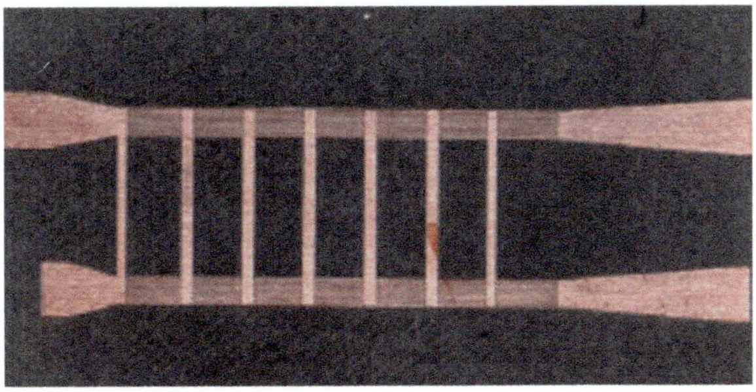

Fig. 8 Wideband phase shifter reported in the literature [9]

In the literature [10], a broadband 4-bit digital phase shifter using interdigital gaps and grounded via holes is proposed.

4 Conclusion

In this paper, the latest research results of microwave phase shifters based on left-/right-handed TLs have been reported at home and abroad, and the latest research results are summarized. The research status of microwave phase shifters based on left-/right-handed TLs is analyzed in depth. The advantages and disadvantages of different methods are compared.

Acknowledgements The paper was supported by the National Natural Science Foundation of China (Grant No. 61701527, 61601499) and the National Natural Science Foundation of Shaanxi Province (Grant No. 2019JQ-583, 2018JQ6023).

References

1. Antoniades, M.A., Eleftheriades, G.V.: A broadband Wilkinson balun using microstrip metamaterial lines. IEEE Antennas Wirel. Propag. Lett. **4**, 209–212 (2005)
2. Dmitry, K., Elena, S., Irina, V.: Broadband digital phase shifter based on switchable right-and left-handed transmission line sections. IEEE Microwave Wirel. Compon. Lett. **16**, 258–260 (2006)
3. Tseng, C.H., Chang, C.L.: Wide-band balun using composite right/ left-handed transmission line. Electron. Lett. **43**, 1154–1155 (2007)
4. Yoo, H., Lee, S.H., Kim, H.: Broadband balun for monolithic microwave integrated circuit application. Microwave Opt. Technol. Lett. **54**, 203–206 (2012)
5. Mao, S.G., Chueh, Y.Z.: Broadband composite right/left-handed coplanar waveguide power splitters with arbitrary phase responses and balun and antenna applications. IEEE Trans. Antennas Propag. **54**, 243–250 (2006)
6. Zou, Y.Z., Lin, Z.L., Ling, T.: A new broadband differential phase shifter fabricated using a novel CRLH structure. J. Zhejiang Univ. Sci. A **8**, 1568–1572 (2007)
7. Siso′, G., Gil, M., Bonache, J.: "Application of metamaterial transmission lines to design of quadrature phase shifters,". Electron. Lett. **43**, pp. 1098–1100 (2007)
8. Cao, W.P., Guo, F., Wang, B.Z.: Design of a broadband balun based on composite right/left handed structure. Microwave Opt. Technol. Lett. **52**, 1310–1313 (2010)
9. Liu, C.J., Menzel, W.: Broadband via-free microstrip balun using metamaterial transmission lines. IEEE Microwave Wirel. Compon. Lett. **18**, 437–439 (2008)
10. Zhu, Q., Gong, C., Xin, H.: Design of high power capacity phase shifter with composite right/ left-handed transmission line. Microwave Opt. Technol. Lett. **54**, 119–124 (2012)

Developing Trends and Recent Research of Dual-Band Planar Antenna

Jian An, Hui-yong Zeng, Bin-feng Zong, Yan Zhao and Juan Bai

Abstract The appearance of dual-band/multi-band antenna cells meets the diverse needs of the modern wireless communication function. Dual-band/multi-band antenna cells have more advantages and better prospect compared with conventional antennas. The latest researches at home and abroad are summarized of dual-band antenna cells. Multiple kinds of research statuses are deeply analyzed. The advantages and disadvantages of different methods are compared. The trend of dual-band/multi-band antenna cells' development is given at the end of the paper.

Keywords Dual-band/multi-band antenna · Planar antenna cells · Wireless communications

1 Introduction

Antenna is the device that emits and receives electromagnetic waves, which is at the forefront of microwave wireless communication and detection system, and its performance has a very important influence on the whole system. With the rapid development of science and technology, the demand for information in the modern society is increasing, which directly promotes the development of communication system to broadband. On the one hand, with the development of electronic technology and the emergence of broadband communications equipment, broadband

J. An
Beijing Space Information Relay Transmission
Technology Research Center, Beijing 102300, China

H. Zeng (✉) · J. Bai
Air and Missile-Defence College, Air Force Engineering University, Xi'an 710051, China
e-mail: hyzeng.1023@163.com

B. Zong
Unit 94710 of the PLA, Wuxi 214000, China

Y. Zhao
Air Force's Military Representative Office, Jilin, China

V. Jain et al. (eds.), *Recent Trends in Intelligent Computing, Communication and Devices*, Advances in Intelligent Systems and Computing 1031,
https://doi.org/10.1007/978-981-13-9406-5_85

antenna and antenna broadband technology are also constantly evolving. On the other hand, in order to expand the system capacity or to achieve multi-mode communication, the actual communication system often works in two or more frequency points; in the case of improper installation of two or more antennas, dual-/multi-frequency antenna was born.

Dual/multi-frequency antenna refers to two or more different frequency bands can meet the performance requirements of the antenna system, specific indicators comprising: VSWR, gain, radiation pattern and efficiency. In fact, most of the broadband antenna and multi-frequency antenna are not much different in nature. When the antenna's multiple resonant frequency fusion occurs, it is considered a broadband antenna. On the contrary, when the resonant frequency is far away, it constitutes a dual-/multi-frequency antenna. One of the advantages of dual-/multi-frequency antenna compared to broadband antenna is to suppress interference of different frequencies.

In the study of dual-/multi-frequency antenna elements, the researchers proposed a variety of methods; traditional methods mainly include branches [1–5], slot-loading method [6–8], short circuit [9–11], and parasitic patch method [12, 13]. And new implementation methods mainly include right-/left-handed transmission lines.

2 Research of Dual-/Multi-band Antenna Units

Branch method is a more intuitive way, equivalent to multiple antennas together and fed by one port. And this method mainly used in printed monopole antenna.

In the Literature [1], a dual-frequency printed monopole antenna operating on a wireless local area network is designed. The two unipolar poles work in different frequency bands and are connected to the central conduction band of the coplanar waveguide, as shown in Fig. 1. Literature [2] designed a double T-type dual-frequency printing monopole antenna, as shown in Fig. 2, began to share a single unipolar microstrip line, and then separated by the length of the different. Other letter-like structures such as G-type, L-type, and F-type [3–5] can achieve dual-frequency work.

Slot-loading method is to remove part of the metal surface of the antenna to change the surface current distribution of the antenna. The introduction of new resonance, in order to achieve the dual/multi-frequency antenna work. This method is mainly used in the microstrip antenna applications. Figure 3 shows a double-slot rectangular microstrip antenna. Adjusting the length of the two slots can change the resonant frequency in a certain range of two operating frequency modes [6]. Figure 4 is a dual-band rectangular microstrip antenna. The π-shaped slot introduces a new mode between the TM10 and TM20 modes to achieve dual-frequency characteristics [7].

The basic principle of the short-circuit method is similar to that of the slot-loading method. The short circuit is carried out at one or more of the microstrip

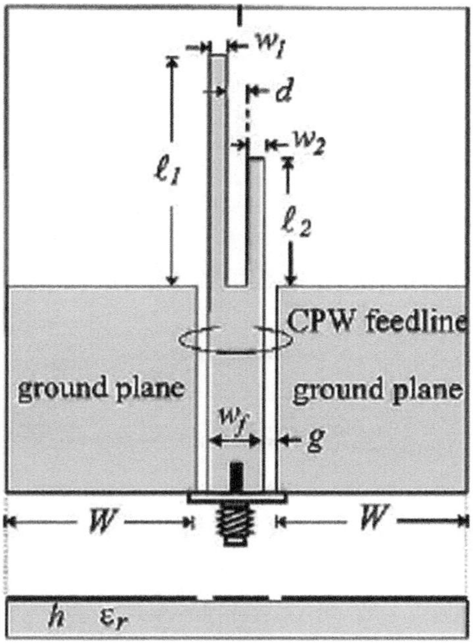

Fig. 1 Literature [1] reported dual-band antenna

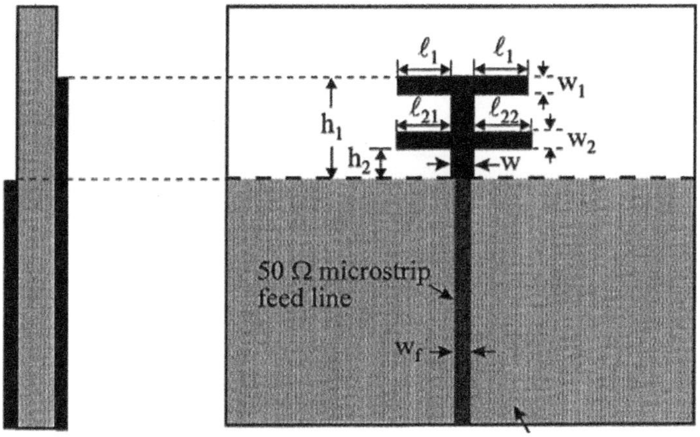

Fig. 2 Literature [2] reported dual-band antenna

patches. The purpose is to change the resonant frequency of the fundamental mode or the high-order mode. As shown in Fig. 5 is a rectangular dual-band microstrip antenna, the short-circuit probe position, and the location of the feed point on the same axis of symmetry. Adjusting the short position of the probe can change the ratio of two frequencies in a certain range.

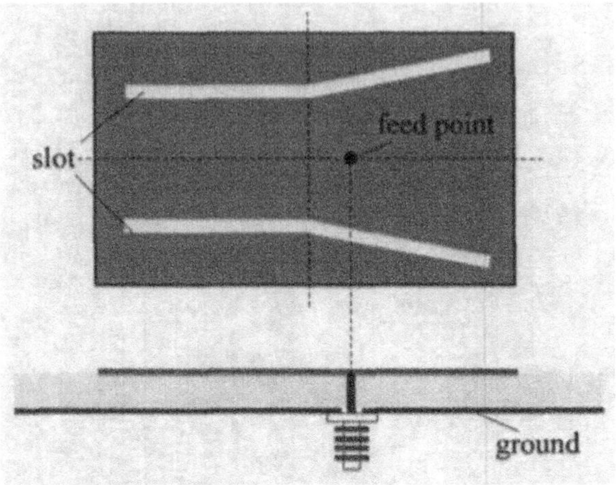

Fig. 3 Literature [6] reported dual-band antenna

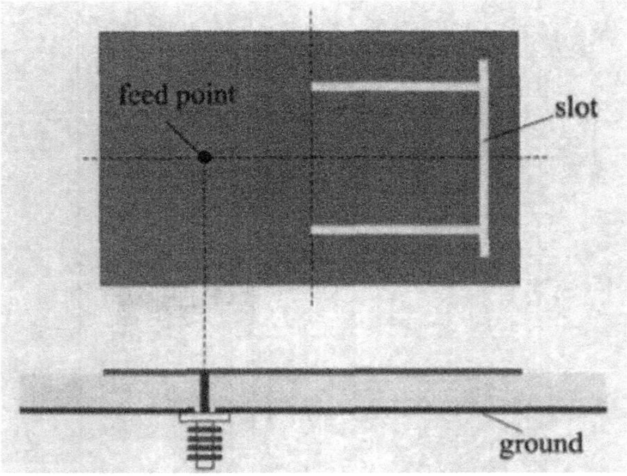

Fig. 4 Literature [7] reported dual-band antenna

Parasitic patch method is to add one or more parasitic patches in the microstrip patch above or around. Parasitic patches produce one or more new resonant so as to achieve dual-/multi-frequency work. Figure 6 is a square microstrip patch antenna, and fractal structures of the parasitic patch are added in its surrounding. The antenna can work in multiple frequency bands [12]. The disadvantage of the parasitic patch method is that it will make the antenna structure more complicated.

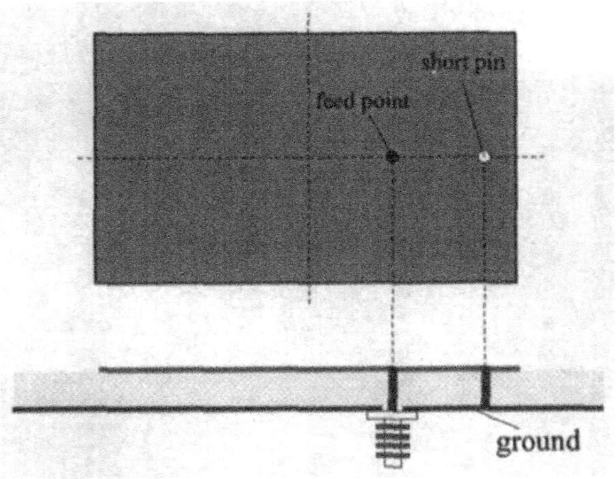

Fig. 5 Literature [9] reported dual-band antenna

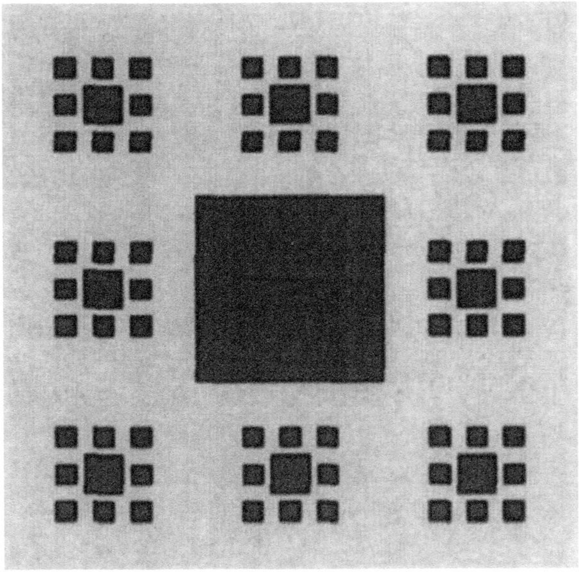

Fig. 6 Literature [12] reported dual-band antenna

3 Conclusion

In this paper, the realization of dual-/multi-frequency antenna cells has been summarized, and the latest research results on dual-/multi-band antenna cells in recent years are summarized. The research status of dual-/multi-band antenna cells in various implementations is summarized. The advantages and disadvantages of different methods are compared.

Acknowledgements The paper was supported by the National Natural Science Foundation of China (Grant No. 61701527, 61601499) and the National Natural Science Foundation of Shaanxi Province (Grant No. 2019JQ-583, 2018JQ6023).

References

1. Chen, H.D., Chen, H.T.: A CPW-fed dual-frequency monopole antenna. IEEE Trans. Antennas Propag. **52**, 978–982 (2004)
2. Kuo, Y.L., Wong, K.L.: Printed double-T monopole antenna for 2.4/5.2 GHz dual-band WLAN operations. IEEE Trans. Antennas Propag. **51**, 2187–2192 (2003)
3. Yeh, S.H., Wong, K.L.: Dual-band F-shaped monopole antenna for 2.4/5.2 GHz WLAN application. In: Proceedings of IEEE Antennas and Propagation Society International Symposium, San Antonio, TX, pp. 72–75 (2002)
4. Lin, Y.H., Chen, H.D., Chen, H.M.: A dual-band printed L-shaped monopole for WLAN applications. Microwave Opt. Technol. Lett. **45**, 214–216 (2003)
5. Kim, T.H., Park, D.C.: CPW-fed compact monopole antenna for dual-band WLAN applications. Electron. Lett. **41**, 291–293 (2005)
6. Lu, J.H.: Dual-frequency operation of rectangular microstrip antenna with bent-slot loading. In: Proceedings of Asia Pacific Microwave Conference, Piscataway, NJ, pp. 33–36 (2000)
7. Chen, H.M.: Single-feed dual-frequency rectangular microstrip antenna with a π-shaped slot. IEE Proc. Microwave Antennas Propag. **148**, 60–64 (2001)
8. Gao, S.C., Li, L.W., Leong, M.S.: FDTD analysis of a slot-loaded meandered rectangular patch antenna for dual-frequency operation. IEE Proc. Microwave Antennas Propag. **148**, 65–71 (2001)
9. Pan, S.C., Wong, K.L.: Design of dual-frequency microstrip antennas using a shorting-pin loading. In: Proceedings of IEEE Antennas and Propagation Society International Symposium, Ailanta, Georagia, pp. 312–315 (1998)
10. Guo, Y.X., Luk, K.M., Lee, K.F.: Dual-band slot-loaded short-circuited patch antenna. In: Proceedings of IEEE Antennas and Propagation Society International Symposium, Salt Lake, UT, pp. 1592–1595 (2000)
11. Pan, S.C., Wong, K.L.: Dual-frequeney triangular microstrip antenna with a shorting pin. IEEE Trans. Antennas Propag. **45**, 1889–1997 (1997)
12. Du, Z., Gong, K.: Analysis of microstrip fractal patch antenna for multi-band communication. Electron. Lett. **37**, 805–806 (2001)
13. Sanad, M.: A compact dual-broadband microstrip antenna having both stacked and planar parasitic elements. In: Proceedings of IEEE Antennas and Propagation Society International Symposium, Piseataway, NJ, pp. 6–9 (1996)

Research on Signal Processing of Mechanical Vibration Based on Time–Frequency Analysis Method

Linsen Du, Hongli Liu, Shuai Li and Zhisheng Dong

Abstract The time–frequency analysis method is applied to the signal processing of mechanical vibration in this paper, in order to analyze and research the rules and characteristics of mechanical vibration. The mechanical is a nonlinear system and mechanical vibration is an unsteady process. Frequency-domain analysis method based on the technology of FFT is insufficient to research on the nonlinear and unsteady system; because, the Fourier transform theory is only suitable for dealing with the linear and time-invariant system. Time–frequency analysis is an effective technology for the processing of unsteady signals. The time–frequency–energy figures obtained from Hilbert–Huang transform and short-time Fourier transform method could show frequency-domain information clearly with the local time-during states, which will describe the dynamic behaviors of the mechanical system accurately. Combining with the characteristics of a mechanical vibration signal, in this paper, the STFT and HHT time–frequency analysis methods are used to compare with analyzing the mechanical vibration signals. Provide the results of signal processing, and compare the advantages and disadvantages of the STFT and HHT time-frequency analysis methods.

Keywords Time–frequency analysis · HHT · STFT

1 Introduction

The basic principle of the traditional Fourier transform is to decompose the original signal approximately into the superposition of a finite number of sinusoidal signals. The amplitude and frequency of each sine wave are fixed, so all kinds of frequency-domain analysis method based on FFT have congenital defects in

L. Du (✉) · H. Liu · S. Li
Research Institute of Highway Ministry of Transport, Beijing 100088, China
e-mail: ls.du@rioh.cn

Z. Dong
Shenyang University of Technology, Shenyang 110870, China

V. Jain et al. (eds.), *Recent Trends in Intelligent Computing, Communication and Devices*, Advances in Intelligent Systems and Computing 1031,
https://doi.org/10.1007/978-981-13-9406-5_86

processing such original signal which is nonlinear and unsteady. The short-time Fourier transform and Hilbert–Huang Transform are important time–frequency analysis methods which could effectively deal with various unsteady signals in practical engineering. The time–frequency–energy graph can provide local time-domain and frequency-domain information which is a great significance to describe the mechanical vibration process accurately.

2 Time–Frequency Analysis Methods

2.1 *The Basic Principle of Hilbert–Huang Transform*

Hilbert–Huang transform (HHT) includes two parts: the empirical mode decomposition [1] (EMD) and the Hilbert spectrum [2] analysis (HSA). The time–frequency–energy graph and the marginal spectrum are very effective methods of analysis, and its spectral distribution has practical physical significance.

2.1.1 Empirical Mode Decomposition

The method of empirical mode decomposition is decomposing the original signal into a set of intrinsic mode functions (IMF) with better Hilbert transformation performance. It means that any signal is represented by a combination of different simple eigenmode oscillation signals. After Hilbert transform, it has a clear meaning of instantaneous amplitude and instantaneous frequency.

(1) Finding all local extreme points of the original signal and all the local maxima (minimum) value points is connected with the cubic spline curve to form the lower (upper) envelope which should envelop all extremum points.
(2) Calculating the average m_1 of the upper and lower envelope, the original data sequence $x(t)$ is subtracted from the average value m_1 to get the new data sequence h_1

$$h_1 = x(t) - m_1 \tag{1}$$

If the h_1 fits IMF conditions, it is considered as the first IMF; if not, it can be used as the original data sequence, and then loop the above process until the new data sequence satisfies IMF conditions. h_1 is isolated from $x(t)$ to get the rest of the sequence r_1.

$$r_1 = x(t) - h_1 \tag{2}$$

r_1 can be as the new original sequence, and then repeat the above step to get the rest of IMF which could be h_n. The decomposition process is ended when the residual r_n satisfies a preset tolerance range (a monotonic function or constant).

2.1.2 Hilbert Spectral Analysis

Hilbert transform was performed for each IMF, and transient frequency was calculated, for any time series $x(t)$, its Hilbert transform $y(t)$ is defined as:

$$y(t) = \frac{1}{\pi} P \int_{-\infty}^{+\infty} \frac{x(\tau)}{t - \tau} \mathrm{d}\tau \tag{3}$$

where P is Cauchy's principal value; the analytic signal $z(t)$ corresponds to $x(t)$ is formulated as follows:

$$z(t) = x(t) + jy(t) = a(t)e^{j\theta(t)} \tag{4}$$

$$a(t) = \left(x^2 + y^2\right)^{1/2};\ \theta(t) = \tan^{-1}\frac{y}{x};\ \omega = -\frac{\mathrm{d}\theta}{\mathrm{d}t} \tag{5}$$

where, $a(t)$ and $\theta(t)$ are instantaneous amplitude and instantaneous phase of the signal $x(t)$, respectively; Instantaneous frequency can be calculated by instantaneous phase calculation.

2.2 *Short-Time Fourier Transform*

The basic principle of short-time Fourier transform (STFT) is: unsteady signal is seemed as a composition of a serious of short-time steady signals based on traditional Fourier transform, added by a window of time-domain for short time. When the window is added, the local spectrum is transformed to a small time spectrum near time t, and then the local spectrum of any position can be obtained by sliding the calculation window, i.e., moving along the entire time axis with the change of time t, and then time–frequency–energy analysis is calculated.

The STFT of certain unsteady signal $s(t)$ is defined as:

$$\mathrm{STFT}_S(t, \omega) = \int_{-\infty}^{+\infty} s(\tau)h(\tau - t)\mathrm{e}^{-i\omega t}\mathrm{d}\tau \tag{6}$$

where $h(t)$ is window function.

For the above formula, parameter t is moved by window function, and then signal s(t) is separated to obtain the signal of the time period near the time t, then the local signal $s_t(\tau)$ is taken by FFT.

$$s(t) = s(\tau)h(\tau - t) \tag{7}$$

3 Signal Processing and Analysis

3.1 Data Processing

By collecting the original signal data of a certain mechanical vibration, the original signal data of the same time period at a different time is intercepted. According to the principle of comparison, the signal data was processed by FFT, HHT, and STFT, respectively, indicating the law of signal processing.

3.2 Processing Method of FFT

The collected data was filtered and the processed time domain diagram was obtained. Through comparative analysis, the data with higher vibration amplitude (Fig. 1) and lower vibration amplitude (Fig. 2) at the same time interval of 2 s are, respectively, intercepted for local amplification processing analysis, followed by FFT processing analysis. The analysis results are as follows:

(1) From the time periods of 132.8–134.8 s and 139.5–141.5 s, the peak value of the mechanical vibration frequency is relatively high around 125 Hz.

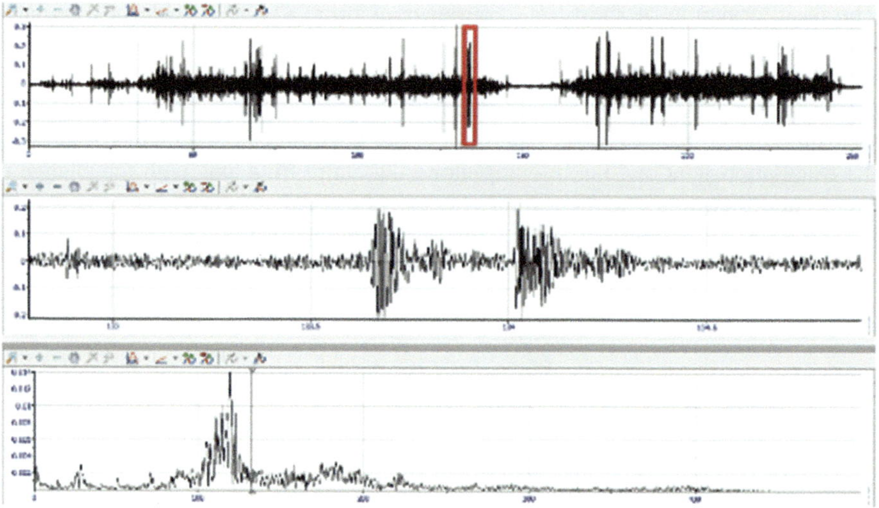

Fig. 1 FFT results of a vibration signal (high amplitude)

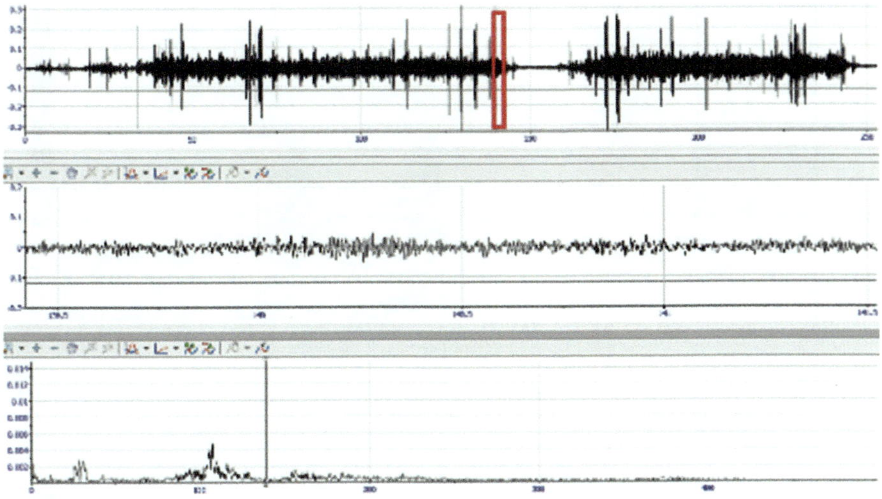

Fig. 2 FFT results of a vibration signal (low amplitude)

(2) The peak values of a certain mechanical vibration frequency are 0.014 and 0.004, respectively, when the vibration frequency is 125 Hz.

3.3 *Processing Method of STFT*

The collected data are, respectively, intercepted for local amplification processing analysis, followed by STFT processing analysis. The analysis results are as follows:

(1) From the time periods of 132.8–134.8 s and 139.5–141.5 s, the relatively higher frequency energy performance of mechanical vibration is about 125 and 25 Hz, respectively.
(2) It can be seen from the STFT time–frequency analysis graph that the relatively higher energy performance of mechanical vibration frequency is around 125 and 25 Hz, respectively. As STFT cannot meet the frequency resolution and time resolution at the same time, the frequency distribution is not clear (Figs. 3 and 4).

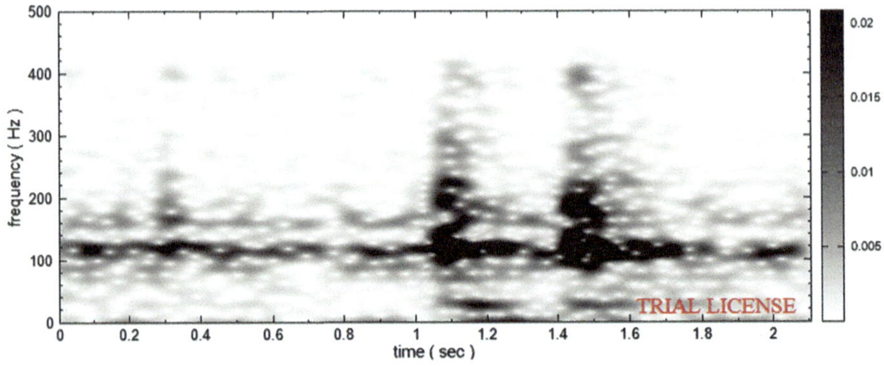

Fig. 3 STFT diagram of local amplification signal (high amplitude)

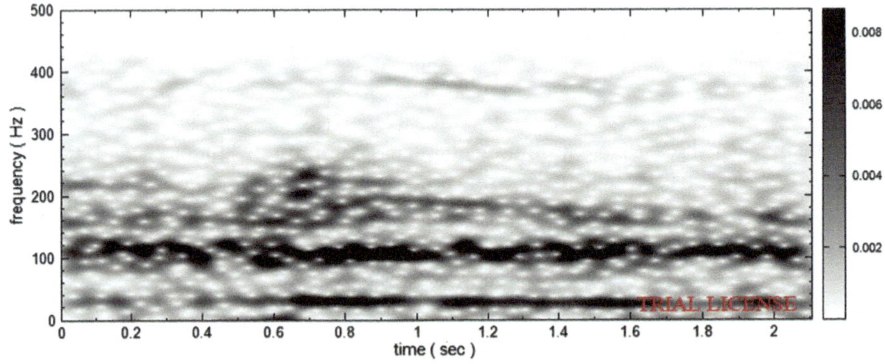

Fig. 4 STFT diagram of local amplified signal (low amplitude)

3.4 Processing Method of HHT

EMD experience decomposition was performed, as shown in Figs. 5 and 6, respectively. Based on EMD experience decomposition, HHT processing analysis was performed, as shown in Figs. 7 and 8, respectively. The analysis results are as follows:

(1) From the time periods of 132.8–134.8 s and 139.5–141.5 s, the relatively higher frequency energy performance of mechanical vibration is about 125 and 25 Hz, respectively.
(2) It can be seen from the HHT time–frequency analysis graph that the relatively higher energy performance of mechanical vibration frequency is around 125 and 25 Hz, respectively.

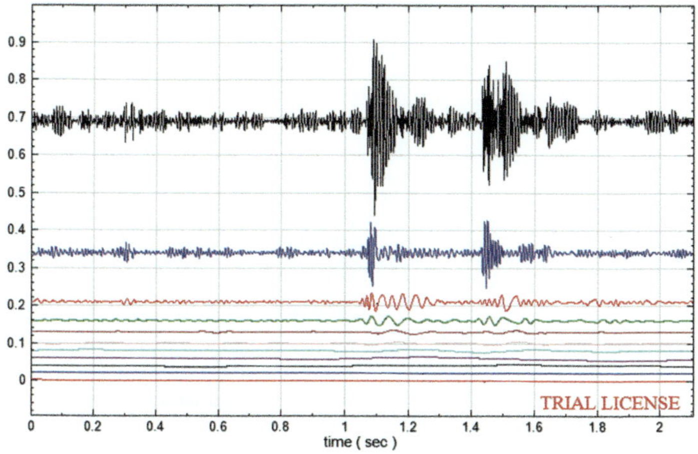

Fig. 5 Empirical mode decomposition diagram of local amplification signal (high amplitude)

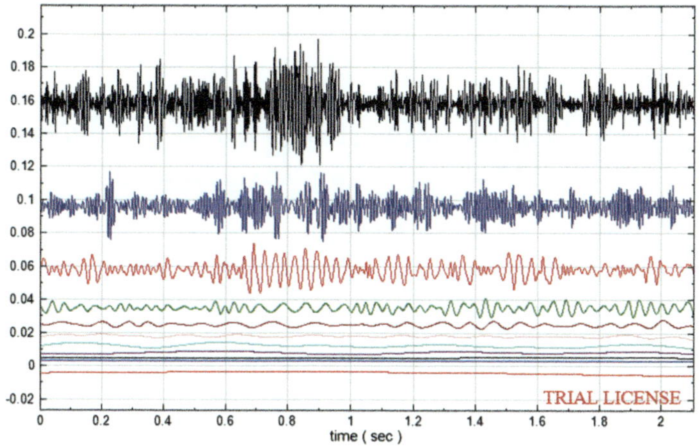

Fig. 6 Empirical mode decomposition diagram of local amplification signal (low amplitude)

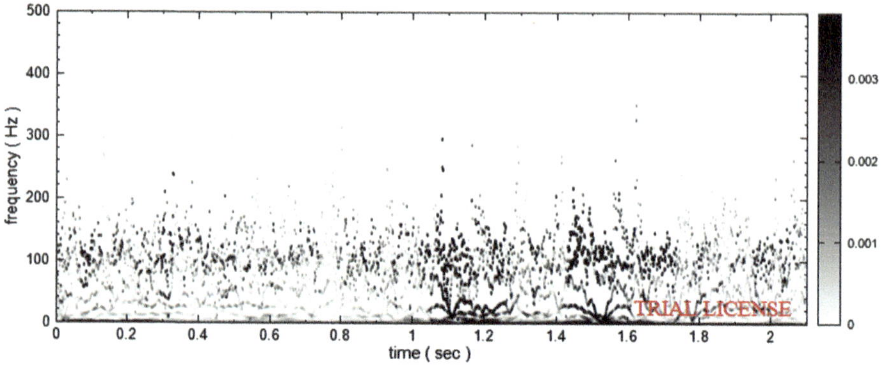

Fig. 7 HHT diagram of local amplification signals (high amplitude)

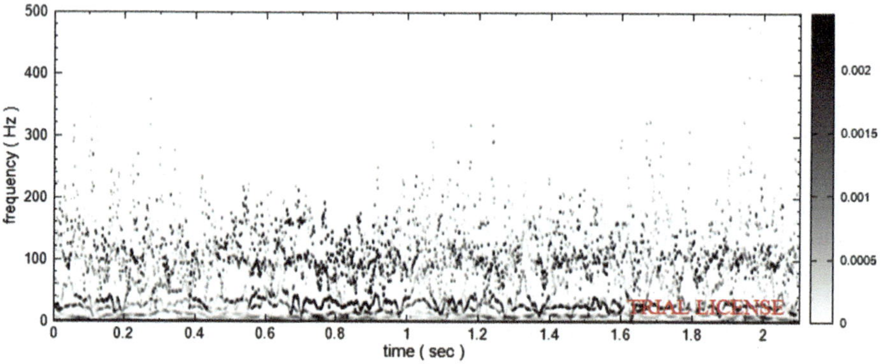

Fig. 8 HHT diagram of local amplification signals (low amplitude)

4 Conclusion

While the traditional Fourier analysis method is suitable for the processing of steady-state mechanical vibration signals, there are congenital defects for the processing of unsteady mechanical vibration signals; Although the STFT time–frequency analysis method is simple to calculate and easy to implement, it has defects of poor time–frequency resolution and cannot resolve the contradiction between time resolution and frequency resolution at the same time. However, when HHT is dealing with unsteady mechanical vibration signals, it first decomposes various eigenwave patterns contained in complex signals through EMD decomposition. HHT transform of signal EMD decomposition results can obtain the amplitude and frequency of the signal at the same time which is the change of characteristics over time and can guarantee time resolution and frequency resolution of problems at the same time. Hilbert-Huang transformation greatly improving the adaptability of unsteady mechanical vibration signal is an effective means which is research on unsteady complex mechanical vibration signal time-frequency characteristics.

References

1. Huang, N.E., Wu, M., Long, S., et al.: A confidence limit for the empirical mode decomposition and Hilbert spectral analysis. Proc. R. Soc. Lond. A **459**(2307), 2317–2345 (2003)
2. Huang, N.E., Shen, Z., Long, S.: A new view of nonlinear water waves: the Hilbert Spectrum. Annu. Rev. Fluid Mech. **31**(1), 417–457 (1999)

Research on Network Security Technology Based on Artificial Intelligence

Lijun Chen, Zhang Yi and Xiaoru Chen

Abstract Maybe you are for hackers or more network security events concerned, thus will increasingly sophisticated hacking techniques as focus on the object, but at the same time, you may need to pay attention to the business behind the operating personnel problems, these problems may be a real threat, and the use of machine learning and artificial intelligence to solve the problems of network security, is the development trend of the present. This paper introduces the necessity of the development of network security technology and the application of artificial intelligence (AI) in solving some problems. This paper also gives a brief overview of some recent advances in network security technology of artificial intelligence and thus looks forward to the application prospect of artificial intelligence in the field of network security (ABI Research in Machine Learning in Cybersecurity to Boost Big Data, Intelligence, and Analytics Spending to $96 Billion by 2021, 2018) [1].

Keyword Privacy security

1 Introduction

Just bought the house, open the phone, decoration information "refresh" immediately; buy tickets through the Web site, then travel insurance promotion. With the "Internet" of daily life, such as shopping, online payment, and access to information, the invasion of privacy security is increasingly frequent, and the "streaking" of citizens' personal information seems to have become the norm.

Since June 1, the cyber security law of the People's Republic of China (hereinafter referred to as the cyber security law) has come into force. The cyber security law has made clear provisions on cyber fraud, the safety of critical information infrastructure, the establishment of cyber security monitoring, early warning, and

L. Chen (✉) · Z. Yi · X. Chen
South China Institute of Software Engineering, Guangzhou University,
Guangzhou 510990, China
e-mail: 372158286@qq.com

V. Jain et al. (eds.), *Recent Trends in Intelligent Computing, Communication and Devices*, Advances in Intelligent Systems and Computing 1031,
https://doi.org/10.1007/978-981-13-9406-5_87

emergency response systems. But even with the law, the "extortion virus" remains a shadow of the industry. How to build a strong cyber security line has become a hot topic at the 2017 China cyber information security summit in Beijing on June 9 [2].

Given the current situation, it is easy to understand why cyber security experts have been working on artificial intelligence (AI) and how it can help alleviate those problems. For example, machine learning (ML), using many of the latest artificial intelligence algorithms, can go a long way toward helping detect malware, making it increasingly difficult to identify and isolate. As malware becomes increasingly adapted to linear, traditional security solutions, ML can learn not only how the malware looks and behaves, but how it evolves. In addition, artificial intelligence system cannot only provide detection, but also take measures to correct certain situations, sort and classify events and threats, and finally liberate technicians from repeated activities.

2 Machine Learning (ML)

The traditional way to prevent network security is to block the malware before it executes, trying to match the code pattern to a known signature. When they fail to do so, however, there is often the little remedy. The execution of malware is hard to stop. The ML algorithm is an attempt to identify malware attacks in real time, combined with artificial intelligence-assisted decision making, machine and network isolation technology to isolate infected computers or entire network segments within milliseconds, providing a way to prevent the spread of malicious code [3].

In any case, one of the most relevant issues in applying AI or ML to cyber security seems to be how to identify useful patterns and how to understand what constitutes a security event when they deviate, how to classify them and act accordingly. From a "normal" perspective, statistical deviations from normal network traffic can yield interesting results, but not all of them necessarily constitute security incidents. A negative test can be just as problematic as a large number of false positives.

3 Supervised ML

Traditional network security has many drawbacks, but in the field of artificial intelligence, supervised machine learning is probably the best preventive measure in network security so far. The huge database contains malware files, which can be further trained on these algorithms to improve the accuracy of detection and reduce the number of false positives and false positives. This applies to spam detection as well, thanks to a large sample database that is expanding with the help of users and Internet service providers who report spam. Compared with the traditional text-based string matching or address blacklist, ML provides a smarter and more efficient approach [5]. Another example of overseeing machine learning for cyber security is the use of circular neural networks (RRN) to separate artificially generated DNS records from those automatically generated by ransomware. In this case, the signature-based blacklist is completely useless, as the number of malicious names grows, changes, and evolves. However, using ML in RRN can yield more accurate results using language analysis [4].

4 Unsupervised ML

While monitoring machine learning may have achieved the best results so far, it is important that due to its inherent limitations, the focus of cyber security seems to be unsupervised ML. In this case, the computer should understand the underlying structure from the data and extrapolate the appropriate output. Unsupervised ML is more subjective and without human expertise to understand them and understand which patterns are really useful, acquired patterns often do not make sense [5]. Some interesting technologies applied in network security are clustering, association rule learning, and dimensionality reduction [6].

Dimensional reduction techniques can be particularly helpful for cyber security. Let us take the example of capturing network traffic. In this case, there may be multiple features: source IP, target IP, protocol, port, payload, MAC address, TTL, routing, and so on. Analyzing all the features of captured data may not be computationally sufficient to solve specific security problems. If we are dealing with real-time analysis, this problem may not be solved at all. However, if the ML algorithm can learn how to reduce the number of related features to a few (feature selection techniques) or group them into a more manageable set (feature extraction) to solve the problem, it will make the problem easier to solve [7].

5 Practical Case—Darktrace

Since the beginning of summer, Beijing has been hit by rainstorms that have not been seen for many years. However, no amount of storms can hinder the progress and efforts of the six cloud partners. With the joint efforts of the six cloud partners, on July 27, 2018, six cloud finally ushered in the most exciting news since its establishment—the release of the whole series of six cloud "AI gene intelligent defense" products. The released security products include four series—industrial security LinSec series, cloud security CSec series, big data security SSA series, and network security NSec series [8].

At present, domestic Internet plus, industrial Internet, and the Internet of things are developing in full swing, and the information security threat caused by this is also highly concerned. Information security threats fall into two categories: known threats and unknown threats. Many information security vendors have provided multiple solutions to known threats. For the prevention of unknown threats, some technologies, such as MTD and honeypot, have emerged, but they cannot solve the problem well. After years of research and experiments, experts in the "superstring" artificial intelligence laboratory in Liufang cloud have applied artificial intelligence technology to the unknown threat defense for the first time, making the security technology self-evolution without relying on a lot of artificial intervention and decision making, so as to solve the problem of unknown threat defense more effectively [9].

The defense against unknown threats by the six Fang Yun artificial intelligence technology is mainly divided into two basic steps:

Step 1 Model training: Before model training, add the blacklist and whitelist detection mechanism into the sample to prevent the model from training based on the "contaminated" sample.

Step 2 Threat detection and abnormal disposal: Four steps are included: blacklist detection, abnormal judgment, abnormal screening, and abnormal disposal.

Since its establishment, Beijing Liufang Cloud Technology Co., Ltd has been devoted to the research of safety technology and products. Chairman Ren reinforced the mission of "connecting all things to safety" and the concept of "technology is the guarantee of safety," pointing out that "technology" is the power source of continuous innovation and development of Liufang cloud!

6 The Foreseeable Future of AI in Cyber Security

Now, despite billions of dollars spent by governments or businesses on cyber security, the number of reports of cyberattacks is growing, and the scale of attacks is growing. In many frontier areas with artificial intelligence (AI) predictive capabilities, security providers, companies, and ourselves can take the upper hand in responding to cyber attacks. Next, let us summarize the key areas of AI cyber defense innovation:

(1) Find and prevent hackers from invading IOT devices

Cisco expects the number of connected devices worldwide to increase to 50 billion by 2020, from the current 15 billion. However, many Internet-connected devices do not have basic security measures due to the limitations of hardware and software resources. Worse, with the release of Mirai's original code, this malware was rampant and can now be used to attack any business or individual. LOT security is one of the most prominent fields in the application of AI technology. AI-based lightweight predictive models can automatically reside and operate on low-compute power devices and can detect and prevent suspicious behavior within the device or network scope in real time.

(2) Prevent malware and files from being executed

File-based cyber attacks are still the main form of cyber attacks. The easiest files to target in this type of cyber attack are executable files (.exe), Acrobat reader (.pdf), and Microsoft Office files. Some start-ups are trying to solve the problem with artificial intelligence. They take advantage of the power of artificial intelligence to find millions of features in every suspicious file, even the slightest code clash. Leaders in developing such file-based AI security systems include Cylance, Deep Instinct and Invincea.

(3) Improve the operation efficiency of the safety operation center

One of the key issues facing the security team is the alarm fatigue caused by the overflow of security alerts received every day. On average, North American businesses need to handle 10,000 security alerts a day. Under certain circumstances, this could make malware a "'leaky net," "even though it has been flagged as a" "suspicious target."' This requires the close coordination of multiple information sources, integrated internal logs, and monitoring systems with external threat intelligence services to automatically classify all events.

This frontier of cyber defense has become a super hot spot because it can help large companies with their own security operations centers deal with cyber threats. Some start-ups are using AI to address the threat, such as Phantom, Jask, StatusToday and CyberLytic.

(4) Quantitative risk

The network risk quantification faced by enterprises is becoming a challenge, mainly because of the lack of historical data and too many variables to consider. Currently, AI technology can process millions of data points while generating predictions that help businesses and online insurers get the most accurate online risk assessments. Some start-ups are doing similar research, including BitSight and security scorecards.

(5) Network traffic anomaly detection

Artificial intelligence can be used to check abnormal network traffic by looking for cross-protocol associations that do not rely on intrusive deep packet inspection to analyze endless metadata correlations between internal and external network traffic. Such start-ups include Vectra Networks, DarkTrace and BluVector.

(6) Detect malicious mobile applications

Ericsson says the number of smartphones worldwide will increase from 2.5 billion today to 6 billion by 2020. Through a study of 100 popular IOS and Android apps, the Arxan study found that 56% of the top IOS apps and all Android apps have been hacked.

Finally, it is important to continue to use data visualization to improve the ability of security analysts to understand broader threats in a shorter amount of time, while reducing workloads and resources. With the booming development of big data, the storage and analysis ability of massive data has become an urgent need for ML to play an important role. However, not only should you be able to analyze it, but you should be able to represent it in a format that you can easily understand at multiple levels of the organization. Therefore, data visualization is one of the areas where ML is most likely to play an important role in the future.

7 Conclusion

In the past few years, technologies such as big data, cloud computing, and artificial intelligence have repeatedly appeared in many forums. In many cases, they do not clearly recognize their importance and do not solve practical problems well. It is well known that when people do not fully understand technology, they usually have two effects: One is that the technology is rejected irrationally (for example, a new operating system), and the other is that if the technology is properly promoted, it is think of it as a panacea for solving all problems (artificial intelligence). It takes a while or even years for the dust to settle, and the market realizes its true potential.

ML's intrusion into network security has forced a shift from active rule-based prevention to reactive real-time detection. Security threats have become so diverse that traditional techniques based on rules inferred from known attacks, i.e., blocking them before they occur, seem to be no longer a viable approach. Many attacks have escaped these mechanisms, causing enormous damage that cannot be prevented once they are activated. The goal of ML is to identify attacks in real time, with little need to interact with people and stop them before they cause serious harm.

We can conclude that artificial intelligence is not currently used to address anticipated network or network security issues. Currently, only one artificial intelligence branch machine learning has been successfully applied to solve a few problems. While Monitor ML offers many interesting and practical solutions, there are still ongoing studies, especially for the use of Monitorless ML, because its ultimate goal is to minimize human intervention when detecting threats [10].

Fund Project No: 2015KTSCX176, CJ201804SYJG201803, SYJG201803.

References

1. ABI Research: Machine Learning in Cybersecurity to Boost Big Data, Intelligence, and Analytics Spending to $96 Billion by 2021. Retrieved March 24, 2018, from https://www.abiresearch.com/press/machine-earning-cybersecurity-boostbig-data-inte/ (2017, January 30)
2. Alghamdi, R.: Hidden Markov Models (HMMs) and security applications. Int. J. Adv. Comput. Sci. Appl. **7**(2) (2016). https://doi.org/10.14569/ijacsa.2016.070205
3. Crosby, S.: Separating fact from fiction: the role of artificial intelligence in cybersecurity. Retrieved March 16, 2018, from https://www.forbes.com/sites/forbestechcouncil/2017/08/21/separating-fact-fromfiction-the-role-of-artificial-intelligence-in-cybersecurity/2/#47045fb329bc (2017, August 21)
4. Darktrace: Technology. Retrieved March 15, 2018, from https://www.darktrace.com/technology (n.d.)

5. Honkela, T.: Multimodally Grounded Translation. Retrieved March 20, 2018, from http://lodel.irevues.inist.fr/tralogy/index.php?id=259&format=print (2011)
6. Kanal, E.: Machine Learning in Cybersecurity. Carnegie Mellon University, Software Engineering Institute. Retrieved March 9, 2018 from https://insights.sei.cmu.edu/sei_blog/2017/06/machine-learning-in-cybersecurity.html (2017, January)
7. Machine Learning Techniques Applied to Cyber Security: Retrieved March 18, 2018, from https://towardsdatascience.com/machine-learning-techniquesapplied-to-cyber-security-d58a8995b7d7 (2017, September 10)
8. Marty, R.: AI and Machine Learning in Cyber Security—Towards Data Science. Retrieved March 16, 2018, from https://towardsdatascience.com/ai-andmachine-learning-in-cyber-security-d6fbee480af0 (2018, January 01)
9. Marty, R.: AI in Cybersecurity: Where We Stand & Where We Need to Go. Retrieved March 12, 2018, from https://www.darkreading.com/threatintelligence/ai-in-cybersecurity-where-we-stand-and-where-we-need-to-go/a/did/1330787 (2018, January 11)
10. McHugh, J.: Testing intrusion detection systems: a critique of the 1998 and 1999 DARPA intrusion detection system evaluations as performed by Lincoln Laboratory. ACM Trans. Inf. Syst. Secur. (TISSEC) **3**(4), 262–294. Retrieved March 12, 2018 (2000)

Outage Performance for Relaying Aided Non-orthogonal Multiple Access

Jinhong Fan and Li He

Abstract Non-orthogonal multiple access is the key technique of the fifth-generation wireless communication. In this paper, we investigate the outage performance of NOMA-based downlink amplify-and-forward half-duplex relaying networks. The closed-form representations of exact outage performance are attained. Simulation results indicate that the concluded result is in excellent agreement with the Monte Carlo simulation.

Keywords Non-orthogonal multiple access (NOMA) · Amplify-and-forward (AF) · Outage probability

1 Introduction

Non-orthogonal multiple access has a higher spectral efficiency than orthogonal multiple access. So it has drawn much attention as multiple access for fifth-generation wireless communication networks [1–5]. Relaying technology can improve throughout and increase the communication range. The outage performance of the amplify-and-forward relaying scheme was investigated in [6]. Cooperative NOMA has drawn much attention because it can improve the NOMA reliability [7–9]. In [7], the users with better channel qualities are selected as a relay in the system. The performance of decode and forward relaying NOMA systems are analyzed [8]. The cooperative NOMA schemes with multiple antennas multiple users and dedicated relays were studied in [9]. In [10], NOMA was used into directed transmission and relay transmission. The overall outage probability of cooperative NOMA system was analyzed. In [11], the authors analyzed the NOMA amplify-and-forward relay of fixed gain in Nakagami-m fading channels. The author in [12] studied outage probability and ergodic sum rate for cooperative

J. Fan (✉) · L. He
School of Information Science and Technology, North China University of Technology, Beijing 100144, China
e-mail: qfjh0922@163.com

V. Jain et al. (eds.), *Recent Trends in Intelligent Computing, Communication and Devices*, Advances in Intelligent Systems and Computing 1031,
https://doi.org/10.1007/978-981-13-9406-5_88

NOMA and attained the exact and asymptomatic closed-form expressions of outage probabilities. In this paper, considering a downlink cooperative NOMA system, the performance is studied.

2 System Model

The downlink cooperative NOMA system model is shown in Fig. 1. In this model, there is one base station S, two mobile users D_1 and D_2, and a relay R. The D_1 directly communicates with the S and is based on the help of R. There is no direct path between the S and D_2 because of long distance between them, so the D_2 needs the assistance of R. We assume that the numbers of all the nodes are one and the nodes operate in a half-duplex mode. The protocol of relay is an AF. This scheme consists of two consecutive same length time slots.

Each channel link is Rayleigh fading. The channel parameters of the links $S \rightarrow D_1$ and $S \rightarrow R$ are denoted by g_1 and g_r with $\Omega_1 = E\{|g_1|^2\}$ and $\Omega_r = E\{|g_r|^2\}$. The channel coefficients between R and D_k is $g_{r,k}$ with $\Omega_{r,k} = E\{|g_{r,k}|^2\}$.

In the first time slot, the S transmits the superimposed signal $x_1 + x_2$ to R and D_1, where x_1 is the signal for D_1 with $P_1 = E\{|x_1|^2\}$, and x_2 is the signal for D_2 with $P_2 = E\{|x_2|^2\}$. Based on the NOMA protocol described in [7], we assume $P_1 < P_2$. The total transmit power for the S is $P_T = P_1 + P_2 = P$. The received signals y_1 at D_1 and y_r at R are given by

$$y_r = g_r(x_1 + x_2) + n_{sr} \tag{1}$$

$$y_1 = g_1(x_1 + x_2) + n_{s1} \tag{2}$$

where n_{s1} and n_{sr} are the complex Gaussian noise at D_1 and R, with zero mean and variance of N_0. D_1 first performs successive interference cancelation (SIC) by

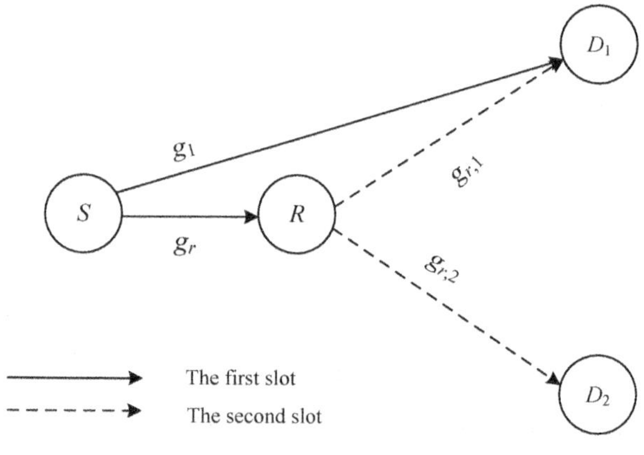

Fig. 1 Downlink cooperative NOMA system

detecting and decoding D_2's information. The received signal-to-interference-and-noise-ratio (SINR) for D_1 is

$$\gamma_{s1\to 2} = \frac{|g_1|^2 a_2 P}{|g_1|^2 a_1 P + N_0} = \frac{|g_1|^2 a_2 \gamma}{a_1 \gamma |g_1|^2 + 1} \tag{3}$$

where $\gamma = P/N_0$ is a transmit signal-to-noise ratio, $a_1 = P_1/P$, $a_2 = P_2/P$ and $a_1 + a_2 = 1$. D_1 can decode its signal with the bellow SINR after D_2's signal is decoded

$$\gamma_{s1} = \frac{|g_1|^2 a_1 P}{N_0} = a_1 \gamma |g_1|^2 \tag{4}$$

In the second time slot, R transmits the signal $x_r = \rho y_r$ to D_1 and D_2 after amplifying it with an amplifying gain $\rho = \sqrt{P_R}/\sqrt{P_T|g_r|^2 + N_0}$, where $P_R = E\{|x_r|^2\}$ denotes the transmit power of the relay and $P_R = P_T = P$. The received signals at D_1 and D_2 are given by

$$y_{r1} = g_{r,1} x_r + n_{r1} = \rho g_{r,1} g_r (x_1 + x_2) + \rho g_{r,1} n_{sr} + n_{r1} \tag{5}$$

$$y_{r2} = g_{r,2} x_r + n_{r2} = \rho g_{r,2} g_r (x_1 + x_2) + \rho g_{r,2} n_{sr} + n_{r2} \tag{6}$$

where n_{r1} and n_{r2} are the complex Gaussian noise at D_1 and D_2, with zero mean and variance of N_0. D_1 first detect D_2's information with the received SINR

$$\begin{aligned}\gamma_{R1\to 2} &= \frac{|g_{r,1}|^2 \rho^2 |g_r|^2 a_2 P}{|g_{r,1}|^2 \rho^2 |g_r|^2 a_1 P + |g_{r,1}|^2 \rho^2 N_0 + N_0} \\ &= \frac{\gamma^2 a_2 |g_{r,1}|^2 |g_r|^2}{\gamma^2 a_1 |g_{r,1}|^2 |g_r|^2 + \gamma\left(|g_{r,1}|^2 + |g_r|^2\right) + 1}\end{aligned} \tag{7}$$

After SIC, the received SINR for D_1 is

$$\gamma_{R1} = \frac{\gamma^2 a_1 |g_{r,1}|^2 |g_r|^2}{\gamma\left(|g_{r,1}|^2 + |g_r|^2\right) + 1} \tag{8}$$

The received SINR for D_2 to detect x_2 is

$$\gamma_{R2} = \frac{|g_{r,2}|^2 \rho^2 |g_r|^2 a_2 P}{|g_{r,2}|^2 \rho^2 |g_r|^2 a_1 P + |g_{r,2}|^2 \rho^2 N_0 + N_0} = \frac{\gamma^2 a_2 |g_{r,2}|^2 |g_r|^2}{\gamma a_1 |g_{r,2}|^2 |g_r|^2 + \gamma\left(|g_{r,2}|^2 + |g_r|^2\right) + 1} \tag{9}$$

3 Outage Probability Analysis

The outage performance of the NOMA downlink cooperative network is evaluated. The exact expressions of the outage probability of the users over Rayleigh fading channels are studied.

The probability density function (PDF) of $|g_i|^2$ is

$$f_{|g_i|^2}(y) = \frac{1}{\Omega_i} \exp\left(-\frac{y}{\Omega_i}\right) \tag{10}$$

The user D_1 combines with observations from S and R by applying selection combining at the second stage. The outage would not occur at D_1 in two cases: In the two slots, D_1 can detect D_2's information and then detect its own information. The outage probability for D_1 is

$$P_{\text{out}}^1 = [1 - P(\gamma_{s1\to 2} \geq \gamma_{\text{th}2}, \gamma_{s1} \geq \gamma_{\text{th}1})] \times [1 - P(\gamma_{R1\to 2} \geq \gamma_{\text{th}2}, \gamma_{R1} \geq \gamma_{\text{th}1})] \tag{11}$$

where $\gamma_{\text{th}1} = 2^{2R_1 - 1}$ and $\gamma_{\text{th}2} = 2^{2R_2 - 1}$, with R_1 and R_2 being the transmission rate at D_1 and D_2.

The outage probability of D_1 is described as follows.

Theorem 1 *The close-form expression for the outage probability of D_1 is*

$$P_{\text{out}}^1 = \left(1 - \exp\left(-\frac{2\theta}{\Omega_1}\right)\right) \left\{1 - \exp\left(-\theta\left(\frac{1}{\Omega_{r,1}} + \frac{1}{\Omega_r}\right)\right)\sqrt{\frac{4\theta(1+\gamma\theta)}{\gamma\Omega_{r,1}\Omega_r}}\, K_1\left(\sqrt{\frac{4\theta(1+\gamma\theta)}{\gamma\Omega_{r,1}\Omega_r}}\right)\right\} \tag{12}$$

where $\theta = \max(\varepsilon, \beta)$, $\varepsilon = \frac{\gamma_{\text{th}2}}{\gamma(a_2 - a_1\gamma_{\text{th}2})}$, $\beta = \frac{\gamma_{\text{th}1}}{a_1\gamma}$, $K_1(\cdot)$ is the first-order modified Bessel function of the second kind.

Proof is seen in Appendix 1.

The outage event of D_2 can be interpreted as it cannot detect its own message at the second slot. D_2's outage probability is

$$P_{\text{out}}^2 = P(\gamma_{R2} < \gamma_{\text{th}2}) \tag{13}$$

The outage probability of D_2 is described by the theorem.

Theorem 2 *The closed-form expression of the outage probability of D_2 is*

$$P_{\text{out}}^2 = 1 - \exp\left(-\varepsilon\left(\frac{1}{\Omega_r} + \frac{1}{\Omega_{r,2}}\right)\right)\sqrt{\frac{4\varepsilon(1+\varepsilon\gamma)}{\Omega_{r,2}\Omega_r\gamma}}\,K_1\left(\sqrt{\frac{4\varepsilon(1+\varepsilon\gamma)}{\Omega_{r,2}\Omega_r\gamma}}\right) \tag{14}$$

Proof is seen in Appendix 2.

4 Numerical Results

For verifying the validity of theory expressions, simulation results are given. Figure 2 shows outage probability for the user D_1 for NOMA. Assumed that the transmission rates are $R_1 = 0.2$, $R_2 = 0.8$, $\Omega_1 = 0.4$, $\Omega_r = 0.5$, and $\Omega_{r,1} = 0.5$. The comparison for the outage probability between the analytical result Eq. (12) and the simulation results is shown in Fig. 2. It is clearly observed that the analytical result Eq. (12) matches the numerical results perfectly.

Figure 3 shows the outage probability of the user D_2 for NOMA with channel estimation error. Assumed that the transmission rates are $R_1 = 0.2$, $R_2 = 0.8$, $\Omega_r = 0.4$, and $\Omega_{r,2} = 0.5$. The comparison for the outage probability between the analytical result Eq. (14) and the simulation results is shown in Fig. 3. It is clearly observed that the analytical result Eq. (14) matches the numerical results perfectly.

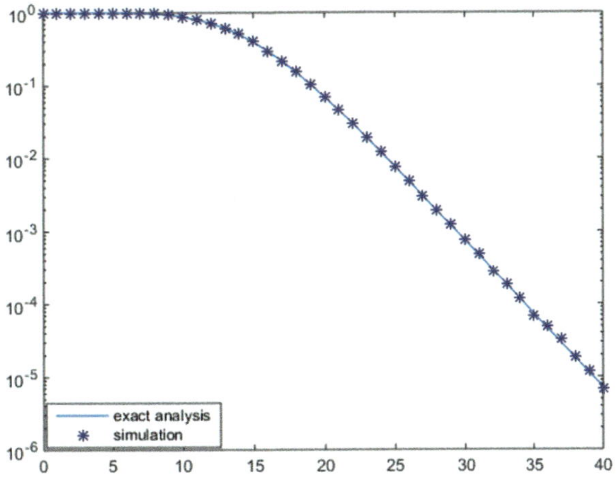

Fig. 2 Outage probability of D_1

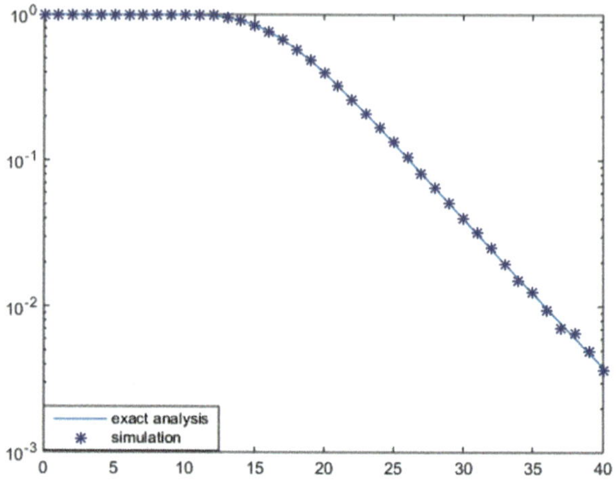

Fig. 3 Outage probability of D_2

5 Conclusion

We study the outage performance of downlink cooperative NOMA in this paper. The exact closed-form expressions of the outage probability are attained. It can provide valuable guidelines for the theoretical basis in the actual communication environment. Results show that our analysis is very close to the exact values.

Acknowledgements This work was supported by Scientific Research Common Program of Beijing Municipal Commission of Education under Grant No. KM201510009008 and Scientific Research Foundation of North China University of Technology.

Appendix 1

Proof of Theorem 1 The outage probability of D_1 is:

$$P_{\text{out}}^{1} = [1 - P(\gamma_{s1\to 2} \geq \gamma_{\text{th2}}, \gamma_{s1} \geq \gamma_{\text{th1}})] \times [1 - P(\gamma_{R1\to 2} \geq \gamma_{\text{th2}}, \gamma_{R1} \geq \gamma_{\text{th1}})]$$

and assumed that $J_1 = 1 - P(\gamma_{s1\to 2} \geq \gamma_{\text{th2}}, \gamma_{s1} \geq \gamma_{\text{th1}})$ and $J_2 = [1 - P(\gamma_{R1\to 2} \geq \gamma_{\text{th2}}, \gamma_{R1} \geq \gamma_{\text{th1}})]$ where J_1 can be solved as

$$
\begin{aligned}
J_1 &= 1 - P(\gamma_{s1\to 2} \geq \gamma_{\text{th2}})P(\gamma_{s1} \geq \gamma_{\text{th1}}) \\
&= 1 - P\left(\frac{|g_1|^2 a_2\gamma}{|g_1|^2 a_1\gamma + 1} \geq \gamma_{\text{th2}}\right)P\left(|g_1|^2 a_1\gamma \geq \gamma_{\text{th1}}\right) \\
&= 1 - P\left(|g_1|^2 \geq \frac{\gamma_{\text{th2}}}{\gamma(a_2 - a_1\gamma_{\text{th2}})} = \varepsilon\right)P\left(|g_1|^2 \geq \frac{\gamma_{\text{th1}}}{a_1\gamma} = \beta\right) \\
&= 1 - P\left[|g_1|^2 \geq \max(\varepsilon, \beta)\underline{\Delta}\theta\right] \\
&= P\left(|g_1|^2 \leq \theta\right) = 1 - \exp\left(-\frac{2\theta}{\Omega_1}\right)
\end{aligned} \tag{15}
$$

Now, J_2 can be solved as

$$
\begin{aligned}
J_2 &= 1 - P(\gamma_{R1\to 2} \geq \gamma_{\text{th2}})P(\gamma_{R1} \geq \gamma_{\text{th1}}) \\
&= 1 - P\left(\frac{a_2\gamma^2|g_{r,1}|^2|g_r|^2}{a_1\gamma^2|g_{r,1}|^2|g_r|^2 + \gamma\left(|g_{r,1}|^2 + |g_r|^2\right) + 1} \geq \gamma_{\text{th2}}\right) \\
&\quad P\left(\frac{a_1\gamma^2|g_{r,1}|^2|g_r|^2}{\gamma\left(|g_{r,1}|^2 + |g_r|^2\right) + 1} \geq \gamma_{\text{th1}}\right) \\
&= 1 - P\left(|g_r|^2 \geq \frac{\theta\left(1 + \gamma|g_{r,1}|^2\right)}{|g_{r,1}|^2 - \theta}, |g_{r,1}|^2 \geq \theta\right) \\
&= 1 - \int_0^\infty \frac{1}{\Omega_{r,1}}\exp\left(-\frac{y}{\Omega_{r,1}}\right)\int_{\frac{\theta(1+\gamma y)}{y-\theta}}^\infty \frac{1}{\Omega_r}\exp\left(-\frac{x}{\Omega_r}\right)\mathrm{d}x\mathrm{d}y \\
&= 1 - \exp\left(-\theta\left(\frac{1}{\Omega_{r,1}} + \frac{1}{\sigma_r^2}\right)\right)\sqrt{\frac{4\theta(1+\gamma\theta)}{\gamma\Omega_{r,1}\Omega_r}}K_1\left(\sqrt{\frac{4\theta(1+\gamma\theta)}{\gamma\Omega_{r,1}\Omega_r}}\right)
\end{aligned} \tag{16}
$$

Equation (16) is obtained with the aid [13] [(3.324.1)]. And substituting (15) and (16) into (11).

The closed-form representation of the outage performance of D_1 is

$$
\begin{aligned}
P_{\text{out}}^1 = &\left(1 - \exp\left(-\frac{2\theta}{\Omega_1}\right)\right) \\
&\left\{1 - \exp\left(-\theta\left(\frac{1}{\Omega_{r,1}} + \frac{1}{\Omega_r}\right)\right)\sqrt{\frac{4\theta(1+\gamma\theta)}{\gamma\Omega_{r,1}\Omega_r}}K_1\left(\sqrt{\frac{4\theta(1+\gamma\theta)}{\gamma\Omega_{r,1}\Omega_r}}\right)\right\}
\end{aligned}
$$

Appendix 2

Proof of Theorem 2 The outage probability of D_2 is:

$$P_{\text{out}}^2 = P(\gamma_{R2} < \gamma_{\text{th2}})$$

and

$$\begin{aligned}
P_{\text{out}}^2 &= P(\gamma_{R2} < \gamma_{\text{th2}}) \\
&= P\left(|g_{r,2}|^2 < \varepsilon\right) + P\left\{|g_r|^2 < (\varepsilon/r)\left(\gamma|g_{r,2}|^2 + 1\right)/\left(|g_{r,2}|^2 - \varepsilon\right), |g_{r,2}|^2 > \varepsilon\right\} \\
&= \int_0^{\varepsilon} f_{|g_{r,2}|^2}(x)\mathrm{d}x + \int_{\varepsilon}^{\infty} f_{|g_{r,2}|^2}(x)\mathrm{d}x \int_0^{\frac{\varepsilon(1+\gamma x)}{\gamma(x-\varepsilon)}} f_{|g_r|^2}(y)\mathrm{d}y \\
&= 1 - \frac{1}{\Omega_{r,2}}\exp\left(-\varepsilon\left(\frac{1}{\Omega_r} + \frac{1}{\Omega_{r,2}}\right)\right)\int_{\varepsilon}^{\infty}\exp\left(-\frac{x-\varepsilon}{\Omega_{r,2}}\right)\exp\left(-\frac{\varepsilon(1+\varepsilon\gamma)}{\Omega_r\gamma(x-\varepsilon)}\right)\mathrm{d}x \\
&= 1 - \frac{1}{\Omega_{r,2}}\exp\left(-\varepsilon\left(\frac{1}{\Omega_r} + \frac{1}{\Omega_{r,2}}\right)\right)\int_0^{\infty}\exp\left(-\frac{y}{\Omega_{r,2}}\right)\exp\left(-\frac{\varepsilon(1+\varepsilon\gamma)}{\Omega_r\gamma y}\right)\mathrm{d}y \\
&= 1 - \frac{1}{\Omega_{r,2}}\exp\left(-\varepsilon\left(\frac{1}{\Omega_r} + \frac{1}{\Omega_{r,2}}\right)\right)\int_0^{\infty}\exp\left(-\frac{4\varepsilon(1+\varepsilon\gamma)}{\frac{\Omega_r\gamma}{4y}} - \frac{y}{\Omega_{r,2}}\right)\mathrm{d}y \\
&= 1 - \exp\left(-\varepsilon\left(\frac{1}{\Omega_r} + \frac{1}{\Omega_{r,2}}\right)\right)\frac{1}{\Omega_{r,2}}\sqrt{\frac{4\varepsilon(1+\varepsilon\gamma)\Omega_{r,2}}{\Omega_r\gamma}}\,K_1\left(\sqrt{\frac{4\varepsilon(1+\varepsilon\gamma)}{\Omega_{r,2}\Omega_r\gamma}}\right) \\
&= 1 - \exp\left(-\varepsilon\left(\frac{1}{\Omega_r} + \frac{1}{\Omega_{r,2}}\right)\right)\sqrt{\frac{4\varepsilon(1+\varepsilon\gamma)}{\Omega_{r,2}\Omega_r\gamma}}\,K_1\left(\sqrt{\frac{4\varepsilon(1+\varepsilon\gamma)}{\Omega_{r,2}\Omega_r\gamma}}\right)
\end{aligned} \tag{17}$$

References

1. Li, Q.C., Niu, H., Papathanassiou, A.T., et al.: 5G network capacity-key elements and technologies. IEEE Veh. Technol. Mag. **9**, 71–78 (2014)
2. Saito, Y., Benjebbour, A., Kishiyama, Y., et al.: System-level performance evaluation of downlink non-orthogonal multiple access (NOMA). In: IEEE International Symposium on Personal Indoor and Mobile Radio Communications, pp. 611–615 (2013)
3. Saito, Y., Benjebbour, A., Kishiyama, Y., et al.: Non-orthogonal multiple access (NOMA) for cellular future radio access. In: IEEE Vehicular Technology Conference, pp. 1–5 (2013)

4. Benjebbour, A., Saito, Y., Kishiyama, Y., et al.: Concept and practical considerations of non-orthogonal multiple access (NOMA). In: Intelligent Signal Processing and Communications Systems, pp. 770–774 (2013)
5. Ding, Z., Yang, Z., Fan, P., et al.: On the performance of non-orthogonal multiple access in 5G systems with randomly deployed users. IEEE Signal Process. Lett. **21**(12), 1501–1505 (2014)
6. Laneman, J.N., Tse, D.N., Wornell, G.W.: Cooperative diversity in wireless networks: efficient protocols and outage behavior. IEEE Trans. Inf. Theory **50**(12), 3062–3080 (2004)
7. Ding, Z., Peng, M., Vincent Poor, H.: Cooperative non-orthogonal multiple-access in 5G systems. IEEE Commun. Lett. **19**(8), 1462–1465 (2015)
8. Liu, H., Ding, Z., Kim, K.J., Kwak, K.S., Poor, H.V.: Decode-and-forward relaying for cooperative NOMA systems with direct links. IEEE Trans. Wireless Commun. https://doi.org/10.1109/twc.2018.2873999
9. Men, J., Ge, J.: Non-orthogonal multiple access for multiple-antenna relaying networks. IEEE Commun. Lett. **19**(10), 1686–1689 (2015)
10. Liang, X., Wu, Y., Ng, D.W.K., Zuo, Y., Jin, S., Zhu, H.: Outage performance for cooperative NOMA transmission with an AF relay. IEEE Commun. Lett. **21**(11), 2428–2431 (2017)
11. Men, J., Ge, J.: Performance analysis of non-orthogonal multiple access in downlink cooperative network. IET Commun. **18**, 2267–2273 (2015)
12. Yue, X., Liu, Y., Kang, S., Nallanathan, A.: Performance analysis of NOMA with fixed gain relaying over Nakagami-m fading channels. IEEE Access **5**, 5445–5454 (2017)
13. Gradshteyn, I.S., Ryzhik, I.M.: Table of Integrals, Series and Products, 7th edn. Academic, San Diego, CA, USA (2007)

Kernel Parameter Optimization of One-Class Classifier and Application in Fault Detection

Haizhen Zhu, Mingqing Xiao, Lina Sun and Xilang Tang

Abstract The one-class classification, different from conventional multi-class or two-class classification, would work well only using samples from the target class, which is suitable for solving problems of fault detection where normal states of system are easy to acquire. In this paper, we utilize Gaussian kernel to construct the one-class classifier where hypersphere works as boundary of normal working dataset of a system. To optimize the performance of the classifier, we adjust two parameters of the Gaussian kernel to optimize the tightness of the hypersphere. A set of experiments are carried out where we select the optimal parameters of the classifier for a banana-shaped dataset. The result shows that the classifier, using selected Gaussian kernel parameters, yields good performance on fault detection.

Keywords One-class classifier · Hypersphere kernel parameters · Fault detection

1 Introduction

Fault detection (FD) is designed to monitor the abnormal status of complex systems [1]. Many systems are trying to add FD technology into a system so as to avoid catastrophic accidents. Thus, the study is receiving overwhelming attention among researchers. Generally, the fault detection is regarded as classification problems for that, intuitively, the status of equipment or systems can be classified into normal ones and abnormal ones. Typically, in machine learning, the classification problems are meant to distribute the datasets into two classes or multi-classes. Relatively speaking, the data of the system is easy to collect under normal operating

H. Zhu (✉) · M. Xiao · X. Tang
ATS Lab, Air Force Engineering University, Xi'an 710038, China
e-mail: Haizhzhu@hotmail.com

X. Tang
e-mail: tangxilang@sina.com

L. Sun
School of Mathematics and Statistics, Xi'an JiaoTong University, Xi'an 710049, China

V. Jain et al. (eds.), *Recent Trends in Intelligent Computing, Communication and Devices*, Advances in Intelligent Systems and Computing 1031,
https://doi.org/10.1007/978-981-13-9406-5_89

conditions, but the data acquisition of the system faulty state may be quite expensive or even impossible. Taking the two-class classifications as an example, if the number of samples of one type of data is much less than the number of samples of another type, the positive and negative samples in the training set are unbalanced [2], which may cause the classifier to be too biased toward a large number of sample categories.

The one-class classification (OCC) is conventionally solved by target density estimation [3]. Afterward, the one-class SVM was proposed by Schölkopf et al. in [4] to solve the problem by constructing a hyperplane that separates the target class and the outliers. Tax and Duin [5] attempted to use a hyperplane to distinguish the target class from the other classes. Also in [6], the outlier generation method was proposed to train the classifier. Additionally, some researchers have been devoting to introducing the OCC method into fault detection and some of them have proposed new kernel methods, which are the main fields and need to be discussed in detail in our future work [7–9]. In this paper, our focus is on the optimization of OCC kernel function parameters and to verify the effectiveness of selection by the experiment on banana-shaped normal dataset.

The reminder of this paper is as follows. The second section would give a brief introduction of different algorithms that can be utilized to solve the classification problem. In the third section, the parameters of the selected kernel are altered to verify their effectiveness. A banana-shaped normal dataset, with artificial generated outliers, is utilized to verify the algorithm. The conclusion is given at the end of this paper.

2 One-Class Classification (OCC)

2.1 Constructing the Hypersphere

One-class classification is designed to solve the problems that we have to distinguish the target class or outliers with only data from the target class. The OCCSVM takes the outliers as an area around the origin in feature space. It is devoted to obtain a hyperplane that can best separate the target and all the other possible classes in the space. However, for the fault detection problem, it is of little possibility that the all the abnormal operating data occupies half side of the feature space, thus making it difficult to obtain such a hyperplane. Typically, target class (i.e., the normal operation state) is surrounded by the outliers, which is distributed in all directions [10]. Therefore, in fault detection, we prefer to utilize the hypersphere to detect the outliers. To obtain an optimized hypersphere, the boundary must be tight enough to fit the target class dataset.

The basic description of constructing a hypersphere to classify the targets is as follows:

The boundary of target class is defined by the two parameters, the center and radius which are represented by O and R. For flexibility, we define a slack variable

ξ. $p_i, i = 1, 2, \cdots N$ is the point in the feature space. To fit the dataset with tightest boundary, the loss function is defined as:

$$L(O, R, \xi) = R^2 + C\sum_i \xi_i \tag{1}$$

where the C is a trade-off parameter of the function. We can tighten the hypersphere by increasing C, when boundary is too slack and vice versa. The target can be achieved by minimizing $L(O, R, \xi)$ under the constraints:

$$\|p_i - O\|^2 \le R^2 + \xi_i \tag{2}$$

The Lagrangian obtained from the function and constraints is as follows:

$$L(O, R, \alpha_i, \beta_i, \xi_i) = R^2 + C\sum_i \xi_i - \sum_i \alpha_i\left\{R^2 + \xi_i - \left(\|p_i\|^2 - 2O \cdot p_i + \|O\|^2\right)\right\} - \sum_i \beta_i \xi_i \tag{3}$$

where the Lagrange multipliers α_i and β_i are greater than zero.

We can get the following constraints by setting the partial derivatives of L to O, R and ξ_i to zero, respectively.

$$\sum_i \alpha_i = 1; O = \sum_i \alpha_i p_i; \text{ and } C = \alpha_i + \beta_i \tag{4}$$

Combining the new constraints and Lagrangian, we can get the updated loss function:

$$L = 1 - \sum_i \alpha_i^2 (p_i \cdot p_i) - \sum_{i \neq j} \alpha_i \alpha_j (p_i \cdot p_j) \tag{5}$$

2.2 *Transformation Using the Kernel Function*

We select the typical Gaussian kernel, and the inner product $(x \cdot y)$ becomes $\exp(-\|x - y\|^2/s^2)$. Then, the loss function is:

$$L = 1 - \sum_i \alpha_i^2 - \sum_{i \neq j} \alpha_i \alpha_j \exp(-\|p_i - p_j\|^2/s^2) \tag{6}$$

In this paper, we are devoted to obtain the tightest boundaries of the hypersphere while maintaining lower error ratios. The error ratio can be calculated by the outer accepted and target class rejected ratio. Assume that the total number of samples in N and the O_+ and T_- represent the number of outer accepted and target class rejected, respectively. Thus, the accuracy is calculated by:

$$\text{Acc} = 1 - \frac{O_+ + T_-}{N} \tag{7}$$

3 Experiments

3.1 Transformation Using the Kernel Function

The kernel parameter s and trade-off parameter C affect shape of hypersphere seriously. The following part of this section would devote to figure out how the parameters affect shape of hypersphere and error of the one-class classifier. In the experiment, we try to separate banana-shaped datasets, which depict the normal working states of a system, from randomly generated outliers, which mimic the fault states of the system, using one-class classification algorithm. Next, we set the parameters to different values to verify their effectiveness.

In the experiment above, we set trade-off parameter C to 0.1 and the kernel parameter s to 8, 4, 2.7, and 2.5, respectively. It is obvious that with the decrease in s, the boundaries vary from loose to tight.

While a larger s may lead to bad performance in detecting fault state. Because, it tries to collect more samples in the boundary in which lots of outliers are included. In Fig. 1d, there is no outlier classified as target class, but it would end with a relatively high false alarm probability for fault detection.

We set the s to 4, and trade-off parameter to 0.01, 0.05, 0.1, and 0.3. We can figure out in Fig. 2 that the larger C is set, the closer the boundary is to a sphere. *B*.

The accuracy of the algorithm was calculated, where we set C to 0.1 and s from 1 to 10. The optimal accuracy reaches 0.9904 when s is 4.2. From Fig. 3, we can see that the accuracy maintains a very high level when the range of s is from 1 to 6 which means that we need to balance the effort to reject the outlier and accept the target class. By setting C we can control the shape of the boundary, which depends on the shape of the data in the experiment. The accuracy of this experiment is illustrated in Fig. 4.

It is obvious that, except for a slight increase in accuracy when C is smaller than 0.1, the accuracy of algorithm decreases with the increase in C. The slack variables are penalized more seriously under larger C. The optimal accuracy, 0.9951, is obtained when we set C to 0.07. For the banana-shaped target class, a relatively small C would give the algorithm necessary flexibility, which enables algorithm to achieve better performance.

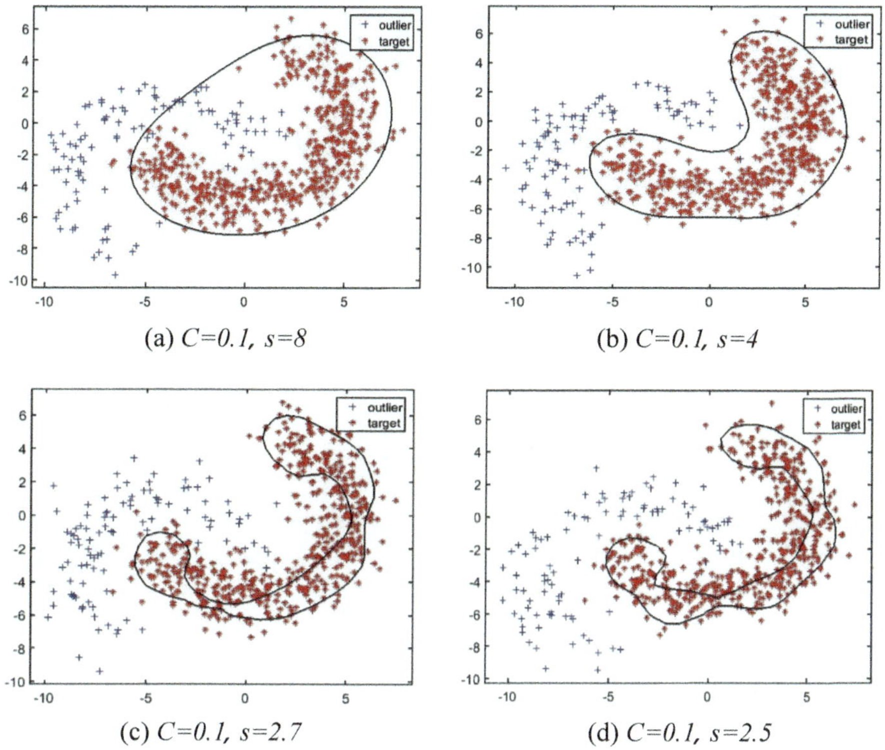

(a) $C=0.1,\ s=8$

(b) $C=0.1,\ s=4$

(c) $C=0.1,\ s=2.7$

(d) $C=0.1,\ s=2.5$

Fig. 1 Performance of the OCC classifier using different *s*s

3.2 *Accuracy of the Classifier*

Tanking both the parameters into account and utilizing the error rate as criteria for evaluating the classifier, we can figure out that for this banana-shaped dataset, the optimized trade-off parameter C and kernel parameter s are (0.12, 5.2) whose corresponding accuracy is 0.9969 (Fig. 5).

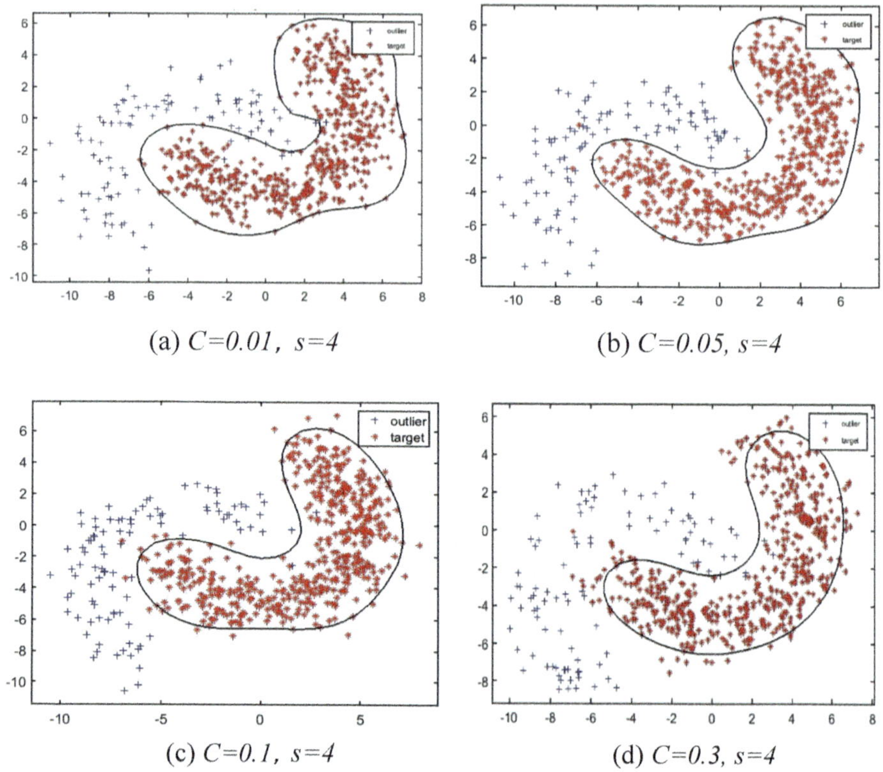

(a) $C=0.01$, $s=4$ (b) $C=0.05$, $s=4$

(c) $C=0.1$, $s=4$ (d) $C=0.3$, $s=4$

Fig. 2 Performance of the OCC classifier using different *C*s

Fig. 3 Accuracy under different *s*s

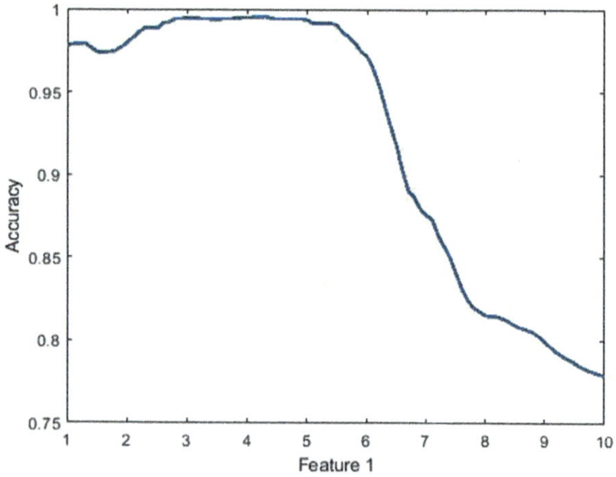

Fig. 4 Accuracy under different *C*s

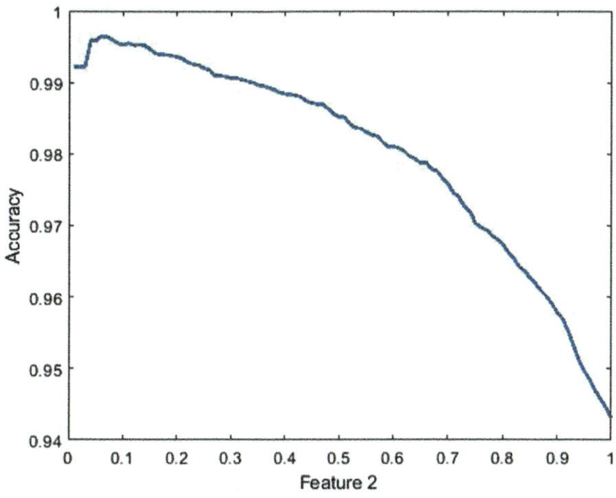

Fig. 5 Accuracy under different *S*s and *C*s

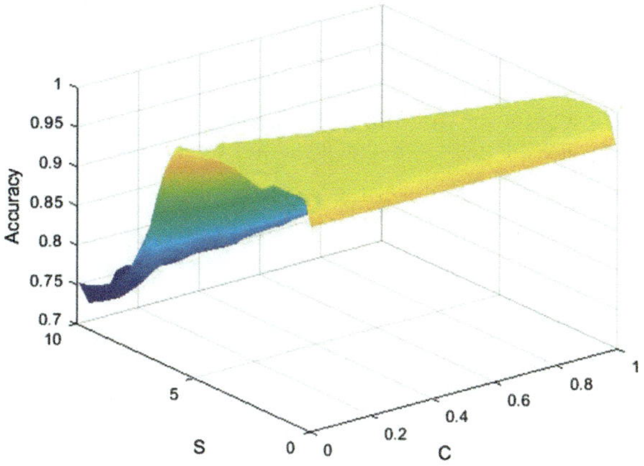

4 Conclusion

In the real-word application, it is quite common that the normal data of a system is abundant, while the fault states are hardly acquired. In this paper, the OCC is utilized to solve the problems in fault detection. The Gaussian kernel is selected to act, and the kernel function of OCC and a hypersphere is constructed to separate the target class from outliers. A banana-shaped normal dataset, with artificial generated outliers, is utilized to verify the algorithm. The effectiveness of two parameters, the C and s, whose function are to adjust the tightness and shape of the hypersphere, are verified through a set of experiments. Moreover, the parameter s affects accuracy of OCC more than the C, and through our experiment, the optimal parameters for the banana-shaped dataset are selected. The outcomes show that the OCC algorithm yields good performance in the experiment and accuracy would reach 0.9969.

Acknowledgements The author would like to thank the referees for their comments, which would be helpful to improve the quality of the current version of paper.

References

1. Rahamathunnisa, U., Chetty, B.C., King, A.C.: A survey on fault detection techniques in different machines—an image processing approach. Int. J. Civil Eng. Technol. **8**(9), 1124–1127 (2017)
2. Wu, S.J., Pham, V.H., Nguyen, T.N.: Two-phase optimization for support vectors and parameter selection of support vector machines: two-class classification. Appl. Soft Comput. **59**, 129–142 (2017)
3. Moya, M.M., Koch, M.W., Hostetler, L.D.: One-class classifier networks for target recognition applications. In: Proceedings World Congress on Neural Networks, vol. III, pp. 797–801 (1993)
4. Scholkopf, B., Williamson, R.C., Smola, A., Shawe-Taylor, J.: SV estimation of a distribution's support. Adv. Neural. Inf. Process. Syst. **41**, 582–588 (2000)
5. Tax, D.M.J., Duin, R.P.W.: Support vector domain description. Pattern Recognit. Lett. **20** (11–13), 1191–1199 (1999)
6. Tax, D.M.J., Duin, R.P.W.: Uniform object generation for optimizing one–class classifiers. J. Mach. Learn. Res. **2**(2), 155–173 (2001)
7. Shin, H.J., Eom, D.H., Kim, S.S.: One-class support vector machines—an application in machine fault detection and classification. Comput. Ind. Eng. **48**(2), 395–408 (2005)
8. Mahadevan, S., Shah, S.L.: Fault detection and diagnosis in process data using one-class support vector machines. J. Process Control **19**(10), 1627–1639 (2009)
9. Xiao, Y., Wang, H., Zhang, L., Xu, W.: Two methods of selecting Gaussian kernel parameters for one-class svm and their application to fault detection. Knowl.-Based Syst. **59**(2), 75–84 (2014)
10. Núñez, V.A.B., Kulkarni, S., Santoso, S., Meléndez, J.: SVM-based classifica-tion methodology for overhead distribution fault events. In: International Conference on Harmonics and Quality of Power, pp. 1–6 (2010)

Development of Single-Wavelength Integrating Nephelometer

Yueqin Wang, Ji Li and Jun Qi

Abstract Atmospheric aerosol refers to a stable mixture of solid particles and liquid particles uniformly dispersed in the atmosphere. The scattering coefficient is one of the important parameters of the optical properties of an aerosol. The integral nephelometer can monitor the scattering coefficient of atmospheric aerosols in real time by measuring the angular integral of light scattering. Based on the similar instruments from abroad, this design has been transformed into light source, signal processing and control, and human–machine interface. A single-wavelength integrated turbidimeter based on integrating sphere light source was developed. The long-term operation data shows that the instrument is stable, reliable, easy to operate, and easy to maintain.

Keywords Integrating nephelometer · Aerosol · Scattering coefficient

1 Introduction

Atmospheric aerosols generally refer to a mixed system composed of the atmosphere and solid and liquid particles suspended in it [1]. The measurement of aerosol scattering and absorption characteristics is the scattering coefficient and absorption coefficient of aerosol. The sum of these two coefficients is the extinction coefficient of aerosol [2]. The integrating nephelometer can monitor the scattering

---

Y. Wang (✉)
Anhui Xinhua University, Hefei 230088, China
e-mail: wyqme@163.com

J. Li
Hefei Institute of Technology Innovation, Chinese Academy of Sciences, Hefei 230088, China

J. Qi
Institute of Applied Technology, Hefei Institution of Physical Science, Chinese Academy of Sciences, Hefei 230088, China

V. Jain et al. (eds.), *Recent Trends in Intelligent Computing, Communication and Devices*, Advances in Intelligent Systems and Computing 1031,
https://doi.org/10.1007/978-981-13-9406-5_90

coefficient of atmospheric aerosols in real time by measuring the angular integral of light scattering.

In the 1950s, foreign researchers researched and invented the first integrating nephelometer and made several improvements in the following decades. The nephelometer technology continued to mature, such as three-wavelength integrating nephelometer from TSI in the USA and M9003 integrating nephelometer from Australian ECOTECH [3]. With the support of the China Meteorological Administration, based on the foreign integrating nephelometer, the light source, signal processing and control, man–machine interface, etc., were modified, and a single-wavelength integrating nephelometer based on the integrating sphere light source was developed.

2 Integrating Nephelometer Measurement Principle

2.1 *Physical Principle*

As the light propagates, its intensity decays, and this attenuation follows the Beer–Lambert law [4]:

$$\frac{I}{I_0} = \exp(-\sigma_{\text{ext}} L) \tag{1}$$

I_0 is the initial intensity of radiation, I is the radiation intensity after passing through a medium with a distance of L, σ_{ext} is the total extinction coefficient,and L is the distance travelled by light. For aerosols, solar radiation is mainly subject to two weakenings when incident on the atmosphere: (1) scattering and absorption of air molecules, and (2) scattering and absorption of aerosol particles.

2.2 *Optical Principle*

Figure 1 reflects the optical principle of integrating nephelometer. The photomultiplier tube can observe scattered light at a solid angle through small holes and apertures. All scattering volumes in the solid angle are illuminated by diffuse light sources. Each piece of scattering volume is:

$$\mathrm{d}V = (R - x)^2 \,\Omega\, \mathrm{d}x \tag{2}$$

According to the geometric relationship:

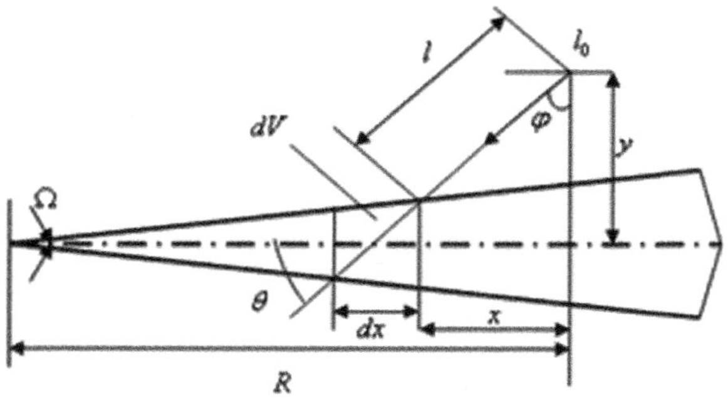

Fig. 1 Optical turbidimeter optical schematic

$$\theta = \frac{\pi}{2} - \varphi \tag{3}$$

$$x = y\cot\theta,\ \mathrm{d}x = y\csc^2\theta\, d\theta,\ l = y\csc\theta,\ \cos\varphi = \sin\theta \tag{4}$$

Then, the scattering volume is:

$$\mathrm{d}V = (R - y\cot\theta)^2\Omega\csc^2\theta\, d\theta \tag{5}$$

Φ is the luminous intensity of the diffuse light source, and the brightness of the φ direction satisfies the cosine relationship, Therefore, the incident light intensity on the scattering volume dV is:

$$I = \Phi\cos\varphi/l^2 = \Phi\sin^3\theta/y^2 \tag{6}$$

The brightness of the scattered light produced by the volume element dV in the solid angle Ω is:

$$dL = I\beta(\theta)dV/(R-x)^2 = (\Phi/y)\,\Omega\,\beta(\theta)\sin\theta d\theta \tag{7}$$

$\beta(\theta)$ is the scattering function [5], and then, the brightness in the unit solid angle received by the photomultiplier tube is:

$$L = \frac{1}{\Omega}\int_{\theta_1}^{\theta_2} dL = \left(\frac{\Phi}{y}\right)\int_{\theta_1}^{\theta_2}\beta(\theta)\sin\theta d\theta \tag{8}$$

Then, the output L of the integrated scatterometer can directly obtain the scattering coefficient σ_{scat}:

$$L = \frac{l_0}{y} \int_0^{\pi} \beta(\theta) \sin\theta d\theta = \frac{l_0}{2\pi y} \sigma_{\text{scat}} \tag{9}$$

2.3 Optical Measurement

During the normal operation of the instrument, three main measurements were taken: shading reference count (C_{sh}), dark count (C_{dark}), and measurement count (C_m). The measurement ratio (MR) is the ratio of the measurement count to the shutter reference count. $\text{MR} = C_m/C_{\text{sh}}$ Since C_{sh} is a known stable value, MR is proportional to the turbidity meter measures the aerosol scattering extinction coefficient as follows:

(1) Calibration of the calibration gas: By controlling the standard gas valve, the standard gas enters the measurement chamber, and the measurement count C_m and the reference count C_{sh} in the open and closed states of the shutter are, respectively, measured, and the ratio of them is calculated to obtain the measurement ratio $MR_s = C_m/C_{\text{sh}}$. (2) Zero gas calibration: The measurement count C'_m and the reference count C'_{sh} are also measured, and their ratio is calculated to obtain the measurement ratio $\text{MR}_z = C'_m/C'_{\text{sh}}$. (3) Draw a calibration curve: The calibration work curve obtained from the above figure is: $y = s \times x + C = 0.0817x + 8.5601$, where S is the slope and C is the intercept. (4) Calculate the scattering coefficient of aerosol particles: For any environmental sample, when the measurement ratio is MR, the aerosol particle scattering coefficient σ_{sp} can be calculated by the following formula (at the time of temperature and pressure, unit: Mm^{-1}):

$$\begin{aligned} \sigma_{\text{sp}} &= \sigma_{\text{scat}} - \sigma_{\text{scat}}(Z) = \frac{(\text{MR} - C \times 10^{-3})}{S \times 10^{-3}} - \sigma_{\text{scat}}(Z) \\ \sigma_{\text{sp}} &= \frac{(\text{MR} - 8.56 \times 10^{-3})}{0.0817 \times 10^{-3}} - 13.6 \end{aligned} \tag{11}$$

(5) The scattering coefficient of the particles converted to standard atmospheric pressure at normal temperature (obtained by PV = nRT) (Fig. 2).

3 Overall Design

The single-wavelength integrating nephelometer consists of a gas path, an optical cavity, and signal processing and control. The gas path includes a sample gas path and a standard gas path. The overall structure block diagram is shown in Fig. 3.

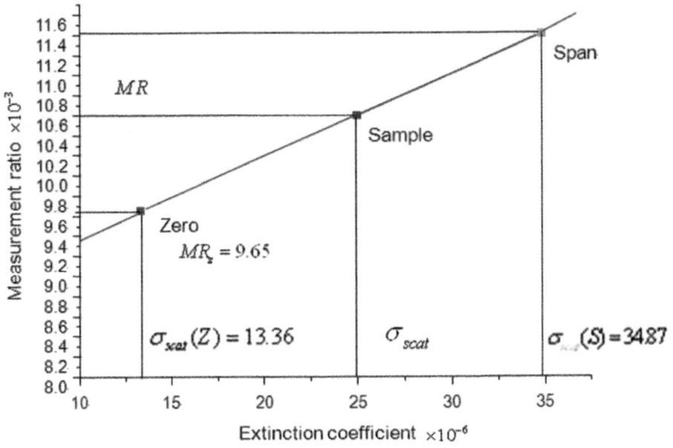

Fig. 2 Calibration curve

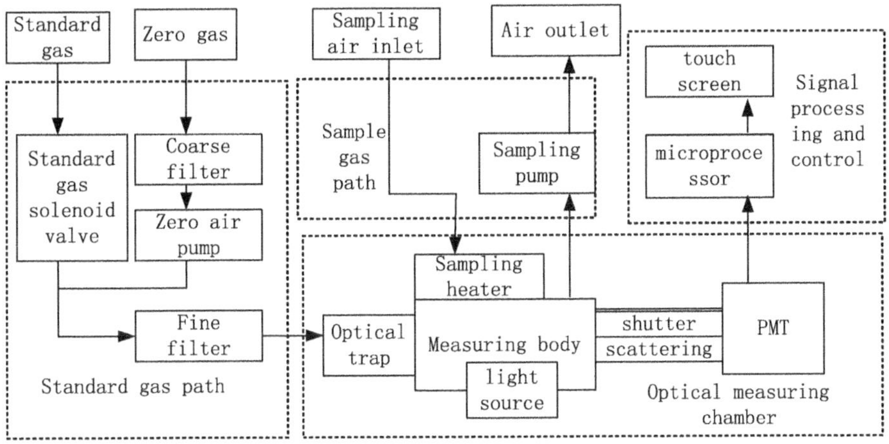

Fig. 3 Block diagram of the overall structure of the integrating nephelometer

3.1 Optical Measuring Chamber

The optical measuring chamber is the optical measuring core of the instrument, and its internal part is mainly composed of an integrating sphere light source, a photomultiplier tube PMT, a flash shutter, an optical trap, and an aperture stop. Its structure is shown in Fig. 4a. When measuring, the integrating sphere light source generates illumination to generate scattered light, the photomultiplier tube PMT receives the sample gas scattered light and performs photoelectric conversion, the scattering shutter provides stable reference light to correct errors caused by light

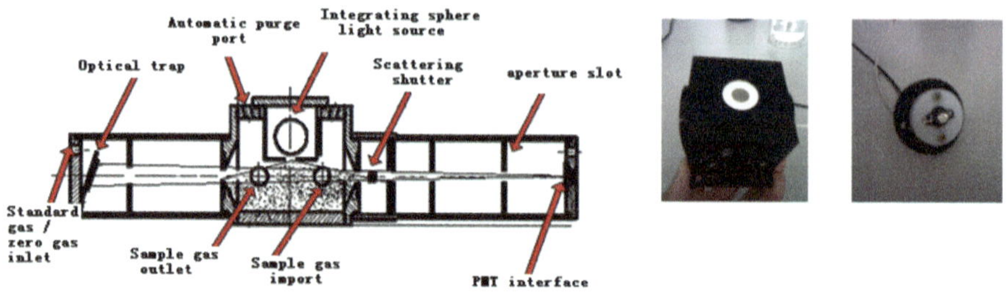

(a) Schematic diagram (b) Physical picture

Fig. 4 Optical measuring chamber

source fluctuation and PMT noise, and the light trap eliminates unwanted the reflected light from the light source and the scattered light from the end of the chamber at the non-detecting end.

It can be known from the measurement principle of the integral turbidity meter that the illumination source only satisfies the calculation formula in the measurement principle with the cosine characteristic, so the cosine characteristic of the light source directly affects the measurement result. The cosine source of the foreign turbidity meter generally adopts a modulated LED array, and the debugging and maintenance of the light source is cumbersome. This design uses an integrating sphere cosine source, and the actual figure is shown in Fig. 4b.

3.2 *Signal Processing and Control*

Signal processing and control is the control core part of the instrument. On the one hand, the signal processing is completed. The single-chip microcomputer sends a timing signal to control the opening and closing of the light source. The photomultiplier tube receives the scattered light and converts the received optical signal into an electrical signal. The CPLD calculates the current time in real time. The number of photons is sent to the MCU for processing, and the final processing data is sent to the touch screen display through the inversion algorithm. On the other hand, the control of the system is completed, including temperature and humidity sensors, pressure sensors, solenoid valves, electromagnets, pumps, zero air pumps, heating sheets, and the like. The signal processing and control block diagram is shown in Fig. 5.

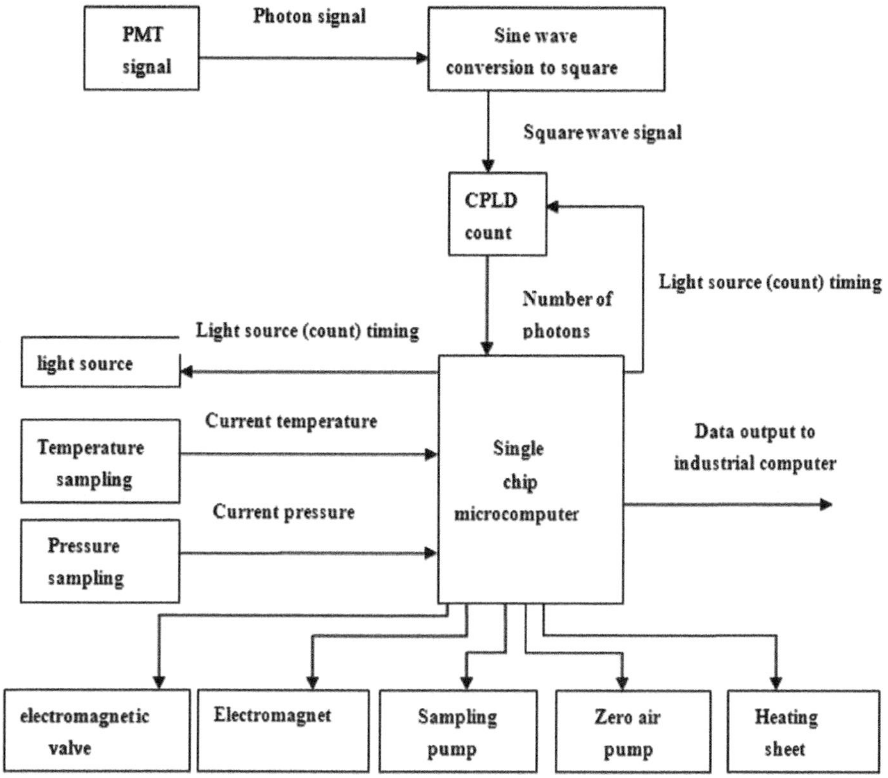

Fig. 5 Signal processing and control block diagram

4 System Test

4.1 Integrating the Ball Source Test

In order to measure whether the integrating sphere light source satisfies the cosine characteristic, a measuring system as shown in Fig. 6 is constructed.

Fig. 6 Cosine characteristics of the source

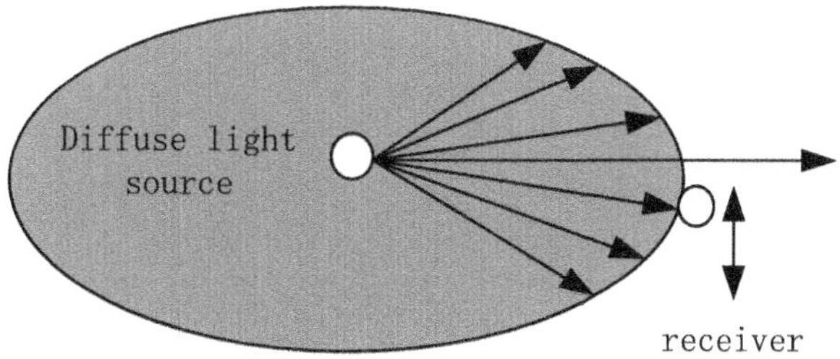

Table 1 Comparison of measured values and theoretical values

Angle (°)	0	5	10	15	20	25	30	35	40
Calculated value	950	946	936	918	893	861	778	778	728
Measurements value	950	942	933	915	890	855	768	768	718
Ratio deviation	1.0	0.996	0.997	0.997	0.997	0.993	0.987	0.987	0.986
%	0.00	0.4	0.3	0.3	0.3	0.7	1.3	1.3	1.4

Angle (°)	45	50	55	60	65	70	75
Calculated value	672	611	545	475	401	325	246
Measurements value	663	605	533	463	390	310	235
Ratio deviation	0.987	0.99	0.978	0.975	0.972	0.953	0.955
%	1.3	1.0	2.2	2.5	2.8	4.7	4.5

(1) On a large plane, select a diffuse source spot to be tested, the spot surface is set up vertically, and the surface normal is parallel to the ground; (2) the receiver and the diffuse spot should be placed on the same plane, the receiver is fixed on the rotating back, and the fixed point of the rotating rod should be perpendicular to and intersect with the spot plane; (3) the opening angle of the receiver on the rotating rod should be larger than the spot diameter; (4) the light source is zero degree normal, and measurement value is recorded at the position of a 5°; (5) move the receiver to the left and right along the circumference, record 5° measurement value for each a, and measure the symmetry left and right; the data repeatability is good.

The measured data is shown in Table 1. The calculated value is calculated according to the cosine formula $I_w = I_0 \cos w$, and I_0 is the standard value. 0° is the observed signal; w is the corner; the measured value is measured according to the cosine formula $I_w = I_0 \cos w$, which deviates from the standard value;

Calculate the calculated and measured values in the above table, as shown in Fig. 7. The measurement results show that: (1) within $0-30^{\circ}$ observation angle, the cosine deviation is less than 1%; (2) within the observation angle $0-70^{\circ}$,

Fig. 7 Curve of calculated and measured values

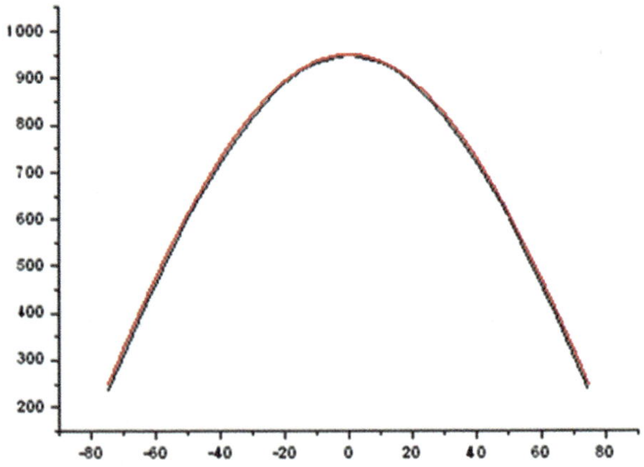

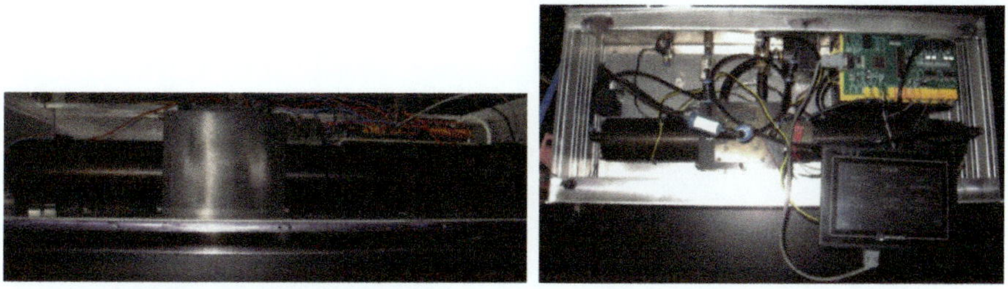

(a) Measurement chamber **(b) Complete instrument**

Fig. 8 Single-wavelength integrating nephelometer

the cosine deviation is less than 5%; it is indicated that the integrating sphere source is a good source of diffusion.

4.2 Prototype Display

At present, the prototype of the domesticated single-wavelength integrated turbidity meter has been developed, and the physical map is shown in Fig. 8. The turbidity meter is running at the Atmospheric Composition Observation Center of the China Meteorological Administration and is compared with similar instruments abroad. The instrument will be further optimized according to the feedback data.

Acknowledgements This work was supported by Research on the Construction of Electronic Information Major Course Group Based on "Integration of theory and practice" (2017jyxm0521), Research on Low Conductivity Fluid Flow Measurement Technology (KJ2018A0596), Analog Circuit (2016gxkcx01), National University Students Innovation and Entrepreneur (201712216017).

References

1. Wang, M., Zhang, R.: Frontier issues in atmospheric aerosol research. Climat. Environ. Res. **1**, 119–124 (2001). (in Chinese)
2. Wang, Z.: Observational study of aerosols and inversion of vertical distribution in semi-arid regions. Lanzhou University (2010). (in Chinese)
3. Jiang, S., Qi, J., Ye, R., et al.: Development of a single-wavelength integrated turbidimeter and several design problems. Analyt. Instr. (5), 5–12 (2015). (in Chinese)
4. Wu, Y., Liu, J., Lu, F., et al.: System design of integral turbidimeter based on LED light source. J. Atmosp. Environ. Optics **7**(005), 370–375 (2012). (in Chinese)
5. Anderson, T.L., Covert, D.S., Marshall, S.F., et al.: Performance characteristics of a high-sensitivity, three-wavelength, total scatter/backscatter nephelometer. J Atmosp Oceanic Technol. **13**(5), 967–986 (1996)

Traffic Analysis of Ad Hoc Network Under Different Communication Conditions

Fang Fang, Chunming Ye and Wei Liu

Abstract To research traffic characteristics of ad hoc network under different communication conditions, a simulation environment of network with AODV protocol is presented. Two networks with different structures are realized to analyze the relationship within traffic, network topology, and communication circumstances. The traffic analysis in experiments illustrates the variations of network throughput and end-to-end delay with the changes of these conditions. Each node also has different traffic features in the process of routing.

Keywords Ad hoc network · Traffic analysis · Network simulation · AODV protocol

1 Introduction

Wireless ad hoc network is widely applied in many fields for its robustness and scalability. Compared with the wired network, the ad hoc network is different in protocols, routing, traffic analysis methods, and so on. The wired network structure is stable and easy to manage. The router, server, and access point in the wired network can supply enough information for understanding the state, diversification of messages, nodes, and network. However, in the wireless ad hoc network, there is no center node to collect messages from the whole network. Analyzing the characteristic of network traffic is a challenging work.

F. Fang
Anhui University of Chinese Medicine, 230037 Hefei, China

C. Ye (✉)
School of Electronic Countermeasure, National University of Defense Technology, 230037 Hefei, China
e-mail: winshepherd@163.com

W. Liu
94789th Unit, 210018 Nanjing, China

V. Jain et al. (eds.), *Recent Trends in Intelligent Computing, Communication and Devices*, Advances in Intelligent Systems and Computing 1031,
https://doi.org/10.1007/978-981-13-9406-5_91

The earlier research on ad hoc network traffic is mainly about the traffic property, which is conditioned by routing protocols and nodes' movement. The relationship between mobility and TCP performance was studied and expected throughput metric was introduced to evaluate network [1]. Some ad hoc network protocols such as DSDV, AODV, and DSR were simulated to compare with their performance on traffic load [2]. These protocols were also analyzed for their influence on throughput, end-to-end delay, and packet delivery fraction in the network [3]. The optimum node density for transferring the maximum number of data packets in AODV network was studied and throughput was measured as the effect [4]. In recent years, it is more likely that the ad hoc and its performance have been discussed in some specific scenarios. New model based on time-varying vehicular speed assumption was proposed to research vehicular ad hoc networks [5]. The information about event-driven warning message propagation was studied to detect the reasonable configuration of messages' TTL time [6]. There are also other researches [7, 8] about traffic features of different kinds of ad hoc network. Most existing studies focus on the performance of the whole network. In fact, each node in the ad hoc network may play a different role during communication and routing process. The diversity of traffic characteristics in each node is important to get the exact state and information about the network. In this research, the traffic in the different node is investigated, and the traffic features are collected to analyze network under the varied circumstance.

2 Protocol and Assessment

2.1 Network Protocol

The protocols in ad hoc network are broadly classified into two types: proactive and reactive. The node with proactive protocol sends messages periodically to collect information and maintain a routing table. This kind of protocol is sensitive to the changes of network structure. However, the periodic messages would burden ad hoc network's overload. The most common proactive protocols include destination sequence distance vector protocol (DSDV) and Optimized Link State Routing (OLSR). The node with reactive protocol only sends messages to establish connection and update routing table when needs to send data. This method is more effective for overload but may cause extra delay at the first beginning of communication. Reactive protocols include Dynamic Source Routing Protocol (DSR), and Ad hoc On-Demand Distance Vector Routing (AODV). We take AODV as an example to study traffic in a single node.

In ad hoc network with AODV protocol, each node within the communication link can hold a routing table. Four kinds of messages are used to generate and refresh the routing table. These messages include neighbor detection (HELLO), route request (RREQ), route reply (RREP), and route error (RERR). HELLO message is sent periodically for the seek of neighbor nodes. RREQ message is sent to find the destination node. RREP message is used to reply RREQ and complete routing table. RERR message is used for notification when error occurs.

2.2 Assessment Indicator

Three assessment indicators are used to analyze traffic in each node, which are network throughput, node throughput, and mean delay. All the indicators in the experiment are normalized for comparison.

(1) Network throughput: We take the number of successfully received messages by all nodes per unit time as the network throughput. The value of throughput indicates the communication state of the whole network.
(2) Node throughput: We take the number of successfully received messages by a single node per unit time as the node throughput. This value only shows the communication state of each node. We combined with network throughput and node throughput to explore the whole and local traffic features in the network.
(3) Mean delay: We take the mean end-to-end delay of all messages in the application layer as the mean delay. Let $D(s)$ denote mean delay. Let rt_i and st_i denote the received time and sent time of the message in node i. N is the number of all nodes in the network. The mean delay is calculated as:

$$D(s) = \frac{1}{N}\sum_{i=0}^{N} (rt_i - st_i) \quad (1)$$

3 Simulation and Model

3.1 Network Structure

There are two scenarios are designed and simulated in experiments. One of the scenarios is communication in the normal state (Scenario 1) and another is nodes communicating when the network topology changes (Scenario 2). Each scenario is tested under two network structures.

(1) net1: Net1 has fewer nodes and communication links. According to the distance between nodes, the links in net1 include 1–2–4–6,1–3–5–7,1–3–8–9–10,4–11–5.
(2) net2: Net2 has more nodes and links. Each node can communicate with its all neighbors. The structures of net1 and net2 are shown in Fig. 1.

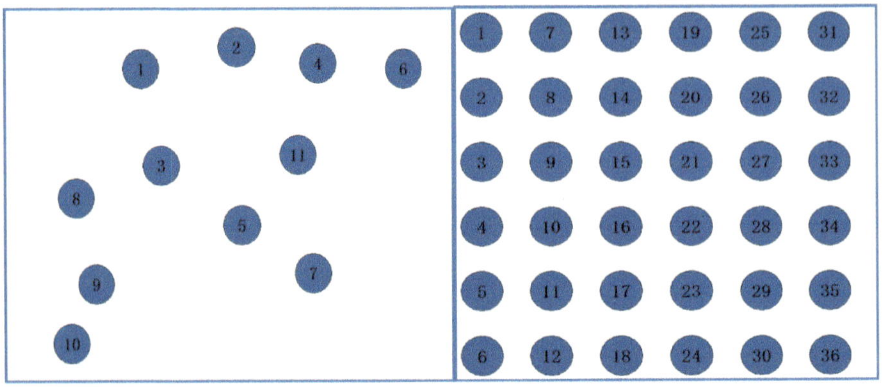

Fig. 1 Network structure left is net1, right is net2

3.2 Traffic Simulation

The simulation program is implemented in Python 3.5. The network includes a routing layer and application layer. All nodes in the network send HELLO, RREQ, and RREP messages based on AODV protocols. Moreover, it would send RERR message when a link breaks. After routing table being completed, the source node will send one data packet per second to destination. The normal communication link in net1 is 6–4–11–5–7. Node 6 is the source node and node 7 is the destination. In net2, the normal link is 1–8–15–22–29–36.

4 Experiment and Results

The simulation includes two scenarios. In scenario 1, the nodes in the network established connection by routing, then send data packets normally. In scenario 2, some nodes between the links would be removed during the communication and the routing table would be rebuilt according to AODV protocol. Each scenario lasts 180s.

4.1 Normal State (Scenario 1)

(1) Traffic of network

First we collect traffic information from two networks in the normal state. The result is shown in Fig. 2. The left graph shows the comparison of network throughput. The right graph shows the comparison of different types of messages' number.

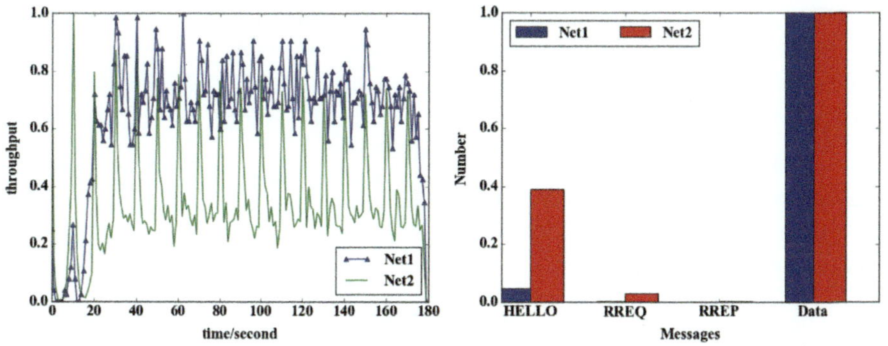

Fig. 2 Throughput and messages of the whole network

The results show that the link has not been established at the beginning, and the throughput is little. After the link been established, throughput becomes much higher. The traffic in net2 has periodic fluctuations for the reason that there are too many HELLO messages in the network. The traffic of net1 appears smoother for the less nodes send HELLO messages.

(2) Traffic of endpoint

We select node 7 in net1 and node 36 in net2 as endpoint to analyze their throughput. The results are shown in Fig. 3. The throughput in the left graph is similar to traffic of the whole network in Fig. 1. However, the throughput of node 36 in net2 doesn't have periodicity. Because the proportion of HELLO messages becomes smaller in node 36. In the right graph, the end-to-end delay curves change little during normal communication.

(3) Traffic of relay node

The traffic characteristics in relay nodes are much the same as that in endpoint. We do not discuss it anymore owing to space reasons.

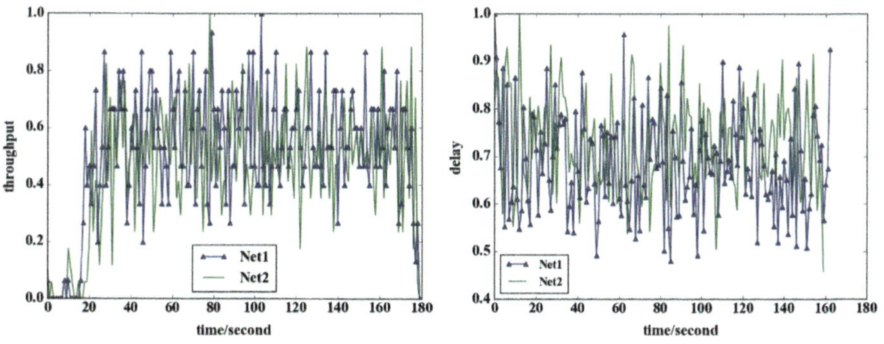

Fig. 3 Throughput and end-to-end delay of endpoints

The results above show that in the normal state the network scale and protocol would greatly affect the traffic features of the whole network. However, in a single node, this influence seems not to be so important.

4.2 Routing Changes (Scenario 2)

(1) Traffic of network

In the next experiments, we remove node 11 from the link in net1 and node 15 from the link in net2 in the middle of normal communication. Then nodes in the link send RERR and other messages to update routing table. The results are shown in Fig. 4. The curves in the left graph show that network throughput has remarkable changes when network rebuilds routing table. The throughput in net1 increases greatly after link recovery because of longer link and more messages in the network.

(2) Traffic of endpoint

The throughput of endpoint node 7 in net1 and node 36 in net2 is illustrated by the right graph in Fig. 4. The traffic in both networks is declined when routing and return after link recovery. The end-to-end delay of these nodes is shown by the left graph in Fig. 5. The delay curves fluctuate at the same time in two networks. The appearance of delay time peak is because many packets are stored in nodes' cache and sent forward after routing table been built.

(3) Traffic of relay node

We choose node 4 in net1 and node 8 in net2 as relay nodes to analyze their traffic. Their throughput is shown in the right graph in Fig. 5. The traffic of these nodes has the same characteristics as the whole network.

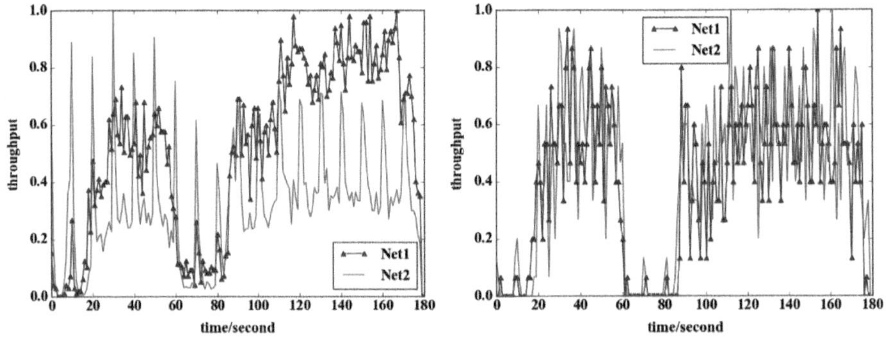

Fig. 4 Throughput and end-to-end delay of network

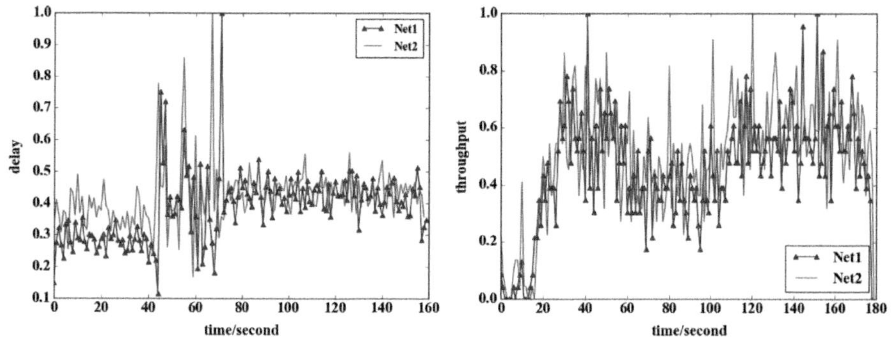

Fig. 5 Throughput of endpoints and relay nodes

5 Conclusion

To analyze traffic characteristics of the ad hoc network, we simulate two networks based on AODV protocol with two kinds of topology and scale. The experiment results indicate that network scale, network structure, protocol, and the position of the node have a different influence on traffic. The difference and relationship of traffic between the whole network and a single node can be used to detect the exact changes in the ad hoc network.

References

1. Holland, G., Vaidya, N.: Analysis of TCP performance over mobile ad hoc networks. Wireless Netw. **8**(2-3), 275–288 (2002)
2. Johansson, P., et al.: Scenario-based performance analysis of routing protocols for mobile ad-hoc networks. In: Proceedings of the 5th Annual ACM/IEEE International Conference on Mobile Computing and Networking. ACM (1999)
3. Zafar, S., Tariq, H., Manzoor, K.: Throughput and delay analysis of AODV, DSDV and DSR routing protocols in mobile ad hoc networks. Int. J. Comput. Netw. Appl. (IJCNA) **3**(2), 1–7 (2016)
4. Royer, E.M., Melliar-Smith, P.M., Moser, L.E.: An analysis of the optimum node density for ad hoc mobile networks. In: IEEE International Conference on Communications, 2001. ICC 2001, vol. 3. IEEE (2001)
5. Zarei, M., Rahmani, A.M., Samimi, H.: Connectivity analysis for dynamic movement of vehicular ad hoc networks. Springer-Verlag, New York, Inc (2017)
6. Zhou, H., Xu, S., Ren, D., et al.: Analysis of event-driven warning message propagation in Vehicular Ad Hoc Networks. Ad Hoc Netw. **55**(C), 87–96 (2017)
7. Varshney, P.K., Agrawal, G.S., Sharma, S.K.: Relative performance analysis of proactive routing protocols in wireless ad hoc networks using varying node density. Inverts J. Sci. Technol. **9**(3), 161–169 (2016)
8. Kuhlmorgen, S., Llatser, I., Festag, A., et al.: Performance evaluation of etsi geonetworking for vehicular ad hoc networks. In: IEEE Vehicular Technology Conference, pp. 1–6 (2015)

Research Progress on Key Technologies of Radar Signal Sorting

Shi-qiang Wang, Caiyun Gao, Qin Zhang, Hui-yong Zeng and Juan Bai

Abstract The complexity and computational complexity of the radar receiver signal processing are mainly concentrated on the signal sorting process. The pre-condition of radar signal sorting is to extract signal features and then to select key features for sorting. This paper discusses several aspects from feature extraction technology and feature selection technology.

Keywords Radar signal sorting · Feature extraction · Feature selection · Kernel space · Rough set theory

1 Introduction

Modern radars are developing in a multi-functional and multi-purpose direction. In order to improve its performance and anti-interference needs, various complex waveform designs are often used to destroy signals structure as much as possible. The signal regularity used for sorting and identification, the adoption of low intercept probability (LPI) technology increases the difficulty of signal classification and deinterleaving, and greatly affects the interception probability of the radar reconnaissance system [1]. The rapid increase in the number of various electronic countermeasures devices, as well as the emergence of extremely complex and dense radar signal environments characterized by the combined application of various working systems and multiple anti-jamming technologies, will make the effectiveness and real-time performance of radar reconnaissance system signal processing hard to guarantee. Based on the above reasons, it can be seen that the complexity and computational complexity of the radar receiver signal processing are mainly concentrated on the signal sorting process. Therefore, this paper

S. Wang · Q. Zhang · H. Zeng (✉) · J. Bai
Air and Missile Defense College, Air Force Engineering University, Xi'an 710051, China
e-mail: 1048768838@qq.com

C. Gao
Basic Department, Air Force Engineering University, Xi'an 710051, China

V. Jain et al. (eds.), *Recent Trends in Intelligent Computing, Communication and Devices*, Advances in Intelligent Systems and Computing 1031,
https://doi.org/10.1007/978-981-13-9406-5_92

provides a theoretical basis for exploring the key technology of pulse deinterleaving technology in modern high-density complex signal environment and exploring a new generation of radar intercept receiver signal sorting processing system.

2 Feature Extraction Technology

Feature extraction refers to a nonlinear mapping (or transformation) from the signal measurement space to the feature space, which realizes the transformation process from the measurement space mode of the huge dimension to the low-dimensional feature space mode [2].

2.1 *Research Directions of Feature Extraction Technology*

Traditional linear transformation methods mainly include principal component analysis (PCA) [3] and linear discrimination analysis (LDA) [4]. The subspace learning of LDA is supervised in order to maximize the ratio of inter-class dispersion (Sb) and intra-class dispersion (Sw) determinants in subspace. LDA assumes that each sample follows a Gaussian distribution and the covariance matrices of the different classes are the same, and all samples are subject to a Gaussian distribution overall.

Due to the existence of non-Gaussian distribution and small samples, feature extraction is also a hot spot in recent years. The work in this area can be divided into the following directions.

(1) A linear feature extraction method for small samples. A typical example of small sample learning is image classification. If the value of all pixel points in the image is directly used as the feature quantity, the dimension of the vector is very high, and the number of samples of each class is very small. A straightforward way to overcome Sw singularity is to regularize the discriminant analysis, making Sw non-singular by matrix smoothing. The Fisherface method uses PCA to reduce the feature dimension from D to $N - M$ (N is the number of samples and M is the number of categories) to make Sw non-singular. However, the reduction of Sw's dimension from D to $N - M$ will result in some loss of authentication information, while down to $N - 1$ dimensions will not be lost. At this time, Sw is still singular, and it needs to extract some features from the zero space of Sw (corresponding to the eigenvalue of 0) [5].
(2) Heteroscedastic discriminant analysis of the covariance matrix in the class. For the case where the category covariance matrix is different, the heteroscedastic discriminant analysis method can obtain better classification performance than LDA. For non-Gaussian distributions or arbitrary distributions, nonparametric

identification analysis is a basic idea for extracting identification features. The resulting method also includes differential analysis based on decision boundaries [6]. The classification projection criterion can also be directly optimized with the classification performance criterion without assuming the parameter probability density, such as minimum classification error (MCE) and mutual information between features and categories [7]. For the case of multimodal distribution for each type of sample, a differential analysis based on mixed Gaussian density can be used [8].

(3) Feature extraction method for maintaining local spatial characteristics. The local retention feature extraction method draws on the idea of manifold learning (ML), in order to maintain the adjacent relationship between sample points in the subspace. The problem with ML is that only the training samples are projected. To extend to the test samples, a parametric model or regression network is needed to represent the projection process. The Local Preserving Projection (LPP) method proposed by He et al. estimates the projection vector by optimizing a local retention criterion, which can be converted into a matrix eigenvalue decomposition problem [9]. Yan et al. proposed a unified framework for feature extraction based on sample proximity analysis, called embedded graph, and proposed a new discriminant analysis method [10].

(4) Nonlinear feature extraction method. Almost all linear feature projection methods can be generalized to the kernel space. Schölkopf et al. first introduced the kernel function to PCA and proposed the Kernel PCA (KPCA) method [11]. Yang conducted an in-depth analysis of the combined PCA dimensionality reduction and FDA feature extraction in kernel space and proposed an effective algorithm [12].

2.2 Feature Extraction Technique Based on Empirical Mode Decomposition

The empirical mode decomposition (EMD) is derived from the time–frequency analysis method proposed by Huang [13, 14] for time-varying, non-stationary signals. The time–frequency (TF) analysis method based on EMD does not need to use the prior knowledge of the signal. The basis function itself is adaptively decomposed from the original signal, which overcomes the above deficiencies.

EMD-based time–frequency analysis consists of two main steps: first, empirical mode decomposition of time series data, decomposition of the intrinsic mode function (IMF) group; then, each eigenmode function is subjected to Hilbert transformation recombination time spectrum analysis. The most critical of these is the empirical mode decomposition method, which essentially smoothes a signal, and the result is that the fluctuations or trends of different scales in the signal are decomposed step by step, resulting in a series of different sequence of data of a feature scale, each sequence being an eigenmode function component. The essence of the EMD method is to decompose the signal into the sum of several IMFs. Different IMFs have different scale features. Extracting these features facilitates a more detailed analysis of signals.

3 Feature Selection Technology

Feature selection refers to the process of selecting some of the most effective features from a set of features to achieve a reduction in the number of feature space dimensions [15]. For radar signal sorting, the feature selection is to select the feature subset that makes the classification ability the strongest from the original feature set, so that the classifier achieves an effect that is similar before the feature selection.

3.1 Research Direction of Feature Selection Technology

In recent years, due to the increasing complexity (the tens of thousands of feature dimensions and the probability density deviating from the Gaussian distribution), new feature selection methods have been proposed and new research hot spots formed [16].

The method of feature selection can be divided into several types according to the degree of interaction between the feature selection process and the classifier, such as filter, wrapper [17], embedded, and hybrid.

(1) Filtered feature selection. Filtered feature selection is completely independent of the classifier, which is also the most common feature selection method. The selection process is small, but often the selected features are not suitable for classification.
(2) Wrapper feature selection. In the wrapper method, the performance of a feature subset is measured using a classifier on the correctness of the validation sample, so that the features selected are more suitable for the classifier. Since many feature subsets are evaluated in the feature selection process (the number of subsets grows exponentially), even with sequential forward search, the amount of wrapper calculation is large, only suitable for cases where the feature dimension is not too high. Another problem with wrapper is that when the training sample is small, it will cause over-fitting, and the generalization performance will be poor.
(3) The embedded method includes the feature selection function in the training process of the classifier, so it is also dependent on the classifier as the wrapper. A classic method is LASSO [18]. Two representative embedded methods have recently been sparse support vector machines [19] and boosting feature selection [20].

(4) Hybrid feature selection combined with different methods to achieve better computational complexity, the trade-off of classification performance is often used when the initial number of features is very large. For example, the method in [21] uses three methods in three stages to reduce the number of features: filtering, clustering, and combined selection. Filtering and wrapper methods are used in combination.

Most of the research work in the field of feature selection has focused on filtering methods. Early work in the field of pattern recognition focused on search strategies [22], and feature subset evaluation criteria mostly adopt distance criteria based on Gaussian density assumptions, such as Fisher criterion and Mahalanobis distance. When the criteria better measure the separability of feature subsets and are relatively stable, simple search strategies can produce good classification performance. Two types of comparative feature evaluation methods are analyzed below: a margin-based approach and a mutual information-based approach.

The basic principle of feature selection is to select the characteristics of the category relating to the exclusion of redundant features. This class correlation and redundancy is usually measured by mutual information (MI). The mutual information between features and categories measures the correlation of features well, while the mutual information between features measures the similarity (redundancy) between them. Therefore, the feature selection method based on mutual information generally follows a pattern in which a feature with the largest cross-information and the smallest mutual information with the previously selected feature is found in the sequential forward search [23].

3.2 Feature Selection Technique Based on Rough Set Theory

Rough set theory is a rigorous mathematical method proposed by Polish scholar Pawlak in 1982 to find rules from data. It has excellent data reasoning performance when dealing with inaccurate, inconsistent, incomplete, and redundant information [24, 25]. An important concept of a rough set is attribute reduction, which refers to finding the smallest set of attributes that can classify objects with the same precision as the original set of attributes and removing the extra attributes to obtain a more robust and concise classification rule.

Zhang et al. used a method of discretization of interval continuous attributes to discretize the eigenvalues firstly and then simplified the feature set by using the attribute reduction method based on discernible matrix and logic operation [26]. The computation time required for this feature selection algorithm, as well as the selected feature set classification performance, is superior to many other selection methods.

The advantage of rough set theory applied to feature selection is that since each feature attribute is inconsistent in distinguishing the categories of data points, the importance of the attributes must be fully considered in the process of clustering. Rough set theory is good at dealing with inaccurate data. The attribute reduction knowledge in the rough set has unique advantages for analyzing the importance of feature attributes. Therefore, the rough set theory is applied to analyze the characteristic attributes of data sets, and different rights are given according to the importance of attributes. Values improve the quality of subsequent clusters.

4 Conclusions

In this paper, we analyze the characteristics of current radar and its transmitted signals and form a conclusion that it is necessary to study the precondition of radar signal sorting in the modern high-density complex signal environment and when exploring a new generation of radar intercept receiver signal sorting processing system. Specifically, this paper presents the latest research progress of feature extraction and feature selection technologies, and thus, this paper provides a theoretical basis for exploring the key technology of radar signal sorting.

Acknowledgements This paper was supported in part by the National Natural Science Foundation of China under the Grant no. 61601499, 61701527, and 61601503.

References

1. Xiong, J., Wang, W.Q., Cui, C., Gao, K.D.: Cognitive FDA-MIMO radar for LPI transmit beamforming. IET J. Mag. **11**(11), 1574–1580 (2017)
2. Murthy, C.A.: Bridging feature selection and extraction: compound feature generation. IEEE Trans. Knowl. Data Eng. **29**(4), 757–770 (2017)
3. Han, N., Song, Y., Song, Z.: Bayesian robust principal component analysis with structured sparse component. Comput. Stat. Data Anal. **109**, 144–158 (2017)
4. Huang, S., Yang, D., Zhou, J., Zhang, X.H.: Graph regularized linear discriminant analysis and its generalization. Pattern Anal. Appl. **18**(3), 639–650 (2015)
5. Chen, L.F., Liao, H.Y.M., Ko, M.T., Lin, J.C., Yu, G.J.: A new LDA-based face recognition system which can solve the small sample size. Pattern Recogn. **33**, 1713–1726 (2000)
6. Zhang, J., Liu, Y.: SVM decision boundary based discriminative subspace induction. Pattern Recogn. **38**(10), 1746–1758 (2005)
7. Hild II, K.E., Erdogmus, D., Tokkola, K., Principe, J.C.: Feature extraction using information-theoretic learning. IEEE Trans. PAMI **28**(9), 1385–1392 (2006)
8. Zhu, M., Matinez, A.M.: Subclass discriminant analysis. IEEE Trans. PAMI **28**(8), 1274–1286 (2006)
9. He, X., Yan, S., Hu, Y., Zhang, H.J.: Learning a locality preserving subspace for visual recognition. In: Proceedings of 9th ICCV, Nice, France, pp. 385–392 (2003)
10. Yan, S., Xu, D., Zhang, B., Zhang, H.J., Yang, Q., Lin, S.: Graph embedding and extension: a general framework for dimensionality reduction. IEEE Trans. PAMI **29**(1), 40–51 (2007)
11. Schölkopf, B., Smola, A., Müller, K.R.: Nonlinear component analysis as a kernel eigenvalue problem. Neural Comput. **10**(5), 1299–1319 (1998)

12. Yang, J., Frangi, A.F., Yang, J.Y., et al.: KPCA plus LDA: a complete kernel Fisher discriminant framework for feature extraction and recognition. IEEE Trans. PAMI **27**(2), 238–244 (2005)
13. Huang, N.E.: The empirical mode decomposition and the Hilbert spectrum for nonlinear and non-stationary time series analysis. J. Proc. R. Soc. Lond. A. **454**, 903–995 (1998)
14. Li, H.G., Hu, Y., Li, F.C., Meng, G.: Succinct and fast empirical mode decomposition. Mech. Syst. Signal Process. **85**, 879–895 (2017)
15. Wang, J., Xu, H.P., Wei, J.M.: Feature selection via vectorizing feature's discriminative information. Lect. Notes Comput. Sci. **9931**(1), 493–505 (2016)
16. Guyon, I., Elisseeff, A.: An introduction to variable and feature selection. J. Mach. Learn. Res. **3**, 1157–1182 (2003)
17. Kohavi, R., Tohu, G.: Wrappers for feature selection. Artif. Intell. **97**(1–2), 273–324 (1997)
18. Tibshirani, R.: Regression selection and shrinkage via the lasso. J. Roy. Stat. Soc. Ser. B **58** (1), 267–288 (1996)
19. Bi, J., Bennett, K., Embrecht, M., Breneman, C., Song, M.: Dimensionality reduction via sparse support vector machines. J. Mach. Learn. Res. **3**, 1229–1243 (2003)
20. Viola, P., Jones, M.J.: Rapid object detection using a boosted cascade of simple features. In: Proceedings of CVPR 2001, Hawaii, vol. 1, pp. 511–518 (2001)
21. Bins, J., Draper, B.A.: Feature selection from huge feature sets. In: Proceedings of 8th ICCV, vol. 2, pp. 159–165 (2001)
22. Kudo, M., Sklansky, J.: Comparison of algorithms that select features for pattern classifiers. Pattern Recogn. **33**(1), 25–41 (2000)
23. Battiti, R.: Using mutual information for selecting features in supervised neural net learning. IEEE Trans. Neural Networks **5**(4), 537–550 (1994)
24. Pawlak, Z.: Rough sets. Int. J. Comput. Inform. Sci. **11**, 341–356 (1982)
25. Jia, X.Y., Shang, L., Zhou, B., Yao, Y.Y.: Generalized attribute reduct in rough set theory. Knowl.-Based Syst. **91**, 204–218 (2016)
26. Zhang, G.X., Hu, L.Z., Jin, W.D.: Discretization of continuous attributes in rough set theory and its application. Lect. Notes Comput. Sci. **3314**, 1020–1026 (2004)

Fault Diagnosis Strategy Optimization Under Unreliable and Multivalued Tests

Yajun Liang, Mingqing Xiao, Xiaofei Wang, Tong Han and Yawei Ge

Abstract In this paper, we propose a method of optimal diagnosis strategy based on Markov decision process with the inference process of fault diagnosis, and the multi-valued outcomes were treated as the independent signals. And, the uncertainty of tests is depicted by the fault detection and false alarm probabilities for the existence of multiple interference sources in real world. The simulation results also illustrate that this method is valid to get the optimal strategy with minimum cost.

Keywords Diagnosis strategy · Multi-valued test · Unreliable tests · Markov decision process

1 Introduction

As a key technique of fault diagnosis, the optimal solution of fault diagnosis strategy with minimum cost and isolation precision has been a research hotspot. Many algorithms have been proposed, for example, greedy algorithm [1], AND/OR graph [2], dynamic programming (DP) [3], genetic algorithm (GA) [4], and so on, which indeed solve the diagnosis strategy problem of perfect and binary tests with great performance.

But for the complex electronic weapon system, the outcomes of tests can be multi-valued and there can be many conditions where a test fails. So, the above algorithms are unsuitable, otherwise, large amount of diagnosis information will be lost and the diagnosis accuracy and efficiency will be reduced inevitably [5]. Moreover, the tests are always unreliable in real systems due to improper setup, operator error, electromagnetic interference, or various environmental conditions [6]. To deal with this, we regard the process of fault diagnosis as a sequential decision problem and model it based on Markov decision process(MDP) which has widely be applied in medical decision making [7], defense of radio networks [8],

---

Y. Liang (✉) · M. Xiao · X. Wang · T. Han · Y. Ge
Air Force Engineering University, Xi'an 710038, China
e-mail: 1214102891@qq.com

V. Jain et al. (eds.), *Recent Trends in Intelligent Computing, Communication and Devices*, Advances in Intelligent Systems and Computing 1031,
https://doi.org/10.1007/978-981-13-9406-5_93

and so on. To simplify this problem, we translate the multi-valued tests into binary tests through measurement region division. The reliability of tests is characterized by the detection and false alarm probabilities. The weight factor of test cost and diagnosis information gain is traduced to acquire different deviation strategies according to actual demands. At last, the simulation results also demonstrate the proposed method is simple and able to achieve the fault diagnosis accuracy and maintenance support in complex environment.

2 Problem Statement

In our work, the repeated test is allowed for multi-outcomes, and the diagnosis strategy optimization problem can be further defined as follows.

1. $\{f_0, f_1, f_2, \ldots, f_m\}(m \geq 0)$ is a finite set of failure sources in a system, where $f_i(0 \leq i \leq m)$ denotes the ith failure source and f_0 denotes the fault-free condition.
2. The a priori probability of each failure source, $p(f_i)$ is known and $\sum_{i=0}^{m} p(f_i) = 1$.
3. $T = \{t_1, t_2, \ldots, t_n\}(1 \leq n)$ is a finite set of tests, where each test $t_j(1 \leq j \leq n)$ can check a subset of failure sources.
4. The reliability of each test $t_j(1 \leq j \leq n)$ is characterized by a probability pair $\{\mathrm{Pd}_j, \mathrm{Pf}_j\}$, where Pd_j and Pf_j are the detection and false alarm probabilities, respectively.
5. The test cost set $C = \{c_1, c_2, \ldots, c_n\}$ measured in terms of time or other economic factors, which corresponding to test set T.

In binary tests, the outcomes are pass or fail. So, the test-fault dependency matrix $B = [b_{ij}]_{m \times n}$ can be obtained as if test t_j is able to monitor failure source f_i, then $b_{ij} = 1$ and 0 otherwise. But in real systems, the measurements are usually multi-valued and different interval values correspond to different system states. For that, the measured values can be divided into intervals. Then, the dependency between each test and fault is further related to the intervals of the measured value. So, each test can be rewritten as $t_j = \{v_{j1}, v_{j2}, \ldots, v_{jr}\}(1 \leq j \leq n, 1 \leq r)$, where v_{jr} represents the rth value region or interval of test t_j. Then multi-valued test-fault dependency matrix $D = [d_{ijr}](1 \leq i \leq m, 1 \leq j \leq n, r \geq 1)$ can be obtained, where $d_{ijr} = 1$ means that test t_j can monitor the failure source f_i and 0 otherwise.

The problem is to get an optimal strategy with minimum cost in long run to isolate the failure source to LRU or SRU accuracy as soon as possible.

3 Diagnosis Model Based on MDP

The fault diagnosis process is considered to be a process of reducing the ambiguity of system fault state, and outcomes of tests at each step are used to induce the possible fault state. The principle of that, named inference engine, is given as

$$\begin{cases} F_{jrf} = \{f_i | d_{ijr} = 1, \forall f_i \in F\} & \text{if } t_{jr} \text{ failed} \\ F_{jrp} = \{f_i | d_{ijr} = 0, \forall f_i \in F\} & \text{if } t_{jr} \text{ passed} \end{cases} \tag{1}$$

where F is the ambiguity group of system fault state before executing test t_{jr}, and the next fault state could be F_{jrf} or F_{jrp} judging from the actual outcomes of test t_{jr}. So, the next fault state is only related to the current fault state and test action being chosen, but has no relationship with the fault states and tests being executed before. This is a typical Markov process. So, the Markov decision process (MDP) is applied to get the optimal strategy in fault diagnosis.

MDP is described by a five-tuple$\{S, A(i), K, p(j|i, a), r(i, a)\}(i, j \in S, a \in A)$, where:

- S is the state space, a set of states which descript all the possible situation of the system be observed at each time step.
- $A(i)$ is a set of workable action s at state i as A is the set of all possible actions.
- K is the set of time steps where decision needs to be made.
- $p(j|i, a)$ denotes the probability function of state transfers from i to j after taking action a, and $\sum_{j \in S} p(j|i, a) = 1$.
- $r(i, a)$ represents the reward or payoff of taking action a at state i.

With theories above, the strategy model based on MDP can be obtained. Consequently, the state space $S = \{F_1, F_2, \ldots, F_q\}(q \geq 1)$ contains all the possible fault state in the diagnosis process. And, the action space $A = T$ is equal to the available test set. Because of the false alarm caused by electromagnetic interference, improper setup, operator error, or environmental conditions, the outcomes of tests are unreliable. In Eq. (1), the transition probability $p(j|i, a)$ is equal to the detection and false alarm probability pair $\{\mathrm{Pd}_j, \mathrm{Pf}_j\}$. Moreover, the reward function can be confirmed considering cost and information gain of each test action, which is given by

$$r(i, a) = r(F, t_j) = \alpha I(F, t_j) - (1 - \alpha)c_j \tag{2}$$

$$I(F, t_j) = -\left(\frac{P(F_{jp})}{P(F)} lb \frac{P(F_{jp})}{P(F)} + \frac{P(F_{jf})}{P(F)} lb \frac{P(F_{jf})}{P(F)}\right) \tag{3}$$

where $F, F_{jP}, F_{jf} \in S$, where F_{jp}, F_{jf} are the subset of F, and lb is the binary logarithm, α is a weight factor which represents the deviation of diagnosis strategy. Also, the cost is negative for we expect bigger return.

For any strategy π, mapped from state space to action space, we can get a specific utility function of summing over all the rewards to estimate the strategy.

The γ-discount $(0 < \gamma < 1)$ criterion is considered in our work. So, the expected total reward of applying strategy π and starting from state F_1 can be calculated as

$$\begin{aligned} V^{\pi}(F_1) &= E^{\pi}\left[r_0 + \sum_{k=1} \gamma^k r_k | s_0 = F_1\right] \\ &= \sum_{t_n \in T} \pi(F_1, t_n)\left[r(F_1, t_n) + \gamma \sum_{F_q \in S} p(F_q|F_1, t_n) V^{\pi}(F_q)\right], \pi \in \Pi \end{aligned} \tag{4}$$

where $E^{\pi}[\cdot]$ is the expectation operation of strategy π, Π is the strategy space, and s_k represents the system state at time step k. So, there is an optimal value function, which is

$$\begin{aligned} V^{\pi^*}(F_1) &= \sup_{\pi \in \Pi} V^{\pi}(F_1), i \in S \\ &= \max_{t_n \in T}\left[r(F_1, t_n) + \gamma \sum_{F_q \in S} P(F_q|F_1, t_n) V^{\pi^*}(F_q)\right] \end{aligned} \tag{5}$$

Then, the optimal strategy π^* can be obtained as

$$\pi^* = \arg\max_{\pi \in \Pi} V^{\pi}(F_1) \tag{6}$$

There are many classic algorithms to get the optimal strategy by solving Eqs. (4) and (5), and we adopt the strategy iteration algorithm in this paper.

4 Simulation Results

In this paper, the proposed method is employed in a practical case of the video amplifier circuit in a tracker to get the optimal fault diagnosis strategy. After reachability analyze, the fault-multivalued test dependency matrix is acquired as shown in Table 1. So, the system has seven possible faults, two tests of multi-valued and two binary tests, where t_2 has two outcomes and t_3 has three outcomes. Then, the fault states of the diagnosis process are obtained according to Eq. (1) as shown in Table 2. So $S = \{F_1 \sim F_{55}\}$ and test action space $A = \{t_1 \sim t_4\}$.

The MDP toolbox is used in MATLAB circumstance to solve the optimal solution in this paper with Eqs. (2) and (3) where $\alpha = 0.50$. We do a comparison of different discount factors. The result is shown in Fig. 1. The best test selections of each fault state are obtained in long run. For simplify, we number the tests and multi-valued tests by natural number, i.e., $v_{21}{=}t_2, v_{22}{=}t_3, \ldots$. And, it is easy to

Table 1 Fault multi-valued test dependency matrix

FT	t_1	t_2		t_3			t_4	$P(f_n)$
	v_1	v_{21}	v_{22}	v_{31}	v_{32}	v_{33}	v_4	
f_0	0	0	0	0	0	0	0	0.30
f_1	1	0	1	0	1	0	0	0.15
f_2	1	0	0	0	1	0	1	0.11
f_3	0	0	0	1	0	0	1	0.05
f_4	0	0	0	0	0	0	1	0.10
f_5	0	1	0	0	0	1	0	0.09
f_6	0	0	1	0	0	1	1	0.08
f_7	1	0	1	1	0	0	0	0.12
Pd_j	0.85	0.87	0.80	0.88	0.92	0.90	0.78	–
Cost (c_j)	1.0	1.2		1.0			1.5	–

Table 2 All the possible fault states of video amplifier circuit

Fault state	Failure source	Fault state	Failure source	Fault state	Failure source	Fault state	Failure source
F_1	$f_0, f_1, f_2, f_3, f_4, f_5, f_6, f_7$	F_2	f_0, f_3, f_4, f_5, f_6	F_3	f_1, f_2, f_7	F_5	f_5
F_4	$f_0, f_1, f_2, f_3, f_4, f_6, f_7$	F_6	f_0, f_2, f_3, f_4, f_5	F_7	f_1, f_6, f_7	F_8	$f_0, f_1, f_2, f_4, f_5, f_6$
F_9	f_3, f_7	F_{10}	$f_0, f_3, f_4, f_5, f_6, f_7$	F_{11}	f_1, f_2	F_{12}	$f_0, f_1, f_2, f_3, f_4, f_7$
F_{13}	f_5, f_6	F_{14}	f_0, f_1, f_5, f_7	F_{15}	f_2, f_3, f_4, f_6	F_{16}	f_3, f_4, f_6
F_{17}	f_3, f_4, f_5	F_{18}	f_6	F_{19}	f_3	F_{20}	f_4, f_5, f_6
F_{21}	f_0, f_3, f_4	F_{22}	f_2	F_{23}	f_1, f_7	F_{24}	f_7
F_{25}	f_0, f_3, f_4, f_6	F_{26}	f_0, f_2, f_3, f_4	F_{27}	f_0, f_1, f_2, f_4, f_6	F_{28}	f_0, f_3, f_4, f_6, f_7
F_{29}	f_0, f_1, f_7	F_{30}	f_0, f_3, f_4, f_5	F_{31}	f_0, f_2, f_4, f_5	F_{32}	f_0, f_5
F_{33}	f_2, f_3, f_4	F_{34}	f_1, f_6	F_{35}	f_6, f_7	F_{36}	f_1
F_{37}	f_0, f_4, f_5, f_6	F_{38}	f_0, f_1, f_2, f_4	F_{39}	f_0, f_1, f_5	F_{40}	f_2, f_4, f_6
F_{41}	f_0, f_3, f_4, f_7	F_{42}	f_0, f_5, f_7	F_{43}	f_0, f_3, f_4	F_{44}	f_4, f_6
F_{45}	f_4, f_5	F_{46}	f_4	F_{47}	f_0, f_4, f_7	F_{48}	f_0, f_2, f_4
F_{49}	f_0	F_{50}	f_0, f_4, f_6	F_{51}	f_0, f_1	F_{52}	f_0, f_7
F_{53}	f_0, f_4, f_5	F_{54}	f_2, f_4	F_{55}	f_0, f_4		

know that there is no difference in strategies with different discounts. With the above results, we can construct the fault diagnosis tree as Fig. 2.

Then, the diagnosis strategies of different weight factors, $\alpha = 0.4, 0.5, 0.6$, are also acquired as shown in Fig. 3, as well as the utility values under them in Fig. 4.

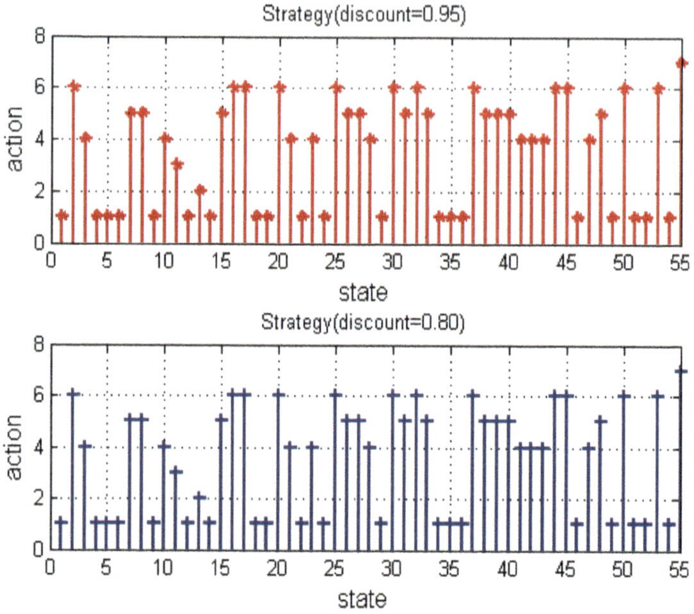

Fig. 1 Optimal strategy map

Fig. 2 Optimal fault diagnosis tree

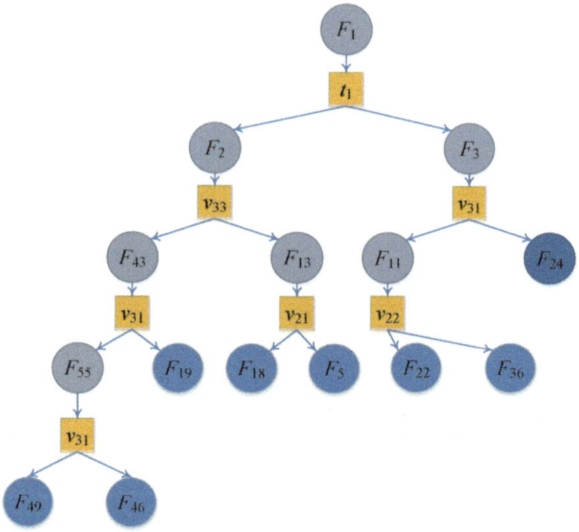

From the above results, different weight factors represent the deviation of diagnosis strategy, seeking more on test cost or the fast diagnosis. So, the weight factor can be set an appropriate value in practice flexibly, to get the optimal diagnosis strategy. Especially, we also can quickly get the optimal strategy starting

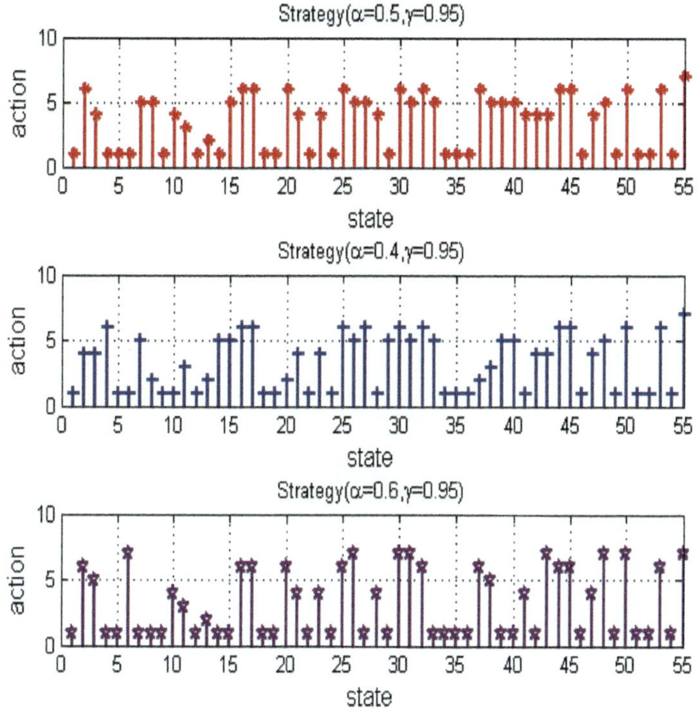

Fig. 3 Optimal strategies under different weights

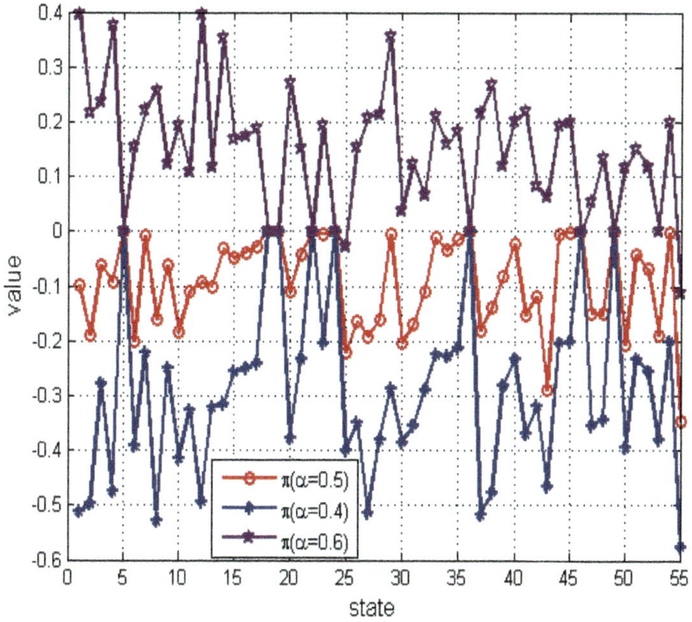

Fig. 4 Optimal utility values

from an arbitrary fault state. This method provides another way to achieve fast fault isolation with minimum cost and saving much valuable time in weapon maintenance support.

5 Summary

This paper provides a novel method to get the optimal fault strategy with unreliable and multi-valued test based on MDP. The validity and efficiency of the proposed method are demonstrated to guide the fault detection and isolation fast in complex and uncertainty environment.

References

1. TU, F., Patfipatik, R.: Rollout strategy for sequential fault diagnosis. IEEE Trans. Syst. Man Cybern. **33**(1), 86–99 (2003)
2. Pattipati, K.R., Alexandridis, M.: Application of heuristic search and information theory to sequential fault diagnosis. IEEE Trans. Syst. Men Cybern. **20**(4), 872–887 (1999)
3. Wei, W.C., Coolen, K., Leus,R.: Sequential testing policies for complex systems under precedence constraints. Expert Syst. Appl. **40**(2), 611–620 (2013)
4. Yu, J.-S., Xu, B., Li, X.-S.: Generation of test strategy for sequential fault diagnosis based on genetic algorithms. Acta Sim. Syst. Sinica **16**(4), 833–836 (2004)
5. Wang, C., Liu, Z., Yang, Z.: Fault diagnosis strategy design of electronic equipment based on multivalue test. Acta Armament. **32**(10), 1287–1291 (2011)
6. Raghavan, V., Shakeri, M., Pattipati, K.: Test sequencing algorithms with unreliable test. IEEE Trans. Syst. man Cybern. Part A Syst. Hum. **29**(4), 347–357 (1999)
7. Bennett, C.C., Hauser, K.: Artificial intelligence framework for simulating clinical decision-making: a Markov decision process approach. Artif. Intell. Med. **57**(1), 9–19 (2013)
8. Wu, Y., Wang, B., Ray Liu, K.J.: Optimal defense against jamming attacks in cognitive radio networks using the markov decision process approach. IEEE Glob. Telecommun. Conf. **45**(2), 1–5 (2014)

Research on Gyro Fault Diagnosis Method Based on Wavelet Packet Decomposition and Multi-class Least Squares Support Vector Machine

Qiang Liu, Jinjin Cheng and Wenhao Guo

Abstract In order to diagnose the faulty gyro, a method of fault diagnosis for the genus gyro based on wavelet packet decomposition and least squares support vector machine is proposed. Four kinds of gyro fault signal models are established by studying the characteristics of the output signals in the gyro fault state. The wavelet packet decomposition is used to perform time–frequency domain decomposition of gyro signals in various states, and the wavelet packet energy entropy value in each frequency band is calculated as a feature vector for characterizing the gyro fault state. The "one-to-many" multi-class least-squares support vector machine is used to identify the feature vectors in one state, and the parameters of LSSVM are optimized by using firefly algorithm and tenfold cross-validation method, respectively. The results show that the two methods are The diagnostic accuracy rate of the gyro fault data is more than 90%, and the calculation speed is fast, and the fault diagnosis of the gyro is successfully realized.

Keywords Fault diagnosis · Gyro · Wavelet packet decomposition · Least squares support vector machine

1 Introduction

In the flight control process of the aircraft, the inertial navigation component is a key component in the aircraft inertial navigation system, and the performance of the inertial navigation component plays an important role in the flight safety of the aircraft. According to relevant statistics, in the fault history data of the inertial navigation system, 60% of the faults are the faults of the group gyro. The gyro is a measuring component used in the inertial navigation component to measure the inertial angular velocity of the aircraft. Once the gyro fails, it will directly affect the

Q. Liu (✉) · J. Cheng · W. Guo
College of Aerospace Engineering, Air Force Engineering University, 710038 Xian, China
e-mail: 353134979@qq.com

V. Jain et al. (eds.), *Recent Trends in Intelligent Computing, Communication and Devices*, Advances in Intelligent Systems and Computing 1031,
https://doi.org/10.1007/978-981-13-9406-5_94

operation of the flight control system. Therefore, it is very important to study the fault diagnosis method of the gyro.

Related studies have shown that when the sensor fails, the signal energy output from different fault modes varies in different frequency bands [1]. Based on this feature, this paper proposes a gyro fault diagnosis method based on wavelet packet decomposition and multi-class least squares support vector machine. The gyro output model in each fault mode is established by analyzing the characteristics of the output signals in each fault mode when the gyro fault occurs. The wavelet packet decomposition is used to decompose the gyro output data in each state, and the wavelet packet energy value of each frequency band signal is calculated as the eigenvector of the gyro signal in each state. Finally, the multi-class least squares support vector machine is used. The feature vector is identified to diagnose the gyro.

2 Fault Diagnosis Model

2.1 Gyro Fault Output Signal Model

When the gyro fails, its failure symptoms generally show an output data failure, which means that there is a big difference between the actual output of the gyro and the real data or even no output [2]. Common gyro failure modes include gyro complete failure, gyro constant drift, gyro constant gain output, and periodic interference faults.

Let y_{out} be the output of the gyro and y_{rout} be the normal output of the gyro. The fault output model under the four failure modes is constructed as follows:

(1) Gyro completely faulty

The fault model is:

$$y_{out}(t) = \begin{cases} y_{rout}(t) & 0 < t < t_f \\ a_i & t_f < t < t_f + \Delta t_f \end{cases} \quad (1)$$

where a_i is a constant, t_f is the fault start time, and Δt_f is the fault duration.

(2) Gyro constant drift failure

The fault model is:

$$y_{out}(t) = y_{rout}(t) + b \quad t_f < t < t_f + \Delta t_f \quad (2)$$

where b is the drift setting.

(3) Gyro constant gain output fault

The fault model is:

$$y_{out}(t) = ky_{rout}(t) \quad t_f < t < t_f + \Delta t_f \tag{3}$$

where k is the gain loss coefficient, $0<k<1$.

(4) Periodic interference fault

The fault model is:

$$y_{out}(t) = \begin{cases} y_{rout}(t) & 0<t<t_f \\ y_{rout}(t) + \text{square}(t) & t_f < t < t_f + \Delta t_f \end{cases} \tag{4}$$

where square (t) is a square wave signal with period T_0.

2.2 *Wavelet Packet Decomposition*

Wavelet packet decomposition [3] is a more elaborate analysis method than wavelet analysis. It divides the frequency band into multiple layers and further decomposes the high-frequency part without wavelet in the wavelet analysis. Compared with wavelet analysis, wavelet packet analysis has higher time–frequency resolution for a wider range of applications.

The structure diagram of the three-layer wavelet packet decomposition is shown in the following Fig. 1:

Where L represents low frequency, H represents high frequency, and the last digit represents the current wavelet packet decomposition layer.

Let $\Omega_{0,0}$ be the vector space of node 0 in the above figure, then the wavelet subvector space $\Omega_{j,n}$ of each node of each layer can be divided into two mutually orthogonal subspaces:

$$\Omega_{j,n} = \Omega_{j+1,2n} + \Omega_{j+1,2n+1} \tag{5}$$

where j is the number of decomposition layers and (j, n) is the node of the jth layer, where $n = 0, 1, \ldots, 2^{j-1}$.

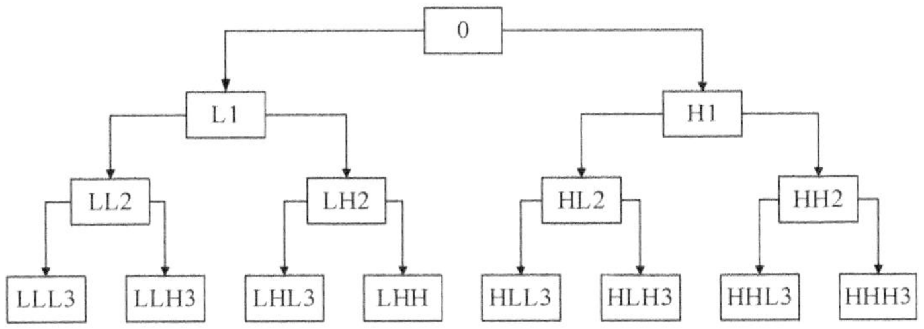

Fig. 1 Three-layer wavelet packet decomposition structure diagram

Let the wavelet function $u_{2n}(t)$ and the orthogonal scaling function $u_n(t)$ satisfy the relationship:

$$u_{2n}(t) = \sqrt{2}\sum_{k \in Z} h(k)u_n(2t - k) \tag{6}$$

$$u_{2n+1}(t) = \sqrt{2}\sum_{k \in Z} g(k)u_n(2t - k) \tag{7}$$

where $g(k)$ is the low-pass filter coefficient and $h(k)$ is the high-pass filter coefficient.

According to the decomposition definition of wavelet decomposition, the energy of the decomposed sub-band signal $x_j^n(t)$ can be calculated by the formula:

$$E_{j,n} = \sum_{k} \left|D_k^{j,n}\right|^2 \tag{8}$$

2.3 *Least Squares Support Vector Machine*

The least squares support vector machine is an intelligent algorithm developed on the basis of support vector machines to solve classification and regression problems. Compared with the traditional SVM, LSSVM introduces the square of the training error as the loss function and replaces the inequality constraint with the equality constraint and directly obtains the analytical solution of the model parameters, avoiding solving the quadratic problem [4].

Set the sample set to $\{(x_i, y_i)\}, i = 1, 2, \ldots, n, x_i \in R^d, y_i \in \{-1, 1\}$. The LSSVM model is:

$$f(x) = \operatorname{sgn}\left[\sum_{i=1}^{n} \alpha_i y_i K(x, x_i) + b\right] \tag{9}$$

3 Experimental Verification

In this paper, the gyro failure data in the alignment process of 100 sets of aircraft inertial navigation components is selected to verify the proposed fault diagnosis method. A total of 50 sets of data were randomly selected as training samples, and the remaining 50 sets of data were used as test samples. Firstly, the eigenvectors are constructed by selecting typical data in five states: gyro normal state, gyro constant gain fault, gyro complete fault, gyro constant drift fault, and gyro periodic signal

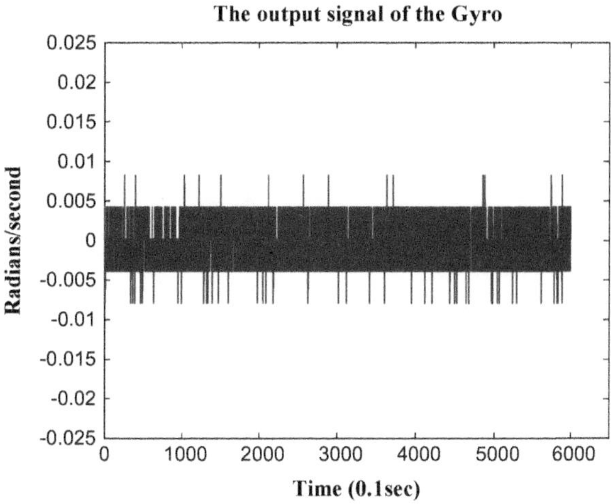

Fig. 2 Gyro normal output historical data

interference fault. The selected characteristic of signal raw data is as shown below (Figs. 2, 3, 4, 5 and 6).

For the original signal, the wavelet energy value in each period is solved every 1s (10 points), and the feature vector is constructed. The results are as follows (Table 1).

It can be seen that the energy distribution of each frequency band in the wavelet packet decomposition under different states is different, indicating that the output data characteristics of the gyro in different states are different.

After acquiring the gyro output fault feature vector, the LSSVM is used to identify the fault pattern of the feature vector.

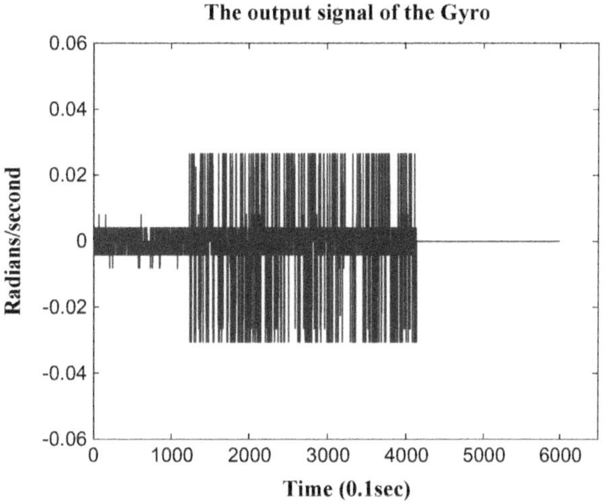

Fig. 3 Gyro complete fault output historical data

Fig. 4 Gyro constant gain fault output historical data

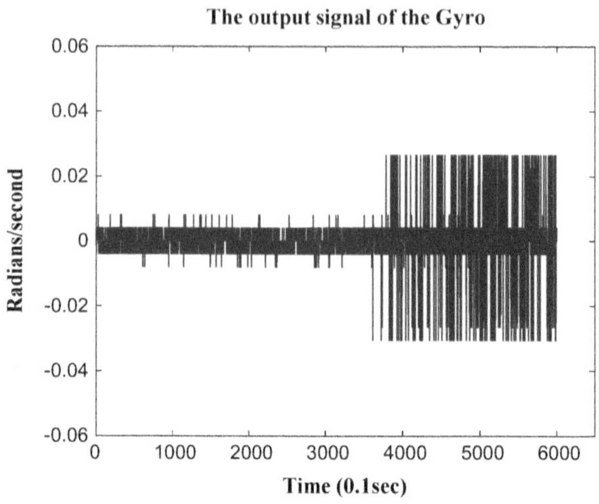

Fig. 5 Gyro drift fault output historical data

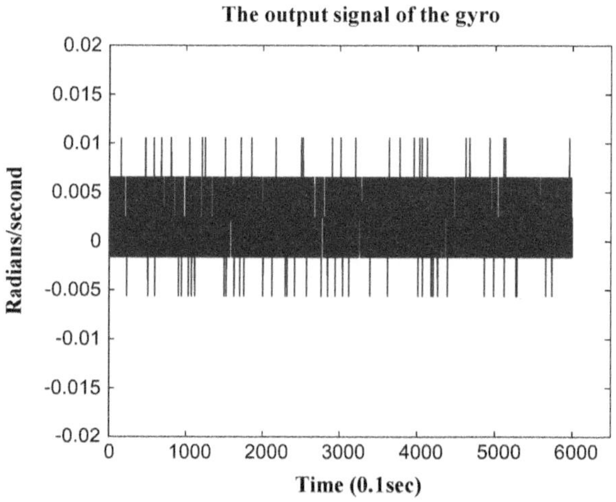

Fig. 6 Gyro periodic signal interference fault output historical data

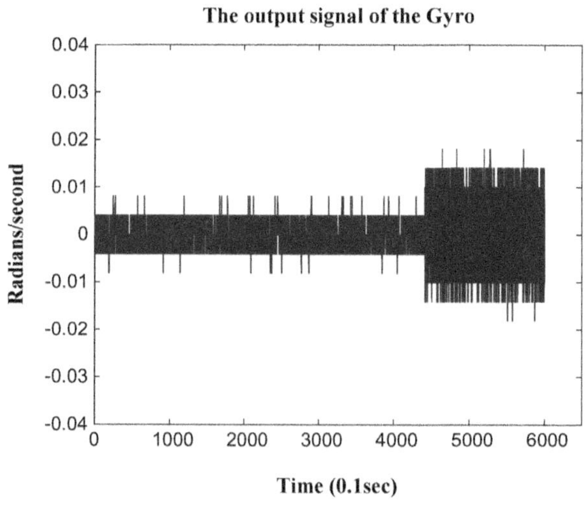

Table 1 Wavelet packet decomposition energy value in each state

Failure mode	Energy entropy value in each frequency band							
	W31	W32	W33	W34	W35	W36	W37	W38
Normal	24.8	8.182	12.8	7.223	16.1	15.0	5.921	9.734
Constant gain fault	27.2	8.403	11.9	7.164	14.7	5.598	15.3	9.499
Complete fault	19	8.779	6.434	5.068	10.4	9.809	5.611	3.862
Drift fault	55.2	4.846	7.298	4.213	9.293	3.462	9.482	6.109
Periodic fault	30.2	8.134	4.415	8.344	12.5	12.5	14.4	9.316

The kernel function used in the LIBSVM designed in this paper is a Gaussian kernel, i.e., $k(x_i, x_k) = \exp(-\frac{\|x_i - x_j\|^2}{2\sigma^2}), \sigma > 0$. The selection of the kernel parameters directly affects the classification performance of the LSSVM. At the same time, the selection of the penalty function C in the objective function also has a great impact on the performance of the LSSVM [6]. To improve the classification performance of LSSVM, this paper uses the firefly algorithm [7] to optimize these two parameters. The fault diagnosis of FA-LSSVM and LSSVM is shown in the following figure (Figs. 7 and 8).

The classification results are shown in the following Table 2.

By analyzing the classification of test samples by FA-LSSVM and LSSVM, it can be seen that the fault diagnosis accuracy rate of FA-LSSVM is higher than LSSVM, and the diagnostic accuracy rates of the two fault diagnosis methods are 96 and 92%, respectively. The FA-LSSVM runs slower than the regular LSSVM. Considering the fault diagnosis correct rate and running speed, the above two

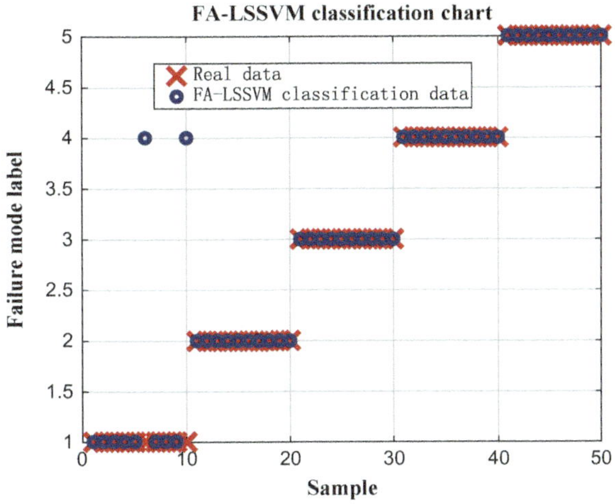

Fig. 7 FA-LISVM classification results

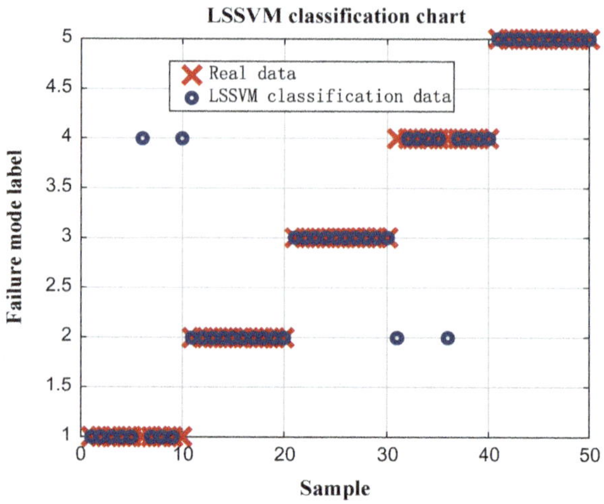

Fig. 8 LISVM classification results

Table 2 Comparison of fault diagnosis results

Method	Parameter optimization method	Final setting parameter value		Error rate (%)	Running speed (s)
		C	σ		
FA-LSSVM	FA	2.287	6.724	4	6.71
LSSVM	10-fold cross-validation	3.62	4.52	8	4.28

methods can complete the fault diagnosis of the gyro with higher accuracy and faster running speed.

4 Conclusion

In this paper, a method for fault diagnosis of genset gyro based on wavelet packet decomposition and multi-class least squares vector machine is proposed. A fault signal model of four fault modes is established for the signal characteristics of the gyro fault signal. The wavelet packet decomposition is used to extract the fault original signal and construct the fault signal feature vector. The results show that the eigenvectors of different fault modes have large differences and the data features are obvious. Finally, the firefly algorithm based on LSSVM and conventional LSSVM is used for feature recognition. The correct rate of fault identification is 96 and 92%, which indicates that the method can effectively diagnose the fault of the gyro.

References

1. Hui, T.E.: The application of wavelet and multi-kernel SVM in UAV sensors fault diagnosis. Nanjing University of Aeronautics and Astronautics (2014). (in Chinese)
2. Zhang, X.: Sensor fault diagnosis for flight control system based on neural network. Nanjing University of Aeronautics and Astronautics (2012). (in Chinese)
3. Liu, Q., Cheng, J., Tan, Y., et al.: A Darlington transistor fault prognostics method based on KPCA and Mahalanobis distance. J. Air Force Eng. Uni. (Nat. Sci.e Edn.) **19**(5), 71–77 (2018)
4. Suykens, J.A.K.: Nonlinear modeling and support vector machines. In: IEEE Instrumentation and Measurement Technology Conference, Budapest, Hungary, pp. 287–294 (2001)
5. Wang, P., Xin, J., Gao, X., et al.: Fault diagnosis of chiller based on independent component analysis and least square support vector machine. J. Beijing Uni. Technol. **43**(11), 1641–1644 (2017)
6. Nie, J., Li, C., Li, W., et al.: Research on the least squares support vector machine optimized by genetic algorithm in the simulation MBR prediction. Conput Eng. Softw. **36**(5), 41–44 (2015)
7. Liu, C., Ye, C.: Firefly algorithm with Chaotic search strategy. J. Syst. Manage. **7**(4), 539–543 (2013)

The Latest Research on Clustering Algorithms Used for Radar Signal Sorting

Shi-qiang Wang, Caiyun Gao, Qin Zhang, Hui-yong Zeng and Juan Bai

Abstract As an important part of electronic intelligence (ELINT) and electronic support measurement (ESM) systems, radar signal sorting directly affects the performance of electronic reconnaissance equipment and is a key technology for decision making. This paper discusses several clustering methods which could be used for radar signal sorting. The discussion includes artificial neural networks, classical clustering algorithm and its improvement, and support vector clustering.

Keywords Radar signal sorting · Clustering algorithm · Artificial neural networks · Support vector clustering · Spectral clustering · Kernel clustering

1 Introduction

Radar signal sorting is a process of separating pulse sequences of different radars from randomly interleaved pulse signal streams and selecting useful signals. It is an important part of electronic intelligence reconnaissance system (ELINT) and electronic support system (ESM). It affects the ability of electronic reconnaissance equipment directly and also affects the subsequent operational decisions. For noncooperative electronic reconnaissance, the intercepted pulse stream generally lacks the necessary prior information, so it is necessary to study the clustering method for sorting the pulse stream lacking a priori information. Clustering refers to grouping data objects into multiple clusters or classes so that the data in the same cluster have higher similarities, while data have larger differences in different clusters [1]. The biggest difference from the classification is that clustering belongs to the unsupervised learning method, and the class to be divided is unknown, so it is especially suitable to solve the separation problem of an interlaced pulse stream.

S. Wang · Q. Zhang · H. Zeng (✉) · J. Bai
Air and Missile Defense College, Air Force Engineering University, Xi'an 710051, China
e-mail: 1048768838@qq.com

C. Gao
Basic Department, Air Force Engineering University, Xi'an 710051, China

V. Jain et al. (eds.), *Recent Trends in Intelligent Computing, Communication and Devices*, Advances in Intelligent Systems and Computing 1031,
https://doi.org/10.1007/978-981-13-9406-5_95

Scholars have proposed a variety of sorting methods based on cluster analysis, such as artificial neural networks [2–8], *K*-means algorithm and its improved algorithm [9–14], hierarchical clustering [15–17], kernel clustering [18–23], and spectral clustering [24–28], but like all multi-parameter methods, the biggest problem for clustering sorting is that it is used for separable categories. The characteristics of clustering do not have good interclass separation ability, which affects the actual use of clustering methods. The emergence of support vector clustering (SVC) [29–32] method solves this problem well. In this paper, we discuss these clustering algorithms mentioned above.

2 Artificial Neural Networks

Considering artificial neural networks, such as self-organizing (SOM) neural networks, radial basis function neural (RBFN) networks, probabilistic neural networks (PNN), etc., with parallel processing and fault tolerance, many documents apply them to radar signals. Sorting, it is expected to use the parallel processing and fault tolerance of artificial neural networks to solve the problem of radar signal sorting in complex environments [3]. In 2001, Xu et al. [4] combined the self-organizing network with PNN to obtain the SOPNN neural network sorting method, which is mainly used for pulse train deinterlacing in a radiation source environment without prior information; Lin [5] selected PW, RF, and DOA as sorting parameters, using Kohonen neural network to sort radar multi-target; in 2006, Guo [6] also selected these parameters and used a Kohonen neural network to achieve pulse sorting; and Kohonen neural network sorting method is different. In 2009, Han et al. [7] used the Kohonen neural network and the mixed characteristics of interpulse parameters and intrapulse parameters to automatically sort the radiation source signals and achieved good results; in 2010, Wang et al. [8] used the Eidos BSB artificial neural network to perform self-associative learning on a large number of radar pulse samples with measurement errors and completed the memory of the pulse mode to realize the sorting function.

However, the neural network-based sorting method has some disadvantages when it is applied to the separation of radar signals from complex systems. For example, self-organizing neural networks require a large network scale when dealing with the identification problem of unknown mode space, and there is no cluster validity verification. When there is a big difference between the actual radiation source environment and the a priori radiation source environment, the false alarm probability is greatly For example, PNN has problems such as prior knowledge of samples, unsatisfactory classification performance for pattern overlap and interleaving, and SOPNN is not ideal for pulse stream sorting under parameter space and parameter interleaving.

3 Classical Clustering Algorithm and its Improvement

3.1 *K-means Clustering Algorithm*

In 1967, MacQueen [10] first proposed the *K*-means clustering algorithm. So far, many clustering tasks have chosen this classic algorithm. In 1998, Huang [11] researched a *K*-modes algorithm which is suitable for clustering objects data in order to overcome the disadvantage that the *K*-means algorithm is only suitable for processing numerical attribute data clustering. In 1999, Huang et al. [12] demonstrated that the algorithm of *K*-modes can only find a local minimum after a finite iteration. In 2001, Chaturvedi et al. [13] proposed a nonparametric clustering method *K*-modes-CGC algorithm for classification attribute data (nominal scale data), similar to the traditional *K*-means algorithm for numerical data (interval scale data). In 2004, Ding [14] researched a consistency-reserved *K*-means algorithm. The algorithm extends the concept of consistency to objects clustering. For any object in a class, its *k*-nearest neighbor must be in this class. The algorithm studies the properties of *k*-nearest neighbor consistency, proposes kNN and kMN consistency coercive and improved algorithms, and proposes an important quality of class *k*-nearest neighbor or *k*-nearest neighbor consistency as data clustering.

3.2 *Hierarchical Aggregation Algorithm*

Hierarchical clustering algorithm, also known as tree clustering algorithm [15], uses data join rules to repeatedly divisive or agglomerative data to form a solution to the clustering problem of hierarchical sequence through a hierarchical architecture. In 2007, Gelbard et al. [16] proposed a new hierarchical aggregation algorithm called the binary-positive method. In the same year, Kumar et al. [17] proposed a new hierarchical clustering algorithm RCOSD based on indiscernible coarse aggregation for continuous data. The key idea of RCOSD is to find a feature set that can capture the continuous information and content information of the data sequence and map these feature sets to an upper approximation space, and apply the constraint similarity approximation technique to obtain the upper approximation of the coarse class cluster, one of which elements can belong to multiple clusters.

3.3 *Kernel Clustering Algorithm*

For the kernel clustering method, its idea is to map the objects in the input space to the high-dimensional feature space using the mapping that is nonlinear and select the suitable kernel function of Mercer which means it is not the inner product of the nonlinear mapping into gather class in feature space [18]. This method is common, and it has a greater promotion than the common clustering method. The advantage of the kernel clustering method is that it increases the probability that the data points are linearly separable through nonlinear mapping, that is, it can better distinguish, extract, and amplify useful features, thereby achieving more accurate clustering, and the algorithm converges faster. In the case of the failure of the classical clustering algorithm, the kernel clustering algorithm often obtains better clustering results [19]. The kernel clustering algorithm can be divided into three forms: support vector clustering (SVC) [20, 21], feature space kernel clustering [22], and input spatial kernel clustering [23].

3.4 *Spectral Clustering Algorithm*

The main idea of the spectral clustering algorithm [24] is that: Firstly, define an affinity matrix which describes the pair's similarity of objects according to a given sample set of data, secondly calculate the eigenvalues and eigenvectors for this matrix, and finally, select suitable feature vectors cluster different data points. The algorithm of spectral clustering is based on the theory of spectral in graph theory [25]. Its nature is transforming the problem of clustering into the optimal partitioning graphs problem. It is the point-to-point algorithm of clustering, which is designed to arbitrary shapes. The clustering on the sample space converges on the global optimal solution, so it has a good application prospect for data clustering. Research scholars at home and abroad have done a lot of research in this direction. Chung [25] gives the classical optimal partitioning function, and Shi [26] proposes the concept of optimal norm partitioning set for the balanced subsets. Kannan [27] and Shi [28], respectively, propose conduction optimization, normalized cut, normalized association optimization, and ratio cut optimization to solve the optimal partition problem of the graph.

4 Support Vector Clustering

As known that support vector clustering (SVC) is a kind of kernel clustering. It means that it is based on a support vector machine (SVM) [29]. It is an unsupervised nonparametric algorithm of clustering developed by Ben-Hur based on the SVDD algorithm [30]. Its basic idea is to use the kernel of Gaussian to map objects in data space into a high-dimensional feature space, then looking for a sphere with the smallest radius that can surround all the objects in feature space, and map the ball back to the data space, and then get the contour set containing all the objects. In this condition, these contours are the boundaries of the cluster. The points enclosed by each closed contour belong to the same cluster [20, 31]. The SVC algorithm is divided into two phases: the training of SVC and the allocation of the cluster. The training of SVC phase includes the determination of the Gaussian kernel width coefficient, the calculation of the kernel matrix, the calculation of the Lagrange multiplier, the selection of the support vector, and the calculation of the feature sphere radius in high-dimensional feature space. And for the clustering allocation stage, it first generates an adjacency matrix and then performs clustering allocation according to the adjacency matrix [32].

The SVC algorithm has two huge advantages: It can generate cluster boundaries of arbitrary shapes; it can analyze and process the noise objects and can separate interlaced clusters. However, the application of SVC to radar signal sorting has the following limitations: The optimization problem solving and the calculation of the adjacency matrix are time consuming, and it is difficult to satisfy the requirements of real time for radar signal sorting. Guoqiang, Xiangyu, Li Zhenxing, etc. [33–36] applied SVC for radar signal sorting, which achieved good results, but it took more time. Therefore, improving the processing speed of the SVC algorithm is of great significance for the real-time sorting of radar signals.

5 Conclusions

Clustering is a process of distinguishing and classifying things according to certain requirements or rules. It is based on a data-driven classification method, that is, an unsupervised learning method. The purpose of clustering is to divide the data set into several classes (or clusters) so that the data in each class are maximally similar, and the data in the same class are not the most different. In this paper, we discuss the latest clustering methods for radar signal sorting, including artificial neural networks, classical clustering algorithm and its improvement, and support vector clustering. It is useful for further research on radar signal sorting.

Acknowledgements This paper was supported in part by the National Natural Science Foundation of China under the Grant No. 61601499, 61701527, and 61601503.

References

1. Aggarwal, C., Han, J., Wang. J.: A framework for projected clustering of high dimensional data streams. In: Proceedings of the 30th VLDB Conference, Toronto, Canada (2004)
2. Ariadna, M., Alberto, S., Benjamin, F.: Classification of radar jammer FM signals using a neural network. Proc. SPIE **10188**, 11–16 (2017)
3. David Wang, C., Thompson, James: An adaptive data sorter based on probabilistic neural networks. IEEE Naecon Dayton Ohio. **3**, 1096–1102 (1991)
4. Xu, X., Zhou, Y.Y., Lu, Q.Z.: Research on real-time deinter leaving technology for radar intercept system. Syst. Eng. Electron. **23**(3), 12–15 (2001)
5. Lin, Z.Y., Liu, G., Dai, G.X.: Application of Kohonen neural network in radar multi-target sorting. J. Military Eng. Uni. **4**(5), 56–59 (2003)
6. Guo, J., Chen, J.W.: A clustering method for processing unknown radar signals. Syst. Eng. Electron. **28**(6), 853–856 (2006)

7. Han, J., He, M.H., Zhu, Y.Q., et al.: A new method for signal sorting of radar emitter based on multi-parameters. Data Acquisition Process. **24**(1), 91–94 (2009)
8. Wang, X.D., Song, M.Z.: Radar pulse sorting method based on Eidos BSB artificial neural network. Modern Electron. Technol. **23**, 6–9 (2010)
9. Xin, F., Hu, X.X., Liu, Y.: Radar signal sorting algorithm of k-means clustering based on data field. In: 3rd IEEE International Conference on Computer and Communications (ICCC), pp. 2262–2266 (2017)
10. Marques, J.P., Wu, Y.F.: Pattern Recognition Concepts, Methods and Applications. 2nd edn., pp. 51–74. Tsinghua University Press, Beijing (2002)
11. Huang, Z.: Extensions to the k-means algorithm for clustering large data sets with categorical values. Data Mining Knowl. Discovery II **2**, 283–304 (1998)
12. Huang, Z., Ma, N.: Fuzzy k-modes algorithm for clustering categorical data. IEEE Trans. Fuzzy Syst. **7**(4), 446–452 (1999)
13. Chaturvedi, A.D., Green, P.E., Carroll, J.D.: K-modes clustering. J. Classif. **18**(1), 35–56 (2001)
14. Ding, C., He, X.: K-nearest-neighbor in data clustering: Incorporating local information into global optimization, pp. 584–589. ACM Press, Nicosia (2004)
15. Fred, A., Leitão, J.: Partitional versus hierarchical clustering using a minimum grammar complexity approach. In: Proceedings of the SSPR&SPR 2000, LNCS vol. 1876, pp. 193–202 (2000)
16. Gelbard, R., Goldman, O., Spiegler, I.: Investigating diversity of clustering methods: an empirical comparison. Data Knowl. Eng. **63**(1), 155–166 (2007)
17. Kumar, P., Krishna, P.R., Bapi, R.S., De, S.K.: Rough clustering of sequential data. Data Knowl. Eng. **3**(2), 183–199 (2007)
18. Cristianini, N., Shawe-Taylor, J.: An introduction to support vector machines and other kernel-based learning methods. Cambridge University Press (2000)
19. Zhang, L., Zhou, W.D., Jiao, L.C.: Nuclear clustering algorithm. Chin. J. Comput. **25**(6), 587–590 (2002)
20. Ben-Hur, A., Horn, D., Siegelmann, H.T.: Support vector clustering. Mach. Learn. Res. **2**, 125–137 (2000)
21. Chiang, J.H., Hao, P.Y.: A new kernel-based fuzzy clustering approach: support vector clustering with cell growing. IEEE Trans. Fuzzy Syst. **11**, 518–527 (2003)
22. Girolami, M.: Mercer kernel-based clustering in the feature space. IEEE Trans. Neural Netw. **13**(3), 780–784 (2002)
23. Zhang, D.Q., Chen, S.C.: Fuzzy clustering using kernel method. In: Proceedings of the 2002 International Conference on Control and Automation, pp. 123–127. Xiamen, China (2002)
24. Cristianini, N., Taylor, J.S.: Kandola, J.S.: Spectral kernel methods for clustering. In NIPS, pp. 649–655 (2001)
25. Chung, F.R.K.: Spectral graph theory (CBMS Regional Conference Series in Mathematics, No. 92). American Mathematical Society (1997, February)
26. Shi, J., Malik, J.: Normalized cuts and image segmentation. IEEE Trans. Pattern Anal. Mach. Intell. (PAMI) (2000)
27. Kannan, R., Vempala, S., Vetta, A.: On clusterings: good, bad, and spectral. In Proceedings of the 41st Annual Symposium on the Foundation of Computer Science, pp. 367–380. IEEE Computer Society (2000, November)
28. Dhillon, I.S., Guan, Y., Kulis, B.: A unified view of kernel k-means, spectral clustering and graph partitioning. Technical Report Technical Report TR-04-25. UTCS (2005)
29. Burges, C.J.C.: A tutorial on support vector machines for pattern recognition. Data Min. Knowl. Disc. **2**(2), 121–167 (1998)
30. Tax, D.M.J., Duin, R.P.W.: Support vector domain description. Pattern Recogn. Lett. **20**(11–13), 1191–1199 (1999)
31. Scholkopf, B., Williamson, R., Smola, A.: Support vector method for novelty detection. Adv. Neural Info. Process. Syst. **12**, 582–588 (2000)

32. Lu, C.K., Jiang, C.Y., Wang, N.S.: A fast algorithm for support vector clustering. J. South China University Technol. **33**(1), 6–9 (2005)
33. Guo, Q., Li, W., Li, H.P.: Application of support vector clustering method in radar signal sorting. Annual Meeting, pp. 237–241 (2005)
34. Guo, Q., Wang, C.H., Li, W.: Radar signal sorting method based on support vector cluster combined type entropy recognition. J. Xi'an Jiaotong University **44**(8), 63–67 (2010)
35. Xiang, W., Tang, J.L.: Radar signal sorting based on improved support vector clustering. Space Sci. Technol. **27**(1), 50–53 (2011)
36. Li, Z.X., Lu, J.S., Zhang, G.Y.: A radar source SVC sorting method with automatic parameter selection. Electron. Info. Technol. **26**(2), 15–20 (2011)

Recent Research and Developing Trends of the Planar Dual-Band Antenna Array

Hui-yong Zeng, Tong Xu, Bin-feng Zong, Yan Zhao and Shi-qiang Wang

Abstract The dual-band antennas and their arrays have many advantages compared with those conventional ones. As far as we know, the dual-band antenna arrays have broad development prospect because they can satisfy the requirement of modern wireless communication diversification. In this paper, the latest researches of dual-band antenna arrays at home and abroad are summarized. Two design methods are deeply analyzed and their research findings are presented. Based on these initial presented findings, the advantages and disadvantages of the different methods are compared. Moreover, the future perspective of the dual-band antenna arrays is given.

Keywords Dual-band antenna · Planar antenna array · Wireless communications

1 Introduction

The so-called dual/multi-band antenna refers to the antenna that provides the required properties in two or more bands. The specific indicators of this kind of antenna include voltage standing wave ratio (VSWR), gain, radiation pattern and efficiency.

During the early ages, the wireless communication systems often operated in a certain band. However, with the development of the communication technology, the single-band systems cannot meet the requirement and the dual/multi-band systems are demanded increasingly. In our daily life, the wireless local area network

H. Zeng · T. Xu · S. Wang (✉)
Air and Missile-Defence College, Air Force Engineering University,
Xi'an 710051, China
e-mail: hyzeng.1023@163.com

B. Zong
Unit 94710 of the PLA, Wuxi 214000, China

Y. Zhao
Air Force's Military Representative Office in Jilin, Jilin, China

V. Jain et al. (eds.), *Recent Trends in Intelligent Computing, Communication and Devices*, Advances in Intelligent Systems and Computing 1031,
https://doi.org/10.1007/978-981-13-9406-5_96

(WLAN) communication system is a typical dual-band system; its lower operation band is 2.4 GHz and the upper one is 5.5 GHz. In practice, the dual/multi-band antenna, which has lower frequency ratio, can be replacement by a wideband antenna. But, the effectiveness of this method is reduced when the frequency ratio becomes larger. Besides, only one antenna can be used in some systems because of the requirement of their miniaturization and low cost. Therefore, dual/multi-band antenna design plays a key role in dual/multi-band system integration [1]. The main advantage of the dual/multi-band antenna is that it can realize the operational properties in two or more antennas.

Generally, the gain of the dual-band antenna is low, which limits its application in some high-gain required systems. Fortunately, the dual-band antenna array is a good way to solve this problem.

For the time being, there are two methods for the planar dual-band antenna array design. One is synthesized by the dual-band antenna elements [2–5], and the other is synthesized by two kinds of single-band antennas [6–8]. Here, the dual-band antenna array, which is synthesized by two independent antenna arrays, is named as composite dual-band antenna array.

2 Antenna Arrays Based on Dual-Band Antenna Element

In order to synthesize the dual-band antenna array, using the dual-band antenna elements is a direct method. Figure 1 illustrates a typical structure of a dual-band antenna array [3]. The reported antenna array has advantages of simple structure,

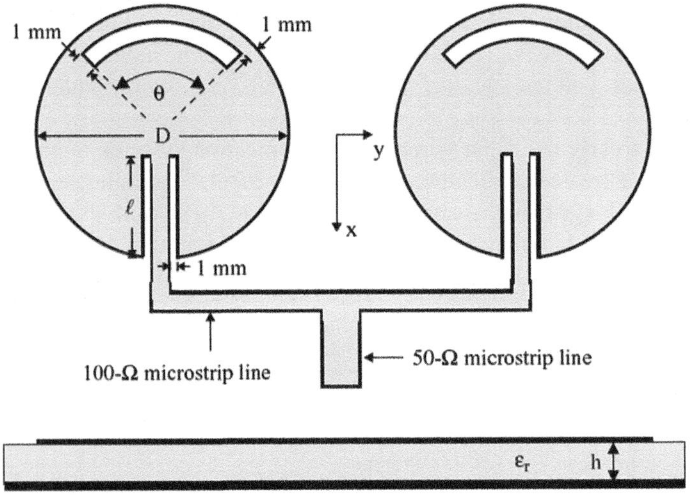

Fig. 1 Dual-band antenna array reported in [3]

low weight, and small size, and it is suitable for applications in good performance required systems such as the base-station radar.

A novel dual-band microstrip planar antenna array, whose photograph was illustrated in Fig. 2, was designed using the direct method. In the antenna design, the inset-fed feed technology is used and good impedance matching is obtained through adjusting the length of the feeding line. Besides, this feed structure is equivalent to load a reactance on the microstrip patch, which results in two zero points and realizes dual band.

However, an obvious drawback of this method is that its effectiveness reduces when the frequency ratio becomes larger. Because the directivity, coupling between the elements and grating lobe should be considered when antenna array synthesized. Usually, the element distance is 0.5λ0–1λ0. As the frequency ratio becomes larger, the distance between the elements should be taken into consideration to avoid strong coupling in lower frequency and avoid grating lobe in the higher one. But the distance is unchanged in finished design, and it is difficult to solve the above problem simultaneously. As a result, the antenna array has worse properties using this method. In short, this method is suitable for the antenna array with a small frequency ratio.

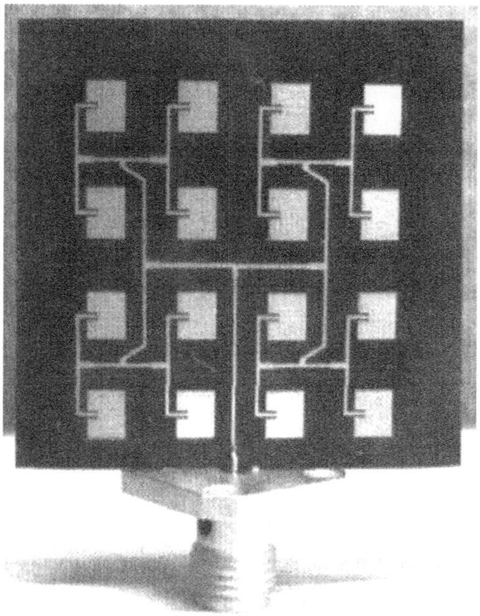

Fig. 2 Dual-band antenna array reported in [4]

3 Composite Dual-Band Antenna Arrays

Another way to design the dual-band antenna arrays is using two kinds of antenna elements with independent control bands. As a result, a dual-band antenna array with large frequency ratio can be designed by using this method. Compared with the direct way using the dual-band element, this one has an additional design flexibility because of the antenna element can be design independently. Besides, based on the property of independent design, it is more convenient to design antenna arrays with broadband, dual-polarized and circularly polarized. Again, the dual-band antenna array, which is synthesized by two independent antenna arrays, is named as composite dual-band antenna array.

To realize the composite dual-band antenna arrays, two methods are used. One is co-aperture method and the other is uncommon aperture method. The first one is synthesized two elements in the same aperture [6–9], and the second one is synthesized two elements in different apertures [10, 11].

Several references have reported the co-aperture method. In [6], a two-layered tri-band antenna array, which illustrates in Fig. 3, is designed for WiMAX application. In this design, two rectangle patches and an *S*-typed patch are used as radiators. It can be seen that these radiators are synthesized in co-aperture at the upper layer of the substrate and the feeding network at the lower substrate. Here, the *S*-typed patch can be used to enhance the work bands. Unfortunately, the measured results indicate that the radiation patterns for the higher frequency at 5.8 GHz degraded because of the grating lobe problem. In [7], a dual-band array antenna and its associated feed networks for 2.4/5.8 GHz wireless local area networks (WLAN) are proposed. Its photograph is illustrated in Fig. 4. The dual-band array is composed of a 2.4 GHz band array formed by 4 × 4 parallel-fed rectangular patches and a 5.8 GHz band array formed by 8 × 8 hybrid-fed double-sided printed dipoles. In this design, the 4 × 4 array and the 8 × 8 array use a common aperture.

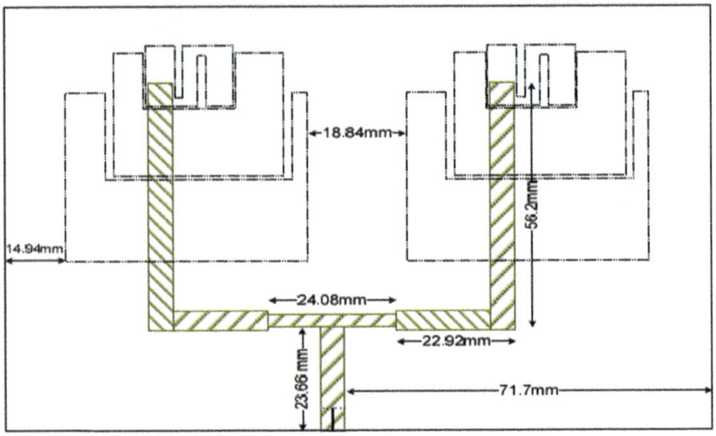

Fig. 3 Dual-band antenna array reported in [6]

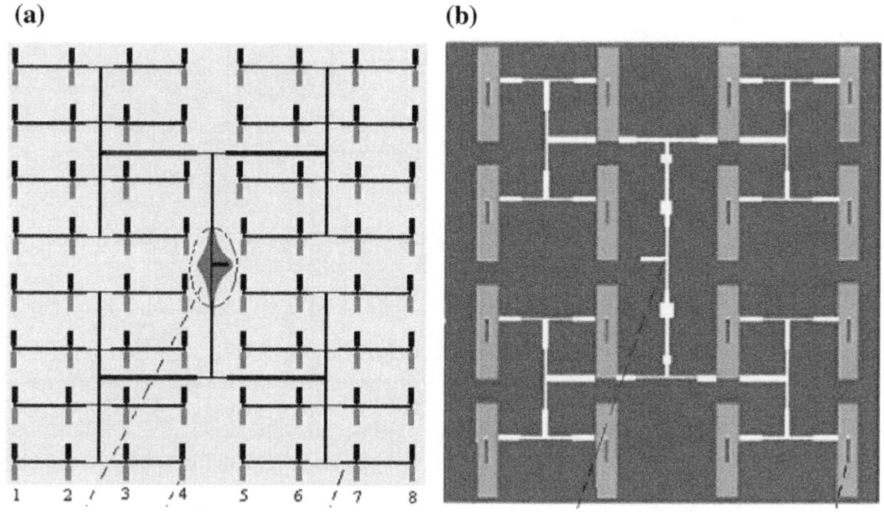

Fig. 4 Dual-band antenna array reported in [7]. **A** 8 × 8 planar dipole; **b** 4 × 4 patch

For the uncommon aperture method, few references have reported. In [11], three types of electromagnetic band gap (EBG) structure are used in a single-port dual-band microstrip antenna array design for 2.4/5.8 GHz. Its photographs are illustrated in Fig. 4. Two photographs in upper are the front and back of the microstrip antenna array and its feeding network. Three photographs in lower are the mushroom EBG, a modified Minkowski EBG, and Sierpinski EBG structures. These EBG structures operate as band rejecters and eliminate the band jamming, thus making two bands operate individually at their own frequencies. The EBG structure in feed line for 2.4 GHz is used to reject the 5.8 GHz band, and the EBG structure in feed line for 5.8 GHz is used to reject the 2.4 GHz band (Fig. 5).

Many particular methods have been used in the design of dual-band antenna array. A single-port planar dual-band antenna is presented in [12], which is with 4 × 2 5.5 GHz and 2 × 2 2.4 GHz array elements. The suspended patches are used to enhance its bandwidth and reduce loss. The topology of the antenna array is depicted in Fig. 6. Its working mechanism can be concluded as follows. A microstrip filter with stopband at 5.5 GHz and passband at 2.4 GHz are employed to feed the 2.4 GHz radiators is designed. This prevents the 2.4 GHz radiators from resonating at 5.5 GHz. So, the antenna array operates at two bands.

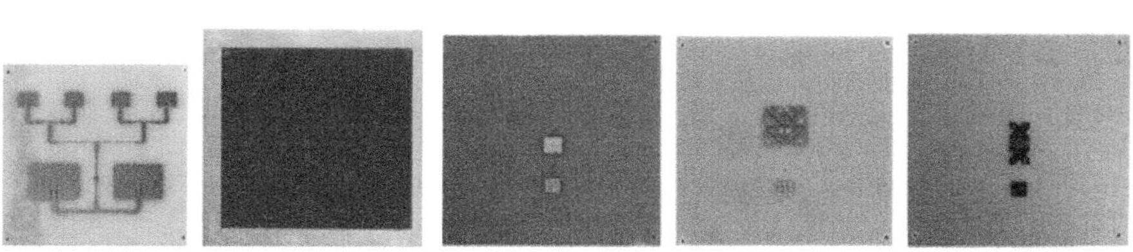

Fig. 5 Dual-band antenna array in [11]

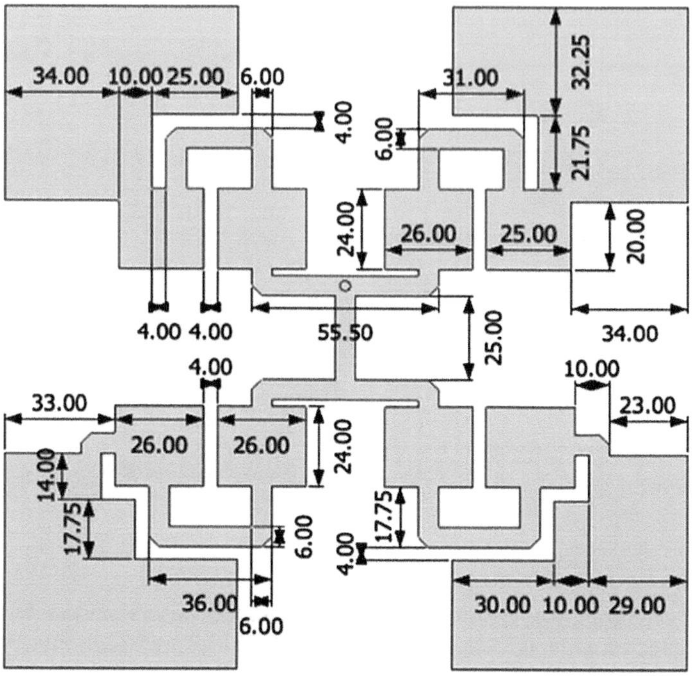

Fig. 6 Dual-band antenna array in [12]

4 Conclusion

In the paper, the design methods of the reported dual-band antenna arrays are summarized. The differences between the dual-band antenna-array-based method and the two single-band antenna-array-based methods are presented. According to the references, a few achievements for the dual-band antenna array design are reported, and it is worthy to do some researches about it.

Acknowledgements The paper was supported by the National Natural Science Foundation of China (Grant No. 61701527, 61601499) and the National Natural Science Foundation of Shaanxi Province (Grant No. 2019JQ-583, 2018JQ6023).

References

1. Ma, H.Q.: Research on wideband antennas and mutiband antennas. A Dissertation Submitted to Xidian University for The Degree of Doctor, Xi'an, China (2009)
2. Pan, S.C., Wong, K.L.: Design of dual-frequency microstrip antennas using a shorting-pin loading. In: Proceedings of IEEE Antennas and Propagation Society International Symposium, pp. 312–315. Ailanta, Georagia (1998)

3. Hsieh, K.B., Wong, K.L.: Inset-microstrip-line-fed dual-frequency circular microstrip antenna and its application to a two-element dual-frequency microstrip array. IEEE Proc. Microwave Trans. Antennas Propag. **147**(10), 359–361 (1999)
4. Fu, J.H., Wu, Q., Zhang, F., Liu, M.: Research on millimeter wave dual-band microstrip planar antenna arrays. Syst. Eng. Electron. **33**(4), 746–749 (2011)
5. Ma, X.L., Wang, Z.S.: The dual-frequency and dual-polarization microstrip antenna for space-borne SAR application. J. Electron. Info. Technol. **24**(12), 1947–1954 (2002)
6. Rahardjo, E.T., Zulkifl, F.Y., Marlena, D.: Multiband microstrip antenna array for WiMAX application. In: Proceedings of Asia Pacific Microwave Conference, pp. 1–4 (2008)
7. He, S.H., Xie, J.D.: Analysis and design of a novel dual-band array antenna with a low profile for 2400/5800 MHz WLAN systems. IEEE Trans. Antennas Propag. **58**(2), 391–396 (2010)
8. Li, C., Lv, X.D.: A L/X dual-frequency co-aperture microstrip array design. In: Proceedings of IEEE Antennas and Propagation Society International Symposium, pp. 795–798 (2005)
9. Shaafi, L.L.: Dual-band dual-polarized perforated microstrip antennas for SAR applications. IEEE Trans. Antennas Propag. **48**(1), 58–66 (2000)
10. Rostan, F., Wiesbeck, W.: Aperture-coupled microstrip patch phased arrays in C-and X-band a contribution to future multi-polarization multi-frequeney SAR Systems. IN: Proceedings of IEEE International Symposium on Phased Array Systems and Technology, pp. 141–146. (1996)
11. Masri, T., Rahim, M.K.A.: Dual-band microstrip antenna array with a combination of mushroom, modified Minkowski and Sierpinski electromagnetic band gap structures. IET Microwave Antennas Propag. **4**(11), 1756–1763 (2010)
12. Toh, W.K., Qing, X.M., Chen, Z.N.: A planar dualband antenna array. IEEE Trans. Antennas Propag. **59**(3), 833–838 (2011)

Design Methods of Wideband Differential Phase Shifters

Hui-yong Zeng, Qin Zhang, Bin-feng Zong, Yan Zhao and Shi-qiang Wang

Abstract Wideband phase shifters are important devices in modern wireless communication, and there are varied methods in previous reports. In this paper, six main methods to design wideband differential phase shifters are concluded. Advantages and disadvantages of every method are analyzed. As a result, clear guidelines are provided to design small planar phase shifters with wideband characteristics for different applications, which is useful to the following designers.

Keywords Differential phase shifters · Wideband · Design methods

1 Introduction

The phase shifter is a two-port network. The controller is a working state in which the DC bias voltage is controlled by the circuit, which causes the phase input and output of the signal to change, thereby achieving phase shifting.

There are many classifications of phase shifters, which are mainly divided into digital and analog. The digital and analog phase shifters mainly distinguish the phase shift between the patterns by the change of the mode. The phase shift of the analog phase shifter is constantly changing. The digital phase shifter is the discrete phase shift value.

We know that the technical difficulty of the phase shifter is that it changes with the frequency. We must ensure that the phase shift in multiple frequency bands is consistent, and more importantly, it requires size miniaturization.

H. Zeng · Q. Zhang · S. Wang (✉)
Air and Missile-Defence College, Air Force Engineering University,
Xi'an 710051, China
e-mail: hyzeng.1023@163.com

B. Zong
Unit 94710 of the PLA, Wuxi 214000, China

Y. Zhao
Air Force's Military Representative Office in Jilin, Jilin, China

V. Jain et al. (eds.), *Recent Trends in Intelligent Computing, Communication and Devices*, Advances in Intelligent Systems and Computing 1031,
https://doi.org/10.1007/978-981-13-9406-5_97

Phased shifters are vital microwave devices in electronic industry applications. Phased shifters are extensively used in phased arrays, beam-forming networks, modulators, and many other wireless communication networks. So, it is meaningful work to design planar phase shifters with wide band. Two methods can be used in design a traditional differential phase shifter. In one case, the phase shift can be realized using two transmission lines with different lengths. The other case is the phase shift can be realized by changing the propagation constant of the transmission with equal length. However, the designed phase shifters operate in narrow bands using these two methods.

2 Design Method of WideBand Phase Shifter

Most of the existing microwave phase shifters are composed of right-hand transmission lines. Its working frequency band is proportional to the length of the transmission line, and there is a strong correlation between the two. At present, two technologies are mainly used for multi-band and miniaturization of phase shifters.

The coupling branch is used to make the phase shifter have multi-band or wideband characteristics. This technology has proved its effectiveness from theory, simulation, testing, etc., but the biggest limitation of this technology is that it is difficult to process, and it is suitable for personal wireless communication that requires mass production, low cost and easy processing. Not very suitable.

The miniaturization is mainly achieved by using a high dielectric constant microwave medium and a process. The high dielectric constant allows the length of the transmission line to be effectively shortened in the operating frequency band, thereby reducing the size of the phase shifter. However, the high value makes the operating frequency range decrease, and it is difficult to perform multi-frequency and wideband in this process.

According to the initial reports, six main methods, which have been used to design wideband phase shifters, can be concluded.

Method 1: *The Schiffman phase shifter and its improvements.* The conventional Schiffman phase shifter employs sections of coupled transmission lines (TL) operating in TEM mode as key elements, but its phase difference is over 10° [1]. The properties of the Schiffman phase shifter are deteriorated because the phase velocity of the even mode is unequal to that of the odd mode. To improve the performances of the Schiffman phase shifter, compensated methods have been used in [2] and [3]. However, this technique cannot improve the phase performance obviously but enhance the band only. In [4], an improved wideband Schiffman phase shifter is presented. In the design, an additional isolated rectangular conductor under the coupled lines acts as a capacitor to decrease the odd-mode impedance and the ground plane under the coupled lines removed to increase the even-mode impedance. As a result, this phase shifter has a small size and good performance compared with the cascading microstrip multi-section coupled-line configuration. Unfortunately, the narrow gap of the coupled line limits its applications in the higher frequency.

Method 2: *based on the branch-line coupler and asymmetric ring-hybrid.* In [5], a wideband phase shifter has been proposed. The phase shifter consists of bias circuits, four switching diodes, and four radial stubs and one Lange coupler. In [6], phase shifter, which is composed of an asymmetric ring-hybrid and reflecting terminations, is presented. But, bandwidths of the phase shifters using this method are narrow than those phase shifters using edge-coupled structures.

Method 3: *multilayer broadside-coupled technique.* In [7], an ultra-wide band phase shifter was designed by exploiting broadside coupling between bottom and top elliptical patches. In the middle layer, an elliptical slot is use as the ground plane. Although the designed phase shifters achieve differential phase stability better than ±3°, the complex structure limits its application.

Method 4: *broadside-coupled microstrip-coplanar waveguide technique.* In [8], compact and planar phase shifters with wideband characteristics are presented. The designed phase shifters achieve differential phase stability better than ±2° in 3–11 GHz. Although the utilization of broadside-coupled microstrip-coplanar waveguide results in easy fabrication, the switched part between the microstrip and coplanar waveguide introduces ground defected structure. So, this method is not a good choice in majority situations.

Method 5: *paralleled open/short stub loaded technique.* In [9], phase shifters with 45°, 90°, and 180° phase shift are designed, and bandwidths of them are about 50%. In [10], a broadband dumb-bell-shaped 45° phase shifter is presented utilizing open-circuit and short-circuit muti-section stubs. The phase shifter has a bandwidth of 2–6 GHz with a maximum phase deviation of ±3.2°. But, the maximum insertion loss of 2.1 dB limits its application in those situations required low insertion loss. In [11], a multiple circular sector structure has been used in phase shifter design. A multi-objective optimization procedure is used to obtain optimum performances, such as low insertion losses, small phase variations in operating bandwidths. The phase shifter has a bandwidth of 2–6 GHz with a maximum phase deviation of ±2.84°, and a maximum insertion loss of 1.06 dB. However, its structure also has complex structure, which is difficult to spread widely.

Method 6: *composite right/left-handed transmission line (CRLH-TL)-based technique.* The phase characteristic of CRLH-TL is nonlinear and it can be used in design phase shifter with broadband. In this method, the most important link is to replace a right-handed transmission line with a CRLH-TL [12]. Gratefully, the phase shifter has a broadband using CRLH-TLs.

3 Conclusion

In this paper, the methods to design phase shifter are summarized. The detail analysis of different methods is presented. As a result, clear guidelines are given for the future designers. In practical, the designers can choose proper method to design the phase shifter according to fabrication, properties, and many other factors. Also, the researches can do some deep research based on the summarized results in this paper.

Acknowledgements The paper was supported by the National Natural-Science Foundation of China (Grant No. 61701527, 61601499) and the National Natural-Science Foundation of Shaanxi Province (Grant No. 2019JQ-583, 2018JQ6023).

References

1. Schiffman, B.: A new class of broadband microwave 90-degree phase shifters. IRE Trans. Microwave Theory **6**(4), 232–237 (1958)
2. Gruszczynski, S., Wincza, K., Sachse, K.: Design of compensated coupled-stripline 3-dB directional couplers, phase shifters, and Magic-T's—Part II: single-section coupled-line circuits. IEEE Trans. Microwave Theory Technique **54**(9), 3501–3507 (2006)
3. Gruszczynski, S., Wincza, K., Sachse, K.: Design of compensated coupled-stripline 3-dB directional couplers, phase shifters, and Magic-T's—Part I: single-section coupled-line circuits. IEEE Trans. Microwave Theory Technique **54**(11), 3986–3994 (2006)
4. Guo, Y., Zhang, Z., Ong, L.: Improved wideband Schiffman phase shifter. IEEE Trans. Microwave Theory Technique **54**(3), 1196–1200 (2006)
5. Kwon, H., Lim, H., Kang, B.: Design of 6–18 GHz wideband phase shifters using radial stubs]. IEEE Microwave Wirel. Compon. Lett. **17**(3), 205–207 (2007)
6. Ahn, H., Wolff, I.: Asymmetric ring-hybrid phase shifters and attenuators. IEEE Trans. Microwave Theory Technique **50**(4), 1146–1155 (2002)
7. Abbosh, A.M.: Ultra-wideband phase shifters. IEEE Trans. Microwave Theory Technique **55**(9), 1935–1941 (2007)
8. Abbosh, A.M.: Broadband fixed phase shifters. IEEE Microwave Wirel. Compon. Lett. **21**(1), 22–24 (2011)
9. Eom, S.Y., Park, H.K.: New switched-network phase shifter with broadband characteristics. Microwave Opt. Technol. Lett. **38**(4), 255–257 (2003)
10. Zheng, S.Y., Yeung, S.H., Chan, W.S., et al.: Improved broadband dumb-bell-shaped phase shifter using multi-section stubs. Electron. Lett. **44**(7), 478–480 (2008)
11. Yeung, S.H., Man, K.F., Chan, W.S.: The multiple circular sectors structures for phase shifter designs. IEEE Trans. Microwave Theory Technique **59**(2), 278–285 (2011)
12. Caloz, C., Itoh, T.: Electromagnetic metamaterials: transmission line theory and microwave applications. Wiley, New Jersey (2006)

Intra-pulse Modulation Feature Analysis for Radar Signals

Shi-qiang Wang, Caiyun Gao, Xingcheng Li, Hui-yong Zeng and Juan Bai

Abstract Radar signal sorting based on intra-pulse parameters is applied in complex electromagnetic environment well. This method provides a new basis and ideas for the sorting and identification of radiation source signals, and also a possible way to improve the current sorting ability of radiation source signals. This paper discusses several aspects from domain analysis, frequency-domain analysis, time–frequency analysis, wavelet analysis, atomic decomposition, high-order statistical analysis, and nonlinear dynamic analysis methods.

Keywords Intra-pulse feature · Time-frequency analysis · Wavelet analysis · Atomic decomposition · High-order statistical · Nonlinear dynamic analysis

1 Introduction

For the problem of signal sorting based on intra-pulse parameters applied in complex electromagnetic environment, after the 1990s, Danielsen proposed the method to solve this by searching and extracting new radar signal features [1]. Intra-pulse characteristic parameters help to reduce the overlapping probability of multi-parameter space, provide a new basis for the sorting and identification of radiation source signals, and also a possible way to improve the current sorting ability of radiation source signals. Intra-pulse intentional modulation includes intra-pulse phase modulation, frequency modulation, amplitude modulation, and hybrid modulation of a combination of three modulations. Common radar signals include linear frequency modulation (LFM), nonlinear frequency modulation (NLFM), two-phase coding (BPSK), four-phase coding (QPSK), and frequency

S. Wang · X. Li · H. Zeng (✉) · J. Bai
Air and Missile Defense College, Air Force Engineering University, Xi'an 710051, China
e-mail: 1048768838@qq.com

C. Gao
Basic Department, Air Force Engineering University, Xi'an 710051, China

V. Jain et al. (eds.), *Recent Trends in Intelligent Computing, Communication and Devices*, Advances in Intelligent Systems and Computing 1031,
https://doi.org/10.1007/978-981-13-9406-5_98

coding (FSK). At present, the general methods of intra-pulse modulation analysis of radar radiation sources include domain analysis methods, frequency-domain analysis methods, time–frequency analysis methods, high-order statistical analysis methods, and nonlinear dynamic analysis methods.

2 Time-Domain and Frequency-Domain Analysis

The time-domain autocorrelation method analyzes the intra-pulse characteristics of the radar signal by calculating the time-domain autocorrelation function of the signal. In view of the fact that the instantaneous autocorrelation encounters some practical problems in engineering applications, such as the sampling frequency in the actual digital intermediate frequency processing is limited by the device, it is impossible to be too high, resulting in the delay interval in the transient autocorrelation algorithm being too small, thus enabling the two calculated points of the correlation are too close, and the correlation of the noise will cause the performance of the algorithm to drop sharply at low signal-to-noise ratio (SNR). Pu et al. [2, 3] adjusted the range of delay interval to four times at the same sampling frequency by adjusting the unambiguous interval of the instantaneous phase so that the anti-noise performance of the algorithm is obtained to a certain extent. The improvement is that by using the unambiguous phase interval and the moving average, the quadratic features of the instantaneous frequency are extracted.

The frequency-domain analysis of the radiation source signal is to study the law of the energy or power of the signal as a function of time, thereby extracting the intra-pulse features that are useful for identifying the sorting. Papers in [4–6] have used time-domain spectrum method, frequency-domain similarity coefficient, high-order similarity coefficient, holder coefficient, symbolic intra-pulse feature extraction, and other methods to analyze the radiation source signal in the frequency domain and according to different principles. Effective new features were extracted separately.

The intra-pulse parameter feature extraction method based on time–frequency distribution mainly focuses on the processing of time–frequency images. In [7], a feature extraction method based on adaptive data-driven window length Wigner time–frequency distribution and feature extraction method based on Choi–Williams time–frequency distribution are proposed. In 2006, Zilberman [8] used CWD distribution to calculate radiation source signal. The time-frequency image is etched and the adaptive threshold binary processing based on recursive expansion is used to determine the center of gravity of the modulation energy, and then the time-frequency image is further processed to extract the binary eigenvector of the radiation source signal; The paper in [9] proposed a feature extraction method based on PWD distribution combined with random transform (RT); in 2010 the paper in [10] proposed a second-order time moment and Wigner based on complex independent component analysis, Wigner-Ville time-frequency distribution (WVD) The feature extraction method of-Hough transform can better identify the

LFM signal; the paper in [11] successively used the fuzzy function to extract the intra-pulse feature information of different signal structures for signal sorting and recognition.

3 Wavelet Analysis and Atomic Decomposition

Wavelet analysis has received extensive attention in the extraction of pattern features without signal reconstruction. In 1992, Delpart [12] proposed a wavelet asymptotic method for intra-pulse instantaneous frequency feature extraction based on the idea of wavelet transform. In [13], the spectrum of radar signal is decomposed by wavelet at different scales, and the low frequency is approximated to wavelet coefficient and high. The energy distribution entropy of the frequency detail wavelet coefficient is taken as the signal characteristic; the time–frequency rearrangement theory is applied to the wavelet spectrum in [14] to obtain the instantaneous parameter characteristics (frequency, phase, and amplitude) of the LFM signal; In 2010, it is provided that a new wavelet base similar to Morlet wavelet is constructed, and the secondary feature extraction of the ridge frequency characteristics of the signal is obtained, and four new features for different modulation modes are obtained [15]. As a generalization of wavelet transform, the wavelet packet also decomposes the high-frequency part, which is a more elaborate analysis method than multi-resolution analysis. In general, feature extraction using wavelet packets refers to the energy of each frequency band as a feature after wavelet packet decomposition [16], which applied the similarity coefficient method to select the extracted features, and selected two optimal feature subsets to identify the radiation source signals by neural network, which can achieve better recognition results. In 2006, the paper in [17] provided that the envelope characteristics of the radar signal are extracted by wavelet, and the four characteristics of the envelope including top drop, rising edge, falling edge, and pulse width are obtained with Morlet. In 2010, the paper in [18] also transforms the wavelet packet into the radiation source signal from different aspects. The application in feature extraction was analyzed.

Multi-parameter atomic decomposition of the radiation source signal can extract its intra-pulse characteristic parameters. Atomic decomposition adaptively approximates signals by sparse representation and has strong adaptability to densely distributed signals with different time–frequency distributions. The application of atomic decomposition to extract the characteristics of radiation source signals mainly focuses on: constructing an appropriate atomic library [19], searching for matching atomic algorithm selection [20], and matching atomic set optimization [21]. In addition, the fractional Fourier transform (FRFT) can perform signal processing on a uniform time–frequency domain, thus describing the time–frequency characteristics of the radiation source signal more completely. In 2009, Si [22] proposed a method based on FRFT for the extraction of eigenvectors of α-envelope curves.

4 High-Order Statistic Analysis

High-order statistic (HOS), meaning the high-order spectral analysis method, can automatically suppress the influence of Gaussian-colored noise on non-Gaussian signals and can suppress the influence of non-Gaussian-colored noise, retaining the amplitude and phase information of the signal, and therefore the source of radiation. Signal recognition and sorting have received extensive attention [23–27]. In [23], the time-domain bi-spectrum features are used to classify the measured radar data and compared with the non-high-order statistical analysis methods. The results show that under the conditions of colored noise and Gaussian noise, the time-domain bi-spectrum feature can be used to obtain good classification performance; in 2003, Yang [24] extracted the fourth-order cumulant of the radar emitter signal as the signal feature and used the Kohonen neural network to complete the separation of the interlaced signal; Sun [25] combined with the higher-order cumulant to give a kind of radar signal sorting method based on blind source separation, which can sort multiple radar signals mixed in Gaussian noise background; Qin [26] used two-spectral features to sort radar radiation source signals in 2011; Han [27] first used HOS to classify different modulation pattern signals, and then used pulse repetition interval (PRI) to match different modulation parameters with the same modulation pattern. The number is further separated, and finally, the effectiveness of the method is verified by simulation.

5 Nonlinear Dynamic Analysis

The nonlinear dynamic analysis method can identify the pulse modulation of the signal by measuring the complexity and irregularity of the signal waveform. For example, the correlation dimension, information dimension, box dimension, entropy, complexity, etc., can all describe the complexity of the signal. The paper in [28–30] has studied the intra-pulse feature extraction of the radiation source signal from these aspects. In 2003, Zhang [28] used the correlation dimension in fractal theory to characterize the complexity and irregularity of the signal, which simplified the design of the classifier. In the same year, the author used the box dimension and information dimension to make the complexity of the signal spectrum shape. Metrics, extracting fractal characteristics of radiation source signals containing information on signal amplitude, frequency and phase variation [29]; literature [30] uses Lempel-Ziv complexity to characterize signal complexity and irregularity, simple calculation by copying and adding. The model describes the signal sequence and takes the number of addition operations required as a measure of the Lempel-Ziv complexity of the sequence. The extracted features have better classification accuracy and noise insensitivity.

Pincus [31] pointed out that approximate entropy (ApEn) is a statistical parameter for measuring the complexity of time series. It only needs shorter data to measure the probability of generating new patterns in the signal and has better anti-interference ability. Therefore, ApEn can be used as a feature to describe the complexity of the radar source signal, and the radiation source signal sequence with different complexities is counted with less data; however, this method has the problem of estimation bias caused by signal self-matching. Richman [32] proposed the sample entropy (SampEn) as the statistical parameter of the time series, which has the same physical meaning and advantages as the approximate entropy and solves the problem of estimation bias and insensitivity to small complexity changes, so it can be considered This parameter is used as the intra-pulse feature of the radiation source signal.

6 Conclusions

It can be seen from the above analysis that domestic and foreign scholars have analyzed and studied the pulse modulation characteristics from different angles, and obtained certain research results, which provided theoretical support for improving the performance of electronic countermeasure equipment. However, some of the feature analysis methods, such as wavelet analysis, atomic decomposition, and time–frequency analysis methods, are computationally complex and cumbersome, and the feature dimensions extracted by high-order statistics analysis methods are too high. These factors are not conducive to engineering applications. In addition, the application of nonlinear dynamics to the processing of radar signals is a new attempt to objectively broaden the research horizon of radar signals, but the research in this area is not yet mature, and the physical meaning of radar signals corresponding to various nonlinear dynamic parameters requires further study.

Acknowledgements This paper was supported in part by the National Natural Science Foundation of China under the Grant no. 61601499, 61701527 and 61601503.

References

1. Danielsen, P.L., Agg, D.A., Burke, N.R.: The application of pattern recognition techniques to ESM data processing. In IEE Colloquium on SP for ESM Systems, 26 Apr: 6/1–6/4 (1988)
2. Pu, Y.W., Jin, W.D., Zhu, M.: Classification of radar emitter signals using cascade feature extractions and hierarchical decision technique. In: ICSP2006 Proceedings (2006)
3. Pu, Y.W., Jin, Y.D., Hu, L.Z.: Classification method of instantaneous frequency derivative characteristics of radar emitter signals. J. Harbin Inst. Technol. **1**, 136–140 (2009)
4. Marian, W., Adam, K., Janusz, D., et al.: The method of regression analysis approach to the specific emitter identification. In: International Conference on MIKON, pp. 491–494 (2006)
5. Janusz, D., Marian, W., Jan, M.: Applying the radiated emission to the specific emitter identification. J. Telecommun. Info. Technol. **2**, 57–60 (2005)

6. Janusz, D., Adam, K., Robert, O.: An application of iterated function system attractor for specific radar source identification. In: 17th International Conference on MIKON, pp. 1–4 (2008)
7. Lundén, J\., Koivunen, V.: Automatic Radar Waveform Recognition. IEEE J. Selected Top. Signal Process. **1**, 124–136 (2007)
8. Zilberman, E.R., Pace, P,E.: Autonomous time-frequency morphological feature extraction algorithm for LPI radar modulation classification. In: ICIP, pp. 2321–2324 (2006)
9. Gulum, T.O., Pace, P.E., Cristi, R.: Extraction of polyphase radar modulation parameters using a wigner-ville distribution and radon transform. In: ICASSP, pp. 1505–1508 (2008)
10. Qiang, G., Ya-jun, L., Chang-hong, W., He-yang, C.: Novel detection method for multi-component LFM signals. In: 2010 First International Conference on Pervasive Computing, Signal Processing and Applications, pp. 759–762 (2010)
11. Pu, Y.W., Wang, J.H., Jin, W.D.: A novel fractional autocorrelation based feature extraction approach for radar emitter signals. In: ICSP2008 Proceedings, pp. 2338–2341 (2008)
12. Delpart, N.: Asymptotic wavelet and gabor analysis: extraction of instantaneous frequencies. IEEE Trans. Info. Theory **38**(3), 644–664 (1992)
13. Chen, T., Jin, W., Chen, Z.: Feature extraction using wavelet transform for radar emitter signals. In: 2009 International Conference on Communications and Mobile Computing, pp. 414–418 (2009)
14. Chen, C.X., He, M.H., Yu, Ci.: A novel method for extraction of in-pulse feature of LFM signal. In: 2010 2nd International Conference on Industrial Mechatronics and Automation, pp. 692–697 (2010)
15. Yu, Z.B., Jin, Y.D., Chen, C.X.: Radar emitter signal recognition based on wavelet ridge frequency cascading characteristics. J. SouthWest JiaoTong Uni. **45**(2), 290–295 (2010)
16. Zhang, G., Jin, W., Hu, L.: Application of wavelet packet transform to signal recognition. In: Proceedings of the 2004 International Conference on Intelligent Mechatronics and Automation Chengdu, China, pp. 542–547 (2004)
17. Zhang, G., Huang, K., Jiang, W., Zhou, Y.: A method for extracting subtle features of radiation sources based on signal envelope. Syst. Eng. Electron. **28**(6), 795–797 (2006)
18. Ren, M.Q., Cai, J.Y., Zhu, Y.Q.: Radar signal feature extraction based on wavelet ridge and high order spectra analysis. In: Radar Conference, 2009 IET International, pp. 1–5 (2009)
19. Lopez-Risueno, G., Grajal, J., Yeste-Ojeda, O.: Atomic decomposition-based radar complex signal interception. IEE Proc. Radar Sonar Navig. **150**(4), 323–331 (2003)
20. Ming, Z., Weidong, J., Yunwei, P., Laizhao, H., et al.: A novel feature extraction approach for radar emitter signals. In: ICSP2008 Proceedings, pp. 2338–2341 (2008)
21. Cheng, J.X., Zhang, G.X.: Time-frequency atom feature extraction method for radar signal source of complex system. J. Xi'an Jiaotong Uni.y **44**(4), 108–113 (2010)
22. Si, X.C., Chai, J.F.: Radar signal feature extraction and automatic classification based on FRFT-Based α-envelope curve. J. Electron. Info. Technol. **31**(8), 1892–1897 (2009)
23. Jouny, I., Moses, R.L., Garber, F.D.: Classification of radar signals using the bispectrum. Int. Conf. Acoustics Speech Signal Process. **5**, 3429–3432 (1991)
24. Yang, Z., Zhu, Y., Xu, G., et al.: Signal sorting of unknown radar radiation sources based on HOS. Electron. Countermeasure **123**(6), 10–13 (2008)
25. Hong, S., Huangbin, A.: A radar signal sorting method based on blind source separation. Modern Radar **28**(3), 47–50 (2006)
26. Qin, K.B., Shen, Q., Wang, J.: A novel method for sorting radar emitter signal based on the bispectrum. In: International Conference on Information Engineering and Computer Science, pp. 1–4 (2009)
27. Han, J., He, M., Cheng, B.: A new method for signal sorting of unknown radar radiation sources. Aerospace Electron. Competition **25**(1), 40–43 (2011)
28. Zhang, G., Jin, W., Hu, L.: Fractal feature extraction of radar emitter signals. In: CEEM' 2003, pp. 161–164. Hangzhou, China (2003)

29. Zhang, G.X., Hu, L.Z., Jin W.D.: Complexity feature extraction of radar emitter signals. In: Asia-Pacific Conference on Environmental Electromagnetics CEEM' 2003, pp. 459–498. Hangzhou, China (2003)
30. Zhang, G.X., Jin, W.D., Hu, L. Z.: Radar emitter signal recognition based on complexity feature. J. Southwest Jiaotong Uni. **12**(2), 116–122 (2004)
31. Pincus, S.M.: Approximate entropy as a measure of system complexity. Pro. Natl. Acad. Sci **88**, 2297–2301 (1991)
32. Richman, J.S., Moorman, J.R.: Physiological time-series analysis using approximate entropy and sample entropy. Am. J. Physiol. Heart Circ. Physiol. **278**(6), H2039–H2049 (2000)

Analysis of the Research Status of Left- and Right-Hended Transmission Lines

Hui-yong Zeng, Xingcheng Li, Bin-feng Zong, Yan Zhao and Lin Geng

Abstract There are two main ways to implement the left and right hended transmission lines: One is to use the lumped parameter components of surface-mount technology (SMT); the other is to use the distributed parameter effect of the integrated transmission line. This paper analyzes the research status of the left- and right-hended transmission lines realized by these two methods.

Keywords Left- and right-hended transmission line · Lumped parameter · Distribution parameter

1 Introduction

Since the electric field, the magnetic field and the wave vector form a left-hended spiral relationship, the left-hended material based on the transmission line structure is called a left-hended transmission line, and for convenience of comparison, the conventional transmission line is called a right-hended transmission line. According to the electromagnetic field theory, the distribution parameter effect occurs when the electromagnetic wave passes through the transmission line because the right-hended effect generated by the parasitic series inductance and the parallel capacitance is inevitable in the actual structure, wherein the parasitic capacitance is generated by the voltage gradient and the parasitic inductance is caused by the flow of current in the direction of metallization and it can be seen that the ideal left-hended trans-

H. Zeng · X. Li · L. Geng (✉)
Air and Missile-Defence College, Air Force Engineering University, Xi'an 710051, China
e-mail: hyzeng.1023@163.com

B. Zong
Unit 94710 of the PLA, Wuxi 214000, China

Y. Zhao
Air Force's Military Representative Office, Jilin, China

V. Jain et al. (eds.), *Recent Trends in Intelligent Computing, Communication and Devices*, Advances in Intelligent Systems and Computing 1031,
https://doi.org/10.1007/978-981-13-9406-5_99

mission line does not exist, so the concept of the left- and right-hended transmission lines has appeared, and many documents are also called composite left- and right-hended transmission lines.

2 Analysis of the Implementation Method of Left- and Right-Hended Transmission Lines

There are no left- and right-hended transmission lines in nature. Its implementation requires artificial design. There are two main implementation methods: One is to use surface-mount technology (SMT) lumped parameter components to achieve [1]; the other is to utilize the distribution of integrated transmission lines. The parameter effect is realized. The most common two-way implementation of the distribution effect of the left- and right-hended transmission lines is the use of interdigitated gaps and short-circuited branches to achieve [2, 3], the typical structure is shown in Fig. 1, and the second is the use of reverse-opening rings. The resonator and its deformation structure are combined with the microstrip gap [4–8], and the typical structure is shown in Fig. 2. The SMT chip components of the lumped parameters can be directly utilized, and the left- and right-hended transmission lines can be realized easily and quickly, but the SMT components only have discrete values and cannot be used for high frequency due to self-resonance so that the left- and right-hended transmission lines based on the SMT components can only work in the lower frequency band; the left- and right-hended transmission lines with distributed parameters have greater application potential due to their simple structure, easy implementation and working in higher-frequency bands. Although the distributed parameter elements can operate at any frequency, they occupy a large size at low frequencies, which is disadvantageous for miniaturization and cost reduction. Therefore, SMT components and distributed components can complement each other. SMT components are generally used at low frequencies, and distributed components are generally used at high frequencies.

Literature [9] proposed a miniaturized ultra-wideband left- and right-hended transmission line structure. On the basis of the traditional left- and right-hended transmission lines based on interdigitated gaps and shorted vias, a grounding stub is added to each side of the interdigitated gap to form a symmetric π-type left- and right-hended transmission line structure, as shown in Fig. 3, which has

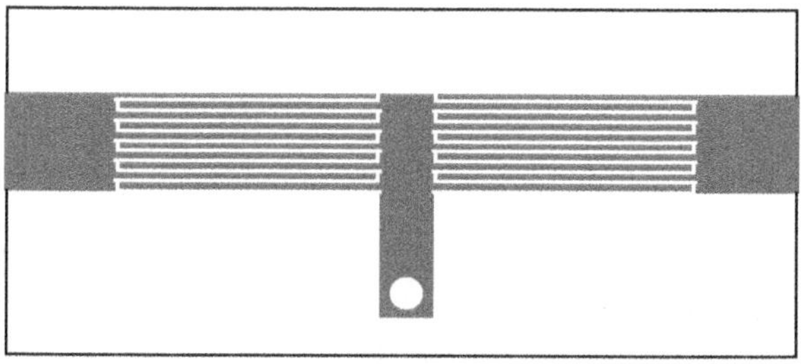

Fig. 1 Intersection gap and short-circuit branch implementation

Fig. 2 Reverse open-loop resonator and microstrip gap implementation

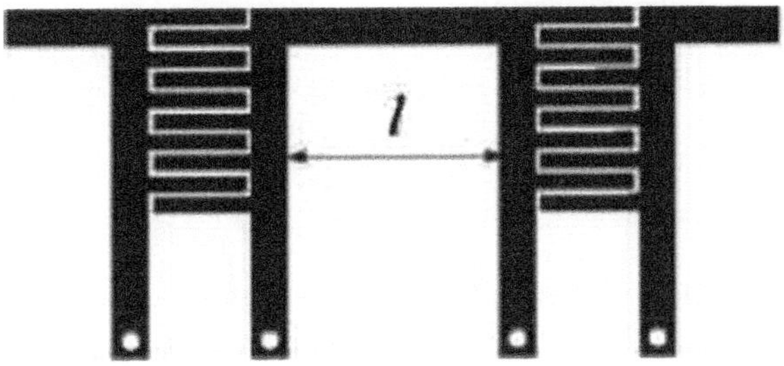

Fig. 3 Left- and right-hended transmission lines reported in [9]

ultra-wideband characteristics; the literature [10] proposes an electric small balance left- and right-hended transmission line based on the Minkowski fractal complementary open-loop resonator, as shown in Fig. 4. It shows that compared with the left- and right-hended transmission lines based on the traditional complementary

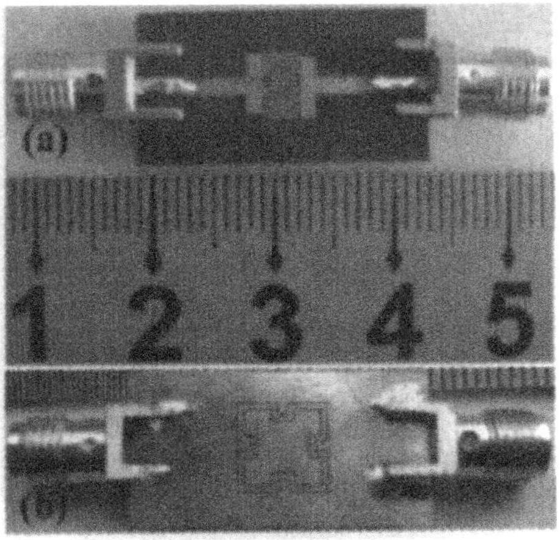

Fig. 4 Left- and right-hended transmission lines reported in [10]

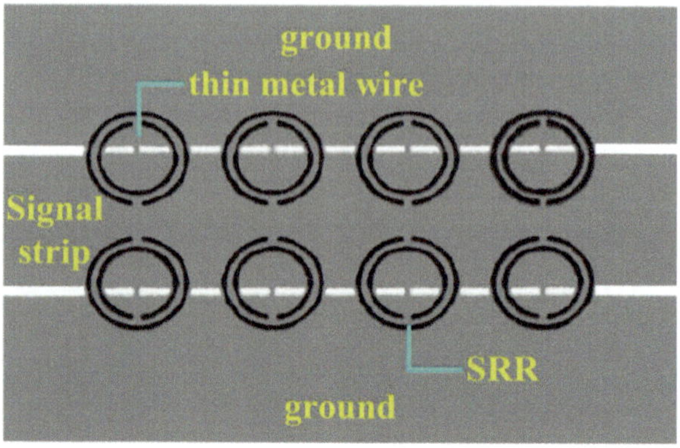

Fig. 5 Left- and right-hended transmission lines reported in [12]

open-loop resonator, the bandwidth of the structure is significantly broadened and the relative bandwidth is 113.7%. The literature [11] proposes a new ultra-wideband left–right-hended transmission line based on coplanar waveguide, according to Bloch periodic circuit theory, from the unit lumped equivalent circuit model, derives the left- and right-hended frequency range and the corresponding phase velocity and attenuation characteristics and extracts the equivalent constitutive parameters of the transmission line. The left- and right-hended parameters of the new structure are independently adjustable. Design and adjustment are convenient. Compared with the common interdigital or slot coupling, the series capacitor formed by surface coupling can effectively suppress high-frequency resonance and widen. The transmission line passband; the literature [12] proposed a left–right-hended coplanar waveguide transmission line based on the open-ring structure, which is symmetrically prepared on the back side of the dielectric plate and facing the coplanar waveguide slot, connecting the coplanar waveguide center signal line and the floor. The thin metal wire is facing the central region of the SRR, as shown in Fig. 5; the [13] uses the SMT component to implement the left- and right-hended transmission lines based on the coplanar waveguide, as shown in Fig. 6.

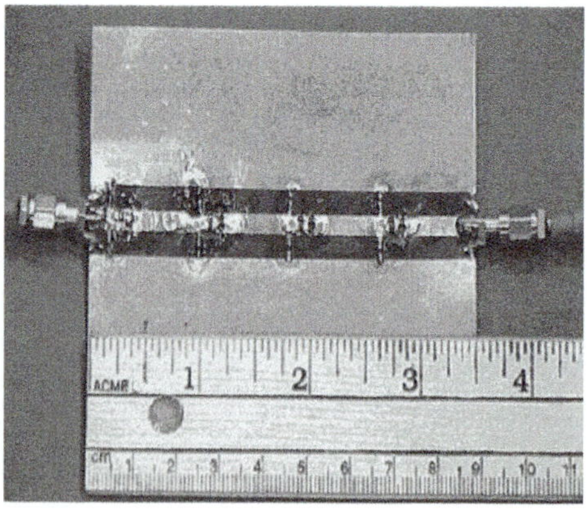

Fig. 6 Left- and right-hended transmission lines reported in [13]

3 Conclusion

The SMT chip components of the lumped parameters can be directly utilized, and the left- and right-hended transmission lines can be realized easily and quickly, but the SMT components only have discrete values and cannot be used for high frequency due to self-resonance; the left- and right-hended transmission lines of the distributed parameters are simple in structure, easy to implement and work in higher-frequency bands with greater application potential. Therefore, SMT components and distributed components can complement each other. This paper compares and analyzes the research status of the two main implementation methods and compares the advantages and disadvantages of the two implementation methods.

Acknowledgements The paper is supported by the National Natural Science Foundation of China (Grant No. 61701527, 61601499) and the National Natural Science Foundation of Shaanxi Province (Grant No. 2019JQ-583, 2018JQ6023).

References

1. Grbic, A., Eleftheriades, G.V.: A backward-wave antenna based on negative refractive index L-C networks. In: Proceedings of IEEE Antennas and Propagation Society International Symposium USNC/URSI National Radio Science Meeting, San Antonio, pp. 340–343 (2002)
2. Sanada, A., Caloz, C., Itoh, T.: Planar distributed structures with negative refractive index. IEEE Trans. Microw. Theory Tech. **52**(4), 1252–1263 (2004)
3. Caloz, C., Itoh, T.: Positive/negative refractive index anisotropic 2-D metamaterials. IEEE Microwave Wirel. Compon. Lett. **13**(12), 547–549 (2003)
4. Zeng, H.Y., Wang, G.M., Yu, Z.W., et al.: Miniaturization of branch-line coupler using composite right/left-hended transmission lines with novel meander-shaped-slots CSSRR. Radioengineering **21**(2), 606–610 (2012)
5. Gil, M., Bonache, J., Gil, I., et al.: Artificial left-hended transmission lines for small size microwave components: application to power dividers. In: Proceedings of European Microwave Conference, Manchester, UK, pp. 1135–1138 (2006)
6. Falcone, F., Lopetegi, T., Laso, M.A.G., et al.: Babinet principle applied to the design of metasurfaces and metamaterials. Phys. Rev. Lett. **93**(19), 197401 (2004)
7. Gil, M., Bonache, J., Selga, J., et al.: Broadband resonant-type metamaterial transmission lines. IEEE Microwave Wirel. Compon. Lett. **17**(2), 97–99 (2007)
8. Bonache, J., Gil, M., Gil, I., et al.: On the electrical characteristics of complementary metamaterial resonators. IEEE Microwave Wirel. Compon. Lett. **16**(10), 543–545 (2006)
9. Fu, G., Zhu, Y., Sun, C., et al.: Design of a miniaturized UWB composite left and right hended transmission line. Sci. Technol. Eng. **11**(29), 7105–7107 (2011)
10. Xu, H., Wang, G., Zhang, C., et al.: Research on composite left and right hended transmission lines based on fractal complementary open-ring resonators. J. Eng. Des. **18**(1), 71–76 (2011)
11. Huang, J., Zhao, Q., Liu, C.: Design and implementation of a new ultra-wideband composite left and right hended transmission line. Chin. J. Radio Sci. **25**(3), 460–465 (2010)

12. Martín, F., Bonache, J., Falcone, F., et al.: Split ring resonator-based left-hended coplanar waveguide. Appl. Phys. Lett. **83**(22), 4652 (2003)
13. Siddiqui, O.F., Mojahedi, M., Eleftheriades, G.V.: Periodically loaded transmission line with effective negative refractive index and negative group velocity. IEEE Trans. Antennas Propag. **51**(10), 2619–2625 (2003)

Application Status of Left- and Right-Handed Transmission Lines in Miniaturized Antenna Units

Hui-yong Zeng, Bin-feng Zong, Yan Zhao, Lujiang Liang and Lin Geng

Abstract For the dual-/multi-frequency characteristics, the wideband phase-shifting characteristics, the miniaturization characteristics, and the zero-/negative-order resonance characteristics of the right- and left-handed transmission lines, based on the miniaturization characteristics of the right- and left-handed transmission lines, the phase compensation of the left- and right-handed structures can be utilized to achieve miniaturization of the antenna. This paper summarizes the application of left- and right-handed transmission lines and "block" left-handed materials in small antenna units.

Keywords Left- and right-handed transmission line · Small antenna unit

1 Introduction

With the development of microwave devices toward integration, multi-functionality, and miniaturization, the size of electronic devices is getting smaller and smaller, and the antennas, as devices for transmitting and receiving electromagnetic wave signals, have to adapt to the development of the times. How to realize the miniaturization of the antenna without affecting the other performances of the antenna is a difficult problem for the antenna designer.

The antenna is the exit and entrance of the wireless communication system for transmitting and receiving electromagnetic wave signals. Its performance is related to not only the performance indicators of the entire wireless communication system, but also an important basis for determining the working mode of the system. As

H. Zeng · L. Liang · L. Geng (✉)
Air and Missile-Defence College, Air Force Engineering University, Xi'an 710051, China
e-mail: hyzeng.1023@163.com

B. Zong
Unit 94710 of the PLA, Wuxi 214000, China

Y. Zhao
Air Force's Military Representative Office, Jilin, China

V. Jain et al. (eds.), *Recent Trends in Intelligent Computing, Communication and Devices*, Advances in Intelligent Systems and Computing 1031,
https://doi.org/10.1007/978-981-13-9406-5_100

microelectronic devices and circuit systems move toward miniaturization, integration, and multi-functionality, the space left for antennas is getting smaller and smaller, which requires antenna designers to design antennas to be smaller and smaller.

In order to save space and enable the antenna to operate in a wide frequency band or in multiple discrete frequency bands, the antenna workers begin to focus on reducing the antenna size, increasing the antenna bandwidth, and achieving multi-frequency radiation of the antenna aspect. How to reduce the size of the antenna and expand the working frequency band of the antenna as much as possible while ensuring good radiation characteristics of the antenna is an arduous and urgent task. However, the size, radiation characteristics, and bandwidth of the antenna are difficult to meet at the same time. When the size of the antenna becomes small, on the one hand, the radiation resistance of the antenna becomes smaller, and the imaginary part of the input impedance of the antenna becomes larger, it is difficult to achieve a good match with the conventional transmission line, so that it is impossible to realize the broadband characteristic of the antenna. On the one hand, the radiation efficiency of the antenna and the radiation area of the antenna are closely related. When the radiation area is reduced, the radiation efficiency of the antenna is inevitably smaller. This requires finding a new way to achieve miniaturization while meeting the requirements of communication systems.

Conventional antenna miniaturization methods mainly include short-circuit loading, selection of high dielectric constant substrates, slotting and slitting, and utilization of lumped components. However, these methods generally sacrifice performance such as gain, efficiency, and bandwidth of the antenna. Based on the miniaturization characteristics of the right- and left-handed transmission lines, the phase compensation of the right- and left-handed structures can be utilized to achieve miniaturization of the antenna.

2 Analysis of the Implementation Method of Small Antenna Unit

Since the electric field, the magnetic field and the wave vector form a left-handed spiral relationship, the left-handed material based on the transmission line structure is called a left-handed transmission line, and for convenience of comparison, the conventional transmission line is called a right-handed transmission line. According to the electromagnetic field theory, the distribution parameter effect occurs when the electromagnetic wave passes through the transmission line, because the right-hand effect generated by the parasitic series inductance and the parallel capacitance is inevitable in the actual structure, wherein the parasitic capacitance is generated by the voltage gradient, and the parasitic inductance is caused by the flow of current in the direction of metallization; it can be seen that the ideal left-handed transmission line does not exist, so the concept of the left- and right-handed transmission lines has appeared, and many documents are also called composite left- and right-handed transmission lines.

In recent years, due to the singular characteristics of metamaterials, it has attracted the attention of scholars in the physical and electromagnetic fields. Compared with traditional metamaterials, the composite left- and right-handed transmission lines in the form of microstrip lines have the characteristics of continuity, low loss, and wide bandwidth, which opens up a completely different idea for the design of microwave radio frequency devices.

In [1], a one-dimensional miniaturized cavity structure based on left- and right-handed media is proposed, which combines the backward wave effect of the left-handed medium with the forward wave effect of the right-handed medium to design a cavity smaller than $\lambda 0/2$, which is applied to The antenna can break through the limitation of the traditional microstrip antenna $\lambda 0/2$ electrical size, and realize the miniaturization of the antenna. The literature [2] uses the left-handed material to make the double-ply resonator cavity, which reduces the cavity thickness to below $\lambda 0/2$. Miniaturized antennas, and the antenna has good directivity; the literature [3] proposed an ultra-small high-directional antenna based on the left-handed material resonator, which consists of two Fabry–Perot cavity reflectors. The thickness of the cavity can reach the order of $\lambda 0/60$; the literature [4] realizes the left-handed characteristic by arranging the periodic flat metal structure on the dielectric substrate and adopts this structure to design a small sub-wavelength cavity antenna and resonance. The thickness of the cavity can also reach $\lambda 0/60$; the literature [5] uses the interdigitated left- and right-handed transmission lines to design a small microstrip antenna operating in the L-band, as shown in Fig. 1, with the same frequency of the traditional rectangular micro. Compared with the antenna, the size of the left- and right-handed microstrip antennas is reduced by 51%, and the gain characteristics are not significantly changed. Since the zero-order resonant frequency is independent of the physical length, the physical length can

Fig. 1 Small antenna reported in [5]

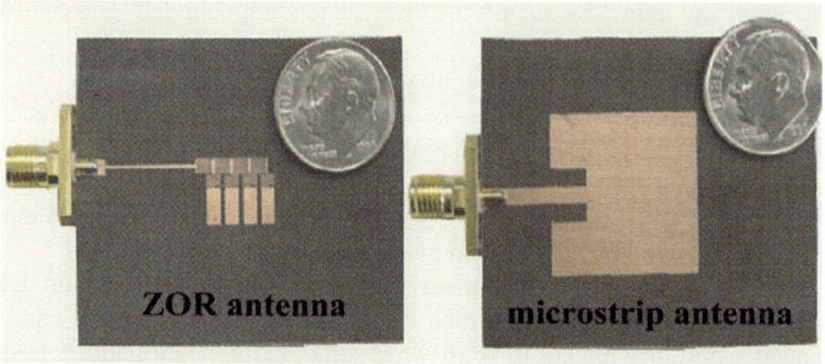

Fig. 2 Small antenna reported in [6]

theoretically be arbitrarily small [6]. A small microstrip antenna is designed using the zero-order resonance characteristics of the left- and right-handed transmission lines, as shown in Fig. 2.

3 Conclusion

Conventional antenna miniaturization methods generally sacrifice performance such as gain, efficiency, and bandwidth of the antenna, while the left- and right-handed transmission lines have dual-/multi-frequency characteristics, wide-band phase-shifting characteristics, miniaturization characteristics, and zero-/negative-order resonance. The miniaturization feature enables the miniaturization of the antenna by utilizing the phase compensation of the left- and right-handed structures. This paper summarizes the application of left- and right-handed transmission lines and "block" left-handed materials in small antenna units.

Acknowledgements The paper was supported by the National Natural Science Foundation of China (Grant No. 61701527, 61601499) and the National Natural Science Foundation of Shaanxi Province (Grant No. 2019JQ-583, 2018JQ6023).

References

1. Engheta, N.: An idea for subwavelength cavity resonators using metamaterials with negative permittivity and permeability. IEEE Antennas Wirel. Propag. Lett. **1**, 10 (2002)
2. Zhou, L., Li, H.Q., Qin, Y.Q., et al.: Directive emissions from subwavelength metamaterial based cavities. Appl. Phys. Lett. **86**, 101101 (2005)
3. Ourir, A., Lustrac, A.D., Lourtioz, J.M.: All metamaterial based subwavelength cavities (λ/60) for ultrathin directive antennas. Appl. Phys. Lett. **88**, 084103 (2006)

4. Ourir, A., Burokur, S.N., Yahiaoui, R., et al.: Directive metamaterial based subwavelength resonant cavity antennas applications for beam steering. C R Phys. **10**(5), 414 (2009)
5. Han, W.: Research on the interdigitated left-handed microstrip antenna. Master's thesis of University of Science and Technology of China, Hefei (2009)
6. Sanada, A., Kimura, M., Awai, I. et al.: A planar zeroth order resonator antenna using left-handed transmission line. In: Proceedings of European Microwave Conference, Amsterdam, The Netherlands (2004)

Research on High-Gain Antenna Unit Based on Left-Handed Materials

Hui-yong Zeng, Yan Zhao, Bin-feng Zong, Juan Bai and Jian An

Abstract The antenna is an important component of the radio system. It is responsible for converting high-frequency current or signal into a magnetic wave propagating in space or replacing the electromagnetic wave propagating in free space into a high-frequency current or signal for processing by the receiver. In recent years, high standards and strict requirements have been put forward on the structure and performance of antennas, and especially bandwidth, gain, and directivity have become the most important performance indicators of antennas. How to design high-gain antennas has become a hot topic. The "block" left-handed material based on the left- and right-handed structure is used as the antenna dielectric substrate or the antenna coating to increase the gain of the antenna. This article summarizes the high-gain antenna elements based on left-handed materials.

Keywords Left-handed material · High-gain antenna unit

1 Introduction

The exploration and development of new materials have always been the unremitting goal and progress of human beings. The left-handed materials different from traditional materials have become one of the most dynamic research fields in the world. The left-handed materials and the traditional right-handed materials are

H. Zeng · J. Bai
Air and Missile-Defence College, Air Force Engineering University, Xi'an 710051, China

Y. Zhao
Air Force's Military Representative Office, Jilin, China

B. Zong
Unit 94710 of the PLA, Wuxi 214000, China

J. An (✉)
Beijing Space Information Relay Transmission Technology Research Center, Beijing 102300, China
e-mail: hyzeng.1023@163.com

V. Jain et al. (eds.), *Recent Trends in Intelligent Computing, Communication and Devices*, Advances in Intelligent Systems and Computing 1031,
https://doi.org/10.1007/978-981-13-9406-5_101

mutually exclusive, many. The nature is complimentary. The left- and right-handed transmission lines, which are important implementations of left-handed materials, have attracted great attention from researchers. The concept of the left- and right-handed transmission lines not only enriches the transmission line theory, but more importantly opens the door for people to freely control the dispersion curve of the transmission line. The design concept of traditional microwave devices.

The first proposal of the left-handed material was that in 1967, Former Soviet Physicist Veselago changed the signs of the dielectric constant ε and the magnetic permeability μ at the same time. The Maxwell equation still holds, and the electric field, the magnetic field, and the wave vector form a left-handed spiral relationship. Veselago named it "left-handed material." Since the material has never been found in nature, this subversive concept has been left unattended; until 1999, Professor Pendry of the Royal College of England first theoretically proved the existence of left-handed materials. Veselago's pioneering work attracted attention. According to Professor Pendry's theory, in 2001, Professor Smith of the Massachusetts Institute of Technology first synthesized the "block" left-handed material. Since then, the left-handed material has gradually become a research in the field of physics and electromagnetics. Hotspots; In 2002, Professor Itoh and Professor Eleftheriades proposed the left-hand material of the transmission line structure at the same time, called the left-right hand transmission line method. The left–right-handed transmission line avoids the resonance structure of the "block" left-handed material, and its transmission response and frequency range can satisfy the microwave. Circuit requirements, with the advantages of low frequency bandwidth and low loss Ease of application in microwave circuits.

As the device for transmitting and receiving electromagnetic waves, the antenna is at the forefront of the microwave wireless communication and detection system. Its performance has a very important impact on the whole system. The function of the feeder system is to ensure the amplitude and phase of each antenna unit in the antenna array as needed. Distribution is the key to affecting the overall performance of the antenna array. At present, the theoretical and applied research of left- and right-handed transmission lines has been carried out in the field of microwave technology, especially in the application of antenna feeder systems. The nonlinear dispersion curve and phase constant of the left- and right-handed transmission lines can be arbitrarily selected in the real number domain, so that they have dual-/multi-frequency characteristics, wideband phase-shifting characteristics, miniaturization characteristics, and zero-/negative-order resonance characteristics which are not possessed by conventional transmission lines. Microwave devices based on left- and right-handed transmission lines have many functions not found in conventional structures. The dual–multi-frequency, wideband, small-scale, and circular-polarized antenna feeder system can be designed by using the left- and right-handed transmission lines, which has important significance and broad application prospects in both military and civilian applications.

2 Analysis of the Implementation Method of High-Gain Antenna Unit

There are many parameters describing the performance of the antenna, such as lobe map, gain, effective aperture, antenna impedance, antenna polarization, etc. Each parameter determines the place where the antenna is suitable for application. However, it has been widely used in many parameters. The pursuit of high antenna gain, because the antenna transmit power is fixed, the higher the gain of the antenna, the farther the electromagnetic wave transmission distance, the wider the coverage of the signal, so scholars at home and abroad have been working on high Research on gain antennas. The currently available high-gain antennas include Yagi antennas, Franklin antennas, parabolic antennas, and high-gain antennas with microstrip structures.

Traditional methods for improving the gain of the antenna are various, such as switching to parabolic antennas and using array antennas and dish antennas, but these antennas are mostly too large, which limits their application. Although the microstrip antenna is small in size, its gain is low, and the radiation direction is susceptible to surface waves. The "block" left-handed material based on the left- and right-hand structure is used as the antenna dielectric substrate or the antenna coating, which can increase the gain of the antenna without increasing the volume.

With the development of antenna technology and the increasingly complex use environment of antennas, the pace of research and improvement of high-gain antennas has never stopped. In the 1920s, the Franklin antenna theory proposed by Franklin received strong attention from everyone, which led to the upsurge of studying high-gain Franklin antennas. It is a common method to use the cavity structure to increase the gain of the antenna. Dielectric resonators increase the efficiency and gain of the antenna by using a vertically coupled feed structure. Another method for improving the antenna gain is the reflective surface method. By adding a reflecting surface, the original omnidirectional antenna is changed into a directional antenna, so that the width of the radiation pattern of the antenna is reduced, thereby achieving the purpose of improving the antenna gain. The famous Cassegrain antenna uses the reflective surface to increase the gain of the antenna.

With the advent of electromagnetic metamaterials, the use of metamaterials in increasing antenna gain is increasing. The literature [1, 2] reported the use of left-handed material flat lens focusing effect to improve the antenna gain, obtained high gain, and achieved a small antenna design; the literature [3] theoretically studied the left-handed material as an antenna. The influence of the cladding on the microstrip antenna. The left-handed material consists of a rectangular split ring and a metal wire. A certain volume of this left-handed material coating is placed in front of the antenna, which can increase the gain of the antenna by 2.8 dB, and the directivity is good; in [4], a new left-handed material structure is constructed by combining a metal split ring structure with a capacitor-loaded metal wire. As a coating of a microstrip antenna, the gain can be significantly improved and the directivity is good; the literature [5] studied. The effect of the left-handed material covering the square-opening ring structure on the performance of the antenna on the microstrip patch antenna is found to increase the gain of the antenna as the number of layers of the left-handed material increases, and the antenna gain of the four-layer structure is reached. 2.12 dB; the literature [6] designed a circular patch antenna based on the left-handed material of the metal split ring structure, the antenna gain increased from 2.02 to 3.51 dB, with better matching performance, and days. Only about half the size of conventional antennas.

3 Conclusion

The antenna is an important component of the radio system. It is responsible for converting high-frequency current or signal into a magnetic wave propagating in space or replacing the electromagnetic wave propagating in free space into a high-frequency current or signal for processing by the receiver.

With the gradual improvement of the degree of social informatization and the increasingly scientific and intelligent life of people, the application of wireless communication technology has been integrated into all aspects of our lives. As a key part of the wireless communication system, the antenna plays an important role in the operation of the entire system. Bandwidth, gain, and directionality have become the most important performance indicators for antennas. How to broaden bandwidth and increase gain has become a hot topic. The "block" left-handed material based on the left- and right-hand structure is used as the antenna dielectric substrate or the antenna coating, which can increase the gain of the antenna without increasing the volume. This paper summarizes the application of left-handed materials in high-gain antenna units and analyzes the advantages and disadvantages of different implementations.

Acknowledgements The paper was supported by the National Natural Science Foundation of China (Grant No. 61701527, 61601499) and the National Natural Science Foundation of Shaanxi Province (Grant No. 2019JQ-583, 2018JQ6023).

References

1. Wu, B.I., Wang, W., Pacheco, J., et al.: A study of using metamaterials as antenna substrate to enhance gain. Prog. Electromagnet. Res. **51**, 295 (2005)
2. Wang, S., Feresidis, A.P., Goussetis, G., et al.: High-gain subwavelength resonant cavity antennas based on metamaterial ground planes. IEEE Proc. Microwave Antennas Propag. **153** (1), 1 (2006)
3. Burokur, S.N., Latrach, M., Toutain, S.: Theoretical investigation of a circular patch antenna in the presence of a left handed medium. IEEE Antennas Wirel. Propag. Lett. **4**, 183 (2005)
4. Rahim, M.K.A., Majid, H.A., Masri, T.: Microstrip antenna incorporated with left-handed metamaterial at 2.7 GHz. In: Proceedings of IEEE Antenna Technology. Santa Monica, CA (2009)

5. Zhao, Y.C., Wan, G.B., Zhao, H.L. et al.: Effects of superstrate with improved SSRRs on the radiation of microstrip antenna. In: Proceedings of IEEE Microwave, Antenna, Propagation and EMC Technologies for Wireless Communications. Beijing, p. 51 (2009)
6. Zani, M.Z.M., Jusoh, M.H., Sulaiman, A.A. et al.: Circular patch antenna on metamaterial. In: Proceedings of Electronic Devices, Systems and Applications, p. 313 (2010)

Analysis of the Implementation Methods of Left- and Right-Hand Transmission Lines in Couplers and Hybrid Rings

Hui-yong Zeng, Jian An, Bin-feng Zong, Yan Zhao and Shi-qiang Wang

Abstract The function of the feeder system is to ensure that the amplitude and phase of each antenna unit in the antenna array are distributed as needed. Therefore, the merits of the feeder system are the key factors affecting the overall performance of the antenna array. Couplers and hybrid rings are important parts of the feeder system. This paper summarizes the application of left- and right-hand transmission lines in couplers and hybrid rings and analyzes the advantages and disadvantages of different implementations.

Keywords Left- and right-hand transmission lines · Coupler · Hybrid ring

1 Introduction

The function of the feeder system is to ensure that the amplitude and phase of each antenna unit in the antenna array are distributed as needed. Therefore, the merits of the feeder system are the key factors affecting the overall performance of the antenna array. The left- and right-hand transmission lines have dual-frequency/multiple frequency characteristics, wideband phase shifting characteristics, miniaturization characteristics, and zero-order/negative-order resonance characteristics, and couplers and hybrid rings are realized based on these characteristics of the right- and left-hand transmission lines.

H. Zeng · S. Wang (✉)
Air and Missile-Defence College, Air Force Engineering University, Xi'an 710051, China
e-mail: hyzeng.1023@163.com

J. An
Beijing Space Information Relay Transmission Technology Research Center, Beijing 102300, China

B. Zong
Unit 94710 of the PLA, Wuxi 214000, China

Y. Zhao
Air Force's Military Representative Office, Jilin, China

V. Jain et al. (eds.), *Recent Trends in Intelligent Computing, Communication and Devices*, Advances in Intelligent Systems and Computing 1031,
https://doi.org/10.1007/978-981-13-9406-5_102

2 Analysis of the Implementation Method

In the design of coupler and hybrid ring using left- and right-hand transmission lines: [1] designed the positive/negative $\lambda 0/4$ short/open route based on the left- and right-hand transmission line units by using the dispersion controllable characteristics and negative length characteristics of the left- and right-hand transmission lines, indicating the corresponding two. The frequencies are independently adjustable, and based on this, a dual-frequency branch line coupler and hybrid ring based on the left- and right-hand transmission lines are designed, as shown in Fig. 1. The literature [2] proposes a new coupling based on the special properties of the left- and right-hand transmission lines. The mechanism theoretically achieves 100% tight coupling. According to this coupling mechanism, symmetric and asymmetric couplers are designed, as shown in Fig. 2; the literature [3] is based on the design of the left- and right-hand transmission lines of reverse open-loop resonators. A small dual-frequency branch line coupler is shown in Fig. 3; the literature [4] proposes a high directional coupler consisting of a left–right-hand transmission line and a right-hand transmission line, where the left–right-hand transmission line is composed of an interdigital capacitor and a parallel short-circuit branch; the literature [5, 6] proposes a left–right-hand transmission line unit based on the coupled line and uses the unit to design a small branch line bridge and a directional coupling with arbitrary coupling degree. The combination of the unit structure and its application design is shown in Fig. 4. The left- and right-hand transmission lines have a simple structure and no grounding via holes is introduced, which has great application potential.

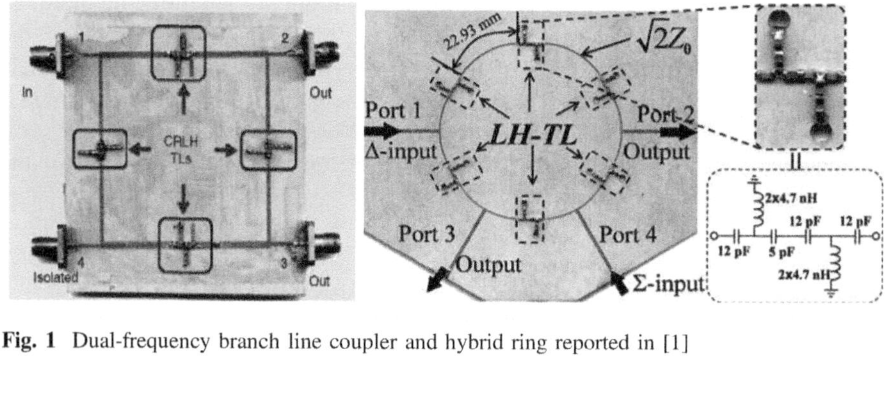

Fig. 1 Dual-frequency branch line coupler and hybrid ring reported in [1]

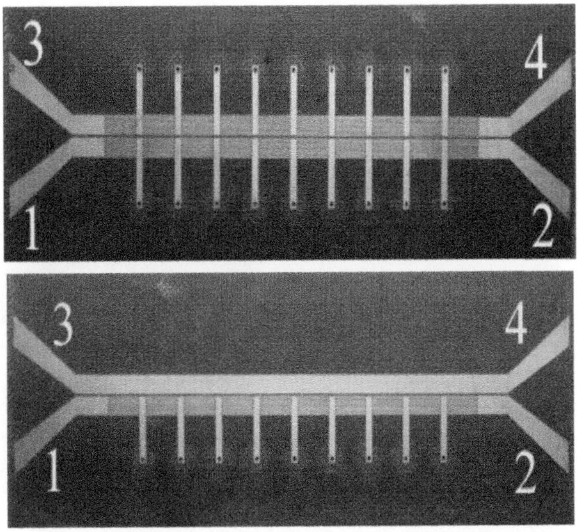

Fig. 2 Symmetric coupler and asymmetric coupler reported in [2]

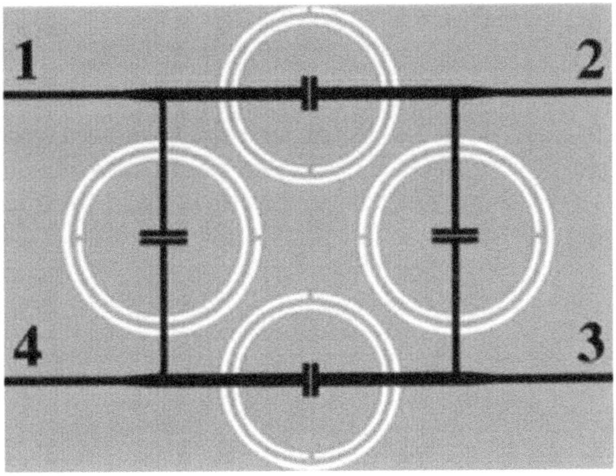

Fig. 3 Branch line coupler reported in [3]

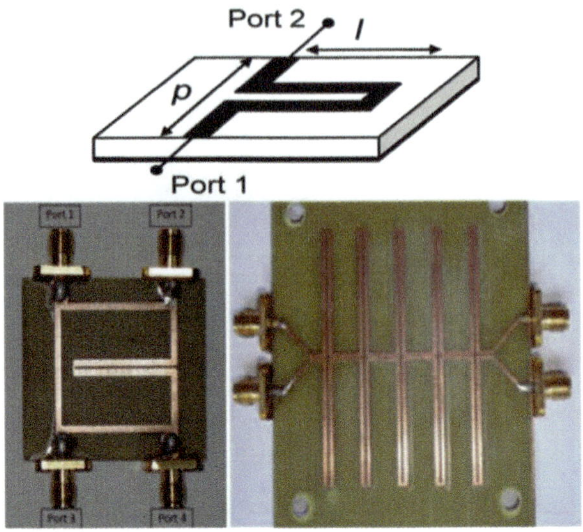

Fig. 4 Directional coupler reported in [6]

3 Conclusion

The function of the feeder system is to ensure that the amplitude and phase of each antenna unit in the antenna array are distributed as needed. Therefore, the merits of the feeder system are the key factors affecting the overall performance of the antenna array. Couplers and hybrid rings are important parts of the antenna's feeder system. This paper summarizes the application of left- and right-hand transmission lines in couplers and hybrid rings and analyzes the advantages and disadvantages of different implementations.

Acknowledgements The paper was supported by the National Natural Science Foundation of China (Grant Nos. 61701527, 61601499) and the National Natural Science Foundation of Shaanxi Province (Grant No. 2018JQ6023).

References

1. Lin, I., Devincentis, M., Caloz, C., et al.: Arbitrary dual-band components using composite right/left-handed transmission lines. IEEE Trans. Microw. Theory Tech. **52**(4), 1142–1149 (2004)
2. Caloz, C., Sanada, A., Itoh, T.: A novel composite right/left-handed coupled-line directional coupler with arbitrary coupling level and broad bandwidth. IEEE Trans. Microw. Theory Tech. **52**(3), 980–992 (2004)
3. Bonache, J., Sisó, G., Gil, M., et al.: Application of composite right/left handed (CRLH) transmission lines based on complementary split ring resonators (CSRRs) to the design of dual-band microwave components. IEEE Microwave Wirel. Compon. Lett. **18**(8), 524–526 (2008)

4. Islam, R., Eleftheriades, G.V.: Printed high-directivity metamaterial MS/NRI coupled-line coupler for signal monitoring applications. IEEE Microwave Wirel. Compon. Lett. **16**(4), 164–166 (2006)
5. Safwat, A.M.E.: Microstrip coupled line composite right/left-handed unit cell. IEEE Microwave Wirel. Compon. Lett. **19**(7), 434–436 (2009)
6. Fouda, A.E., Safwat, A.M.E., Hadia, E.H.: On the applications of the coupled-line composite right/left-handed unit cell. IEEE Trans. Microw. Theory Tech. **58**(6), 1584–1591 (2010)

Comparison of Radar Signal Sorting Method Between Single and Multi-parameter Based on

Shi-qiang Wang, Caiyun Gao, Xingcheng Li, Hui-yong Zeng and Juan Bai

Abstract With the increasing complexity of radar waveform design, traditional radar signals such as fixed-frequency, jagged, and jitter are becoming less and less common in modern radar design, and complex repeat frequency patterns such as agile and random repeat frequency become the current radar. Therefore, the sorting algorithm based on PRI has been difficult to adapt to the radar signal sorting of complex systems. In addition, the PRI-based sorting algorithm also has the problems of slow speed and poor sorting effect. In contrast, multi-parameter-based sorting method could achieve better results.

Keywords Radar signal sorting · Time of arrival (TOA) · Pulse width (PW) · Carrier frequency (RF) · PRI transform · Support vector clustering

1 Introduction

In general, the parameter used in the inter-pulse routine parameter signal sorting is information extracted from each pulse, including TOA, RF, PA, PW, and DOA and other parameters. Signal sorting systems generally rely on single parameters or multiple parameters for sorting. The most common ones rely on a single pulse parameter known as TOA, the pulse repetition interval (PRI) is obtained by TOA, and then, different algorithms are used for sorting. The most used ESM system is sequence search, accumulating difference histogram, and timing difference histogram sorting method [1, 2], and PRI transform algorithm and other improved algorithms are also applied [3].

With the emergence of a large number of complex system radars, the traditional five parameters can be varied in various parameter domains, even overlapping each

S. Wang · X. Li · H. Zeng (✉) · J. Bai
Air and Missile Defense College, Air Force Engineering University, Xi'an 710051, China
e-mail: 1048768838@qq.com

C. Gao
Basic Department, Air Force Engineering University, Xi'an 710051, China

V. Jain et al. (eds.), *Recent Trends in Intelligent Computing, Communication and Devices*, Advances in Intelligent Systems and Computing 1031,
https://doi.org/10.1007/978-981-13-9406-5_103

other, which causes the signal sorting to be extremely difficult. The multi-parameter-based signal sorting can achieve a relatively good sorting effect, generally adopting the unsupervised learning method [4] to achieve multi-parameter signal sorting. The papers in [5–7] have studied and discussed the multi-parameter radar signal method from different aspects.

2 PRI-Transform-Based Sorting Method

The method of sorting by single pulse parameter is mainly performed by transforming the TOA to obtain the PRI of the pulse stream of the radiation source and then performing further sorting. First, let the receiver have an intermediate frequency of 30 MHz, a bandwidth of 20 MHz, and an ADC sampling frequency of 150 MHz. Simulate the TOA of the starting pulse of a radiation source with a uniformly distributed random number from 0 to the total interception time T_{int}. When the total intercept time is $T_{\text{int}} = 50$ ms, a typical experiment (SNR = 15 dB) generates a total of 476 pulses (Rd1–Rd6), and the details of the generated pulse and residual pulse are given in Table 1.

PRI transform is a complex-valued autocorrelation integral algorithm based on pulse arrival time series. The algorithm transforms the TOA difference of the pulse sequence into a PRI spectrum, and the PRI value corresponding to the pulse sequence can be estimated from the peak position [8, 9]. The PRI transform algorithm has received extensive attention due to its ability to suppress harmonics and pulse jitter very well [10–13]. The results of PRI transform simulation are shown in Fig. 1.

As can be seen from Fig. 1, the PRI transformation method is basically invalid when sorting. Only a small number of pulse signals in the two radars of Rd2 and Rd5 can be separated, and the residual identification fails, resulting in more false alarms. The reason for this phenomenon is that the PRI modulation rules of the six radar signals to be sorted are more complicated.

Table 1 Simulation of the loss of generated pulses

Radar source	Rd1	Rd2	Rd3	Rd4	Rd5	Rd6	Total	Loss rate
Generate pulse	152	59	65	112	32	152	476	9.24%
Residual pulse	138	54	59	101	29	138	432	

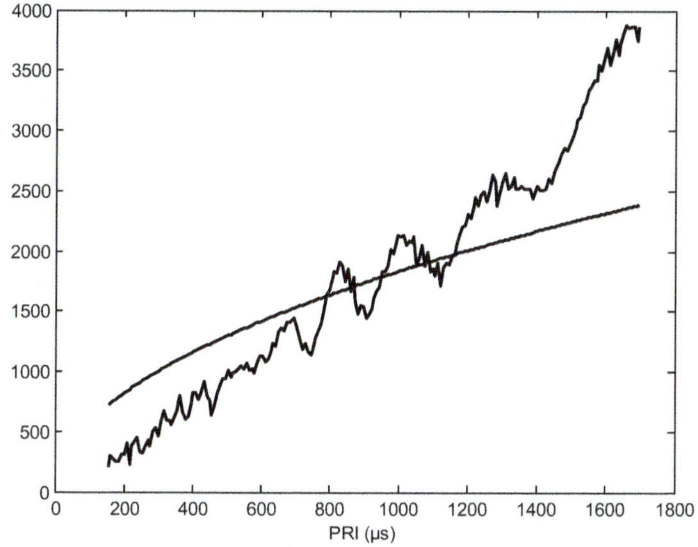

Fig. 1 Schematic diagram of PRI transform sorting

3 Multi-parameter-Based Sorting Method

Take [RF, PW] as the characteristic parameter. The following uses K-means and SE-MSVC sorting method to compare and explain. It is proved that the feature vector is not conducive to signal sorting, and the superiority of SE-MSVC method [14] is verified. The feature vector is first normalized, and the absence of a missed pulse is set. The results of three typical K-means clustering are shown in Fig. 2.

In Fig. 2, $C1$ represents the first cluster generated by the algorithm, $C2$, and so on, and the symbol "×" indicates the cluster center. It can be seen from the figure that even if the number of clusters is specified, the K-means clustering method often aggregates Rd5 and Rd6 into one class. Correspondingly, Rd1 is split into two

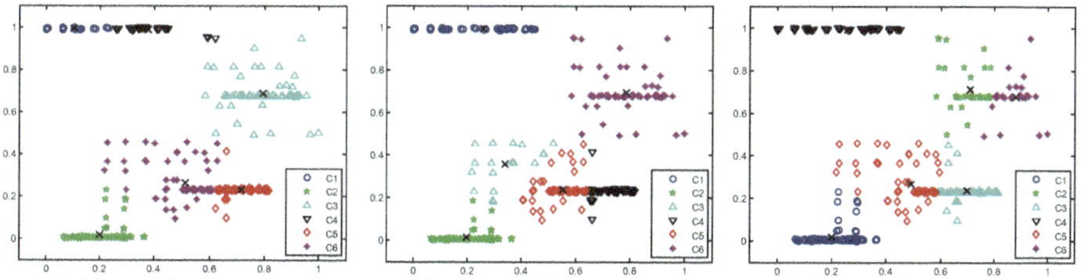

(*a*) Clustering results 1 (*b*) Clustering results 2 (*c*) Clustering results 3

Fig. 2 Typical K-means clustering results

categories, and it can be seen that other clusters are also varying degrees of error clustering.

For the SE-MSVC method, the penalty factor C, the Gaussian kernel width q, and the validity verification index SE parameters play a crucial role in the validity of the clustering results. The process of adjusting C and q by SE is explained by a typical sorting process. First, let $C = 1$ and calculate the initial value $q = 0.607$ according to the formula (1), and the initial q determined by the formula (1) means that the value of the kernel function formed by all the data points is large so that all the data forms only one cluster, and $C = 1$ is ensured. The boundary support vector, i.e., the outlier, is not formed at this time.

$$q = \frac{1}{\max_{ij} \left\| g_i - g_j \right\|^2} \tag{1}$$

Then, run the MCMSVC algorithm according to the parameters q to obtain an intermediate clustering result. At this time, the number of support vectors increases sharply. Therefore, the C value is reduced heuristically according to the $p = 1/(NC)$ [15], where the number of samples is represented as $N = 432$, and p indicates the proportion of the boundary support vector allowed. After reducing the C value, continue to execute the SVC algorithm according to the last algorithm running with q, and determine whether the number of SVs in the clustering result increases sharply or whether the cluster formed by the single sample vector is included; if both conditions are false, according to the modified cone surface clustering identifier (MCCL) algorithm, the outliers in the clustering result are processed, and the SE value is calculated from the processed clustering result. If the SE is not the maximum value, the value q is continuously increased. The algorithm is executed, and the parameters are adjusted according to the validity verification index SE.

Figure 3a shows the clustering results at the time of the parameter (the MCCL algorithm has not been used to process the outliers at this time), where the smooth curve represents the contour, the symbol "O" represents the support vector, and the remaining symbols represent vectors of different directions and different forms. The symbols represent different cluster classes. It can be seen that the number of clusters containing most samples splits into multiple meaningful clusters. The abnormal value is processed according to the MCCL algorithm, and the SE value is 0.7089. At the same time, Fig. 3a also shows that the parameters play a role in smoothing the cluster boundary. The final cluster distribution is shown in Fig. 3b.

It can be seen from Fig. 3b that due to the radiation sources Rd1 and Rd2, the RF-PW characteristics of Rd5 and Rd6 overlap severely and the separability is poor, so even using the SE-MSVC method with better clustering performance, it is still difficult to effectively distinguish the above four sources. Conversely, SE-MSVC can increase the probability of linearly separable overlapping features by nonlinear mapping. Therefore, even if the RF-PW feature of Rd2 is located inside the Rd4 feature, the radiation source Rd4 completely covers part of the characteristic samples of Rd2. The two can be effectively distinguished. It should be noted

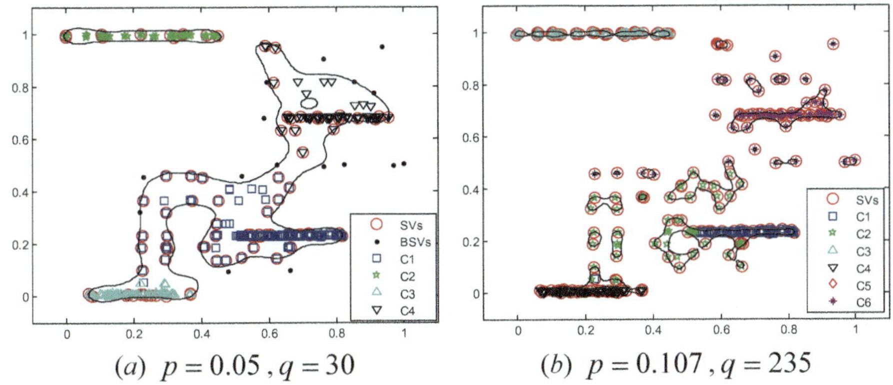

Fig. 3 Cluster distribution map under different parameters

that the result shown in Fig. 3b is the clustering situation after the abnormal value is processed by the MCCL algorithm.

In the clustering process, the penalty factor C (described by p), the Gaussian kernel width q, the SE index value, and the cluster number value change with the number of iterations as shown in Fig. 4. The values q in the figure take the natural logarithm, p is magnified 10 times, and other parameters are compared.

In Fig. 4, "p value" and "q value" represent the values describing the parameters, respectively, and "Index" and "Cluster" represent the SE index value and the number of clusters, respectively.

It can be seen from Fig. 4 that with the increase of the parameters p and q, the best clustering effect is obtained when the SE index reaches the maximum value of 2.3672, and the number of clusters is consistent with the number of radiation sources. The corresponding parameters are $p = 0.107$ and $\ln q = 5.461$, and the

Fig. 4 Trend charts of various parameters

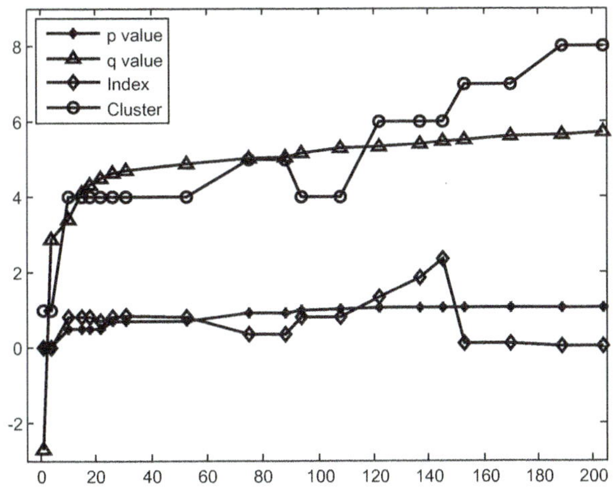

Table 2 Comparison of sorting results between two clustering methods

Method	Sorting result	Rd1	Rd2	Rd3	Rd4	Rd5	Rd6
		$n_1 = 138$	$n_2 = 54$	$n_3 = 59$	$n_4 = 101$	$n_5 = 29$	$n_6 = 51$
K-means	Correct numbers	127	19	30	81	13	10
	Misselected numbers	11	35	29	20	16	41
	Accuracy rate	64.81%					
SE-MSVC	Correct numbers	119	32	59	83	3	51
	Misselected numbers	19	22	0	18	26	0
	Accuracy rate	**80.32%**					

number of iteration times is 145. Then, with the increase of the value q, the value of the validity verification index decreases sharply, indicating that the inter-class aggregation and inter-class separation between the clusters determined by the algorithm deteriorate after the value q is increased again. Therefore, the SE index value is used to cluster the algorithm. MCMSVC is able to correctly give clustering results.

The K-means algorithm is compared with the SE-MSVC sorting method. And the statistical average results of the 20 sorting are shown in Table 2. Among them, the parameter and the adjustment step are different in each sorting experiment. The misselected pulse refers to the pulse belonging to the current radiation source and is distributed to other radiation sources, indicating the actual pulse number of the first type of radiation source signal.

It can be seen from Table 2 that the sorting effect obtained by the K-means algorithm is not ideal, and the average sorting accuracy is only 64.81%, and the pulse generated by the radiation source Rd6 is almost completely misselected, which will cause a larger probability of missed police. The SE-MSVC sorting method can obtain 80.32% sorting accuracy, and the effect is better. Except that the radiation sources Rd2 and Rd5 are selected with more pulses, the correct sorting pulse of other signal sources exceeds 80%.

4 Conclusions

In this paper, we analysis the radar signal sorting methods based on single parameter and multi-parameter. It can be considered that the single-parameter-based sorting algorithm has the problems of slow speed, poor sorting effect on incomplete data and contaminated pulse parameters, and inability to process large amounts of complex data. However, multi-parameter-based sorting method could achieve better sorting results. So, it is necessary to research better sorting methods based on multi-parameter with new technology.

Acknowledgements This paper was supported in part by the National Natural Science Foundation of China under the Grant No. 61601499, 61701527, and 61601503.

References

1. Liu, X., Si, X.: A new method for radar signal sorting. Syst. Eng. Electron. **32**(1), 53–56 (2010)
2. Wang J., Lei P., Yang D. et al.: A novel deinterleaving algorithm of radar pulse signal based on DSP. In: IEEE International Symposium on Industrial Electronics (ISIE 2009), Seoul, Korea, 5–8 July, pp. 1899–1903 (2009)
3. Fuhua, F., Xuezhong Y.: Improved method for deinterleaving radar pulse trains with stagger PRI from dense pulse series. In: 2nd International Conference on Signal Processing Systems (ICSPS), Dalian, China, 5–7 July, pp. V3-250–V3-253 (2010)
4. Wang, B., Chen, Q., Wang, C.: Clustering-based frequency hopping signal sorting. J. Beijing Univ. Posts Telecom **32**(2), 80–84 (2009)
5. Wilkinson, D.R., Watson, A.: Use of metric techniques in ESM data processing. IEE Proc. Part F Radar Sig. Process. **132**(7), 121–125 (1985)
6. Hassan, H.E.: A new algorithm for radar emitter recognition. In: Proceedings of the 3rd ISISPA, vol. 2, pp. 1097–1101 (2003)
7. Jin, D., Wen, Z., Li, H.: Application of DBSCAN algorithm in radar full pulse signal sorting. Electron. Countermeasure **2**, 19–22 (2011)
8. Nelson, D.J.: Special purpose correlation functions for improved signal detection and parameter estimation. In: International Conference on Acoustics, Speech, and Signal Processing (ICASSP'93), pp. 73–76 (1993)
9. Nishinuchi, K., Kobayashi, M.: Improved algorithm for estimating pulse repetition intervals. IEEE Trans. Aerosp. Electron. Syst. **36**, 407–421 (2000)
10. Han, J., He, M.-H., Yan, W.-J. et al.: Radar signal sorting based on PRI transform and wavelet transform. J. Microcomput. **23**(9), 164–166 (2007)
11. Zou, S., Zhang, Q., Yan, X.: Radar signal sorting based on PRI transform. Comput. Simul. **23** (6), 41–44 (2006)
12. Chang, W., Wang, J., Wu, P.: Efficient PRI transform algorithm for DSP parallel processing. Aerosp. Electron. Front **24**(1), 50–53 (2008)
13. Han, J., He, M., Cheng, B.: An improved PRI transform algorithm. Electron. Warfare **124**, 25–29 (2009)
14. Wang, Z.L., Zhang, D.F., Bi, D.Y., et al.: Multiple-parameter radar signal sorting using support vector clustering and similitude entropy index. Circ. Syst. Sig. Process. **33**(6), 1985–1996 (2014)
15. Ben-Hur, A., Horn, D., Siegelmann, H.T., et al.: Support vector clustering. J. Mach. Learn. Res. **2**, 125–137 (2001)

Radar Signal Sorting Based on Core Cluster Support Vector Clustering

Shi-qiang Wang, Caiyun Gao, Tong Xu, Hui-yong Zeng and Juan Bai

Abstract With the increasing complexity of the modern electronic countermeasure environment, radar signal sorting faces serious challenges in ensuring effectiveness. At present, the conventional inter-pulse parameters or intra-pulse features are used, or the two directly form a vector to achieve signal sorting, but they all have certain problems. This paper proposes a radar signal sorting technology based on core cluster support vector clustering (CCSVC), which not only overcomes the existing technology or limited application scope, or computational complexity, or is not conducive to engineering implementation.

Keywords Radar signal sorting · Clustering algorithm · Core cluster · Support vector clustering · Inter-pulse parameters · Intra-Pulse parameters

1 Introduction

Signal sorting refers to the separation of different signals from randomly interlaced source signal streams based on intercepted radar characteristic parameters, arrival time, and position data. It is the premise and basis for different radar types, attributes, use identification, and threat degree judgment, so it directly affects the performance of radar reconnaissance equipment and is related to subsequent operational decisions [1]. However, with the rapid increase in the number of various electronic countermeasure devices, the signal density of the electromagnetic threat environment has reached millions of orders, and modern radars are moving toward multi-functional and multi-purpose directions. A radar may have multiple working states and variety of systems, and in order to improve their performance and anti-interference needs, often use a variety of complex waveform design to

S. Wang · T. Xu · H. Zeng (✉) · J. Bai
Air and Missile Defense College, Air Force Engineering University, Xi'an 710051, China
e-mail: 1048768838@qq.com

C. Gao
Basic Department, Air Force Engineering University, Xi'an 710051, China

V. Jain et al. (eds.), *Recent Trends in Intelligent Computing, Communication and Devices*, Advances in Intelligent Systems and Computing 1031,
https://doi.org/10.1007/978-981-13-9406-5_104

minimize the signal regularity used in signal sorting and identification, coupled with low intercept probability (LPI) technology [2], these all put new and higher requirements on the real-time, accuracy and reliability of signal sorting.

2 Traditional Sorting Method

So far, the researchers have proposed the following three broad categories of methods to achieve radar signal sorting.

The first method is signal sorting using inter-pulse parameters [3–7]; signal sorting by this method is usually fast and easy to implement. However, since only inter-pulse parameters are used, it is not well adapted to the radar waveform design. The more complicated it is, the traditional radar signals such as fixed-frequency, jagged, and jitter are becoming less and less common in modern radar design and cannot be used in traditional five parameters (TOA, RF, PA, PW, and DOA). The complex system radar signal sorting can be changed in each parameter domain and even overlap each other.

The second method is the signal sorting using intra-pulse characteristics [8–14]; the intra-pulse characteristic parameters help to reduce the overlapping probability of multi-parameter space, provide a new basis and ideas for the sorting and identification of radiation source signals, and also improve the current possible way and idea of the signal sorting ability of the radiation source. However, due to the use of complex analysis methods, it is easy to produce problems such as high computational complexity and difficulty in implementation.

The third method is to use the inter-pulse parameters and intra-pulse features to directly form the feature vector [15]; this method reduces or partially reduces the cross-pulse or intra-pulmonary characteristics by combining inter-pulse parameters and intra-pulse parameters. Stacking can achieve better results, but this method requires in-pulse feature extraction for all radar signals, so the processing invisibly adds to the processing burden of the electronic support system, so this design is not easy to implement.

The use of inter-pulse parameters for signal sorting is usually fast and easy to implement. However, due to the use of inter-pulse parameters, it is not well adapted to radar waveform design and becomes more and more complicated. Radar signals with traditional repetition frequency of fixed, jagged, jitter, etc. are becoming less and less common in modern radar design and cannot be used in traditional five-parameter (TOA, RF, PA, PW, and DOA) complex system radar signals that can be varied in various parameter domains and even overlap each other for sorting. The use of intra-pulse features for signal sorting can reduce the overlapping probability of multi-parameter space, can provide a new basis and ideas for the sorting and identification of radiation source signals, and is also a possible way to improve the sorting ability of current radiation source signals. However, due to the use of complex analysis methods, it is easy to produce problems such as high computational complexity and difficulty in implementation. A sorting method that directly forms a feature vector using inter-pulse parameters and intra-pulmonary features can achieve better results by combining inter-pulse parameters and intra-pulmonary parameters to completely reduce or partially reduce the overlap caused by the use of inter-pulse or intra-pulmonary features. However, this method requires in-pulse feature extraction for all radar signals so that the processing invisibly adds to the processing burden of the electronic support system, and therefore, this design is not easy to implement.

3 Cluster-Based Support Vector Clustering Sorting

Firstly, we define the concept of the core cluster; that is, keep the points outside the optimal hypersphere as much as possible in the original SVC algorithm, which is called Bounded Support Vectors (BSVs), so as to make as many points on the optimal hypersphere as possible, that is, Support Vectors (SVs) are converted into BSVs, and finally the core clusters of the inter-pulse features are retained, so as to ensure the misselection pulse generated during the sorting is ensured to the greatest extent, and the feature clusters are guaranteed to have the most Aggregation within the best class and separation between classes. Therefore, a cluster can be defined as a cluster class that satisfies the following formula:

$$\mathrm{SE} = \max\{\mathrm{SE}_c, 2 \leq c \leq N - 1\} \quad (1)$$

where SE denotes a similar entropy index, indicating that different numbers of clusters are obtained by the SVC clustering algorithm.

This cluster-based support vector clustering is called core cluster support vector clustering (CCSVC) sorting technology. The technique first uses clustering pre-sorting of interleaved signals by conventional parameters, and then the intra-pulse data corresponding to the missing pulse is selected; the intra-pulse modulation features which are favorable for signal sorting are extracted. Then, the feature selection algorithm is used to select the key features of the radar emitter signal, and the clustering support vector clustering technique is used to select the in-pulse. The characteristics are clustered and sorted, and finally, the two sorting results are combined to complete the final sorting.

In core cluster support vector clustering (CCSVC) sorting, the main steps can be explained as follows:

Step 1 Clustering and sorting are performed on the radiation source signals characterized by inter-pulse parameters according to the RSVC algorithm;
Step 2 According to the missing pulse generated in Step 1, the pulse data that needs to be processed later is launched;
Step 3 After the feature extraction performed on the intra-pulse data in Step 2, select features and form a feature vector;

Step 4 Using MCCL-based support vector clustering (SVC) algorithm combined with SE index sorting method, cluster the intra-pulse features obtained in Step 3;
Step 5 Combine the core cluster generated in Step 1 with the sorting result generated in Step 4 to complete the sorting of the radiation source signal.

The SVC algorithm can not only cluster the data distribution of arbitrary shapes, but also eliminate the outliers after the relaxation amount is added, which ensures the rationality and robustness of the sorting effect. This shows that the SVC algorithm has certain anti-noise ability. The anti-noise performance of the extracted and selected feature vectors has been theoretically analyzed and experimentally verified. Combining the above two points, the use of nuclear cluster support vector clustering CCSVC for signal sorting theoretically also has the anti-noise ability of both.

The effectiveness and anti-noise performance of the CCSVC sorting algorithm are further examined below. The following parameters are used for simulation experiments. The simulated signals are conventional radar signal (CW), linear frequency modulated radar signal (LFM), nonlinear frequency modulated radar signal (NLFM), and two. Phase-coded radar signal (BPSK), four-phase coded radar signal (QPSK) and frequency-coded radar signal (FSK). At this time, the interception time is changed from each interval to the simulation data of the interception time. The noise ratio is changed from 0 to 20 dB and changes every 4 dB.

Using RSVC to pre-sort the inter-pulse parameters, and further sorting the missing pulses by SE-MSVC, the final statistical results obtained by CCSVC sorting can be obtained, as shown in Fig. 1.

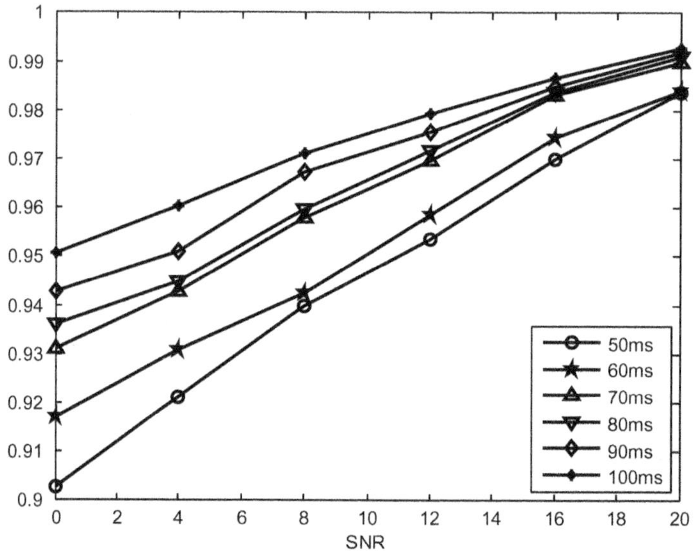

Fig. 1 CCSVC algorithm sorting results

Figure 1 shows that the CCSVC method has better noise immunity. When the signal-to-noise ratio is 0 dB, the average sorting accuracy rate corresponding to different interception times is over 90%. With the improvement of SNR, sorting performance has also been gradually upgraded, and its correct rate is close to 100%.

4 Conclusions

In this paper, a radar signal sorting technique based on kernel cluster support vector clustering (CCSVC) is proposed. This technique firstly uses the conventional parameters to perform cluster pre-sorting on the interleaved signals and then selects the intra-pulse data corresponding to the missing pulses. The in-pulse modulation features which are beneficial to signal sorting are extracted. Then, the feature selection algorithm is used to select the key features of the radar emitter signal, and the proposed cluster-based support vector clustering method is used to cluster the selected features. Finally, the two sorting results are combined to complete the final sorting. The technology not only overcomes the problems of existing technologies or limited application scope, or computational complexity; is not conducive to engineering implementation; and has higher sorting accuracy rate than traditional sorting methods.

Acknowledgements This paper was supported in part by the National Natural Science Foundation of China under the Grant No. 61601499, 61701527, and 61601503.

References

1. Wiley, R.G.: ELINT: The Interception and Analysis of Radar Signals, 2nd edn. Artech House, Boston, MA (2006)
2. Zhigang, Li: Research on Radar Signal Sorting and Recognition Technology of Inter-pulse Waveform Transformation. Harbin Engineering University, Harbin (2007)
3. Wilkinson, D.R., Watson, A.: Use of metric techniques in ESM data processing. IEE Proc. Part F Radar Sig. Process. **132**(7), 121–125 (1985)
4. Wang, J., Lei, P., Yang, D. et al.: A novel deinterleaving algorithm of radar pulse signal based on DSP. In: IEEE International Symposium on Industrial Electronics (ISIE 2009), Seoul, Korea, 5–8 July, pp. 1899–1903 (2009)
5. Fuhua, F., Xuezhong, Y.: Improved method for deinterleaving radar pulse trains with stagger PRI from dense pulse series. In: 2nd International Conference on Signal Processing Systems (ICSPS), Dalian, China, 5–7 July, pp. V3-250–V3-253 (2010)
6. Campbell, J.W, Saperstein, S.: Signal Recognition in Complex Radar Environments. Watkins-Johnson Tech. Notes, November/December **3**(6) (1976)
7. Hassan, H.E.: A new algorithm for radar emitter recognition. In: Proceedings of the 3rd International Symposium on Image and Signal Processing and Analysis, vol. 2, pp. 1097–1101 (2003)

8. Danielsen, P.L., Agg, D.A., Burke, N.R.: The application of pattern recognition techniques to ESM data processing. In: IEE Colloquium on Signal Processing for ESM Systems, 26 Apr, pp. 6/1–6/4 (1988)
9. Pu, Y., Jin, W., Zhu, M. et al.: Classification of radar emitter signals using cascade feature extractions and hierarchical decision technique. In: ICSP2006 Proceedings (2006)
10. Pu, Y., Jin, Y., Hu, L.: Classification of radiation source signals based on quadratic feature extraction of instantaneous frequency. J. SouthWest JiaoTong Univ. **42**(3), 373–379 (2007)
11. Marian, W., Adam, K., Janusz, D. et al.: The method of regression analysis approach to the specific emitter identification. In: International Conference on MIKON, pp. 491–494 (2006)
12. Janusz, D., Marian, W., Jan, M.: Applying the radiated emission to the specific emitter identification. J. Telecommun. Inf. Technol. **2**, 57–60 (2005)
13. Janusz, D., Adam, K., Robert, O.: An application of iterated function system attractor for specific radar source identification. In: 17th International Conference on MIKON 2008, pp. 1–4 (2008)
14. Chen, T., Jin, W.: Feature extraction of radar emitter signals based on symbolic time serials analysis. In: Proceedings of the 2007 International Conference on Wavelet Analysis and Pattern Recognition, Beijing, China, 2–4 Nov, pp. 1277–1282 (2007)
15. Chen, T.: Research on Signal Sorting Technology of Radar Emitter Based on Intra-pulse Characteristics. Southwest Jiaotong University, Chengdu (2010)

Radar Signal Unintentional Modulation Feature and Clustering Sorting Methods

Shi-qiang Wang, Hui-yong Zeng, Tong Xu, Caiyun Gao and Juan Bai

Abstract With the complexity of the radar system, radar signal sorting is facing an increasingly severe form. Therefore, the research on radar emitter signal sorting model and algorithm based on intentional modulation feature and clustering in the current stage has important theoretical and practical significance. This paper studies the meaning, research methods of intentional modulation features and clustering sorting method.

Keywords Radar signal sorting · Clustering algorithm · Unintentional modulation feature · Fingerprint feature · Intra-pulse subtle feature

1 Introduction

As one of the key technologies for signal processing of electronic reconnaissance equipment, the research on the theory and technology realization of radar signal sorting has been highly valued by various countries. Relevant research literature indicates that domestic and foreign research scholars have carried out signal sorting technology for radiation source signals. Lots of research work has also achieved a lot of results, such as sorting methods based on unintentional modulation feature, clustering.

The unintentional modulation feature in the pulse, that is, the "fingerprint" feature characterizing the individual characteristics of the radar signal, is inherently attached to the signal by the radar using a certain type of modulator and is basically difficult to completely eliminate. In general, it is an undesirable variety of parasitic modulations due to devices such as launch tubes and high-voltage power supplies [1]. The emergence of complex system radars makes the parameters of the

S. Wang · H. Zeng · T. Xu · J. Bai
Air and Missile Defense College, Air Force Engineering University, Xi'an 710051, China

C. Gao (✉)
Basic Department, Air Force Engineering University, Xi'an 710051, China
e-mail: 1048768838@qq.com

V. Jain et al. (eds.), *Recent Trends in Intelligent Computing, Communication and Devices*, Advances in Intelligent Systems and Computing 1031,
https://doi.org/10.1007/978-981-13-9406-5_105

conventional five-parameters overlap each other. In addition, for non-cooperative electronic reconnaissance, the intercepted pulse stream generally lacks the necessary prior information, and it is impossible to know for sure the number of categories of intercepted signals, so it is necessary to study the unsupervised learning method of multi-parameters of the radiation source signals lacking a priori information.

2 Unintentional Modulation Feature

The intra-pulse subtle feature of the radar, that is, the unintentional modulation feature, is the individual characteristic of the radiation source. In 2004, Kawalec [2] pointed out that the key to individual radiation source identification (SEI) is to extract the unintentional modulation of the signal, and the author will frequency-modulate the radar pulse signal. Features are used as subtle features in the vein, and linear discriminant analysis (LDA) and principal component analysis (PCA) methods are introduced to analyze these features. In 1993, Langley [3] proposed the concept of unintentional frequency modulation (UMOF) as an important feature of specific radiation source identification; in addition, Kawalec and Langley also used FM characteristics as fingerprints for signal classification and performed with actual data. The verification is very representative; the unintentional amplitude modulation (UMOA) feature can be extracted from the video envelope of the signal; Zhang [4] studied this. The literature [5] estimates the power spectrum of the fundamental, second harmonic and third harmonic according to the autocorrelation operation, estimates the power of each harmonic according to the power spectrum and uses the power ratio as the harmonic power constraint feature vector.

Marian [6] obtained the transformed measurement points by performing a fixed matrix transformation on the two-dimensional Euclidean space of the intra-pulse signal samples, performed regression analysis on the transformed measurement points to obtain the joint coordinates of the feature points and identified these joint coordinates as the classification. A new feature of the type of radiation source signal is extracted similarly. Janusz [7] first uses the second regression analysis method to analyze the transformed measurement points and obtain some feature points, and then constructs a global measurement function (GMF) with Lagrange polynomial form based on these feature points. The closed plane area and arc length determined by the generalized sampling function are used as the characteristics of the radiation source signal. Finally, the extracted features are used to classify the signals of the same type of radiation source, and a better classification effect is obtained. Meanwhile, the paper in [8] considers that the generalized sampling function (GSF) is constructed according to the affine mapping (AM) stacked function system (IFS) for feature extraction.

The paper [9] based on the principle of the transmitter extracts subtle features for individual identification of the radiation source; in addition, Chen [10] uses the signal subtle information contained in the bispectrum and bispectral Gaussian noise subtle feature extraction is performed with features such as small clutter influence; Wang [11] analyzed the performance of kernel point sorting and optimization on the whole plane of the fuzzy function and used the fuzzy function slice with frequency offset close to zero (including zero) as the radiation source, which are the main representative characteristics of the signal. This method can obtain a sparse feature set, which can help to find the robust individual characteristics of the radiation source, which can be applied to the unintentional modulation recognition of stationary and moving radiation sources. At the same time, the author extracts the representative slice of the fuzzy function as the subtle feature of the motion radar. Thus, the stable individual characteristics of the motion radar radiation source signal are retained [12].

Intra-pulse subtle feature analysis plays an irreplaceable role in sorting and identifying individual radiation sources. However, as an unconventional parameter, subtle features are inherent features added to the signal by the radar using a certain modulator. The characteristics are often reflected in the nuances of similar signals, so a larger sampling bandwidth is required to capture this feature information. For example, most unintentional frequency modulation occurs at the leading and trailing edges of the pulse oscillator signal, which indicates that a high sampling rate is required to capture this characteristic information. Therefore, intra-pulse subtle features are not a conventional de-interleaving or recognition tool.

3 Clustering Sorting Methods

As mentioned, the large number of complex system radars causes the parameters of the conventional five parameters to overlap each other. In addition, for non-cooperative electronic reconnaissance, the intercepted pulse stream does not possess the necessary prior information and the exact number of categories of intercepted signals, it is necessary to study the unsupervised learning method for sorting multiple parameters of the radiation source signal lacking a priori information. The unsupervised learning method, that is, the clustering method, refers to the classification of samples without reference to the classification target. It has been widely used in many problems, so it has received widespread attention in signal sorting. Considering artificial neural networks, such as self-organizing (SOM) neural networks, probabilistic neural networks (PNN), radial basis function neural networks (RBFN), etc., with parallel processing and fault tolerance, many documents apply them to radar signals. Signal sorting is expected to use the parallel processing and fault tolerance of artificial neural networks to solve the problem of radar signal sorting in complex environments [13]. In 2001, Xu [14] combined the self-organizing network with PNN to obtain the SOPNN sorting method, which is mainly used for pulse train de-interleaving in a radiation source environment without prior information; Lin [15] selected PW, RF and DOA as sorting parameters, using Kohonen neural network to sort radar multi-target; in 2009, Han [16] used the Kohonen neural network and the mixed characteristics of interpulse parameters and intra-pulse parameters to automatically sort the radiation source signals and achieved good results; in 2010, Wang [17] used the Eidos BSB artificial neural network to perform self-associative learning on a large number of radar pulse samples with measurement errors and completed the memory of the pulse mode to realize the sorting function.

However, the neural-network-based sorting method has some disadvantages when it is applied to the separation of radar signals from complex systems. For example, self-organizing neural networks require a large network scale when dealing with the identification problem of unknown mode space, and there is no cluster validity verification. When there is a big difference between the actual radiation source environment and the a priori radiation source environment, the false alarm probability is greatly high. For example, PNN has problems such as prior knowledge of samples, unsatisfactory classification performance for pattern overlap and interleaving, and SOPNN is not ideal for pulse stream sorting under parameter space disconnection [18].

In 1992, Chandra [19] considered a two-stage sorting method suitable for frequency-agile signals. The method first clustered the DOA and RF parameters and then used the PRI and PW parameters to recalculate the averaged values; in 2003, Eric [20] proposed an online classification fuzzy mode rearrangement method for the actual collection of pulse signal sets and achieved a better classification effect; in the same year, Mao [21] proposed a support-based vector analysis of radar signal sorting methods. The method firstly uses DOA, RF and PW to cluster the pulse signal stream and then uses the support vector analysis method to further sort the pulsed stream after clustering. In 2004, Xu [22] discussed the passive positioning conditions in a single station. A quadratic clustering method is proposed in the same year. Zhang Wanjun used K-means clustering to sort unconventional radar signals with similar parameters and overlapping each other [23]; afterward, Zhu Zhengwei studied multiple parts with intra-pulse modulation characteristics. Clustering and sorting of phased array radar signal interleaved pulse flow [24]; in 2008, Zhang [25] introduced the two potential functions of pseudo-gravity field and pseudo-nuclear force field by using the concept of field and potential in physics. The function describes the spatial distribution of the data and finally uses the obtained spatial distribution characteristics to achieve signal clustering; the literature [26] studied the sorting method based on fuzzy clustering algorithm; Chen [27] studied a radar signal sorting algorithm based on kernel fuzzy clustering. In the same year, Zhao [28] applied the optimization algorithm to signal sorting and improved ant colony clustering method. In 2011, Chen [29] proposed a sorting method based on gray relational hierarchical clustering algorithm, introducing gray correlation measure to measure the similarity between intra-pulse feature samples, and realizing clustering sorting of radar emitter signals. Jin [30] used the density-based clustering algorithm (DBSCAN) to cluster the feature vectors composed of DOA, RF and PRI to obtain the sorting results; Lee [31] also sorted based on density clustering. The method was studied. In addition, granular clustering based on non-uniform [32], subspace clustering method and support vector clustering (SVC) algorithm in radar signal sorting also is studied in [33, 34].

As with all multi-parameter methods, the biggest problem for clustering sorting is that the features used for clustering do not have good interclass separation ability, which affects the actual use of clustering methods. Support vector clustering (SVC) is one of the most effective unsupervised classification methods. It has two significant advantages: It can generate cluster boundaries of arbitrary shapes; it can analyze noise data points and can separate interlaced clusters. However, the application of SVC to radar signal sorting has the following limitations: The optimization problem solving and the calculation of the adjacency matrix are time consuming, and it is difficult to meet the real-time requirements of radar signal sorting. Guo and Wang [33, 34] applied SVC for radar signal sorting, which achieved good results, but it took more time. Therefore, improving the processing speed of the SVC algorithm is of great significance for the real-time sorting of radar signals.

4 Conclusions

The intra-pulse subtle feature of the radar, that is, the unintentional modulation feature, is the individual characteristic of the radiation source. This feature not only achieves the effect of the signal recognition of the radiation source, but also can be used for signal sorting of the radar radiation source. Clustering refers to grouping data objects into multiple classes or clusters so that objects in the same cluster have higher similarity, and objects in different clusters have larger differences. The biggest difference of classification is that clustering belongs to unsupervised learning method, and the class to be divided is unknown, so it is especially suitable to solve the separation problem of interlaced pulse stream.

Acknowledgements This paper was supported in part by the National Natural Science Foundation of China under the Grant No. 61601499, 61701527 and 61601503.

References

1. Chai, J.: Research on sorting and identification technology of radar signals in complex environment. Harbin Engineering University, Harbin (2009)
2. Kawalec, A., Owczarek, R.: Radar emitter recognition using intrapulse data. In: Proceedings of 15th International Conference on Microwaves, Radar and Wireless Communications, vol. 2, pp. 435–438 (2004)
3. Langley, L.E.: Specific emitter identification (SEI) and classical Parameter fusion technology. In: WESCON'93, Conference Record, pp. 377–381 (1993)
4. Zhang, G., Huang, K., Jiang, W., Zhou, Y.: A method for extracting subtle features of radiation sources based on signal envelope. Syst. Eng. Electron. **28**(6), 795–797 (2006)
5. Yu, Z., Chen, C., Jin, W., et al.: Feature extraction of radar emitter harmonic power constraint based on nonlinear characters of the amplifier. In: 2nd International Congress on Image and Signal Processing, pp. 1–4 (2009)

6. Marian, W., Adam, K., Janusz, D., et al.: The method of regression analysis approach to the specific emitter identification. In: International Conference on Microwaves, Radar & Wireless Communications, pp. 491–494 (2006)
7. Janusz, D., Marian, W., Jan, M.: Applying the radiated emission to the specific emitter identification. J. Telecommun. Inf. Technol. **2**, 57–60 (2005)
8. Janusz, D., Adam, K., Robert, O.: An application of iterated function system attractor for specific radar source identification. In: 17th International Conference on Microwaves, Radar and Wireless Communications, pp. 1–4 (2008)
9. Kawalec, A., Owezarek, R.: Radar emitter recognition using intrapulse data. In: 15th International Conference on Microwaves, Radar and Wireless Communications, vol. 2, pp. 435–438 (2004)
10. Chen, T.W, Jin, W.D., Li, J.: Feature extraction using surrounding-line integral bispectrum for radar emitter signal. In: 2008 IEEE International Joint Conference on Neural Networks (IEEE World Congress on Computational Intelligence), pp. 294–298 (2008)
11. Wang, L., Ji, H.: Optimizing zero-slice feature of ambiguity function for radar emitter identification. In: 2009 7th International Conference on Information, Communications and Signal Processing (ICICS), pp. 1–4 (2009)
12. Wang, L., Ji, H., Shi, Y.: Radiation source identification of motion radar based on representative slice of fuzzy function. Syst. Eng. Electron. **32**(8), 1630–1634 (2010)
13. Wang, C.D., Thompson, J.: An adaptive data sorter based on probabilistic neural networks. In: Proceedings of the IEEE 1991 National Aerospace and Electronics Conference NAECON 1991, Dayton, Ohio, vol. 3, pp. 1096–1102 (1991)
14. Xu, X., Yiyu, Z., Lu, Q.: Research on real-time signal sorting processing technology of radar interception system. Syst. Eng. Electron. **23**(3), 12–15 (2001)
15. Lin, Z., Liu, G., Dai, G.: Application of Kohonen neural network in radar multi-target sorting. J. Mil. Eng. Univ. **4**(5), 56–59 (2003)
16. Han, J., He, M.H., Zhu, Y.Q.: A new method for signal sorting of radar emitter based on multi-parameters. Data Acquisition Process. **24**(1), 91–94 (2009)
17. Wang, X., Song, M.: Radar pulse sorting method based on Eidos BSB artificial neural network. Mod. Electron. Technol. **23**, 6–9 (2010)
18. Guo, Q.: Theoretical study on signal sorting of unknown radar emitters in complex environments. Harbin Engineering University, Harbin (2007)
19. Chandra, V., Bajpai, R.C.: ESM Data Processing parametric deinterleaving approach. technology enabling tomorrow: computers, communications and automation towards the 21st century. In: IEEE Region 10 International Conference (TENCON), vol. 1, pp. 26–30 (1992)
20. Eric, G., Yvon, S., Pierre, L.: A pattern reordering approach based on ambiguity detection for online category learning. IEEE Trans. Pattern Anal. Mach. Intell. **25**(4), 524–528 (2003)
21. Mao, W., Zhu, Y., Wang, J.: Application of support vector analysis in radar signal sorting. J. Air Force Radar Acad. **17**(3), 19–21 + 24 (2003)
22. Xu, D., Jiang, W., Yiyu, Z.: Secondary clustering method for pulse sorting of radiation source. Aerosp. Electron. Front **20**(3), 26–29 (2004)
23. Zhang, W., Fan, Y., Tan, Y.: Application of clustering method in radar signal sorting. Radar Sci. Technol. **2**(4), 219–223 (2004)
24. Zhu, Z.: Clustering and sorting method of radar signals. Electron. Countermeasure **6**, 6–10 (2004)
25. Zhang, H.C., Yan, H.L., Gong, L.L.: A new clustering method for unknown radar radiation source. Comput. Eng. Appl. **44**(27), 200–202 (2008)
26. He, H., Xu, J., Xu, Z.: Fuzzy clustering and sorting method for radar signals. J. Aerosp. Comput. Technol. **38**(5), 21–24 (2008)
27. Chen, B., Luo, L., Zhao, G.: Radar signal sorting algorithm based on kernel fuzzy clustering. Ship Electron. Countermeasure **32**(4), 76–79 (2009)
28. Zhao, G., Luo, L., Chen, B.: Research on radar signal sorting algorithm based on improved ant colony clustering. Electron. Inf. Countermeasure Technol. 24(2), 27–30 + 40 (2009)

29. Chen T.W., Jin Y.D., L.J.: Clustering and sorting algorithm for radar emitter signals. J. Circ. Syst. **16**(3), 56–61 (2011)
30. Jin, D., Wen, Z., Li, H.: Application of DBSCAN algorithm in radar full pulse signal sorting. Electron. Countermeasure **2**, 19–22 (2011)
31. Lee, D.W., Han, J.W., Lee, W.D.: Adaptive radar pulses clustering based on density cluster window. In: ITC-CSCC: International Technical Conference on Circuits Systems, Computers and Communications, pp. 1377–1380 (2010)
32. Yue, H., Luo, J., Lü, J., et al.: Non-uniform granularity clustering method for radar signals. Fire Command Control **33**(8), 24–26 (2008)
33. Guo, Q., Li, W., Li, H.: Application of support vector clustering method in radar signal sorting. In: 2005 Annual Meeting, pp. 237–241 (2005)
34. Wang, Z.L., Zhang, D.F., Bi, D.Y., et al.: Multiple-parameter radar signal sorting using support vector clustering and similitude entropy index. Circ. Syst. Signal Process. **33**(6), 1985–1996 (2014)

Research on Product Preference Image Measurement Based on the Visual Neurocognitive Mechanism

Chen Yang, Lin Li and Chen Zhi-ang

Abstract In order to obtain user's preferred image more accurately in the process of innovative design of perceptual products, eye movement combined with EEG measurement technology is used to measure the user's preferred image of products. Eight earpieces that met the user's preference were used as the starting stimulus and five pairs of mutually sensitive sentimental image adjective pairs as the detecting stimulus. Based on the eye movement behavior data and EEG signals of the subjects, the participants' preferred image orientation of earpieces was analyzed. The results show that the image words of "Complex-Concise," "Retro-Modern," "Smart-Heavy," and "Lively-Serious" made the subject's average fixation duration shorter than other words. In addition, the power of α waves captured by the left frontal channel (F3) was significantly lower than the average power of the α waves captured by the right frontal channel (F4) when the subject viewed the image words of "Concise," "Gorgeous," "Modern," "Smart," and "Lively." It shows that user's preferred image of perceptual products can be acquired more scientifically and effectively by using the combined physiological measurement technology of eye tracking and EEG, and reference can be provided for design applications based on user's preferred image in the jewelry.

Keywords Eye-tracking technology · EEG technology · Perceptual image · Preferred image

---

C. Yang
Key Laboratory of Advanced Manufacturing Technology, Ministry of Education, Guizhou University, Guizhou, China

L. Li (✉) · C. Zhi-ang
School of Mechanical Engineering, Guizhou University, Guizhou, China
e-mail: linlisongbai@163.com

V. Jain et al. (eds.), *Recent Trends in Intelligent Computing, Communication and Devices*, Advances in Intelligent Systems and Computing 1031,
https://doi.org/10.1007/978-981-13-9406-5_106

1 Introduction

The perceptual image of the product is the most direct feeling that the product gives us. The image can be identified because it expressed the form that reflects certain cultural and social background characteristics [1]. Since a user's expression of product image reflects his/her perceptual preference for the product, understanding consumer's preferences for image is a key for companies to manufacture products that satisfy user's psychological expectations and promote consumption, while can help designers improve design ability. In the existing product image studies, psychometric methods such as semantic difference method are widely used to obtain a perceptual image. It is difficult to identify the real and accurate mental state of user by using a single psychometric method because the expression of preferred image is vague and incomplete. Along with the development of neuroscience and brain science, the physiological measurement techniques that scientific instruments are used to measure and record user's psychological reactions have been increasingly applied in the fields of marketing and product design. From the perspective of the visual neurocognitive mechanism, Khushaba et al. studied the formation of user's preferences, the formulation of purchasing decisions, the product evaluation, and design application [2–4]. In view of the fact that it is difficult to fully express preferences when people are asked, physiological measurement methods have been used to "deliberately" reveal hidden information of consumer's preferred image from the origin of human cognition and transform it into physiological activity information [5–7]. It is an inevitable trend for in-depth implementation of user-driven design in the future.

Vision is the main channel to obtain product information, which directly affects user's preferences and purchase intentions [8]. A large number of studies used eye-tracking method to acquire user's perceptual preferences [9, 10].

EEG is a useful tool for studying the relationship between brain activity and emotional state [11, 12], so consumer's EEG can be used to judge preferences. Researchers believe that the medial prefrontal lobe mainly characterizes consumer's preferences and value assessments of product and has a function of purchasing prediction, while the orbitofrontal lobe mainly characterizes consumer's pleasant experience and purchase intentions [13–15].

In addition, the combination of eye-tracking and EEG technology can further understand the mechanism of individual cognitive activities [16]. Yang et al. proposed a method of affective pictures classification based on information about the user's experience collected by eye movement and EEG signals [7]. Khushaba et al. used the two technologies to study the physiological decision-making process of consumers and observed and evaluated the cortical activities in different brain regions when they make preference selections for product [5]. Li et al. established a relationship model among the experimental indexes of eye movement, the experimental indexes of EEG, and PAD emotional state values [17]. Tang et al. proposed a cognitive evaluation method for aesthetic experience of the mobile design combining the two technologies [18].

These studies have laid a foundation for the measurement and evaluation of perceptual information. With the deepening of user-driven design, acquiring user's preferred image is an important premise for perceptual product design, which makes it possible to design products closer to the user's appeals. Therefore, based on the theory of cognitive neuroscience and cognitive psychology, the cognition of the preferred image was reflected by eye–brain data from the perspective of visual neurocognitive mechanism, and the deep hidden design information from user was explored and revealed from the source of cognition, which provides reliable data and scientific basis for designers to obtain user's preferred image more accurately.

2 Measurement of User's Preferred Image for Silver Earpieces with Eye-Tracking and EEG Technology

Under the guidance of cognitive neuroscience theory and visual perception principle, the changes of eye movement and EEG when user made a simple matching were explored to obtain user's preferences for the samples, and image adjectives were used as the representations of preferences. The matching situation was designed as user's judgments of the image adjectives that they thought could describe the product in accordance with their preferences. The image adjectives that fit their preferred samples were the preferred image words. Taking the silver earpieces as an example, the user's preferred product samples and image words were made into test materials. The subject's eye movement data were recorded by SMI eye tracker, and the EEG signals were recorded by Emotiv EPOC+.

2.1 *Preparation of Stimuli Materials*

Firstly, eight product pictures conforming to user's preferences were used as priming stimuli in this experiment, as shown in Fig. 1. The product pictures were imported into the picture viewing module.

Perceptual adjectives about silver earpieces were collected. Manual screening, clustering, questionnaire, and other methods were used, and the number of pairs of perceptual image words was reduced to five pairs, as shown in Fig. 2.

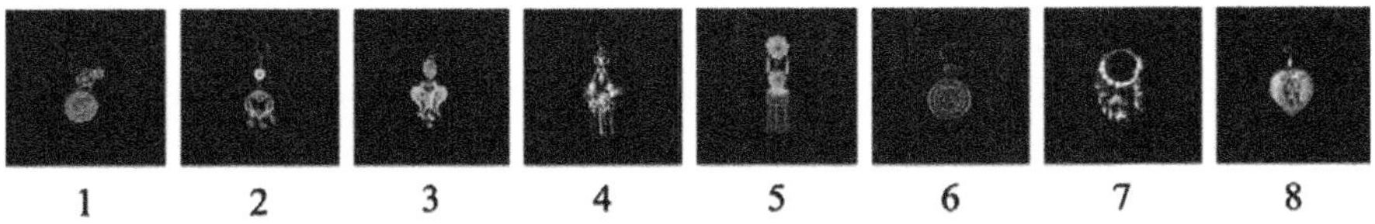

Fig. 1 Samples conform to user's preferences

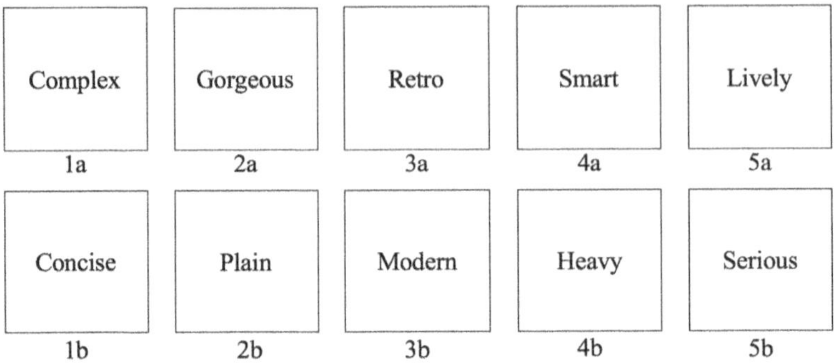

Fig. 2 Image word pairs

The image-word pictures were imported into the picture viewing module as detecting stimuli.

2.2 *Measurement Experiment of User's Preferred Image with Eye-Tracking and EEG Technology*

2.2.1 Participants

Eighteen consumers of silver ornaments were selected. The ratio of male to female was 1:1. All subjects were between the ages of 20 and 40 (the mean age is 27).

2.2.2 Experimental Equipment

Wireless multi-channel EEG system Emotiv EPOC+ was used to collect EEG data. It consists of fourteen channels with two additional reference electrodes located behind the ears. Figure 3 shows the distribution of electrode positions. Meanwhile, RED desktop eye tracker from SMI Company was used to collect eye movement data. With e-prime, a psychological programming software, the two devices were synchronously measured.

2.2.3 Experimental Procedures

(1) Subject enters the laboratory, registers personal information, and takes a rest to calm down.
(2) Experimenter introduces the process of eye tracking combined with EEG experiment and precautions to the subject.

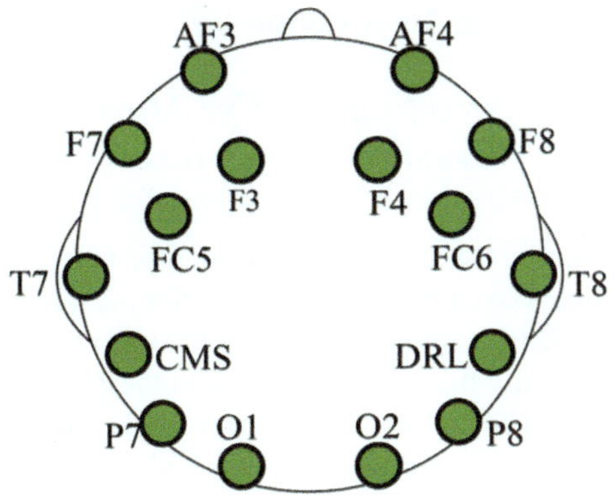

Fig. 3 14 channel electrode profile

(3) Subject wears and calibrates the instruments.
(4) Experimenter starts the experimental procedure, and the subject reads the instructions.
(5) The pre-experiment starts running; the subject observes the pre-experimental materials and then uses left or right key on the mouse to decide whether the image words can describe the product samples. The left key indicates "match" and the right key indicates "no match."
(6) The experiment starts running; the operation mode is the same as that of the pre-experiment. The experimenter controls and observes the collection of physiological data.
(7) At the end of the experiment, instruments stop recording.

The experimental procedures are shown in Fig. 4.

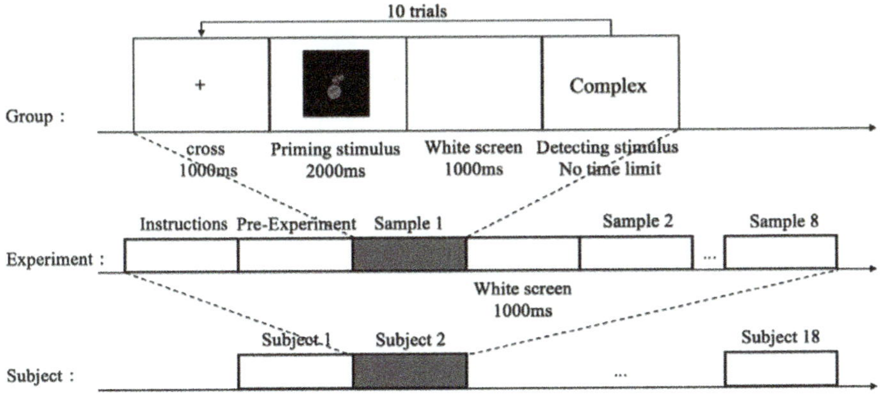

Fig. 4 The experimental procedures

3 Data Processing

3.1 *Selecting the Eye Movement and EEG Measurement Indexes*

There is a significant correlation between average fixation duration (AFD) in the eye movement index and preference [19]. When the product image matching result is definite, the AFD is shorter; when the product image matching result is fuzzy, the thinking time is prolonged and the AFD is significantly increased [20, 21]. So, AFD was selected as the eye movement index to analyze the user's preferences.

For brain cognition, cerebral cortex is mainly divided into the frontal, parietal, occipital, and temporal lobe, and each has a different functional division of labor. Among them, the frontal lobe is an important area of processing cognitive information [22]. Therefore, F3 and F4 were selected as the main channels to obtain brain cognitive information. The common rhythm waves in EEG are delta (δ), theta (θ), alpha (α), beta (β), and gamma (γ). α is thought to be involved in the processing of emotions [15]. The left and right hemispheres of human brain differ in structure and function. They lead different types of emotion processing, while the right hemisphere is mainly responsible for the processing of negative emotions and the left hemisphere is mainly responsible for the processing of positive emotions [23]. The relationship between α waves and emotions is mainly reflected in the asymmetry between the left and right hemispheres of the frontal region, which is often measured by the frontal α asymmetry index(FαAI), that is, the power of α waves captured by F4 minus the power of α waves captured by F3 [24]. Therefore, the FαAI was selected as the EEG index to characterize the preferred image.

3.2 *Processing the Eye Movement and EEG Measurement Indexes*

The collected eye movement data were analyzed by BeGaze software.

The collected EEG data were processed as following steps: (1) preprocessing: The doped artifacts in the collected EEG signals were removed by filtering and independent component analysis; (2) EEG data segmentation: From 500 ms before the emergence of each image as the detecting stimulus to 1000 ms after its emergence, a total of 4500 ms was used as an analysis period; (3) feature extraction: The power spectral density was calculated, Fourier transform was used to map the original EEG signal to α band, and α band power spectrum was calculated. EEG power spectrum was obtained from the integration of EEG power spectrum density (PSD) [25]:

$$\mathrm{PS}_\alpha = \sum_{w1}^{w2} p(w) * d. \tag{1}$$

d: Frequency Resolution, $d \frac{1}{4.096\,\mathrm{S}}$
PS_α The power of α waves
$w1$, $w2$: The upper and lower limits of frequency band
FαAI [17] = Power of α waves captured by F4—power of α waves captured by F3.

4 Results

4.1 Eye Movement Data

The AFD is as shown in Table 1. When the product image matching results are definite, that is, the subject confirms that those image words match (not match) the descriptions of the product, the AFD of is shorter. When the product image matching results are fuzzy or there are no obvious affective preferences, the AFD is longer.

The experimental results show that when the subject made matching judgments on the five pairs of image words, Group 1, 3, 4, and 5 evoke shorter AFD, and the results indicate that most subjects are more efficient in matching "1a–1b," "3a–3b," "4a–4b," "5a–5b"; that is, most subjects make it clear whether these image words are consistent with the description of the product, but the positive and negative emotions of subjects cannot be distinguished. The AFD of the Group 2 is significantly longer than that of the other groups, indicating that subjects do not have obvious emotional bias toward the matching of "2a–2b" and product; more cognitive processing resources need to be allocated to judge it. Since the user's preferences in implicit cognitive process are composed of preference tendency and preference degree, eye movement data alone cannot draw conclusions about preference tendency, which needs to be further confirmed with EEG data.

Table 1 AFD of five image-word pairs (eight samples of silver earpieces)

Image word	1		2		3		4		5	
	1a	1b	2a	2b	3a	3b	4a	4b	5a	5b
AFD (ms)	557.32	549.02	723.16	714.50	593.51	590.55	543.81	522.42	556.86	590.40

Table 2 FαAI a of five image-word pairs (eight samples of silver earpieces)

Image word	1		2		3		4		5	
	1a	1b	2a	2b	3a	3b	4a	4b	5a	5b
FαAI	−0.011	0.053	0.037	−0.012	−0.023	0.038	0.055	−0.017	0.058	−0.025

4.2 EEG Data

The preference-related characteristic variables of EEG generated by the subject in making judgments are used to calculate the changes of EEG spectral activities. The FαAI can be used to judge the emotional state of subjects and obtain the user's preferences from emotions. The greater the FαAI, the more the image words fits user's preferences.

The FαAI during the matching experiment is shown in Table 2; when subjects make matching judgments for the five image-word pairs, "1b," "2a," "3b," "4a," and "5a" than "1a," "2b," "3a," "4b," and "5b" induce greater FαAI, indicating that subjects are more likely to generate positive emotions when judging whether the former image words can describe the preferred samples, which further verify that the former image words are more consistent with user's cognitive preferences.

5 Conclusion

By observing the subjective evaluation results, eye movement data, and EEG signals recorded during the stimulation process, from the visual cognitive neurological perspective of eye tracking combined with EEG technology, consumer's preferred image words of product are obtained. For silver earpieces in this study, eye movement results show that users have obvious emotional bias toward image words such as "Complex-Concise," "Retro-Modern," "Smart-Heavy," and "Lively-Serious" and have certain cognitive ambiguity toward "Gorgeous-Plain." EEG results show that "Concise," "Gorgeous," "Modern," "Smart," and "Lively" are more likely to induce positive emotions. However, as a result of eye movement measurement, it is known that user's emotional preferences for the matching of "Gorgeous" and silver earpieces are less obvious; that is, user's preferred image for silver earpieces is "Concise," "Modern," "Smart," and "Lively." Eye tracking combined with EEG technology can be used as an effective method to measure user's perceptual preferences for image. The results of the research can lay a foundation for measurement of perceptual information and its manifestation from the perspective of the visual cognitive neural mechanism, a scientific and effective measurement method of user's preferred image for product design based on user's cognitive preferences is provided, and a reference is provided for the researches on the design applications based on user's preferred image for silver earpieces.

There are still some limitations in this study. Because EEG measurement requires rigorous and elaborate experimental design, mouse keys and people's operating habits, extracted EEG duration, subjects' educational level and understanding ability and other factors will affect the experimental results, so more comprehensive experimental design is needed to improve the accuracy of the experimental results.

6 Funding

This project is supported by the National Natural Science Foundation of China [Grant No. 51465007 and 51865003].

Acknowledgements I would like to thank the National Natural Science Foundation of China for providing financial support for this research. Thanks to the Human Body Experimental Committee for approval of this experiment.

References

1. Khushaba, R.N., Wise, C., Kodagoda, S.: Consumer neuroscience: assessing the brain response to marketing stimuli using electroencephalogram (EEG) and eye tracking. Expert Syst. Appl. **40**(9), 3803–3812 (2013)
2. Kenning, P.H., Plassmann, H.: How neuroscience can inform consumer research? IEEE Trans. Neural Syst. Rehabil. Eng. **16**(6), 532–538 (2008)
3. Yang, M.Q., Lin, L., Milekic, S.: Affective image classification based on user eye movement and EEG experience information. Interact. Comput. **30**(5), 417–432 (2018)
4. Moshagen, M., Thielsch, M.: Facets of visual aesthetics. Int. J. Hum Comput Stud. **68**(10), 689–709 (2010)
5. Kostyra, E., Wasiak-Zys, G., Rambuszek, M.: Determining the sensory characteristics, associated emotions and degree of liking of the visual attributes of smoked ham. A multifaceted study. LWT-Food Sci. Technol. **65**, 246–253 (2016)
6. Colombo, B., Laddaga, S., Antonietti, A.: Psychology and design. The influence of the environment's representation over emotion and cognition. An ET study on Ikea design. Procedia Manuf. **3**(6), 2259–2266 (2015)
7. Lee, Y.Y., Hsieh, S.: Classifying different emotional states by means of EEG-based functional connectivity patterns. Plos One **9**(4), e95415 (2014)
8. Bos, D.O.: EEG-based emotion recognition. The influence of visual and auditory stimuli. Emotion **1359**, 667–670 (2012)
9. Guo, F., Ding, Y., Wang, T.: Applying event related potentials to evaluate user preferences toward smartphone form design. Int. J. Ind. Ergon. **54**, 57–64 (2016)
10. Khushaba, R.N., Greenacre, L., Kodagoda, S.: Choice modeling and the brain: a study on the Electroencephalogram (EEG) of preference. Expert Syst. Appl. **39**(16), 12378–12388 (2012)
11. Briesemeister, B.B., Tamm, S., Heine, A.: Approach the good, withdraw from the bad—a review on frontal alpha asymmetry measures in applied psychological research. Psychology **4**(3), 261–267 (2013)
12. Gao, X.Q., Wang, Y.Y., Ge, L.Z.: The combination of eye movement technique and EEG technique—a new approach to cognitive research. Ergonomics **11**(1), 36–38 (2005)

13. Li, S., Zhuang, X.X., Liu, W.L.: Study on emotion measurement method of EEG and eye movement technology fusion. IE&M **12**(19), 144–148 (2014)
14. Tang, B.B., Guo, G., Wang, K.: Combined with eye movement and EEG, the auto industry design user experience selection. Comput. Integr. Manuf. Syst. **6**(21), 1449–1459 (2015)
15. Jantathai, S., Danner, L., Joech, M.: Gazing behavior, choice and color of food: does gazing behavior predict choice? Food Res. Int. **54**(2), 1621–1626 (2013)
16. Yang, C., Chen, C., Tang, Z.C.: Research on product image reasoning model based on EEG. J. Mech. Eng. **12** (2017)
17. Guo, F., Qu, Q.X., Zhang, X.Y.: Study on the relationship between user eye movement behavior and web design elements. IE&M **19**(5), 129–139 (2014)
18. Costa, T., Rognoni, E., Galati, D.: EEG phase synchronization during emotional response to positive and negative film stimuli. Neurosci. Lett. **406**(3), 159–164 (2006)
19. Davidson, R. J.: Cerebral asymmetry, emotion, and affective style. In: Davidson, R.J., Hugdahl, K. (eds.) Brain Asymmetry, pp. 361–387. The MIT Press, Cambridge, MA, US (1995)
20. Coan, J.A., Allen, J.J.B.: Frontal EEG asymmetry as a moderator and mediator of emotion. Biol. Psychol. **67**(1), 7–49 (2004)
21. Chen, M., Wang, H.Y., Xue, C.Q.: Product image-semantic matching evaluation based on event-related potentials. J. Southeast Univ. **44**(1), 58–62 (2014)

Research on Theoretical Line Loss Calculation Analysis and Loss Reduction Measures of Main Network Based on Multiple Factors

Weiru Wang, Xincong Shi, Mengzan Li, Xueting Cheng, Xinyuan Liu, Chengjun Huo and Jun Pi

Abstract Line loss is an important economic and technological index of power grid, which has been paid more and more attention by power grid enterprises. It is necessary to analyze the comprehensive factors affecting the line loss rate of the power grid and carry out relevant theoretical research on line loss. Firstly, the main factors affecting the theoretical line losses of transmission lines and transformers are analyzed in this paper. Taking two different operation modes of Yanhuai UHVDC in Shanxi power grid as examples, the theoretical line losses are measured and calculated. The loss reduction measures are proposed from the aspects of power grid operation mode and grid structure, which provide the scientific technical basis for the company in power grid planning and energy-saving renovation.

Keywords Theoretical line loss · Loss reduction measures · Fixed loss · Variable loss

1 Introduction

Line loss is an important economic and technical index of power grid assessment. It not only reflects the rationality of a power grid structure and operation mode, but also reflects the planning and design, production technology and operation management level of power grid [1]. At present, as an important assessment index for electric benchmarking, line loss has been paid more and more attention by power grid enterprises. It is necessary to analyze the comprehensive factors affecting the

W. Wang (✉) · M. Li · X. Cheng · X. Liu · J. Pi
State Grid Shanxi Electric Power Research Institute, Taiyuan, China
e-mail: hdwangweiru@163.com

X. Shi
State Grid Shanxi Electric Power Company Lingchuan Power Supply Company, Jincheng, China

C. Huo
State Grid Shanxi Electric Power Company, Taiyuan, China

V. Jain et al. (eds.), *Recent Trends in Intelligent Computing, Communication and Devices*, Advances in Intelligent Systems and Computing 1031,
https://doi.org/10.1007/978-981-13-9406-5_107

line loss rate of the power grid and carry out relevant theoretical research on line loss.

Power loss can be divided into three parts: fixed loss, variable loss and other losses. Fixed loss mainly includes iron loss of transformer and voltage coil loss of meter. Variable loss mainly includes line loss and copper loss of the transformer. Other losses include station power loss, management loss and other losses caused by unknown factors in power grid [2]. Power grid is a very complex system composed of a large number of transmission, transformation and distribution equipment. There are many factors affecting line loss. Power grid structure, voltage level, operation mode, load distribution and new energy access capacity and form will all have an impact on power grid loss.

It is very necessary and urgent to strengthen the application research of theoretical calculation results to diagnose and analyze the line loss of power grid. For this reason, electric power enterprises have formulated technical standards for the theoretical calculation of line loss, carried out the theoretical calculation of line loss, found weak links in the power grid and formulated effective loss reduction measures to provide the technical basis for power grid planning and energy-saving transformation. Firstly, this paper analyzes the main factors affecting the theoretical line loss of power grid, and combining with the actual situation of Shanxi power grid, chooses two typical operation modes of Yanhuai UHVDC before and after commissioning to calculate and analyze the theoretical line loss of Shanxi main network. And then, it puts forward corresponding loss reduction measures, which provide the scientific basis for technology for the company in power grid planning and energy-saving transformation.

2 Influencing Factors Analysis of Theoretical Line Loss [3]

The theoretical line loss of the main network of power grid is mainly composed of line loss, transformer loss, station power loss and other losses. The line loss and transformer loss account for the largest proportion, generally about 90% [4]. Therefore, this paper mainly makes a detailed analysis of transmission line loss and transformer loss.

2.1 Loss Analysis of Transmission Line

When the conductor passes through the current, it will produce active power loss and reactive power loss. The product of power loss and time is electric energy loss. Active power loss is almost the only concern in actual production and operation. The following is a detailed analysis of power loss.

When the line is running, the transmission power P can be expressed in Formula (1):

$$P = \sqrt{3}UI\cos\varphi\ (\text{kW}) \tag{1}$$

Among them,

U	The line head end voltage, kV
I	The line head end current, A
$\cos\varphi$	Average power factor

According to Joule's law, the power loss ΔP on the resistance R of a line when it is running is

$$\Delta P = 3I_{\text{rms}}^2 R \times 10^{-3}\ (\text{kW}) \tag{2}$$

Among them, Irms, the effective current value, can be expressed by the average current I_{av} and the shape coefficient K. And the line resistance R can be expressed by the resistance r_0 per kilometer and line length as follows:

$$I_{\text{rms}} = K \cdot I_{\text{av}} \tag{3}$$

Therefore,

$$\Delta P = 3K^2 I_{\text{av}}^2 r_0 L \times 10^{-3}\ (\text{kW}) \tag{4}$$

According to Formula (1)–(4), the formula for calculating the line loss rate with the line head power P as a variable is derived as shown in Formula (5):

$$\frac{\Delta P}{P}\% = \frac{K^2 \cdot P \cdot r_0 \cdot L}{10 \cdot U^2 \cdot \cos\phi^2}\% \tag{5}$$

According to Formula (5), the loss rate of transmission lines is mainly affected by the following factors:

① Line length. In principle, the longer the line, the higher the line loss.
② Wire material. In the selection of conductors, resistivity directly affects the loss of lines. Different conductor materials have different loss rates.
③ Voltage level. The higher the voltage level of the transmission line, the lower the line loss rate.
④ The influence of load distribution. With uneven load distribution and large voltage drop, the line loss rate is increased.
⑤ The influence of power factor. The variation of power factor directly affects the theoretical line loss. The larger the power factor, the lower the line loss rate.

2.2 Loss Analysis of Transformer

Transformer loss is composed of no-load loss P_0 and load loss P_R. No-load loss mainly refers to the iron loss of transformer. After the transformer is put into operation, the iron loss is basically unchanged, which is called fixed loss [5, 6]. Load loss mainly refers to the copper loss of transformer. The effect of fluctuation coefficient K_T of load loss on the results should be considered in calculation. The relationship between the load loss fluctuation coefficient K_T and the shape factor K is as follows:

$$K_T = K^2 \tag{6}$$

The formula for calculating active load loss of transformer is

$$\Delta P_R = P_K \frac{K \cdot I_{av}}{I_N} = P_K K^2 \beta^2 = K_T \beta^2 P_K \tag{7}$$

Among them,

ΔP_R Active load loss of transformer, kW
P_K Rated load power loss of transformer, kW
P Average output active power of transformer, kW
S_N Rated capacity of transformer, kVA
S Average output apparent power of transformer, kVA
I_{av} Average output current of transformer, A
I_N Rated current of transformer, A
β Average load factor of transformer, the calculation formula is

$$\beta = \frac{I_{\text{av}}}{I_N} = \frac{S}{S_N} \tag{8}$$

Therefore, the loss rate of the transformer can be expressed as

$$\frac{\Delta P}{P}\% = \frac{P_0 + P_R}{P}\% = \frac{P_0 + K_T \cdot \beta^2 \cdot P_K}{\beta \cdot S_N \cdot \cos\phi}\% \tag{9}$$

It can be seen that once the transformer is put into operation, the no-load loss P_0, load loss fluctuation coefficient KT and the rated load power loss K of transformer are basically unchanged. The average load coefficient β of the transformer directly determines the loss rate of the transformer.

3 Case Study

The simulation calculation uses the network version theoretical line loss calculation software developed by Hohai University. The calculation method of the main network line loss is power flow method. In the calculation process, the grid model needs to be built, and the basic parameters and operation data of the power grid also to be input.

The main power flow of Shanxi power grid presents the trend pattern of transmission from North to South. Yanhuai UHVDC is connected to the northern Shanxi power grid. After Yanhuai UHVDC and its supporting projects are put into operation, with the increase of large-scale outgoing power supply, great changes have taken place in the distribution of power flow in the main network of Shanxi power grid. And the distribution of power flow has a direct impact on the loss rate of the power grid.

In order to make the data comparable, this paper chooses two days before and after Yanhuai UHVDC put into operation as the database, when the load and grid structure of Shanxi power grid are equivalent. After Yanhuai UHVDC put into operation, the total outgoing power of the whole network increased by 81 percentage points.

3.1 Influence of Yanhuai UHVDC Operation on Theoretical Line Loss of Shanxi 500 kV Power Grid

After Yanhuai UHVDC put into operation, large-scale increase in power delivery in northern Shanxi power grid has greatly affected the start-up mode of 500 kV power plants and the distribution of 500 kV power flow in Shanxi power grid. The distribution of power flow in Shanxi power grid is relatively more balanced.

The power supply of Shanxi 500 kV power grid increased by 37,375.8 MWh, 12.72 percentage point increasing over the same period last year. However, due to the increase of external power supply by 78,368.5 MWh, the actual power supply of Shanxi 500 kV decreased by 40,992.7 MWh, affecting the total loss of electricity reduced by 687.0 MWh, down by 19.97 percentage points from the same period last year. It affects the reduction of line loss rate of this layer. The theoretical line loss calculation results of the two representative days in Shanxi 500 kV power grid are shown in Table 1.

As the grid structure remains unchanged, the iron loss is basically the same. The transmission power of main transformers in 500 kV Yantong, Minghaihu and Lvliang substations has been reduced to varying degrees, affecting the reduction of copper loss in the 500 kV main network. Consequently, the total loss rate of the transformer is reduced.

After Yanhuai UHVDC put into operation, the transmission power of the main passageway of 500 kV North to South power transmission has been greatly

Table 1 Calculation results of representative days of 500 kV grid

	Power supply volume (MWh)	Loss of electricity(MWh)						Power loss rate (%)	Copper and iron loss ratio
		Loss of line	Copper loss	Iron loss	Station power loss	Other losses	Summation		
After the operation of Yanhuai UHVDC	331,103.2	2213.2	79.2	208.7	38.9	214.0	2754.0	0.83	0.38
Before the operation of Yanhuai UHVDC	293,727.4	2874.2	108.0	206.9	39.4	212.5	3441.0	1.17	0.52
Year-on-year change	37,375.8	−661	−28.8	1.8	−0.5	1.5	−687.0	−0.34	−0.14
Year-on-year percentage	12.72	−23.00	−26.67	0.87	−1.27	0.71	−19.97	/	/

Table 2 500 kV main transmission line loss situation table

Sequence number	Line name	After the operation of Yanhuai UHVDC		Before the operation of Yanhuai UHVDC	
		Transmission power (MWh)	Line loss rate (%)	Transmission power (MWh)	Line loss rate (%)
1	Shenyan II line	5728.61	0.29	21,308.34	0.79
2	Shenyan I line	5415.37	0.22	20,131.33	0.61
3	Xinhou I line	14,738.16	0.34	29,140.26	0.68
4	Xinhou II line	14,987.68	0.32	29,651.04	0.64
5	Shuoyun line	7108.19	0.41	17,658.59	0.52
6	Lvji line	12,271.98	0.67	19,600.66	1.03
7	Lvhuo line	3231.21	0.15	12,436.71	0.35

reduced, which affects the total line loss reduced by 23 percentage points, and the loss rate of 500 kV line is reduced.

The loss of 500 kV main transmission lines on the two representative days is shown in Table 2.

3.2 *Influence of Yanhuai UHVDC Operation on Theoretical Line Loss of 220 KV Line in the Vicinity of UHVDC*

After Yanhuai UHVDC was put into operation, the power flow of 220 kV lines near Yanhuai UHVDC increased and the transmission power increased, which increased the corresponding line loss and line loss rate. The detailed calculation results are shown in Table 3.

4 Loss-Reducing Measures

Under the premise of ensuring the safe operation of the power grid, the following measures are recommended to reduce the loss of main network:

(1) According to the law of load variation, adjust and optimize the operation mode of power grid in time, improve the distribution of power flow, avoid circuitous power supply and reduce the proportion of heavy-load and light-load

Table 3 Loss of 220 kV line in the vicinity of Huaihuai UHVDC

Sequence number	Line name	After the operation of Yanhuai UHVDC		Before the operation of Yanhuai UHVDC	
		Transmission power (MWh)	Line loss rate (%)	Transmission power (MWh)	Line loss rate (%)
1	Hushui II line	2858.52	0.55	675.01	0.16
2	Hushui I line	2682.44	0.52	632.07	0.15
3	Huxiang II line	3134.89	0.31	2123.59	0.13
4	Huxiang I line	3147.07	0.30	2122.73	0.12
5	Huyu II line	4927.65	0.77	3601.05	0.60
6	Huyu I line	5821.08	0.66	4254.72	0.51
7	Shuoxiang II line	3083.85	0.62	1030.02	0.22
8	Shuoxiang I line	3193.97	0.70	1082.78	0.24

transmission and transformation equipment, so that the power grid is in an economic operation state.

(2) On the premise of satisfying the security and stability of power grid and transmission plan, priority should be given to increasing the transmission power of transmission channels with low line loss rate, so as to minimize the loss of transmission channels from North to South.

(3) Strengthen the management of reactive power and voltage in each station, increase the power factor as much as possible and control the voltage at each voltage pilot node close to the upper limit of the voltage curve, which can not only improve the stability level of the system, but also reduce the loss of the line.

5 Conclusions

In this paper, the main factors affecting theoretical line loss are discussed. And the influencing factors of transmission line and transformer theoretical line loss are analyzed in depth. It is pointed out that the main factors include power flow distribution, voltage level, transformer load rate, etc. This paper calculates the line loss rate of Shanxi main network before and after Yanhuai UHVDC operation and analyzes the main reasons affecting the line loss rate. Finally, the loss reduction measures and suggestions are put forward from the aspects of power grid operation mode, reactive power and voltage management, equipment operation economy and so on, which can provide decision-making reference for power grid planning and energy-saving transformation.

References

1. Yu, Z.: Power loss in power grid. China Electric Power Press, Beijing (2002)
2. Niu, Y.: Application and case study of power loss reduction and energy conservation technology. China Electric Power Press, Beijing (2013)
3. Li, J.: Brief discussion on management measures of line loss in rural power grid. China Sci. Technol. Inf. **22**, 4 (2005)
4. DL/T 686—1999. Guide of calculation of grid energy loss
5. Liu, L., Niu, Y.: Operation loss and energy saving of power transformer. China Electric Power Press, Beijing (2018)
6. Ma, Y., Ding, Y., Zhao, S., et al.: Line loss calculation and analysis of several typical modes of Qinghai main network. China Sci. Technol. Inf. **36**, 1 (2017)

Industry 4.0 and Applications

Theory and Practice: Workers' Quality Promotion by Labor and Skill Competitions in New Era

Shuling Li, Shufen Wang and Hui Yang

Abstract In this new era, the different forms of labor and skill competitions with diversified contents not only become an important tool for improving quality and efficiency, but also play a significant role in comprehensively enhancing workers' quality. Relying on practical experience, this thesis analyzes the value and mechanism of labor and skill competition in driving the team construction of industrial workers and attempts to establish the theoretical framework as well as application suggestions for the promotion mechanism of labor and skill competition in promoting of workers' quality.

Keywords Labor and skill competition · Workers' quality · Promotion

1 Overview About Labor Competition

As a feature of Chinese enterprises, labor competition had been existed before the founding of New China. Originally being a specific activity that stimulates workers' enthusiasm through competition, it gradually evolved into the diversified and conventional excellence selections directing at technological innovation, energy conservation, consumption reduction and quality improvement.

Under the encouragement of labor union system, labor competition has become an important tool to encourage workers in a state-owned corporation, enhance productivity, improve product quality and complete difficult tasks in the past decades of economic construction. It also turned into the significant means to promote workers' quality. Labor and skill competitions have continued to infiltrate non-state-owned enterprises and SMEs during recent years, which have also

S. Li (✉)
Capital University of Economics and Business, Beijing, China
e-mail: 103146706@qq.com

S. Wang · H. Yang
China University of Labor Relations, Beijing, China

V. Jain et al. (eds.), *Recent Trends in Intelligent Computing, Communication and Devices*, Advances in Intelligent Systems and Computing 1031,
https://doi.org/10.1007/978-981-13-9406-5_108

facilitated personnel management, technological upgrading and innovative production of these enterprises.

In the new era of information technology, the form of labor and skills competitions is presenting regular, diversified, and also it presents long-term and stable tendency. It is more and more popular with workers and enterprises. It also plays an important role in motivating employees to practice their skills and learn knowledge. Currently, the Chinese government put forward the development idea of promoting industrial upgrading and quality improvement via industrial restructuring so as to turn China from a manufacturing power into an intelligent producer. The foundation of these series of changes depends on a superior workforce team. According to the appeal of *Reform Plan for the Construction of Industrial Worker Teams in the New Era*, the construction of industrial workers is significant and urgent. Labor competition is an important approach for the construction of industrial workers.

Supported by the National Labor Union's regional labor competition performance evaluation project, the research group conducted an in-depth study on some representative enterprises in Area A and Area B during July–August, 2016. Meanwhile, 2205 questionnaires were randomly distributed to the technical workers in these two regions with 1486 valid questionnaires, and 513 questionnaires were issued to competition organizers with 441 valid questionnaires. In addition, the research group also summarized the labor and skills competitions is widespread and deep in the economic development area through the survey on multiple foreign and private enterprises conducted by Yizhuang Economic Development Zone in Beijing, as well as in-depth interviews to more than 30 technical experts. In July 2018, the survey group came back to Area A, further confirmed the value of labor and skill competitions in the new period in promoting workers' quality, also it is much helpful to workers' professional ethics and career development opportunity, etc. Many real cases show that because of all kinds of labor and skill competitions, workers have the platform to show their professional skill and motivate their passion. A number of excellent labor models and artisans came out, and they have not only become the demonstration of others, but also promoted more people to make progress.

2 Basic Features System of Labor and Skill Competition in New Era

The labor and skill competition in the new era has its own characteristics. The most typical feature is that it has already integrated with employees to achieve practice and study in competition. The promotion of workers' quality in such kind of competition mode is subtle, lasting and comprehensive.

2.1 Diversified Forms

In the new era, the labor unions at all levels join hands with enterprises and institutions to develop various types of labor and skill competitions; including the competition focusing on long-term performance, such as "youth position expert"; those paying attention to short-term effects, such as various types of project competitions; group competitions such as "Worker Pioneer, Ankang Cup, Energy Saving and Emissions Reduction Month," etc.; various skill competitions directing at individuals; "job contests and appraisal competition" inside enterprises; production labor competitions for corporate operating progress; QC group activities, rational advice activities, technological innovations and technical collaborations for the improvement of corporate technology, quality and management; "technical contest" and "five small" invention-creation activities. In short, all collective activities containing contents of comparison, evaluation, preference and awards can all be included in the scope of "labor and skill competition." Table 1 presents the types and proportions of major labor competitions in the two demonstration areas.

Table 1 Proportions of different types of labor competitions participated by workers in the two demonstration areas (data source: research group reports; same as the follows)

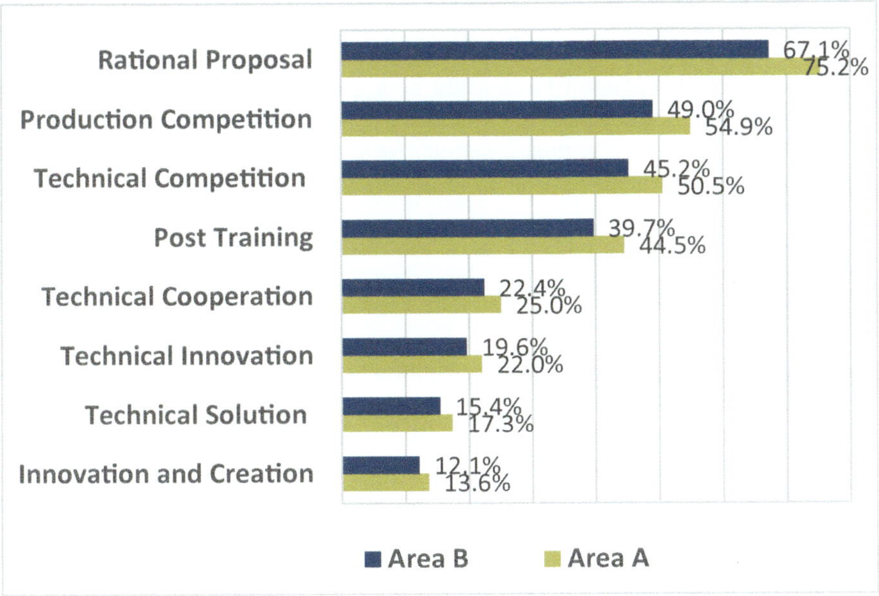

2.2 Extensive Coverage

Based on original labor competition, the labor and skills competition in the new era especially added the concept of skills so that it can cover many fields, industries, regions and technical jobs. Firstly, more and more workers are participating in competitions. Only 616 respondents of the 2205 questionnaires in the labor and skill competition demonstration area expressed that they had never participated in any competition, merely accounting for 27.9%, most of whom are new employees within six months of working. Second, the industries involved are extensive. It contains workers in traditional commercial, service and manufacturing industries and attracts emerging industries, new jobs and new skills, such as computer programming, animation design and software testing. Most of the corporate internal job contests will cover the whole staffs. For example, Company JEG holds no less than 200 job-related skill competitions each year with 80% of workshops and teams participating in on-the-job training activities 1. Thirdly, the types of units involved are extensive. The backbones of labor and skill competition are expanding from large- and medium-sized state-owned enterprises to foreign companies, private enterprises and a large number of emerging SMEs.

2.3 Higher Participation Initiative

The benefits of competition have made both companies and workers pay more attention to it and become passive participants. The staffs tend to regard it as a glory to join in the industrial, regional, national or even worldwide competitions, and many enterprises are actively organized and they also cherish the opportunity.

2.4 Multiple Targets

From the organizer's point of view, in addition to improving labor productivity and work efficiency, the goal of holding labor and skill competitions also includes enhancing quality, lowering costs, saving energy, reducing pollution and promoting workers' quality. This is both the requirement of economic society and staffs as well as enterprises' own development needs. From the participants' perspective, the labor and skill competition will bring various expected benefits, including enhancing technology, increasing income, job promotion, honor awards, career development opportunity, etc.

2.5 *Innovative Contents*

Innovation is the important content and basic feature of the labor competition in the new era. That is to say, no matter whether it is innovation themed, the form of competition shall keep pace with times to meet the individual development in contemporary business. Especially for technology enterprises, innovation is the main purpose of enterprises.

3 Mechanism of Workers' Quality Promotion by Labor and Skill Competition in New Era

Labor and skill competitions promote workers' quality; there is a set of specific measures generalized by Area B: "Training—contest—competition—promoting." This is a four-in-one quality enhancement mode, that is, through more than 100,000 frontline workers have grown up in the multi-form, multi-level, multi-industry and multi-channel technical training as well as activities, so that they eventually have the possibility of career promotion. Specifically, the role of each link is shown in Table 2.

Table 2 Model of promoting workers' quality by each link in labor and skill competition

Link	Tool	Content	Result	Description
Before competition	Training (or self-study)	A lot of knowledge learning	Increased relevant knowledge reserve	Demand of competition itself
	Guidance from senior workers	High-intensity and difficult training	Quickly enhanced skills	Common method for organizing competition
In competition	Competition process	Match between players	Acquired different techniques and methods by making up for weakness	Players' special achievements
	Competition itself	motivate personal potential	Created new skill proficiency	
After competition	Experience exchange meeting	Summary and demonstration	Popularized technical experience	Expansion of competition value
	Personal reflection	Reflection and accumulation	Consolidated skills	Real personal benefits
	Incentive measures	Awards in different categories	Devoted to know learning and growth enthusiastically	Powerful guarantee of competition continuity
Post training	Labor competition based on the post	Competition integrated into daily work	Promoted overall skill improvement	

4 Constructing Effective System to Promote Workers' Quality

4.1 Goal Incentive Mechanism

Goal is participants' motive force and organizer's baton since a scientific goal positioning will help exert the role of competition. The key points of setting goals are as follows:

First. Set goal clearly and specifically. Especially for quality promotion, the objectives that participants shall achieve need to be proposed in some quantifiable positions, e.g., the specific decrease of error rate and increase of qualification rate through job training, etc.

Second. Set a reasonable goal that is in line with workers' objective reality.

Third. Take into account the overall, classification and stratification of goals. It is necessary to set sub-types and sub-level targets for different groups so as to really play a leading role and encouragement.

Fourth. Combine organizational goals with personal goals.

Trade union of Area B has positioned the labor and skill competition as allowing employees to "learning from practice, training from learning, contesting from training and innovating from contesting" through competitions, which gradually formed a new pattern of staff skill competitions in non-state-owned enterprises that the area organizes general projects, enterprises hold different projects, one theme for every year and a unique model for specific enterprise.

4.2 Training Support Mechanism

Except for being essential for labor and skill competitions, pre-competition training also is the basic way to promote workers' quality. During the 12th Five-Year Plan, more than 220,000 employees in Area B received skill training, and over 15,000 were promoted their technical level. In practice, each skill competition will be accompanied by the skill training and exercise of thousands or even tens of thousands' employees, which resolve many workers' skills barriers, moral education or innovative ideas in a relatively short period of time with forming a positive atmosphere of "comparing, learning and catching up." [1]

In particular, training support should be given to migrant workers, many of whom lack systematic skills and technology learning. For this reason, enterprises may encourage migrant workers to take part in the competition and stimulate their enthusiasm to learn technology by giving full training.

There are many approaches for training in the era of the Internet, e.g., based on "post rival competition, team confront and elite election," Company JEG carries out full-coverage and multi-level activities of "learning techniques, training skills and improving technique." They organize training activities from multiple channels, including online training, practical teaching of shifts, small technical class and top-ranking "apprentice training" activities [2].

4.3 Appraisal Promotion Mechanism

The fairness, openness and generality of the process of electing labor and skill competition players will directly affect staffs' enthusiasm. A reasonable election mechanism will generate a large number of first-line skilled talents and enable the participants at each technical level to fully reflect their right of development in posts, teams and workshops. Therefore, when selecting contestants, it is required to respect their willingness, pay attention to objective facts, publicize sufficiently and achieve equal opportunities. Besides, the appraisal criteria shall focus on actual efficacy and value of technical skills and inventions.

4.4 Incentive Mechanism

Incentive mechanism is critical to the smooth proceeding of labor and skill competitions and promotion of workers' quality. This mechanism includes target incentives; winners' material and spiritual reward systems; title promotion system; welfare incentives such as recuperation, etc. It can be found from the survey that there are many similarities in employees' expectation of incentives as shown in Tables 3 and 4.

The survey shows that the interviewed employees are more concerned about the satisfaction of spiritual level and career development ability, and material incentives still occupy important positions. According to the principle that the unsatisfied need is the motivation of future actions, incentives measures shall base on different emphases. The effect and use of specific incentive methods are shown in Table 5.

5 Summary

There are tremendous energies among the workers and the masses. At the dawn of the age of knowledge, information and intelligence, it is urgent to quickly improve workers' skills and qualities. Innovative competitions and labor skill competitions based on positions and the incentives rotted in career development will greatly promote the development of ordinary workers.

Table 3 Attraction of different types of benefits to basic level staffs in competition—interviewed employees in area B

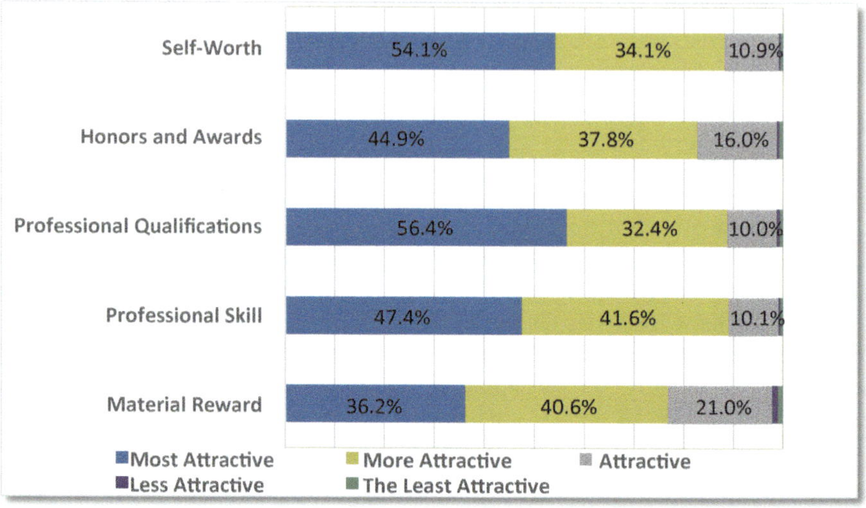

Table 4 Attraction of different types of benefits to basic level staffs in competition—interviewed employees in area A

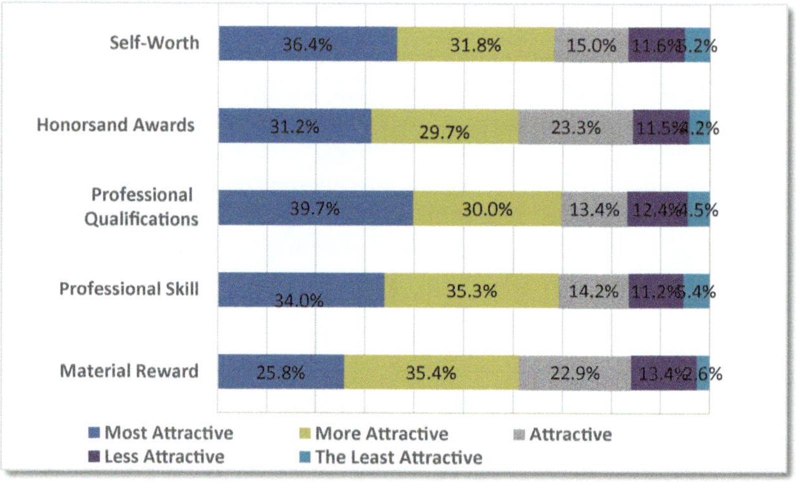

Table 5 Incentive mechanism and effect

Incentive type	Incentive method	Incentive effect	Matters need attention
Material incentive	Bonuses, items or similar materialized rewards	Most people are looking forward to this with good incentive effect	The bonus shall be attractive enough
Spiritual incentive	Naming the innovation achievement following the inventor; honor titles of model worker, excellent craftsman and high-skilled personnel, etc.	The best way to identify personal values with long-term effect	It is recommended to match with publicity
Title promotion	The winners pass the certification of technicians or senior technicians, and the technical level is advanced	The most positive affirmation of participants' technical skills with longer-term benefits and incentive effects	Using the procedure strictly to achieve openness and transparency
Career development	Transferring dispatched workers to formal worker, etc., and giving the winners further training, advanced studies, higher education, special creative studios, etc.	The shift of employment status has given employees a stronger sense of belonging, and the cultivation of occupational potential is also very popular	Good at discovering and supporting potential employees because they play stronger demonstration and driving effects
Welfare incentive	Settling down, recuperating, traveling, etc.	A local residence has great incentive for outstanding contestants, and other benefits are also popular	It will be more popular by expanding the reward scope

1. Technical experts and model workers are trained from the competition. Almost all model workers and technical experts who are selected every year by the union system and human society department are the strong performers in competitions. The competition makes these talents stand out and create more open stages for them.
2. Ordinary employees grow through competitions. The survey results show that the proportion of grassroots workers who believe that corporate labor competition is effective in quality promotion, respectively, is 77.8% (Area B) and 73.7% (Area A).
3. Workers' creativity is released during the competition. Innovative competitions allow a large number of rational proposals, inventions, techniques and operation laws accumulated in frontline operations have swarmed. For example, a young software engineer from a private company in Shanghai is deeply attracted to the research and development after accidentally participating in the technical innovation competition among workers. He achieved a dozen of patents within just a few years.

By means of strengthening the mechanism construction and promoting the effects via mechanism certainly will enable labor and skill competitions to efficiently facilitate the construction of industrial workers.

This paper is supported by the project *Research about the evaluation of labor competition of labor union* of China University of Labor Relations.

References

1. Liu haibin.: The national model labor competition exchange meeting for promoting regional development was highly praised for its experience in the construction of bin hai new area. http://www.bhxqgh.com/Mem/NewsShow.aspx?id=4661.2014.11.12
2. Trade Union of Hebei province.: New model of innovate labor competition. New type of work group-trade union of Jizhong energy fengfeng group innovation No136 work competition experience, Trade Union Information, 2017.13

Modeling of Assembly System Complexity and Its Application for Planning Horizon Problem

Fei He, Kang Shen and Ning Guo

Abstract For a more understanding about manufacturing system complexity, assembly system and its complexities are focused. The hierarchy of the assembly system is analyzed, and three kinds of primary complexities are classified and analyzed facing to different system layers. Thereinto, the static complexity is utilized to define the complexity of the manufacturing process and resources. Process complexity is defined to describe the complexity of production sequence which is defined by production planning. Then, the control complexity is the information which needs to describe the complexity of control methods. Then the process complexity is classified as static, dynamic, and real-time dynamic process complexities, which are used to describe the complexity for ideal, actual, and real-time production processes, respectively. The process complexities are applied to research the production planning effectiveness, and a method for appraising the initial planning and calculating the maximum planning horizon is presented. A case study for a weekly engine production planning is performed, and a four-day planning horizon is calculated and suggested to direct an actual engine production situation.

Keywords Complexity · Assembly system · Entropy measuring · Planning effectiveness

1 Introduction

Usually, manufacturers are needed to improve the flexibility, reliability, and responsiveness capacity to meeting the demands of market [1]. To coping with the problems, the manufacturing systems are becoming more and more complex and contain a large number of resources such as facilities, tools, materials, and employees.

F. He · K. Shen · N. Guo (✉)
Department of Industrial Engineering, School of Mechanical Engineering, Nanjing University of Science and Technology, Nanjing, China
e-mail: gnvara@qq.com

V. Jain et al. (eds.), *Recent Trends in Intelligent Computing, Communication and Devices*, Advances in Intelligent Systems and Computing 1031,
https://doi.org/10.1007/978-981-13-9406-5_109

Recently, complexity has been defined in an analytical form. According to Frizelle [2–4] and Deshmukh et al. [5], the manufacturing complexity is always classified into static complexity and dynamic complexity. Static complexity is arising from the impact the product structure has on the resources that will produce it, and the dynamic complexity is determined by the operational behavior from direct observations of the process, in particular on how queues behave.

Zhang and Efstathiou [6] calculate the information-based complexity of mass customization manufacturing system and compare the complexity under various inventory management strategies based on the complexity measurement. Kuzgunkaya and ElMaraghy [7] assess the manufacturing system's structure complexity; the presented metric would be helpful in selecting the least complex manufacturing system configuration that meets the requirements. Liu et al. [8] establish the system-level complexity flow model to obtain the complexity source of welding system, and complexity source sensitivity indices are proposed to identify key station and key equipment that contribute most to the complexity.

In this chapter, we analyze and classify the complexity of the assembly system based on three hierarchies' analysis of system. Two kinds of process complexities are researched and measured, respectively. Then, the diversity between static and dynamic process complexities is utilized to estimate the effectiveness of the initial planning.

2 The Complexity Classification of Assembly System

2.1 Hierarchies of Assembly System

Hierarchy is one of the typical characteristic of the system based on general system theory, and an analysis for the production line with three layers is described in Fig. 1.

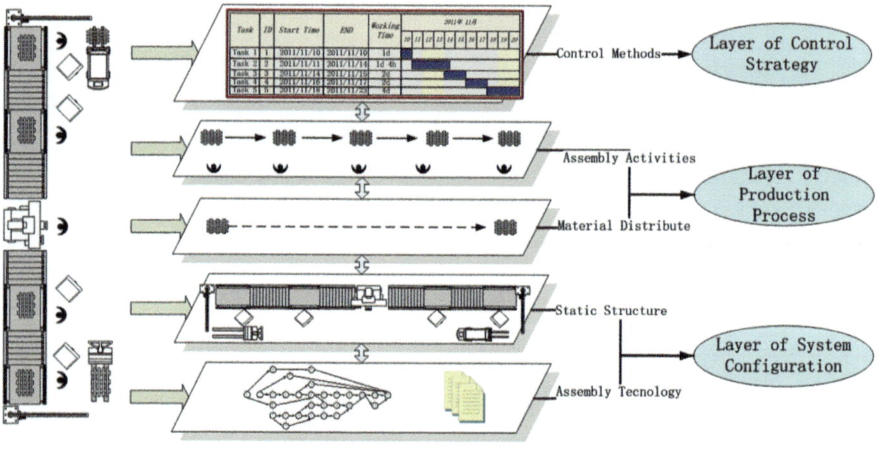

Fig. 1 Layers of assembly manufacturing system

Layer 1: Layer of system configuration

System configuration always focuses on the relationship between all kinds of resources in production line, which are used to implement the production tasks. It is always used to calculate the gathered complexity of equipments, operation stations, fixtures, and storage racks.

Layer 2: Layer of production process

Production process focuses on the execution of assembly works defined by the planning. Assembly works involve all the production tasks in every station of production line, and it can be defined as the moving activities of work-in-process from the station to the next. Besides the assembly works, lots of work to ensure the production such as maintenance, fixture, facility switching can also be seen as the assembly activities.

Layer 3: Layer of control strategy

The control strategy layer is defined as how the assembly work is organized and optimized. It contains the management work and control methods such as mast production plan, daytime planning, and shop-floor scheduling.

2.2 *Classification of the Complexity*

By the analysis of the hierarchies of assembly system, the complexity of assembly system can be more classified as follows:

a. **Static complexity**

Assembly technology complexity: It concerns about the operation technologies of every task and the relationship between them. The complexity can be separated as two parts—one is the information of the quality of tasks and their sequence, and another one is decided by all the elements of technology such as equipment, tool, and parameter.

Static structure complexity: It is defined as the function of the resources of the production line and its configuration. The static structure complexity is only used to describe the complexity of the resources themselves and does not consider the relationship between the resources and the assembly work.

b. **Process complexity**

Assembly process complexity: It is defined to measure the complexity of batches of products to be assembled. The parameters such as operation time, set-up time, maintenance time, switch time, failure rate, and qualified rate must be considered and calculated for the complexity.

Material distribution complexity: It is defined to describe the complex of the distribution of parts and components while the products are assembled in the production line.

c. **Control complexity**

The control complexity can be classified by the planning-making methods such as scheduling complexity, planning complexity, and mast planning complexity.

3 Resolving and Measuring of Process Complexity

3.1 Two Characteristics of Process Complexity

The process complexity is defined to measure the information of the state of pre-planned activities executed by the manufacturing resource. So the processing time of every task and their sequences are the most important parameters for measuring the complexity. Figure 2 shows the relationship between static and dynamic process complexities.

Static process complexity (SPC) is primarily used to measure the information about the expected state of the predefined assembly process without anomalous event. This complexity can be measured based on the information contained in initial planning.

Dynamic process complexity (DPC) This complexity is defined to measure all the predictable and unpredictable information for the system operation. Besides the normal procedure of every assembly task, the unplanned changes such as equipment failure, operation delay, and urgent tasks must also be observed and calculated.

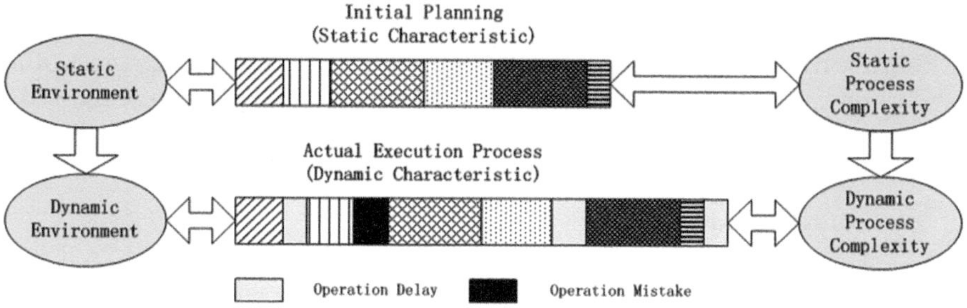

Fig. 2 Relationship between static and dynamic process complexities

3.2 *Measuring of Static Process Complexity*

a. **Measure with task execution time**

Because the execution of every task should not regard the previous tasks in certain time especially under a push assembly model, every task can be seemed as independent. We first propose the complexity measurement just considering the execution time of every task. According to the measurement of information entropy, the static process complexity *H*(*S*) can be defined as:

$$H(S) = -\sum_{i=1}^{m} p_i \log p_i \tag{1}$$

$$p_i = t_i / \sum_{i=1}^{m} t_i \tag{2}$$

$$\sum_{i}^{m} p_i = 1 \tag{3}$$

where m = number of tasks, t_i = the execution time of the task i, and p_i means the rate of task i incidence in the whole process.

b. **Measure with several parameters**

According to Eq. (1), the tasks' execution time is the only parameter to calculate complexity. But under an actual manufacturing environment, other parameters such as production cost, operation space, and operators' skill level may influence the whole production process, and they should also be considered to calculate the complexity. So the probability distribution p_i to measure the complexity may convert into a multidimensional probability distribution $p_{ij\ldots n}$, and $p_{ij\ldots n} = p_i \cdot j \cdot \ldots \cdot n$

Where n presents the total number of parameters that influence the process.

So Eq. (1) can be also presented as

$$\begin{aligned} H(S_{ij\cdots n}) &= -\sum_{i=1}^{m} p_{ij\ldots n} \log p_{ij\ldots n} \\ &= H(S_i) + H(S_j) + \cdots + H(S_n) + H(S_j|S_i) + \cdots + H(S_n|S_i S_j \ldots) \end{aligned} \tag{4}$$

If the parameters are all independent, the complexity can be calculated by:

$$H(S_{ij\cdots n}) = \sum_{i=1}^{n} H(S_i) \tag{5}$$

3.3 Measuring of Dynamic Process Complexity

The dynamic process complexity concerns the practical state of the production process. As the production planning is executing, two kinds of events may occur. One is the predecided events by the plan, another is the abnormal events which have not been anticipated (such as equipment failure and operation delay). These two kinds of events should be considered to measure the dynamic process complexity.

Corresponding to Eq. (1), the dynamic process complexity *H*(*D*) with one parameter can be presented as:

$$H(D) = -\sum_{j=1}^{m'} p_j \log p_j = -\sum_{i=1}^{m} p'_i \log p'_i - \sum_{k=1}^{m'-m} p_k \log p_k \quad (6)$$

$$p'_i = \frac{t_i}{\sum_{i=1}^{m} t_i + \sum_{k=1}^{m'-m} t_k} \quad (7)$$

$$p_k = \frac{t_k}{\sum_{i=1}^{m} t_i + \sum_{k=1}^{m'-m} t_k} \quad (8)$$

where m' = number of actual activities to achieve the plan, t_k = the execution time of the unplanned activity k, p_j = probability of any task *j* in whole actual process, p'_i = probability of planned task i in actual process, p_k = probability of unplanned activity *k* in actual process, $-\sum_{i=1}^{m} p'_i \log p'_i$ presents the complexity caused by the planned tasks, and $-\sum_{k=1}^{m'-m} p_k \log p_k$ presents the complexity caused by the unplanned activities.

3.4 Real-Time Dynamic Process Complexity

The real-time dynamic process complexity (RDPC) is proposed to describe the actual changing of the system state. The real-time dynamic process complexity can also be measured by Eq. (8) when all the unexecuted tasks are implemented under an ideal environment. So according to Eq. (8), the measure of the complexity at time *t* can be presented as:

$$
\begin{aligned}
H(D_{\mathrm{RT}\_t}) &= H(D_i') + H(S_i') \\
&= -\sum_{i=1}^{m_t} p_i \log p_i + \left(-\sum_{j=1}^{m_t'} p_j \log p_j - \sum_{k=1}^{m_t''-m_t-m_t'} p_k \log p_k \right)
\end{aligned} \tag{9}
$$

where $H(D_i')$ = the dynamic complexity caused by the incurred unplanned activities at time t, $H(St)'$ = the static complexity caused by the planned tasks, m_t'' = number of practical activities to achieve the plan at time t, m_t = number of incurred unplanned activities at time t, m_t' = numbers of incurred planned tasks at time t, p_i = probability of incurred unplanned activity i, p_j = probability of incurred planned task j, and p_k = probability of non-occurred planned task k.

4 Appraising the Production Planning Effectiveness

4.1 Appraising of Initial Planning

According to the classification of process complexity, the transformation of the static and dynamic process complexities is mainly affected by two reasons—one of which is dynamic process complexity increase rate and the other is static complexity decrease rate. So the total variation of the process complexity can be calculated by the difference of the two values. The method to appraise the effectiveness of the initial planning is presented as follows.

Firstly, the increase of the dynamic assembly process complexity at time t can be given as:

$$
H(\Delta D\_t) = H(D_{\mathrm{RD}\_t}) - H(S_t') = -\sum_{i=1}^{m_t} p_i \log p_i \tag{10}
$$

Secondly, the decrease of the static assembly process complexity primarily caused by the occurrence of the unplanned event at time t can be defined as:

$$
\begin{aligned}
H(\Delta S\_t) &= H(S) - \left(H(D_{\mathrm{RD}\_t}) - \left(-\sum_{i=1}^{m_t} p_i \log p_i \right) \right) \\
&= H(S) - \left(-\sum_{j=1}^{m_t'} p_j \log p_j - \sum_{k=1}^{m_t''-m_t-m_t'} p_k \log p_k \right)
\end{aligned} \tag{11}
$$

Thirdly, the difference of the variation of the static and dynamic process complexities which can be called as process complexity change rate (PCCR) is as follows:

$$\Delta h(t) = H(\Delta D\_t) - H(\Delta S\_t) \tag{12}$$

Then, the average remaining SAPC at time t is:

$$H(S_{\text{remain}})_{\text{avg}} = \frac{H(D_{\text{RD}}\_t) - \left(-\sum_{i=1}^{m_t} p_i \log p_i - \sum_{j=1}^{m'_t} p_j \log p_j \right)}{M - M'} = \frac{-\sum_{k=1}^{m''_t - m_t - m'_t} p_k \log p_k}{\Delta M} \tag{13}$$

If $\Delta h(t) \geq H(S_{\text{remain}})_{\text{avg}}$, the initial plan does not suit the manufacturing environment anymore, and a new plan should be programmed.

4.2 The Maximum Planning Horizon

The maximum planning horizon is the period that the initial static process complexity can support the average complexity consumption rate, and the horizon is a statistic value which can be used as a reference for the future production by the long-term observation. The method for the maximum planning horizon can be given as follows.

Firstly, based on Eq. (11), the final increase of the dynamic process complexity is defined as:

$$H(\Delta D)_{\text{final}} = H(D) - (-P'_S \log P'_S) = -\sum_{i}^{m'} p'_D \log p'_D \tag{14}$$

The average increase of the DAPC is

$$H(\Delta D_{\text{avg}}) = \frac{H(\Delta D)_{\text{final}}}{\Delta T_{\text{final}}} = \frac{H(\Delta D)_{\text{final}}}{T_{\text{final}} - T} \tag{15}$$

where T_{final} = the actual final completion time, T = the anticipated completion time

Then, the final decrease of the static process complexity is presented as

$$H(\Delta S)_{\text{final}} = H(S) - P'_S \log P'_S \tag{16}$$

The average decrease of the SAPC is

$$H(\Delta S_{\text{avg}}) = H(\Delta S)_{\text{final}}/T \tag{17}$$

So, the maximum planning horizon can be given as:

$$T_{\text{MAX}} = \frac{H(S)}{H(\Delta D_{\text{avg}}) - H(\Delta S_{\text{avg}})} = \frac{H(S)}{\frac{H(\Delta D)_{\text{final}}}{T_{\text{final}} - T} - \frac{H(\Delta S)_{\text{final}}}{T}} \tag{18}$$

Within T_{MAX}, the increase of the dynamic process complexity in unit time is still less than the decrease of the static process complexity. Once beyond the time limit, the variation of the dynamic characteristics cannot be covered by the decrease of the static process complexity.

The uncertainty factor which may occur from the implementation of the plan is disposed according to the maximum plan horizon, then the stability and flexibility of the planning increase.

5 Appraising the Production Planning Effectiveness

In this section, we suppose a system that involves one-engine mixed-model assembly line, five types of engines that can be intermixed to be assembled in the assembly line, and the products that are produced batch to batch.

Suppose the five products is P1–P5. Figure 3a is showing a daily planning sample, where M means the equipment maintenance, J1–J5, respectively, present the assembly period of the P1–P5, and S1–S5 present the equipment adjusting time of every work. Because of the uncertainty, the actual implementation of the planning sample is shown in Fig. 3b. Moreover, if the situation such as the J3 arises, it can be substituted as Fig. 3c. The execution time of every task is shown in Table 1.

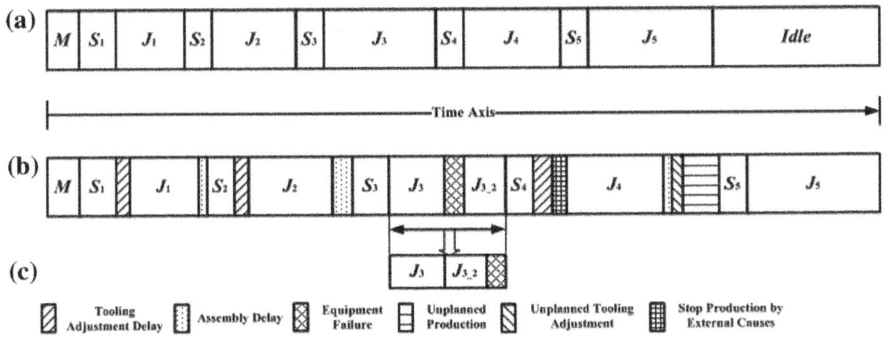

Fig. 3 Example of ideal and actual production processes

Table 1 Measure of static assembly process complexity

State	Activity	Proposed time (h)	Probability p_i	$-p_i \log p_i$
1	Equipment maintenance	2	0.0185	0.1065
2	Tooling adjustment S1	3	0.0278	0.1437
3	Assembly J1	16	0.1481	0.4081
4	Tooling adjustment S2	5	0.0463	0.2052
5	Assembly J2	18	0.1667	0.4308
6	Tooling adjustment S3	4	0.037	0.1761
7	Assembly J3	25	0.2315	0.4887
8	Tooling adjustment S4	5	0.0463	0.2052
9	Assembly J4	12	0.1111	0.3522
10	Tooling adjustment S5	3	0.0278	0.1436
11	Assembly J5	15	0.1389	0.3956
	Aggregate	108	1	3.0557 bit

a. **Measure of initial static process complexity**

The static process complexity of Fig. 3a is measured as shown in Table 1.

So the value of static process complexity is 3.0557 bit based on Eq. (1).

b. **Compare PCCR with Average Remaining static process complexity**

As shown in Fig. 3b, several kinds of disturbance affect the execution process of the initial planning. Here we propose the calculation of the real-time dynamic process complexity at the time 6, for example. Table 2 presents all values of process complexity change rate in the every unit time.

Before calculating the real-time dynamic process complexity at time 6, the supposed entire process after time 6 is shown as in Fig. 4. In Fig. 4, the forward 6 h which contain M, S1, and D1 are the incurred events.

$$H(D_{\mathrm{RT}\_6}) = -\sum_{i=1}^{m_t} p_i \log p_i + \left(-\sum_{j=1}^{m'_i} p_j \log p_j - \sum_{k=1}^{m_i^* - m_i - m_i^*} p_k \log p_k\right) \tag{19}$$
$$= 0.2485 + 0.0621 + 2.7924 = 3.1030\,\text{bit}$$

$$H(\Delta D\_6) = 0.0621\,\text{bit} \tag{20}$$

$$H(\Delta S\_6) = H(S) - 0.2485 - 2.7924 = 0.0148\,\text{bit} \tag{21}$$

$$\Delta h(t) = H(\Delta D\_t) - H(\Delta S\_t) = 0.0473\,\text{bit} \tag{22}$$

$$H(S_{\text{remain}})_{\text{avg}} = \frac{2.7924}{9} = 0.3102\,\text{bit} \tag{23}$$

So if $H(S_{\text{remain}})_{\text{arg}} > \Delta h(t)$, the initial planning can still be executed.

Table 2 Values of PCCR and average remaining static process complexity at every time

No	Interval	Duration	RDPC	Average remaining SPC	No	Interval	Duration	RDPC	Average remaining SPC
1	$t = 0$	$0 \leq t \leq 5$	3.0557	0.2778	14	$t = 91$		3.4559	0.2787
2	$t = 6$	$6 < t \leq 22$	3.103	0.3102	15	$t = 92$		3.4798	0.2774
3	$t = 23$	$23 < t \leq 28$	3.1494	0.2968	16	$t = 93$		3.5196	0.2761
4	$t = 29$		3.1953	0.3089	17	$t = 94$	$94 < t \leq 106$	3.5428	0.2748
5	$t = 30$	$30 < t \leq 48$	3.2225	0.3075	18	$t = 107$		3.5816	0.2481
6	$t = 49$		3.267	0.2867	19	$t = 108$		3.62	0.2470
7	$t = 50$	$50 < t \leq 79$	3.2933	0.2854	20	t = 109		3.6578	0.2458
8	$t = 80$		3.3367	0.2643	21	$t = 110$		3.6796	0.2447
9	$t = 81$		3.3622	0.2630	22	$t = 111$		3.6952	0.2436
10	$t = 82$		3.3809	0.2617	23	$t = 112$		3.7069	0.2425
11	$t = 83$		3.3953	0.2604	24	$t = 113$		3.7157	0.2414
12	$t = 84$		3.4064	0.2591	25	$t = 114$	$114 < t \leq 132$	3.7222	0.2403
13	$t = 85$	$85 < t \leq 90$	3.4151	0.2578					

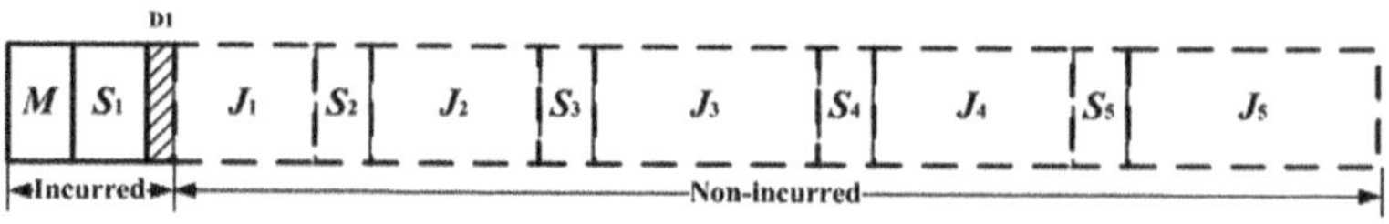

Fig. 4 Supposed entire process after time 6

As shown in Table 2, we can see, at the time 80 h, that the PCCR is bigger than average remaining static process complexity, and the initial planning should be refreshed at this time.

c. **Solving the maximum planning horizon**

The dynamic assembly process complexity is measured in Table 3. Finally, the dynamic process complexity is 3.7222 bit in which the complexity caused by the planned events decreases to 2.737 bit, and the complexity caused by the uncontrolled events increases to 0.9852 bit.

Table 3 Measure of dynamic assembly process complexity

State	Event	Attribute	Duration	Probability	Ctr H	Un Ctr H
1	Equipment maintenance	Ctr	2	0.0152	0.0916	
2	S1	Ctr	3	0.0227	0.1241	
3	S1 delay	Un Ctr	1	0.0076		0.0534
4	J1	Ctr	16	0.1212	0.369	
5	J1 delay	Un Ctr	1	0.0076		0.0534
6	S2	Ctr	5	0.0379	0.1789	
7	S2 delay	Un Ctr	2	0.0152		0.0916
8	J2	Ctr	18	0.1364	0.392	
9	J2 delay	Un Ctr	2	0.0152		0.0916
10	S3	Ctr	4	0.0303	0.1529	
11	J3-J3′	Ctr	25	0.1894	0.4546	
12	Equipment failure	Un Ctr	6	0.0456		0.2027
13	S4	Ctr	5	0.0379	0.1789	
14	S4 delay	Un Ctr	2	0.0152		0.0916
15	Stop production	Un Ctr	2	0.0152		0.0916
16	J4	Ctr	12	0.0909	0.3145	
17	J4 delay	Un Ctr	1	0.0076		0.0534
18	Unplanned adjustment	Un Ctr	1	0.0076		0.0534
19	Unplanned assembly	Un Ctr	6	0.0455		0.2027
20	S5	Ctr	3	0.0227	0.1241	
21	J5	Ctr	15	0.1136	0.3565	
	Aggregate		132	1	2.737 bit	0.9852 bit

The average increase of the dynamic process complexity is:

$$H(\Delta D_{\mathrm{avg}}) = \frac{H(\Delta D)_{\mathrm{final}}}{\Delta T} = \frac{H(\Delta D)_{\mathrm{final}}}{T_{\mathrm{final}} - T} \frac{0.9852}{24} = 0.04105\,\mathrm{bit} \tag{24}$$

The average decrease of the static process complexity is:

$$H(\Delta S_{\mathrm{avg}}) = \frac{H(\Delta S)_{\mathrm{final}}}{T} = \frac{3.0557 - 2.737}{108} = 0.002951\,\mathrm{bit} \tag{25}$$

So the maximum planning horizon is:

$$T_{\mathrm{MAX}} = \frac{H(S)}{H(\Delta D_{\mathrm{avg}}) - H(\Delta S_{\mathrm{avg}})} = \frac{3.0557}{0.04105 - 0.002951} = 80.2\,\mathrm{h} \tag{26}$$

It means that the period of the planning formulation is 80.2 h, and it is close to the time calculated by the comparison between PCCR and average remaining static process complexity.

6 Avenues for Future Work

Although the process complexity has been classified and measured, the concept of assembly process we research here is still rough. Beside the process complexity, the complexity characteristics containing the system configuration and the control strategy have not been researched in detail. So an important future work is to discuss and measure all kinds of subcomplexities and their relationship to expand an entire assembly system complexity model.

Acknowledgements This research work is supported by the National Natural Science Foundation of China, under Grant No. 51575280.

References

1. Shuiabi, E., Thomson, V., Bhuiyan, N.: Entropy as a measure of operational flexibility. Eur. J. Oper. Res. **165**, 696–707 (2005)
2. Frizelle, G., Wookcock, E.: Measuring complexity as an aid to developing operational strategy. Int. J. Oper. Prod. Manag. **15**(5), 26–39 (1994)
3. Frizelle, G., Suhov, Y.M.: An entropic measurement of queueing behaviour in a class of manufacturing operations. Royal Soc. **457**, 1579–1601 (2001)
4. Sivadasan, S., Efstathiou, J., Frizelle, G., Schrin, J., Calinescu, A.: An information-theoretic methodology for measuring the operational complexity of supplier-customer systems. Int. J. Oper. Prod. Manag. **22**(1), 80–102 (2002)

5. Deshmukh, A.V., Talavage, J.J., Barash, M.M.: Complexity in manufacturing systems, Part 1: Analysis of static complexity. IIE Trans. **30**(4), 35–44 (1998)
6. Zhang, T., Efstathiou, J.: The complexity of mass customization systems under different inventory strategies. Int. J. Comput. Integr. Manuf. **19**(5), 423–433 (2006)
7. Kuzgunkaya, o, ElMaraghy, H.A.: Assessing the structure complexity of manufacturing systems configurations. Int. J. Flex. Manuf. Syst. **18**(2), 145–171 (2006)
8. Liu, H., Xu, K., Pan, Z.: Modeling and application of mixed model assembly system complexity introduced by auto-body personalization. Int. J. Adv. Manuf. Technol., 1–12 (2015)

Robust Sliding Mode Control of Ship Based on Neural Network Under Uncertain Conditions

Renqiang Wang, Keyin Miao, Yue Zhao, Hua Deng, Jianming Sun and Jiabao Du

Abstract A robust sliding mode control algorithm of ship based on neural network under uncertain conditions was designed. The algorithm could effectively solve the problem of ship motion control under model uncertainty and external disturbance. Based on the nonlinear response motion mathematical model of the ship, the RBF neural network was used to effectively approximate the ship system function and external disturbance. Then, Lyapunov stability theory and backstepping method were used to design the controller of ship motion. The simulation results verified that the control algorithm tracked the set signal well and the controller had good robustness.

Keywords Ship · Control · Uncertainty · RBF neural network · Robust · Lyapunov

1 Introduction

The problem of uncertainty [1] must be fully considered and resolved for robust controller designed of ship motion. In the process of controller designed, many parameters need to be adaptively estimated [2], the fuzzy system [3] was used to approximate the ship state system continuous function, and then, the pole configuration adaptive fuzzy controller was designed. The defect of the method is that the design parameters are too much, and as the fuzzy rules increase, the adaptive estimation parameters increase in multiples. On the basis of the literature [4], the fuzzy system and the neural network which have an ability to approximate arbitrary continuous functions with arbitrary precision were used to approximate the uncertain continuous function in the system. The advantage of this strategy is that the number of design parameters is greatly reduced. In the literature [5–7], the parameters of the ship model are uncertain, so, the RBF neural network was used to

R. Wang (✉) · K. Miao · Y. Zhao · H. Deng · J. Sun · J. Du
College of Navigation, Jiangsu Maritime Institute, Nanjing 211170, China
e-mail: wangrenqiang2009@126.com

V. Jain et al. (eds.), *Recent Trends in Intelligent Computing, Communication and Devices*, Advances in Intelligent Systems and Computing 1031,
https://doi.org/10.1007/978-981-13-9406-5_110

approximate the uncertain term, and the control algorithm based on the network weight adaptive law was designed.

To solve the problem of ship heading control under model uncertain conditions, this paper proposed a design scheme of ship heading sliding mode robust controller based on RBF neural network and backstepping algorithm. The controller made the state signal index in the system progressively stable; that is, the ship tracked the set heading with an arbitrarily small error so as to achieve a consistent final bound.

2 Ship Nonlinear Motion Model with Uncertain Conditions

The following nonlinear motion of mathematical model of ship was considered [8]:

$$\ddot{\psi} + \frac{1}{T}H\left(\dot{\psi}\right) = \frac{K}{T}\delta \tag{1}$$

where $H\left(\dot{\psi}\right) = a_1\dot{\psi} + a_2\dot{\psi}^3 + a_3\dot{\psi}^5 + \cdots$.

Considering the slow change of the state of the ship, the nonlinear function of the ship is generally taken as $H\left(\dot{\psi}\right) = \alpha\dot{\psi} + \beta\dot{\psi}^3$.

The nonlinear motion equation of the ship's heading dynamic system can be expressed as:

$$\begin{cases} \dot{\psi} = r \\ \dot{r} = -\dfrac{K_0}{T_0}\left(\alpha r + \beta r^3\right) + \dfrac{K_0}{T_0}\delta + d(t) \end{cases} \tag{2}$$

According to engineering practice, $d(t)$ is usually bounded interference; that is, $|d(t)| \leq \rho$, $\rho > 0$, it is an unknown constant.

Definitions: $x_1 = \psi$, $x_2 = r$, and $u = \delta$; Eq. (2) can be simplified.

$$\begin{cases} \dot{x}_1 = x_2 \\ \dot{x}_2 = f(\bar{x}_2) + bu + d(t) \end{cases} \tag{3}$$

where $\bar{x}_2 = [x_1, x_2]$, $f(\bar{x}_2) = -\frac{1}{T}H(x_2)$, $b = \frac{K}{T}$.

3 Ship Motion Controller Was Designed Based on RBF Neural Network

3.1 RBF Neural Network

RBF network algorithm is $h_{nn}(x) = \omega^T \phi(x)$, x is neural network input, $\omega = [\omega_1, \omega_2, \ldots, \omega_l] \in R^l$ is weight matrix between input layer and output layer, $l > 1$ is the number of neural network nodes, $\phi_i(x) = [\phi_1(x), \phi_2(x), \ldots, \phi_l(x)]^T$, and usually, $\phi_i(x)$ is a Gaussian function, i.e.,

$$\phi_i(x) = \exp\left[\frac{-(x-\mu_i)^T(x-\mu_i)}{\eta_i^2}\right], \quad i = 1, 2, \ldots l \tag{4}$$

where $\mu_i = [\mu_1, \mu_2, \ldots, \mu_l]^T$ is center vector, and η_i is the width of Gaussian function.

It is shown that RBF network can approximate any continuous function with the precision value on compact set, $x \in \Omega_x \subset R^q$, i.e.,

$$h(x) = \omega^{*^T} \phi(x) + \varepsilon^*, \ \forall x \in \Omega_x \tag{5}$$

where ω^* is the ideal weight, and ε^* is the reconstruction error.

Symbol $\hat{\omega}$ is used to represent the estimated value of ω^*. And the estimation error is defined as $\tilde{\omega}_i = \hat{\omega}_i - \omega_i^*$, $i = 1, 2, \cdots, n$. Therefore, $f(\bar{x}_2)$ can be approximated as shown in Eq. (9):

$$f(\bar{x}_2) = \omega^T \phi(\bar{x}_2) + \varepsilon \tag{6}$$

3.2 Ship Robust Intelligent Controller Designed

Considering heading error $z_1 = x_1 - y_d$, it is defined $z_2 = x_2 - \sigma_2$; therefore,

$$\dot{z}_1 = z_2 + \sigma_2 - \dot{y}_d \tag{7}$$

Further, the following virtual control law is selected:

$$\sigma_2 = -k_1 z_1 + \dot{y}_d \tag{8}$$

where k_1 was a design parameter.

From Eqs. (7) and (8), Eq. (9) can be derived:

$$\dot{z}_1 = -k_1 z_1 + z_2 \tag{9}$$

The Lyapunov function is constructed as follows:

$$V_1 = \frac{1}{2} z_1^2 \tag{10}$$

Equation (11) is obtained by differentiating Eq. (10).

$$\dot{V}_1 = z_1 \dot{z}_1 = z_1(-k_1 z_1 + z_2) = -k_1 z_1^2 + z_1 z_2 \tag{11}$$

It is considered that $z_2 = x_2 - \sigma_2$; Eq. (12) can be obtained as:

$$\dot{z}_2 = \dot{x}_2 - \dot{\sigma}_2 = f(\bar{x}_2) + bu + d(t) - \dot{\sigma}_2 \tag{12}$$

Furthermore, Eq. (13) is obtained as:

$$\dot{\sigma}_2 = -k_1 \dot{z}_1 + \ddot{y}_d = -k_1(\dot{x}_1 - \dot{y}_d) + \ddot{y}_d = -k_1 x_2 + k_1 \dot{y}_d + \ddot{y}_d \tag{13}$$

Equation (14) is obtained by bringing Eq. (13) into Eq. (12).

$$\dot{z}_2 = f(\bar{x}_2) + bu + d(t) - \dot{\sigma}_2 = bu + f(\bar{x}_2) + d(t) - (-k_1 x_2 + k_1 \dot{y}_d + \ddot{y}_d) \tag{14}$$

Considering that the system control law u containing the gain term b, so, the virtual control law υ_u is designed instead of bu, and it is described as follows:

$$\upsilon_u = -k_2 z_2 - \Phi^2(x) z_2 - z_1 - k_1 x_2 + k_1 \dot{y}_d + \ddot{y}_d \tag{15}$$

where k_2 and $\Phi(x)$ are intelligent damping terms.

High-frequency gain b and control law υ_u are improved by Eq. (16).

$$u = \hat{\theta} \upsilon_u = \hat{\theta}\left[-k_2 z_2 - \Phi^2(x) z_2 - z_1 - k_1 x_2 + k_1 \dot{y}_d + \ddot{y}_d\right] \tag{16}$$

where $\hat{\theta}$ is the estimated value of $\theta = \frac{1}{b}$, $\dot{\hat{\theta}} = -\gamma \upsilon_u z_2$, γ is a constant.

3.3 Lyapunov Stability Analysis

Make the following settings: $\tilde{\theta}$ was the difference between θ and $\hat{\theta}$, which was the estimated value of θ, i.e.,

$$\tilde{\theta} = \hat{\theta} - \theta \tag{17}$$

Equation (18) is obtained by combining Eq. (16) with Eq. (17).

$$\begin{aligned}\dot{z}_2 &= b\hat{\theta}\upsilon_u + f(\bar{x}_2) + d(t) - (-k_1x_2 + k_1\dot{y}_d + \ddot{y}_d) \\ &= b\theta\upsilon_u + f(\bar{x}_2) + d(t) - (-k_1x_2 + k_1\dot{y}_d + \ddot{y}_d) + b\tilde{\theta}\upsilon_u \\ &= \upsilon_u + f(\bar{x}_2) + d(t) - (-k_1x_2 + k_1\dot{y}_d + \ddot{y}_d) + b\upsilon_u\tilde{\theta}\end{aligned} \tag{18}$$

The second Lyapunov function is constructed as follows:

$$V_2 = \frac{1}{2}z_1^2 + \frac{1}{2}z_2^2 + \frac{b}{2\gamma}\tilde{\theta}^2 \tag{19}$$

Equation (20) is obtained by differentiating Eq. (19):

$$\begin{aligned}\dot{V}_2 &= -k_1z_1^2 + z_1z_2 + z_2\dot{z}_z + \tilde{\theta}\dot{\hat{\theta}} \\ &= -k_1z_1^2 + z_2\left[z_1 + \upsilon_u + f(\bar{x}_2) + d(t) - (-k_1x_2 + k_1\dot{y}_d + \ddot{y}_d) + b\upsilon_u\tilde{\theta}\right] - \tilde{\theta}b\upsilon_u z_2 \\ &= -k_1z_1^2 + z_2[z_1 + \upsilon_u + f(\bar{x}_2) + d(t) - (-k_1x_2 + k_1\dot{y}_d + \ddot{y}_d)] \\ &= -k_1z_1^2 - k_2z_2^2 + z_2\left[-\Phi^2(x)z_2 + f(\bar{x}_2) + d(t)\right]\end{aligned} \tag{20}$$

Further, RBF network is used to approximate $f(\bar{x}_2)$ and $d(t)$ as follows:

$$f(\bar{x}_2) + d(t) = \omega^T\phi(\bar{x}_2) + \varepsilon + d(t) \le \omega^T\phi(\bar{x}_2) + \bar{\varepsilon} + \rho \le \bar{\omega}\Phi(\bar{x}_2) \tag{21}$$

where $\bar{\omega} = \max\{\omega_i, \bar{\varepsilon}, \rho\}, i = 1, 2, \ldots, n$. $\Phi(\bar{x}_2) = \|\phi(\bar{x}_2)\|_2 + 1$.

According to Young's inequality, Eq. (22) can be derived as follows:

$$\begin{aligned}\dot{V}_2 &\le -k_1z_1^2 - k_2z_2^2 + z_2\left[-\Phi^2(x)z_2 + \bar{\omega}\Phi(\bar{x}_2)\right] \\ &\le -k_1z_1^2 - k_2z_2^2 - \Phi^2(\bar{x}_2)z_2^2 + \frac{\bar{\omega}^2}{4} + \Phi^2(\bar{x}_2)z_2^2 \\ &= -k_1z_1^2 - k_2z_2^2 + \frac{\bar{\omega}^2}{4} \\ &\le -AV_2 + B\end{aligned} \tag{22}$$

where $A = \min\{k_1, k_2\}$,$B = \frac{\bar{\omega}^2}{4}$. It is possible to ensure that all signals of entire closed-loop system are bounded by adjusting parameters k_1 and k_2.

4 Simulation

The simulation of training teaching ship "Yulong" [7] was carried out to verify the algorithm described above. The initial conditions of the system simulation were set as follows: the initial heading was 0, interference signal was

$d(t) = 0.5 + 2\sin(0.3t)$, command signal $\psi_r(t) = 15(\mathrm{sgn}(\sin(5\pi t/1000)) + 1)$ with the range of 0°–30°, $k_1 = 0.15$, $k_2 = 0.15$, $\gamma = 7.5$.

Figure 1 shows the heading of the ship in consideration of the interference. It could be seen the controller tracked preset heading without overshoot and overcame external interference. It is indicated that the control algorithm has better robustness. And Fig. 2 shows the ship's control rudder angle curve when considering interference. The trajectory had a certain amount of jitter, but it was not violent in a reasonable range.

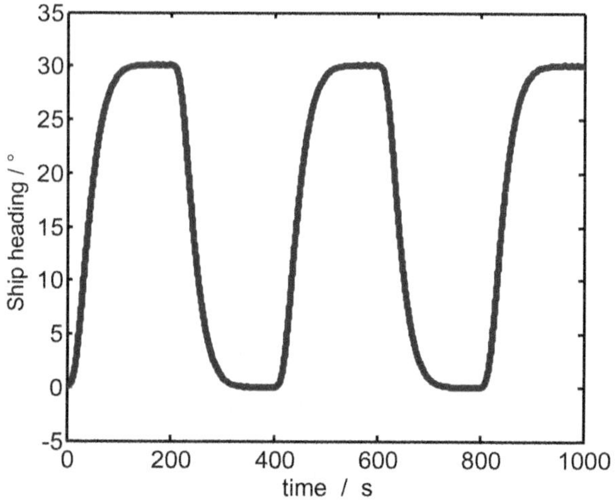

Fig. 1 Heading of the ship

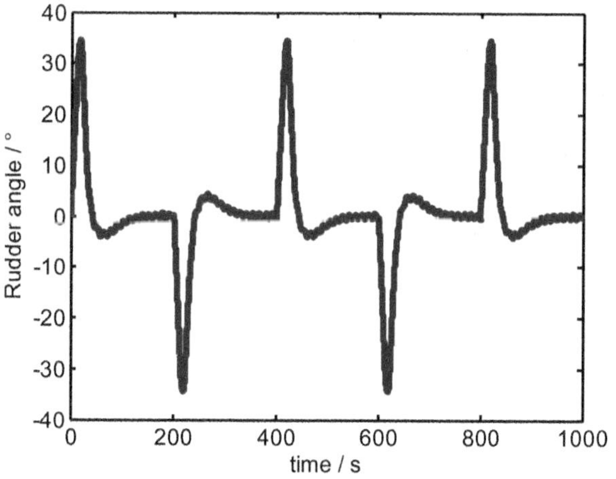

Fig. 2 Output of rudder angle

5 Conclusion

To solve the problem of ship motion control under model uncertainty and external disturbance, RBF neural network was used to effectively approximate the ship system function and external disturbance, and then, Lyapunov stability theory and backstepping method were used to design a robust sliding mode controller of ship. The simulation verified the algorithm tracked the set signal well with good robustness.

Acknowledgements This work was supported by the Natural Science Research Project of Universities in Jiangsu Province under Grant No. 18KJB580003.

References

1. Kumarawadu, S., Kumara, K.: On the speed control for automated surface vessel operation. In: Proceedings of the 2007 Third International Conference on Information and Automation for Sustainability, pp. 135–140, Melbourne, Australia (2007)
2. Wang, R., Zhao, Y., Miao, K.: Application of neural network minimum parameter learning algorithm in ship's heading tracking control. In: 9th International Symposium on Computational Intelligence and Design, vol. 12, pp. 135–139 (2016)
3. Fan, Y., Sun, X., Wang, G., Guo, C.: On fuzzy self-adaptive PID control for USV course. In: 34th Chinese Control Conference, vol. 7, pp. 8472–8478 (2015)
4. Hu, G., Zhou, Y.: Application of fuzzy neural network in ship course control. Appl. Mech. Mater. **135**(1511), 309–315 (2012)
5. Li, T.-S., Yang, Y.-S., Hong, B.-G., Qin, Y.-X.: Robust adaptive fuzzy design for ships track-keeping control. Control Theory Appl. **3**(4), 445–448 (2007)
6. Hu, G.S., Zhou, Y.B., Xiao, H.R.: Application of fuzzy neural network in ship course control. Appl. Mech. Mater. **13**(10), 309–315 (2012)
7. Wang, X., Li, T., Luo, W.: Direct adaptive neural network control for a class of ship course uncertain discrete-time nonlinear systems. Mar. Eng. Front. **3**(1), 42–48 (2013)
8. Wang, Y., Guo, C., Sun, F.: Dynamic neural fuzzified adaptive control of ship course with parametric modelling uncertainties. Int. J. Model. Identif. Control **13**, 251–258 (2015)

Research on Dynamic Security Access Control Technology Based on Resource Attributes

Zhimin He, Lin Peng, Min Xu, Gang Wang, Hai Yu and Zhansheng Hou

Abstract After research and analysis, this program studies the dynamic security access control technology for resource attributes to realize the security and reliability of cross-domain resource access in the new generation of dispatch control system, including the effective control of data access between the monitoring system and the analysis decision center and between services.

Keywords Resource · Attributes · Dynamic security · Access control

1 Introduction

1.1 Background

The development of China's power grid has entered the stage of smart grid with high integration of power flow, information flow, and business flow. The power grid dispatch control system and communication network are the "brain" and "nerve center" of the smart grid, which manage and control the reliable operation of the power grid. The information security risk of cyberspace can form a deadly threat to smart grid entities through the destruction of grid dispatch control systems and communication networks. Currently, network security has become an important part of national security. The central government set up a network security and informatization leading group and proposed that "no national security without network security." The concept of network security has evolved from traditional information system security protection to cyberspace confrontation. In fact, the network has become the fifth battle space after land, sea, sky, and space. At the

Z. He (✉) · L. Peng · M. Xu · G. Wang · H. Yu · Z. Hou
Global Energy Interconnection Research Institute, Nanri Road 8, Nanjing 210003, China
e-mail: 825097034@qq.com

Z. He
State Grid Key Laboratory of Information & Network Security, Beiqijia, Beijing 100000, China

V. Jain et al. (eds.), *Recent Trends in Intelligent Computing, Communication and Devices*, Advances in Intelligent Systems and Computing 1031,
https://doi.org/10.1007/978-981-13-9406-5_111

international level, there has been a national-level game or even a partial cyber war around the "network-making rights."

1.2 Significance

This project studies the dynamic security access control technology for resource attributes to realize the security and reliability of cross-domain resource access in the new generation of dispatch control system, including the effective control of data access between the monitoring system and the analysis decision center and between services.

1.3 Overseas Research Levels

Research on access control [1] began in the 1960s, and the discretionary access control (DAC) model was based on the access matrix model proposed by Lampson in 1971. It was standardized by Conway, Maxwell, and Morgan. The latter access control model is widely used in UNIX-like operating systems.

In 1976, Denning et al. proposed the lattice model theory, namely the mandatory access control (MAC) model, and applied the theory to implement an access control system. The system determines whether the subject is allowed to access the object by assigning different security attributes to the subject and the object. Because the model has high intensity confidentiality, it is widely used in the military field.

In 1992, David Ferraiolo and Rick Kuhn of the National Institute of Standards and Technology of the United States proposed the role-based access control (RBAC) model framework, which facilitated the authorization management by introducing roles to separate users from permissions. In the process of continuous evolution, RBAC made the NIST RBAC reference model a national standard under the impetus of NIST in 2004. Many improvements such as Fuzzy RBAC, TRBAC, GTRBAC, GEO-RBAC, and Ex-RBAC have been developed based on the RBAC classic model.

Due to the importance of access control [2] in the security field and the importance attached by various industries, the application environment of access control is complicated, and a model cannot meet all the requirements. Therefore, in addition to RBAC, there are many access control models such as task-based access control (TBAC) model and behavior-based access control (ABAC) model in the field of access control models.

1.4 Domestic Research Levels

In China, Fan Yundong et al. proposed a cloud computing access control model based on the trust value evaluation. By dynamically calculating the comprehensive trust value of user behavior, the model determines its trust level, enabling users to dynamically obtain access rights and improve the security of cloud computing. Wang Yuding et al. proposed a cloud computing data access control scheme DACPCC with access rights. On the basis of the traditional CP-ABE encryption system, the privilege control key is set to encrypt the data in the cloud, and the data provider controls the privilege. The choice of key to control access to the data. Yu Yang et al. proposed an attribute-based cloud manufacturing collaborative platform access control model [3]. Firstly, the access request and policy decision are formalized, and then the policy decision case of the proposed model is given. The model not only enables dynamic and fine-grained authorization of collaborative services, but also improves the flexibility, scalability, and security of the cloud manufacturing collaboration platform.

China, with the development of smart grid, for more in-depth study of access control, intelligent scheduling control system in which the full use of domestic security Linux operating system, which uses mandatory access control and role access control to achieve the operating system level security protection; at the same time introduced The security label technology establishes a unified identity role management system within the scope of the national dispatching system and implements the role-based secure access of the service provider to the requester according to the access control principle of read/write [4].

2 Solutions and Key Technologies

In the new generation of dispatch control system, the resource distribution is dynamic; that is, the location of the resource is dynamically changed, and the usage scenario is open; that is, the user is not fixedly using or accessing the system and resources of the domain in which it is located, and there is cross-domain access. In the case, the existing security system is difficult to adapt to the security requirements of the new system. Therefore, it is necessary to study dynamic security access control technology for resource attributes.

2.1 Method of Constructing Resource Attribute Set

The use of attributes as the smallest authorization unit replaces the traditional access control model based on identity-based authorization, which meets the new requirements of resource access control in an open network environment. The access control strategy, model, and implementation mechanism are all associated with the user's attribute set. The attribute set is used to uniformly model the subject, object, authority, and environment attributes, and the entity features are described in all aspects from different perspectives.

(1) Entity attribute discovery mechanism

Using the probabilistic model-based learning method, using the preset extraction rules and existing dictionaries, first extract different entities, concepts, and attributes, and count the number of occurrences of an attribute and the given attribute when given a concept. The number of occurrences of a concept, then the Bayesian probability model is used to score the attributes to associate the attribute values with the corresponding concepts, so as to achieve the mining of the optimal attributes. Under the condition of satisfying the independent and complete constraints, the calculation is lower. Complexity is to get the best possible set of query properties.

(2) Attribute permission association

The bottom-up method is used to automatically filter and generate candidate roles based on the existing "user-privilege" correspondence. The collection makes the Cartesian product of the "user-role" and "role-permission" obtained as close as possible to the original "user-privilege" relationship. The role-based state output algorithm is used to generate the role set and the complete role state, including the assignment relationship between the user and the role, the assignment relationship between the role and the authority, and the hierarchical relationship between the roles.

(3) Attribute set for user groups

Since the identity information of the user is represented as a set of attributes, the user group consisting of two or more users may also have the same attribute set, so that the attribute set may be used to represent a single user or a user group composed of multiple users. Therefore, the attribute set can be flexibly adjusted by a comprehensive description of the user identity information, so that it represents a single user or a user group.

(4) Description of access control policy

Extensible Access Control Markup Language (XACML) description method, the extensible, platform-independent, and high-security access control policy description language, realizes the subject and object in access control, context information, accurate description of authorization policies, and access control permission definitions for sensitive information such as network services and digital rights provide an important technical platform for the construction of access models and the preparation of access control policies.

2.2 Transfer Mapping Technique for Resource Attributes

The transfer mapping technology of resource attributes in the new generation of dispatch control system is studied. When the cross-domain access is implemented, the user attributes are converted into the target domain attributes according to the conversion rules to ensure reliable access of users to cross-domain resources (Fig. 1).

By verifying the registration information of the subject and the registration information of the resource in the identity confirmation module, it is confirmed whether both parties are registered users. If it is legal, proceed to the next step, and redirect the principal and resources, allocate a memory space, and prepare for negotiation. Otherwise, the service is denied.

The registration information of the subject registration information and the resource is transmitted to the attribute provisioning negotiation service module, and the attribute provisioning negotiation service module establishes the subject attribute controller and the resource attribute controller. The attribute controller establishes a link with the subject and the resource attribute certificate through the registration information and returns the rejected service if the resource or the subject attribute certificate does not match the registration information. Otherwise, proceed to the next step.

First, the resource requester and the resource provider provide the minimum attribute that they think can be matched and the attribute set that the other party needs to provide. The attribute controller converts the data that can be identified in the attribute matching relationship table and matches the attribute matching negotiation service module. If both parties are not satisfied, the attributes that are not successfully matched are passed from the attribute controller to the resource requester and the resource provider. Then, on the basis of this, the next allocation is performed, the number of iterations is limited, and finally, the attribute matching result set of both parties is obtained.

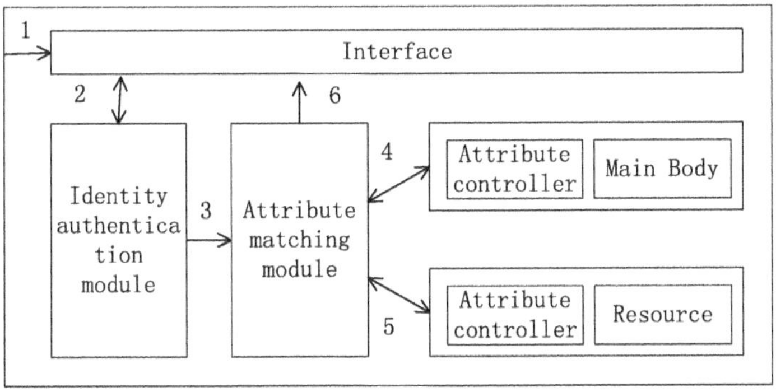

Fig. 1 Attribute matching

2.3 Resource Access Control Technology

This part of the research is based on the access control technology of resources in the new generation of dispatch control system, constructs a trust-based access authority adjustment mechanism, realizes the dynamic adjustment of resource access rights, and regulates the access of users of the new generation of dispatch control system users. The overall framework of multi-domain access control based on trust is shown in Fig. 2.

The cloud user first obtains the corresponding role through the role management center, and then interacts with the authentication and authorization center to submit the user ID, password, role, and information of the resource to be accessed, and requests authorization access. If the resource requested by the user is in the local domain, the local domain access control policy is adopted. If the external domain resource is requested to be accessed, the cross-domain access control policy is adopted to perform rights allocation and management.

The main way to introduce trust into role-based access control is to use trust as a basic attribute of cloud users and cloud services or cloud resources. The access control authentication, authorization, and trust management in the local domain are handled by the authentication and authorization center (AAC), while the cross-domain access control and trust management require the master authentication and authorization center (MAAC) and AAC work together.

The main difference between trust-based multi-domain access control and the traditional access control mechanism is that when the users access in the local domain and across domains, two access control policies are adopted. In the multi-domain access control model based on trust, when the user logs in, the user's identity is first verified. For the trusted user, whether or not to authorize is determined according to its identity. At the same time, the user behavior trust level in the model reflects the credibility of user behavior, so that authorization is no longer a static mechanism based solely on identity trust, but a dynamic mechanism based on the combination of identity trust and behavior trust. Therefore, the model implements a combination of user identity trust and behavioral trust.

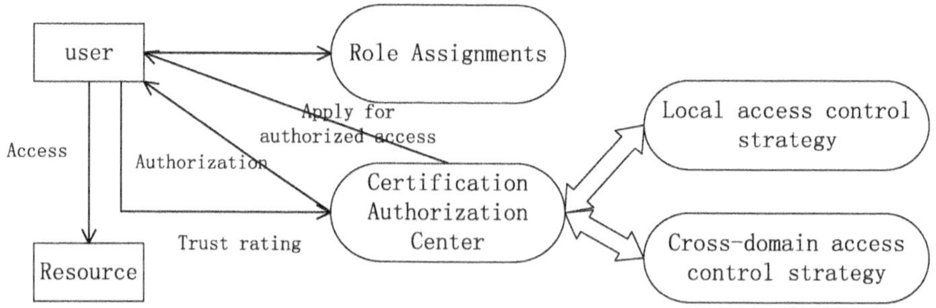

Fig. 2 Trust-based access control

3 Conclusion

This project studies the dynamic security access control technology for resource attributes. Firstly, it studies the construction method of resource attribute set in the new generation scheduling control system and uses access control description language to express the access control strategy to ensure that the attribute set is easy to construct and expand. Secondly, the new research A transfers mapping technology for resource attributes in a generation of dispatch control systems, which implements cross-domain access and securely converts user attributes into target domain attributes according to conversion rules to ensure reliable access to cross-domain resources. Finally, the research is based on a new generation of dispatch control systems. The access control technology of resources is built, and the access authority adjustment mechanism based on trust is constructed to realize the dynamic adjustment of resource access rights and to standardize the access of resources of users of the new generation dispatch control system. This mechanism simplifies the control of user authorization and access policies, increases the flexibility and fine-grained-ness of access authorization and access control policy expression, and enhances the flexibility and scalability of the access control mechanism.

Acknowledgements This work was financially supported by the science and technology project to State Grid Corporation "Research on Key Technologies of Supporting Platform under Physical Distribution and Logic Unified Architecture". I would like to express my heartfelt gratitude to all the scholars who are quoted in this paper, they gave lots of useful advices during the process of studying this topic research and also provided enthusiastic help in the process of typesetting and writing thesis!

References

1. Fang, L., Yin, L., Guo, Y., Fang, B.: A survey of key technologies in attribute-based access control scheme. Chin. J. Comput. **40**(7), 1690–1697 (2017)
2. Li, J.: Design and Implementation on Security Access Control System. Beijing University of Posts and Telecommunications (2011)
3. Pen, W., Liu, X., Guo, H., Song, C.: Research on trust based access control in cross-domain. Appl. Res. Comput. **33**(6), 1791–1796 (2016)
4. Bie, Y., Lin, G., Trust-based access control strategy in multi-domain of cloud computing. Inf. Secur., 39–52 (2012)

Design of Remote Condition Monitoring System for Armored Vehicles Based on Beidou

Ming Chen, Xiaoming Zhang, Dongxiang Zhou and Yantao Wang

Abstract Based on the actual demand of armored vehicle remote condition monitoring, and the contradiction between conventional remote communication methods and the army's confidentiality provisions, this paper presents a remote condition monitoring system for armored vehicles, taking Beidou satellite navigation system as location and remote communication method, and it realizes the unification of remote monitoring and communication security and stability. The hardware and software design schemes of the system are introduced in detail. The experimental results show that the system has stable communication and accurate test results and has realized the remote condition monitoring of armored vehicle engine.

Keywords Beidou · Short message · Remote condition monitoring

1 Introduction

With the gradual advancement of military informationization and digitalization, it is an inevitable trend for management departments to establish real-time equipment condition monitoring system for the convenience of networking, automation and online equipment management. At present, there are many systems for remote monitoring of vehicle condition in the market; however, the remote communication methods chosen by these systems are mostly GPRS, WiFi, and other network signals. There is a risk of information being stolen easily, and it is highly dependent on communication base station construction. In isolated areas where there are few base stations, the weak signal connection is usually unstable, which is difficult to meet the security and stability requirements of military equipment remote communication. Because of the above situation, this paper designs a remote

M. Chen · X. Zhang (✉) · D. Zhou · Y. Wang
Department of Vehicle Engineering, Academy of Army Armored Force, Beijing, China
e-mail: 13661366835@139.com

V. Jain et al. (eds.), *Recent Trends in Intelligent Computing, Communication and Devices*, Advances in Intelligent Systems and Computing 1031,
https://doi.org/10.1007/978-981-13-9406-5_112

monitoring system for armored vehicles with Beidou as the positioning and remote communication mode.

Beidou Satellite Navigation System (Beidou System for short) is a global satellite navigation system built and operated independently by China. Compared with the other three major satellite navigation systems in the world—GPS, GLONASS, and Galileo—the Beidou system has integrated navigation and communication capabilities, not only has the functions of location, navigation, and timing, but also enables 120 short messages of Chinese characters/times to communicate remotely [1]. The remote monitoring system based on Beidou introduced in this paper is precisely relying on this short message communication function to realize the remote transmission of information.

2 The Overall Architecture of Remote Monitoring System

The remote monitoring system is mainly composed of vehicle terminal and monitoring center. The vehicle terminal and the monitoring center can transmit data through Beidou short message module, and the system frame diagram is shown in Fig. 1.

The vehicle terminal collects the equipment parameters in real time by a large number of sensors deployed on armored vehicles. After signal conditioning and A/D conversion, the collected signals are input into the MCU, which then stores the data and the location information into the built-in hard disk,and extract the

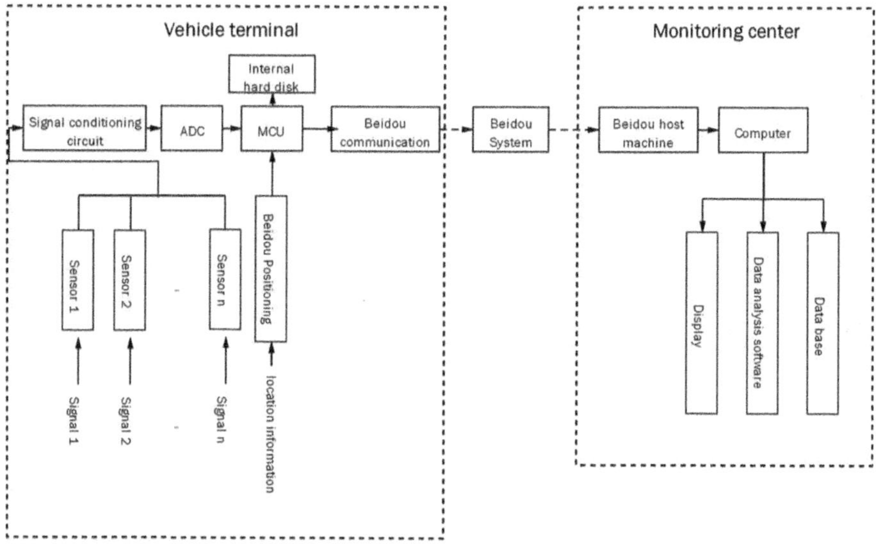

Fig. 1 Architecture of an armored vehicle remote monitoring system based on Beidou

eigenvalues of the monitored parameters at the same time. The extracted feature values and vehicle location information are transmitted to the monitoring center through the Beidou short message function. The Beidou command machine of the monitoring center receives the state information of the armored vehicle and transmits these data to the computer. Then the computer will analyze these data and display the analysis results, so that the equipment management department can timely understand the real-time status of the equipment.

3 Hardware Design

The system hardware mainly includes vehicle terminal and Beidou host machine.

3.1 Hardware Design of Vehicle Terminal

The design of vehicle terminal must first determine the parameters to be collected. The terminal mainly collects two types of vehicle parameters: one is the basic operating parameters of general equipment chassis, mainly including engine working time, driving mileage, working time of each gear, and the running track of the vehicle. Another class of parameters is engine performance parameters, including engine output shaft speed, engine output shaft torque, starting current, starting voltage, fuel flow, cylinder pressure, cylinder head vibration, and exhaust pressure.

In order to achieve accurate acquisition and timely processing of the parameters to be collected, the terminal uses STM32F207 chip as MCU and two ADS1278 chips as A/D converters. The STM32F207 [2] chip has a wide range of voltage support, it can still work normally under the condition of voltage instability, and its operation is powerful, so that it can achieve good control of data acquisition and timely processing of data. ADS1278 [3] has the characteristics of fast conversion speed and high precision, which can ensure the accuracy and timeliness of data acquisition.

Beidou positioning and short message function are realized by BDT data transmission baseboard with BDM910 module produced by Beidou Star Navigation Technology Co., Ltd. The BDM910 module supports both Radio determination satellite service (RDSS) and radio navigation satellite system (RNSS) functions. Because of Beidou, RDSS can send and receive short message information, it can also locate, but its positioning accuracy is low, power consumption is large, and speed is slow. RNSS has no communication function, but its positioning accuracy is high, speed is fast, and the energy consumption is low [4]. The terminal uses the RNSS mode of the BDM910 module to locate and RDSS for short message communication. BDM910 module obtains the position information of equipment from Beidou system, transmits the position information to MCU through RS232

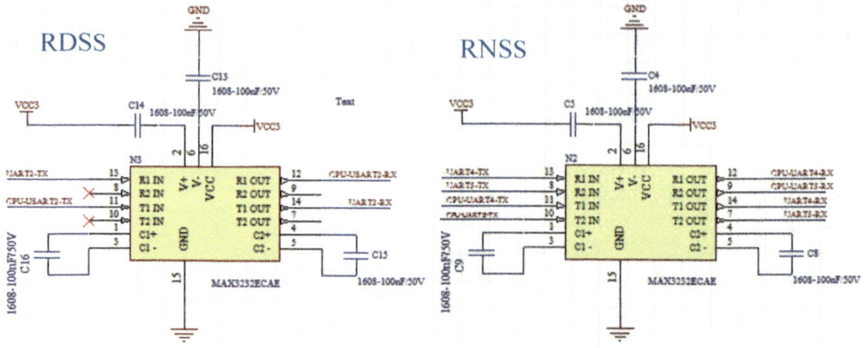

Fig. 2 Communication circuit diagram of the Beidou data transmission baseboard and MCU

interface on the baseboard, and obtains short message edited by MCU through another RS232 serial port on the baseboard. The communication circuit between the data transmission baseboard and the MCU is shown in Fig. 2.

3.2 *Beidou Host Machine*

Beidou host machine adopts BD-1 command user machine (type 100) which is designed by Beijing BDStar Navigation Co., Ltd. The BD-1 command user machine is a special user machine designed to facilitate group networking and dispatching command. Besides having the function of location, communication and time service of the ordinary user, the model can also supervise the subordinate users. The machine can monitor the location and communication information of the subordinate users, and send the broadcast information to them; it is widely used in national defense, public security, border protection, transportation, disaster relief, emergency, and other fields. Applying it to the remote monitoring system of armored vehicles can effectively realize the command and dispatching monitoring of the monitoring center to the equipment owned by it.

4 Software Design

The system software mainly includes vehicle terminal and monitoring center part.

4.1 Design of Vehicle Terminal Software

The software design of the terminal is written in C language and compiled by Keil C compiler. The communication frequency of the Beidou communication card used by the terminal is once a minute, and the terminal software uses the time interval between two adjacent short messages as the working cycle to perform cyclic work. At first, the system will initialize the program after boot, then the real-time parameters collecting of armored vehicle engine will start, and a working cycle begins. The terminal will store the collected data into the internal memory. When a working cycle finishes, the processor will extract the eigenvalues of these parameters in the cycle, editing these eigenvalues in short message and send them to the monitoring center, then a new working cycle begins. The software flowchart is shown in Fig. 3.

The method of extracting the eigenvalues of the monitored parameters will be formulated according to the use of parameters, the basic operating parameters are only used for recording equipment utilization condition, and the extraction method of these eigenvalues is relatively simple. The working time of engine, mileage, and working time of the each gear will extract the accumulated value, equipment location takes latest position information, and the engine performance parameters are used to characterize the engine health condition; the extraction of eigenvalues needs to reflect the current performance of the engine. The rotational speed and torque will be combined to calculate the output power of the engine, and that will reflect the working ability of the engine, the starting current and the starting voltage will be combined to calculate the total work done by the electronic motor during starting the engine, and that will be used to reflect cylinder tightness, the fuel flow

Fig. 3 Vehicle terminal software flowchart

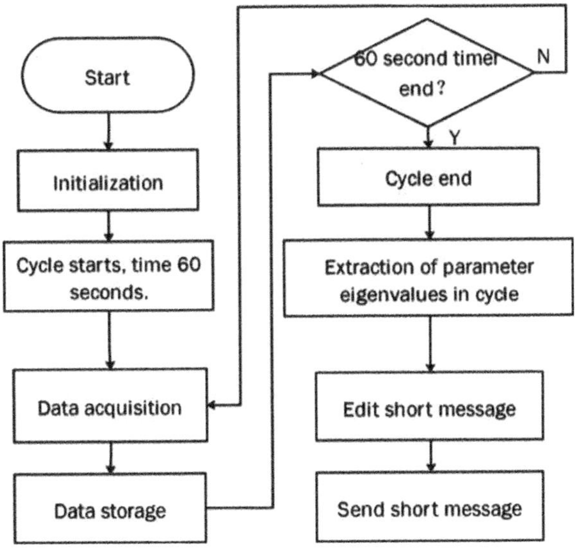

Table 1 Parameter eigenvalue extraction method

Monitoring parameters		Eigenvalue extraction	Purpose
Basic operating parameters	Working time of engine	Σ	Acquire basic operation parameters of equipment
	Mileage	Σ	
	Working time of the each gear	Σ	
	Equipment positioning	Latest positioning information	
Engine performance parameters	Rotational speed (n)	$T \times n/9550$	Calculate output power
	Torque (T)		
	Starting current (I)	$U \cdot I \cdot t$	Reflect cylinder tightness
	Starting voltage (U)		
	Fuel flow (q)	Σq	Reflect engine economic performance
	Pressure in cylinder (P)	$P_{\max}$	Judge the position of the oil supply advance angle
	Cylinder head vibration	Vibration severity	Monitoring supernormal vibration
	Exhaust gas temperature	Average value	Evaluate engine heat load

will extract the accumulated value to reflect engine economic performance; the maximum value of cylinder pressure extraction is used to judge whether the position of the oil supply advance angle is abnormal,the vibration signal of the cylinder head will be used to calculate the vibration intensity to monitor whether the engine is vibrating abnormally, and the average value of exhaust gas temperature will be calculated to evaluate engine heat load. The extraction method of each parameter eigenvalue is shown in Table 1.

4.2 Design of Monitoring Center Software

Monitoring center is based on the realization of equipment management departments to implement convenient and efficient remote monitoring and management of their equipment. The software mainly includes five modules: system login, equipment management, equipment health condition query, history record, and system exit. Among them, the equipment management module is mainly used to manage the basic information of their equipment, including the newly installed

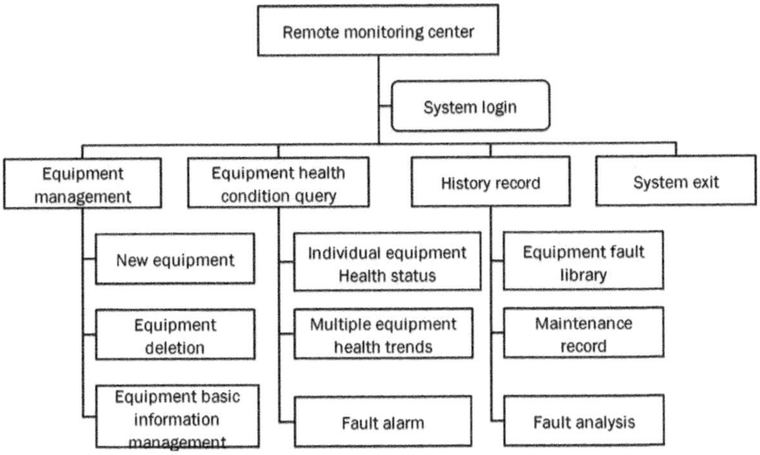

Fig. 4 Software design of monitoring center

equipment entry, the deletion of decommissioning equipment, and the management of basic equipment information. The equipment health condition query module supports individual equipment health status query, multiple equipment health situation query, and fault alarm functions when fault signs appear. The history record module includes equipment fault library, maintenance record, fault analysis, etc. It is convenient for the equipment management and R&D (research and development) department to analyze the parts with a high failure rate and provides a reference for the use or upgrading of equipment in the next step. The design of the monitoring center software structure is shown in Fig. 4.

5 Conclusion

This paper presents the design of armored vehicle remote monitoring system based on Beidou. The system adopts Beidou as positioning and communication mode to realize the remote monitoring of armored vehicles by the equipment management department; at the same time, it satisfies the army's requirements for confidentiality of equipment information. After testing, the system can timely and accurately grasp the health status and location information of the armored vehicles monitored. Using this, the system can help the equipment management department easily grasp the current number and quality of the vehicles, consumption, location and other information, and provide a great convenience for equipment management.

Taking Beidou short message as the communication mode in this paper, compared with other remote communication methods, it has the following advantages: (1) strong secrecy, because the Beidou system is completely developed by our country and operates independently, it can guarantee the information security fundamentally and prevent the communication content from being intercepted by the enemy under the conditions of war; (2) high stability, Beidou communication is equivalent to building a communication base station in space, the possibility of communication interruption caused by the destruction of the base station is lower than other communication methods under the conditions of war; (3) wide coverage area. Considering that military operations may take place anywhere, military communications must be accessible in any area. Because Beidou is satellite communication, satellite signals can still be covered even in deserted and uninhabited areas, which ensures that communication can be achieved anywhere. This is unparalleled in any other ways of remote communication.

It can be foreseen that with the further improvement of the Beidou system, this remote monitoring system will be able to play a greater role in the army equipment management.

References

1. China Satellite Navigation System Management Office. Report on the Construction and Development of Beidou Satellite Navigation System [EB/OL]. 23 May 2018. http://www.beidou.gov.cn/zt/dhnh/djzgwxdhxsnh/nhdt9/201805/t20180523_14729.html
2. Li, J.: Research and implementation of STM32F207 based industrial core board. Hunan University (2012)
3. Gao, X., Chen, F.: High-precision signal acquisition system based on ADS 1278. Electron. Technol. **44** (08), 47–48 + 43 (2015)
4. Wang, Y., Zhou, Y., Guo, W., et al.: Design of ARV recovery control system based on Beidou positioning communication. Meas. Control Technol. **37**(04), 120–124 + 134 (2018)

Development and Implementation of Small Industrial Robot Arm Based on Arduino

Ye Tian, Kun Zhang, Yuming Zhang, Fanghong Bi and Jun Yang

Abstract The control system of this manipulator consists of Arduino development board and STM32 development board. The main function of Arduino development board is to realize the control of the sensor on the arm. STM32 is mainly used to control the motion of the steering gear. The US-100 ultrasonic the position recognition module is responsible for sensing the distance information. The combination of the ultrasonic position module, the Arduino development board, the STM32 development board, and the mechanical arm enables the robot arm to realize the multi-degree of freedom of the robot arm by sensing the distance information of the grasped object. This small industrial robotic arm makes a meaningful exploration of the application and popularization of robotic arms in the production lines of small and medium-sized enterprises and small and medium-sized products.

Keywords Arduino · STM32 · Small mechanical arm · Ultrasonic ranging

1 Introduction

With the introduction of the "Made in China 2025" strategy, the industrial robot industry has developed rapidly. According to data released by the China Robot Industry Alliance [1], China has become the world's largest robot consumer market for many years [2]. However, due to the high cost and complicated operation of large industrial robot arms, it is difficult to promote in small and medium-sized enterprises. This small robotic arm uses the open-source hardware Arduino

Y. Tian · K. Zhang · Y. Zhang · F. Bi · J. Yang (✉)
School of Information Science and Engineering, Yunnan University, Kunming, China
e-mail: 646531201@qq.com

V. Jain et al. (eds.), *Recent Trends in Intelligent Computing, Communication and Devices*, Advances in Intelligent Systems and Computing 1031,
https://doi.org/10.1007/978-981-13-9406-5_113

development board and the STM32 development board as the control core [3]. Through the rich online packaging function library of the Arduino open-source platform [4], various modular professional functions can be connected according to the actual needs of users, reducing development costs. At the same time, save a lot of development time [5].

2 Overall Design of Small Industrial Robotic Arms

The system consists of Arduino UNO development board, STM32 development board, US-100 ultrasonic ranging module, small industrial steering gear, base, bracket, hand grip, and power supply [6]. The Arduino UNO development board is responsible for the control of the US-100 sensor, and the STM32 development board is responsible for controlling the six industrial steering gears of the robot arm [7]. The overall structure of the system is shown in Fig. 1.

2.1 Design of Small Industrial Robot Arm Control System

This small industrial six-degree-of-freedom robotic arm consists of the US-100 ultrasonic intelligent sensing module, the control module based on the Arduino control board, the robot arm, and the power module [8]. The control system is as follows: The intelligent robot arm uses the Arduino control board as the core controller to connect to the US-100 ultrasonic sensor module [9], STM32 control board, and steering gear through the DuPont line [10]. The control program developed by the Arduino IDE integrated development environment acts on the servo drive module. The steering gear drive module drives the various steering gears of the robot arm for an accurate motion to form a closed control system. Finally, through system debugging, accurate operation control of the robot arm with multiple degrees of freedom is realized.

The system uses ultrasonic sensors for distance measurement. When the user is at a certain distance from the sensor, the mechanical arm can be triggered to work. The US-100 ultrasonic ranging module has non-contact ranging function in the range of 2 cm–3 m. The module supply voltage is 5 V, the working current is 3.8 mA, it supports analog voltage output, and the operation is stable and reliable. This module according to the different application scenarios can be set to different range (maximum measuring distance of 1 and 3 m), respectively; when the range pin is floating, the range is 3 m. The US-100 converts the measured distance into an analog voltage output that is proportional to the measured distance [11].

Ranging works:

After the module is energized, the system is first to determine distance from the base of the foot, and to set a different range based on the input level. When the

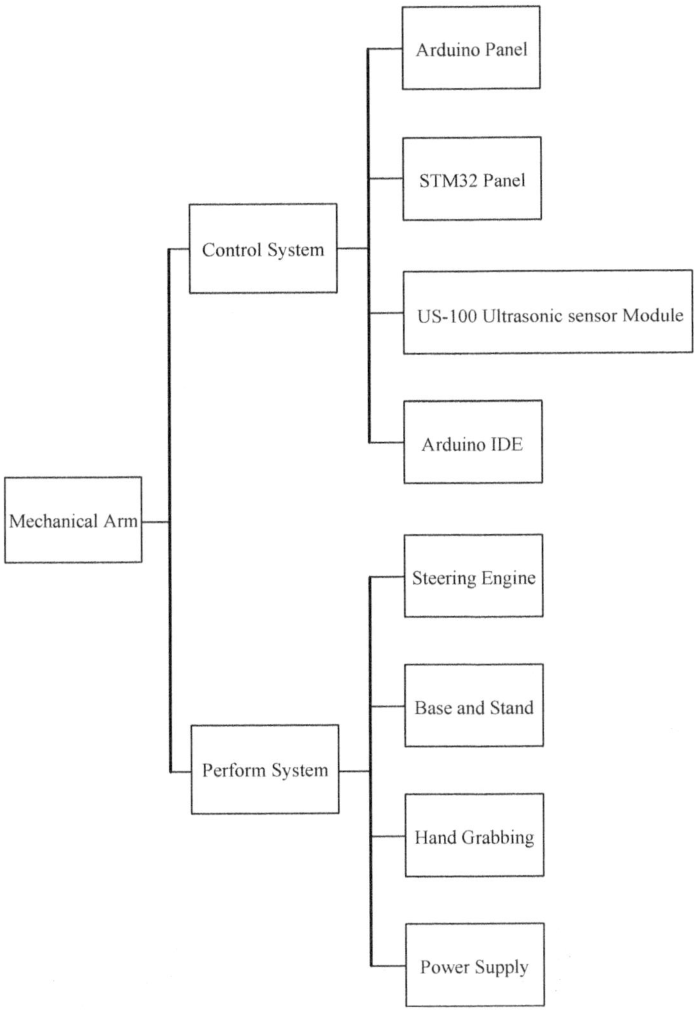

Fig. 1 Overall structure of the system

range leads to a higher power, the range is 3 m, and when the range is low voltage, the range is 1 m. Then, the system begins to measure the distance continuously, and at the same time, the result of the measurement is output in out pin through the analog voltage. When the distance changes, the analog voltage changes with it.

The analog voltage is proportional to the measured distance, and the output range of the analog voltage is 0–V_{cc} [12].

1. When the system measures 1 m, it measures the distance: $L = 1024 * V_{out}/V_{cc}$ (mm). When the output voltage is 0 V and the corresponding distance is 0 m, the output V_{cc} corresponds to 1.024 m.

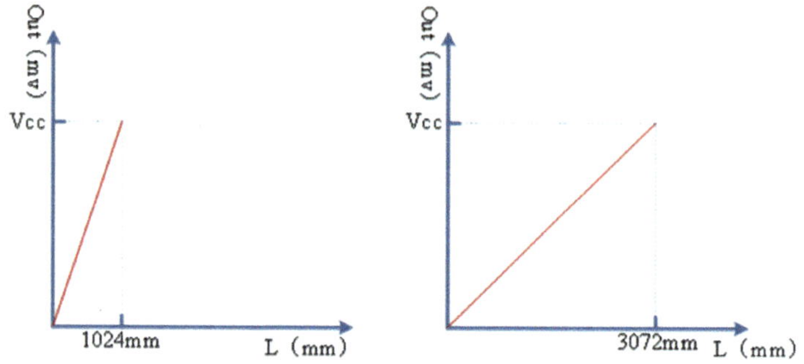

Fig. 2 Relationship between measured distance and output voltage

2. When the system measures 3 m, it measures the distance: $L = 3096 * V_{out}/V_{cc}$ (mm). When the output voltage is 0 V and the corresponding distance is 0 m, the output V_{cc} corresponds to 3.072 m [13].

The relationship between the measured distance and the output voltage is shown in Fig. 2.

2.2 *Steering Engine Control*

The main component of the design of the steering gear is the servo motor. The rotor stops before the signal comes in; the rotor moves immediately after the signal arrives. So we can input different signals to the steering gear to control its rotation to different angles [14].

The servo receives the PWM signal. When the signal enters the internal circuit, a bias voltage is generated, which triggers the motor to move the potentiometer through the reduction gear. When the voltage difference is zero, the motor stops, thereby achieving the servo effect. Simply put, the servo is given a specific PWM signal, and the servo can be rotated to the specified position.

There are three wires on the servo, which are GND, VCC and SIG, and the PWM input to the servo signal line.

3 System Implementation

This ultrasonic ranging sensing arm contains a 3000 mA power supply that supplies all the components needed for the STM32 servo control board, steering gear, and Arduino control panel. After the power is turned on, the ultrasonic distance sensor and the robot arm work as follows: The ultrasonic sensor receives different distance information and transmits the signal to the Arduino control panel [15]. When the received distance information is within the valid range, the control board passes the burning program. Judging the range of the detection distance, issuing different commands to the STM32 control panel according to the interval, the STM32 control board calls the corresponding action group stored therein, thereby controlling the different angles of the steering gears of different parts so that the mechanical arm makes various action. The motor drive module here uses PWM signal to control the speed and forward and reverse of the DC motor so that the software execution flow of the six servos in the six intelligent induction robot is shown in Fig. 3.

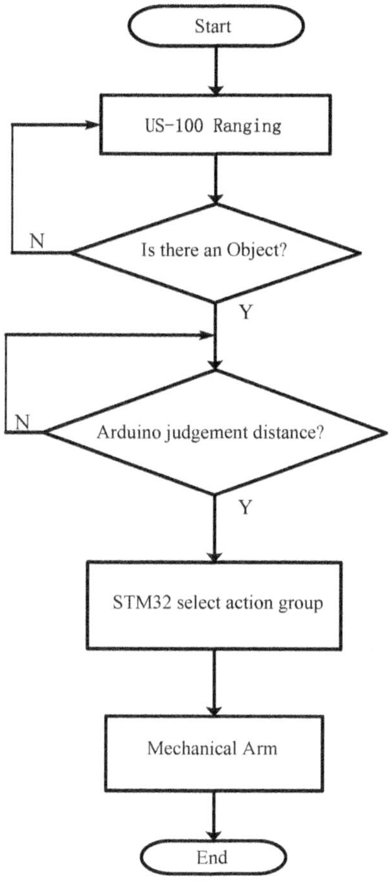

Fig. 3 Intelligent robot arm system implementation flowchart

4 Summary

This article is based on the Arduino control board, STM32 control board, and US-100 ultrasonic ranging module design of the small industrial robot arm hardware system. The program design is developed by Arduino IDE, which has the advantages of high modularity, low cost, and simple use. The US-100 ultrasonic ranging module used in this system has its own temperature compensation function, which solves the problem that the accuracy of the ranging accuracy of the ordinary ultrasonic ranging module is affected by temperature and can realize accurate ranging under different temperature conditions. In the later stage, various types of sensor modules can be replaced, which can realize various operations such as voice recognition and grabbing, voice control and grabbing and make useful explorations for the application of the robot arm in small and medium-sized enterprises.

References

1. Du, J., Zhang, X., Zhao, Y.: Design of wireless synchronous manipulator based on STM32. Dev. Innov. Mach. Electr. Prod. **28**(4), 40–42 (2015)
2. Ning, C., Liu, F., Sun, Z.: Design of multi-degree-of-freedom manipulator based on STM32. Electron. World (9), pp. 143–143 (2017)
3. D'Ausilio, A.: Arduino: a low-cost multipurpose lab equipment. Behav. Res. Methods **44**(2), 305–313 (2012)
4. Chen, C., Zhang, Z.: Application and development prospect of Arduino. Light Ind. Technol. (5) (2018)
5. Margolis, M.: Arduino Cookbook. Oreilly Vlg Gmbh & Co (2018)
6. Perez, I., Cuevas, J.Á.H., Galindo, J.T.E.: Design and construction of a desktop AC susceptometer using an Arduino and a Bluetooth for serial interface. Eur. J. Phys. **39**(3) (2018)
7. Li, H.: Key points for improvement of ultrasonic sensor technology. Electron. Technol. Softw. Eng. (12) (2018)
8. Zhang, H.: Design of intelligent car system based on wireless communication. Electron. Qual. **1**, 19–20 (2018)
9. Chen, J.: Design and implementation of sonar robot sonar ranging function based on Arduino technology. Comput. Knowl. Technol. (14) (2018)
10. Hong, J., Wu, Y., Ouyang, C., et al.: Based on STM32 Bluetooth ultrasonic ranging. Sci. Technol. **5**, 75 (2017)
11. Xiao, X., Liu, C., Yu, P.: Application research of ultrasonic sensor based on Arduino control board. Big Technol. (17) (2017)
12. Tan, Y., Shi, G., Wang, X. et al.: Design and implementation of digital steering gear control system based on STM32. Sci. Technol. Econ. Guide (15) (2018)
13. Yao, Q., Wang, Y.: Design of Multi-channel PWM Steering Controller Based on STC15F2K60S2. Softw. Guide (6) (2018)
14. Jiang, H., Liu, S., Zhang, B: Inverse kinematics analysis of a six-degree-of-freedom modular manipulator. J. Zhejiang Univ. (Engineering Science) **44**(7), 1348–1354 (2010)
15. Gan, Y., Dai, X.: Review of research on coordinated control of multiple manipulators. Control Decis. **28**(3), 321–333 (2013)

Study on Adaptive Cruise Internal Model Control Strategy of Truck

Jingjing Fan, Wenbo Chu and Li Wang

Abstract In order to resolve the difficulties of the truck, adaptive cruise control (ACC) technology, including how to extract radar targets and how to improve the control performance of the complex system. This paper presents a two-input/two-output time-delay system, idealizes to a first-order time-delay form, and an internal model PID (IMPID) control strategy is proposed. The simulation analysis and validation of the control strategy are carried out by Trucksim. The simulation results show that the IMPID control strategy can successfully realize the main functions and realize decoupling, the dynamic performance is excellent, and response is positive.

Keywords Adaptive cruise control · Internal model control · Trucksim

1 Introduction

The vehicle, especially large goods trucks, operates on the highway, and the adaptive cruise control (ACC) system continuously scans the front road of the vehicle through the car sensor (radar) installed in the front of the vehicle. When the distance from the front vehicle is small, the ACC control unit can properly brake the wheels and reduce the output power of the engine so that the truck stays at a safe distance from the vehicle in front. ACC system can effectively reduce the driver's operating load and improve the safety of the truck.

The current ACC control strategies are divided into the following categories: (1) rule-based control strategy; (2) the control strategy based on PID; (3) various intelligent control strategies (mainly including fuzzy control, neural network, ant

J. Fan (✉) · L. Wang
Beijing Key Lab of Urban Intelligent Traffic Control, North China University of Technology, Beijing 100144, China
e-mail: fjj419@126.com

W. Chu
China Intelligent and Connected Vehicles Research Institute Co., Ltd, Beijing 100176, China

V. Jain et al. (eds.), *Recent Trends in Intelligent Computing, Communication and Devices*, Advances in Intelligent Systems and Computing 1031,
https://doi.org/10.1007/978-981-13-9406-5_114

algorithm, etc.). Among them, the energy management strategy based on rules and PID is the most widely used because of the advantages of good real time and high reliability. Other control strategies based on some optimization theories must also be refined into improved rule-based control strategies to be downloaded to the real-time operation of the controller [1–6].

In this paper, based on the analysis of the characteristics of the vehicle adaptive cruise system and the establishment of a dual input-output system, a system model is established. Through decoupling and simplification of the system as a first-order delay, an internal model PID control strategy is derived, and the control parameters are optimized. In the trucksim environment, the simulation test results show that the internal model control strategy is effective and the vehicle operation is stable, and the response is timely.

2 Structure of ACC System

The ACC system studied in this paper is used in trucks, and the structure of the system is shown in Fig. 1. The input of the corresponding controller is the relative speed and relative distance. The actuator is the electronic throttle opening and the brake pressure and constitutes a typical double-input and double-output system (Table 1).

3 Internal Model Control Strategy

The whole vehicle is simplified as a generalized transfer function $G_p(s)$, which $G(s)$ is a process transfer function matrix and $K(s)$ is the transfer function matrix of the process decoupling, and $G_m(s)$ is a model of the generalized process.

The transfer function matrix of the whole decoupling internal model control system is:

Fig. 1 Structure of truck ACC system

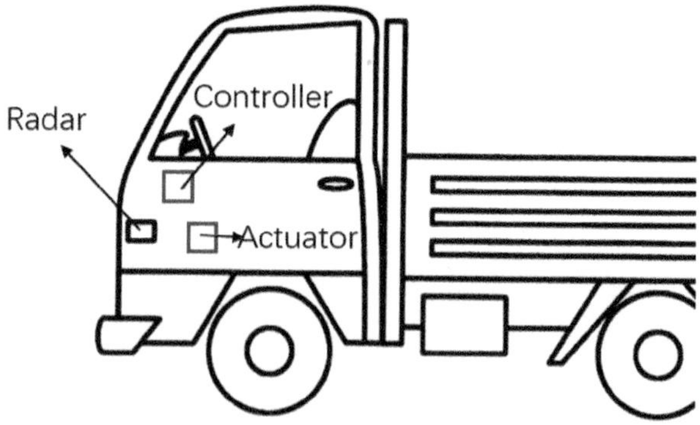

Table 1 Main parameters

Parameters	Values
Curb weight (m/kg)	5000
Truck power (p/kW)	175
Transmission	Seven-step manual
Driving form	Rear wheel drive
Wheel (r/m)	0.51

$$H(s) = G(s)K(s)Q(s)[I + (G(s)K(s) - G_m(s))Q(s)]^{-1}$$

Consider dual-input dual-output process:

$$G(s) = \begin{bmatrix} g_{11}(s) & g_{12}(s) \\ g_{21}(s) & g_{22}(s) \end{bmatrix} = \begin{bmatrix} \frac{k_{11}e^{-L_{11}s}}{T_{11}s+1} & \frac{k_{12}e^{-L_{12}s}}{T_{12}s+1} \\ \frac{k_{21}e^{-L_{21}s}}{T_{21}s+1} & \frac{k_{22}e^{-L_{22}s}}{T_{22}s+1} \end{bmatrix}$$

In the formula, $g_{ij}(s)$ is each channel transfer function; k_{ij}, T_{ij}, L_{ij} are the gain, time constant, and time delay of each channel.

$$K(s) = G^*(s) = \begin{bmatrix} \frac{k_{22}e^{-L_{22}s}}{T_{22}s+1} & -\frac{k_{12}e^{-L_{12}s}}{T_{12}s+1} \\ -\frac{k_{21}e^{-L_{21}s}}{T_{21}s+1} & \frac{k_{11}e^{-L_{11}s}}{T_{11}s+1} \end{bmatrix}$$

$$G_p(s) = G_m(s) = I \det G(s) = [A - B]\begin{bmatrix} 1 & 0 \\ 0 & 1 \end{bmatrix}$$

In the formula, $A = k_{11}k_{22}e^{-(L_{11}+L_{22})s}/[(T_{11}(s)+1)(T_{22}(s)+1)]$, $B = k_{12}k_{21}$ $e^{-(L_{12}+L_{21})s}/[(T_{12}(s)+1)(T_{21}(s)+1)]$.

The minimum time is multiplied by the absolute value of the error integral criterion to reduce the order of $\det(G(s))$, and the transfer function matrix of the feedback controller is obtained:

$$C(s) = Q(s)\left[I - G_p(s)Q(s)\right]^{-1} = c(s)I$$

$$c(s) = \frac{q(s)}{1 - g_p(s)q(s)} = \frac{Ts+1}{k(\alpha s + 1 - e^{-Ls})}$$

Into the form of the PID controller, so: $c(s) = T/[k(\alpha + L)] + 1/[k(\alpha + L)s]$,so:

$$K_p = T/[k(\alpha + L)]$$

In the formula, K_p is the proportional gain coefficient of PID control.

Thus, get the control results of the accelerator pedal and brake pedal:

$$\alpha = \begin{cases} (V_{x\_set} - V_x)p_\alpha & D > D_{set} \\ 0 & D \leq D_{set} \end{cases}$$

$$\beta = (D_{set} - D)p_\beta$$

The radar signal is transformed according to the distance of *X*- and *Y*-direction, and the straight line distance and angle are calculated.

$$D = \sqrt{\Delta x^2 + \Delta y^2}$$

$$\delta = \arctan(\Delta x, \Delta y)$$

4 Test and Result Analysis Based on Trucksim

Based on the Trucksim developed truck system and test conditions, the test conditions in Fig. 2 (Fig. 3).

The test trajectory is shown as follows:

Figure 4 shows the speed following condition, as can be seen from the diagram that the speed can follow the front car well, and not too aggressive, in the corners can also be good detection of the front car to avoid a collision.

Fig. 2 Test conditions include the two trucks

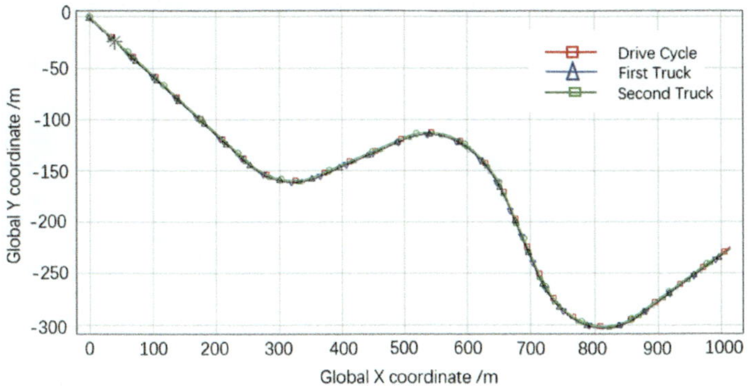

Fig. 3 Test trajectory

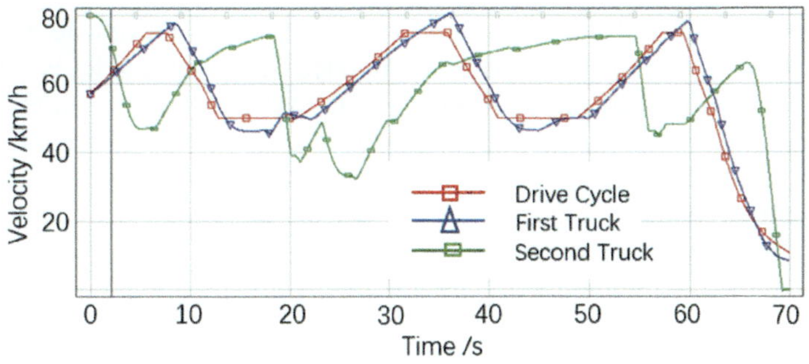

Fig. 4 Speed following condition

Figures 5 and 6 show the control results of throttle opening and brake pressure, which can be seen in the effective control decoupling, and the change of the control amount is relatively smooth, can make the vehicle control process smooth, and achieve better fuel economy.

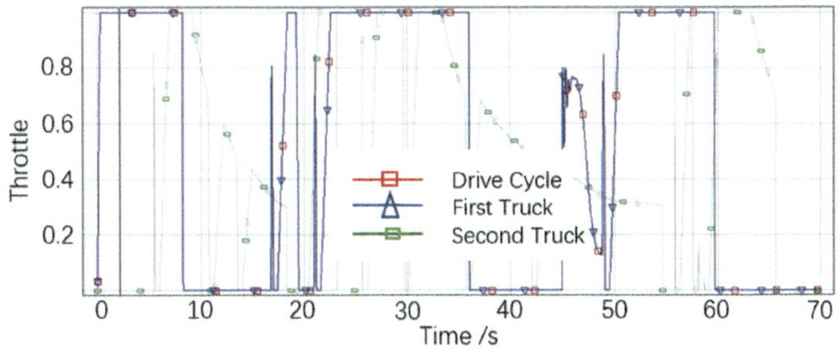

Fig. 5 Control result of throttle opening

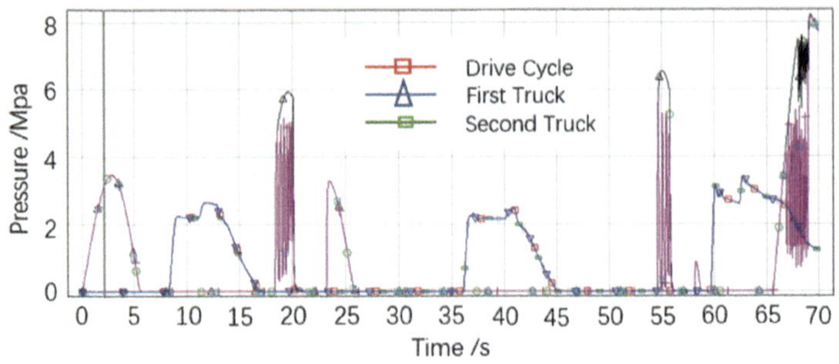

Fig. 6 Control result of brake pressure

5 Conclusion

In this paper, based on the analysis of the characteristics of the vehicle ACC system and the establishment of dual input-output system, a system model is established. Through decoupling and simplification of the system as a first-order delay, an internal model PID control strategy is derived, and the control parameters are optimized. In the Trucksim environment, the simulation test is carried out under the typical running condition of the vehicle under the target cycle condition, and the following conclusions are obtained:

(1) By analyzing the characteristics of radar signal and the characteristics of the system, a system model of double input and dual output is established, and the system decoupling is carried out. By simplifying the calculation as a first-order delay, an internal model PID control strategy is obtained.
(2) The test platform and test conditions are developed based on Trucksim, and the fast optimization and verification of the control strategy can be realized. The internal model control strategy is validated by the platform. The results show that the control strategy can achieve better control function and the control quantity decoupling is effective, and the control quantity is smooth, which is conducive to the stability of vehicle control and the improvement of fuel economy.

Acknowledgements This research was supported by the Beijing New Star Project Interdisciplinary Science and Technology (XXJC201709).

References

1. Mun, H., Kim, B.: A study on the V2V-communication-based AEB system for high-speed driving under various road surface conditions. Springer Berlin Heidelberg **354**, 247–252 (2016)
2. Lin, M.,Yoon, Y., Kim, B.: A study and analysis for calculating the brake-application time of AEB systems considering the gradient. Int. J. Control Automation **8** (2015)

3. Zhang, D., Li, K., Wang, J.: A curving ACC system with coordination control of longitudinal car-following and lateral stability. Vehicle Syst. Dyn. **50**(7), 1085–1102 (2012)
4. Luo, Y., Cao, K., Da, Yi, Chu, W., Li, K.: A novel hierarchical global chassis control system for distributed electric vehicles. SAE Int. J. Passenger Cars Electron. Electri. Syst. **7**(2), 313–327 (2014)
5. Li, K., Chen, T., Luo, Y., Wang, J.: Intelligent environment-friendly vehicles: concept and case studies. IEEE Trans. Intell. Transp. Syst. **13**(1), 318–328 (2012)
6. Luo, Y., Zhu, T., Wan, S., Zhang, S., Li, K.: Optimal charging scheduling for large-scale EV (electric vehicle) deployment based on the interaction of the smart-grid and intelligent-transport systems. Energy **97**, 359–368 (2016)
7. Luo, Y., Chen, T., Zhang, S., Li, K.: Intelligent hybrid electric vehicle ACC with coordinated control of tracking ability, fuel economy, and ride comfort. IEEE Trans. Intell. Transp. Syst. **16**(4), 2303–2308 (2015)

Online Method for Measuring the Air Preheater Leakage Rate of Metallurgical Gas Boiler

Yalan Ye, Hongming Wang, Xiang An and Wenhao Jiang

Abstract In order to make full use of surplus gas, metallurgical gas boilers have been widely applied in steel mills, especially in China, India, and other Asian countries. The leakage of air preheater is one of the common faults of metallurgical gas boiler. It is helpful for the steady operation of gas boiler to grasp the air preheater leakage rate in real time. However, the current methods for measuring the air leakage rate cannot be used in online monitoring for metallurgical gas boiler. According to the characteristics of metallurgical gas and the definition of air leakage rate, this paper deduced an online method for measuring the air preheater leakage rate, which is suitable for metallurgical gas boiler and has good practicability. Finally, the online model was verified by experimental data. The results show that the proposed model has high accuracy and can meet the engineering requirement.

Keywords Air preheater · Leakage rate · Metallurgical gas · Boiler · Online measuring

1 Introduction

Different kinds of by-product metallurgical gases are produced in the smelting process of iron and steel enterprises, including blast furnace gas, converter gas, and coke oven gas. Normally, the steel production process cannot completely consume all the gas. Many steel mills use gas boiler to balance the residual gas, especially in Asian countries such as China, India, and Indonesia. The types of gas boiler include pure blast furnace gas boiler, pure converter gas boiler, pure coke oven gas boiler, and multifuel gas boiler [1–8].

Y. Ye (✉) · H. Wang · X. An
Jiangsu Maritime Institute, Nanjing 211170, China
e-mail: yyl_jessica@sina.com

W. Jiang
MCC Huatian Engineering & Technology Corporation, Nanjing 210000, China

V. Jain et al. (eds.), *Recent Trends in Intelligent Computing, Communication and Devices*, Advances in Intelligent Systems and Computing 1031,
https://doi.org/10.1007/978-981-13-9406-5_115

Air preheater is one of the main heating surfaces in the tail of boiler, and its running condition will directly affect the stable operation of the boiler. For gas boiler, especially for the boiler with high sulfur content, the air preheater is prone to leak, which will undoubtedly bring adverse effects on the safety and economic operation of boiler. Therefore, it is necessary to monitor the air preheater leakage rate online in order to grasp the air preheater running condition in real time.

Currently, the air preheater leakage rate is mainly measured and calculated according to China's national standard *Performance test code for utility boiler* (standard number: GB/T 10184) [9] and American's national standard ASME PTC 4.3 *Air Heaters* [10]. The methods include accurate method and rough method. However, the accurate measuring method needs complete gas composition data, but most steel plants cannot provide gas composition online for gas boiler at present. And the rough estimate method is only suitable for coal-fired boiler, but not suitable for metallurgical gas boiler. Therefore, the current methods cannot be directly used in online measurement of the air preheater leakage rate of metallurgical gas boiler.

Based on the definition of air preheater leakage rate and combined with the fuel characteristics of metallurgical gas, an online model for measuring and calculating the air preheater leakage rate was presented in this paper, which is suitable for metallurgical gas boiler. This model can make online analysis on the air preheater leakage rate in metallurgical gas boiler, and it has certain practical significance.

2 Definition of Air Preheater Leakage Rate

The air preheater leakage rate is the ratio of air mass leaking into the flue gas side to total flue gas mass at the flue inlet, which can be calculated as follows:

$$A_L = \frac{\Delta m_a}{m'_{fg}} \times 100 \tag{1}$$

where A_L is the air leakage rate, Δm_a is the air mass leaking into the flue gas side of air preheater, and m'_{fg} is the total flue gas mass at the flue inlet of air preheater.

3 Current Method for Measuring Air Preheater Leakage Rate

3.1 Measuring and Calculating Model

3.1.1 Accurate Model

The air mass leaking into the flue gas side of air preheater (Δm_a) is equal to the difference between the flue gas mass at the inlet and outlet of air preheater. Therefore, formula (1) can be rewritten as follows:

$$A_L = \frac{m''_{fg} - m'_{fg}}{m'_{fg}} \times 100 \tag{2}$$

where m''_{fg} is the total flue gas mass at the outlet of air preheater.

Based on the above definition, the following formula was provided in China's GB/T 10184-2015 *Performance test code for utility boiler*:

$$A_L = \frac{(\alpha'' - \alpha')V_{a.\text{th}}\rho_a}{\alpha' V_{a.\text{th}}\rho_\text{a} + (V_{fg.\text{th}}\rho_{fg} - V_{a.\text{th}}\rho_a)} \times 100 \tag{3}$$

where α' and α'' are the excess air coefficients at the inlet and outlet of air preheater, respectively. $V_{a.\text{th}}$ is the theoretical air volume needed for combustion of unit volume gas. $V_{fg.\text{th}}$ is the theoretical flue gas volume produced by combustion of unit volume gas. ρ_a is the air density.

The model in American's ASME PTC 4.3 *Air Heaters* is not exactly the same as China's GB/T 10184-2015, but is similar. Limited to space, this article does not describe the ASME model in detail.

3.1.2 Rough Model

In GB/T 10184-2015, a rough model for estimating the air leakage rate of air preheater is put forward as follows:

$$A_L = \frac{\alpha'' - \alpha'}{\alpha'} \times 90 \tag{4}$$

Obviously, the above rough model is much simpler than the accurate model.

3.2 *Difficulties in Applying the Current Model to Online Measuring*

3.2.1 Difficulties in Applying the Accurate Model to Online Measuring

As can be seen from formula (3), if accurate measuring method is applied, it is necessary to measure the excess air coefficient at the inlet of air preheater, the excess air coefficient at the outlet of air preheater, as well as the air density. Besides, the theoretical air volume and the theoretical flue gas volume corresponding to unit volume gas should be calculated through the fuel composition. Thus, the following data need to be collected:

- Gas composition;
- Atmospheric pressure, atmospheric temperature, and atmospheric humidity;
- Flue gas composition at the inlet of air preheater and flue gas composition at the outlet of air preheater.

The obtainment of the above parameters is feasible through experimental sampling and analysis, but it is difficult to realize online measurement. Due to the limitation of field conditions, most steel plants cannot provide gas composition analyzer for gas boiler. Therefore, the accurate calculation method cannot be directly used to measure the air preheater leakage rate.

3.2.2 Difficulties in Applying the Rough Model to Online Measuring

The conversion coefficient 90 in formula (4) is an empirical value, which is the average result calculated according to the characteristics of most kinds of coal. It is suitable for coal-fired boiler, and the result has high accuracy. However, the conversion coefficient 90 is not applicable to metallurgical gas boiler for the following reasons:

On the one hand, the composition of metallurgical gas is different from that of coal, and the corresponding conversion coefficient cannot be taken as the coefficient of coal.

On the other hand, the coefficient of different metallurgical gases is very different. After calculation, we can get the following conclusion: For blast furnace gas boiler, the coefficient is between 38 and 48; for converter gas boiler, the coefficient is between 58 and 66; and for coke oven gas boiler, the coefficient is between 95 and 98.

Thus, the rough model shown in formula (4) cannot be used in metallurgical gas boiler.

4 Online Calculation Model for Air Preheater Leakage Rate of Metallurgical Gas Boiler

4.1 New Calculation Formula for Air Preheater Leakage Rate

In order to solve the problem that the current measuring method is not suitable for online measurement of air preheater leakage rate in metallurgical gas boiler, a new calculation model is derived from the definition of air preheater leakage rate, which is easy to simplify and is introduced thoroughly in the following text.

4.1.1 Formula for the Air Mass Leaking into the Flue Gas Side of Air Preheater

In the air leakage rate definition formula (1), the air mass leaking into the flue gas side of air preheater Δm_a can be calculated as follows:

$$\Delta m_a = 1.293(1 + h_{a.ab})(\alpha'' - \alpha')V_{a.d.\mathrm{th}} \tag{5}$$

where $h_{a.ab}$ is the absolute humidity of air. $V_{a.d.\mathrm{th}}$ is the theoretical dry air volume needed for combustion of unit volume gas.

The excess air coefficient can be calculated by the following formula:

$$\alpha = \frac{21 + \lambda \varphi_{\mathrm{fg}}(\mathrm{O}_2)}{21 - \varphi_{\mathrm{fg}}(\mathrm{O}_2)} \tag{6}$$

where $\varphi_{fg}(\mathrm{O}_2)$ is the volume content of O_2 in dry flue gas. λ is the dimensionless factor related to fuel composition. It can be calculated as follows:

$$\lambda = \frac{V_{fg.d.\mathrm{th}} - V_{a.d.\mathrm{th}}}{V_{a.d.\mathrm{th}}} \tag{7}$$

where $V_{\mathrm{fg.d.th}}$ is the theoretical dry flue gas volume generated by combustion of unit volume gas.

Theoretical dry air volume $V_{a.d.\mathrm{th}}$ and theoretical dry flue gas volume $V_{fg.d.\mathrm{th}}$ can be calculated according to formula (8) and formula (9), respectively:

$$V_{a.d.\mathrm{th}} = \frac{1}{21}\left[0.5\,\varphi(\mathrm{H}_2) + 0.5\,\varphi(\mathrm{CO}) + \sum\left(m + \frac{n}{4}\right)\varphi(\mathrm{C}_m\mathrm{H}_n) + 1.5\varphi(\mathrm{H}_2\mathrm{S}) - \varphi(\mathrm{O}_2)\right] \tag{8}$$

$$V_{fg.d.\mathrm{th}} = \frac{\varphi(\mathrm{CO}_2) + \varphi(\mathrm{CO}) + \sum m\varphi(\mathrm{C}_m\mathrm{H}_n) + \varphi(\mathrm{H}_2\mathrm{S})}{100} + 0.79V_{a.d.\mathrm{th}} + \frac{\varphi(\mathrm{N}_2)}{100} \tag{9}$$

where $\varphi(\mathrm{CO})$, $\varphi(\mathrm{H}_2)$, $\varphi(\mathrm{C}_m\mathrm{H}_n)$, $\varphi(\mathrm{H}_2\mathrm{S})$, $\varphi(\mathrm{CO}_2)$, $\varphi(\mathrm{N}_2)$, and $\varphi(\mathrm{O}_2)$ are volume fractions of CO, H_2, $\mathrm{C}_m\mathrm{H}_n$, $\mathrm{H}_2\mathrm{S}$, CO_2, N_2, and O_2 in gas, respectively.

4.1.2 Formula for Flue Gas Mass at the Inlet of Air Preheater

Flue gas mass at the inlet of air preheater m'_{fg} can be calculated by the following formula:

$$m'_{fg} = \rho_{gas} + h_{gas} + 1.293(1 + h_{a.ab})\alpha' V_{a.d.\mathrm{th}} \tag{10}$$

where h_{gas} is the moisture content of gas and ρ_{gas} is the dry gas density that can be calculated as follows:

$$\begin{aligned}\rho_{gas} = {} & 0.0125\varphi(\mathrm{CO}) + 0.0009\varphi(\mathrm{H_2}) + \sum(0.0054m + 0.00045n)\varphi(\mathrm{C}_m\mathrm{H}_n) \\ & + 0.0152\varphi(\mathrm{H_2S}) + 0.0196\varphi(\mathrm{CO_2}) + 0.0125\varphi(\mathrm{N_2}) + 0.0143\varphi(\mathrm{O_2})\end{aligned} \tag{11}$$

4.1.3 New Calculation Formula for the Air Preheater Leakage Rate

By combining new formulas for air mass and flue gas mass as above, formula (1) can evolve into:

$$A_L = \frac{\Delta m_a}{m'_{fg}} \times 100 = \frac{1.293(1 + h_{a.ab})(\alpha'' - \alpha')V_{a.d.\mathrm{th}}}{\rho_{gas} + h_{gas} + 1.293(1 + h_{a.ab})\alpha' V_{a.d.\mathrm{th}}} \times 100 \tag{12}$$

Further evolution of formula (12) can be obtained as follows:

$$A_L = \frac{\alpha'' - \alpha'}{\alpha' + \frac{\rho_{gas} + h_{gas}}{1.293(1 + h_{a.ab})V_{a.d.\mathrm{th}}}} \times 100 \tag{13}$$

We make the following setting:

$$b = \frac{\rho_{gas} + h_{gas}}{1.293(1 + h_{a.ab})V_{a.d.\mathrm{th}}} \tag{14}$$

Then, formula (13) can be changed as follows:

$$A_\mathrm{L} = \frac{\alpha'' - \alpha'}{\alpha' + b} \times 100 \tag{15}$$

This is a new calculation formula for the air preheater leakage rate in metallurgical gas boiler.

4.2 *Simplified Processing of the New Calculation Formula*

It can be seen from formula (15) that the obtainment of excess air coefficient α (including α' and α'') and parameter b is the key link to calculate the air preheater leakage rate. The excess air coefficient α can be obtained according to formula (6) by the volume content of O_2 in dry flue gas $\varphi_{fg}(O_2)$, theoretical dry flue gas volume $V_{fg.d.\mathrm{th}}$, and theoretical dry air volume $V_{a.d.\mathrm{th}}$. The parameter b can be obtained according to formula (14) through the gas density ρ_{gas}, gas moisture content h_{gas}, air absolute humidity $h_{a.ab}$, and theoretical dry air volume $V_{a.d.\mathrm{th}}$. Therefore, there are six key variables to solve the air preheater leakage rate, and the specific processing methods are as follows:

- The volume content of O_2 in dry flue gas $\varphi_{fg}(O_2)$

  It can be obtained by online measurement of oxygen meter.

- The theoretical dry flue gas volume $V_{fg.d.\mathrm{th}}$

  It can be calculated by online measurement of gas calorific value $Q_{\mathrm{net}.d}$:
  For blast furnace gas, $V_{fg.d.\mathrm{th}} = 1.470 \times 10^{-4} Q_{\mathrm{net}.d} + 1$.
  For converter gas, $V_{fg.d.\mathrm{th}} = 1.449 \times 10^{-4} Q_{\mathrm{net}.d} + 1$.
  For coke oven gas, $V_{fg.d.\mathrm{th}} = 1.607 \times 10^{-4} Q_{\mathrm{net}.d} + 1$.

- The theoretical dry air volume $V_{a.d.\mathrm{th}}$

  It also can be calculated by online measurement of gas calorific value $Q_{\mathrm{net}.d}$:
  For blast furnace gas, $V_{fg.d.\mathrm{th}} = 1.955 \times 10^{-4} Q_{\mathrm{net}.d}$.
  For converter gas, $V_{fg.d.\mathrm{th}} = 1.858 \times 10^{-4} Q_{\mathrm{net}.d}$.
  For coke oven gas, $V_{fg.d.\mathrm{th}} = 1.107 \times 10^{-4} Q_{\mathrm{net}.d} + 2.381$.

- The gas density ρ_{gas}

  For blast furnace gas, it can be simplified to a fixed value 1.359 kg/m^3.
  For converter gas, it can be simplified to a fixed value 1.367 kg/m^3.
  For coke oven gas, it can be simplified to a fixed value 0.468 kg/m^3.

- The gas moisture content h_{gas}

  It can be considered as saturated gaseous water vapor and can be simplified as 0.05 kg/m^3.

- The air absolute humidity $h_{a.ab}$

  It can be simplified to a fixed value 0.01 kg/kg.

5 Online Measuring Specific Scheme

5.1 Measurement Parameters Required for Online Measuring of the Air Leakage Rate

It can be seen from the previous section that the simplified calculation model of air leakage rate is greatly simplified and the measurement parameters required are greatly reduced. Ultimately, the required measurement parameters are as follows:

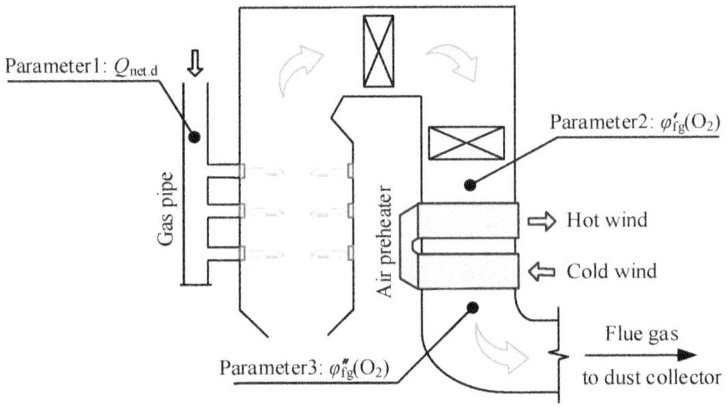

Fig. 1 Layout diagram of measurement parameters in metallurgical gas boiler

- Parameter 1: Gas calorific value $Q_{\text{net}.d}$,
- Parameter 2: The volume content of O_2 in dry flue gas at the inlet of air preheater $\varphi'_{fg}(O_2)$, and
- Parameter 3: The volume content of O_2 in dry flue gas at the outlet of air preheater $\varphi''_{fg}(O_2)$.

The measuring locations of the above measurement parameters in metallurgical gas boiler are shown in Fig. 1.

5.2 *Online Measuring Steps*

Combined with the simplified treatment, the air preheater leakage rate in metallurgical gas boiler can be measured online according to the following steps:

Step 1: Obtain the oxygen content in flue gas at the inlet of air preheater $\varphi'_{fg}(O_2)$, oxygen content in flue gas at the outlet of air preheater $\varphi''_{fg}(O_2)$, and lower calorific value of gas $Q_{\text{net}.d}$ through online measurement.

Step 2: According to the simplified treatment formula provided in the previous section, calculate $V_{fg.d.\text{th}}$ and $V_{a.d.\text{th}}$ through $Q_{\text{net}.d}$.

Step 3: Calculate the dimensionless factor λ according to formula (7) through $V_{fg.d.\text{th}}$ and $V_{a.d.\text{th}}$.

Step 4: Calculate the excess air coefficient at the inlet of air preheater α' according to formula (6) through $\varphi'_{fg}(O_2)$ and λ. Calculate the excess air coefficient at the outlet of air preheater α'' according to formula (6) through $\varphi''_{fg}(O_2)$ and λ.

Step 5: Based on the simplified values in the previous section, calculate the parameter b according to formula (14).

Step 6: Calculate the air preheater leakage rate A_L according to formula (15).

6 Model Verification

In order to verify the reliability of the online measuring model proposed in this paper, the test data of blast furnace gas boiler in A steel plant and coke oven gas boiler in B steel plant are taken as the samples. The air preheater leakage rate is calculated according to the accurate method and the online method, respectively. The original data are given in Table 1, and the calculated results are given in Table 2.

It can be seen from Table 2 that under all operating conditions, the maximum and minimum errors of the proposed online model for measuring air preheater leakage rate were 0.65 and 0.29%, respectively. And the mean measuring error was only 1.38%. Therefore, the proposed online model is reasonable and feasible, which can meet the requirements of engineering.

Table 1 Original data

Parameter	Blast furnace gas boiler		Coke oven gas boiler	
	Operating condition 1	Operating condition 2	Operating condition 1	Operating condition 2
Gas composition $\varphi(H_2)$/%	3.0	1.6	58.6	57.1
Gas composition $\varphi(CO)$/%	21.6	23.5	7.2	8.5
Gas composition $\varphi(CH_4)$/%	0.4	0.4	26.2	25.1
Gas composition $\varphi(C_mH_n)$/%	0.0	0.0	2.0	3.3
Gas composition $\varphi(N_2)$/%	53.2	56.2	2.2	2.4
Gas composition $\varphi(CO_2)$/%	21.8	18.3	3.6	3.1
Gas composition $\varphi(O_2)$/%	0.0	0.0	0.2	0.5
Gas calorific value $Q_{net.d}$/kJ m^{-3}	3197	3286	17,832	18,213
Gas moisture content h_{gas}/kg m^{-3}	0.05	0.04	0.05	0.06
Air absolute humidity $h_{a.ab}$/kg kg^{-1}	0.013	0.011	0.092	0.095
O_2 content in dry flue gas at the inlet of air preheater $\varphi'_{fg}(O_2)$/%	2.75	2.83	3.65	3.87
O_2 content in dry flue gas at the outlet of air preheater $\varphi''_{fg}(O_2)$/%	3.18	3.29	4.17	4.42

Table 2 Calculated results of the air leakage rate

Parameter	Blast furnace gas boiler		Coke oven gas boiler	
	Operating condition 1	Operating condition 2	Operating condition 1	Operating condition 2
Accurate measuring result of A_L/%	2.108	2.303	2.620	2.822
Online measuring result of A_L/%	2.114	2.306	2.603	2.813
Relative error of online measurement/%	0.29	0.14	0.65	0.30

7 Conclusions

Air leakage is one of the common faults of air preheater, which will bring adverse effect to not only safety but also economic performance of the metallurgical gas boiler. Therefore, it is necessary to monitor the air preheater leakage rate online, in order to grasp the air preheater leakage condition in real time.

In the current methods for measuring the air preheater leakage rate, the accurate method requires complete gas composition analysis, but almost all gas boilers are not equipped with gas composition online analyzer at present. In addition, the rough method is not suitable for metallurgical gas boiler. Therefore, the existing methods cannot be directly used to measure the air preheater leakage rate of metallurgical gas boiler online.

In this paper, an air leakage rate online measuring method for metallurgical gas boiler was proposed, which was proved feasible for the online analysis on air preheater leakage rate of metallurgical gas boiler. The new method has certain practical value.

Acknowledgements This work was supported by the Natural Science Foundation of the Jiangsu Higher Education Institutions of China (Grant No. 17KJD470001) and sponsored by Qing Lan Project of the Jiangsu Higher Education Institutions of China.

References

1. Chen, S.L., Sun, T.F.: Transformation of the 220 t/h boiler from burning coking gas to BF gas at Nanjing Steel. Metallur. Power. **12**, 19–21 (2014)
2. Tao, H.W., Fu, N.W.: Application of 400 t/h superpressure gas-fired boiler in Ningbo Iron & Steel Co., Ltd. Metallur. Power. **6**, 53–56 (2012)
3. Wang, H.J.: The practice of all combustion converter gas in a 35t/h boiler. Metallur. Power. **34**(1), 42–46 (2015)
4. Yu, Q.S.: Design of a 75 t/h gas boiler. Jiangxi Energy. **2**, 68–70 (2012)
5. Wang, H.T., Cheng, Y.L., Yang, T.Z.: Research of 220 t/h high temperature and pressure plant boiler with combusting pure BFG. Energy Conservation Technol. **23**(5), 422–425 (2005)
6. Ren, X.H., Fan, Y.J., Lu, S.Q.: The application of coke oven gas in power boiler. Energy Conserv. Environ. Prot. **6**, 20–21 (2001)
7. Zhang, H.Q., Wang, H.J., Hu, L.Y.: Practice of 75 t/h boiler burning blast furnace gas and converter gas. Energy Conserv. **10**, 50–52 (2010)
8. Liu, T., Liu, Y., Pang, S.: Combustion characteristic study of a 170t/h three gases co-fired boiler. Energy for Metallur. Ind. **34**(1), 42–46 (2015)
9. General Administration of Quality Supervision, Inspection and Quarantine of the People's Republic of China: GB/T 10184-2015 Performance test code for utility boiler. China Standards Press (2016)
10. The American Society of Mechanical Engineers: ASME PTC 4.3-1968 (Supplement to Performance Test Code for Steam Generating Units, PTC4.1) (1968)

A Design of Cyber-Physical System Architecture for Smart City

Xinghua Xia, Changxiao Liu, Hongcheng Wang and Zhonghua Han

Abstract To promote the development of smart city, cyber-physical system (CPS) is intended to be used in modern smart city information management system. As unity of computation and physical process, CPS can connect the virtual space and the physical world to create a real network interconnection world of physical objects. In this paper, we study architecture design for CPS at first and emphasize the service function for information management system of smart city. According to service-oriented architecture framework of CPS, we present a physical architecture model of CPS for smart city. There are five layers in the architecture framework, i.e., perceptual connection layer, network layer, convergence layer, cognition layer, and configuration layer. Moreover, comparative studies are discussed.

Keywords Smart city · Cyber-physical system (CPS) · Architecture design · Service-oriented

1 Introduction

Since the first concept of smart city was proposed by IBM in 2008, constructions of smart city have been received more and more attention of the government departments at all levels. More than 250 cities have been put forward the construction of smart city in China, which have been supported by local government departments [1, 2]. Initially, smart city is put forward by IBM. According to its definition, smart city should sense, analyze, and integrate the key information of core systems in running cities. Rapid, reliable, and intelligent response to all kinds of needs during city management can be provided which is dependent on its intelligent services.

Intelligent services of smart city rely on perception layer of the Internet of things, which can provide more information sources, more complex data types, and more

X. Xia (✉) · C. Liu · H. Wang · Z. Han
Shenyang Jianzhu University, Shenyang 110168, China
e-mail: xxh8787@sjzu.edu.cn

V. Jain et al. (eds.), *Recent Trends in Intelligent Computing, Communication and Devices*, Advances in Intelligent Systems and Computing 1031,
https://doi.org/10.1007/978-981-13-9406-5_116

massive data volume. Thus, it needs to construct a unified data access and pooling mechanism. Cyber-physical System (CPS) is another challenging proposition after cloud computing and the Internet of things. It can be seen as an extension of the Internet of things and combined with the core technologies of cloud computing.

During designing architecture of smart city, there are many surveys and suggestions presented. In general, there are four perspectives of architectures about smart city, i.e., architectural layers [3, 4], service-oriented architecture [5], event-driven architecture [6, 7], and Internet of things (IoT) [8, 9]. Moreover, there are also perspectives of combined architectures, which manage to integrate characteristics and technologies.

Architectural designing is also the basic element among CPS researches, since the function, performance, compatibility, and flexibility of CPS are all embodied in the architectures [10]. Wang et al. proposed a service-oriented architecture framework of CPS [11]. Chen et al. analyzed the control and features of CPS and presented a three-layer architecture frame [12]. Lee proposed unified five-level architecture for Industry 4.0-based manufacturing systems [13]. Wan et al. presented a novel architecture for integrating vehicular cyber-physical systems and mobile cloud computing [14]. Recent years, several researches have focused on smart city architecture design [15, 16].

It can be seen that CPS is a deep embedded real-time system, whose architecture is very complex. The main characteristics of CPS include virtual fusion, high degree of autonomy, heterogeneity, real time, intelligent computation, security, flexibility, and predictability. With these advantages, problems raised by rapid urban development would be solved effectively by advanced technologies of smart city combined with CPS.

2 CPS Architecture Design for Smart City

2.1 CPS Architecture

CPS is a complex integration of computation with physical process. Massive embedded computers existed in information world should provide the ability of controlling the physical process in real time. While, cybernetic process will affect computation with feedback loops. These newborn properties are not presented in existed computing and networking abstractions yet. Generally, CPS architecture should possess two characteristics:

(1) Flexibility: CPS architecture should support high-level flexibility. Various CPS components in the system can be added or removed freely. In order to adapt varied operation conditions and applications, the architecture of the system should be self-organized and optimized dynamically.
(2) Heterogeneous: CPS architecture should support various quality of services (QoS) requirements throughout every level of CPS. The heterogeneous is

embodied various components existed in CPS, i.e., different types of physical entities, various computing modules, different technologies of networks, varied operating system, and diverse system models.

3 A CPS Architecture Design for Smart City

Smart city information service system is a complex system which is composed of various urban components, massive volume data, and heterogeneous processing platform and diversified business systems. With unified infrastructure and loosely coupled service structure, a globally unified information service platform can be constructed, which can realize information sharing and system interconnection. Thus, the performance of reutilization efficiency and interoperability of task, service, and data can be improved.

In our work, we focus on the service function of smart city and present a design of CPS architecture framework for smart city which combined with service-oriented architecture as shown in Fig. 1.

There are five layers in our proposed architecture framework, i.e., perceptual connection layer, network layer, convergence layer, cognition layer, and configuration layer.

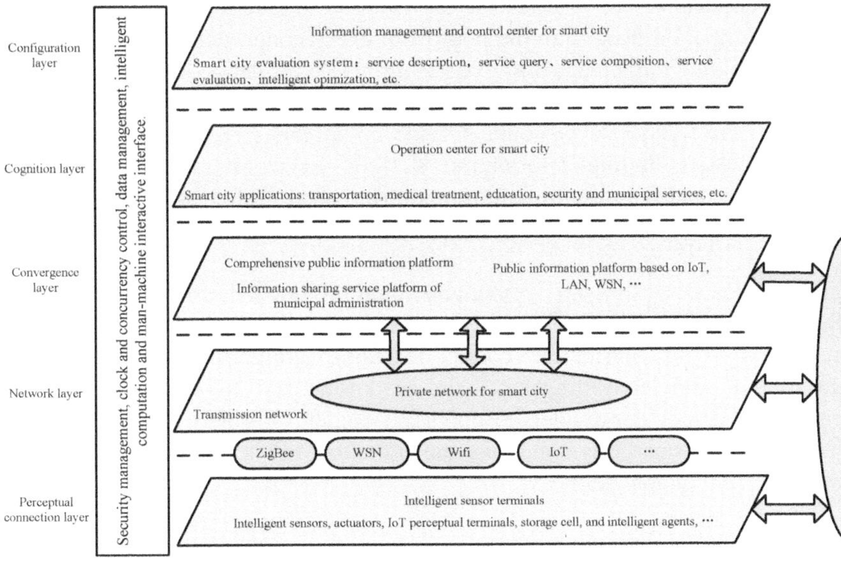

Fig. 1 Physical architecture framework of CPS for smart city

(1) Perceptual connection layer

During designing architecture of smart city, it should provide function to collect all kinds of information by sensors and intelligent terminals. Thus, acquiring accurate and reliable data effectively from real physical world makes CPS application feasible. Generally, information might be directly obtained with all kinds of sensors from various controllers and actuators or obtained by diverse smart city running systems. Various and vast physical components should be included in perceptual connection layer, i.e., intelligent sensors, actuators, perceptual terminals of Internet of things, storage cell, and intelligent agents. Data stream generated from perceptual connection layer could be transferred to network layer by various means, such as Zigbee, WSN, Wi-fi, IoT, and LAN.

(2) Network layer

Meaningful information should be inferred depending on effective data management and analysis. There are several methods which have been implemented on information conversion for network layer. Simultaneously, network layer should also provide capabilities of description and intelligent analysis heterogeneous data generated by heterogeneous nodes, effective localization caused by node mobility, coverage of perceptual abilities, and network congestion caused by massive data transmission, etc. Moreover, the second layer of CPS architecture should possess the ability of prediction and diagnoses, which could bring self-awareness to smart city.

(3) Convergence layer

In CPS architecture, convergence layer will play the role as information pool. Massive information flows into the pool from every connected unit. Thus, specific analysis should be implemented to extract intrinsic information. Convergence layer should provide self-comparison and self-judgment abilities for CPS system. Moreover, convergence layer could provide comprehensive information platform for smart city, including information sharing service platform of municipal administration, and public information platform based on IoT and WSN.

(4) Cognition layer

Cognition layer acts as monitor center and could provide a thorough knowledge for CPS system, including task analysis, task scheduling, task execution, task monitoring, which could contribute to correct decision-making. Proper infographics are necessary to completely transfer acquired knowledge to users or operators. Cognition layer could provide the operation center for smart city, which could cover all areas of smart city including transportation, medical treatment, education, security, and municipal services.

(5) Configuration layer

Configuration layer acts as central supervisory control. Through configuration layer being implemented, information from cyber space is feedback to real physical space. The reliability and correctness of decision-making in real physical space could be ensured accordingly. Configuration layer should provide functions, such as service description, service query, service composition, and service evaluation which could act as an information management and control center for smart city.

Moreover, it should be noticed that security management, clock and concurrency control, data management, intelligent computation, and man–machine interactive interface should exist in the whole operating system of smart city based on CPS, which could be the guarantee for normal operation of a CPS system.

During smart city system running process, long-distance information transmission between communication nodes can be completed through telecommunication network, wireless sensor network, broadcast network, Internet, etc., and a broader connectivity will be achieved. Accessibility, reliability, safety, and security of communication would directly affect the implementation of the subsystems of smart city. Special information transmission is mainly transmitted through dedicated communication network, since there are stricter requirements of reliability, safety, security, and real time during the information transferred. In this case, public networks will not be used unless there are incomplete conditions or special condition requirements.

Smart city is a new emerged operation mode, which will make changes in city management, economic development, and daily life in all walks of life. Thus, it seems very important of related policy construction, which includes safety guarantee, standard specification, evaluation system, policy, and regulations. These are also essential contents of informationalized construction for smart city. Besides, it might also emphasize the vitality of economic development and its supporting industries.

4 Comparative Studies

In this section, some comparative studies are employed. Some architectures which have been implemented on smart city designing are discussed. The main characteristics of architecture, including data acquisition, transmission, convergence, cognition, and configuration, are extracted for comparative study, as given in Table 1. In our opinion, it is also important of implementing convergence layer, which heterogonous information from various connected components of smart city

Table 1 Comparison of architectures in literature

Literature	Data acquisition	Data transmission	Convergence	Cognition	Configuration
Vilajosana [3]	Yes	Yes	No	Yes	Yes
Su [4]	Yes	Yes	No	Yes	No
Anthopoulos [5]	Yes	Yes	No	Yes	No
Our proposed	Yes	Yes	Yes	Yes	Yes

Table 2 Comparison of existing city and smart city based on CPS

Data source	Existing city	Smart city based on CPS
Sensor	Sensors and fault detection	Self-awareness
Actuator	Condition-based monitoring and diagnostics	Self-prediction and self-comparison
Operating system	Event-driven system	Service-oriented system. Self-configuration and self-organization

system could be proper processed. Configuration layer could make components of smart city with self-configure and self-adaptive abilities.

Implementing CPS in smart city could provide several advantages over several aspects as given in Table 2. There are massive perceptive sensors and actuators distributed in smart city. CPS could provide abilities of self-awareness, self-prediction, and self-comparison of smart sensors and intelligent actuators. For smart city operating system, aggregated knowledge could provide abilities of self-configuration and self-organization, which will make changes in city management, economic development, and daily life in all walks of life.

5 Conclusions

In our work, we focus on research of smart city architecture based on CPS, and the aim is to explore a new development model for the construction of smart city in China. Smart city information system based on CPS would make the urban operation system more intelligent, which is dependent on perception, network, intelligence, and visualization of urban operating system. Smart city based on CPS would be more self-adaptive and self-adjustable.

Acknowledgements Our work is supported by Liaoning Scientific Enterprise Public Welfare Research Fund: Research of key technologies of smart city information system based on CPS (20170048).

References

1. Zhang, N., Chen, X., Song, G.: Key issues of smart city development in China: an empirical study. Urban Develop. Stud. **22**(6), 22–33 (2015)
2. Dagn, A., Wang, D., Jun, L., He, J.: Progress and trends of smart city development in China. Geometics World **22**(4), 1–7 (2015)
3. Vilajosana, I., Llosa, J., Martinez, B., et al.: Bootstrapping smart cities through a self-sustainable model based on big data flows. IEEE Commun. Mag. **51**(6), 128–134 (2013)
4. Su, K., Li, J., Fu, H.: Smart city and the applications. In: International Conference on Electronics, Communications and Control. IEEE, pp. 1028–1031 (2011)

5. Anthopoulos, L., Fitsilis, P.: From digital to ubiquitous cities: defining a common architecture for urban development. In: Sixth International Conference on Intelligent Environments. IEEE Computer Society, pp. 301–306 (2010)
6. Filipponi, L., Vitaletti, A., G. Landi, G., et al.: Smart city: an event driven architecture for monitoring public spaces with heterogeneous sensors. In: Fourth International Conference on Sensor Technologies and Applications. IEEE Computer Society, pp. 281–286 (2010)
7. Hasan, S., Riain, S., Curry, E.: Approximate semantic matching of heterogeneous events. In: ACM International Conference on Distributed Event-Based Systems, ACM, pp. 252–263 (2012)
8. Samaras, C., Vakali, A., Giatsoglou, M., et al.: Requirements and architecture design principles for a smart city experiment with sensor and social networks integration. In: Panhellenic Conference on Informatics, pp. 327–334 (2013)
9. Horng, G.: The adaptive recommendation mechanism for distributed parking service in smart city. Wireless Pers. Commun. **80**(1), 394–413 (2015)
10. Ta, Y., Goddard, S., Perez, L.: A prototype architecture for cyber-physical systems. Acm Sigbed Rev. **5**(1), 1–2 (2008)
11. Wang, X., Chen, L., Huang, H., et al.: A service-oriented architecture framework of cyber-physical systems. J. Comput. Res. Develop. **47**, 299–303 (2010)
12. Chen, L., Wang, X., Deng, S.: Cyber-physical system architecture design. Comput. Sci. **38**(5), 295–300 (2011)
13. Lee, J., Bagheri, B., Kao, H.: A Cyber-Physical Systems architecture for industry4.0-based manufacturing systems. Manufact. Lett. **3**, 18–23 (2015)
14. Wan, J., Zhang, D., Lin, K., et al.: VCMIA: a novel architecture for integrating vehicular cyber-physical systems and mobile cloud computing. Mobile Netw. Appl. **19**(2), 153–160 (2014)
15. Li, Y., Dai, W., Ming, Z., et al.: Privacy protection for preventing data over-collection in smart city. IEEE Trans. on Comput. **65**(5), 1339–1350 (2016)
16. Paganelli, F., Turchi, S., Giuli, D.: A web of things framework for restful applications and its experimentation in a smart city. IEEE Syst. J. **10**(4), 1412–1423 (2017)

Effect of the Engine Working Condition on the Vehicle Emissions Based on Real-World Driving

Zhilei Ma, Chao He, Xueyuan Liu, Jiaqiang Li, Ming Liu and Heng Wei

Abstract Road emission test is for heavy-duty diesel trucks, retrofitted compressed natural gas cars, and light gasoline cars. To study the relationship between engine speed, load, and emission, the relationship between engine speed, engine emissions per cylinder per working cycle and emission ratio is put forward. The results showed that the NO_x, CO, and CO_2 emission rates of diesel trucks, natural gas cars, and gasoline cars increase with the increase of engine speed and load. The engine emissions per cylinder per working cycle increases with the increase of engine load. The trend of CO emission rate of heavy diesel truck is not obvious with engine condition.

Keywords Vehicle · Engine condition · Load · PEMS · Emission

1 Introduction

By the end of 2017, China's car ownership reached 217 million, an 11.9% increase compared with that of 2016. In order to reduce the environmental impact of vehicle emission pollutants, China issued a national six emission regulations for light and heavy-duty vehicles. On the basis of further reducing the emission limits of each pollutant, the requirements of vehicle road emission test are proposed, and the limit value of road emission is stipulated [1, 2]. In addition, Chinese oil consumption is increasing, and the external dependence has risen to 65.4% [3]. In order to reduce the consumption of petroleum energy, natural gas is increasingly used as vehicle fuel to reduce the consumption of petrol and diesel [4, 5].

Z. Ma · C. He (✉) · X. Liu · J. Li · M. Liu
School of Vehicle and Transportation, Southwest Forestry University, Kunming 650224, China
e-mail: hcsmile@aliyun.com

H. Wei
Key Laboratory of Vehicle Environmental Protection and Safety of Yunnan Institutions of Higher Education, Kunming 650224, China

V. Jain et al. (eds.), *Recent Trends in Intelligent Computing, Communication and Devices*, Advances in Intelligent Systems and Computing 1031,
https://doi.org/10.1007/978-981-13-9406-5_117

The emission of light and heavy vehicles is significantly different from that of laboratory type certification in the real road driving [6]. The emission factor of CO, HC, and NO_x is reduced with the increase of vehicle speed while the light-duty vehicle is driving at medium and low speed within 70 km/h [7, 8]. The emission factor of CO, HC, and NO_x is reduced with the increase of vehicle speed when the heavy vehicle travels under the medium and low speed within the 50 km/h [9, 10]. The emission of CO and NO_x increases with the increase of acceleration [11]. Increased vehicle load will generally increase the emission factor of CO, HC, and NO_x pollutants [12, 13]. The engine operating condition is closely related to speed, acceleration, and load, but the relationship between engine operating condition and emission is less. Referring to the real road test requirements of the national six emission regulations, this paper carries on the road test to a heavy diesel truck, a modified CNG sedan, and a light petrol sedan, and studies the relationship between NO_x, CO, CO_2 emissions and engine speeds and loads.

2 Materials and Methods

2.1 Test Vehicles and Equipment

In Kunming, three vehicles were tested for actual road emissions. The vehicle 1 is a heavy-duty diesel truck, with a displacement of 3.8 L and a national four emission standard. Car 2 for light-duty gasoline car is modified after the combustion of compressed natural gas dual-fuel vehicles, displacement 1.6 L, certified by the country four emission standards. Car 3 is for light petrol sedan, displacement 1.6 L, and the national five emission standards.

The portable emission measurement system (PEMS) by sensors (USA, SEMTECH-Ecostar) was used to investigate the vehicle exhaust pollutants. The equipment is mainly composed of fuel economy analysis module (FEM), nitrogen oxides analysis module (NO_x), air flowmeter, and power distribution module (PDM). The FEM module uses the non-dispersive infrared analyzer (NDIR) to test CO and CO_2 emissions, and the NO_x module is used to test NO and NO_2 emissions with non-dispersive UV analysis method. In addition, the use of GPS is to determine the location of the test vehicle; the use of meteorological stations is to collect temperature, humidity, and atmospheric pressure information. Using the J1939 CAN communication box to connect the vehicle onboard diagnostic interface (OBD), the vehicle and engine operating parameters can be captured.

2.2 Test Conditions

Referring to GB17691-2018 "pollutant emission limitation and measurement method for heavy duty diesel vehicles" and GB18352.6-2016 "pollutant emission limitation and measurement method for light vehicle (China phase sixth)" practical road traffic pollutant emission measurement method to plan test Road.

Car 1 belongs to the N2 type heavy diesel vehicles, so the average speed of urban road driving cycle should be in the range of 15–30 km/h, suburban road driving cycle in the range of 45–70 km/h, and high-speed road driving cycle greater than 70 km/h. The testing distance of the urban, suburban, high-speed road accounted for 45%, 25%, and 30%, respectively, allowing for a ± 5% deviation. Car 2 and 3 belong to the M1 class light vehicle, so the urban road driving speed is less than 60 km/h, and the average speed should be in the range of 15–40 km/h. The suburban roads travel between the 60–90 km/h—the speed of the road is greater than 90 km/h—cover 90–110 km/h the speed range of the city, suburban, high-speed road accounted for 34%, 33% and 33%, and allow the error of ±10%, but the proportion of urban road can not be less than 29%.

3 Test Results and Analysis

3.1 The Relationship Between Engine Speed and Emission Rate

The variable p is introduced to indicate the emission of one pollutant per working cycle per cylinder of the engine, for subsequent analysis of the relationship between the emission rate and engine speed, emission factors, and driving gear.

The relationship between the emission rate and engine speed is as following:

$$m = (p \times n \times i)/(30\tau) \quad (1)$$

m is a pollutant discharge rate (mg/s), p is the engine per cylinder per working cycle of a pollutant emissions (mg), n is the engine speed (rpm), i is the number of cylinders, and for the engine punch.

In the whole road test process, the car 1 engine is mainly in the range of 700–2400 rpm, the car 2 engine mainly in the range of 700–3600 rpm, and the car 3 engine in the range of 800–3400 rpm. The speed of each vehicle is divided into four intervals to investigate the variation of emission rate under different rotational speed range. Because the vehicle 1 and the car 2, 3 speed range difference is large, for easy observation, the speed range is divided, as shown in Table 1.

According to the size of the engine load, the engine load is divided into five intervals, and the partitioning method is shown in Table 2.

Table 1 Segments of the engine speed range

Interval	Car 1 speed range/rpm	Car 2, 3 rpm range/rpm
S1	$n \leq 1000$	$n \leq 1000$
S2	$1000 < n \leq 1500$	$1000 < n \leq 2000$
S3	$1500 < n \leq 2000$	$2000 < n \leq 3000$
S4	$n > 2000$	$n > 3000$

Table 2 Segments of the engine load zone

Interval	N1	N2	N3	N4	N5
Load	$1 \leq 20$	$20 < 1 \leq 40$	$40 < 1 \leq 60$	$60 < 1 \leq 80$	$1 > 80$

The engine speed and load distribution are shown in Fig. 1. It can be seen that with the engine speed increase, the engine workload increased and the proportion of large load increased.

The relationship between engine speed and NO_x emission rate is shown in Fig. 2. The NO_x emission rate of three vehicles increases with the increase of engine speed. Among them, the NO_x emission rate of the Car 1 and 2 is higher, mainly due to the high excess air coefficient and low emission standards.

The relationship between engine speed and CO emission rate is shown in Fig. 3. Car 2, Car 3 CO emission rates of Car 2 and 3 increases with the engine speed

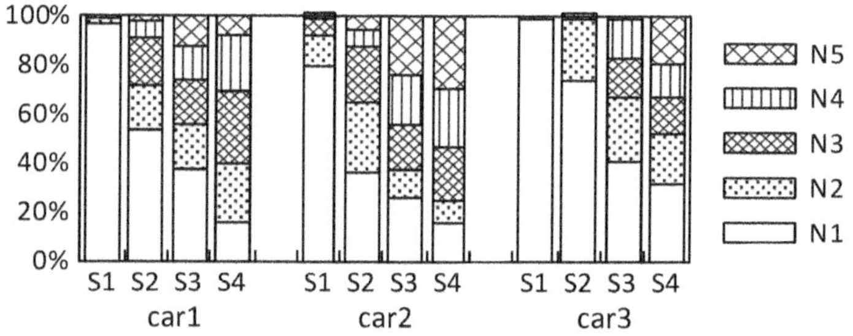

Fig. 1 Engine speed and load distribution

Fig. 2 Relationship between rotational speed and NO_x emission rate

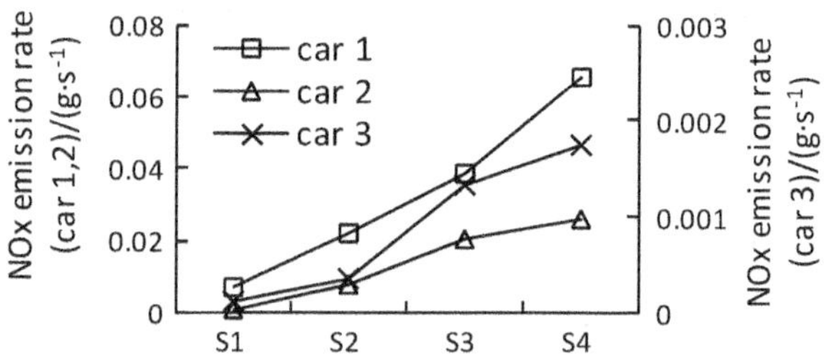

Fig. 3 Relationship between rotational speed and CO emission rate

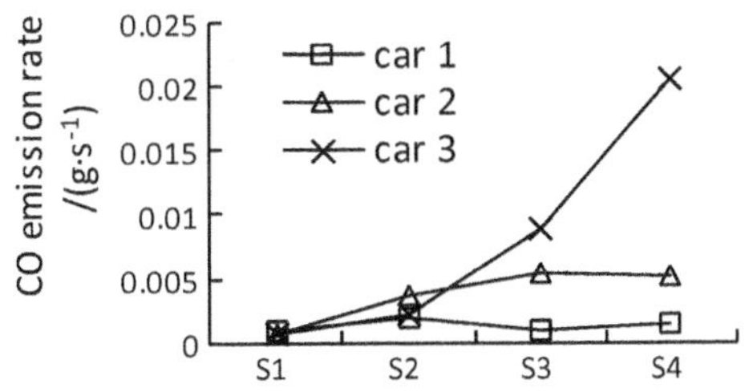

Fig. 4 Relationship between rotational speed and CO_2 emission rate

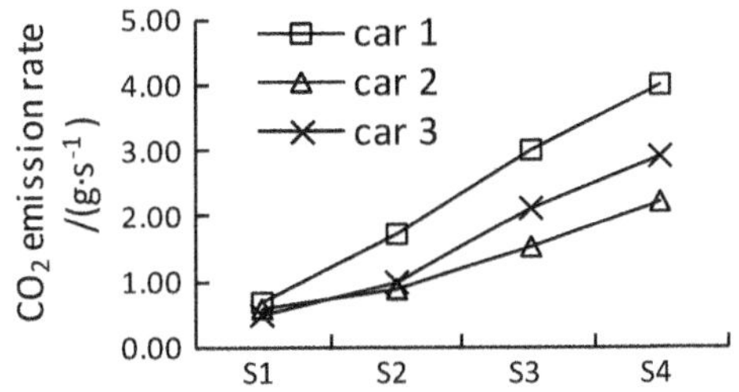

increase, but the CO emission rates of Car 1 with the engine speed change trend is not obvious, mainly because of the higher excessive air coefficient of the car 1 for diesel vehicles.

The relationship between engine speed and CO_2 emission rate is shown in Fig. 4. The CO_2 emission rate of three vehicles increases with the increase of engine speed. Vehicle 1 is a heavy-duty truck with a higher engine displacement and therefore a higher CO_2 emission rate. The vehicle 2 uses natural gas, the CO_2 emission rate is low, manifests the fuel gas has the certain energy-saving function.

When the number of cylinders and the number of engine strokes are determined, the NO_x and CO_2 emission rate increases with the increase of engine speeds.

3.2 The Relationship Between Engine Load and Emission Rate

In the engine S4 speed range, each load occupies a certain proportion. In order to study the change rule of each working cycle of engine per cylinder, the variable p, that is, the relationship between load and discharge rate of engine speed in S4 interval, is analyzed.

Figure 5 shows the relationship between the engine load and the NO_x emission rate. At a certain speed range, the NO_x emission rate of three vehicles increases with the increase of engine load. It can be seen that when the engine speed range is certain, as the load increases, the NO_x per cylinder per working cycle discharge p increases.

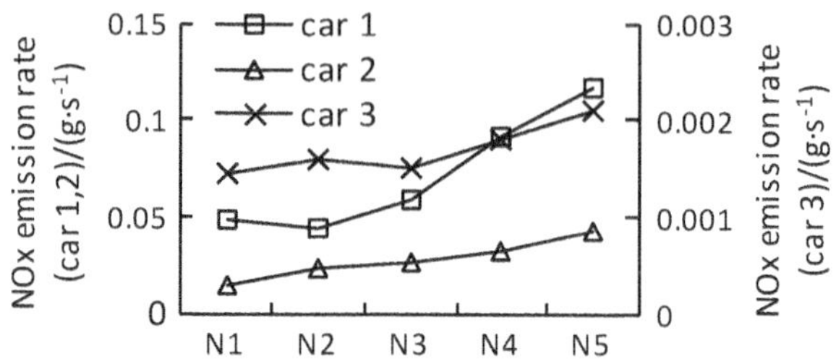

Fig. 5 Relationship between load and NO_x emission rate

Figure 6 shows the relationship between the engine load and the CO emission rate. When the speed range is certain, the CO emission rate of vehicle 2 and car 3 is increased with the increase of engine load, and the CO per cylinder per working cycle emission p increases as the load increases. However, the trend of CO emission rate of vehicle 1 is not obvious with the change of engine loads. Because the vehicle 1 is a diesel vehicle, the excess air coefficient is higher.

Figure 7 shows the relationship between the engine load and the CO_2 emission rate. At a certain speed range, the CO_2 emission rate of three vehicles increases with the increase of engine load. It can be seen that, when the engine speed range is certain, with the load rising, CO_2 per cylinder per working cycle emissions p increases.

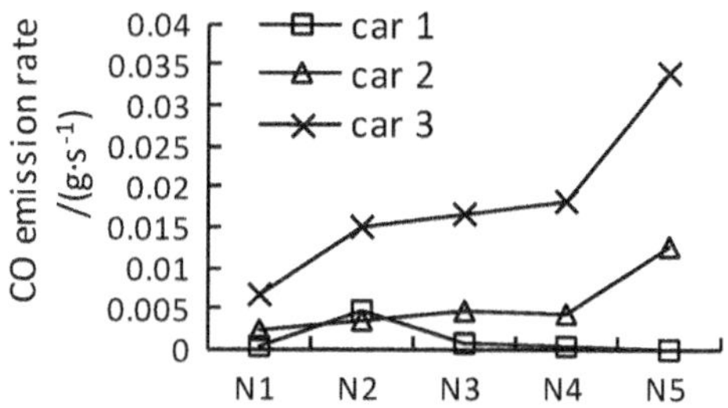

Fig. 6 Relationship between load and CO emission rate

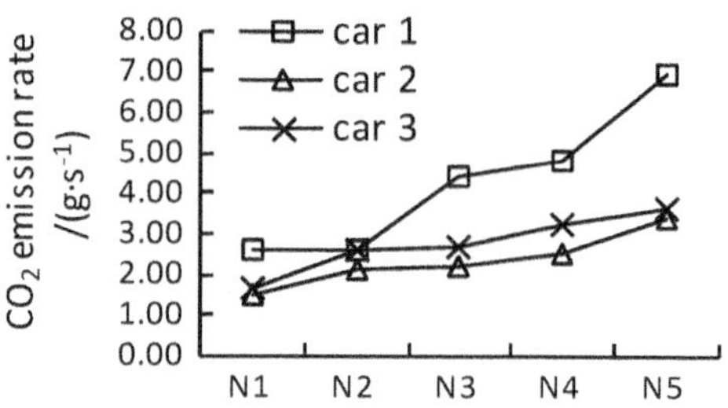

Fig. 7 Relationship between load and CO_2 emission rate

4 Conclusion

(a) As the engine speed rises, the engine load increases and the proportion of heavy load increases.
(b) The emission rate of NO_x, CO, and CO_2 of gasoline cars and natural gas cars increases with the increase of engine speed and load, and the discharge of pollutants per working cycle increases with the increase of engine load.
(c) The NO_x and CO_2 emission rate of heavy-duty diesel trucks increases with the increase of engine speed and load. The trend of CO emission rate varying with engine operating condition is not obvious.

Acknowledgements The authors gratefully acknowledge financial support from "Science and Technology Project of Yunnan Province (No. 2017FD108)," "National Natural Science Foundation of China (No. 51266015)," and "Agricultural Joint Project of Yunnan Province [No. 2017FG001(-010)]."

References

1. Ministry of Environmental Protection.: GB18352.6–2016 Light Vehicle Pollutant Emission Limit and Measurement Method (China sixth stage). China Environmental Science Press, Beijing (2016)
2. Department of Ecological Environment: GB17691-2018 Heavy Duty Diesel Vehicle Pollutant Emission Limit and Measurement Method (China sixth stage). China Environmental Science Press, Beijing (2018)
3. Peng, Y., Dong, X.: China Oil and Gas Industry Development Analysis and Prospect Report Blue Book (2016–2017). China Petrochemical Press, Beijing (2017)
4. Hao, H., Liu, Z., Zhao, F. et al.: Natural gas as vehicle fuel in china: a review. Renew. Sustain. Energy Rev. **62**(5), 521–533 (2016)
5. Spearrin, R.M., Triolo, R.: Natural gas-based transportation in the USA: economic evaluation and policy implications based on MARKAL modeling. Energy Res. **38**, 1879–1888 (2014)
6. Gersan, Ding, Y., Yin, H.: Research status of vehicle emission testing system. J. Automot. Saf. Energy Conserv. **8**(2), 111–121 (2017)
7. Gersan, Wang, A., Wang, M. et al.: PEMS for the actual road gas emission test of urban vehicles. J. Automot. Saf. Energy Conserv. **1**(2), 141–145 (2010)
8. Qiang, M., Huang, W.: Comparison of emission of hybrid electric car and traditional gasoline car based on PEMS test. Environ. Eng. **34**(4), 166–171 (2016)
9. Wu, X., Peng, M., Fang, R., et al.: Emission test analysis of diesel buses based on PEMS. Environ. Sci. Technol. **35**(1), 146–149 (2012)
10. Wang, A., Ge, Y., Tan, J., et al.: On-road pollutant emission and fuel consumption characteristics of buses in Beijing. J. Environ. Sci. **23**(3), 419–426 (2011)
11. Fu, B., Yang, Z., Yin, H., et al.: Study on the emission characteristics of the light-duty gasoline vehicle in actual driving. Autom. Eng. **39**(4), 376–380 (2017)
12. Yao, Z., Wang, Q., Zhang, Y., et al.: Study on the effect of load on the emission of heavy diesel vehicles. Environ. Pollut. Control **34**(3), 63–67 (2012)
13. Wang, Y., Ji, Z., Yin, H., et al.: Influence factors of pollutant emission factor measurement for heavy diesel vehicles. Environ. Sci. Res. **27**(3), 232–238 (2014)

Self-calibration Method for Two DOF Cable-Driven Joint Module

Tianjiang Zheng, Yi Wang, Guilin Yang, Wenjun Shen, Zaojun Fang and Kaisheng Yang

Abstract A modular cable-driven robot consists of a number of consecutively connected cable-driven joint modules. It has the advantages of lightweight structure and compliant behavior so that it can perform intrinsically safe human–robot interactions. However, a modular cable-driven robot normally has a low positioning accuracy due to the manufacturing and assembly errors in each of the cable-driven joint modules. In this work, a self-calibration method is proposed for the 2-DOF cable-driven universal joint module by utilizing its actuation redundancy. Different from the conventional self-calibration method where the absolute cable lengths need to be measured, the proposed error model is formulated based on the relative lengths changes of the diving cables. The effectiveness of the self-calibration method is verified through computer simulations.

Keywords Cable-driven robot · Self-calibration · Kinematics · Identification

1 Introduction

The cable-driven robot has been attracted much attentions due to their advantages of lightweight, large workspace, and intrinsically safe properties. The cable-driven robot has been applied in many areas such as large workpiece detection [1], pick and place applications [2], and robots for rehabilitation [3]. The structure design and kinematics analysis are of key importance to build this kind of robot. The cable-driven robot is usually designed modularly and each joint module has two or three degrees of freedom (DOF).

Y. Wang · W. Shen
University of Chinese Academy of Sciences, Beijing, China

T. Zheng · Y. Wang · G. Yang (✉) · W. Shen · Z. Fang · K. Yang
Zhejiang Key Laboratory of Robotics and Intelligent Manufacturing Equipment Technology, Ningbo, China
e-mail: glyang@nimte.ac.cn

V. Jain et al. (eds.), *Recent Trends in Intelligent Computing, Communication and Devices*, Advances in Intelligent Systems and Computing 1031,
https://doi.org/10.1007/978-981-13-9406-5_118

Normally, there are two reasons that will resulting the kinematic errors at the robot joint end point, the first reason is due to the mechanic manufacturing errors, the other reason is come from the mechanical assembly. Therefore, the identification and compensation of the kinematic errors are very important for the robot control and robot trajectory design. In previous efforts, the geometric errors calibration method has been studied by Guilin Yang et al. who developed the calibration algorithm [4] for serial-linked robots and parallel robots [5, 6]. Chen et al. [7] and Mustafa et al. [8] finished studying on the calibration of cable-driven robot. In Chen's works, it need to measure the angles of the joint, and the angle sensors are embedded in their robot joint at the universal joint of center shut. Mustafa's method is proposed for their designed 7 DOF robot, and they need to measure the length between the mounting points at moving platform. Lau [10] has proposed a calibration method to find the initial pose of the cable robot using only the relative length. However, in his method, the model is considered to be ideal, and the geometric error is not taken onto consideration. Jing et al. [9] adopted the machine learning method for robot calibration; however, this method needs a mount of calculation and training. Other recent works related to the robot calibration can be found in Wang [11] and Tao et al. [12]. Either the above-mentioned calibrations need extra sensors or they cannot be used for the cable-driven robot joint module. To keep the force closure property, the cable-driven robot joint is usually designed as a redundant system, and there always exist one or more extra redundant cables. The redundant cables can be used for the self-calibration.

In this paper, we proposed a simple and explicit method to calibrate the gross geometric errors for 2-DOF cable-driven robot joint module; in this method, the only used parameters are the relative cable length which are able to record by driven motor's encoders. In this method, the first step is to measure the relative cable length and compute its nominal value, then built the errors transmission differential model, and calculate the cable length errors at the redundant cables; the error parameters at mounting points are then identified by using the least square method

2 Kinematic Model of the Cable-Driven Robot Joint

The cable-driven robot joint shown in Fig. 1 is composed of two platforms and a center shut. The moving platform and the base platform, the center shut has been designed as a universal joint, the moving platform and base platform are connected by four stainless steel cables.

The kinematic diagram of the proposed two DOF cable-driven joint is shown in Fig. 2. Three frames are attached to the joint module. Frame $\{B\}$ is attached to the geometric center of the base platform and frame $\{P\}$ is attached to the geometric center of the moving platform. Between the base platform and the moving platform, there is a frame $\{U\}$ attached to the cross shaft. The relative position from the base

Fig. 1 Cable-driven joint

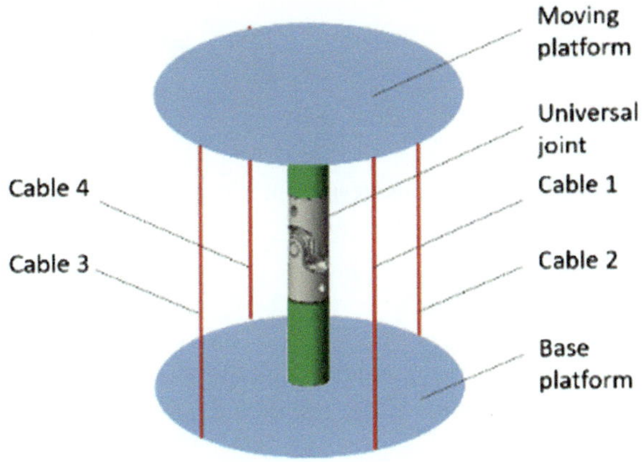

Fig. 2 Kinematics diagram of the 2-DOF joint module

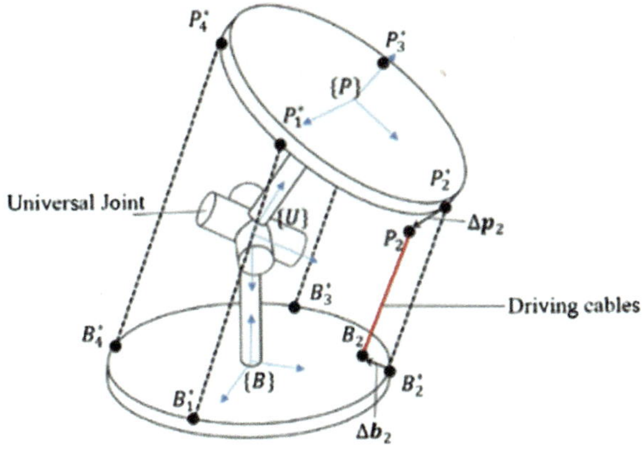

platform to moving platform is represented by a 4×4 matrix ${}^{B}T_{P}$ which is an element of SE(3). By applying the product of exponentials formula, ${}^{B}T_{P}$ is given by:

$$ {}^{B}T_{p} = {}^{B}T_{U0} e^{\hat{\xi}_1 \theta_1} e^{\hat{\xi}_2 \theta_2} {}^{U}T_{P0} \tag{1} $$

where θ_1 and θ_2 are the angular position of the universal joint; $\hat{\xi}_1$ and $\hat{\xi}_2$ are the rotation axes of the universal joint in twist coordinate system; ${}^{B}T_{U0}$ and ${}^{B}T_{P0}$ are the transformation matrices between $\{B\}$ and $\{U\}$, $\{U\}$ and $\{P\}$. In this paper, we only considered the cable mounting positioning errors of the platforms, these errors will be represented in the frames of $\{B_i^*\}$ respect to $\{B\}$ and $\{P_i^*\}$ respect to $\{P\}$; To calibrate these errors, normally, the cable length of the ith cable is determined by:

$$l_i^2 = \left(P_i^{\{B\}} - B_i^{\{B\}}\right)^{\mathrm{T}} \left(P_i^{\{B\}} - B_i^{\{B\}}\right) \tag{2}$$

where $\begin{bmatrix} P_i^{\{B\}} & 1 \end{bmatrix}^{\mathrm{T}} = {}^{B}T_P(\theta_1, \theta_2) \begin{bmatrix} P_i^{\{P\}} & 1 \end{bmatrix}^{\mathrm{T}} B_i^{\{B\}} = B_i^* + \Delta \boldsymbol{b}_i$. By differentiating the Eq. (2) and considering the differentiation of the Eq. (1), it gives:

$$\begin{bmatrix} \mathrm{d}P_i^{\{B\}} \\ 0 \end{bmatrix} = \left(\mathrm{Ad}_{({}^{B}T_{U0})}\xi_1\right)^{\wedge} \begin{bmatrix} P_i^{\{B\}} \\ 1 \end{bmatrix} \mathrm{d}\theta_1 + \left(\mathrm{Ad}_{({}^{B}T_{U0} e^{\hat{\xi}_1 \theta_1})}\xi_2\right)^{\wedge} \begin{bmatrix} P_i^{\{B\}} \\ 1 \end{bmatrix} \mathrm{d}\theta_2 + \begin{bmatrix} \mathrm{Rd}P_i^{\{P\}} \\ 0 \end{bmatrix} \tag{3}$$

where $\mathrm{Ad}_{(M)}$ is the adjoint transformation matrix M. Combining Eqs. (2) and (4), the differentiation of the cable length is given by:

$$\mathrm{d}l_i = \mathbb{A}_i \left\{ \mathrm{Rd}P_i^{\{P\}} - \mathrm{d}B_i^{\{B\}} + \mathbb{C}_i \mathrm{d}\theta_1 + \mathbb{H}_i \mathrm{d}\theta_2 \right\} \tag{4}$$

where $\mathbb{A}_i = \frac{\left(P_i^{\{B\}} - B_i^{\{B\}}\right)^{\mathrm{T}}}{P_i^{\{B\}} - B_i^{\{B\}}}$ $\mathbb{C}_i = \mathrm{Row}_{1-3}\left\{ \left(\mathrm{Ad}_{({}^{B}T_{U0})}\xi_1\right)^{\wedge} \begin{bmatrix} P_i^{\{B\}} \\ 1 \end{bmatrix} \right\}$, $\mathbb{H}_i = \mathrm{Row}_{1-3}$ $\left\{ \left(\mathrm{Ad}_{(e^{\hat{\xi}_1 \theta_1} {}^{B}T_{U0})}\xi_2\right)^{\wedge} \begin{bmatrix} P_i^{\{B\}} \\ 1 \end{bmatrix} \right\}$, and $\mathrm{Row}_{1-3}(M)$ is a function that returns the first 3 rows of a matrix M. By utilizing the Eq. (4), the forward kinematics and inverse kinematics can be solved by using numerical method.

3 Self-calibration Model of the Cable-Driven Robot

In the presented robot joint module, there are four cables mounted for each joint. Two cables are enough to find the pose of the robot end effectors while the other two cables are used for the self-calibration. For example, l_1 and l_2 are used for driving the universal joint that will cause a joint displacement $[\theta_1, \theta_2]^{\mathrm{T}}$ and the other two cables $[l_3, l_4]^T$ will also have a displacement. If the model is accurate, the measured relative cable lengths $[l_3, l_4]^{\mathrm{T}}$ will be equal to their nominal value. However, due to the existence of errors which is shown in Fig. 2, the angles θ_1 and θ_2 will have errors inevitably. Then, these errors will result in the errors of cable $[l_3, l_4]^{\mathrm{T}}$, which are $[\mathrm{d}l_3, \mathrm{d}l_4]^{\mathrm{T}}$. There are three steps to deduct the self-calibration error model. Firstly, the relationship between $[\mathrm{d}l_3, \mathrm{d}l_4]^{\mathrm{T}}$ and $[\mathrm{d}\boldsymbol{b}_3^T, \mathrm{d}\boldsymbol{b}_4^T, \mathrm{d}\boldsymbol{p}_3^T, \mathrm{d}\boldsymbol{p}_4^T, \mathrm{d}\theta_1, \mathrm{d}\theta_2]^{\mathrm{T}}$ is given by:

$$\begin{bmatrix} \mathrm{d}\Delta l_3 \\ \mathrm{d}\Delta l_4 \end{bmatrix} = (J_a - J_{a0}) \begin{bmatrix} \mathrm{d}\boldsymbol{b}_3 \\ \mathrm{d}\boldsymbol{p}_3 \\ \mathrm{d}\boldsymbol{b}_4 \\ \mathrm{d}\boldsymbol{p}_4 \end{bmatrix} + \mathbb{D}_{34} \begin{bmatrix} \mathrm{d}\theta_1 \\ \mathrm{d}\theta_2 \end{bmatrix} \tag{5}$$

where $J_a = \begin{bmatrix} -\mathbb{A}_3 & \mathbb{A}_3 R & 0 & 0 \\ 0 & 0 & -\mathbb{A}_4 & \mathbb{A}_4 R \end{bmatrix}$, $\mathbb{D}_{34} = \begin{bmatrix} \mathbb{A}_3\mathbb{C}_3 & \mathbb{A}_3\mathbb{H}_3 \\ \mathbb{A}_4\mathbb{C}_4 & \mathbb{A}_4\mathbb{H}_4 \end{bmatrix}$ $\mathrm{d}\Delta l_i = \mathrm{d}(l_i - l_{i0}) = \mathrm{d}l_i - \mathrm{d}l_{i0}$ $J_{a0} = J_a|_{\theta_1=0,\theta_2=0}$.

The second step is to analyze the relationship between $[\mathrm{d}\theta_1, \mathrm{d}\theta_2]^{\mathrm{T}}$ and $\left[\mathrm{d}\boldsymbol{b}_1^T, \mathrm{d}\boldsymbol{p}_1^T, \mathrm{d}\boldsymbol{b}_2^T, \mathrm{d}\boldsymbol{p}_2^T\right]^{\mathrm{T}}$. We assume that the cables l_1 and l_2 are driven cables, this mean that the initial errors of l_1 and l_2 are zeros, then the differentiation equation of l_1 and l_2 is given by:

$$\mathbb{D}_{12}\begin{bmatrix} \Delta\theta_1 \\ \Delta\theta_2 \end{bmatrix} = -(J_b - J_{b0})\begin{bmatrix} \Delta\boldsymbol{b}_1 \\ \Delta\boldsymbol{p}_1 \\ \Delta\boldsymbol{b}_2 \\ \Delta\boldsymbol{p}_2 \end{bmatrix} \tag{6}$$

The last step is to find the relationships between the cable length errors $[\mathrm{d}\Delta l_3, \mathrm{d}\Delta l_4]^{\mathrm{T}}$ and the error parameters $\left[\mathrm{d}\boldsymbol{b}_1^T, \mathrm{d}\boldsymbol{p}_1^T, \mathrm{d}\boldsymbol{b}_2^T, \mathrm{d}\boldsymbol{p}_2^T, \mathrm{d}\boldsymbol{b}_3^T, \mathrm{d}\boldsymbol{b}_4^T, \mathrm{d}\boldsymbol{p}_3^T, \mathrm{d}\boldsymbol{p}_4^T\right]^{\mathrm{T}}$. Combining the Eqs. (5) and(6), the relationship is then given by:

$$\begin{bmatrix} \mathrm{d}\Delta l_3 \\ \mathrm{d}\Delta l_4 \end{bmatrix} = -\mathbb{D}_{34}\mathbb{D}_{12}^{-1}(J_b - J_{b0})\begin{bmatrix} \mathrm{d}\boldsymbol{b}_1 \\ \mathrm{d}\boldsymbol{p}_1 \\ \mathrm{d}\boldsymbol{b}_2 \\ \mathrm{d}\boldsymbol{p}_2 \end{bmatrix} + (J_a - J_{a0})\begin{bmatrix} \mathrm{d}\boldsymbol{b}_3 \\ \mathrm{d}\boldsymbol{p}_3 \\ \mathrm{d}\boldsymbol{b}_4 \\ \mathrm{d}\boldsymbol{p}_4 \end{bmatrix} \tag{7}$$

The Eq. (7) can be written into matrix format which is given by:

$$Y = JX \tag{8}$$

where $Y = \begin{bmatrix} \mathrm{d}\Delta l_3 \\ \mathrm{d}\Delta l_4 \end{bmatrix}$ is the cable length error; $J = \left[-\mathbb{D}_{34}\mathbb{D}_{12}^{-1}(J_b - J_{b0}) \;\; (J_a - J_{a0})\right]$ is the calibration Jacobian; $X = \left[\mathrm{d}\boldsymbol{b}_1^T, \mathrm{d}\boldsymbol{p}_1^T, \ldots, \mathrm{d}\boldsymbol{b}_4^t, \mathrm{d}\boldsymbol{p}_4^T\right]^{\mathrm{T}}$ is the error variable.

For the ith test, the Eq. (14) becomes $Y_i = J_i X$ and combines all the pose tested equations, and it gives:

$$\mathbb{Y} = \mathbb{J}X \tag{9}$$

where $\mathbb{Y} = \begin{bmatrix} Y_1^T & Y_2^T & \cdots & Y_n^T \end{bmatrix}^{\mathrm{T}}$; $\mathbb{J} = \begin{bmatrix} J_1^T & J_2^T & \cdots & J_n^T \end{bmatrix}^{\mathrm{T}}$; $n \geq 12$ is the number of test positions. Based on this equation, the geometric error can be identified by using iterative method. The flowing chart based on this proposed method shown in Fig. 3.

Fig. 3 The flowing chart of the method

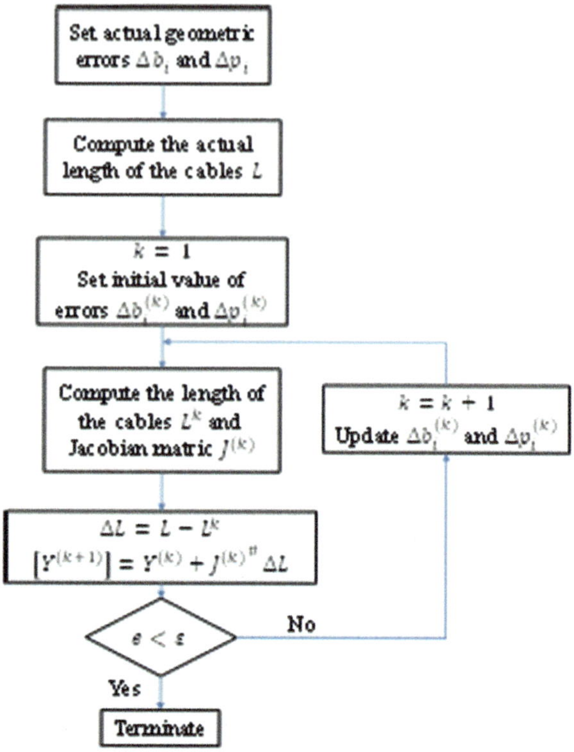

Table 1 The geometric parameters of the simulation

$B_1^{*\{B\}} = P_1^{*\{P\}} = [100 \quad 0 \quad 0]^T$
$B_2^{*\{B\}} = P_2^{*\{P\}} = [0 \quad 100 \quad 0]^T$
$B_1^{*\{B\}} = P_1^{*\{P\}} = [-100 \quad 0 \quad 0]^T$
$B_1^{*\{B\}} = P_1^{*\{P\}} = [0 \quad -100 \quad 0]^T$
$[\Delta b_1^T \quad \Delta p_1^T]^T = [0 \quad 0 \quad -0.5 \quad 2 \quad -1 \quad 0]^T$
$[\Delta b_2^T \quad \Delta p_2^T]^T = [1 \quad -0.5 \quad 0.1 \quad -0.5 \quad 0.5 \quad 2]^T$
$[\Delta b_3^T \quad \Delta p_3^T]^T = [0.2 \quad 0 \quad 0.5 \quad 1 \quad 1 \quad -5]^T$
$[\Delta b_1^T \quad \Delta p_1^T]^T = [-1 \quad 1 \quad 0 \quad 2 \quad 3 \quad 1]^T$

4 Simulation Results

In order to verify the proposed method, a simulation example for the calibration of a single joint module is conducted. In the simulation, the parameters shown in Fig. 2 are set as in the Table 1, where the units of the parameters are set by mm. In the simulation, 30 sets of pose have been selected for the self-calibration.

The simulation results are shown in Fig. 4. From Fig. 4a, it is shown that the second norm of the cable length error is $\|Y\| = 117.107\,\text{mm}$ while after the calibration, the estimated cable length error is decreased to $\|Y\| = 6.871 \times 10^{-3}\,\text{mm}$. Form Fig. 4b, it is shown that the second norm of the joint angle error is $\|\Delta\theta^*\| = 0.063$ while after the calibration, the estimated error is decreased to $\|\Delta\theta\| = 2.3756 \times 10^{-7}$.

The result has shown that the algorithm is effective. In Fig. 5, a uniformly distributed random noise with a range of 10^{-3} mm is added to the relative cable length. Applying the same process, the iteration also converges as shown in Fig. 5. Form Fig. 5a, it is shown that the second norm of the cable length error is $\|Y\| = 117.10827\,\text{mm}$ while after the calibration, the estimated cable length error is decreased to $\|Y\| = 0.02945\,\text{mm}$. Form Fig. 5b, it is shown that the second norm of the joint angle error is $\|\Delta\theta^*\| = 0.06365$ while after the calibration, the estimated error is decreased to $\|\Delta\theta\| = 1.387 \times 10^{-4}$.

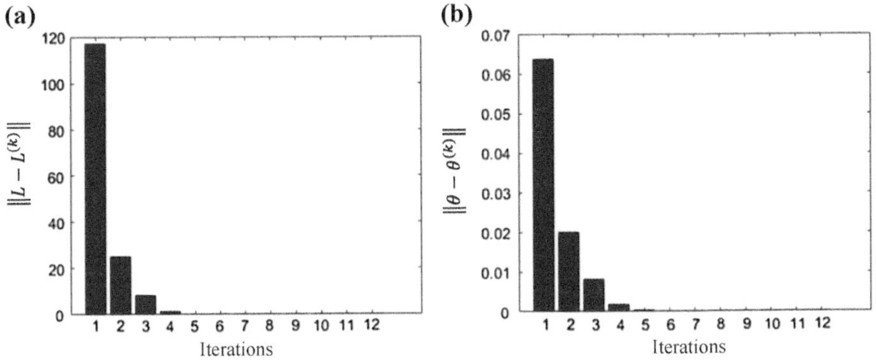

Fig. 4 Simulation results without noise: **a** The cable length error; **b** The joint angle error

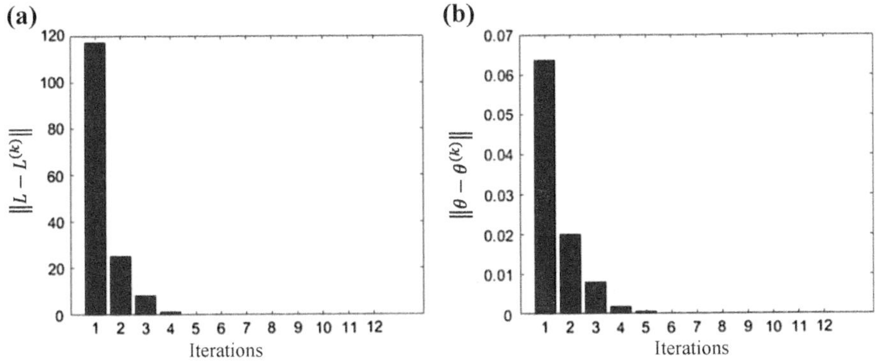

Fig. 5 Simulation results with noise: **a** The cable length error; **b** The joint angle error

5 Conclusion

This paper presented a self-calibration method for 2-DOF cable-driven robots. The novel self-calibration method is developed for the cable-driven robot with redundant driving cables. With the self-calibration method, the geometric errors can be calibrated by using only the information of the relative cable length. To find the model errors, this paper first built the differential error model of the redundant cables due to their nominal value might not equal to their actual value because of the mounting errors. Then we identified the geometrical mounting errors by utilizing the presented error model and the model has been solved by using least square method. A simulation is conducted to verify the proposed algorithm. The result shown that the proposed method can effectively identify the geometrical errors even with system noise. The estimated cable length error is from 117.10827 decreased to 0.02945 mm and the joint angle error decreases from 0.063 to 2.3756×10^{-7} after 12 iterations. Future work will focus on the calibration of the geometric error for the entire cable-driven robot arm.

Acknowledgements This research is supported by National Natural Science Foundation of China (Project code: 51705510 and 51475448) and Pre-research Project of Equipment Development Department of PRC Central Military Commission (61409230101), and Qianjiang Talent Project (QJD1602033).

References

1. Lafourcade, P., Llibre, M.: Design of a parallel wire-driven manipulator for wind tunnels. In: Proceedings Workshop on Fundamental Issues and Future Research Directions for Parallel Mechanisms and Manipulators, Quebec City, Canada
2. Kawamura, S., Choe, W., Tanaka, S.: Development of an ultrahigh speed robot FALCON using wire drive system. In: Proceedings IEEE International Conference on Robotics and Automation, Nagoya, Japan (1995)
3. Suzuki, Y., Kuwahara, H., Ohnishi, K.: Development and verification of tendon-driven rotary actuator for haptics with flexible actuators and a PE line. In: Proceedings IEEE International Workshop on Advanced Motion Control, Nagaoka, Japan (2010)
4. Chen, I.M., Yang, G.L., Tan, C.T., Yeo, S.H.: Local POE model for robot kinematic calibration. Mech. Mach. Theory **36**(11–12), 1215–1239 (2001)
5. Yang, G., Chen, I., Yeo, S.H., Lim, W.K.: Simultaneous base and tool calibration for self-calibrated parallel robots. Robotica **20**(04) (2002)
6. Tao, P.Y., Mustafa, K.S., Yang, G., Tomizuka, M.: Robot work cell calibration and error compensation. In: Handbook of Manufacturing Engineering and Technology. SpringerLink (2015)
7. Chen, Q., Chen, W., Yang, G., Liu, R.: An integrated two-level self-calibration method for a cable-driven humanoid arm. IEEE Trans. Autom. Sci. Eng. **10**(2), 380–391 (2013)
8. Mustafa, S.K., Yang, G., Yeo, S.H., Lin, W.: Kinematic calibration of a 7-DOF self-calibrated modular cable-driven robotic arm. In: IEEE International Conference on Robotics and Automation ICRA, pp. 1288 (2008)

9. Jing, W., Tao, P.Y., Yang, G., Shimada, K.: Calibration of industry robots with consideration of loading effects using product-of-exponential and gaussian process. In: IEEE International Conference on Robotics and Automation, Stockholm, Sweden (2016)
10. Lau, D.: Initial length and pose calibration for cable-driven parallel robots with relative length feedback. In: Initial Length and Pose Calibration for Cable-Driven Parallel Robots with Relative Length Feedback, pp. 140–151 (2017)
11. Wang, H., Gao, T., Kinugawa, J., Kosuge, K.: Finding measurement configurations for accurate robot calibration: validation with a cable-driven robot. IEEE Trans. Rob. **33**(5), 1156–1169 (2017)
12. Tao, P.Y, Mustafa, K.S. et al.: Robot Work Cell Calibration and Error Compensation. SpringerLink (2015)

Study on Discontinuous Lane Recognition Method Based on Multi-threshold Fusion

Xu Tao and Chu Wenbo

Abstract For lane recognition involved in the research of intelligent vehicles, especially for discontinuous lane recognition, higher requirements are put forward for recognition methods. This chapter proposes a solution based on multi-threshold fusion after analyzing the limitations of the single-threshold algorithm. On the basis of the theoretical analysis, a discontinuous lane recognition method, which combines camera distortion correction, white balance algorithm, HSL color space threshold, gradient threshold, gradient magnitude threshold, gradient direction threshold, perspective transformation, reverse perspective transformation, and quadric curve fitting, is designed and carried out. The experimental results show that the lane recognition method based on multi-threshold fusion can accurately identify discontinuous lanes and can provide an accurate boundary for intelligent vehicle path planning.

Keywords Multi-threshold fusion · Lane recognition · Machine vision

1 Introduction

With the rapid development of intelligent connected vehicles in recent years, more and more high requirements have been put forward for the vehicle sensor system. Lane recognition as a part of the vehicle sensor system, especially the recognition of discontinuous lane, can provide intelligent vehicle's path planning with the accurate boundary, so the lane detection technology in the intelligent vehicle sensor system is particularly important. But at the same time, it also puts forward higher requirements for lane recognition methods.

X. Tao (✉)
Beihang University, Beijing 100191, China
e-mail: justin_tao@126.com

C. Wenbo
China Intelligent and Connected Vehicles Research Institute Co., Ltd., Beijing 100176, China

V. Jain et al. (eds.), *Recent Trends in Intelligent Computing, Communication and Devices*, Advances in Intelligent Systems and Computing 1031,
https://doi.org/10.1007/978-981-13-9406-5_119

References [1–6] all adopted the single-threshold method to detect and identify lane lines, so the results are relatively limited. After a detailed analysis of the limitations of single-threshold method, this chapter designs an algorithm based on multi-threshold fusion, which combines camera distortion correction, white balance algorithm, HSL color space threshold, gradient threshold, gradient magnitude threshold, gradient direction threshold, perspective transformation, reverse perspective transformation, and quadric curve fitting for discontinuous lane recognition method design. The experimental results show that the lane recognition method based on multi-threshold fusion can accurately identify discontinuous lanes and provide an accurate boundary for intelligent vehicle path planning.

2 Structure of Lane Recognition System

The structure of the discontinuous lane recognition system based on multi-threshold fusion is shown in Fig. 1.

In Fig. 1, due to the internal camera parameters, the images captured by cameras have a certain degree of distortion. Therefore, the internal parameters of the camera should be obtained by calibrating the camera, and the image distortion should be corrected. At the same time, the colors of the objects will change due to the colors of the rays, and the images captured under different rays will have different color temperatures. Therefore, the color temperatures of the pictures should be corrected through white balance to restore the real colors. Then, the final binary image is

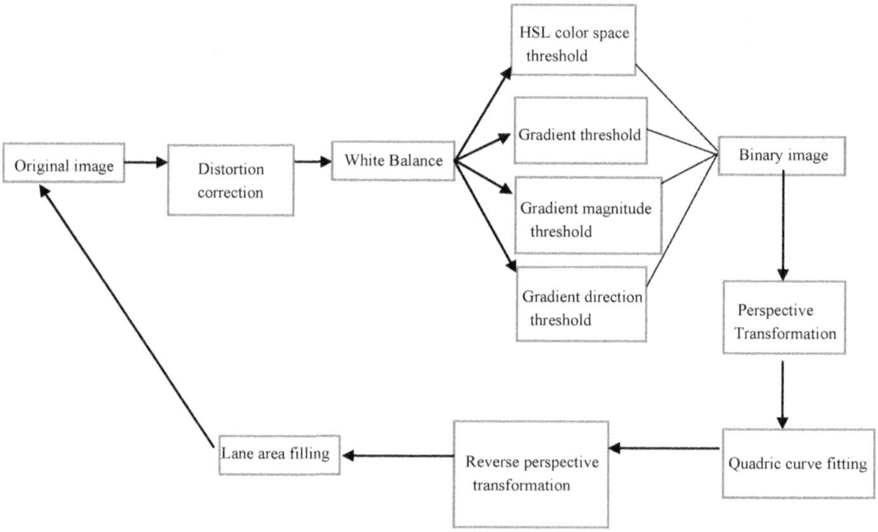

Fig. 1 Structure of discontinuous lane recognition system

obtained by HSL color space threshold, gradient threshold, gradient magnitude threshold, and gradient direction threshold. Convert the final binary image by perspective transformation; then, the lane lines will account for the majority of the image. By means of quadric curve fitting, the equation of lane line is obtained. And the lane line area is mapped to the original image through reverse perspective transformation, and finally, the lane is marked in the original image.

3 Method Design

3.1 Distortion Correction

Generally, the image captured by the camera has a degree of distortion. Due to the characteristics of the lens itself, rays will bend at the edge of the camera lens to a certain degree, which is called radial distortion. Radial distortion can be corrected by the following formula:

$$\begin{aligned} x_{\text{corr}} &= x_{\text{dis}}\left(1 + k_1 r^2 + k_2 r^4 + k_3 r^6\right) \\ y_{\text{corr}} &= y_{\text{dis}}\left(1 + k_1 r^2 + k_2 r^4 + k_3 r^6\right) \end{aligned} \tag{1}$$

where x_{dis} and y_{dis} represent the distortion coordinates, x_{corr} and y_{corr} represent the correction coordinates, k_1, k_2, and k_3 represent the radial distortion parameters.

By taking multiple photographs of the checkerboard, the pixel positions of the black and white points in the checkerboard were found by OpenCV. At the same time, by constructing the relative positions of these diagonal points, two vector lists, img_points and obj_points were obtained, respectively. And the camera distortion coefficient was obtained through OpenCV. Finally, the image was corrected. The correction of image distortion is shown in Fig. 2. The image on the left side is the original image, and the image on the right side is undistorted image. It can be seen that after calibration, the image captured by the camera is closer to the real situation, and the distorted lines caused by distortion are corrected.

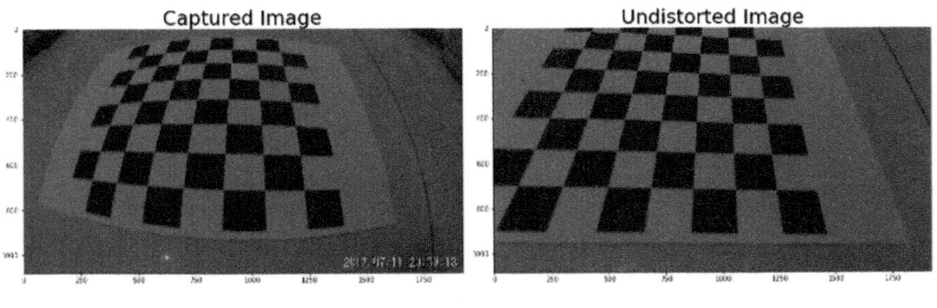

Fig. 2 Correction of image distortion

3.2 White Balance

First, *r*, *g*, and *b* components were separated from the image, and the mean values of *r*, *g*, and *b* components were obtained, respectively, as mean_r, mean_g, and mean_b. The correction coefficients KB, KG, and KR of the *r*, *g*, and *b* components were obtained according to Eq. (2).

$$\begin{aligned} \mathrm{KB} &= (\mathrm{mean\_}r + \mathrm{mean\_}g + \mathrm{mean\_}b)/(3 * \mathrm{mean\_}b) \\ \mathrm{KG} &= (\mathrm{mean\_}r + \mathrm{mean\_}g + \mathrm{mean\_}b)/(3 * \mathrm{mean\_}g) \\ \mathrm{KR} &= (\mathrm{mean\_}r + \mathrm{mean\_}g + \mathrm{mean\_}b)/(3 * \mathrm{mean\_}r) \end{aligned} \tag{2}$$

After the correction coefficients were obtained, multiply the original *r*, *g*, and *b* components by the correction coefficients KB, KG, and KR, respectively. If the value is greater than 255, the value is directly assigned to 255. The result of white balance is shown in Fig. 3, where the image before white balance is on the left side and the image after white balance is on the right side.

3.3 Combined Thresholds' Design

The lane is actually the edge of the image, which can be extracted by edge detection. The purpose of edge detection is to mark out the points with obvious changes in brightness in the digital image. Usually, the edge detection method based on search is adopted, that is, the edge strength is calculated in the form of the first derivative (gradient mode), then the local direction (gradient direction) of the edge is estimated, and the local maximum value of gradient mode can be found. In this chapter, the edge is calculated by using the Sobel operator. The operator consists of longitudinal and transverse directions and convolutes the image in plane to obtain the longitudinal and transverse approximate brightness values. Assuming that A is the original image, G_x and G_y are the images detected by transverse and

Fig. 3 Result of white balance

longitudinal edges, respectively, and the calculated values of G_x and G_y are obtained by the Sobel operator in Eq. (3):

$$\begin{gathered} G_x = \begin{pmatrix} -1 & 0 & 1 \\ -2 & 0 & 2 \\ -1 & 0 & 1 \end{pmatrix} * A \\ G_x = \begin{pmatrix} -1 & 0 & 1 \\ -2 & 0 & 2 \\ -1 & 0 & 1 \end{pmatrix} * A \end{gathered} \tag{3}$$

The calculation method of gradient amplitude is shown in Eq. (4):

$$G = \sqrt{G_x^2 + G_y^2} \tag{4}$$

The calculation method of gradient direction is shown in Eq. (5):

$$\theta = \arctan\left(\frac{G_y}{G_x}\right) \tag{5}$$

In the actual operation, a three-order Sobel operator was adopted, and the x and y gradient thresholds were selected as (25, 200). The gradient amplitude thresholds were selected as (30, 100). The gradient direction thresholds were selected as (0.8, 1.2).

Then, the image was converted from RGB space to HSL space. HSL color mode is a color standard in the industry. It obtains all kinds of colors by changing the three-color channels of hue (H), saturation (S), and lightness (L) and superimposing them. This standard includes almost all colors that human vision can perceive. And it is one of the most widely used color systems. Combined with the relevant features of the lane image, the threshold of the S channel was selected as (140, 255), and the threshold of the L channel was selected as (120, 255), so that it can achieve a better binarization effect.

By selecting the above thresholds, the obtained binary image is shown in Fig. 4. The image on the left side is the original image, and the image on the right side is the image after binarization. It can be seen that the pickup of the lane counter in the image after binarization is better than that in the original image.

3.4 *Perspective Transformation and Reverse Perspective Transformation*

The essence of perspective transformation is to project the image onto a new visual plane. Its general transformation formula is shown in Eq. (6):

Fig. 4 Binarization result with combined threshold

$$[x'\ y'\ w'] = [u\, v\, w] * \begin{bmatrix} a_{11} & a_{12} & a_{13} \\ a_{21} & a_{22} & a_{23} \\ a_{31} & a_{32} & a_{33} \end{bmatrix} \tag{6}$$

$(u\, v)$ are the original image coordinates, and $(x'/w'\ y'/w')$ are the image coordinates after the transformation.

Given the coordinates of the four pixels corresponding to the perspective transformation, the perspective transformation matrix can be obtained. The inverse process of perspective transformation is reverse perspective transformation.

The effect of perspective transformation is shown in the picture on the right side of Fig. 5.

3.5 *Quadric Curve Fitting of Lane*

The histogram was firstly obtained from the lower half of the bird's-eye view (the right part of Fig. 5), and the result is obtained as shown in Fig. 6.

The lane is fitted according to the following procedures:

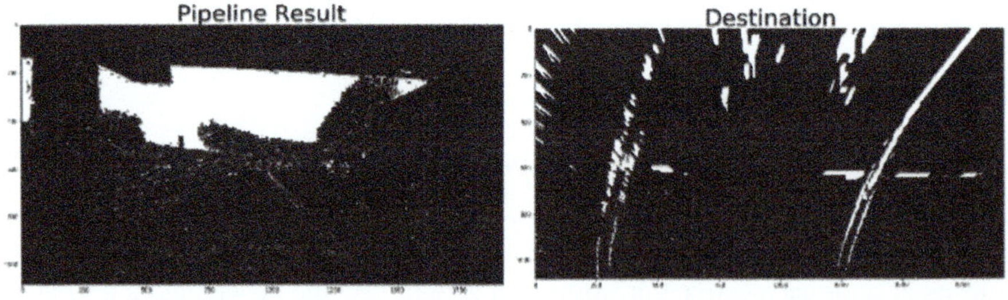

Fig. 5 Result of perspective transformation

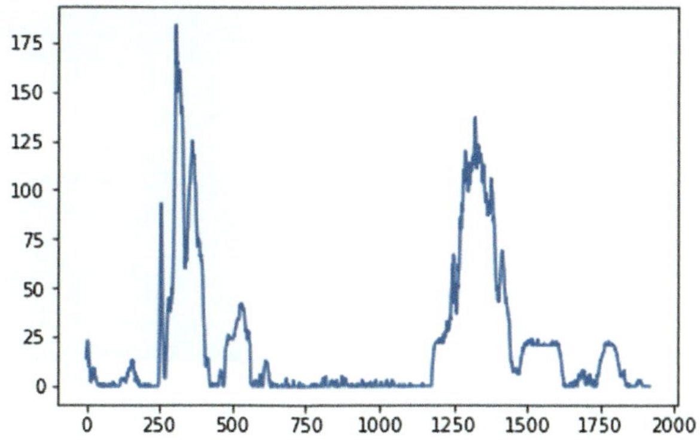

Fig. 6 Histogram of bird's-eye view

① Create a search window at the bottom of the image with a height of one-ninth of the image. Divide the search window into two halves—the left side and the right side.
② Find the pixel coordinates of the maximum value through the histogram, and draw the rectangular frame using the edge variable parameters in the area.
③ Statistics all nonzero pixels in the rectangular frame. If the number of the nonzero pixels exceeds the set threshold, obtain the center of gravity of the rectangle and take it as the center of the next search window.
④ Use the quadratic curve to fit the center of gravity of all the rectangles in the image of the left and right halves, so the lane equation is obtained.

The lane fitting result is shown in Fig. 7, where the dotted line is the lane obtained by fitting.

By reverse perspective transformation, the result was mapped to the original image, and the result of lane marking in the original image can be obtained, as shown in Fig. 8.

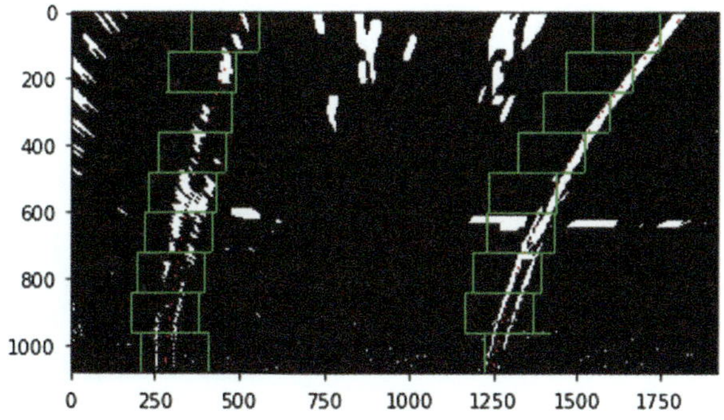

Fig. 7 Lane fitting results of bird's-eye view

Fig. 8 Result of lane marking in original image

4 Verification

In order to verify the effect of the designed algorithm on the recognition of discontinuous lane, the single-side discontinuous lane and the two-side discontinuous lane were selected for verification, as shown in Figs. 9 and 10, respectively. The results show that the algorithm is effective in recognizing discontinuous lanes and meets the design requirements.

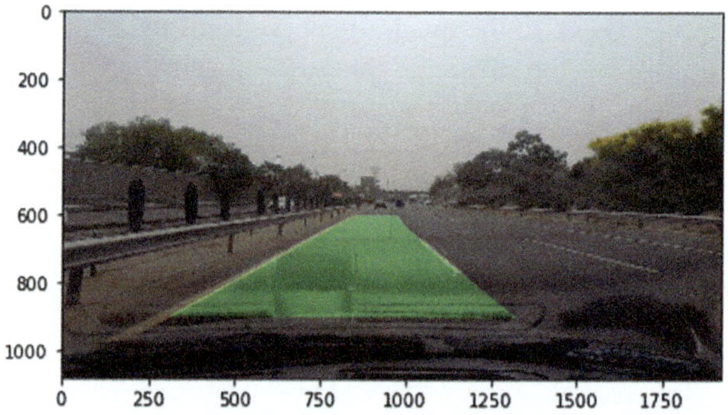

Fig. 9 Recognition effect of discontinuous lane on single side

Fig. 10 Recognition effect of discontinuous lane on double sides

5 Conclusion

This chapter designs an algorithm based on multiple threshold fusion, and the algorithm combines camera distortion correction, white balance algorithm, HSL color space threshold, gradient threshold, gradient magnitude threshold, gradient direction threshold, perspective transformation, reverse perspective transformation, and quadric curve fitting. The validity of the algorithm is verified by experiments under real conditions, and the conclusions are the following:

(1) The algorithm can accurately identify discontinuous lanes and provide accurate boundary information for intelligent vehicle path planning.
(2) By the combination of multiple thresholds, the robustness of lane recognition is greatly improved.

References

1. Yuehua, Cao, Wenguang, Luo, Hongli, Lan, Xiaodong, Zhao: Study on recognition algorithm of lane line on road in complicated environment. Mod. Electron. Techn. **40**(14), 109–113 (2017)
2. Geheng, Chen, Xiaoxu, Pan, Zuohui, Hou: Preceding vehicle detection algorithm based on recognition and multi-characteristics. Sci. Technol. Eng. **16**(15), 245–250 (2016)
3. Wang, Y., Fan X., Liu J., Pang Z.: Lane line recognition using region division on structured roads. J. Comput. Appl. **35**(09):2687–2691+2700 (2015)
4. Chen, Y., Yan, H., Xin, L.: A study on adaptive canny edge detection and improvement in lane recognition. J. Transp. Inf. Saf. **30**(01), 148–151 (2012)
5. Fan, C., Di, S., Hou, L., Shi, X.: Linear model based lane mark identification algorithm. Appl. Res. Comput. **29**(01):326–328+332 (2012)
6. Yu, H., Zhang, W.: Lane detection and tracking based on dynamic region of interest. Ind. Instrum. Autom. **05**, 103–106 (2009)

Research and Application of Multi-source Information Management for Electric Power Emergency

Xiyuan Xu, Zhen Yu, Yongsheng Men and Yusong Guo

Abstract In order to meet the actual needs of power emergency events, contact, and multi-channel interaction, this chapter designs a method of multi-source information management of power emergency fusion based on communication technology. And introduces the method to develop a set of power emergency command and support the work of the multi-source information management system. This method realizes the real-time and two-way interaction of telephone, fax, short message, voice, video and other multi-source information between the incident scene and the emergency command center. It could break the traditional emergency information exchange, emergency command and emergency response of the space and time limit, make up for the lack of effective emergency command information acquisition and transmission means.

Keywords Power · Unified communication · Mobile emergency · Emergency command · Multi-source information

1 Introduction

As the public infrastructure of national economy and people's livelihood, power system bears the important responsibility of providing reliable power supply for the society. In recent years, natural disasters, such as earthquakes, typhoons, snow, and ice, have occurred frequently in China, causing serious damage to the power system. In addition, terrorist attacks and other accidents also affect the safety of the power system. In this context, power enterprises have established emergency command centers to deal with all kinds of emergencies and carry out rapid and

X. Xu (✉) · Z. Yu · Y. Men · Y. Guo
Global Energy Interconnection Research Institute Co., Ltd., 102209 Beijing, China
e-mail: waxxy2009@126.com

X. Xu
State Grid Laboratory of Electric Power Communication Network Technology, 102209 Beijing, China

V. Jain et al. (eds.), *Recent Trends in Intelligent Computing, Communication and Devices*, Advances in Intelligent Systems and Computing 1031,
https://doi.org/10.1007/978-981-13-9406-5_120

efficient rescue and disaster relief. In the process of dealing with emergencies, emergency command centers at all levels need to know the scene of emergencies timely and accurately, so as to ensure that sufficient information can be provided to assist the relevant personnel to make reasonable decisions [1, 2].

Nowadays, with the continuous development of information technology and the increasing of information acquisition technology, there are more and more terminals which can be used to collect information on the spot of power emergencies. Many kinds of on-site information acquisition terminals bring more abundant on-site information and more timely on-site reports, but also bring a series of problems. The main problem is that a large number of videos, pictures, texts, and language information generated by various terminals cannot be managed effectively and uniformly due to the different ways of production and communication channels, which brings some difficulties to information management.[3–8].

2 Multi-source Information Classification for Power Emergency

After the occurrence of power emergencies, the most important task for power emergency command is to quickly and accurately grasp the situation on the spot. The information collected by various field information acquisition terminals and returned to the power emergency command center is multi-source information of power emergencies (hereinafter referred to as "multi-source information"). For the multi-source information of power emergencies, there are three main classification methods: classification by business type, classification by manifestation, and classification by collection terminal.

(1) By business type:

According to business types, multi-source information can be divided into basic information, damage information, weather information, and other field information.

(2) By manifestation type

According to the manifestation, multi-source information can be divided into text information, picture information, voice information, and video information.

(3) By collection terminal type

According to the acquisition terminal, multi-source information can be divided into mobile terminal collect information, cell phone collect information, telephone collect information, fax collect information, PC collect information.

The three classification methods of multi-source information are interrelated and should not be treated separately. A variety of acquisition terminals collect basic information, damage information, weather information, and other field information as text information, picture information, voice information, and video information, and then transmit emergency command center through the corresponding Internet,

	Mobile terminal	Cellphone	Telephone	FAX	PC
Text	basic information damage information weather information other field information	damage information weather information other field information	damage information other field information	damage information weather information other field information	basic information damage information weather information other field information
Picture	damage information other field information	damage information other field information	Unable to collect	damage information other field information	damage information weather information other field information
Voice	damage information other field information	damage information other field information	damage information other field information	Unable to collect	damage information weather information other field information
Video	damage information other field information	Unable to collect	Unable to collect	Unable to collect	damage information other field information

Fig. 1 Relation graph of information classification

mobile communication network, public telephone exchange network, and other transmission channels. The diagram of the three classification methods is shown in Fig. 1.

3 Present Situation of Information Unified Communication Technology

3.1 Unified Communication

With the continuous development of information technology, a new communication technology has emerged, that is, converged communication technology. The so-called convergent communication includes the convergence of multiple communication networks as well as the convergence of multiple terminals. It integrates Internet technology and communication technology, and integrates multiple communication networks on one platform to realize many application services such as telephone, fax, data transmission, audio and video conferences, call center, and instant messaging. Fusion communication can realize the unification of all kinds of communication and user experience, adapt to the communication needs of different industries and even different enterprises, and combine with enterprise applications, such as integration with OA systems, mail and office software, and third-party applications [9, 10].

3.2 Application Status of Unified Communication Technology in Other Industries

In view of the superiority of convergent communication and the future development of convergent communication, some government agencies and large state-owned enterprises such as China Mobile and China Telecom have adopted the solution of convergent communication in their daily work. [11] At the same time, in view of the limited funds, the convergent communication technology can simplify the user wiring, facilitate users' use, provide a unified user experience, and provide various communication services to users, which can meet most of the internal and external communication needs. Fusion communication technology has also been applied in the majority of small and medium-sized enterprises [12].

3.3 Application Status of Unified Communication Technology in Electric Power Industry

As a new and excellent technology, integrated communication technology has also been applied in the power industry, such as the application of power communication network in Tianjin Power Baodi Power Supply Company of State Grid [13] and the application of integrated communication platform in Zhengzhou Power Supply Company of State Grid [14].

Although integrated communication technology has been applied in the power industry, it has not been applied in the field of emergency command in the power industry. The emergence of integrated communication technology provides a direction for solving the problem that the data collected by various terminals on the spot of power emergencies cannot be effectively and uniformly managed.

4 Research on Multi-source Information Fusion Communication Technology for Power Emergency

4.1 Business Needs

At present, the way to deal with multi-source information is to collect information from various acquisition terminals and transmit it to different communication systems through their respective transmission channels. The information acquired by various communication systems is collected manually or semi-automatically by relevant personnel, and then submitted to emergency commanders for assistant decision making. This method has some problems, such as poor timeliness of information acquisition, large workload of information integration, and decentralized information storage, which cannot meet the needs of emergency command for multi-source information management. It is urgent for a unified management of multi-source information and system to assist in emergency command.

4.2 Key Technology Research

In order to build a multi-source information fusion system for power emergencies, the following three aspects of technology need to be studied.

(1) Information transmission:

After collecting field information by various acquisition terminals, the information will be transmitted to emergency command centers or related personnel through various transmission modes. Because of the different transmission modes, the information format of the emergency command center or related personnel is different. The unified management of multi-source information should complete the unified identification of the information transmitted by various transmission modes. To solve this problem, the fusion and unified identification of multi-source information can be completed based on fusion communication technology in the process of information transmission.

(2) Information storage:

The multi-source information transmission methods and forms obtained by emergency commanders are quite different, and they are in a relatively disordered state. The unified management of multi-source information should complete the standardized storage of information. To solve this problem, we can adopt matrix storage technology, that is, video information, voice information, picture information, and text information. Multi-source information is associated with emergencies, respectively. Finally, we can realize matrix storage of information according to event and video, voice, picture, and text.

(3) Visualization technology:

The multi-source information obtained by emergency command and management personnel is relatively rich. In order to make better use of multi-source information and better assist emergency commanders in decision making, more visualization methods should be provided to realize the diversity display of multi-source information. To solve this problem, four kinds of information display technologies, namely audiovisual display, chart display, list display, and map display, can be used to display the diversity of multi-source information.

5 Design of Multi-source Information Fusion System for Power Emergency

5.1 System Architecture Design

The system architecture design is completed around four main processes: multi-terminal acquisition, fusion communication, matrix storage, and diversified display. The system architecture is shown in Fig. 2.

The system collects on-site information of power emergencies through mobile emergency terminal, mobile phone, fixed telephone, fax machine, and computer terminals, obtains multi-source information, and transmits it to the fusion application service through various transmission channels; the fusion communication service converts and fuses multi-source information, and obtains the computer's capability. The information related to power emergencies is identified and transmitted to the information network of power enterprises. The information related to power emergencies is stored in matrix and provided with display pages for display of multi-source information.

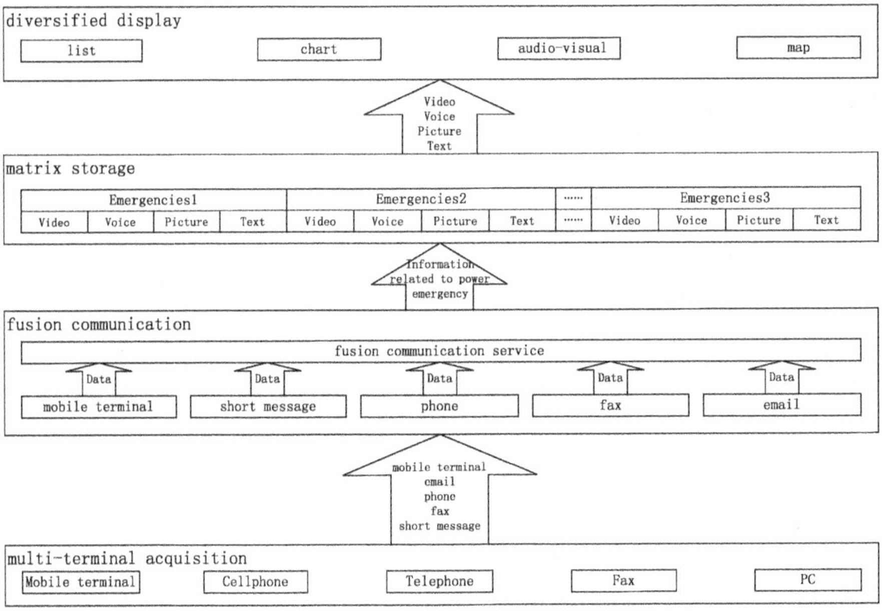

Fig. 2 System architecture diagram

5.2 System Application Architecture

The design of system application architecture is based on the overall architecture. The system is designed to be physical layer, service layer, data layer, and application layer.

The system application architecture is shown in Fig. 3.

(1) Physical layer:

As the basic carrier of the whole system and the physical support of the management method, the physical layer mainly includes mobile terminal, cell phone, telephone, fax, PC and other acquisition terminals, network equipment, short message platform, server, and other necessary hardware equipment.

Physical layer is mainly responsible for information collection in management methods, and at the same time, it works with service layer to complete information transmission in management methods. After collecting the field information of power emergencies, the physical layer transmits the information to the service layer through a variety of transmission channels.

(2) Service layer:

As the technical core of the whole system, the service layer mainly includes the fusion communication service which integrates various communication modes through the fusion communication technology.

The service layer links the preceding and the following, integrates various communication modes, and is responsible for the information transmission in the management method with the physical layer. The information related to electric power emergencies is received and processed to form the information related to electric power emergencies, and the information is transmitted to the data layer.

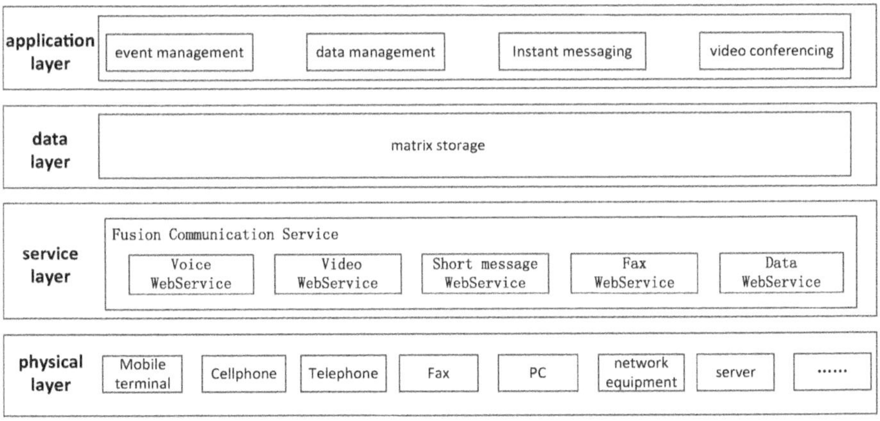

Fig. 3 System application architecture

(3) Data layer:

Data layer, as the data warehouse of the whole system, mainly includes data storage.

The data layer is mainly responsible for information storage in the management method. The data layer classifies and stores the information related to the received power emergencies in a matrix manner, and at the same time provides data support for the application layer.

(4) Application layer:

Application layer is the application core of the whole system. The first three levels are the support of this level and serve for this level.

The application layer is mainly responsible for completing the information display in the management method. On the basis of data storage in the data layer, the application layer centralizes the management and classifies the multi-source information of power emergencies. At the same time, it can interact with the scene of emergencies in real time, which provides technical support for the multi-source information management of power emergencies and the emergency command of power emergencies.

5.3 System Function Realization

Based on the above design ideas, in order to achieve the above management methods, the system realizes the basic functions of integration management, sending, receiving, and displaying of short message, picture information, video information, voice information, and fax information, and realizes the event management function, so that all information in the system can be associated with emergencies. The system realizes the functions of timely communication and audio-video consultation between the acquisition terminals and between the acquisition terminals and the command center.

The realization of the above functions meets the demand of power enterprises for timely and accurate grasp of multi-source information in emergency command, ensures timely and accurate information exchange between emergency scene and emergency command center, and enables emergency commanders to intuitively and accurately understand the feedback information of emergency scene.

6 System Application

Multi-source information fusion system for power emergency has been applied in State Grid Corporation of China. Take the State Grid Corporation of China against the typhoon 201617 Megi as an example.

In the whole anti-Taiwan process, the multi-source information fusion system of power emergency provides unified services to emergency commanders to assist decision making after fusing basic information of power grid, damage information of power grid, weather information, emergency repair site information, and so on.

(1) Basic information of power grid:

After the typhoon, according to the impact range predicted by the Central Meteorological Observatory, the system will obtain and integrate the relevant provincial power grid basic information into the system through the power system information Intranet to provide emergency commanders.

(2) Damaged information in power grid:

The typhoon landed on the coast of Hui'an County, Quanzhou, Fujian Province, at 4:40 a.m. on the 28th. After the typhoon landed, the system will continue to integrate the damaged information of power grid acquired by means of power system information Intranet, mobile phone short message, and e-mail into the system and provide it to emergency commanders.

(3) Weather information:

After the typhoon is generated, the system integrates weather-related information such as weather forecast information, typhoon path information, and satellite cloud image information released by meteorological departments into the system and provides them to emergency commanders.

(4) Rush repair site information:

After the typhoon passed, the information obtained by mobile terminal, power system information Intranet, mobile phone short message, and e-mail will be integrated into the system and provided to emergency commanders.

7 Conclusion

The multi-source information management method and multi-source information fusion system described in this chapter have been initially used in emergency command of State Grid Corporation of China, and achieved good results.

(1) Multi-source information management method for power emergencies:

The multi-source information management method of power emergencies based on fusion communication technology realizes the unified management and flexible display of field information of power emergencies and provides key support for objective decision making and flexible command of power emergencies.

(2) Multi-source information fusion system for power emergencies:

The multi-source information fusion system for power emergencies is deployed in the information network environment of power enterprises. Real-time acquisition and interaction of all kinds of emergency information on the scene of emergencies are fully realized through various communication modes, and voice communication, video conference, and instant information exchange between the scene of emergencies and the emergency command center are realized. In the unified emergency decision making and command work, break through the time and space constraints, efficient and rapid research and decision making and emergency disposal work. Its advantage lies not only in the integrated use of pictures, videos, voice, and other information on the scene of emergencies, but also in the integration of existing isolated communication networks to form an emergency command system centered on emergencies, which makes the emergency command work more efficient.

In the process of on-site disposal of unexpected events, the efficient upload and delivery of on-site information and command orders makes the on-site disposal more efficient and orderly and makes the State Grid Corporation obtain huge indirect economic benefits as well as huge social benefits.

The method and system described in this chapter provide an important support for the emergency command work of electric power and have a strong reference value for improving the emergency system of electric power enterprises.

Acknowledgements Supported by Science and Technology projects of State Grid Corporation of China. Any opinions, findings, and conclusions or recommendations expressed in this material are those of the authors and do not necessarily reflect the views of the sponsors.

References

1. Chen, X.: Research and Construction of Power Grid Emergency Platform. China Electric Power Publishing House (2010, December)
2. Liu, C., Xu, X.: Research on assessment of electric power emergency command center. Telecommun. Sci. **9**, 172–175 (2011)
3. Tian, S., Chen, X., Zhu, C., et al.: Study on electric power emergency management platform. Power Syst. Technol. **1**, 30–34 (2008)
4. Xu, X., Cao, J., Qu, X., et al.: Intelligent construction of basic support system of electric power emergency command center. Electrotechnical Appl. **S1**, 269–271 (2013)
5. Li, T., Wu, B., Pan, L., et al.: Research on information access method of emergency command center. Electric Power Inf. Technol. **11**, 38–41 (2012)
6. Xu, X., Cao, J., Men, Y., et al.: Application of local TV access system based on VOD technology in electric power emergency command center. Electrotechnical Appl. **S2**, 299–302 (2013)
7. Zhang, J., Pan, Y., Wang, Q., et al.: Research on application of soft switch in power grid emergency command. Telecommun. Electric Power Syst. **31**(8), 14–16 (2010)
8. Ma, X., Guo, Y., Li, L., et al.: Design of power emergency communication system based on mobile internet. Electric Power Inf. Commun. Technol. **7**, 17–22 (2016)

9. Li, J.: The integration of communication internet plus age. Appl. Electron. Techn. (z1), 70–73 (2015)
10. Zheng, J., Cao X., Zhao, L., et al.: Research and design of unified communication system based on IMS architecture. Electric Power Inf. Commun. Technol. (8), 78–81 (2016)
11. Gan, X.: Research on application scheme of full IP converged communication industry. Commun. Inf. Technol. (1), 43–46 (2012)
12. Song, J., Li, S.: Research on the scheme of small and medium sized enterprises based on IP communication network system. Netw. Secur. Technol. Appl (3), 57–57 (2015)
13. Tian, J., Yu, J., Liu, B.: Application of communication technology based on IP in power communication network. Tianjin Electric Power Technol. (3), 1–3 (2011)
14. Hui, Wu: Application of converged communication technology in electric power industry. Electric Power Inf. Commun. Technol. **09**(9), 85–88 (2011)

Optimization Design of Natural Gas Pipeline Based on a Hybrid Intelligent Algorithm

Yongtu Liang, Jianqin Zheng, Bohong Wang, Taicheng Zheng and Ning Xu

Abstract Natural gas pipeline is an important link between natural gas exploitation and utilization. To coordinate the domestic and foreign natural gas resources and satisfy the needs of various regions' economic development, China plans to continue accelerating the construction of natural gas pipeline network during the 13th five-year plan period. However, the problems of irrational pipeline construction and unexpected high cost of investment come out in the early stage. To solve these problems, this paper proposes a hybrid intelligent algorithm for optimization design. Firstly, the pipeline throughput is determined based on predicting the natural gas demand in the region with grayscale coupled neural network model, and then a mathematical model is established for natural gas pipeline system design, using particle swarm optimization (PSO) algorithm coupled with simulated annealing (SA) to solve this problem. A real gas pipeline in China is taken as an example to demonstrate the effectiveness of SA-PSO compared with traditional PSO algorithm.

Keywords Natural gas pipeline · Pipeline throughput · Optimization design · Mathematical model · SA-PSO algorithm

1 Introduction

In recent years, the state has emphasized green development and optimized energy structure [1, 2]. Natural gas is an indispensable pillar of the energy economy for the future development, and its main mode of transportation is pipeline transportation. For the construction of a pipeline, the first thing is the pipeline throughput, which is determined by the demand for natural gas consumption. Therefore, accurately

Y. Liang (✉) · J. Zheng · B. Wang · T. Zheng · N. Xu
National Engineering Laboratory for Pipeline Safety/MOE Key Laboratory of Petroleum Engineering/Beijing Key Laboratory of Urban Oil and Gas Distribution Technology, China University of Petroleum-Beijing, Fuxue Road no. 18, 102249 Changping District, Beijing, People's Republic of China
e-mail: liangyt21st@163.com

V. Jain et al. (eds.), *Recent Trends in Intelligent Computing, Communication and Devices*, Advances in Intelligent Systems and Computing 1031,
https://doi.org/10.1007/978-981-13-9406-5_121

predicting the natural gas demand in an area is conducive to the rational construction of the pipeline. Various methods for predicting natural gas demand are introduced in previous work [3], and grayscale coupled neural network model is effective. Minor improvements in parameters of pipeline system design and operating conditions can greatly reduce construction cost and operating cost [4].

At present, the steps of commonly used pipeline design method include designing a variety of schemes under the requirements of the transmission volume, evaluating the design scheme through pipeline simulation software or economic comparison, and determining the final construction scheme. At the same time, many scholars have applied intelligent algorithms to optimize pipeline design based on a mathematical model. Sanaye et al. [5] proposed an optimization model which minimized the sum of investment and operating costs, solving by genetic algorithm. Hamedi et al. [6] introduced a mixed integer nonlinear programming model to minimize direct or indirect distribution cost. Üster et al. [7] considered multi-period construction and pipeline network extension for the construction of gas pipeline network and established a mixed integer nonlinear programming model with the minimum investment and operating cost as the objective function, which was solved by online solver. Marketa et al. [8] considered the diversity of gas sources, and a multi-period mixed integer nonlinear programming model was established to optimize the design of the gas pipeline network based on the mass and energy balance equations. Chebouba [9] aimed at the steady flow of the gas pipeline, the operation scheme of the pipeline system is optimized with the minimum energy consumption as the objective function, and the ant colony algorithm is used to solve the problem. The solution result was better than the traditional descent method and dynamic programming. Tran et al. [10] proposed four MILP models for optimal line pack planning to compensate for the fluctuation of gas demand under scenarios. Mikolajkova et al. [11] proposed a mathematical model which considered daily variations in the energy demand, gas flow rate, pressures in the pipes, and compressor duties. Alves et al. [12] proposed a two-objective optimization model which minimized the transportation fare and maximized the transported gas volume to support the decision of regulatory authorities. Alem et al. [13] considered linear, branch, and looped topology network configurations and established a two-objective optimization model which minimized power consumption and maximized gas delivery flow rate. Amir et al. [14] proposed a two-objective optimization model which simultaneously maximized gas delivery flow and line pack, and minimized operating cost including sum of fuel consumption and carbon dioxide emission cost. Although the pipeline design is optimized, it is often caught in the local optimum due to the specification of the pipeline diameter and wall thickness. The combination of intelligent algorithms is used less in this research, but may avoid local optimization.

In this work, considering the uncertainty of natural gas demand, the influencing factors affecting natural gas demand are analyzed. The grayscale coupled neural network model is established to predict the demand for the future in the city, as an important indicator for determining pipeline throughput. By analyzing the total cost compositions of the natural gas pipeline system, a model of process parameter optimization design is established to minimize annual total cost. Finally, SA coupled with PSO is used to solve the model to avoid the local optimum.

The remainder of this paper is organized as follows. Section 2 introduces the grayscale coupled neural network model for predicting the natural gas consumption, and PSO algorithm coupled with SA is also shown in Sect. 2. A mathematical model is established for pipeline optimization design to minimize the annual total cost, and the constraints of safety and economics are considered in Sect. 3. In Sect. 4, a real gas pipeline is taken as an example to verify the effectiveness of the method. Section 5 is the conclusion of this work.

2 Methodology

2.1 Grayscale Coupled Neural Network Model

Grayscale model prediction is a method for predicting systems with uncertainties. It identifies the degree of difference between the development factors of system factors, generates and processes the original data to find the law of system variation, which can generate a data sequence with strong regularity, and then establishes a corresponding differential equation model to predict the future development of things. The basic idea is to use the original data to form the original sequence (0) and to generate the sequence (1) by the cumulative generation, which can weaken the randomness of the original data and make it exhibit obvious characteristic laws. A model of a differential equation type, that is, a GM model, is created for generating the transformed sequence (1). The most widely used gray prediction model is the GM (1,1) model for a variable of the series prediction and the first-order differential.

The neural network model consists of three layers: the input layer, the hidden layer, and the output layer [15]. The core is to continuously adjust the network parameters (weights and bias) by transmitting errors backward while correcting errors to achieve or approximate the system. For the learning process of neural networks, it is roughly divided into the following steps:

(1) Initializing parameters, including weights, offsets, network layer structure, activation functions;
(2) Cyclic calculations, including forward propagation, calculation of errors and backpropagation, adjustment of parameters;
(3) Return to the final neural network model.

In the actual prediction, if the uncertainty factors are relatively higher in the prediction object, it is difficult to obtain a better prediction result by using only a single prediction model. Combining neural networks to predict based on grayscale model is currently a popular predictive method. In this work, GM (1,1) predicts each variable, the predicted values are used as the input of the neural network, and the output of the neural network is used to determine the pipeline throughput.

2.2 SA-PSO Algorithm

Particle swarm optimization (PSO) is an evolutionary computation technique proposed by Eberhart and Dr. Kennedy in 1995 [16], which is derived from the behavioral study of predation of birds. The operators of the algorithm in this work are briefly described as follows:

(1) Initialize the particle swarm: Several individuals should be randomly generated to form this first generation, and the number of individuals is called the group size. In this paper, the dimension of each individual is the number of the optimization variables.
(2) Evaluate fitness: The fitness function is the basis of the continuous evolutionary search of the algorithm, and the fitness values of each are calculated and evaluated. Introducing a penalty function for constraints, when the individuals do not satisfy the constraints, the fitness is large and thus easier to eliminate.
(3) Find the two 'extreme values': In each iteration, the individual updates itself by tracking two 'extreme values,' namely individual optimal value and optimal group value. In this paper, the least annual cost of the pipeline is the optimal value for individual and group, respectively.
(4) Update the position and velocity: To form the next generation, after finding the two 'extreme values,' velocity and position of each are updated during iterations according to (1) and (2):

$$v_i(t+1) = w \times v_i + c_1 \times r_1 \times (p_i(t) - x_i(t)) + c_2 \times r_2 \times \left(p_g(t) - x_i(t)\right) \tag{1}$$

$$x_i(t+1) = x_i(t) + v_i(t+1) \tag{2}$$

where v_i and x_i are the velocity and position of particle i, respectively, w is the inertia weight, c_1 and c_2 are the respective acceleration parameters and represent two random numbers in the range [0, 1], p_i is the best position for a particle i, and p_g denotes the best position in the group at the iteration.

(5) End until convergence: The best individual will be selected with constant iteration, and until the maximum iteration is satisfied, the least annual cost of the natural gas pipeline is optimized in this work.

The idea of simulated annealing (SA) is based on the principle of solid annealing [17]. Starting from a higher temperature, this temperature is called the initial temperature, accompanied by the continuous decline of temperature parameters, the solution in the algorithm. Stable, however, it is possible that such a stable solution is a local optimal solution. At this time, the simulated annealing algorithm will jump out such a local optimal solution with a certain probability to find the global optimal solution of the objective function.

The PSO algorithm is applied to the solution of the model, and the dimension of the particle corresponds to the number of the optimization variables. Since D and δ satisfy the standard specification constraints, the PSO algorithm should set the closest actual standard pipe diameter and wall thickness after each update of the particles in the process of particle regeneration, so D and δ are easy to fall into a certain pipeline diameter and wall thickness during the update process, resulting in local minimum. Based on this, the idea of SA is coupled into the optimization process. The PSO algorithm particles are not always updated in the direction of large fitness, but based on the SA, a certain probability is allowed to update in the direction of small fitness. In this way, it is possible to jump out of a fixed value during the particle update process, and D and δ are easy to spread over the standard set to avoid falling into local optimum. The algorithm framework is shown in Fig. 1.

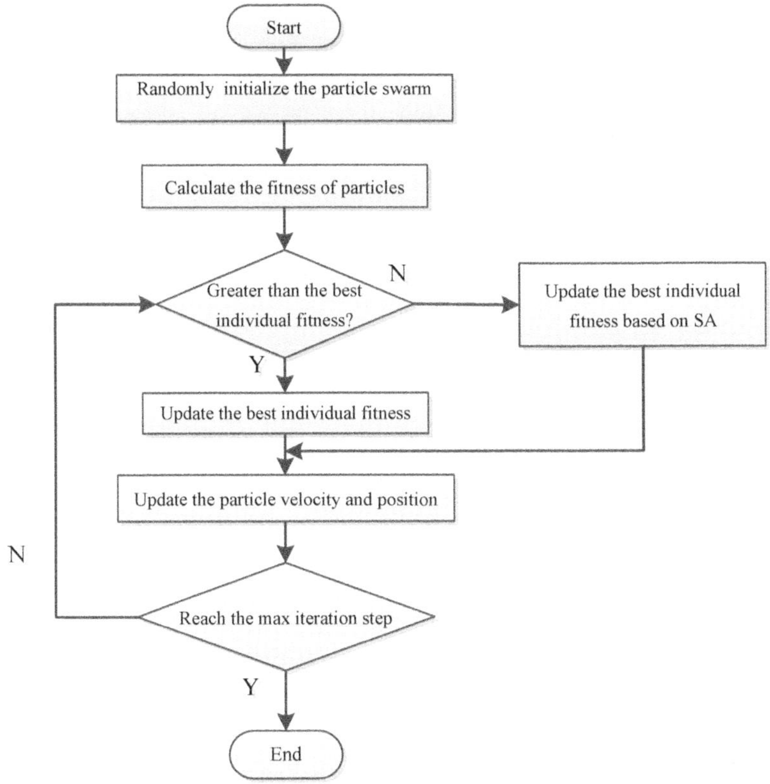

Fig. 1 Flowchart for the proposed algorithm

3 Mathematical Model

The total cost of the life of the gas pipeline is composed of five parts: pipeline construction cost, pipeline maintenance cost, compressor construction cost, compressor operation cost, and compressor maintenance cost. Pipeline construction cost and compressor construction cost are fixed investments during the construction period, which are one-time expenditures, while the other three expenses are expenditures year after year. Considering the time value of funds, the construction period costs are converted to the standard rate of return to each year of the operation period, and the total cost per year is calculated. On the basis of analyzing the total cost of the gas pipeline system, the mathematical model of the optimization design of the natural gas pipeline system is established with the minimum annual total cost as the objective function.

3.1 Objective Function

$$\min f = C1 + C2 + C3 + C4 + C5 \tag{3}$$

Objective function (3) minimizes the annual total cost of the pipeline and has five parts. The first part $C1$ is the annual pipeline construction cost, and $C2$ is the annual compressor station construction cost. The second part $C3$ is the annual compressor station operating cost. And the third part $C4$ and $C5$ are the annual maintenance cost of compressor station and pipeline, respectively.

$$C1 = \{a1 \times D + [a2 \times \delta \times (D - \delta) \times \pi \times \rho \times 10^{-3} + a3 \times D \times \pi] \times (1 + a4)\} \times L \times I \tag{4}$$

$$I = i \times (1 + i)^t / ((1 + i)^t - 1) \tag{5}$$

Equation (4) describes that the construction cost of a natural gas pipeline is related to the pipeline diameter and the wall thickness. Equation (5) describes the relationship between the annual discount factor and the benchmark yield and operating period.

$$C2 = (b0 + b1 \times N) \times n \times I \tag{6}$$

$$N = \frac{1}{\eta} \frac{4k}{k - 1} \mathrm{QZT}(\varepsilon^{\frac{k-1}{k}} - 1) \tag{7}$$

Equation (6) describes that the construction cost of the compressor station is related to the number of seats in the compressor stations and its power. Equation (7) is the formula for calculating the power of the compressors.

$$C3 = 10^{-4} \times \frac{n \times N \times tt \times 24 \times 3600}{\eta_c H_d} \times W \tag{8}$$

Compressor station drive mode for gas pipeline system is considered as fuel drive, and Eq. (8) is the formula for calculating operating cost, related to the operating time and the price of natural gas on site.

$$C4 = \alpha C2 \tag{9}$$

$$C5 = \beta C1 \tag{10}$$

Equations (9) and (10) describe the maintenance cost of compressor station and pipeline, which are related to the construction cost of the compressor station and pipeline, respectively.

3.2 *Constraints*

$$\varepsilon_{\min} \leq \varepsilon \leq \varepsilon_{\max} \tag{11}$$

$$P_{\min} \leq P_s \leq P_{\max} \tag{12}$$

$$P_{\min} \leq P_d \leq P_{\max} \tag{13}$$

These are boundary condition constraints. In order to ensure the safety of the pipeline operation, the pressure ratio of the compressors, and the inlet and outlet pressures are in the safe range, which are shown in constraints (11), (12), and (13), respectively.

$$D \in DD \tag{14}$$

$$\delta \in \delta\delta \tag{15}$$

These are pipeline specification constraints. The gas pipeline diameter and wall thickness of the gas pipeline are manufactured according to a certain standard series, and constraints (14) and (15) ensure the pipeline specification standard.

$$\delta \geq \frac{P_d D}{2\sigma F \phi} + r \tag{16}$$

This is pipeline strength constraint. Constraint (16) is the minimum wall thickness, considering that the operating pressure of the natural gas pipeline must meet the pipeline strength requirements at the highest pressure and the occurrence of pipeline corrosion.

$$D/\delta \leq M \tag{17}$$

This is pipeline stability constraint. Constraint (17) shows that in order to prevent cross-sectional instability, the ratio of pipe diameter to wall thickness is limited for buried pipelines.

$$P_d^2 - P_s^2 = CL_0Q^2 \tag{18}$$

$$P_s^2 - P_o^2 = C(L - n \times L_0)Q^2 \tag{19}$$

$$C = \frac{\lambda Z \Delta_* T}{C_0^2 D^5} \tag{20}$$

These are hydraulic constraints. Considering the constraints of pipeline inlet and outlet pressure and station spacing, constraints (18), (19), and (20) describe the pressure drop equation between stations and the end pressure loss of the pipeline.

$$0.9(L - n \times L_0) \leq \frac{P_s^2 - P_o^2}{2CQ^2} \leq 1.1(L - n \times L_0) \tag{21}$$

This is end gas storage constraint. Considering the end gas storage, constraint (21) ensures that the end length is limited to 0.9–1.1 times the optimal end length.

4 Case Study

4.1 Pipeline Throughput Determination

The factors that affect the demand for natural gas considered by the grayscale coupled neural network model are the GDP, weather (temperature), and regional population of the region. Considering the increase in natural gas demand in the region and the stable production constraints of pipeline gas sources, the predicted demand for the fifth year of the region is an important indicator for determining pipeline output. The length of the pipeline is initially designed to be 1000 km. By checking the Web site of the National Bureau of Statistics of the People's Republic of China (http://www.stats.gov.cn/), the basic data is obtained.

Firstly, the grayscale model is used to predict the predicted value of GDP, temperature, and population for the fifth year, and as the input of the neural network, the demand for natural gas in the region in the fifth year is predicted to determine the pipeline throughput. The number of input layer nodes in the neural network model is 3, and the number of output layer nodes is 1. The number of hidden layer nodes is determined by empirical formula and trial and error, and it turns out to be 5 finally. It turns out that the pipeline throughput is determined as 26 billion cubic meters per year.

4.2 Results and Discussion

According to the above analysis, the mathematical model is solved by PSO algorithm and SA-PSO algorithm, respectively. In order to ensure the stability of the algorithm and solve the optimal solution, ten calculations are performed for each of these two algorithms. The number of initial particles is 100, and the number of iterations is 50. The convergence curves are shown in Fig. 2.

Comprehensive consideration of convergence stability and the fitness values, it can be concluded from Fig. 2 that the results of SA-PSO algorithm are better than that of PSO algorithm. And the comparison of the solution results is shown in Table 1.

By analyzing the solution results, the annual cost of the plan given through SA-PSO algorithm is RMB ¥2.430 billion, which saves approximately 0.188 billion yuan, 7.2% compared with the PSO algorithm. According to the analysis of Fig. 2a and b, we can find that in the iterative process, although the solution result is close to 2.600 billion yuan, SA-PSO can jump out of the local optimal solution more easily and find a smaller value while PSO sometimes falls into a local minimum and it is hard to find smaller values. Based on the above analysis, it can be concluded that the SA-PSO algorithm is superior to the traditional PSO algorithm in solving the natural gas pipeline system model established in this paper.

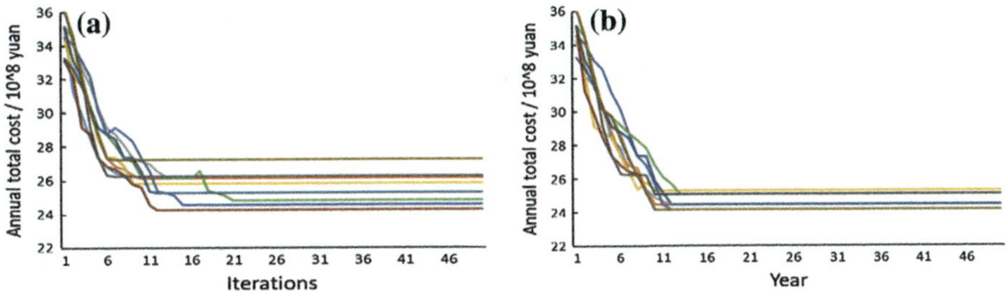

Fig. 2 Convergence curves for **a** PSO algorithm and **b** SA-PSO algorithm

Table 1 Comparison of the solution results of these two algorithms

	The annual total cost ($\times 10^8$ yuan)	Diameter (mm)	Wall thickness (mm)	Pressure ratio	Outlet pressure (MPa)
PSO	26.18	1219	20.1	1.49	9.45
SA-PSO	24.30	1219	21.8	1.45	9.60

5 Conclusion

This study proposed a hybrid intelligent algorithm for optimization design of natural gas pipeline. Considering the development of the region, this work comprehensively analyzes the actual natural gas consumption and introduces a grayscale coupled neural network model to predict the demand and determine the pipeline throughput. Taking construction cost, operating cost, and maintenance cost into consideration, a mathematical model is established to optimize the pipeline parameters, contributing to the minimal annual total cost. In order to avoid falling into local optimum, SA-PSO is proposed to traverse the entire pipeline specification sets easily. Finally, a real gas pipeline is taken to verify the effectiveness of the method.

References

1. Lu, S., Wang, J., Shang, Y., et al.: Potential assessment of optimizing energy structure in the city of carbon intensity target. Appl. Energy **194**, 765–773 (2016)
2. Wang, Z., Zhu, Y., Zhu, Y., et al.: Energy structure change and carbon emission trends in China. Energy **115**(369–377) (2016)
3. Zheng, J., Wang, B., Zhang, H., Liang, Y.: Research progress in natural gas demand forecasting. J. Petrochemical Univ. **31**(04), 1–6 (2018)
4. Wang, B., Yuan, M., Zhang, H., et al.: An MILP model for optimal design of multi-period natural gas transmission network. Chem. Eng. Res. Des. **129**(1221–1231) (2018)
5. Sanaye, S., Mahmoudimehr, J.: Optimal design of a natural gas transmission network layout. Chem. Eng. Res. Des. **91**(12), 2465–2476 (2013)
6. Hamedi, M., Zanjirani Farahani R, Husseini, M.M., et al.: A distribution planning model for natural gas supply chain: a case study. Energy Policy **37**(3), 799–812 (2009)
7. Üster, H., Dilaveroğlu, Ş.: Optimization for design and operation of natural gas transmission networks. Appl. Energy **133**(133), 56–69 (2014)
8. Mikolajková, M., Haikarainen, C., Saxén, H., et al.: Optimization of a natural gas distribution network with potential future extensions. Energy **125**, 848–859 (2017)
9. Chebouba, A., Yalaoui, F., Smati, A., et al.: Optimization of natural gas pipeline transportation using ant colony optimization. Comput. Oper. Res. **36**(6), 1916–1923 (2009)
10. Tran, T.H., French, S., Ashman, R., et al.: Linepack planning models for gas transmission network under uncertainty. Eur. J. Oper. Res. **268**(2), 688–702 (2018)
11. Mikolajková, M., Saxén, H., Pettersson, F.: Optimization of the cost of compression in the Finnish natural gas pipeline. In: Gernaey K.V., Huusom J.K., Gani, R (eds) Computer Aided Chemical Engineering, pp. 2249–2254. Elsevier (2015)
12. Alves, F.D.S, Souzaj, N.M.D., Costa, A.L.H.: Multi-objective design optimization of natural gas transmission networks. Comput. Chem. Eng. **93**, (212–220) (2016)
13. Demissie, A., Zhu, W., Belachew, C.T.: A multi-objective optimization model for gas pipeline operations. Comput. Chem. Eng. 100, 94–103 (2017)
14. Alinia Kashani, A.H., Molaei, R.: Techno-economical and environmental optimization of natural gas network operation. Chem. Eng. Res. Des. 92(11): 2106–22 (2014)
15. Dayhoff, J.E.: Neural Network Architectures: An Introduction. Van Nostrand Reinhold Co., (1990)

16. Kennedy, J., Eberhart, R.: Particle swarm optimization. In: Proceedings of the Fourth IEEE International Conference on Neural Networks, 1995 (2002)
17. Laarhoven P.J.M.V., Aarts, E.H.L.: Simulated annealing: theory and applications. D. Reidel Publishing Company (1987)

A Load-Shedding Technique Based on the Measurement Project Definition

Mario José Diván and María Laura Sánchez Reynoso

Abstract The real-time data processing is becoming a key aspect in relation to the Internet of things (IoT) applications. The IoT is characterized by the heterogeneity of the devices, and for that reason, the data providing rate of each one is variable and unpredictable. Because the data arriving rate from the data sources could exceed the data processing rate, the use of the load-shedding techniques is necessary. The metadata-guided processing strategy is a real-time data processing schema which the project definitions are based on a framework. Here, a new load-shedding technique based on the measurement project definition is introduced. This allows balancing between the data variability and the priority, retaining the important data based on the expert's knowledge from the project definition and the variations of the data series related to each metric.

Keywords Data stream · Load shedding · Measurement · Project definition · IoT

1 Introduction

From the definition of the data-driven decision making outlined by Provost and Fawcet [1], it is possible to define the information-driven decision making such as the process in which each decision avoids the intuition and it is based on those data that simultaneously are consistent, true, opportune, and interesting. This is a key asset in the current economy which continuously is oriented to the real-time decision making [2].

In this sense, the data stream applications are oriented to process the data such as they are and without any change, which implies a challenge in terms of the online processing and its available resources [3]. Thus, the data stream processing should solve the processing demand just with the available resources located in the processor.

M. J. Diván (✉) · M. L. Sánchez Reynoso
Economy and Law School, National University of La Pampa, La Pampa, Argentina
e-mail: mjdivan@eco.unlpam.edu.ar

V. Jain et al. (eds.), *Recent Trends in Intelligent Computing, Communication and Devices*, Advances in Intelligent Systems and Computing 1031,
https://doi.org/10.1007/978-981-13-9406-5_122

Because the data stream processing takes the data such as they are, the data sources are heterogeneous, and they are out of the control of the processor [4]. The Internet of things (IoT) allows opening a wide field of applications related to the data stream processing in which the measurement process has a key role for knowing the current state related to an entity under monitoring [5]. Thus, the measurement process is substantiated by measurement and evaluation (M & E) frameworks for making possible the comparability of the results, the process repeatability, and its extensibility [6].

The metadata-guided processing strategy (known as PAbMM) [7] is a real-time data processing engine based on the Apache Storm Topology (http://storm.apache.org). It is able to make the metadata-driven data processing supported by an M & E framework with the aim to automatize the measurement process from the heterogeneous data sources.

Because the data sources are heterogeneous, each one has its data providing rate. Thus, the data providing rate could exceed the data processing rate making necessary the use of the load-shedding techniques for selective discarding of data, minimizing the impact on the final results [8]. As the main contribution, the incorporation of a metadata-driven load-shedding technique based on the M&E project definition at PAbMM is introduced. This is useful because originally the load shedding just was oriented to the data itself without any kind of discrimination around the measurement project and its criticality. Now, PAbMM can discard the measures based on the project definition, balancing between its associated criticality, and the data variation.

The article is structured around five sections. Section 2 introduces some related works. Section 3 synthetically describes the metadata-guided processing strategy and the role related to the project definition. Section 4 presents the metadata-driven load-shedding technique. Section 5 outlines the conclusions and future works.

2 Related Works

In [9], a distributed data stream engine named THEMIS is introduced, which incorporates the supporting for the distributed load-shedding techniques. In this scenario, each data processing site keeps the independence in relation to the local resources, and for that reason, the load-shedding strategy is distributed and locally managed by each site (i.e., it has autonomy). In addition, the source information content (SIC) metric is introduced for quantifying the contribution of each data in the expected results. Thus, from the contribution to the results, the tuples from the data sources are retained or not. However, it manages the data in terms of contribution to the final results without considering the meaning of the data itself, and its relative importance. That is to say, in case of an outpatient monitoring of a person, the heart frequency is essential in detriment of the environmental temperature (e.g., related to the entity context), and for that reason, in the case of saturation in the processing capacity, it is necessary to preserve the heart frequency measures in place of the environmental temperature measures.

In [10], a treatment of the load-shedding techniques is made in relation to query scheduling for prioritizing the continuous queries (CQ) when the processing capacity is near to be committed. It is interesting because the adaptive load manager named ALoMa discriminates the CQ based on its importance, and the ALoMa behavior is aligned with the query scheduler, fostering the mutual synergy. However, given a critical CQ, the data are processed without information related to its meaning or importance.

In [11], the load-shedding technique is load-aware which gives origin to the load-aware shedding (LAS) strategy. In this kind of strategy, a runtime cost model is used for measuring and monitoring the execution times related to each tuple. Thus, using the runtime collected information is possible to determine when the data stream processor is near to be saturated and to apply the selective discard on the data. In this way, the data discarding is applied in a direct way on the tuple, without discriminating the meaning or importance related to the data in its context.

3 IoT and the Automatization of the Measurement Process

The metadata-guided data processing strategy is a topology mounted on Apache Storm. It is specialized in active monitoring of entities. In PAbMM, before starting the processing of the measures, the measurement project must be formally defined following a measurement framework. Thus, the project definition contains the entity to be monitored, the attributes to be quantified by the metrics, the metric definition, and the associated sensors, among other aspects. The project definition could be interchanged between systems by mean of the CINCAMI/Project Definition (PD) schema, which has an open-source library with the aim of fostering their use and application [12].

Because the data sources (i.e., sensors) are heterogeneous (e.g., a set of devices in IoT), a measurement interchange schema is necessary for the data homogenization. In this way, PAbMM uses a measurement interchange schema named CINCAMIMIS for fostering the data interoperability between the data sources and the processors [6]. Even, an open-source library is available for fostering its use and application illustrated for the IoT context [13].

In the processing architecture, all the measures from the data sensors are collected and communicated through a component named "measurement adapter" (MA). The measurement adapter is located on the mobile devices, and it is responsible for translating the measures from the sensors (expressed in the original data format) to the measurement interchange schema (i.e., CINCAMIMIS).

Figure 1 synthesizes the high level of CINCAMIMIS. On the one hand, the "A" represents that the lower tags could be multiple and without a given order (e.g., each *measurementItemSet* could have a set of *measurementItem* tag) in any order. On the other hand, the "S" represents that the lower tags appear just one time and must keep a given order (e.g., first the *idEntity* tag, second the *Measurement* tag, and third the *context* tag). The CINCAMIMIS stream identifies the measurement adapter responsible for the message generation (*dsAdapterID* in Fig. 1), the data source related to each measure (*dataSourceID* in Fig. 1), a footprint for verifying the message integrity (*footprint* in Fig. 1), the entity ID related to the measure located under the Measurement tag, jointly with the measures related to the context, under the *context* tag.

In this way, PAbMM could be characterized through five main processes: (i) *Configuration and Startup (CS)*: On the one hand, it defines the association between sensors and the measurement adapter by mean of the measurement project definition. On the other hand, the data processors load and start up the data structures and buffers following the project definition, reaching ready to receive the measurement stream from each measurement adapter; (ii) *Collecting and Adapting (CA)*: It is responsible for the data collecting from each data source related to the measurement adapter, the translating to the measurement interchange schema of them and its sending to the data gatherer; (iii) *Data Gathering (DG)*: It receives the data streams from the measurement adapters and communicates it to the analysis and smoothing process jointly with the decision-making process. In addition, it monitors the data processing load, and it makes the data replication to the external targets when it is requested; (iv) *Analysis and Smoothing (AS)*: It performs the online statistical analysis on the data streams. In case some abnormal situation is detected in comparison with the project definition, it throws an alarm to the decision maker for its evaluation; and (v) *Decision Making (DM)*: It monitors the typified situations using incremental classifiers (i.e., Hoeffding trees), sending the alarms when a situation is detected. In addition, it evaluates the received alarms from the classifiers and the statistical analysis, using an organizational memory for making the recommendations when it is necessary.

The metadata-driven load-shedding technique carries forward the selective discarding inside the data gathering process in PAbMM (see Fig. 2), based on the measurement project definition. The project definition incorporates the knowledge from the domain's experts, who establish the data retaining priority in terms of the aim of each measurement project (i.e., establishing the weighting for each entity and its associated attributes).

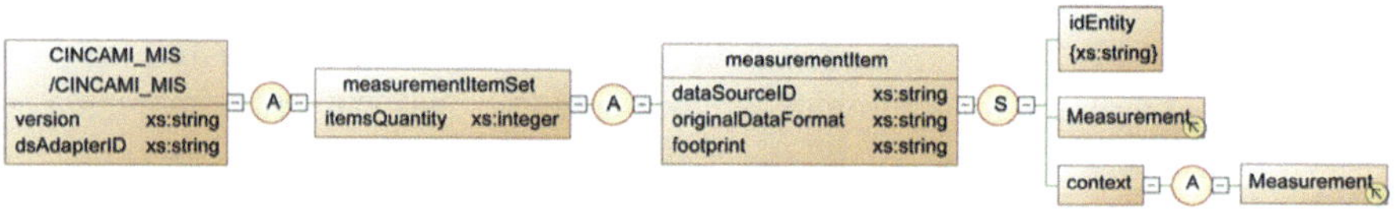

Fig. 1 High level of the CINCAMI/measurement interchange schema, version 2

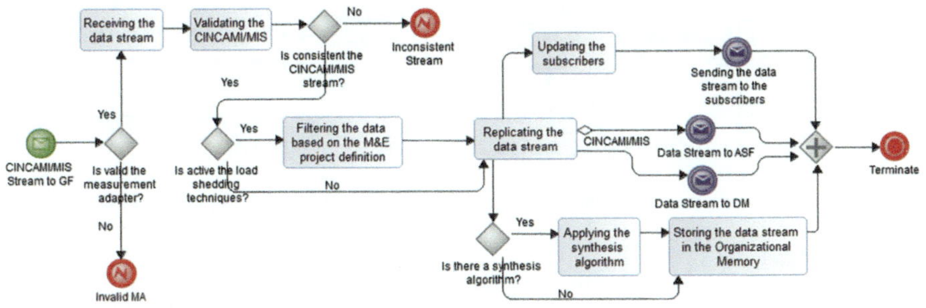

Fig. 2 Data gathering process using BPMN notation

4 A Metadata-Driven Load-Shedding Technique

A load-shedding technique should determine *when* it is applied, the way in which the data are discarding (i.e., *How*), *where* is performed, and on *what* data specifically [10]. As it is possible to see in Fig. 2, the load-shedding technique is applied to the gathering function when the data are received (i.e., *where*). In addition, the data to be analyzed by the load-shedding technique are related to each measurement stream characterized through a logical window under the form of a CINCAMI/MIS message (i.e., *what data*). For determining the total load (TL) and based on [10], the cumulative load by measurement adapter (MA) is computed using the formula 1.

$$\mathrm{TL} = \sum_{m=1}^{n} \mathrm{MA}_m = \sum_{m=1}^{n} \left(\sum_{E=1}^{n'} r_m^E * \mathrm{load}_m^E \right) \quad (1)$$

where:

- r_m^E: It symbolizes the data arriving rate related to the entity "*E*" from the measurement adapter "*m*".
- load_m^E: It is the cumulative processing time for the entity "*E*" and measurement adapter "*m*", in all the operations from the AS to the DM (e.g., replication and data synthesis).

When TL exceeds the established threshold as a parameter in PAbMM, the load shedding is automatically enabled in the DG. When the load shedding is activated, each CINCAMIMIS message is translated in a chained hash table where the first level is in function of the *idEntity* tag (see Figs. 1 and 3), while for each *idEntity*, the second level is in function of the *idMetric* tag (specified inside the *Measurement* tag; see Figs. 1 and 3). The first level related to the entity contains its weighting jointly the associated hash tables for its associated attributes. The second level for a given entity contains the weighting for each attribute jointly with the received measures (see Fig. 3).

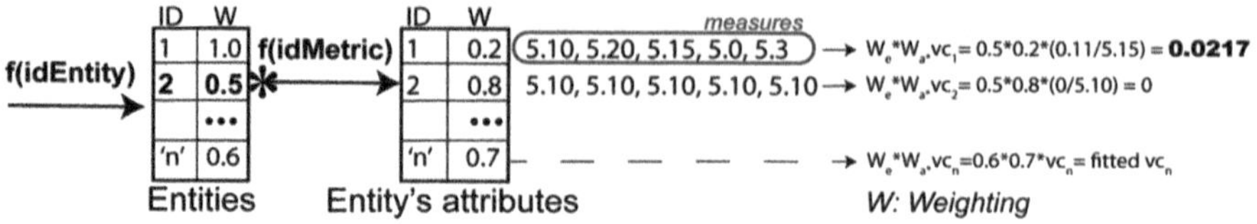

Fig. 3 An automatic scoring for the load shedding based on the metadata and data series

The variation coefficient (VC) is computed as the relation between the standard deviation and the arithmetic mean [14], and it is dimensionless. While the weighting of the entities and their associated attributes indicates its relative importance (established by the domain's experts), the VC analyzes the variability of the measures. Thus, the product among the entity's weighting (W_e), attribute weighting (W_a), and the CV gives a value which can be globally ordered in descendent fashion for being taken as the retaining policy. That is to say, the calculus previously indicated is used like an automatic scoring based on the project definition (the weighting related to the entities and their attributes) and the variation of the data series. It is important to highlight that PAbMM is specifically oriented to the measurement projects, and in the case of the attribute monitoring with data processing saturation, the key aspect is to analyze and act on the variations and not when the measures continue converging to a given value. For example, Fig. 3 has the attributes 1 and 2 for the entity 2. The attribute 1 is lesser important (0.2) than the attribute 2 (0.8); however analyzing the measures for each one, the attribute 2 does not present any variation while the attribute 1 does. For this reason, attribute 1 is located upper than attribute 1 in the retaining data political associated with the automatic scoring.

Finally, using the automatic scoring, the priority data are processed (the high values in the scoring) organized by entity's attributes, discarding all the others until the TL falls under the specified parameter in PAbMM, disabling the load-shedding techniques.

5 Conclusions

A metadata-driven load-shedding technique based on the M & E project definition was introduced. It consists of two chained hash tables with weightings and the measures of the associated attributes for obtaining the automatic scoring. Thus, in PAbMM the load-shedding technique is enabled when the total load exceeds the reference parameter, acting on each CINCAMI/MIS message (such as a logical window) inside the data gathering process. The proposed approach allows computing an automatic scoring using the weighting for entities and the associated attributes jointly with the variability of each data series. Thus, the discarding is oriented to the less important data with low (or null) data variability, which is a key in a measurement process for avoiding affecting the results. As future work, the effect of the outliers in this approach will be analyzed.

References

1. Provost, F., Fawcett, T.: Data science and its relationship to big data and data-driven decision making. Big Data **1**(1), 51–59 (2013)
2. Mohan, L., Potnis, D.: Real-time decision-making to serve the unbanked poor in the developing world. In: Proceedings of the 2017 ACM SIGMIS Conference on Computers and People Research, Bangalore (2017)
3. Morales, G., Bifet, A.: SAMOA: scalable advanced massive online analysis. J. Mach. Learn. Res. **16**, 149–153 (2015)
4. Jankov, D., Sikdar, S., Mukherjee, R., Teymourian, K. Jermaine, C.: Real-time high-performance anomaly detection over data streams: grand challenge. In: 11th ACM International Conference on Distributed and Event-based Systems, Barcelona, Spain (2017)
5. Saatkamp, K., Breitenbucher, U., Leymann, F., Wurster, M.: Generic driver injection for automated IoT application deployments. In: 19th International Conference on Information Integration and Web-based Applications & Services, Salzburg, Austria (2017)
6. Diván, M., Martín, M.: Towards a consistent measurement stream processing from heterogeneous data sources. Int. J. Electri. Comput. Eng. (IJECE) **7**(6), 3164–3175 (2017)
7. Diván, M., Sánchez Reynoso, M.: Behavioural similarity analysis for supporting the recommendation in PAbMM. In: 1st International Conference on Infocom Technologies and Unmanned Systems (ICTUS), Dubai (2017)
8. Querzoni, L., Rivetti, N.: Data streaming and its application to stream processing: tutorial. In: 11th ACM International Conference on Distributed and Event-based Systems, Barcelona, Spain (2017)
9. Kalyvianaki, E., Fiscato, M., Salonidis, T. Pietzuch, P.: THEMIS: fairness in federated stream processing under overload. In: 2016 International Conference on Management of Data, San Francisco, California, USA (2016)
10. Pham, T., Chrysanthis, P., Labrinidis, A.: Avoiding class warfare: managing continuous queries with differentiated classes of service. VLDB J.—Int. J. Very Large Data Bases **25**(2), 197–221 (2016)
11. Rivetti, N., Busnel, Y., Querzoni, L.: Load-aware shedding in stream processing systems. In 10th ACM International Conference on Distributed and Event-based Systems, Irvine, California (2016)
12. Diván, M., Sánchez Reynoso, M.: Fostering the interoperability of the measurement and evaluation project definitions in PAbMM. In: 7nd International Conference on Realiability, Infocom Technologies and Optimization (ICRITO'2018), Noida (2018)
13. Diván, M.: Applying the real-time monitoring based on wireless sensor networks: the Bajo Giuliani Project. In: 7th International Conference on Reliability, Infocom Technologies and Optimization (ICRITO'2018), Noida, India (2018)
14. James, J., Witten, D., Hastie, T., Tibshirani, R.: An introduction to statistical learning with applications in R, 8th edn. Springer Science+Business Media, New York (2017)

Design of Test Platform of Connected-Autonomous Vehicles and Transportation Electrification

Hossam A. Gabbar, Abul Hasan Fahad and Ahmed M. Othman

Abstract This paper presents a technical review on potential test beds in order to design an integrated testing and verification platform of connected and autonomous vehicle and transportation electrification (CAVTE) at Automotive Center of Excellence (ACE) facility within UOIT. In collaboration with Canadian Standards Association (CSA), CAVTE will test, verify, certify, and lead the standardization activity of modern and innovative vehicular and electrification technologies such as autonomous vehicles, wireless charging, energy storage systems, V2V and V2I connectivity, and cyber-physical security. CAVTE will support auto manufacturers, infrastructure service providers, utilities and government agencies with its industry-leading facilities. Within the ACE at UOIT, CAVTE can facilitate testing in different operating and weather conditions which will enable realistic and comprehensive results and will enable faster commercialization and adoption of innovative transportation technologies.

Keywords Testing · Validation · Connected and automated vehicles · Transportation electrification · Battery management system · EV chargers and supply equipment · Electro-magnetic compatibility · Powertrain · Vehicular communication · Smart grid interoperability · Cybersecurity

H. A. Gabbar (✉) · A. H. Fahad · A. M. Othman
Faculty of Energy Systems and Nuclear Sciences, University of Ontario Institute of Technology (UOIT), Oshawa, Canada
e-mail: hossam.gabbar@uoit.ca

A. M. Othman
Department of Electric Power & Machine, Faculty of Engineering, Zagazig University, Zagazig, Egypt

V. Jain et al. (eds.), *Recent Trends in Intelligent Computing, Communication and Devices*, Advances in Intelligent Systems and Computing 1031,
https://doi.org/10.1007/978-981-13-9406-5_123

1 Introduction

For mass adoption of electric vehicles (EVs), it is crucial that manufactured vehicle and associated vehicular systems are tested and certified by appropriate standardization. To facilitate this task, there are several automotive Standard Development Organizations (SDOs) such as IEC, SAE, ISO, UL LLC, CSA Group, SAC from China, and AISC from India [1]. In the case of USA and Canada, Fig. 1 represents the technical committee structure and IEC/ISO level activities [1].

After the standards, the next step is the development of testing process, which ensures products adhere to the relevant standardized requirement. Passing these requirements lead the way to certification. There are several steps in a testing process in the automotive industry [2]. The first step is planning phase and it consists of resource allocation and role assignments for performing the tests. The second step is analysis and design phase where test cases are specified and simulation models are produced. Thirdly, in implementation and execution phase, tests are conducted and necessary data as per plan are gathered. The fourth step is evaluation of exit-criteria and reporting comprises of the test results evaluation. Control phase is associated with second, third, and fourth steps. Control step is the comparison of current test activities against test plan, and if a step is not satisfactory, it results in intervention. The workflow can be shown in a figure (Fig. 2).

The current paper describes the technical research work currently underway in order to plan and develop an integrated testing and verification platform of autonomous vehicle and transportation electrification (CAVTE) at Automotive Center of Excellence (ACE) facility at UOIT. The platform will evaluate functions and capabilities of connected and autonomous transportation systems and support the development and use of Canadian and International standards, as well Certification activities. Subsequent sections of this paper will describe the findings from the current literature regarding the construction of testing setups in order to

Fig. 1 Electric vehicle committee structure

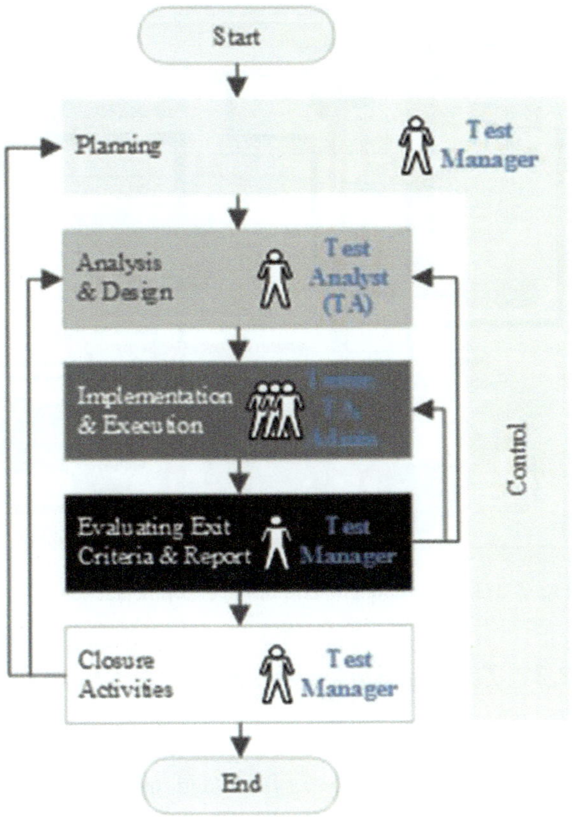

Fig. 2 Steps of test process

test the following components: battery and battery management systems, electric vehicle chargers and supply equipment, powertrain, electro-magnetic compatibility (EMC), autonomous electric vehicles, intra-vehicular communication, EV-smart grid interoperability (V2G), and cybersecurity. Lastly, an integrated test platform is proposed.

2 Testing EV Batteries and Battery Management System

2.1 Battery Tests

Essentially, there are two types of tests for battery: (1) performance tests and (2) safety tests [3]. Performance tests evaluate battery performance in operational conditions. Safety tests are conducted in order to evaluate a battery under internal and/or external failure conditions. In order to perform such tests, design of an experimental setup can be found in [4]. Figure 3 describes the hardware configuration of the setup.

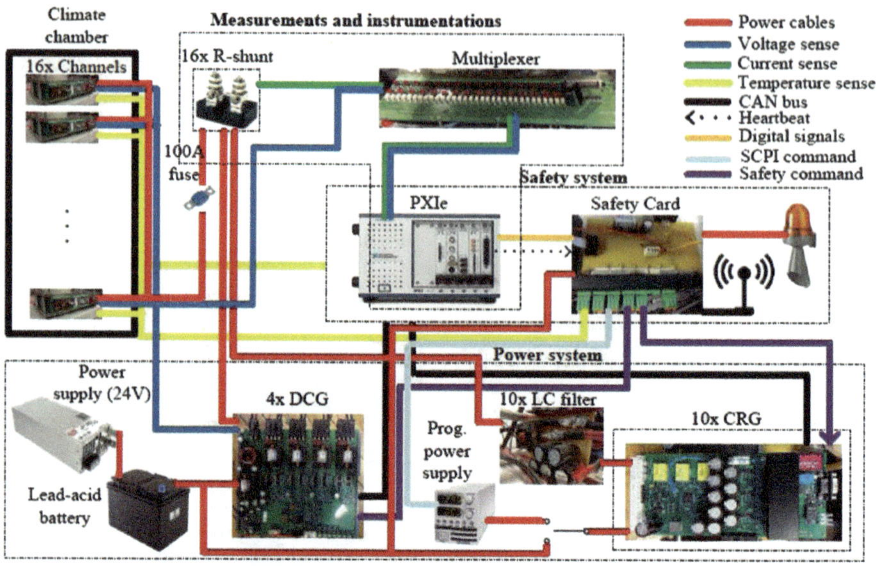

Fig. 3 Hardware configuration of the experimental setup

2.2 *Testing Battery Management System (BMS)*

Battery management systems (BMS) are responsible for monitoring, controlling, and protecting the battery system and the single cells [5]. Many of the final tests are performed on a complete battery system and the result depends on the correct performance of the BMS. Each BMS has a considerable number of in- and output signals. Typical interfaces are shown in Fig. 4. The tests to be conducted on a BMS system can be grouped in the following categories: environmental tests, functional and safety tests, and electrical tests. Detail procedure of each test is outlined in [5].

3 Testing Electric Vehicle Chargers and Supply Equipment

Various ongoing standardization activities for charging infrastructure on conductive and wireless power transfer technologies are given in [6]. To fulfill testing requirement presented by the standards, a potential setup was found in [7] (Fig. 5).

In this paper, a bidirectional charger testing setup was proposed. The setup included a power system simulator and bidirectional controllable power conditioner connected to EV battery test bed. Communication and control are done by a smart interface controller.

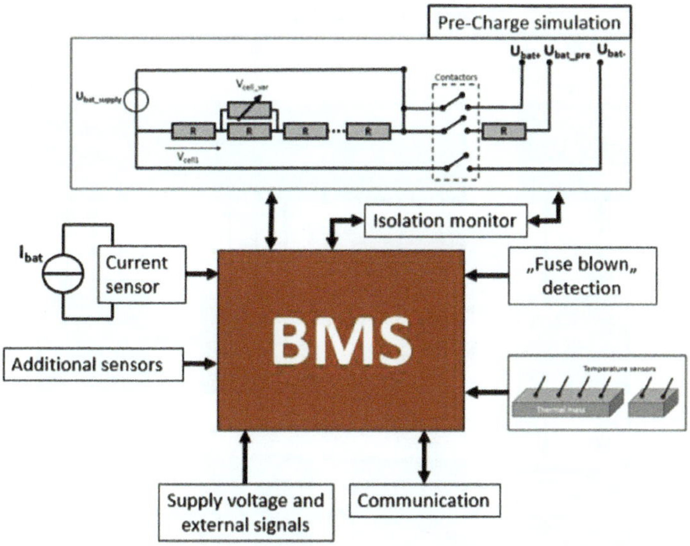

Fig. 4 BMS interface and possible test setup

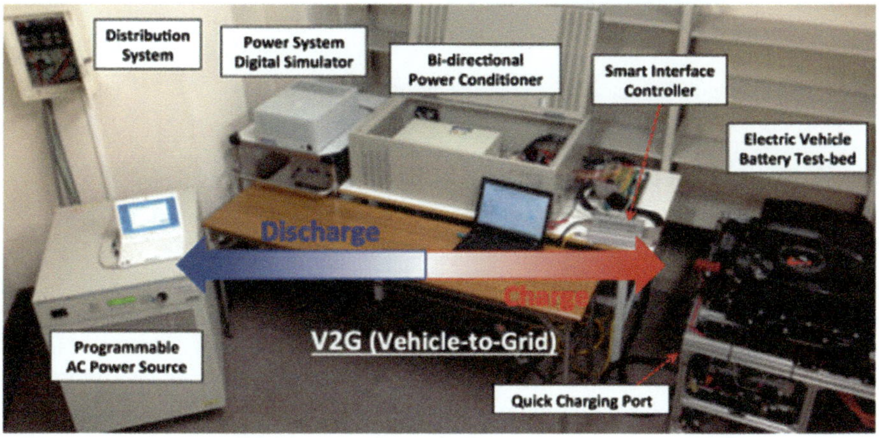

Fig. 5 Experimental setup of EV and charging system

4 Testing Electric Vehicle Powertrain

SAE outlines standard testing methods for electrified vehicle powertrain in its "Vehicle Power Test for Electrified Powertrains" document (SAE J2908_201709). To fulfill the testing requirement, a testing bench was outlined in [8]. There were twelve separate components in the reported test bench and communication between these modules were completed via CAN Bus. Figure 6 describes the test bench setup.

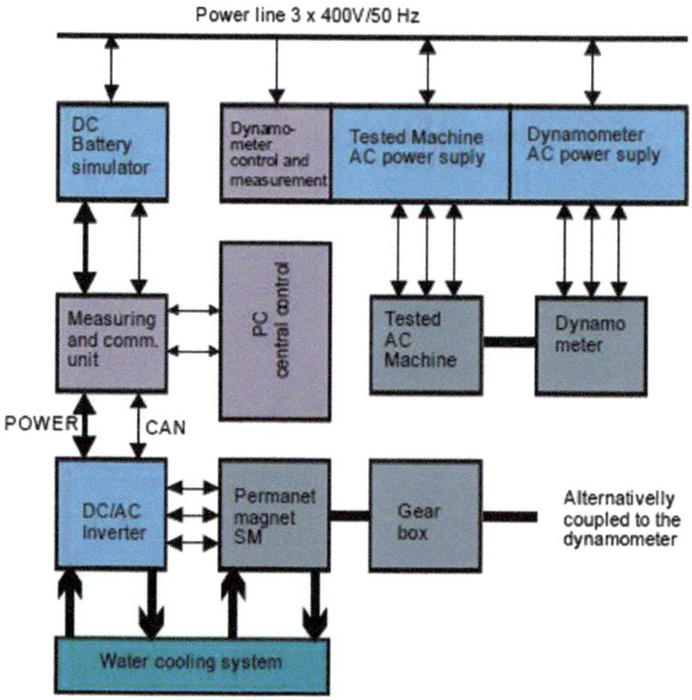

Fig. 6 Block diagram of automotive electric powertrain testing bench

5 Testing Electro-Magnetic Compatibility (EMC)

The electric drive system is a new component in modern passenger vehicles which comprises of high-voltage power source, high-frequency switching high-voltage converter, high-power electric motor, and shielded or unshielded high-power cables. Additionally, a vehicle also has low-voltage electronics. So, electro-magnetic interference between these two systems must be carefully investigated. Also, EMC is a big issue in wireless power transfer [9]. So, standardization activity is underway to address this situation. A typical EMC test setup is proposed in [10]. It contains shielded anechoic chamber and can simulate test conditions of drive, brake and recuperation. The setup is given in Fig. 7.

6 Testing of Autonomous Electric Vehicles

A method to test the autonomous vehicle step-by-step from the simulation environment to the public road is proposed [11]. The workflow is described in Fig. 8. Autonomous vehicle collects data on the surroundings through distributed sensors

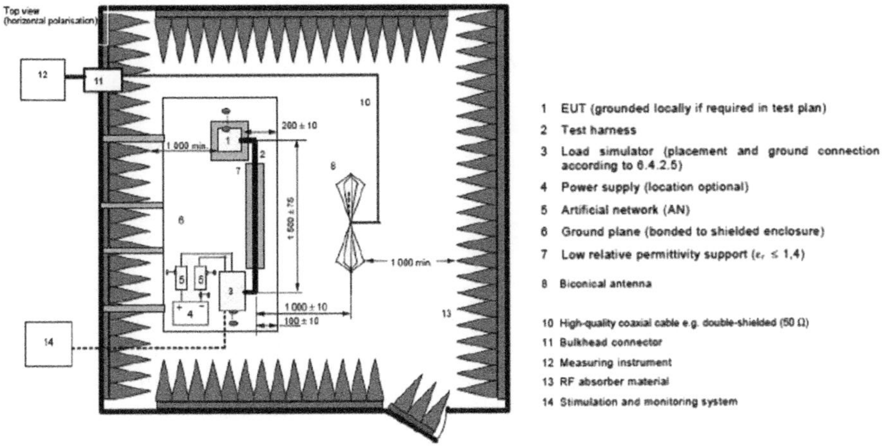

Fig. 7 Test setup in chamber (CISPR25 ed.3)

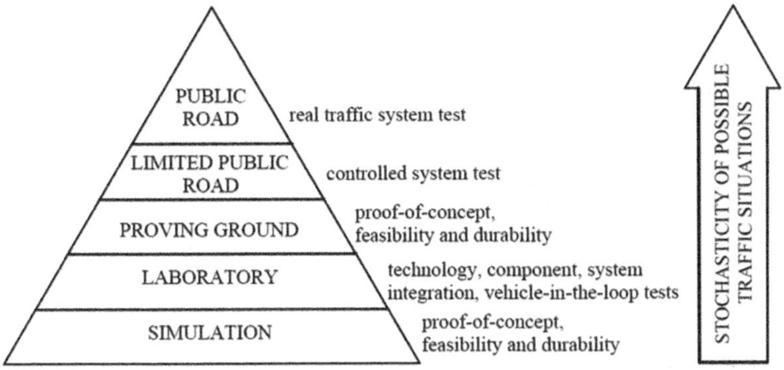

Fig. 8 Autonomous vehicle testing and validation pyramid

(cameras, radar, LIDAR, and GPS), performs sensor-fusion processing algorithms to build a model, and then take a moving decision based on that environment model. For an autonomous vehicle, taking traffic decision is not an easy task as there can be infinite number of situations in road condition. Hence, authors believe the framework described will be a good fit for testing and certifying autonomous vehicles.

Another workflow is proposed in [12]. The algorithm derives autonomous driving test case from traffic observations. In this process, road observations are processed to detect maneuvers, maneuver sequences, and corresponding vehicle behavior. Figure 9 describes the proposed workflow.

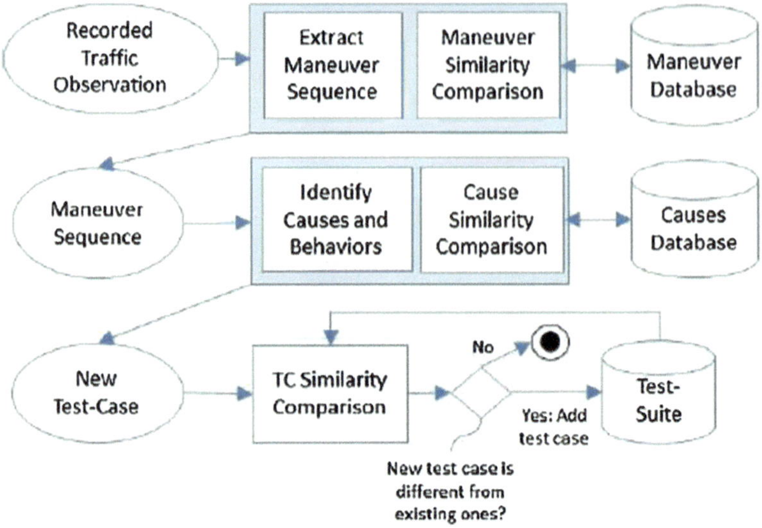

Fig. 9 Process of test case generation

7 Testing EV Intra-vehicular Communication

Three separate SAE protocol classes have been published for intra-vehicular communication. They are Class A (scope: actuators and sensors interconnection, bit rate ~ 10 kbps), Class B (scope: non-real-time control and ECU-to-ECU communication, bit rate >125 kbps), and Class C (scope: real-time and critical applications) [13]. Also there are more advanced protocols supporting high-speed communication, such as FlexRay and MOST. Figure 10 shows timeline of automotive network architecture.

Wang [13] proposes an automotive network test bed with the aim to explore an automotive network design that integrates Ethernet, ZigBee, and Wi-Fi with the legacy CAN in an implementation of a prototype platform for automotive networks. Figure 11 describes the complete setup.

8 Testing of EV-Smart Grid Interoperability (V2G)

Kotsakis [14] describes The Smart Grid Interoperability laboratory (SGILab) of the JRC is a testing facility on the interoperability of smart grid systems. This facility is dedicated to the assessment of interoperability aspects between smart grids and electric vehicles and the study of integration aspects of microgrids. Figure 12 gives a simplified schematic diagram of the laboratory layout.

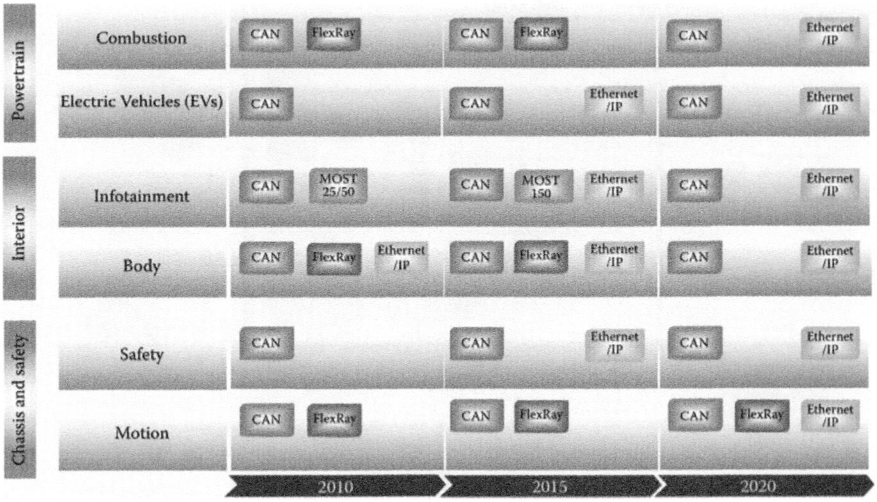

Fig. 10 Timeline of automotive network architecture

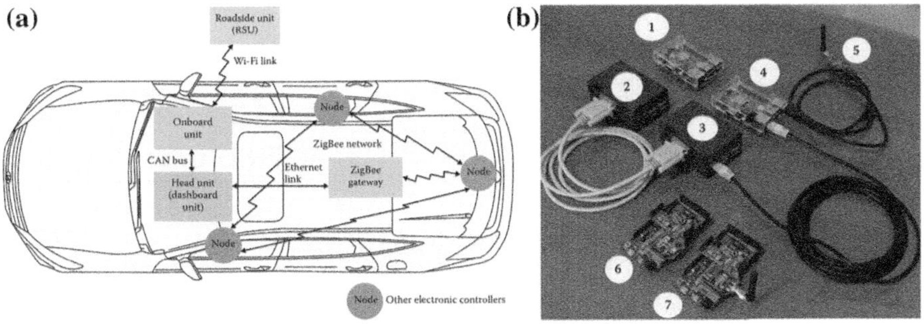

Fig. 11 **a** Block diagram of the architecture of the automotive network test bed platform. **b** An implementation of a minimum viable test bed for automotive communication using Raspberry Pi and Arduino boards

Gowri [15] presents a test setup as shown in Fig. 13. This test bed was implemented at Pacific Northwest National Laboratory. The objectives were testing and validating vehicle-grid communication standards. This test bed supported SAE with standardization requirement definition and development work of SAE J2847 and J2931 documents.

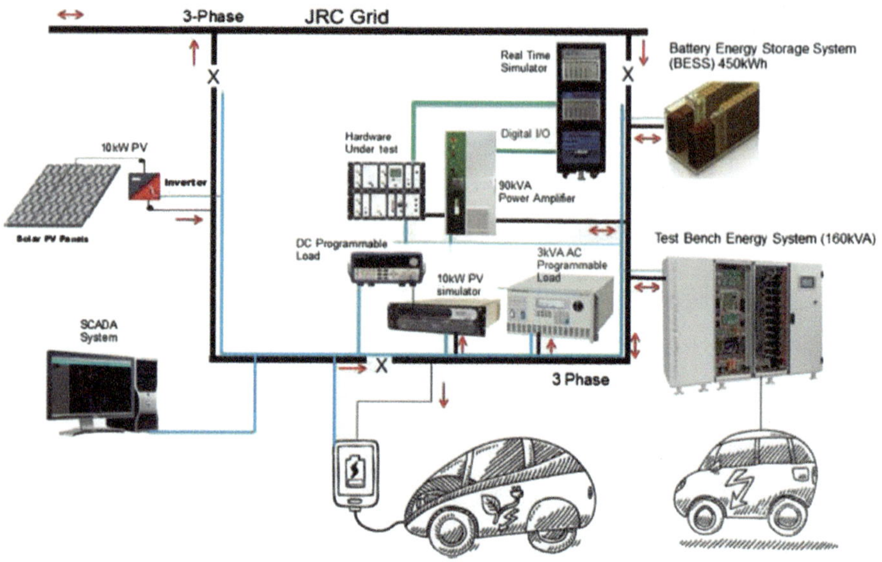

Fig. 12 Simplified physical structure of the SGILab in Ispra, IT

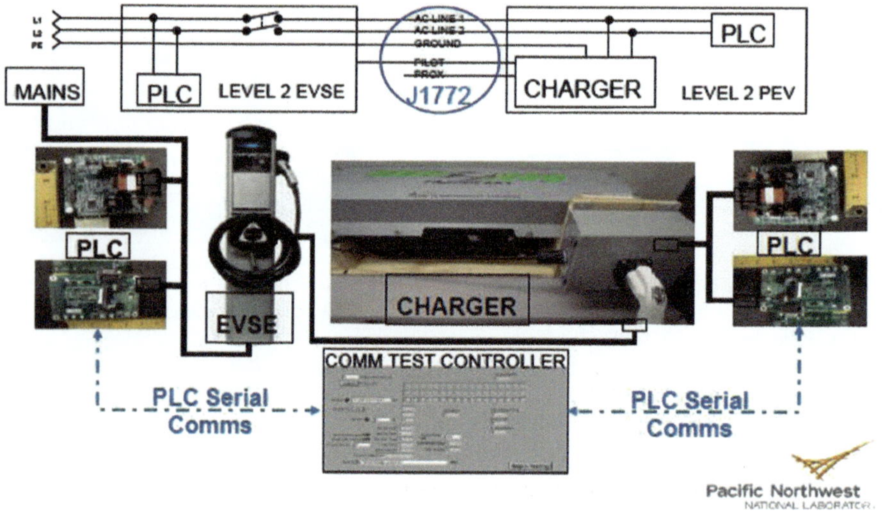

Fig. 13 Functional test schematic

9 Testing EV Cybersecurity

Within the EV, controller area network (CAN) is used for data transmission between distributed electronic control units (ECUs). Fowler [16] reports the CAN bus to be vulnerable to non-encrypted direct access via OBD (on-board diagnostic) port. To study and address this vulnerability, a cybersecurity test bed was proposed. A vehicle simulation case was configured and in the OBD port, a Bluetooth-enabled dongle was connected. An unwanted CAN message was injected by unauthorized pairing to turn the headlights on and off (undesirable behavior). Thus, the security threat scenario was implemented and it was concluded that more complex tests can be done using the same test bed.

10 A Novel Integrated Test Bed

The primary aim of CAVTE will be to support the research and innovation relating to the Development of Canadian Standards and Certification process on the basis of key performance indicators (KPIs) such as safety, reliability, cost, human interactions, and environmental impacts. CAVTE facility will enable Canadian Standards Association (CSA) to evaluate different technologies and systems and conduct different tests accordingly. The target test facility will be connected via wireless sensor network and IoT with layers of cybersecurity to monitor and control the target CAVTE setup at UOIT via remote access and operation by CSA team. Figure 14 shows the layout of CAVTE setup.

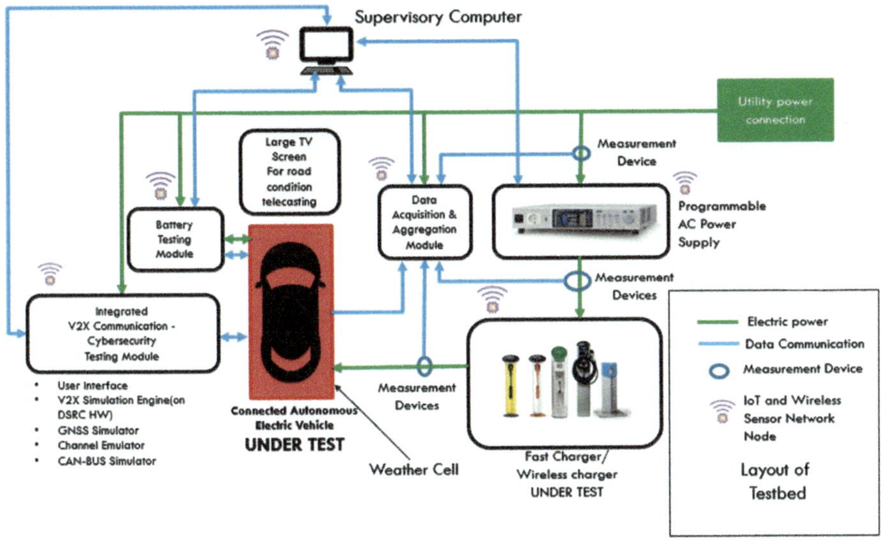

Fig. 14 Integrated test platform schematic

11 Conclusion

This paper presents a technical review of potential automotive testbeds to be developed in UOIT and finally proposes a novel integrated test platform. The study was conducted to contribute to the design of integrated testing and verification platform of connected and autonomous vehicle and transportation electrification (CAVTE) ACE facility at UOIT. This center will bolster the R&D activities on Automotive Testing and Verification in Canada. Activities at this center will contribute to mass adoption of electric vehicles and intelligent transportation solutions.

References

1. Green, J., Hartman, B., Glowacki, P.: A system-based view of the standards and certification landscape for electric vehicles. World Electr. Veh. J. **8**, 564–575 (2016)
2. Schulze, T., Stavesand, J.E.: Hardware-in-the-loop test process for modern E/E systems. In: Gühmann, C., Riese, J., von Rüden, K. (eds.) Simulation and Testing for Vehicle Technology. Springer, Cham (2016)
3. Castillo, E.C.: 18—Standards for electric vehicle batteries and associated testing procedures. In: Scrosati, B., Garche, J., Tillmetz, W. (eds.) Woodhead Publishing Series in Energy, Advances in Battery Technologies for Electric Vehicles, pp. 469–494. Woodhead Publishing (2015)
4. Soares, R., Bessman, A., Wallmark, O., et al.: An experimental setup with alternating current capability for evaluating large lithium-ion battery cells. Batteries **4**, 38 (2018)
5. Stadler, A., Klink, T.: D2.5—Development of reliability test procedures for EV BMS (2015)
6. Bablo, J.: Electric vehicle infrastructure standardization. World Electr. Veh. J. **8**, 576–586 (2016)
7. Ota, Y., Taniguchi, H., Baba, J., Yokoyama, A.: Implementation of autonomous distributed V2G to electric vehicle and DC charging system. Electr. Power Syst. Res. **120**, 177–183 (2015)
8. Mindl, P., Mňuk, P., Čřfovský, Z., Haubert, T.: EV drives testing and measurement system. In: International Conference on Electrical Drives and Power Electronics (EDPE), pp. 328–332. Tatranska Lomnica, Slovakia (2015)
9. Obayashi, S., Tsukahara, H.: EMC issues on wireless power transfer, pp. 601–604. International Symposium on Electromagnetic Compatibility, Tokyo, Japan (2014)
10. Tsang, J. (2012) Presentation for IEEE HK EMC Society
11. Zsolt, S., Nyerges, A., Hamar, Z., Hesz, M.: Technical specification methodology for an automotive proving ground dedicated to connected and automated vehicles. Periodica Polytech. Trans. Eng. **45**(3), 168–174 (2017)
12. Wolschke, C., Kuhn, T., Rombach, D., Liggesmeyer, P.: Observation based creation of minimal test suites for autonomous vehicles. In: 2017 IEEE International Symposium on Software Reliability Engineering Workshops (ISSREW), pp. 294–301. Toulouse, France (2017)
13. Wang, Y., Tian, D., Sheng, Z., Jian, W.: Connected Vehicle Systems. CRC Press, Boca Raton (2017)
14. Kotsakis, E., Fulli, G., Masera, M.: Smart Grid Interoperability lab at the joint research centre (JRC) of the European Commission: Towards a European platform for real time simulation 2016 AEIT International Annual Conference (AEIT), pp. 1–6. Capri, Italy (2016)
15. Gowri, K.: Testing and validation of vehicle to grid communication standards (2011)
16. Fowler, D.S., Cheah, M., Shaikh, SA., Bryans, J.: Towards a testbed for automotive cybersecurity. In: 2017 IEEE International Conference on Software Testing, Verification and Validation (ICST), pp. 540–541. Tokyo, Japan (2017)

Resilient Micro Energy Grids for Nuclear Power Plants During Normal and Emergency Operations

Hossam A. Gabbar and Muhammad R. Abdussami

Abstract This paper represents the integration of a nuclear power plant (NPP) with the Micro Energy Grid (MEG). Traditionally, MEG is supported by only renewable energy sources. But in this paper, we focused to connect the NPP with MEG and tried to observe the current and voltage characteristic in both normal and emergency conditions. To confirm the stability and the reliability of the system, the current and voltage level must be regained after occurrence of the fault in the system. By conducting MATLAB Simulink simulation, we have shown the voltage and current scenario of the load side during the normal and abnormal condition. The simulation results ensure the resiliency of the MEG after abnormal condition without affecting the consumer level. The results also confirm the regularity of the power flow to the load (electrical) side during normal as well as fault condition.

Keywords Micro energy grid · Nuclear power plant · Resilient micro energy grid

1 Introduction

Though the whole world is moving towards adopting renewable energies, nuclear power has more advantages over the renewable energy sources considering the needs of human being and global climate changes. A sudden increase in energy demand by the locality cannot be fulfilled by the green energy sources as the probability of getting energy by utilizing renewable sources is quite unpredictable. If the wind is calm, energy demand cannot be mitigated by wind energy; or if there is a gloomy midday, the load demand cannot be fulfilled by solar power as well.

H. A. Gabbar (✉) · M. R. Abdussami
Faculty of Energy Systems and Nuclear Science, University of Ontario Institute of Technology (UOIT), 2000 Simcoe Street North, L1H7K4 ON Oshawa, Canada
e-mail: Hossam.Gabbar@uoit.ca

H. A. Gabbar
Faculty of Engineering and Applied Science, University of Ontario Institute of Technology (UOIT), 2000 Simcoe Street North, L1H7K4 ON Oshawa, Canada

V. Jain et al. (eds.), *Recent Trends in Intelligent Computing, Communication and Devices*, Advances in Intelligent Systems and Computing 1031,
https://doi.org/10.1007/978-981-13-9406-5_124

In this case, the nuclear power plant (NPP) is a better approach for the solution of power demand [1].

As fossil fuel storage is running out rapidly, the renewable energy sources (solar, wind, and hydro) are incorporated as fuel in the MEG. But due to the lack of continuous availability of renewable energy sources, they hamper the grid reliability [2]. In addition, the number of natural gas-fired MEG is also doubled by the last decade. Due to easy means of natural gas transportation through pipeline and less carbon emission from burning natural gas, the natural gas-fired MEG is adopted all around the world. Natural gas-fired MEG has also better grid reliability than others [2, 3].

Because of the availability of nuclear fuel from different countries, the safer form of energy source than other energy sources, and no environmental pollution from NPP by greenhouse gases, nuclear power is called 'Sustainable Development' technology. The construction cost of the NPP is comparatively higher than coal, oil, and gas-based power plant; however, once the NPP is built, it is less expensive because of less fuel cost and improved technology. Uranium, the most popular nuclear fuel, is found widely in both sea water and earth which is the key to the keen deployment of nuclear power plant [4].

The integration of NPP into an existing MEG is quite challenging. It may cause the MEG to an unstable condition. The proper reinforcement must be put into the NPP when an NPP will be incorporated with an existing MEG. By doing a simulation in a software, named power systems analysis framework (PSAF), it can be concluded that the transmission line from bus to bus is germinated and the transmission line from NPP to load is trebled to make the whole network stable after integrating NPP with the MEG. For simplicity, the network was assumed as a seven-bus generic system [5].

MEG can be defined as a native energy grid which can operate with or without associating to a centralized energy grid [6]. MEG increases energy efficiency; reduces the transmission loss; minimizes the drawback, high costs, and operational difficulties of traditional energy grid [7].

Uniqueness, diversity, controllability, interactivity, and independence are the main features of MEG. The energy storage systems are installed in the MEG to confirm the stability and the reliablity of the energy grid. The energy storage system supplies energy to grid, as well to load, in case of emergencies. The MEG can operate both at the grid-connected mode as well as at the islanded mode. Bidirectional power supply is available between MEG and the traditional grid which also provides great support for the traditional grid, in case of a blackout. MEG ensures a certain maximum demand for a particular region or locality [8].

A number of hazards, such as lack of distributed energy resources (DER) in MEG, power system faults, grid failure, feeder circuit disconnection, failure of DC to AC inverters, fault in cooling system, cooling overload, thermal overload, fault in heating systems, loss of electric and gas boiler, earthquake, water flood, and lightning can happen in MEG. Depending on the level and probability of hazards, specific protection measures must be taken at MEG [9].

Three solutions can be proposed for the resilient operation of MEG during an emergency. Firstly, electric vehicles (EVs) can be used as an emergency electrical power supply source where the EV and the residential buildings will be connected in the bidirectional mode for supplying electrical power. Secondly, Distributed Generation (DG) can be used in case of grid blackout where the renewable energy sources will be used in DG technology. In this method, storage systems are not mandatory; the DG technology will come into action when the energy grid goes to shut down mode. Thirdly, a Software-defined Network (SDN) framework is proposed where on-grid energy storages, EV, and energy storage systems from nearest locality will be interconnected to control, coordinate, and supply the electric energy to the intended area at an emergency. This scenario is named as 'Virtual Power Plant (VPP)' [10].

2 Micro Energy Grid (MEG)

MEG is powered by the distributed energy resources (DER), e.g. solar, wind, hydro, and other renewable energies. The idea of MEG compels the researchers to focus on 'Distributed Generation (DG)' which is the key concept of MEG. MEG reduces the dependency of the consumer on the centralized power grid and increases the power reliability at the customer end. Due to the integration of renewable energy into the power grid and the switching to the power electronics devices, power level fluctuation and voltage transiency may occur in the power system. To diminish this kind of interruption, information and communication technologies are being used in MEG nowadays [11]. A comparison between the Traditional Energy Grid (TEG) and Micro Energy Grid (MEG) is shown in the Table 1.

3 Electrical System in Nuclear Power Plant

The number and types of electrical equipment vary from NPP to NPP. The electrical energy is one of the most crucial things; this is used to maintain the reliability of the nuclear power plant. Micro energy grid, connected with NPP, must be designed in such a way so that the equipment of the NPP has enough power supply. Based on the usage frequency of electrical equipment in NPP and the duration of power outage, the power of a nuclear power plant is classified into four categories [12]:

3.1 Class-1 Power

Class-1 power is used to activate the most sensitive equipment that should be uninterruptable in the NPP, all the time. The outage of Class-1 power causes system blackout. A partial list of the equipment supported by Class-1 power is enlisted in the Table 2 [12].

Table 1 A comparison between TEG and MEG

Traditional energy grid (TEG)	Micro energy grid (MEG)
Electrical power is generated at the generating station	Electrical power is generated at the distribution end
The main fuel of the conventional grid is fossil fuel, e.g. coal, gas	Renewable energy sources (wind, hydro, biomass, solar, etc.) are the primary fuel for MEG
The existing traditional energy grids are rated as megawatt as well as Gigawatt unit	Usually, the MEG is rated as megawatt
The Energy grid is centralized	The infrastructure of the energy grid is grown up where there is existence of locality and availability of renewable energy sources
Power transmission cost is very high	Transmission and distribution cost is low
A transmission and distribution systems are designed separately and some specific rules are followed to design the systems	Due to less transmission line, transmission and distribution system are very simple. But, as some of the DC energy sources (e.g. solar) are integrated into MEG, some extra protection measures are taken here
As the power is transmitted from generating station to customer end by the very long transmission line, the transmission power loss is high	Due to the short transmission line, the transmission loss is small
A blackout may occur in the conventional energy grid as the demand is always higher than the supply	As the power demand is quite comparable with the generation, MEG is more reliable
Since fossil fuel (e.g. coal and gas) are the main source of producing electricity in TEG, the rate of carbon emission is high	As renewable sources are used in MEG, the carbon emission is lower in power generation

Table 2 Example of Class-1 equipment

Class-1 Power
Class-2 inverters
DC seal oil pumps for generator
DC lube oil pump for turbine generator bearings
Turbine trip circuits
Turbine turning gear
DC stator cooling pumps
Control and protection systems for station electrical distribution systems

3.2 Class-2 Power

The Class-2 power sources are only available in AC form. The unavailability of Class-2 will shut down the nuclear reactor. The electrical equipment operated by Class-2 power can sustain the power interruption in order of milliseconds. Some electrical equipment powered by Class-2 power are shown in the Table 3 [12].

3.3 Class-3 Power

Class-3 power predominantly maintains the fuel cooling at the time of the reactor shutdown. It also provides backup support to the Class-4 power. A list of equipment supported by power-3 is given in the Table 4 [12].

3.4 Class-4 Power

The electrical systems supported by Class-4 power can accept infinite power interruption. Class-4 power is supplied to the whole NPP during the normal operating condition. Some electrical equipment in NPP operated by Class-4 are listed in the Table 5 [12].

Table 3 Example of Class-2 equipment

Class-2 Power
Digital control computers
Reactor regulation instrumentation
Process valves (Electrical)
Auxiliary oil pumps on the turbine and generator
Emergency lighting system

Table 4 Example of Class-3 equipment

Class-3 Power
Auxiliary boiler feed and condensation extraction pumps
Shutdown system cooling pumps
Heat transport feed pumps
Class-1 power rectifiers
Emergency core coolant injection pumps
Instrument air compressors
End shield cooling, moderator circulating, fire water and service water pumps

Table 5 Example of Class-4 equipment

Class-4 Power
Main boiler feed and main heat transport circulating pumps
Condenser cooling water pumps
Generator excitation
Heating and ventilation systems
Normal lighting

4 Design Methodology

To ensure the resiliency of the MEG for integrating NPP, a model was developed consisting of MEG, electrical/household load, wind farm, NPP, and ESS. The model was developed and the simulation was conducted in MATLAB Simulink. In the beginning, the NPP and the wind farm were integrated to the MEG and they collectively supplied electricity to the household load. The ESS was initially kept separated by a circuit breaker. After a certain time, a three-phase fault was made at NPP and the NPP was disconnected from the whole system. After the fault occurred at the NPP side, the circuit breaker was tripped and the MEG was isolated from the household load. Within a few times, a signal transferred to another circuit breaker and it triped; this forced the ESS to come into the action. In this mode, the MEG (except wind farm) and the NPP were disconnected from the whole system, whereas the wind farm and the ESS continued to supply electricity to the household load. Figure 1 shows the MATLAB Simulink model of the system.

Some assumptions were made to the system equipment block for simplicity. The ratings and the specifications of the equipment blocks are discussed below:

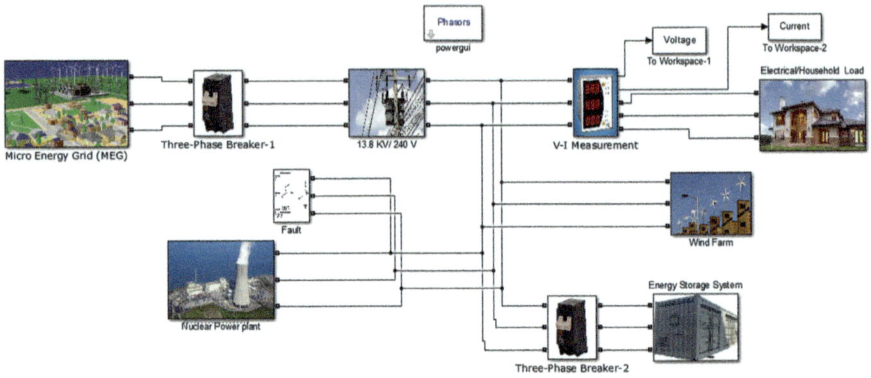

Fig. 1 Simulink model of the sytem

4.1 Micro Energy Grid (MEG)

The following parameters were set for emulating the micro energy grid:

Phase-to-phase r.m.s voltage: 13.8 kV
Frequency: 60 Hz
Phase-to-phase base voltage: 13.8 kV
Base power: 100 MVA
Generator Type: Swing
Quality factor: 7.

4.2 Electrical/Household Load

Though there are industrial, agricultural, and other different types of load available, we considered here a typical household load. The household load consists of several numbers of houses. The rating of a typical household load is as follows:

Nominal voltage (phase-to-phase): 240 V
Frequency: 60 Hz
Active power: 15 MW
Reactive power (Inductive): 3.5 MVAR
Reactive power (Capacitive): 875 KVAR.

4.3 Wind Farm

The wind farm, selected here, consists of three wind turbines, having rating of 2×2 MW. The wind power was transmitted to the MEG form 25 km long by a transmission line. Then, a transformer, rated as 25 kV/240 V, was used to connect the wind farm with the MEG.

Total generated power: 12 MW
Frequency: 60 Hz
Turbine wind speed: 9 m/s

4.4 Nuclear Power Plant (NPP)

Section 2 shows the equipment that is used in an NPP. As the implementation of all the equipment is quite difficult in MATLAB Simulink, we have chosen only some of the essential equipment for simulation. For simplicity, we took the main generator, unit service transformer (UST), main output transformer (MOT), motor load, emergency energy storage system, high resistive load, LC load, and some protective devices. The emergency energy storage system was used for the abnormal or faulty case of inside the NPP. As any kind of fault was not considered inside the NPP, the emergency energy storage system was out of operation throughout the whole simulation time. Some reasonable assumptions were made for the rating of the equipment.

Capacity (main generator): 100 MVA
Generator type: PQ
Output terminal voltage (main generator): 22 kV
Rated output power (main generator): 728 MW
Power factor: 0.9 (lagging)
Frequency: 60 Hz
Main output transformer rating: 22/500 kV
Unit service transformer rating: 22/11.6 kV
Emergency power system generator: Initially disconnected
Motor load: 2250 H.P.

4.5 *Energy Storage System (ESS)*

For ESS, a diesel generator was assumed that was connected to the load by a circuit breaker. The circuit breaker was initially opened. It came into operation after occurring the fault at the NPP side and disconnecting of the MEG from the load. The parameter of the ESS is as follows:

Energy storage system type: Diesel generator
Generator type: PV
Rotor type: Salient-pole
Frequency: 60 Hz
Nominal Power: 100 MVA.

5 Simulation Results

In Fig. 1, at t = 0.20 s, a three-phase fault (A-B-C) occurred at the NPP side. Then the NPP was disconnected and the circuit breaker-1 was tripped to isolate the MEG from the electrical load. At t = 0.26 s, the ESS came into action (by closing the circuit breaker-2) and started supplying electricity to the electric load along with the wind farm. As the rated voltage of the MEG was 13.8 kV, a step-down transformer was used to connect the MEG with the household load. The phasor analysis was

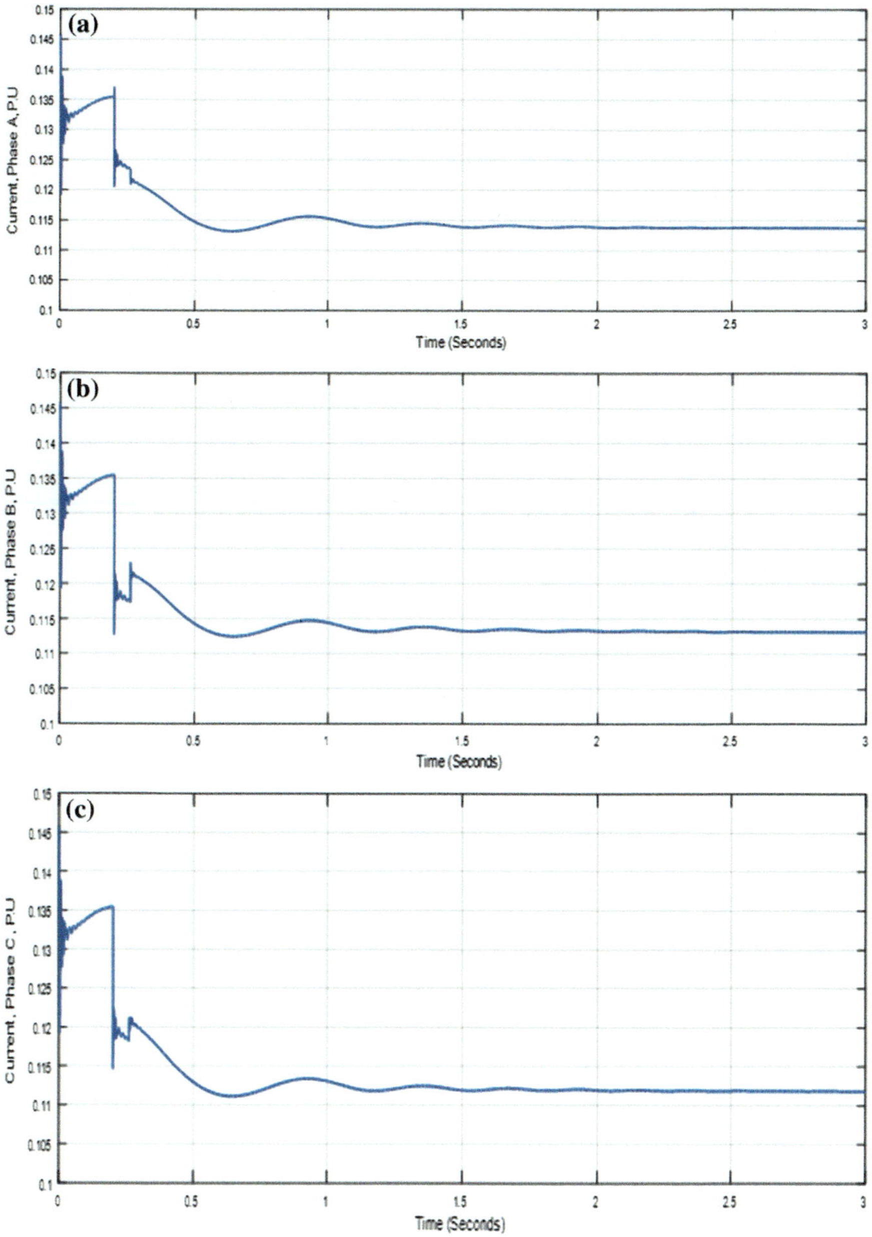

Fig. 2 **a** Current level at household load side before and after the fault at phase A. **b** Current level at household load side before and after the fault at phase B. **c** Current level at household load side before and after the fault at phase C

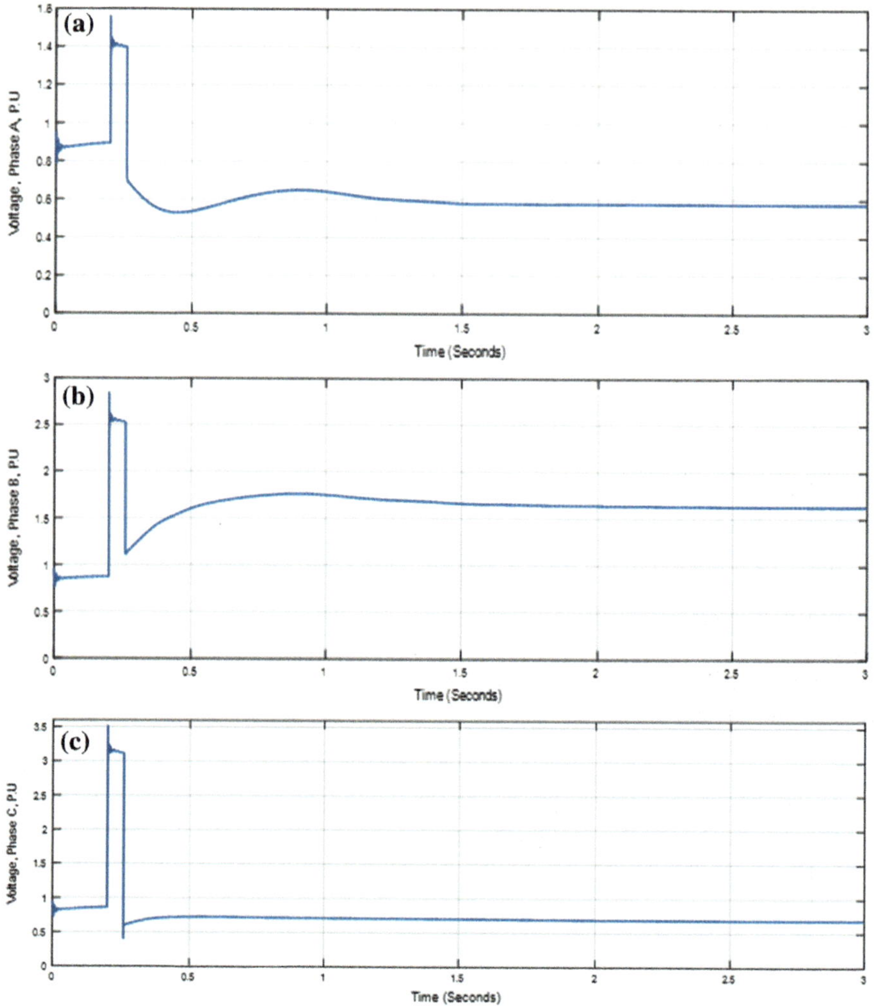

Fig. 3 **a** Voltage level at household load side before and after the fault at phase A. **b** Voltage level at household load side before and after the fault at phase B. **c** Voltage level at household load side before and after the fault at phase C

carried out for the whole model. The three-phase voltage and current were measured during the normal and abnormal conditions. We took only the magnitude of the current and voltage to plot the graph for both conditions (normal and abnormal). Figures 2 and 3 depict that electricity is supplied to the load continuously before and after the fault occurred at the NPP side. In both figures, the voltage level and current level were recuperated after clearing the fault. Both the figures also ascertain the continuation of the power flow to the load (electric). The simulation was run for a total 3.00 s. The whole scenario confirms the resiliency of the MEG integrated with NPP for both the normal and abnormal conditions. After clearing the fault, a comprehensive measure must be taken at the load side to make the power flow smoother. The whole calculations are shown here in the per-unit system.

6 Conclusions

From the conventional definition, the technologies that are used in MEG, are renewable sources. By the above model and simulation results, it is proved that the NPP, along with the renewable sources, can be integrated into MEG; however, some extra care must be taken to control the voltage and the frequency level of the MEG. As the integration of the NPP into MEG is a complex idea, a number of step-by-step protective measures also must be taken to ensure the continuous flow of power supply at the consumer side.

References

1. Can we replace all coal, oil, gas and nuclear power plants with carbon-free electricity? https://www.quora.com/ (Access on: 27 Sept 2018)
2. Gabbar, H.A., Bower, L., Pandya, D., Agarwal, A., Tomal, M.U., Islam, F.R.: Resilient micro energy grids with gas-power and renewable technologies. In: The 2nd IEEE Conference on Power Engineering and Renewable Energy, ICPERE 2014, pp. 1–6 (2014)
3. MIT Energy Initiative, http://energy.mit.edu/news/grid-reliability-and-the-role-of-natural-gas/ , accessed on 03 Aug 2018
4. The Importance of Nuclear Energy in the Global Economy, Silverio Henríquez A, January (2018)
5. Rashed, M.R.H., Zaman, M., Mobassear-ul-Islam, Md. Farsid Raihan: An analysis on the required reinforcement for embedding a nuclear power plant in a generic power system. In: 4th International Conference on Advances in Electrical Engineering (ICAEE), Dhaka, Bangladesh, pp. 597–600 (2017 September 28–30)
6. Department of Energy, How Microgrids Work, https://www.energy.gov/articles/how-microgrids work (Accessed on: 09 Aug 2018)
7. Driesen, J., Katiraei, F.: Design for distributed energy resources. Power and Energy Mag. IEEE, 6(3), 30–40 (2008)
8. Zhou, X., Guo, T., Ma, Y.: An overview on microgrid technology. In: International Conference on Mechatronics and Automation, August 2–5, Beijing, China, pp. 76–81 (2015)
9. Gabbar, Hossam A., Koraz, Yahya: Risk assessment of micro energy grid protection layers. Energies **10**, 1176 (2017)
10. Maharjan, S., Zhang, Y., Gjessing, S., Ulleberg, Ø., Eliassen, F.: Providing microgrid resilience during emergencies using distributed energy resources. 2015 IEEE Globecom Workshops (GC Wkshps)
11. ul Hassan, S., ul Abideen, Z., Izhar, T.: Advanced control techniques for micro-grids power quality improvement. In: Asian Conference on Energy, Power and Transportation Electrification (ACEPT) (2017)
12. Jiang, J.: Electrical Systems, Book, Chapter 11 (Available at: http://www.nuceng.ca/)

Research on APP Icon Based on Logo Design

Wang Xueying and Zhang Bingjian

Abstract The design of logo with simple, refined, and highly summarized graphic design content assumes an important concern as it intends to convey a wealth of contents. There exists an essence of the expression of the characteristics of things. APP start icon is the third-party application graphic symbol, which performs the role of the transfer of APP information. The success or failure of its design will directly affect the user interaction with the APP process. With an increasing number of APPs being developed, the start icon represents diverse user requirements. Hence, there exists a higher requirement for the design of the APP boot icon. The APP is based on the similarities and differences between the icon, logo design, and the app launcher icons research, analysis of the app promoter icon, and the sign of the relationship. Based the logo design, the commencement of study the current APP will iconic, visual and interaction to provide reference to support. Through this study of the APP startup icon art style, the future development trend of APP start icon design can be developed.

Keywords APP startup icon design · The relationship between APP and start flag icon · Design performance

1 Introduction

Symbol is a configuration that embodies the core features of things and communicates the information of the subject. It is a symbol that converts the essential features and values of the subject into figures and characters. An APP with a very good design is simple, easy to identify, rich in connotation, simple to convey information, and convey to the viewer with accurate and refined creative graphics. It enables the viewer to produce association and imagination, rich meaning and vivid signs can arouse the attention of the viewer.

W. Xueying (✉) · Z. Bingjian
Wuxi Institute of Art & Technology, Jiangsu, China
e-mail: 942652240@qq.com

V. Jain et al. (eds.), *Recent Trends in Intelligent Computing, Communication and Devices*, Advances in Intelligent Systems and Computing 1031,
https://doi.org/10.1007/978-981-13-9406-5_125

In a simple sense, a characteristic symbol distinguishing something from others can be called a "sign." In a narrow sense, signs mainly refer to symbols that convey information with simple, obvious, understandable figures or characters as principal elements. According to the different contents and usage of signs, they can be classified into three categories: trademarks, logos, and public signs.

1.1 An Overview of the APP Startup Icon

APP the acronym stands for. Application Program. In the current scenario APP mostly refers to the application of third-party smartphones [with reference to Wikipedia]. Mobile phone software is installed on mobile phones. At the moment, famous application stores have Apple's APP Store, Google's Google Play Store, Microsoft's Marketplace, etc. Icon is a graphic symbol with a special meaning. It has a clear indication of representation. After the appearance of a computer, more signs are expressed in the computer interface, such as program, data, switch button, and state display. The APP icon includes the boot icon and the tool icon. The start icon is the image of an APP (Fig. 1), which has the features of simplicity and refinement, high recognition, and both creative and interesting features. The minimalist design of the tool icon (Fig. 2) requires a designer's superior modeling s'sand generalization ability.

The APP startup icon is the identification of the application with clear expression function. In general, it is the icon of the application. In recent years, the rapid popularization of mobile Internet has led to the new development of communication technology. More and more information is transferred from the PC side to the mobile terminal, and the people's life has been changed greatly. In the APP of smartphone, APP startup icon is the most intuitive expression of application. It is equivalent to the logo of the enterprise, conveying the idea and purpose of the enterprise. The APP startup icon, as the user's media for a certain application of mobile devices, is to guide users to download and use it. It plays the role of passing APP content and using APP, so it is particularly important to design a concise, refined, and beautiful APP startup icon.

With the passage of time, the society has been progressing, and the logo design has gradually changed from simple graphics to symbols with aesthetic and information connotation. Nowadays, various cultural activities and economic talks are also the visual core of the purpose and concept of the activities, as well as an important part of the economy, culture, and development. A landmark historical background is not covered here. The background of APP startup icon is mainly studied and analyzed.

The appearance of smartphones has made a great change in the design of the icon of the mobile interface. It is no longer a simple icon of "information," "setting," "address book," and so on. It has developed into a wide variety of APP graphics, such as video, social, shopping, and games.

Fig. 1 iOS 7 part APP startup icon

Fig. 2 Tool icon (part of the Taobao APP)

Fig. 3 Application icons on early mobile screen

Early mobile phone function is relatively simple, the screen is black and white blue, the application icon is relatively monotonous, mostly in the daily life of the decorative graphics, icons style simple, such as Fig. 3 "time," "magnifying mirror" and other icons, function at a glance, simple and not fancy.

In January 2000, NTT DoCoMo of Japan launched D502 i[, the largest telecom operator in Japan, NTT DoCoMo, which was established in August 14, 1991. The first color screen mobile phone has been colored since then, and the icon has become more hierarchical and personalized (Fig. 4). However, the style of the icon

Fig. 4 The first color screen mobile phone

Fig. 5 2007 screen

of the mobile phone has not changed in a normal sequence or continues to be the simplest style before, only on the basis of the previous period a few colorful ornaments and modern sense.

In 2002, 3G technology was used in foreign markets. China used 3G technology only 5 years later than foreign countries. Although 3G entered China relatively late, it greatly facilitates people's lives and changes their habits in subtle ways. At this time, the icon of the mobile phone is novel and unique, and there are many kinds, including social interaction and video games (Fig. 5).

Fig. 6 iOS 5 system partially quasi-materialized APP boot icon

In June 2007, the first Apple mobile phone iPhone (Fig. 6) appeared in the USA. This iPhone not only has the function of communication and SMS, but also includes the functions of mail, network, chat, and game. It widens the function of the mobile phone and opens a new era of the development of the mobile phone. At this time, the APP startup icon also changed a lot. On the basis of keeping the APP boot icon indicating function, the iPhone APP startup icon is more creative and highly textured and brings a more vivid and fashionable experience to the users. The upgrading of mobile devices and its powerful functions has brought important development opportunities to APP, and also let us see the rich variety of APP startup icons. Especially in the environment dominated by Apple iOS system and Android system, APP has gradually become a very important development direction. The APP startup icon, as the application entry of APP, can directly guide users to download and use.

Since the launch of iOS 7, the design of APP startup icon has undergone the trend change from materialization to flattening (Figs. 7, 8). With the progress of technology and the increase of users' needs, only when considering the APP's own information and functional characteristics, we can make excellent APP startup icons.

Fig. 7 The first iPhone screen

2 Symbiosis and Development of APP Startup Icon and Logo

By sorting out the case of APP startup icon and logo design, the author sums up the graphic design, text expression, and color positioning, focusing on the integration and interconnection of the design elements, graphic allegory, and design style of the APP startup icon and logo design [1].

The graphic design of the symbol is to generalize the content highly, to extract the main features of things, to convey information, and to express the main image

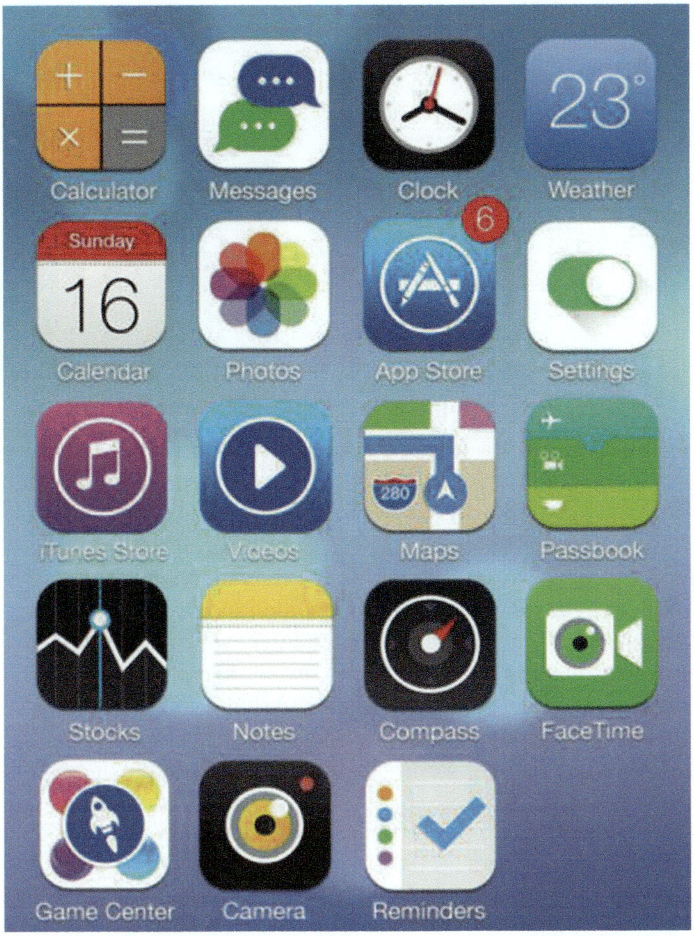

Fig. 8 iOS 7 flat system part APP startup icon

Fig. 9 Camel signs

Fig. 10 Where to travel APP startup icon

of things with a concise tone, so that the logo has a unique visual charm and artistic appreciation value [2]. The same is true of the APP startup icon, and the main elements of the logo and the APP start icon are graphics, all of which are made up of the main content of the main figure and also have close connections in the graphic design techniques, such as the use of figurative and abstract graphics.

The figurative figure is a graphic representation of the specific images of nature, such as the design of the American camel logo, which uses a figurative camel (Fig. 9) to express the brand's adventurous, brave, and persistent spirit. The camel not only means comfortable outdoor equipment, but also tells the consumer of the camel at the spiritual level. Find spirit, encourage consumers to challenge themselves in the open air, find the world beautiful, seek adventure and happiness. APP application "where to travel" start icon is also a camel (Fig. 10), to find the meaning of the discovery of the world, the expression of APP to provide users with "travel, in touch" service ideas.

APP startup icon and logo are also closely related to abstract graphic design. Different from the figurative figure, the abstract figure itself does not have to have the meaning of pictograph. It only expresses the meaning with the symbols of abstract modeling, such as the design and connection of abstract graphics. The "Nike" trademark is a fluent tick (Fig. 11), representing the speed and explosive force, and the affirmative meaning of the tick, and the sign itself is expressed. The affirmation and encouragement to the user. Similar to this APP startup icon, such as "graphics tick in the APP startup icon of the Do it.im Pro" is the meaning of the work's affirmation and life (Fig. 12), allowing people to see this icon to feel the processing and induction functions of the APP program.

Concrete graphics show the main characteristics of things with their intuition. They are clear and easy to understand. They can shorten the distance from the viewer psychologically and create a sense of closeness. Abstract graphics show some complex and difficult features by symbolism. It can express the essential features of things more with association and imagination and achieve the purpose of transmitting information by symbol or APP startup icon.

Fig. 11 Nike logo

Fig. 12 Doit.im Pro APP icon

Text creativity is also the design subject of logo and APP startup icon, and the designed text can convey more abundant information. Many symbols like the use of the name of the company as a sign, the appropriate word change can make the mark become eye-catching, beautiful, more able to show its theme characteristics, so that people have association and memory. Robinson [Robinson: Peng Burton's assistant and partner accountant, a classicist calligrapher. Named "Coca-Cola," the English name "Coca-Cola" (Fig. 13). He thinks "two capitals will look good together," so "Coca-Cola" is used; "Coca" is the spice of the cocoa leaves, and "Cola" is the ingredient extracted from the cocoa fruit. His logo of "Coca-Cola" has been used for more than 100 years. McDonald's takes its first letter "M" as the design subject (Fig. 14). The Latin alphabet is the most common letter in the world. It is the universal language of international circulation. The letter "M" is like the opening gate of two fans. It symbolizes joy and delicacy, and it is rich in meaning. The APP startup icon Path is a social application for sharing between close friends (Fig. 15).

Fig. 13 Logo of Coca-Cola

Fig. 14 McDonald's logo

Fig. 15 Path APP startup icon

The visual design of icons directly uses the initials P of Path as the design element, presenting a concise effect, deforming P, and making the alphabet P more friendly. The text itself is the carrier of transmission of information, plus the rich creative design, which is more likely to arouse the viewer's interest and attention, making the APP start icon or logo more visually attractive.

The same is in the design between the square, the APP start icon and logo in the elements, graphic allegorical, design style, there are some similarities, but there are differences between them, I think the difference between the two, it is beneficial for us to better master the overall performance of the APP start map design.

3 Differences in Graphic Boundaries

The form and means of modern logo design are rich and diverse, as far as its shape is concerned, the conventional square, garden, triangle, rectangle, ellipse, diamond, polygon and so on. As for the irregular shape, it is more variable and flexible according to the needs of the graphics. And the logo design is not limited by the specific size (except party a specific requirements), regardless of the horizontal, vertical, oblique, large, small, even three-dimensional such as relief, a special style of a circular carving are the signs of the performance. The most important difference between APP startup icon and logo design is the difference in shape and size specification. The design of the APP startup icon is relatively standard, such as the APP startup icon for Apple iOS system. The designer icons must be in the rounded rectangle and have a standard size (Fig. 16).

iPhone 6 Plus

- App Icon - 180 x 180px
- AppStore Icon - 1024 x 1024px
- Spotlight Icon - 120 x 120px
- Settings Icon - 87 x 87px

Fig. 16 iPhone 6 plus icon design size

4 Difference Between the Application Carriers

As far as the application carrier is concerned, the difference between the APP startup icon and the logo is that using the mobile digital device as a fixed medium (Fig. 17), and using the functional characteristics of the digital device multimedia, some APP startup icons can display the icon features in the form of animation to

Fig. 17 Different mobile digital devices

Fig. 18 APP icon launcher with the data changing

enhance the creative design of the icons [3]. For example, Apple's mobile calendar APP (Fig. 18), its icons change every day with the changes of days and weeks, which greatly facilitates people's life. In addition, the display of APP startup icon is inseparable from digital mobile devices. It is an interactive port for users to enter APP and needs to be attached to digital devices as the medium.

The sign is more static than its own characteristics. As far as the APP startup icon is different from the carrier of the logo design, the logo can be used in a series of print media, such as all kinds of office things and public relations gifts, and can also be used in various kinds of carriers, such as transportation, clothing, neon lights, and buildings, with hundreds of application categories. Therefore, there is a significant difference between the APP startup icon and the logo on the application carrier.

5 Different Ways of User Experience

The APP startup icon is displayed in the digital media, its display medium determines its particularity, and people need to click to get the interaction with the third-party application, which determines its interactive experience. For example, if we want to enter the APP program (Fig. 19), we need to click the WeChat APP startup icon to get into the program, and then, the APP startup icon is equivalent to a button, which is quite different from the logo [4]. Symbol is the symbol of transmitting information. People only receive information simply. Most of them are set in the people's sight (Fig. 20) in the form of awakening and do not have to have physical interaction with the sign itself.

From the time of the birth of the logo and the APP startup icon, the development background of the logo is far earlier than the APP startup icon, and there are many scholars at home and abroad to study logo design, and the logo has a relatively mature design method. Based on the logo study of the APP startup icon design, it is not difficult to find that the two are compatible with each other in the elements, graphic allegory, and design style. Because the design rules of the APP start icon are not perfect at present, the above analysis can be used as a good paving for the visual design of the APP startup icon [5].

Fig. 19 Finger click into the APP icon

APP startup icons and symbols are the design between the square inches. Both need to convey the information briefly. The similarity of information transfer helps the designer to draw on the graphic, text, and color design of the logo when it is

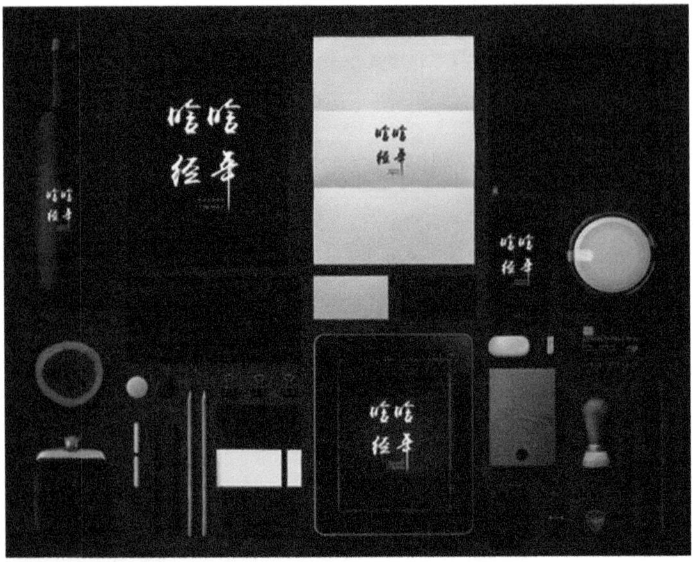

Fig. 20 Logo application

designed for the APP startup icon, thus helping the APP start the icon to shrink. Short exploration time, better development of APP startup icon, to a certain extent, logo design provides inspiration and reference for APP startup icon.

6 APP Startup Icon Is the New Carrier of Logo Design

Compared with the traditional logo design, APP startup icon design has more interactive functions. A sign is a static form of presentation, used to convey information and stand on outdoor or graphic albums. Some of the symbols are presented on the mobile end with the APP boot icon, making the logo more interactive, which is a sign of a new carrier on the mobile device [6].

At the same time, the development of science and technology also expands the carrier medium of the symbol. The symbol is displayed in the form of APP icon in front of us; there are two situations. One is based on the existing logo and designs the APP start map that is consistent with the symbol graphics according to the specified size of the APP startup icon to expand its spread on different media. Strength (Fig. 21) to facilitate user identification. The other is to design a brand new logo for mobile devices to meet the needs of mobile interface design. Based on the above theory, it can be said that the logo has a new development on mobile devices, broadened the scope of the symbol application media, better service for the public, and then promoted the development of the APP startup icon.

Fig. 21 APP and store display of ICBC bank logo

7 Conclusions

With the continuous improvement of mobile devices such as smartphones, the APP startup icon has also been widely developed. Because of the short development time and fast speed of the APP startup icon, its design performance is not very mature. Designers lack experience in the design of APP startup icons. Many times are still in the imitative state of the logo, ignoring the needs of the user interaction experience, and the APP startup icon appears to be incompatible with information transmission and cognitive experience. In accordance with the premise of modern people's aesthetic, this paper is based on the logo design, from the similarities and differences between the APP startup icon and the logo design, and the design of the APP startup icon is combed and summarized.

Through the analysis of the relationship between APP startup icon and logo design, it is concluded that there are differences and connections between them. In terms of the relationship between them, their elements are exactly the same, and the graphic meaning comes from one continuous line, and the design style is interlinked. The method of displaying rich meaning in a concise form provides an inspiration for the visual design of the APP startup icon, making the APP startup icon optimized in many aspects, such as information transmission and user identification. It also improves the visual tension and artistic affinity of the APP startup icon. As far as the difference is concerned, the difference between application carrier and user experience is an important manifestation of the sign from tradition to interaction. The application of APP startup icon in mobile media provides a new direction for logo design and interactive display. It not only changes the people's living habits imperceptibly, but also has a positive effect on the practice and development of the logo.

The coexistence of APP startup icon vision and interaction can not only make the information of the APP startup icons fully displayed, thus enhance the expression of the APP startup icon information, but also enable the APP to start the icon to meet the user's needs, and then make the APP startup icon bring an interesting aesthetic and smooth user experience to the audience. Whether the physical simulation design or the flattening of the new design, the APP startup icon is developing toward a more artistic and more distant direction. Advanced intelligent human–machine interaction will lead the APP launch icon into the next new era.

Acknowledgements The scene of the entrance seems to be still in sight. The three years of graduate study are coming to an end. In the past three years, I would like to express my sincere gratitude to my teachers and friends who care for me.

First, I want to thank my tutor, Professor Xie Yansong. Teacher Xie gave me great guidance and help in setting topics, modifying outlines, and writing papers. During my three-year postgraduate study in Nanyi, I was impressed by Mr. Xie's seriousness and strictness, which benefited me a lot. Secondly, I would like to thank Mr. Tong Fang and Mr. Lu Yi for their help in the literature. Especially in the guidance of topic selection, they have given me new ideas and understanding. Finally, I would like to thank my friends and classmates again for their constant concern and encouragement during the three-year study.

This paper is a summary of my three-year study, and my own views on art and design knowledge. There are still some defects and shortcomings in this paper. Please criticize and correct them.

References

1. Xun, L.: New media art. Shanghai Jiao Tong University Press, Shanghai (2011)
2. Leshan, L.: Interface design. Science Press, Beijing (2004)
3. Yansong, X.: Logo design. Shanghai, Shanghai Fine Arts Publishing House (2007)
4. Shneiderman, B., Plasisant, C.: Designing the user interface–strategies for effective human—computer interaction. Publishing House of Electronics Industry (2010)
5. Encent's User Research and Experience Design Department.: Beside you, design for you: Tencent's user experience design. Publishing House of Electronics Industry (2013)
6. Zhao, C.: Interface design—from man-machine relationship to interpersonal relationship. Decoration (2002)

Laser Radar Application in Vehicle Detection Under Traffic Environment

Bixiang Li and Lan Fang

Abstract According to the real-time laser radar is strong, wide detection depth, the advantages of little influence of the environment, put forward complex environment real-time vehicle detection method based on laser radar. Proposed method the HDL-64E-64 laser radar scanning complex traffic environment, perception by a laser radar system to get the lidar data, using 3D laser point cloud projection transform the lidar data to the body coordinate system, and establish the data grid structure. By gradient method for cutting the lattice structure of the lidar data, index figure obstacles markers in the cutting block. Box reduction method is used to detect the obstacles the shape, volume, speed, and coordinates, using cascade classifier detection accuracy control in the more than 0.9 mm. At last, by experimental verification, the proposed method has the advantages of low complexity and high efficiency.

Keywords Laser radar · A complex environment · Real-Time vehicle detection · Grid structure · Box of reduction

1 Introduction

In the high-tech field, robots can complete navigation, reconnaissance, detection, analysis, and other tasks; in people's daily life, robots can assist in driving navigation, avoid obstacles, and provide solutions for driving safety problems. The robot cannot realize its function without the sensor system, which uses the sensor system to extract detection data through data processing to obtain environmental information, and analysis of environmental information to obtain decision-making scheme. In real-time vehicle detection, machine vision system is often used for road surface and road marking detection. Laser sensors have high detection, classification, and

B. Li (✉) · L. Fang
Department of Information Engineering, City College of Wuhan University of Science and Technology, Wuhan 430083, China
e-mail: lbx153@163.com

V. Jain et al. (eds.), *Recent Trends in Intelligent Computing, Communication and Devices*, Advances in Intelligent Systems and Computing 1031,
https://doi.org/10.1007/978-981-13-9406-5_126

tracking accuracy for obstacles. Distance sensors are suitable for safe distance detection without chromaticity information. GPS is generally used in dynamic detection and three-dimensional modeling.

Laser radar is a technical clustering mechanical system of laser sensors and distance sensors. It has the advantages of wide detection depth and small interference from environment. It can shape and classify the outline of obstacles and is suitable for real-time vehicle detection in complex environment. Obstacle perception is one of the most important functions of laser radar. By cutting the laser radar data to restore the real-time outline of obstacles, it can provide vehicle driving suggestions and enhance the level of vehicle information and safety performance. According to the real-time advantages of laser radar, a real-time vehicle detection method in complex environment based on laser radar is proposed. The detection principle is introduced. The complexity and effectiveness of the proposed method are verified by experiments.

2 Real-Time Vehicle Detection in Complex Environment of Laser Radar

2.1 Principle of Laser Radar Detection

If the laser beam touches the object after it is emitted, the feedback reflected light and scattered light will be sent to the laser receiver. The scanner will use the time of laser feedback with the distance between object and laser radar which is calculated by [1]. The angle between the object and the laser radar is calculated by using the angle information of the laser beam transmitting position to sense the environment. These are the basic detection principles of laser radar. Compared with machine vision system [2], both systems have excellent real-time performance, but the laser radar's perception and anti-jamming ability to the environment are more suitable for real-time vehicle detection in complex environment.

HDL-64E-64 laser radar is a multithreaded intelligent mobile robot detection system. Scanning lines are evenly distributed on the upper and lower ends of the laser radar, and the upper and lower ends each has a laser receiver incident lens to facilitate the collection of laser beams [3]. HDL-64E-64 laser radar has a horizontal scanning range of 360°, a vertical scanning range of 30°, a scanning range from 30 to 150 m, a laser beam reflectance of 80%, a minimum scanning angle of 0.1°, a minimum ranging accuracy of 0.01 m, which are able to work normally in cloudy, rainy, foggy weather and other bad sight.

The position of HDL-64E-64 laser radar on the vehicle is the top of the skylight with the best view. Before real-time vehicle detection in complex environment, two coordinate systems need to be created, namely the laser radar perceptual coordinate system and the body coordinate system. The sketch of the coordinate system is represented in Fig. 1. The placement angles of the two coordinate systems are the

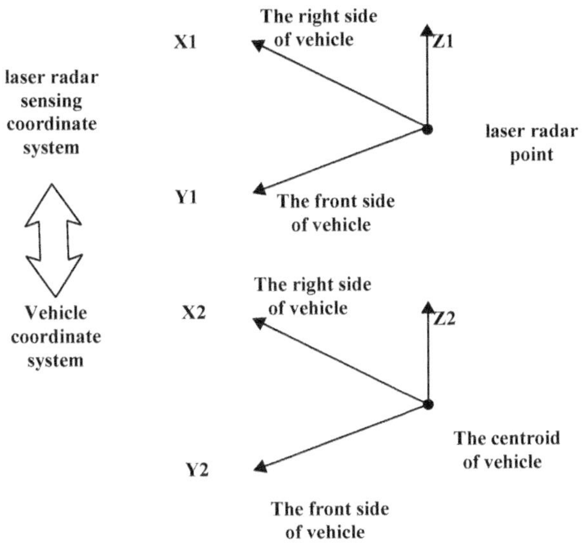

Fig. 1 Sketch map of vehicle coordinate system

same. The Y-axis is the front of the vehicle, and the X-axis is the right side. The coordinate points can be transformed into each other.

HDL-64E-64 laser radar transmits laser beam scanning vehicle and laser radar data in complex environment. The data corresponds to the laser radar sensing coordinate system. Radar data is transmitted by user datagram protocol [4]. The data packet specification contains ten data cutting blocks. Each cutting block has 90 bytes. The volume is relatively large. With the body coordinate system as the standard, the laser radar data is converted into standard binary data, which reduces the complexity of real-time vehicle detection method in complex environment based on laser radar.

2.2 *Three-Dimensional Laser Point Cloud Projection*

The real-time vehicle detection method in complex environment based on laser radar uses 3D laser point cloud projection to transform vehicle and complex environment's laser radar data and creates a grid structure with cutting block area of G. M and N represent the number of long and wide cutting blocks in the grid structure. The laser radar data is displayed as three-dimensional coordinates in the laser radar sensing coordinate system, keeping the Z-axis coordinate points unchanged. Let my and mx be the distance between the laser radar sensing coordinate system and the Y-axis and X-axis of the vehicle body coordinate system, and be responsible for converting all negative coordinates into positive coordinates. The laser radar data coordinates are represented by (x_1, y_1, z_1). The corresponding coordinates in the body coordinate system are set to (x_2, y_2, z_2). The conversion formula is as follows:

$$x_2 = \frac{x_1 + m_x}{G} \tag{1}$$

$$y_2 = \frac{y_1 + m_y}{G} \tag{2}$$

$$z_2 = z_1 \tag{3}$$

In the grid structure of the proposed method, each cutting block represents one laser radar data, and there is no correlation between the data in different cutting blocks, which is represented by the uncorrelated three-dimensional laser point cloud segmentation data. However, the proposed method cannot directly use the segmentation data in the reconstruction and classification of obstacle contours, so it is necessary to segment the data. The section data is put into the grid structure to record the coordinates of the body of the laser point cloud and the laser intensity data [5]. The projection of the 3D laser point cloud is based on hash mapping [6]. The projection relation of the laser radar data can be quickly shaped, and the related data of the vehicle and the complex environment can be clustered by the grid structure. Restore obstacles and classify them.

2.3 *Laser Radar Data Cutting*

In the process of coordinate system transformation, *Z*-axis coordinate points remain unchanged. After projection of 3D laser point cloud, the grid structure clears the *Z*-axis coordinate points and transforms the laser radar data into two-dimensional coordinates. This method can reduce the calculation, stabilize the coordinate values, and improve the real-time detection success rate. However, this method also has defects. If the distribution of the laser radar data is not dense, the raster structure will not cover all the data. Moreover, in complex environment, the laser radar data may contain non-target road surface data, interfering with the real-time detection accuracy, which is not conducive to the restoration and classification of the obstacle contour. In order to achieve local optimum of laser point cloud in grid structure, gradient method is used to cut laser radar data for real-time vehicle detection in complex environment based on laser radar.

HDL-64E-64 laser radar revolving scan line collects laser radar data. The scan line is circular on the road surface of the vehicle. The unit vector direction is outward and perpendicular to the laser beam.

$$\tan\theta = \frac{\Delta_z}{\text{distance}} \tag{4}$$

In the formula, θ is the descriptive parameter, Δ_z is the coordinate difference of *Z*-axis between scanning lines, and distance is the vertical distance between *X*-axis and *Y*-axis between scanning lines. If there are obstacles in the vehicle, the distance and Δ_z will be reduced. Under reasonable circumstances, the gradient value $\tan\theta$ will not change or change in a small range. A gradient threshold is proposed. When the gradient value changes beyond this threshold, the laser radar data at the coincidence point of the scanning line is cut into twice the depth of the original cutting block.

2.4 Classification of Obstacles in Complex Environment

In complex environment, HDL-64E-64 laser radar uses cascade classifier to classify obstacles. It is composed of three classifier nodes in series. All obstacle laser radar data points form a set of sub-windows. The detection result is positive through three classifier nodes. The set of sub-windows is classified into obstacles, and the test results are deleted as negative sets (Note: The positive set contains most of the target detection data points, while the negative set usually contains useless data such as flowers, trees, fences, and road signs).

3 Confirmatory Experiments

In the complex environment of real life, Table 1 is the test result statistics table of vehicle real-time detection in the campus with narrow roads and more flowers and trees.

In Table 1, the best detection results of effective detection distance, obstacle removal rate, and vehicle response time are given by the proposed method. It shows that the proposed method can effectively detect vehicles in real time under the premise of low complexity in complex environment.

Table 1 Statistical table of test results in complex campus environment

	Effective detection distance/m	Obstacle removal rate/%	Vehicle response time/s
Laser radar	983	96.75	0.25
GPS	961	92.18	0.74
ALV	915	92.18	0.59

4 Conclusions

In this chapter, a real-time vehicle detection method based on HDL-64E-64 laser radar in complex environment is proposed, which collects, transforms, cuts, restores, and classifies the data between vehicle and complex environments in turn, and detects the correlation between vehicle and complex environments in real time. Compared with the real-time vehicle detection method based on GPS and ALV, this method can play the best detection level in complex environment.

Acknowledgements This work was supported by two funds: 1. City College of Wuhan University of Science and Technology, Teaching and Research Project: Research and Practice of Computer Network Course Group Practice Teaching System Reform Based on Cloud Computing (Project No: 2018CYZDJY007). 2. Innovation Entrepreneurship Project: Promote student employment plans with reform and practice teaching systems and innovation in science and technology competitions (Project No: 13).

References

1. July I, lasted for days, Tests A O. Environmental perception and sensor data fusion for unmanned ground vehicle. Mathe. Probl. Eng. (4), 1–12 (2013)
2. Nicholls, K.W., Corr, H.F.J., Stewart, C.L., et al.: A ground-based radar for measuring vertical strain rates and time-varying basal melt rates in ice sheets and shelves. J. Glaciol. **61**(230), 1079–1087 (2015)
3. Wang, S., Dai, X., Ning, X.U., et al (2017). Overview on environment perception technology for unmanned ground vehicle. J. Changchun Uni. Sci. Technol (2017)
4. Zhao, T.T., Chen, W.P., Yi-Fei, G.U., et al.: Design of automatic cruise unmanned ground vehicle based on environmental perception. Radio Eng (2017)
5. Liu, X.: Capacity impacts and optimal geometry of automated cars' surface parking facilities. 1–3 (2013)
6. Anusha, S.P., Sharma, A., Vanajakshi, L., et al.: Model-based approach for queue and delay estimation at signalized intersections with erroneous automated data. J. Trans. Eng. **142**(5), 04016013 (2016)

Research on Ranging Algorithm Based on the Fringe Projection

Li Chang and Gege Huang

Abstract Aiming at the problem of the existing positioning and ranging methods depended on the external conditions such as the number of sensors, artificial markers, and the ground, this paper proposes a positioning method based on fringe projection. Firstly, the fringe image of the reference surface and the fringe image of the background are captured, respectively, and the centerlines of the fringes are extracted by using the ZS (Zhang-Suen) thinning algorithm. Secondly, the width of the fringe at any point on the reference surface is determined according to the principle of minimum Euclidean distance. Then, the projection model is established, and the distance between the reference surface and the projector is determined. Finally, the three-dimensional reconstruction technique is used to obtain the three-dimensional topography of the measured object, and the distance between the measured object and the projector is calculated. The experimental results show that the ranging algorithm based on fringe projection is feasible, and the mean square error of the ranging result is 0.0434 m.

Keywords Grating projection · ZS thinning · Projection ranging · Three-dimensional reconstruction

1 Introduction

For the robot perception system, positioning is the basis for completing various tasks. At present, the existing ranging methods mainly include the monocular method and the binocular method. Although the binocular ranging method has three-dimensional measurement capability, the system structure is complicated [1, 2].

L. Chang (✉) · G. Huang
School of Information Science and Engineering, Shenyang University of Technology, Shenyang 110870, China
e-mail: changlianli@163.com

V. Jain et al. (eds.), *Recent Trends in Intelligent Computing, Communication and Devices*, Advances in Intelligent Systems and Computing 1031,
https://doi.org/10.1007/978-981-13-9406-5_127

Aiming at the problem of the existing positioning and ranging methods depended on the external conditions such as the number of sensors, artificial markers, and the ground, this paper proposes a ranging method based on fringe projection and three-dimensional reconstruction technology, which can not only acquires the three-dimensional shape of the measured object, but also achieves distance measurement. It provides more environmental information for the service robot and assists the service robot in accomplishing various tasks better [1, 2].

2 Spatial Ranging Principle

As shown in Fig. 1, the distance L from the reference surface to the projector is measured based on the fringe width, and then the height S of the measured object is obtained according to the three-dimensional topography reconstruction technique. So the distance between the object and the projector called L_p can be expressed as:

$$L_{\mathrm{p}} = L - S \tag{1}$$

2.1 The Ranging Algorithm Principle Based on the Fringe Width

The reference surface fringe image and the background fringe image are collected, and then the gray value of each corresponding pixel is subtracted to determine the approximate region of the reference surface. The region is extracted from the reference surface fringe image, and binarization is performed to obtain a deformed fringe binary image. The width of the gating fringe is obtained by ZS parallel thinning algorithm [3–5].

The projection model is shown in Fig. 2. The optical axis of the projector is perpendicular to the reference surface. O_p is the optical center of the projector, P is the projection center, 2α is the horizontal projection angle, D is the width of the projection screen, and L is the vertical projection distance.

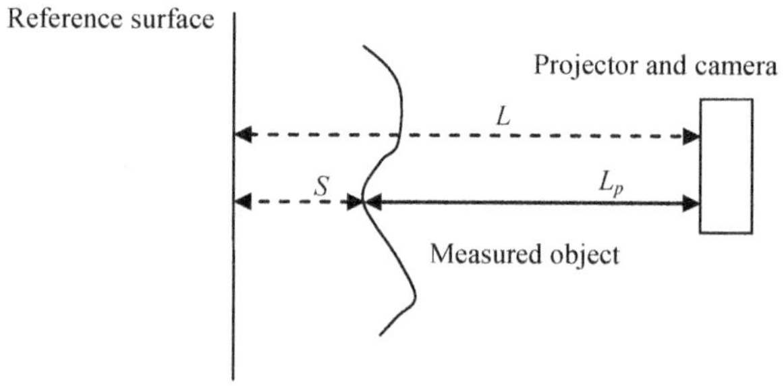

Fig. 1 Schematic diagram of the ranging method based on three-dimensional reconstruction

Fig. 2 Projection model

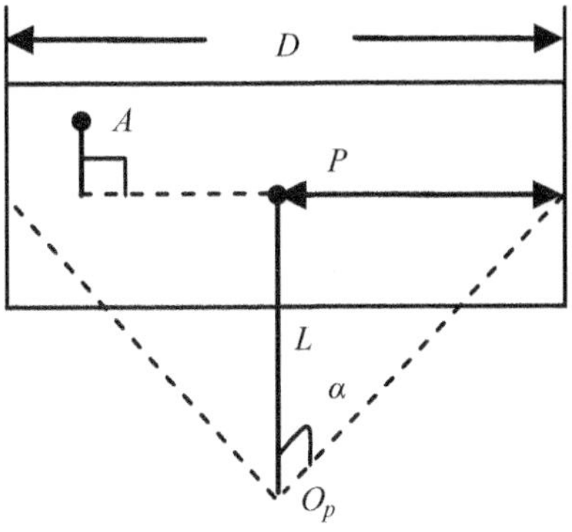

According to the projection principle, the relationship between the gating fringe width W and the projection distance L can be expressed as:

$$L = \frac{m}{2T \tan \alpha} W \tag{2}$$

where m is the resolution of the projector, and T is the period of the grating fringe.

2.2 Three-Dimensional Reconstruction

When the surface of the measured object itself is not flat, the width of the fringe will be irregularly changed. The mathematical model between the height and the phase is established, and the height value of any point S on the surface of the measured object can be calculated by the true phase value:

$$S = \frac{L_0(\phi_A - \phi_B)}{(\phi_A - \phi_B) + 2\pi D_0/\lambda_0} \tag{3}$$

As shown in Fig. 3, Φ_B is the phase value of point B on the reference surface, and Φ_A is the phase value of the surface of the measured object, λ_0 is the pitch of the grating, $L_0 = O_pO$, $D_0 = O_CO_p$ [6, 7].

Fig. 3 Schematic diagram of the grating projection measurement system

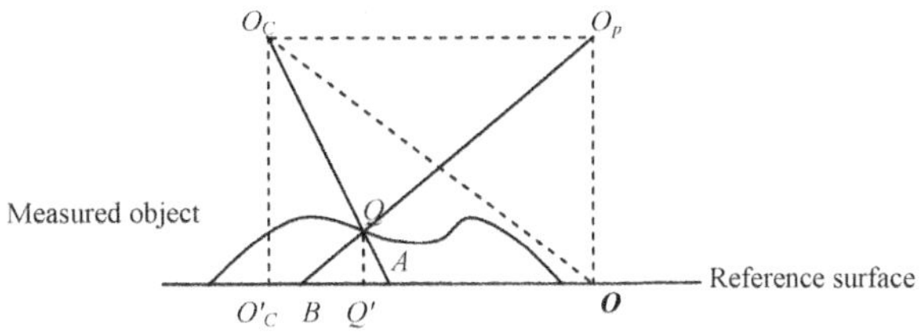

3 Experimental Process and Results

The experimental system consists of a projector, a camera, and a computer. The experimental site is the indoor space of a family, and the size is about 5 m × 5 m × 3 m. The projector and the camera are installed on the roof. When the projector projects the fringe vertically downwards, the maximum vertical projection distance is about 3 m. There is also a certain distance between the measured object and the roof in the room. Therefore, the measurement range is determined between 1 and 3 m.

Since the resolution of the projector is 1280 × 720, the grating fringes with the period of 32, 40, 64, or 128 are generally selected in order to ensure that the fringes in the projection screen are full period. However, in this experiment, when the fringe period is less than 64, the fringes are relatively dense and the fringe width changes little with the change of the distance. Conversely, when the fringe period is greater than 64, the error of the detection result is larger for the smaller measured object and even the width of the fringes on the object surface cannot be judged. At the same time, if the measured object has the fringe pattern features, the fringe with a different period may be selected for projection. Therefore, this article will choose a period of 64 grating fringes for projection.

Figures 4 and 5 are the fringe image I_1 of the reference surface and the fringe image I_0 of the background, respectively. I_1 and I_0 are used for difference and binarization, and the deformation fringe binary image I of the reference surface is obtained, as shown in Fig. 6.

After ZS thinning, the centerline of the fringes as shown in Fig. 7 is obtained. In this experiment, the center point of the deformation fringe image is considered as the measured object point of the reference surface, as shown by the green marking point in Fig. 7. The yellow and red marks in Fig. 7 are the two ends of the fringe width calculation. The abscissa is equal to the object point, and the ordinate is determined by the center line of the fringe. The Euclidean distances at the two ends

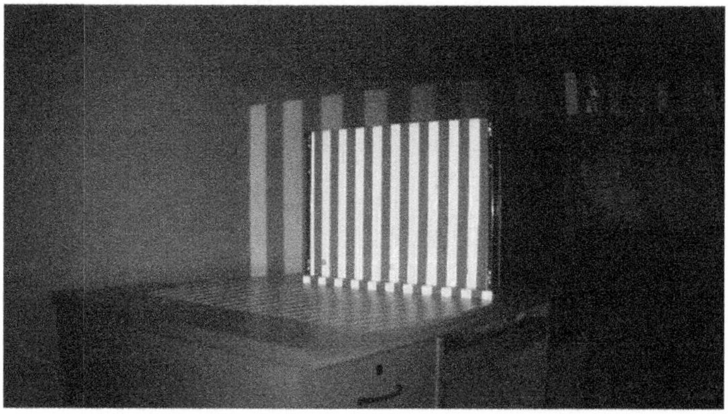

Fig. 4 Fringe image of the reference surface

Fig. 5 Fringe image of the background

Fig. 6 Deformation fringe binary images

are taken as the fringe width. In turn, the distance between the reference surface and the projector is determined based on the projection model.

An insulated kettle with a diameter of 0.14 m is used as the measured object and placed between the projector and the reference surface for three-dimensional topography reconstruction experiments. Four fringe images of the measured object are collected, as shown in Fig. 8.

The four-step phase shift method is used to obtain the wrapped phase diagram of the deformed fringe. The phase unwrapping is performed through the least square method to obtain the true phase as shown in Fig. 9, and the three-dimensional topography of the measured object can be seen. Finally, the height of each point on the object is calculated according to formula (3), as shown in Fig. 10.

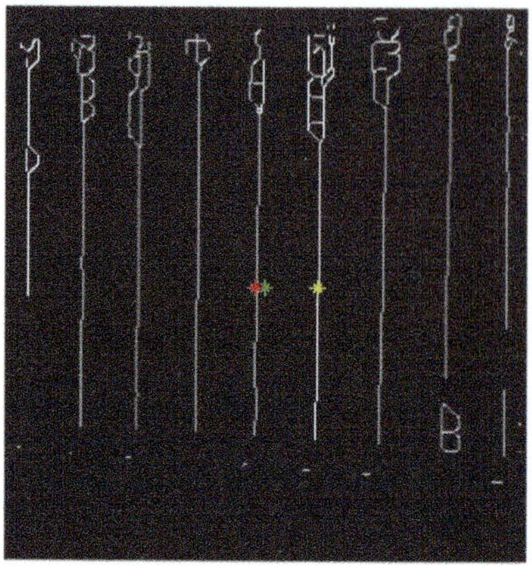

Fig. 7 Thinning results

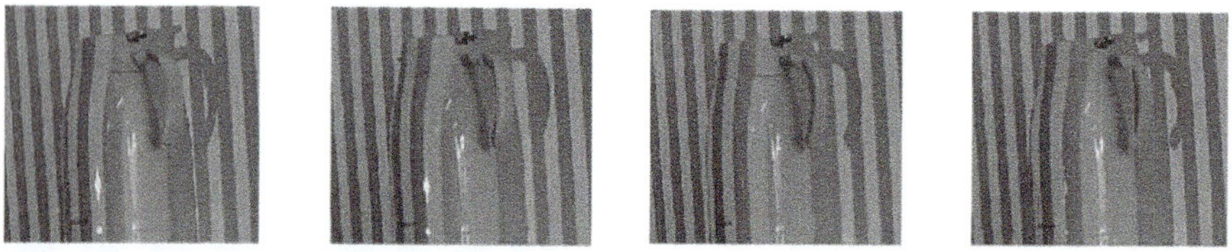

(a) Phase shift is 0 (b) Phase shift is $\pi/2$ (c) Phase shift is π (d) Phase shift is $3\pi/2$

Fig. 8 Fringe image of the measured object

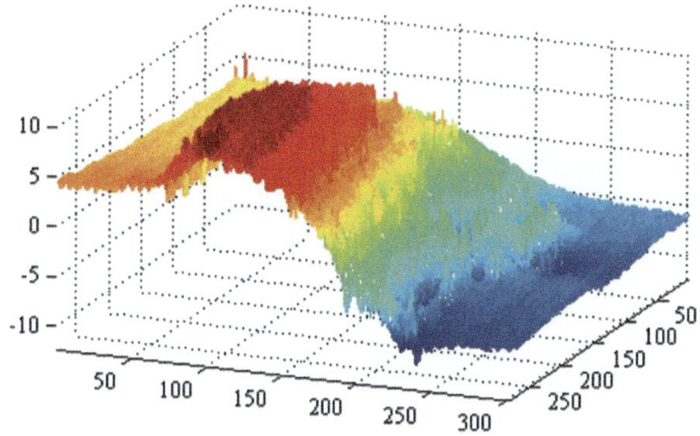

Fig. 9 Phase unwrapping result

In Fig. 10, the height of the measured object can be clearly obtained. As the color deepens, the height becomes higher and higher. The shape of the measured object is not flat, but the height measured in the experiment is almost flat. Its main reason is that the camera has a certain angle and the measured object being parallel to the reference surface. The height of the blue area is negative, which is caused by

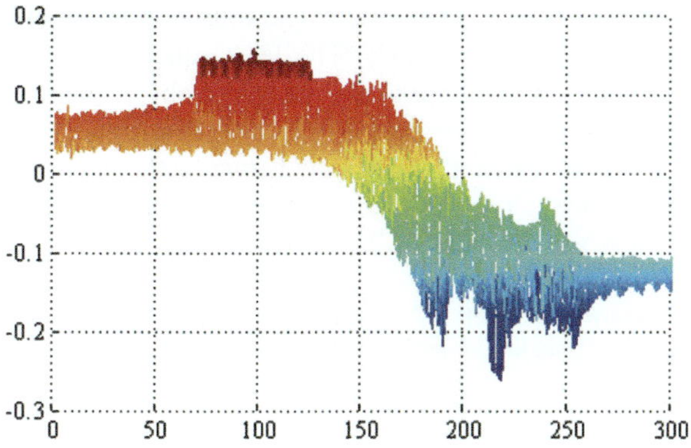

Fig. 10 Height of the measured object

Table 1 Experimental results

Real distance (m)	Measured distance (m)
1.0	1.0653
1.2	1.2312
1.4	1.4376
1.6	1.5952
1.8	1.7938
2.0	2.0011
2.2	2.2195
2.4	2.4490
2.6	2.6887

the shadow. In the ranging algorithm, the blue area can be ignored, and the height of the location of the measured object can be recorded.

Moving the position of the measured object, the distance between the projector and the reference surface and the height of the measured object are measured. The distance between the measured object and the projector is obtained by the difference of the distance and height. The experimental results are shown in Table 1. When the distance is more than 2.6 m, the width of the detected fringe is almost constant due to the external light intensity and other factors, so that the measured value of the distance is inaccurate. The error analysis of the experimental results shows that the mean square error of this ranging method is 0.0434 m.

4 Conclusion

A ranging method combining fringe projection and three-dimensional reconstruction is proposed. The width of the fringe is calculated by the ZS thinning algorithm. The relationship between the projection distance and the fringe width is determined according to the projection model. The distance between the reference surface and the projector is determined. The four-step phase shift method is used to obtain the wrapped phase diagram. The least squares phase unwrapping method is used to obtain the true phase of the measured object. Then the three-dimensional shape and the height of the measured object are determined according to the grating projection principle. Then the distance between the measured object and the projector is determined. The mean squared error of the experimental ranging result is 0.0434 m, which verifies the feasibility of this method.

Acknowledgements The paper is supported by the Science and Technology Foundation of Shenyang under the project F16-205-1-11.

References

1. Jiang, J., Tu, D.: Human-robot collaboration navigation of service robots for the elderly and disabled in an intelligent space. CAAI Trans. Intell. Syst. **9**(5), 560–568 (2014)
2. Li, H.: Three-Dimensional Distance Determining and Positioning Based on Binocular Vision. South China University of Technology (2012)
3. Stoykova, E., Berberova, N., Park, J.S., et al.: Pattern projection profilometry with sinusoidal gratings under coherent illumination. 3D Res. **4**(1), 1–9 (2013)
4. Escribano, C., Giraldo, A., Sastre, M.A.: Digitally continuous multivalued functions, morphological operations and thinning algorithms. J. Math. Imaging Vis. (42), 76–91 (2012)
5. Ben Boudaoud, L., Solaiman, B., Tari, A.: A modified ZS thinning algorithm by a hybrid approach. Vis. Comput. 1–18 (2017)
6. Da, F., Gai, S.: Grating Projection Three-Dimensional Precision Measurement. Science Press (2011)
7. Shang, Z., Li, W., Dong, M., et al.: 3D shape measurement system based on fringe projection in 4-step phase shifting. J. Appl. Optics **36**(4), 584–589 (2015)

Author Index

V. Jain et al. (eds.), *Recent Trends in Intelligent Computing, Communication and Devices*, Advances in Intelligent Systems and Computing 1031,
https://doi.org/10.1007/978-981-13-9406-5